DESIGN OF MACHINERY

AN INTRODUCTION TO THE SYNTHESIS AND ANALYSIS OF MECHANISMS AND MACHINES

Second Edition

New Media Version

McGraw-Hill Series in Mechanical Engineering

Jack P. Holman, *Southern Methodist University*
John R. Lloyd, *Michigan State University*
Consulting Editors

Anderson: *Computational Fluid Dynamics: The Basics with Applications*
Anderson: *Modern Compressible Flow: With Historical Perspective*
Arora: *Introduction to Optimum Design*
Borman and Ragland: *Combustion Engineering*
Cengel: *Heat Transfer: A Practical Approach*
Cengel: *Introduction to Thermodynamics and Heat Transfer*
Cengel and Boles: *Thermodynamics: An Engineering Approach*
Cengel and Turner: *Fundamentals of Thermal-Fluid Sciences*
Culp: *Principles of Energy Conversion*
Dieter: *Engineering Design: A Materials and Processing Approach*
Doebelin: *Engineering Experimentation: Planning, Execution, Reporting*
Driels: *Linear Controls Systems Engineering*
Edwards and McKee: *Fundamentals of Mechanical Component Design*
Gibson: *Principles of Composite Material Mechanics*
Hamrock: *Fundamentals of Fluid Film Lubrication*
Hamrock, Jacobson, and Schmid: *Fundamentals of Machine Elements*
Heywood: *Internal Combustion Engine Fundamentals*
Histand and Alciatore: *Introduction to Mechatronics and Measurement Systems*
Holman: *Experimental Methods for Engineers*
Jaluria: *Design and Optimization of Thermal Systems*
Kays and Crawford: *Convective Heat and Mass Transfer*
Kelly: *Fundamentals of Mechanical Vibrations*
Martin: *Kinematics and Dynamics of Machines*
Mattingly: *Elements of Gas Turbine Propulsion*
Modest: *Radiative Heat Transfer*
Norton: *Design of Machinery: An Introduction to the Synthesis and Analysis of Mechanisms and Machines*
Oosthuizien and Carscallen: *Compressible Fluid Flow*
Oosthuizien and Naylor: *Introduction to Convective Heat Transfer Analysis*
Reddy: *An Introduction to the Finite Element Method*
Rosenberg and Karnopp: *Introduction to Physical Systems Dynamics*
Schlichting: *Boundary-Layer Theory*
Shames: *Mechanics of Fluids*
Shigley and Mischke: *Mechanical Engineering Design*
Shigley and Uicker: *Theory of Machines and Mechanisms*
Stoeker: *Design of Thermal Systems*
Stoeker and Jones: *Refrigeration and Air Conditioning*
Turns: *An Introduction to Combustion: Concepts and Applications*
Ullman: *The Mechanical Design Process*
Wark: *Advanced Thermodynamics for Engineers*
Wark and Richards: *Thermodynamics*
White: *Fluid Mechanics*
White: *Viscous Flow*
Zeid: *CAD/CAM Theory and Practice*

DESIGN OF MACHINERY

AN INTRODUCTION TO THE SYNTHESIS AND ANALYSIS OF MECHANISMS AND MACHINES

Second Edition

New Media Version

Robert L. Norton

Worcester Polytechnic Institute

Worcester, Massachusetts

Boston   Burr Ridge, IL   Dubuque, IA   Madison, WI   New York   San Francisco   St. Louis
Bangkok   Bogotá   Caracas   Lisbon   London   Madrid
Mexico City   Milan   New Delhi   Seoul   Singapore   Sydney   Taipei   Toronto

McGraw-Hill

A Division of The McGraw·Hill Companies

DESIGN OF MACHINERY: An Introduction to the Synthesis and Analysis of Mechanisms and Machines

Published by McGraw-Hill, an imprint of The McGraw-Hill Companies, Inc., 1221 Avenue of the Americas, New York, NY, 10020. Copyright © 2001, 1999, 1992 by The McGraw-Hill Companies, Inc. All rights reserved. No part of this publication may be reproduced or distributed in any form or by any means, or stored in a database or retrieval system, without the prior written consent of The McGraw-Hill Companies, Inc., including, but not limited to, in any network or other electronic storage or transmission, or broadcast for distance learning. Some ancillaries, including electronic and print components, may not be available to customers outside the United States.

This book is printed on acid-free paper.

1 2 3 4 5 6 7 8 9 0 QPF/QPF 0 9 8 7 6 5 4 3 2 1 0

ISBN 0-07-237960-X (text only)
ISBN 0-07-242351-X (set)

Vice president and editorial director: Kevin T. Kane
Publisher: Thomas Casson
Senior sponsoring editor: Jonathan W. Plant
Developmental editor: Debra D. Matteson
Marketing manager: John T. Wannemacher
Project manager: Christina Thornton-Villagomez
Production supervisor: Rose Hepburn
Supplement coordinator: Mark Mattson
Cover design: Gino Cieslik
Book design: Wanda Siedlecka
Printer: Quebecor Printing Book Group/Fairfield

Cover photo: Viper cutaway courtesy of the DaimlerChrysler Corporation, Auburn Hills, MI.

All text, drawings, and equations in this book were prepared and typeset electronically, by the author, on a *Macintosh*® computer using *Freehand*®, *MathType*®, and *Pagemaker*® desktop publishing software. The body text was set in Times Roman, and headings set in Avant Garde. Printer's film color separations were made on a laser typesetter directly from the author's disks. All *clip art* illustrations are courtesy of *Dubl-Click Software Inc.*, 22521 Styles St., Woodland Hills CA 91367, reprinted from their *Industrial Revolution* and *Old Earth Almanac* series with their permission (and with the author's thanks).

Library of Congress Cataloging-in-Publication Data

Norton, Robert L.
        Design of machinery: an introduction to the synthesis and analysis of mechanisms and machines / Robert L. Norton – 2nd ed., new media edition
            p.   cm. —(McGraw-Hill series in mechanical engineering)
        Includes bibliographical references and index.
        ISBN 0-07-237960-X
        1. Machinery—Design.   2. Machinery, Kinematics.   3. Machinery, Dynamics of.
        I. Title.   II. Series.
TJ175.N58   2001   00-032898
        621.8'15—dc21

The book's web page is at http://www.mhhe.com
The author's web page is at http://me.wpi.edu/norton.htm

ABOUT THE AUTHOR

Robert L. Norton earned undergraduate degrees in both mechanical engineering and industrial technology at Northeastern University and an MS in engineering design at Tufts University. He is a registered professional engineer in Massachusetts. He has extensive industrial experience in engineering design and manufacturing and many years experience teaching mechanical engineering, engineering design, computer science, and related subjects at Northeastern University, Tufts University, and Worcester Polytechnic Institute.

At Polaroid Corporation for 10 years, he designed cameras, related mechanisms, and high-speed automated machinery. He spent three years at Jet Spray Cooler Inc., designing food-handling machinery and products. For five years he helped develop artificial-heart and noninvasive assisted-circulation (counterpulsation) devices at the Tufts New England Medical Center and Boston City Hospital. Since leaving industry to join academia, he has continued as an independent consultant on engineering projects ranging from disposable medical products to high-speed production machinery. He holds 13 U.S. patents.

Norton has been on the faculty of Worcester Polytechnic Institute since 1981 and is currently professor of mechanical engineering, the Russell G. Searle Distinguished Instructor, and head of the design group in that department. He teaches undergraduate and graduate courses in mechanical engineering with emphasis on design, kinematics, vibrations, and dynamics of machinery.

He is the author of numerous technical papers and journal articles covering kinematics, dynamics of machinery, cam design and manufacturing, computers in education, and engineering education and of the text *Machine Design: An Integrated Approach*. He is a Fellow of the American Society of Mechanical Engineers and a member of the Society of Automotive Engineers. Rumors about the transplantation of a Pentium microprocessor into his brain are decidedly untrue (though he could use some additional RAM). As for the unobtainium[*] ring, well, that's another story.

---

[*] See Index.

This book is dedicated to the memory of my father,

Harry J. Norton, Sr.

who sparked a young boy's interest in engineering;

to the memory of my mother,

Kathryn W. Norton

who made it all possible;

to my wife,

Nancy Norton

who provides unflagging patience and support;

and to my children,

Robert, Mary, and Thomas,

who make it all worthwhile.

CONTENTS

PREFACE
to the New Media Version of the Second Edition

The medium is the message.
MARSHALL MCLUHAN

This *New Media Version* of the second edition has been enhanced by the addition of new software on the attached CD-ROM. The Working Model 4.0 2D Homework Edition is still included free of charge on the CD-ROM. In addition, Professor Shih-Liang (Sid) Wang of North Carolina A&T has created the package *Mechanism Simulation in a Multimedia Environment* containing 43 new *Working Model* files based on the book's figures and 6 new *Matlab*® models for kinematic analysis and animation.

In combination with the 20 *Working Model* files previously supplied, these now provide 69 models that bring the text's figures to life with animation, graphs, and numerical output. For each of Professor Wang's simulations, a video file of the mechanism can be played independently of the *Working Model* program, or the student can open, run, modify, interact with, save, print and create new *Working Model* simulation files for any assignment with the provided *Working Model* program. Microsoft Internet Explorer is used to navigate among hyperlinked HTML files that contain text, picture, video, *Matlab*, and *Working Model* files.

Some *Matlab* files supplied will analyze fourbar, slider crank, and inverted slider crank linkages and animate their motion. Other *Matlab* files calculate the tooth profile of an involute spur gear, show the geometric generation of an involute and the motion of an elliptic trammel. *Matlab* source code is provided. The *Matlab* program is not. Extensive comments are provided within each *Matlab* file identifying the equations used from the text by number. The student can modify these models for other applications.

The supplied student versions of the author-written programs, FOURBAR, FIVEBAR, SIXBAR, SLIDER, DYNACAM, ENGINE, and MATRIX have all been revised, enhanced, and improved. Most now allow Fourier transforms of their variables to be calculated.

Also included is the *FE Exam Interactive Review for Kinematics and Applied Dynamics* by E. Anderson and J. Hashemi. This is a set of interactive review quizzes.

This revision of the second edition has also allowed the text to be updated to match changes made to some problem statements during the creation of the solutions manual. All known errors in the text have been corrected and many suggestions for improvement from users also have been incorporated in this revision. If you find any other errors, please email the author at *norton@wpi.edu*. Errata as discovered, and other book information, will be posted on the author's web site at http://me.wpi.edu/norton.htm.

The author would like to express his appreciation to Professor Sid Wang for his efforts in creating the *Working Model* and *Matlab* files. Professor Thomas A. Cook's herculean effort in the creation of the 1 200-page solutions manual and its *Mathcad*® files is also greatly appreciated. Professors M. Corley of Louisiana Tech, R. Devashier of U. Evansville, K. Gupta of U. Illinois-Chicago, J. Steffen of Valparaiso University, and D. Walcerz of York College all provided useful suggestions or corrections.

Robert L. Norton
Norfolk, Mass.
May, 2000

PREFACE

to the Second Edition

*Why is it we never have time to do
it right the first time, but always
seem to have time to do it over?*
ANONYMOUS

The second edition has been revised based on feedback from a large number of users of the book. In general, the material in many chapters has been updated to reflect the latest research findings in the literature. Over 250 problem sets have been added, more than doubling the total number of problems. Some design projects have been added also. All the illustrations have been redrawn, enhanced, and improved.

Coverage of the design process in Chapter 1 has been expanded. The discussions of the Grashof condition and rotatability criteria in Chapter 2 have been strengthened and that of electric motors expanded. A section on the optimum design of approximate straight line linkages has been added to Chapter 3. A discussion of circuits and branches in linkages and a section on the Newton-Raphson method of solution have been added to Chapter 4. A discussion of other methods for analytical and computational solutions to the position synthesis problem has been added to Chapter 5. This reflects the latest publications on this subject and is accompanied by an extensive bibliography.

The chapters formerly devoted to explanations of the accompanying software (old Chapters 8 and 16) have been eliminated. Instead, a new Appendix A has been added to describe the programs FOURBAR, FIVEBAR, SIXBAR, SLIDER, DYNACAM, ENGINE, and MATRIX that are on the attached CD-ROM. These programs have been completely re-written as *Windows* applications and are much improved. A student version of the sim-ulation program *Working Model* by *Knowledge Revolution,* compatible with both *Macintosh* and *Windows* computers, is also included on CD-ROM along with 20 models of mechanisms from the book done in that package. A user's manual for *Working Model* is also on the CD-ROM.

Chapter 8 on cam design (formerly 9) has been shortened without reducing the scope of its coverage. Chapter 9 on gear trains (formerly 10) has been significantly expanded and enhanced, especially in respect to the design of compound and epicyclic trains and their efficiency. Chapter 10 on dynamics fundamentals has been augmented with mate-rial formerly in Chapter 17 to give a more coherent treatment of dynamic modeling. Chapter 12 on balancing (formerly 13) has been expanded to include discussion of mo-ment balancing of linkages.

Robert L. Norton
Mattapoisett, Mass.
August, 1997

The author would like to express his appreciation to all the users and reviewers who have made suggestions for improvement and pointed out errors, especially those who responded to the survey about the first edition. There are too many to list here, so rather than risk of-fense by omitting anyone, let me simply extend my sincerest thanks to you all for your efforts.

PREFACE
to the First Edition

When I hear, I forget
When I see, I remember
When I do, I understand
ANCIENT CHINESE PROVERB

This text is intended for the kinematics and dynamics of machinery topics which are often given as a single course, or two-course sequence, in the junior year of most mechanical engineering programs. The usual prerequisites are first courses in statics, dynamics and calculus. Usually, the first semester, or portion, is devoted to kinematics, and the second to dynamics of machinery. These courses are ideal vehicles for introducing the mechanical engineering student to the process of design, since mechanisms tend to be intuitive for the typical mechanical engineering student to visualize and create.

While this text attempts to be thorough and complete on the topics of analysis, it also emphasizes the synthesis and design aspects of the subject to a greater degree than most texts in print on these subjects. Also, it emphasizes the use of computer-aided engineering as an approach to the design and analysis of this class of problems by providing software that can enhance student understanding. While the mathematical level of this text is aimed at second- or third-year university students, it is presented *de novo* and should be understandable to the technical school student as well.

Part I of this text is suitable for a one-semester or one-term course in kinematics. Part II is suitable for a one-semester or one-term course in dynamics of machinery. Alternatively, both topic areas can be covered in one semester with less emphasis on some of the topics covered in the text.

The writing and style of presentation in the text is designed to be clear, informal, and easy to read. Many example problems and solution techniques are presented and spelled out in detail, both verbally and graphically. All the illustrations are done with computer-drawing or drafting programs. Some scanned photographic images are also included. The entire text, including equations and artwork, is printed directly from computer disk by laser typesetting for maximum clarity and quality. Many suggested readings are provided in the bibliography. Short problems, and where appropriate, many longer, unstructured design project assignments are provided at the ends of chapters. These projects provide an opportunity for the students *to do and understand*.

The author's approach to these courses and this text is based on over 35 years' experience in mechanical engineering design, both in industry and as a consultant. He has taught these subjects since 1967, both in evening school to practicing engineers and in day school to younger students. His approach to the course has evolved a great deal in that time, from a traditional approach, emphasizing graphical analysis of many structured problems, through emphasis on algebraic methods as computers became available, through requiring students to write their own computer programs, to the current state described above.

The one constant throughout has been the attempt to convey the art of the design process to the students in order to prepare them to cope with *real* engineering problems in practice. Thus, the author has always promoted design within these courses. Only recently, however, has technology provided a means to more effectively accomplish this goal, in the form of the graphics microcomputer. This text attempts to be an improvement over those currently available by providing up-to-date methods and techniques for analysis and synthesis which take full advantage of the graphics microcomputer, and by emphasizing design as well as analysis. The text also provides a more complete, modern, and thorough treatment of cam design than existing texts in print on the subject.

The author has written seven interactive, student-friendly computer programs for the design and analysis of mechanisms and machines. These programs are designed to enhance the student's understanding of the basic concepts in these courses while simultaneously allowing more comprehensive and realistic problem and project assignments to be done in the limited time available than could ever be done with manual solution techniques, whether graphical or algebraic. Unstructured, realistic design problems which have many valid solutions are assigned. Synthesis and analysis are equally emphasized. The analysis methods presented are up to date, using vector equations and matrix techniques wherever applicable. Manual graphical analysis methods are de-emphasized. The graphics output from the computer programs allows the student to see the results of variation of parameters rapidly and accurately and reinforces learning.

These computer programs are distributed on CD-ROM with this book, which also contains instructions for their use on any IBM compatible, Windows 3.1 or Windows 95/98/NT capable computer. Programs SLIDER, FOURBAR, FIVEBAR and SIXBAR analyze the kinematics and dynamics of those types of linkages. Program DYNACAM allows the design and dynamic analysis of cam-follower systems. Program ENGINE analyzes the slider-crank linkage as used in the internal combustion engine and provides a complete dynamic analysis of single and multicylinder engine inline, V, and W configurations, allowing the mechanical dynamic design of engines to be done. Program MATRIX is a general purpose linear equation system solver.

All these programs, except MATRIX, provide dynamic, graphical animation of the designed devices. The reader is strongly urged to make use of these programs in order to investigate the results of variation of parameters in these kinematic devices. The programs are designed to enhance and augment the text rather than be a substitute for it. The converse is also true. Many solutions to the book's examples and to the problem sets are provided on the CD-ROM as files to be read into these programs. Most of these solutions can be animated on the computer screen for a better demonstration of the concept than is possible on the printed page. The instructor and students are both encouraged to take advantage of the computer programs provided. Instructions for their use are in Appendix A.

The author's intention is that synthesis topics be introduced first to allow the students to work on some simple design tasks early in the term while still mastering the analysis topics. Though this is not the "traditional" approach to the teaching of this material, the author believes that it is a superior method to that of initial concentration on detailed analysis of mechanisms for which the student has no concept of origin or purpose.

Chapters 1 and 2 are introductory. Those instructors wishing to teach analysis before synthesis can leave Chapters 3 and 5 on linkage synthesis for later consumption. Chapters 4, 6, and 7 on position, velocity, and acceleration analysis are sequential and build upon each other. In fact, some of the problem sets are common among these three chapters so that students can use their position solutions to find velocities and then later use both to find the accelerations in the same linkages. Chapter 8 on cams is more extensive and complete than that of other kinematics texts and takes a design approach. Chapter 9 on gear trains is introductory. The dynamic force treatment in Part II uses matrix methods for the solution of the system simultaneous equations. Graphical force analysis is not emphasized. Chapter 10 presents an introduction to dynamic systems modeling. Chapter 11 deals with force analysis of linkages. Balancing of rotating machinery and linkages is covered in Chapter 12. Chapters 13 and 14 use the internal combustion engine as an example to pull together many dynamic concepts in a design context. Chapter 15 presents an introduction to dynamic systems modeling and uses the cam-follower system as the example. Chapters 3, 8, 11, 13, and 14 provide open ended project problems as well as structured problem sets. The assignment and execution of unstructured project problems can greatly enhance the student's understanding of the concepts as described by the proverb in the epigraph to this preface.

ACKNOWLEDGMENTS The sources of photographs and other nonoriginal art used in the text are acknowledged in the captions and opposite the title page, but the author would also like to express his thanks for the cooperation of all those individuals and companies who generously made these items available. The author would also like to thank those who reviewed various sections of the first edition of the text and who made many useful suggestions for improvement. Mr. John Titus of the University of Minnesota reviewed Chapter 5 on analytical synthesis and Mr. Dennis Klipp of Klipp Engineering, Waterville, Maine, reviewed Chapter 8 on cam design. Professor William J. Crochetiere and Mr. Homer Eckhardt of Tufts University, Medford, Mass., reviewed Chapter 15. Mr. Eckhardt and Professor Crochetiere of Tufts, and Professor Charles Warren of the University of Alabama taught from and reviewed Part I. Professor Holly K. Ault of Worcester Polytechnic Institute thoroughly reviewed the entire text while teaching from the pre-publication, class-test versions of the complete book. Professor Michael Keefe of the University of Delaware provided many helpful comments. Sincere thanks also go to the large number of undergraduate students and graduate teaching assistants who caught many typos and errors in the text and in the programs while using the pre-publication versions. Since the book's first printing, Profs. D. Cronin, K. Gupta, P. Jensen, and Mr. R. Jantz have written to point out errors or make suggestions which I have incorporated and for which I thank them. The author takes full responsibility for any errors that may remain and invites from all readers their criticisms, suggestions for improvement, and identification of errors in the text or programs, so that both can be improved in future versions. Contact _norton@wpi.edu_.

Robert L. Norton
Mattapoisett, Mass.
August, 1991

PART I

*Take to Kinematics. It will repay you. It is
more fecund than geometry;
it adds a fourth dimension to space.*

CHEBYSCHEV TO SYLVESTER, 1873

KINEMATICS OF
MECHANISMS

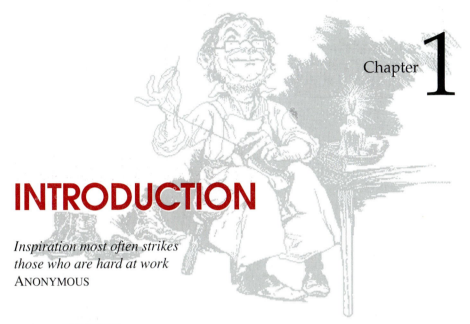

INTRODUCTION

*Inspiration most often strikes
those who are hard at work*
ANONYMOUS

1.0   PURPOSE

In this text we will explore the topics of **kinematics** and **dynamics of machinery** in respect to the **synthesis of mechanisms** in order to accomplish desired motions or tasks,
and also the **analysis of mechanisms** in order to determine their rigid-body dynamic
behavior. These topics are fundamental to the broader subject of **machine design**. On
the premise that we cannot analyze anything until it has been synthesized into existence,
we will first explore the topic of **synthesis of mechanisms**. Then we will investigate
techniques of **analysis of mechanisms**. All this will be directed toward developing your
ability to design viable mechanism solutions to real, unstructured engineering problems
by using a **design process**. We will begin with careful definitions of the terms used in
these topics.

1.1   KINEMATICS AND KINETICS

KINEMATICS   *The study of motion without regard to forces.*

KINETICS   *The study of forces on systems in motion.*

These two concepts are really *not* physically separable. We arbitrarily separate them for
instructional reasons in engineering education. It is also valid in engineering design
practice to first consider the desired kinematic motions and their consequences, and then
subsequently investigate the kinetic forces associated with those motions. The student
should realize that the division between **kinematics** and **kinetics** is quite arbitrary and
is done largely for convenience. One cannot design most dynamic mechanical systems
without taking both topics into thorough consideration. It is quite logical to consider
them in the order listed since, from Newton's second law, $\mathbf{F} = m\mathbf{a}$, one typically needs to

know the **accelerations** (**a**) in order to compute the dynamic **forces** (**F**) due to the motion of the system's **mass** (*m*). There are also many situations in which the applied forces are known and the resultant accelerations are to be found.

One principal aim of **kinematics** is to create (design) the desired motions of the subject mechanical parts and then mathematically compute the positions, velocities, and accelerations which those motions will create on the parts. Since, for most earthbound mechanical systems, the mass remains essentially constant with time, defining the accelerations as a function of time then also defines the dynamic forces as a function of time. **Stresses**, in turn, will be a function of both applied and inertial (*m*a) forces. Since engineering design is charged with creating systems which will not fail during their expected service life, the goal is to keep stresses within acceptable limits for the materials chosen and the environmental conditions encountered. This obviously requires that all system forces be defined and kept within desired limits. In machinery which moves (the only interesting kind), the largest forces encountered are often those due to the dynamics of the machine itself. These dynamic forces are proportional to acceleration, which brings us back to kinematics, the foundation of mechanical design. Very basic and early decisions in the design process involving kinematic principles can be crucial to the success of any mechanical design. A design which has poor kinematics will prove troublesome and perform badly.

1.2   MECHANISMS AND MACHINES

A **mechanism** is a device which transforms motion to some desirable pattern and typically develops very low forces and transmits little power. A **machine** typically contains mechanisms which are designed to provide significant forces and transmit significant power.[1] Some examples of common mechanisms are a pencil sharpener, a camera shutter, an analog clock, a folding chair, an adjustable desk lamp, and an umbrella. Some examples of machines which possess motions similar to the mechanisms listed above are a food blender, a bank vault door, an automobile transmission, a bulldozer, a robot, and an amusement park ride. There is no clear-cut dividing line between mechanisms and machines. They differ in degree rather than in kind. If the forces or energy levels within the device are significant, it is considered a machine; if not, it is considered a mechanism. A useful working **definition of a mechanism** is *A system of elements arranged to transmit* **motion** *in a predetermined fashion.* This can be converted to a definition of a **machine** by adding the words **and energy** after **motion**.

A mechanism

A machine

Mechanisms, if lightly loaded and run at slow speeds, can sometimes be treated strictly as kinematic devices; that is, they can be analyzed kinematically without regard to forces. Machines (and mechanisms running at higher speeds), on the other hand, must first be treated as mechanisms, a kinematic analysis of their velocities and accelerations must be done, and then they must be subsequently analyzed as dynamic systems in which their static and dynamic forces due to those accelerations are analyzed using the principles of kinetics. **Part I** of this text deals with **Kinematics of Mechanisms**, and **Part II** with **Dynamics of Machinery**. The techniques of mechanism synthesis presented in Part I are applicable to the design of both mechanisms and machines, since in each case some collection of moveable members must be created to provide and control the desired motions and geometry.

1.3   A BRIEF HISTORY OF KINEMATICS

Machines and mechanisms have been devised by people since the dawn of history. The ancient Egyptians devised primitive machines to accomplish the building of the pyramids and other monuments. Though the wheel and pulley (on an axle) were not known to the Old Kingdom Egyptians, they made use of the lever, the inclined plane (or wedge), and probably the log roller. The origin of the wheel and axle is not definitively known. Its first appearance seems to have been in Mesopotamia about 3000 to 4000 B.C.

A great deal of design effort was spent from early times on the problem of timekeeping as more sophisticated clockworks were devised. Much early machine design was directed toward military applications (catapults, wall scaling apparatus, etc.). The term **civil engineering** was later coined to differentiate civilian from military applications of technology. **Mechanical engineering** had its beginnings in machine design as the inventions of the industrial revolution required more complicated and sophisticated solutions to motion control problems. **James Watt** (1736-1819) probably deserves the title of first kinematician for his synthesis of a straight-line linkage (see Figure 3-29a on p. 121) to guide the very long stroke pistons in the then new steam engines. Since the planer was yet to be invented (in 1817), no means then existed to machine a long, straight guide to serve as a crosshead in the steam engine. Watt was certainly the first on record to recognize the value of the motions of the coupler link in the fourbar linkage. **Oliver Evans** (1755-1819) an early American inventor, also designed a straight-line linkage for a steam engine. **Euler** (1707-1783) was a contemporary of Watt, though they apparently never met. Euler presented an analytical treatment of mechanisms in his *Mechanica sive Motus Scienta Analytice Exposita* (1736-1742), which included the concept that planar motion is composed of two independent components, namely, translation of a point and rotation of the body about that point. Euler also suggested the separation of the problem of dynamic analysis into the "geometrical" and the "mechanical" in order to simplify the determination of the system's dynamics. Two of his contemporaries, **d'Alembert** and **Kant**, also proposed similar ideas. This is the origin of our division of the topic into kinematics and kinetics as described above.

In the early 1800s, L'Ecole Polytechnic in Paris, France, was the repository of engineering expertise. **Lagrange** and **Fourier** were among its faculty. One of its founders was **Gaspard Monge** (1746-1818), inventor of descriptive geometry (which incidentally was kept as a military secret by the French government for 30 years because of its value in planning fortifications). Monge created a course in elements of machines and set about the task of classifying all mechanisms and machines known to mankind! His colleague, **Hachette**, completed the work in 1806 and published it as what was probably the first mechanism text in 1811. **Andre Marie Ampere** (1775-1836), also a professor at L'Ecole Polytechnic, set about the formidable task of classifying "all human knowledge." In his *Essai sur la Philosophie des Sciences*, he was the first to use the term "**cinematique**," from the Greek word for motion,[*] to describe the study of motion without regard to forces, and suggested that "this science ought to include all that can be said with respect to motion in its different kinds, independently of the forces by which it is produced." His term was later anglicized to *kinematics* and germanized to *kinematik*.

Robert Willis (1800-1875) wrote the text *Principles of Mechanism* in 1841 while a professor of natural philosophy at the University of Cambridge, England. He attempted to systematize the task of mechanism synthesis. He counted five ways of obtaining rel-

* Ampere is quoted as writing "(The science of mechanisms) must therefore not define a machine, as has usually been done, as an instrument by the help of which the direction and intensity of a given *force* can be altered, but as an instrument by the help of which the direction and *velocity* of a given motion can be altered. To this science . . . I have given the name Kinematics from Κινμα—motion." in Maunder, L. (1979). "Theory and Practice." *Proc. 5th World Cong. on Theory of Mechanisms and Machines*, Montreal, p. 1.

ative motion between input and output links: rolling contact, sliding contact, linkages, wrapping connectors (belts, chains), and tackle (rope or chain hoists). **Franz Reuleaux** (1829-1905), published *Theoretische Kinematik* in 1875. Many of his ideas are still current and useful. **Alexander Kennedy** (1847-1928) translated Reuleaux into English in 1876. This text became the foundation of modern kinematics and is still in print! (See bibliography at end of chapter.) He provided us with the concept of a kinematic pair (joint), whose shape and interaction define the type of motion transmitted between elements in the mechanism. Reuleaux defined six basic mechanical components: the link, the wheel, the cam, the screw, the ratchet, and the belt. He also defined "higher" and "lower" pairs, higher having line or point contact (as in a roller or ball bearing) and lower having surface contact (as in pin joints). Reuleaux is generally considered the father of modern kinematics and is responsible for the symbolic notation of skeletal, generic linkages used in all modern kinematics texts.

In this century, prior to World War II, most theoretical work in kinematics was done in Europe, especially in Germany. Few research results were available in English. In the United States, kinematics was largely ignored until the 1940s, when **A. E. R. De-Jonge** wrote "What Is Wrong with 'Kinematics' and 'Mechanisms'?,"[2] which called upon the U.S. mechanical engineering education establishment to pay attention to the European accomplishments in this field. Since then, much new work has been done, especially in kinematic synthesis, by American and European engineers and researchers such as **J. Denavit**, **A. Erdman**, **F. Freudenstein**, **A. S. Hall**, **R. Hartenberg**, **R. Kaufman**, **B. Roth**, **G. Sandor**, and **A. Soni**, (all of the U.S.) and **K. Hain** (of Germany). Since the fall of the "iron curtain" much original work done by Soviet Russian kinematicians has become available in the United States, such as that by **Artobolevsky**.[3] Many U.S. researchers have applied the computer to solve previously intractable problems, both of analysis and synthesis, making practical use of many of the theories of their predecessors.[4] This text will make much use of the availability of computers to allow more efficient analysis and synthesis of solutions to machine design problems. Several computer programs are included with this book for your use.

1.4   APPLICATIONS OF KINEMATICS

One of the first tasks in solving any machine design problem is to determine the kinematic configuration(s) needed to provide the desired motions. Force and stress analyses typically cannot be done until the kinematic issues have been resolved. This text addresses the design of kinematic devices such as linkages, cams, and gears. Each of these terms will be fully defined in succeeding chapters, but it may be useful to show some examples of kinematic applications in this introductory chapter. You probably have used many of these systems without giving any thought to their kinematics.

Virtually any machine or device that moves contains one or more kinematic elements such as linkages, cams, gears, belts, chains. Your bicycle is a simple example of a kinematic system that contains a chain drive to provide torque multiplication and simple cable-operated linkages for braking. An automobile contains many more examples of kinematic devices. Its steering system, wheel suspensions, and piston-engine all contain linkages; the engine's valves are opened by cams; and the transmission is full of gears. Even the windshield wipers are linkage-driven. Figure 1-1a shows a spatial linkage used to control the rear wheel movement of a modern automobile over bumps.

(*a*) Spatial linkage rear suspension   (*b*) Utility tractor with backhoe   (*c*) Linkage-driven exercise mechanism
       Courtesy of Daimler Benz Co.   *Courtesy of John Deere Co.*   *Courtesy of ICON Health & Fitness, Inc.*

FIGURE 1-1

Examples of kinematic devices in general use

Construction equipment such as tractors, cranes, and backhoes all use linkages extensively in their design. Figure 1-1b shows a small backhoe that is a linkage driven by hydraulic cylinders. Another application using linkages is that of exercise equipment as shown in Figure 1-1c. The examples in Figure 1-1 are all of consumer goods which you may encounter in your daily travels. Many other kinematic examples occur in the realm of producer goods—machines used to make the many consumer products that we use. You are less likely to encounter these outside of a factory environment. Once you become familiar with the terms and principles of kinematics, you will no longer be able to look at any machine or product without seeing its kinematic aspects.

1.5   THE DESIGN PROCESS

Design, Invention, Creativity

These are all familiar terms but may mean different things to different people. These terms can encompass a wide range of activities from styling the newest look in clothing, to creating impressive architecture, to engineering a machine for the manufacture of facial tissues. **Engineering design**, which we are concerned with here, embodies all three of these activities as well as many others. The word **design** is derived from the Latin **designare**, which means *"to designate, or mark out."* Webster's gives several definitions, the most applicable being *"to outline, plot, or plan, as action or work. . . to conceive, invent – contrive."* **Engineering design** has been defined as ". . . *the process of applying the various techniques and scientific principles for the purpose of defining a device, a process or a system in sufficient detail to permit its realization . . . Design may be simple or enormously complex, easy or difficult, mathematical or nonmathematical; it may involve a trivial problem or one of great importance."* **Design** is a universal constituent of engineering practice. But the complexity of engineering subjects usually re-

TABLE 1-1
A Design Process

1   Identification of
    Need

2   Background
    Research

3   Goal Statement

4   Performance
    Specifications

5   Ideation and
    Invention

6   Analysis

7   Selection

8   Detailed Design

9   Prototyping and
    Testing

10 Production

Blank paper syndrome

quires that the student be served with a collection of **structured**, **set-piece problems** designed to elucidate a particular concept or concepts related to the particular topic. These textbook problems typically take the form of *"given A, B, C, and D, find E."* Unfortunately, real-life engineering problems are almost never so structured. Real design problems more often take the form of *"What we need is a framus to stuff this widget into that hole within the time allocated to the transfer of this other gizmo."* The new engineering graduate will search in vain among his or her textbooks for much guidance to solve such a problem. This **unstructured problem** statement usually leads to what is commonly called "**blank paper syndrome**." Engineers often find themselves staring at a blank sheet of paper pondering how to begin solving such an ill-defined problem.

Much of engineering education deals with topics of **analysis**, which means *to decompose, to take apart, to resolve into its constituent parts.* This is quite necessary. The engineer must know how to analyze systems of various types, mechanical, electrical, thermal, or fluid. Analysis requires a thorough understanding of both the appropriate mathematical techniques and the fundamental physics of the system's function. But, before any system can be analyzed, it must exist, and a blank sheet of paper provides little substance for analysis. Thus the first step in any engineering design exercise is that of **synthesis**, which means *putting together.*

The design engineer, in practice, regardless of discipline, continuously faces the challenge of *structuring the unstructured problem.* Inevitably, the problem as posed to the engineer is ill-defined and incomplete. Before any attempt can be made to *analyze the situation* he or she must first carefully define the problem, using an engineering approach, to ensure that any proposed solution will solve the right problem. Many examples exist of excellent engineering solutions which were ultimately rejected because they solved the wrong problem, i.e., a different one than the client really had.

Much research has been devoted to the definition of various "design processes" intended to provide means to structure the unstructured problem and lead to a viable solution. Some of these processes present dozens of steps, others only a few. The one presented in Table 1-1 contains 10 steps and has, in the author's experience, proven successful in over 30 years of practice in engineering design.

ITERATION   Before discussing each of these steps in detail it is necessary to point out that this is not a process in which one proceeds from step one through ten in a linear fashion. Rather it is, by its nature, an iterative process in which progress is made haltingly, two steps forward and one step back. It is inherently *circular.* To **iterate** means *to repeat, to return to a previous state.* If, for example, your apparently great idea, upon analysis, turns out to violate the second law of thermodynamics, you can return to the ideation step and get a better idea! Or, if necessary, you can return to an earlier step in the process, perhaps the background research, and learn more about the problem. With the understanding that the actual execution of the process involves iteration, for simplicity, we will now discuss each step in the order listed in Table 1-1.

Identification of Need

This first step is often done for you by someone, boss or client, saying "What we need is . . ." Typically this statement will be brief and lacking in detail. It will fall far short of providing you with a structured problem statement. For example, the problem statement might be "We need a better lawn mower."

Background Research

This is the most important phase in the process, and is unfortunately often the most neglected. The term research, used in this context, should *not* conjure up visions of white-coated scientists mixing concoctions in test tubes. Rather this is research of a more mundane sort, gathering background information on the relevant physics, chemistry, or other aspects of the problem. Also it is desirable to find out if this, or a similar problem, has been solved before. There is no point in reinventing the wheel. If you are lucky enough to find a ready-made solution on the market, it will no doubt be more economical to purchase it than to build your own. Most likely this will not be the case, but you may learn a great deal about the problem to be solved by investigating the existing "art" associated with similar technologies and products. The patent literature and technical publications in the subject area are obvious sources of information and are accessible via the worldwide web. Clearly, if you find that the solution exists and is covered by a patent still in force, you have only a few ethical choices: buy the patentee's existing solution, design something which does not conflict with the patent, or drop the project. It is very important that sufficient energy and time be expended on this research and preparation phase of the process in order to avoid the embarrassment of concocting a great solution to the wrong problem. Most inexperienced (and some experienced) engineers give too little attention to this phase and jump too quickly into the ideation and invention stage of the process. *This must be avoided!* You must discipline yourself to *not* try to solve the problem before thoroughly preparing yourself to do so.

Identifying the need

Goal Statement

Once the background of the problem area as originally stated is fully understood, you will be ready to recast that problem into a more coherent goal statement. This new problem statement should have three characteristics. It should be concise, be general, and be uncolored by any terms which predict a solution. It should be couched in terms of **functional visualization**, *meaning to visualize its function*, rather than any particular embodiment. For example, if the original statement of need was *"Design a Better Lawn Mower,"* after research into the myriad of ways to cut grass that have been devised over the ages, the wise designer might restate the goal as **"Design a Means to Shorten Grass."** The original problem statement has a built-in trap in the form of the *colored* words "lawn mower." For most people, this phrase will conjure up a vision of something with whirring blades and a noisy engine. For the **ideation** phase to be most successful, it is necessary to avoid such images and to state the problem generally, clearly, and concisely. As an exercise, list 10 ways to shorten grass. Most of them would not occur to you had you been asked for 10 better lawn mower designs. You should use **functional visualization** to avoid unnecessarily limiting your creativity!

Reinventing the wheel

Performance Specifications *

When the background is understood, and the goal clearly stated, you are ready to formulate a set of performance specifications. These should **not** be design specifications. The difference is that **performance specifications** define **what** *the system must do*, while **design specifications** define **how** *it must do it*. At this stage of the design process it is unwise to attempt to specify *how* the goal is to be accomplished. That is left for the **ideation** phase. The purpose of the performance specifications is to carefully define and

Grass shorteners

* Orson Welles, famous author and filmmaker, once said, *"The enemy of art is the absence of limitations."* We can paraphrase that as *The enemy of design is the absence of specifications.*

**Performance
Specifications**

Lorem
Ipsum
Dolor amet
Euismod
Volutpat
Laoreet
Adipiscing

TABLE 1-2
**Performance Specifi-
cations**

1 Device to have self-
contained power
supply.

2 Device to be
corrosion resistant.

3 Device to cost less
than $100.00.

4 Device to emit < 80
dB sound intensity
at 50 feet.

5 Device to shorten
1/4 acre of grass
per hour.

6 etc . . . etc.

TABLE 1-3
The Creative Process

5a Idea Generation

5b Frustration

5c Incubation

5d Eureka!

---

constrain the problem so that it both *can be solved* and *can be shown to have been solved* after the fact. A sample set of performance specifications for our "grass shortener" is shown in Table 1-2.

Note that these specifications constrain the design without overly restricting the engineer's design freedom. It would be inappropriate to require a gasoline engine for specification 1, since other possibilities exist which will provide the desired mobility. Likewise, to demand stainless steel for all components in specification 2 would be unwise, since corrosion resistance can be obtained by other, less-expensive means. In short, the performance specifications serve to define the problem in as complete and as general a manner as possible, and they serve as a contractual definition of what is to be accomplished. The finished design can be tested for compliance with the specifications.

Ideation and Invention

This step is full of both fun and frustration. This phase is potentially the most satisfying to most designers, but it is also the most difficult. A great deal of research has been done to explore the phenomenon of "**creativity**." It is, most agree, a common human trait. It is certainly exhibited to a very high degree by all young children. The rate and degree of development that occurs in the human from birth through the first few years of life certainly requires some innate creativity. Some have claimed that our methods of Western education tend to stifle children's natural creativity by encouraging conformity and restricting individuality. From "coloring within the lines" in kindergarten to imitating the textbook's writing patterns in later grades, individuality is suppressed in favor of a socializing conformity. This is perhaps necessary to avoid anarchy but probably does have the effect of reducing the individual's ability to think creatively. Some claim that creativity can be taught, some that it is only inherited. No hard evidence exists for either theory. It is probably true that one's lost or suppressed creativity can be rekindled. Other studies suggest that most everyone underutilizes his or her potential creative abilities. You can enhance your creativity through various techniques.

CREATIVE PROCESS   Many techniques have been developed to enhance or inspire creative problem solving. In fact, just as design processes have been defined, so has the *creative process* shown in Table 1-3. This creative process can be thought of as a subset of the design process and to exist within it. The ideation and invention step can thus be broken down into these four substeps.

IDEA GENERATION   is the most difficult of these steps. Even very creative people have difficulty in inventing "on demand." Many techniques have been suggested to improve the yield of ideas. The most important technique is that of *deferred judgment*, which means that your criticality should be temporarily suspended. Do not try to judge the quality of your ideas at this stage. That will be taken care of later, in the **analysis** phase. The goal here is to obtain as large a *quantity* of potential designs as possible. Even superficially ridiculous suggestions should be welcomed, as they may trigger new insights and suggest other more realistic and practical solutions.

BRAINSTORMING   is a technique for which some claim great success in generating creative solutions. This technique requires a group, preferably 6 to 15 people, and attempts to circumvent the largest barrier to creativity, which is *fear of ridicule*. Most people, when in a group, will not suggest their real thoughts on a subject, for fear of be-

ing laughed at. Brainstorming's rules require that no one is allowed to make fun of or criticize anyone's suggestions, no matter how ridiculous. One participant acts as "scribe" and is duty bound to record all suggestions, no matter how apparently silly. When done properly, this technique can be fun and can sometimes result in a "feeding frenzy" of ideas which build upon each other. Large quantities of ideas can be generated in a short time. Judgment on their quality is deferred to a later time.

Brainstorming

When working alone, other techniques are necessary. **Analogies** and **inversion** are often useful. Attempt to draw analogies between the problem at hand and other physical contexts. If it is a mechanical problem, convert it by analogy to a fluid or electrical one. Inversion turns the problem inside out. For example, consider what you want moved to be stationary and vice versa. Insights often follow. Another useful aid to creativity is the use of **synonyms**. Define the action verb in the problem statement, and then list as many synonyms for that verb as possible. For example:

Problem statement: Move this object from point A to point B.

The action verb is "move." Some synonyms are push, pull, slip, slide, shove, throw, eject, jump, spill.

Frustration

By whatever means, the aim in this **ideation** step is to generate a large number of ideas without particular regard to quality. But, at some point, your "mental well" will go dry. You will have then reached the step in the creative process called **frustration**. It is time to leave the problem and do something else for a time. While your conscious mind is occupied with other concerns, your subconscious mind will still be hard at work on the problem. This is the step called **incubation.** Suddenly, at a quite unexpected time and place, an idea will pop into your consciousness, and it will seem to be the obvious and "right" solution to the problem . . . **Eureka!** Most likely, later analysis will discover some flaw in this solution. If so, back up and **iterate!** More ideation, perhaps more research, and possibly even a redefinition of the problem may be necessary.

In "Unlocking Human Creativity"[5] Wallen describes three requirements for creative insight:

- *Fascination with a problem.*

- *Saturation with the facts, technical ideas, data, and the background of the problem.*

- *A period of reorganization.*

The first of these provides the motivation to solve the problem. The second is the background research step described above. The period of reorganization refers to the frustration phase when your subconscious works on the problem. Wallen[5] reports that testimony from creative people tells us that in this period of reorganization they have no conscious concern with the particular problem and that the moment of insight frequently appears in the midst of relaxation or sleep. So to enhance your creativity, saturate yourself in the problem and related background material. Then relax and let your subconscious do the hard work!

Analysis

Once you are at this stage, you have structured the problem, at least temporarily, and can now apply more sophisticated analysis techniques to examine the performance of the

Eureka!

design in the **analysis phase** of the design process. (Some of these analysis methods will be discussed in detail in the following chapters.) Further iteration will be required as problems are discovered from the analysis. Repetition of as many earlier steps in the design process as necessary must be done to ensure the success of the design.

Selection

When the technical analysis indicates that you have some potentially viable designs, the best one available must be **selected** for **detailed design, prototyping,** and **testing.** The selection process usually involves a comparative analysis of the available design solutions. A **decision matrix** sometimes helps to identify the best solution by forcing you to consider a variety of factors in a systematic way. A decision matrix for our better grass shortener is shown in Figure 1-2. Each design occupies a row in the matrix. The columns are assigned categories in which the designs are to be judged, such as cost, ease of use, efficiency, performance, reliability, and any others you deem appropriate to the particular problem. Each category is then assigned a **weighting factor**, which measures its relative importance. For example, reliability may be a more important criterion to the user than cost, or vice versa. You as the design engineer have to exercise your judgment as to the selection and weighting of these categories. The body of the matrix is then filled with numbers which rank each design on a convenient scale, such as 1 to 10, in each of the categories. Note that this is ultimately a *subjective ranking* on your part. You must examine the designs and decide on a score for each. The scores are then multiplied by the weighting factors (which are usually chosen so as to sum to a convenient number such as 1) and the products summed for each design. The weighted scores then give a ranking of designs. Be cautious in applying these results. Remember the source and subjectivity of your scores and the weighting factors! There is a temptation to put more faith in these results than is justified. After all, they look impressive! They can even be taken out to several decimal places! (But they shouldn't be.) The real value of a decision

	Cost	Safety	Performance	Reliability	RANK
Weighting Factor	.35	.30	.15	.20	1.0
Design 1	3 / 1.05	6 / 1.80	4 / .60	9 / 1.80	5.3
Design 2	4 / 1.40	2 / .60	7 / 1.05	2 / .40	3.5
Design 3	1 / .35	9 / 2.70	4 / .60	5 / 1.00	4.7
Design 4	9 / 3.15	1 / .30	6 / .90	7 / 1.40	5.8
Design 5	7 / 2.45	4 / 1.20	2 / .30	6 / 1.20	5.2

FIGURE 1-2

A decision matrix

matrix is that it breaks the problem into more tractable pieces and forces you to think about the relative value of each design in many categories. You can then make a more informed decision as to the "best" design.

Detailed Design

This step usually includes the creation of a complete set of assembly and detail drawings or **computer-aided design** (CAD) part files, for *each and every part* used in the design. Each detail drawing must specify all the dimensions and the material specifications necessary to make that part. From these drawings (or CAD files) a prototype test model (or models) must be constructed for physical testing. Most likely the tests will discover more flaws, requiring further **iteration**.

Prototyping and Testing

MODELS   Ultimately, one cannot be sure of the correctness or viability of any design until it is built and tested. This usually involves the construction of a prototype physical model. A mathematical model, while very useful, can never be as complete and accurate a representation of the actual physical system as a physical model, due to the need to make simplifying assumptions. Prototypes are often very expensive but may be the most economical way to prove a design, short of building the actual, full-scale device. Prototypes can take many forms, from working scale models to full-size, but simplified, representations of the concept. Scale models introduce their own complications in regard to proper scaling of the physical parameters. For example, volume of material varies as the cube of linear dimensions, but surface area varies as the square. Heat transfer to the environment may be proportional to surface area, while heat generation may be proportional to volume. So linear scaling of a system, either up or down, may lead to behavior different from that of the full-scale system. One must exercise caution in scaling physical models. You will find as you begin to design linkage mechanisms that a **simple cardboard model** of your chosen link lengths, coupled together with thumbtacks for pivots, will tell you a great deal about the quality and character of the mechanism's motions. You should get into the habit of making such simple articulated models for all your linkage designs.

TESTING   of the model or prototype may range from simply actuating it and observing its function to attaching extensive instrumentation to accurately measure displacements, velocities, accelerations, forces, temperatures, and other parameters. Tests may need to be done under controlled environmental conditions such as high or low temperature or humidity. The microcomputer has made it possible to measure many phenomena more accurately and inexpensively than could be done before.

Production

Finally, with enough time, money, and perseverance, the design will be ready for production. This might consist of the manufacture of a single final version of the design, but more likely will mean making thousands or even millions of your widget. The danger, expense, and embarrassment of finding flaws in your design after making large quantities of defective devices should inspire you to use the greatest care in the earlier steps of the design process to ensure that it is properly engineered.

The **design process** is widely used in engineering. Engineering is usually defined in terms of what an engineer does, but engineering can also be defined in terms of *how* the engineer does what he or she does. **Engineering** is *as much a method, an approach, a process, a state of mind for problem solving, as it is an activity*. The engineering approach is that of thoroughness, attention to detail, and consideration of all the possibilities. While it may seem a contradiction in terms to emphasize "attention to detail" while extolling the virtues of open-minded, freewheeling, creative thinking, it is not. The two activities are not only compatible, they are symbiotic. It ultimately does no good to have creative, original ideas if you do not, or cannot, carry out the execution of those ideas and "reduce them to practice." To do this you must discipline yourself to suffer the nitty-gritty, nettlesome, tiresome details which are so necessary to the completion of any one phase of the creative design process. For example, to do a creditable job in the design of anything, you must *completely* define the problem. If you leave out some detail of the problem definition, you will end up solving the wrong problem. Likewise, you must *thoroughly* research the background information relevant to the problem. You must *exhaustively* pursue conceptual potential solutions to your problem. You must then *extensively* analyze these concepts for validity. And, finally, you must *detail* your chosen design down to the last nut and bolt to be confident it will work. If you wish to be a good designer and engineer, you must discipline yourself to do things thoroughly and in a logical, orderly manner, even while thinking great creative thoughts and iterating to a solution. Both attributes, creativity and attention to detail, are necessary for success in engineering design.

1.6   OTHER APPROACHES TO DESIGN

In recent years, an increased effort has been directed toward a better understanding of design methodology and the design process. Design methodology is the study of the process of designing. One goal of this research is to define the design process in sufficient detail to allow it to be encoded in a form amenable to execution in a computer, using "artificial intelligence" (AI).

Dixon[6] defines a design as a *state of information* which may be in any of several forms:

> . . . words, graphics, electronic data, and/or others. It may be partial or complete. It ranges from a small amount of highly abstract information early in the design process to a very large amount of detailed information later in the process sufficient to perform manufacturing. It may include, but is not limited to, information about size and shape, function, materials, marketing, simulated performance, manufacturing processes, tolerances, and more. Indeed, any and all information relevant to the physical or economic life of a designed object is part of its design.

He goes on to describe several generalized states of information such as the *requirements* state which is analogous to our **performance specifications**. Information about the physical concept is referred to as the *conceptual* state of information and is analogous to our **ideation** phase. His *feature configuration* and *parametric* states of information are similar in concept to our **detailed design** phase. Dixon then defines a design process as:

> The series of activities by which the information about the designed object is changed from one information state to another.

Axiomatic Design

N. P. Suh[7] suggests an *axiomatic approach* to design in which there are four domains: **customer** domain, **functional** domain, **physical** domain, and the **process** domain. These represent a range from "what" to "how," i.e., from a state of defining what the customer wants through determining the functions required and the needed physical embodiment, to how a process will achieve the desired end. He defines two axioms that need to be satisfied to accomplish this:

1 Maintain the independence of the functional requirements.

2 Minimize the information content.

The first of these refers to the need to create a complete and nondependent set of performance specifications. The second indicates that the best design solution will have the lowest information content (i.e., the least complexity). Others have earlier referred to this second idea as *KISS,* which stands, somewhat crudely, for "*keep it simple, stupid*."

The implementation of both Dixon's and Suh's approaches to the design process is somewhat complicated. The interested reader is referred to the literature cited in the bibliography to this chapter for more complete information.

1.7 MULTIPLE SOLUTIONS

Note that by the nature of the design process, there is **not** any **one** correct answer or solution to any design problem. Unlike the structured "engineering textbook" problems, which most students are used to, there is no right answer "in the back of the book" for any real design problem.* There are as many potential solutions as there are designers willing to attempt them. Some solutions will be better than others, but many will work. Some will not! There is no "one right answer" in design engineering, which is what makes it interesting. The only way to determine the relative merits of various potential design solutions is by thorough analysis, which usually will include physical testing of constructed prototypes. Because this is a very expensive process, it is desirable to do as much analysis on paper, or in the computer, as possible before actually building the device. Where feasible, mathematical models of the design, or parts of the design, should be created. These may take many forms, depending on the type of physical system involved. In the design of mechanisms and machines it is usually possible to write the equations for the rigid-body dynamics of the system, and solve them in "closed form" with (or without) a computer. Accounting for the elastic deformations of the members of the mechanism or machine usually requires more complicated approaches using **finite difference** techniques or the **finite element method** (FEM).

1.8 HUMAN FACTORS ENGINEERING

With few exceptions, all machines are designed to be used by humans. Even robots must be programmed by a human. **Human factors engineering** is the study of the human-machine interaction and is defined as *an applied science that coordinates the design of devices, systems, and physical working conditions with the capacities and requirements of the worker.* The machine designer must be aware of this subject and design devices to "fit the man" rather than expect the man to adapt to fit the machine. The term **ergonom-**

* A student once commented that "*Life is an odd-numbered problem.*" This (slow) author had to ask for an explanation, which was: "*The answer is not in the back of the book.*"

1

ics is synonymous with *human factors engineering*. We often see reference to the good or bad ergonomics of an automobile interior or a household appliance. A machine designed with poor ergonomics will be uncomfortable and tiring to use and may even be dangerous. (Have you programmed your VCR lately, or set its clock?)

There is a wealth of human factors data available in the literature. Some references are noted in the bibliography. The type of information which might be needed for a machine design problem ranges from dimensions of the human body and their distribution among the population by age and gender, to the ability of the human body to withstand accelerations in various directions, to typical strengths and force generating ability in various positions. Obviously, if you are designing a device that will be controlled by a human (a grass shortener, perhaps), you need to know how much force the user can exert with hands held in various positions, what the user's reach is, and how much noise the ears can stand without damage. If your device will carry the user on it, you need data on the limits of acceleration which the body can tolerate. Data on all these topics exist. Much of it was developed by the government which regularly tests the ability of military personnel to withstand extreme environmental conditions. Part of the background research of any machine design problem should include some investigation of human factors.

1.9   THE ENGINEERING REPORT

Communication of your ideas and results is a very important aspect of engineering. Many engineering students picture themselves in professional practice spending most of their time doing calculations of a nature similar to those they have done as students. Fortunately, this is seldom the case, as it would be very boring. Actually, engineers spend the largest percentage of their time communicating with others, either orally or in writing. Engineers write proposals and technical reports, give presentations, and interact with support personnel and managers. When your design is done, it is usually necessary to present the results to your client, peers, or employer. The usual form of presentation is a formal engineering report. Thus, it is very important for the engineering student to develop his or her communication skills. *You may be the cleverest person in the world, but no one will know that if you cannot communicate your ideas clearly and concisely.* In fact, if you cannot explain what you have done, you probably don't understand it yourself. To give you some experience in this important skill, the design project assignments in later chapters are intended to be written up in formal engineering reports. Information on the writing of engineering reports can be found in the suggested readings in the bibliography at the end of this chapter.

1.10   UNITS

There are several systems of units used in engineering. The most common in the United States are the **U.S. foot-pound-second (fps) system**, the **U.S. inch-pound-second (ips) system**, and the **System International (SI)**. All systems are created from the choice of three of the quantities in the general expression of Newton's second law

$$F = \frac{ml}{t^2}$$

(1.1a)

where F is force, m is mass, l is length, and t is time. The units for any three of these variables can be chosen and the other is then derived in terms of the chosen units. The three chosen units are called *base units*, and the remaining one is then a *derived unit*.

Most of the confusion that surrounds the conversion of computations between either one of the U.S. systems and the SI system is due to the fact that the SI system uses a different set of base units than the U.S. systems. Both U.S. systems choose ***force***, *length*, and *time* as the base units. Mass is then a derived unit in the U.S. systems, and they are referred to as *gravitational systems* because the value of mass is dependent on the local gravitational constant. The SI system chooses ***mass**, length,* and *time* as the base units and force is the derived unit. SI is then referred to as an *absolute system* since the mass is a base unit whose value is not dependent on local gravity.

The **U.S. foot-pound-second (fps)** system requires that all lengths be measured in feet (ft), forces in pounds (lb), and time in seconds (sec). Mass is then derived from Newton's law as

$$m = \frac{Ft^2}{l} \qquad (1.1b)$$

and the units are:

Pounds seconds squared per **foot** (lb-sec^2/ft) = **slugs**

The **U.S. inch-pound-second (ips)** system requires that all lengths be measured in inches (in), forces in pounds (lb), and time in seconds (sec). Mass is still derived from Newton's law, equation 1.1b, but the units are now:

Pounds seconds squared per **inch** (lb-sec^2/in) = **blobs**

This mass unit is not slugs! It is worth twelve slugs or one blob.[*]

Weight is defined as the force exerted on an object by gravity. Probably the most common units error that students make is to mix up these two unit systems (**fps** and **ips**) when converting weight units (which are pounds force) to mass units. Note that the gravitational acceleration constant (g) on earth at sea level is approximately 32.2 **feet** per second squared which is equivalent to 386 **inches** per second squared. The relationship between mass and weight is:

Mass = weight / gravitational acceleration

$$m = \frac{W}{g} \qquad (1.2)$$

It should be obvious that, if you measure all your lengths in **inches** and then use $g = 32.2$ **feet**/sec^2 to compute mass, you will have an error of a *factor of 12* in your results. This is a serious error, large enough to crash the airplane you designed. Even worse off is the student who neglects to convert weight to mass *at all* in his calculations. He will have an error of either 32.2 or 386 in his results. This is enough to sink the ship!

To even further add to the student's confusion about units is the common use of the unit of **pounds mass** (lb$_m$). This unit is often used in fluid dynamics and thermodynamics and comes about through the use of a slightly different form of Newton's equation:

[*] It is unfortunate that the mass unit in the **ips** system has never officially been given a name such as the term *slug* used for mass in the **fps** system. The author boldly suggests (with tongue only slightly in cheek) that this unit of mass in the **ips** system be called a *blob* (bl) to distinguish it more clearly from the *slug* (sl), and to help the student avoid some of the common units errors listed above.

Twelve slugs = one blob.

Blob does not sound any sillier than slug, is easy to remember, implies mass, and has a convenient abbreviation (bl) which is an anagram for the abbreviation for pound (lb). Besides, if you have ever seen a garden slug, you know it looks just like a "*little blob.*"

$$F = \frac{ma}{g_c} \tag{1.3}$$

where m = mass in lb_m, a = acceleration and g_c = the gravitational constant.

The value of the **mass** of an object measured in **pounds mass** (lb_m) is *numerically equal* to its **weight** in **pounds force** (lb_f). However the student *must remember to divide* the value of m in lb_m by g_c when substituting into this form of Newton's equation. Thus the lb_m will be divided either by 32.2 or by 386 when calculating the dynamic force. The result will be the same as when the mass is expressed in either slugs or blobs in the $F = ma$ form of the equation. Remember that in round numbers at sea level on earth:

$$1\ \text{lb}_m = 1\ \text{lb}_f \qquad\qquad 1\ \text{slug} = 32.2\ \text{lb}_f \qquad\qquad 1\ \text{blob} = 386\ \text{lb}_f$$

The **SI** system requires that lengths be measured in meters (m), mass in kilograms (kg), and time in seconds (sec). This is sometimes also referred to as the **mks** system. Force is derived from Newton's law, equation 1.1b and the units are:

$$\text{kilogram-meters per second}^2\ (\text{kg-m/sec}^2) = \text{newtons}$$

Thus in the SI system there are distinct names for mass and force which helps alleviate confusion. When converting between SI and U.S. systems, be alert to the fact that mass converts from kilograms (kg) to either slugs (sl) or blobs (bl), and force converts from newtons (N) to pounds (lb). The gravitational constant (g) in the SI system is approximately 9.81 m/sec^2.

The principal system of units used in this textbook will be the U.S. **ips** system. Most machine design in the United States is still done in this system. Table 1-4 shows some of the variables used in this text and their units. The inside front cover contains a table of conversion factors between the U.S. and SI systems.

The student is cautioned to always check the units in any equation written for a problem solution, whether in school or in professional practice after graduation. If properly written, an equation should cancel all units across the equal sign. If it does not, then you can be *absolutely sure it is **incorrect.*** Unfortunately, a unit balance in an equation does not guarantee that it is correct, as many other errors are possible. Always double-check your results. You might save a life.

1.11   WHAT'S TO COME

In this text we will explore the topic of **machine design** in respect to the **synthesis of mechanisms** in order to accomplish desired motions or tasks, and also the analysis of these mechanisms in order to determine their rigid-body dynamic behavior. On the premise that we cannot analyze anything until it has been synthesized into existence, we will first explore the topic of synthesis of mechanisms. Then we will investigate the analysis of those and other mechanisms for their kinematic behavior. Finally, in Part II we will deal with the **dynamic analysis** of the forces and torques generated by these moving machines. These topics cover the essence of the early stages of a design project. Once the kinematics and kinetics of a design have been determined, most of the conceptual design will have been accomplished. What then remains is **detailed design**—sizing the parts against failure. The topic of *detailed design* is discussed in other texts such as reference [8].

Table 1-4   Variables and Units
Base Units in Boldface – Abbreviations in ( )

Variable	Symbol	ips unit	fps unit	SI unit
Force	F	**pounds (lb)**	**pounds (lb)**	newtons (N)
Length	l	**inches (in)**	**feet (ft)**	**meters (m)**
Time	t	**seconds (sec)**	**seconds (sec)**	**seconds (sec)**
Mass	m	lb–sec^2 / in (bl)	lb–sec^2 / ft (sl)	**kilograms (kg)**
Weight	W	pounds (lb)	pounds (lb)	newtons (N)
Velocity	v	in / sec	ft / sec	m / sec
Acceleration	a	in / sec^2	ft / sec^2	m / sec^2
Jerk	j	in / sec^3	ft / sec^3	m / sec^3
Angle	θ	degrees (deg)	degrees (deg)	degrees (deg)
Angle	θ	radians (rad)	radians (rad)	radians (rad)
Angular velocity	ω	rad / sec	rad / sec	rad / sec
Angular acceleration	α	rad / sec^2	rad / sec^2	rad / sec^2
Angular jerk	φ	rad / sec^3	rad / sec^3	rad / sec^3
Torque	T	lb–in	lb–ft	N–m
Mass moment of inertia	I	lb–in–sec^2	lb–ft–sec^2	N–m–sec^2
Energy	E	in–lb	ft–lb	joules
Power	P	in–lb / sec	ft–lb / sec	watts
Volume	V	in^3	ft^3	m^3
Weight density	γ	lb / in^3	lb / ft^3	N / m^3
Mass density	ρ	bl / in^3	sl / ft^3	kg / m^3

1.12   REFERENCES

1   **Rosenauer, N., and A. H. Willis**. (1967). *Kinematics of Mechanisms*. Dover Publications: New York, p. 275 ff.

2   **de Jonge, A. E. R.** (1942). "What Is Wrong with 'Kinematics' and 'Mechanisms'?" *Mechanical Engineering*, **64**(April), pp. 273-278.

3   **Artobolevsky, I. I.** (1975). *Mechanisms in Modern Engineering Design*. N. Weinstein, translator. Vols. 1 - 5. MIR Publishers: Moscow.

4   **Erdman, A. E.**, ed. (1993). *Modern Kinematics: Developments in the Last Forty Years*. Wiley Series in Design Engineering, John Wiley & Sons: New York.

5   **Wallen, R. W.** (1957). "Unlocking Human Creativity." *Proc. of Fourth Conference on Mechanisms*, Purdue University, pp. 2-8.

6   **Dixon, J. R.** (1995). "Knowledge Based Systems for Design." *Journal of Mechanical Design*, **117b**(2), p. 11.

7   **Suh, N. P.** (1995). "Axiomatic Design of Mechanical Systems." *Journal of Mechanical Design*, **117b**(2), p. 2.

8   **Norton, R. L.** (1996). *Machine Design: An Integrated Approach*. Prentice-Hall: Upper Saddle River, NJ.

1.13 BIBLIOGRAPHY

*For additional information on the **history of kinematics**, the following are recommended:*

Artobolevsky, I. I. (1976). "Past Present and Future of the Theory of Machines and Mechanisms." *Mechanism and Machine Theory*, **11**, pp. 353-361.

Brown, H. T. (1869). *Five Hundred and Seven Mechanical Movements*. Brown, Coombs & Co.: New York, republished by USM Corporation, Beverly, MA., 1970.

de Jonge, A. E. R. (1942). "What Is Wrong with 'Kinematics' and 'Mechanisms'?" *Mechanical Engineering*, **64**(April), pp. 273-278.

de Jonge, A. E. R. (1943). "A Brief Account of Modern Kinematics." *Transactions of the ASME*, pp. 663-683.

Erdman, A. E., ed. (1993). *Modern Kinematics: Developments in the Last Forty Years*. Wiley Series in Design Engineering, John Wiley & Sons: New York.

Ferguson, E. S. (1962). "Kinematics of Mechanisms from the Time of Watt." *United States National Museum Bulletin*, **228**(27), pp. 185-230.

Freudenstein, F. (1959). "Trends in the Kinematics of Mechanisms." *Applied Mechanics Reviews*, **12**(9), September, pp. 587-590.

Hartenberg, R. S., and J. Denavit. (1964). *Kinematic Synthesis of Linkages*. McGraw-Hill: New York, pp. 1-27.

Nolle, H. (1974). "Linkage Coupler Curve Synthesis: A Historical Review - II. Developments after 1875." *Mechanism and Machine Theory*, 9, pp. 325 - 348.

Nolle, H. (1974). "Linkage Coupler Curve Synthesis: A Historical Review -I. Developments up to 1875." *Mechanism and Machine Theory*, 9, pp. 147-168.

Nolle, H. (1975). "Linkage Coupler Curve Synthesis: A Historical Review - III. Spatial Synthesis and Optimization." *Mechanism and Machine Theory*, 10, pp. 41-55.

Reuleaux, F. (1963). *The Kinematics of Machinery*, A. B. W. Kennedy, translator. Dover Publications: New York, pp. 29-55.

Strandh, S. (1979). *A History of the Machine*. A&W Publishers: New York.

*For additional information on **creativity and the design process**, the following are recommended:*

Alger, J. R. M., and C. V. Hays. (1964). *Creative Synthesis in Design*. Prentice-Hall: Upper Saddle River, NJ.

Allen, M. S. (1962). *Morphological Creativity*. Prentice-Hall: Upper Saddle River, NJ.

Altschuller, G. (1984). *Creativity as an Exact Science*. Gordon and Breach: New York.

Buhl, H. R. (1960). *Creative Engineering Design*. Iowa State University Press: Ames, IA.

Dixon, J. R., and C. Poli. (1995). *Engineering Design and Design for Manufacturing—A Structured Approach*. Field Stone Publishers: Conway, MA.

Fey, V., et al. (1994). "Application of the Theory of Inventive Problem Solving to Design and Manufacturing Systems." *CIRP Annals*, **43**(1), pp. 107-110.

Gordon, W. J. J. (1962). *Synectics*. Harper & Row: New York.

Haefele, W. J. (1962). *Creativity and Innovation*. Van Nostrand Reinhold: New York.

Harrisberger, L. (1982). *Engineersmanship*. Brooks/Cole: Monterey, CA.

Osborn, A. F. (1963). *Applied Imagination*. Scribners: New York.

Pleuthner, W. (1956). "Brainstorming." *Machine Design*, January 12, 1956.

Suh, N. P. (1990). *The Principles of Design*. Oxford University Press: New York.

Taylor, C. W. (1964). *Widening Horizons in Creativity*. John Wiley & Sons: New York.

Von Fange, E. K. (1959). *Professional Creativity*. Prentice-Hall: Upper Saddle River, NJ.

For additional information on **Human Factors***, the following are recommended:*

Bailey, R. W. (1982). *Human Performance Engineering: A Guide for System Designers*. Prentice-Hall: Upper Saddle River, NJ.

Burgess, W. R. (1986). *Designing for Humans: The Human Factor in Engineering*. Petrocelli Books.

Clark, T. S., and E. N. Corlett. (1984). *The Ergonomics of Workspaces and Machines*. Taylor and Francis.

Huchinson, R. D. (1981). *New Horizons for Human Factors in Design*. McGraw-Hill: New York.

McCormick, D. J. (1964). *Human Factors Engineering*. McGraw-Hill: New York.

Osborne, D. J. (1987). *Ergonomics at Work*. John Wiley & Sons: New York.

Pheasant, S. (1986). *Bodyspace: Anthropometry, Ergonomics & Design*. Taylor and Francis.

Salvendy, G. (1987). *Handbook of Human Factors*. John Wiley & Sons: New York.

Sanders, M. S. (1987). *Human Factors in Engineering and Design*. McGraw-Hill: New York.

Woodson, W. E. (1981). *Human Factors Design Handbook*. McGraw-Hill: New York.

For additional information on **writing engineering reports***, the following are recommended:*

Barrass, R. (1978). *Scientists Must Write*. John Wiley & Sons: New York.

Crouch, W. G., and R. L. Zetler. (1964). *A Guide to Technical Writing*. The Ronald Press: New York.

Davis, D. S. (1963). *Elements of Engineering Reports*. Chemical Publishing Co.: New York.

Gray, D. E. (1963). *So You Have to Write a Technical Report*. Information Resources Press: Washington, D.C.

Michaelson, H. B. (1982). *How to Write and Publish Engineering Papers and Reports*. ISI Press: Philadelphia, PA.

Nelson, J. R. (1952). *Writing the Technical Report*. McGraw-Hill: New York.

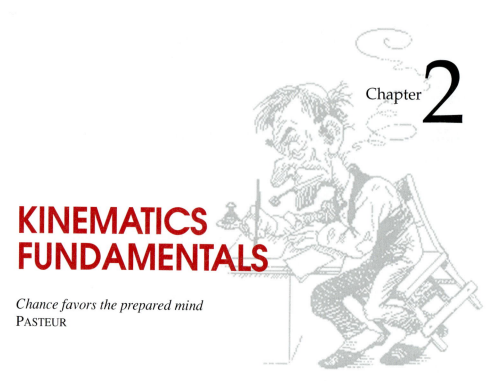

KINEMATICS FUNDAMENTALS

Chance favors the prepared mind
PASTEUR

2.0   INTRODUCTION

This chapter will present definitions of a number of terms and concepts fundamental to the synthesis and analysis of mechanisms. It will also present some very simple but powerful analysis tools which are useful in the synthesis of mechanisms.

2.1   DEGREES OF FREEDOM (*DOF*)

Any mechanical system can be classified according to the number of **degrees of freedom** (*DOF*) which it possesses. The system's *DOF* is equal to *the number of independent parameters (measurements) which are needed to uniquely define its position in space at any instant of time*. Note that *DOF* is defined with respect to a selected frame of reference. Figure 2-1 shows a pencil lying on a flat piece of paper with an *x, y* coordinate system added. If we constrain this pencil to always remain in the plane of the paper, three parameters (*DOF*) are required to completely define the position of the pencil on the paper, two linear coordinates (*x, y*) to define the position of any one point on the pencil and one angular coordinate (θ) to define the angle of the pencil with respect to the axes. The minimum number of measurements needed to define its position are shown in the figure as *x, y,* and θ. This system of the pencil in a plane then has **three** *DOF*. Note that the particular parameters chosen to define its position are not unique. Any alternate set of three parameters could be used. There is an infinity of sets of parameters possible, but in this case there must be three parameters per set, **such as two lengths and an angle**, to define the system's position because *a rigid body in plane motion has three DOF*.

Chapter 2

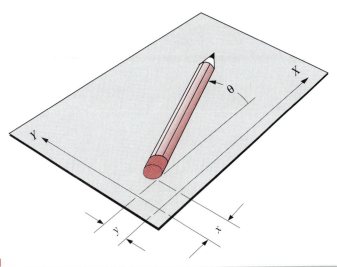

FIGURE 2-1

A rigid body in a plane has three *DOF*

Now allow the pencil to exist in a three-dimensional world. Hold it above your desktop and move it about. You now will need six parameters to define its **six** *DOF*. One possible set of parameters which could be used are **three lengths**, (x, y, z), plus **three angles** (θ, ϕ, ρ). *Any rigid body in three-space has six degrees of freedom.* Try to identify these six *DOF* by moving your pencil or pen with respect to your desktop.

The pencil in these examples represents a **rigid body**, or **link**, which for purposes of kinematic analysis we will assume to be incapable of deformation. This is merely a convenient fiction to allow us to more easily define the gross motions of the body. We can later superpose any deformations due to external or inertial loads onto our kinematic motions to obtain a more complete and accurate picture of the body's behavior. But remember, we are typically facing a *blank sheet of paper* at the beginning stage of the design process. We cannot determine deformations of a body until we define its size, shape, material properties, and loadings. Thus, at this stage we will assume, for purposes of initial kinematic synthesis and analysis, that *our kinematic bodies are rigid and massless.*

2.2 TYPES OF MOTION

A rigid body free to move within a reference frame will, in the general case, have **complex motion**, which is a simultaneous combination of **rotation** and **translation**. In three-dimensional space, there may be rotation about any axis (any skew axis or one of the three principal axes) and also simultaneous translation which can be resolved into components along three axes. In a plane, or two-dimensional space, complex motion becomes a combination of simultaneous rotation about one axis (perpendicular to the plane) and also translation resolved into components along two axes in the plane. For simplicity, we will limit our present discussions to the case of **planar (2-D) kinematic systems**. We will define these terms as follows for our purposes, in planar motion:

Pure rotation

the body possesses one point (center of rotation) which has no motion with respect to the "stationary" frame of reference. All other points on the body describe arcs about that center. A reference line drawn on the body through the center changes only its angular orientation.

Pure translation

all points on the body describe parallel (curvilinear or rectilinear) paths. A reference line drawn on the body changes its linear position but does not change its angular orientation.

Complex motion

a simultaneous combination of rotation and translation. Any reference line drawn on the body will change both its linear position and its angular orientation. Points on the body will travel nonparallel paths, and there will be, at every instant, a center of rotation, which will continuously change location.

Translation and **rotation** represent independent motions of the body. Each can exist without the other. If we define a 2-D coordinate system as shown in Figure 2-1, the x and y terms represent the translation components of motion, and the θ term represents the rotation component.

2.3   LINKS, JOINTS, AND KINEMATIC CHAINS

We will begin our exploration of the kinematics of mechanisms with an investigation of the subject of **linkage design**. Linkages are the basic building blocks of all mechanisms. We will show in later chapters that all common forms of mechanisms (cams, gears, belts, chains) are in fact variations on a common theme of linkages. Linkages are made up of links and joints.

A **link**, as shown in Figure 2-2, is an (assumed) rigid body which possesses at least two **nodes** which are *points for attachment to other links.*

Binary link   *- one with two nodes.*

Ternary link   *- one with three nodes.*

Quaternary link   *- one with four nodes.*

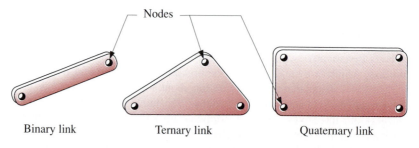

Binary link   Ternary link   Quaternary link

FIGURE 2-2

Links of different order

A **joint** is *a connection between two or more links (at their nodes), which allows some motion, or potential motion, between the connected links.* **Joints** (also called **kinematic pairs**) can be classified in several ways:

1   By the type of contact between the elements, line, point, or surface.

2   By the number of degrees of freedom allowed at the joint.

3   By the type of physical closure of the joint: either **force** or **form** closed.

4   By the number of links joined (order of the joint).

Reuleaux [1] coined the term **lower pair** to describe joints with surface contact (as with a pin surrounded by a hole) and the term **higher pair** to describe joints with point or line contact. However, if there is any clearance between pin and hole (as there must be for motion), so-called surface contact in the pin joint actually becomes line contact, as the pin contacts only one "side" of the hole. Likewise, at a microscopic level, a block sliding on a flat surface actually has contact only at discrete points, which are the tops of the surfaces' asperities. The main practical advantage of lower pairs over higher pairs is their better ability to trap lubricant between their enveloping surfaces. This is especially true for the rotating pin joint. The lubricant is more easily squeezed out of a higher pair, nonenveloping joint. As a result, the pin joint is preferred for low wear and long life, even over its lower pair cousin, the prismatic or slider joint.

Figure 2-3a shows the six possible lower pairs, their degrees of freedom, and their one-letter symbols. The revolute (R) and the prismatic (P) pairs are the only lower pairs usable in a planar mechanism. The screw (H), cylindric (C), spherical, and flat (F) lower pairs are all combinations of the revolute and/or prismatic pairs and are used in spatial (3-D) mechanisms. The R and P pairs are the basic building blocks of all other pairs which are combinations of those two as shown in Table 2-1.

A more useful means to classify joints (pairs) is by the number of degrees of freedom that they allow between the two elements joined. Figure 2-3 also shows examples of both one- and two-freedom joints commonly found in planar mechanisms. Figure 2-3b shows two forms of a planar, **one-freedom** joint (or pair), namely, a rotating pin joint (R) and a translating slider joint (P). These are also referred to as **full joints** (i.e., full = 1 *DOF*) and are **lower pairs**. The pin joint allows one rotational *DOF*, and the slider joint allows one translational *DOF* between the joined links. These are both contained within (and each is a limiting case of) another common, one-freedom joint, the screw and nut (Figure 2-3a). Motion of either the nut or the screw with respect to the other results in helical motion. If the helix angle is made zero, the nut rotates without advancing and it becomes the pin joint. If the helix angle is made 90 degrees, the nut will translate along the axis of the screw, and it becomes the slider joint.

Figure 2-3c shows examples of two-freedom joints (higher pairs) which simultaneously allow two independent, relative motions, namely translation and rotation, between the joined links. Paradoxically, this **two-freedom joint** is sometimes referred to as a "**half joint**," with its two freedoms placed in the denominator. The **half joint** is also called a **roll-slide joint** because it allows both rolling and sliding. A spherical, or ball-and-socket joint (Figure 2-3a), is an example of a three-freedom joint, which allows three independent angular motions between the two links joined. This *ball joint* would typically be used in a three-dimensional mechanism, one example being the ball joints in an automotive suspension system.

TABLE 2-1
The Six Lower Pairs

Name (Symbol)	DOF	Contains
Revolute (R)	1	R
Prismatic (P)	1	P
Helical (H)	1	RP
Cylindric (C)	2	RP
Spherical (S)	3	RRR
Planar (F)	3	RPP

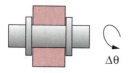

Revolute (R) joint—1 *DOF*

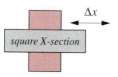

Prismatic (P) joint—1 *DOF*

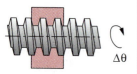

Helical (H) joint—1 *DOF*

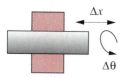

Cylindric (C) joint—2 *DOF*

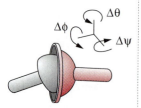

Spherical (S) joint—3 *DOF*

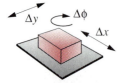

Planar (F) joint—3 *DOF*

(*a*) The six lower pairs

Rotating full pin (R) joint (form closed)

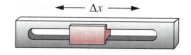

Translating full slider (P) joint (form closed)

(*b*) Full joints - 1 *DOF* (lower pairs)

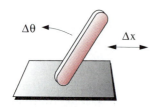

Link against plane (force closed)

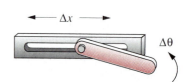

Pin in slot (form closed)

(*c*) Roll-slide (half or RP) joints - 2 *DOF* (higher pairs)

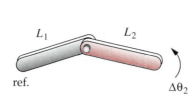

First order pin joint - one *DOF*
(two links joined)

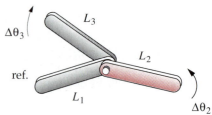

Second order pin joint - two *DOF*
(three links joined)

(*d*) The order of a joint is one less than the number of links joined

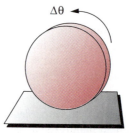

May roll, slide, or roll-slide, depending on friction

(*e*) Planar pure-roll (R), pure-slide (P), or roll-slide (RP) joint - 1 or 2 *DOF* (higher pair)

FIGURE 2-3

Joints (pairs) of various types

2

A joint with more than one freedom may also be a **higher pair** as shown in Figure 2-3c. Full joints (lower pairs) and half joints (higher pairs) are both used in planar (2-D), and in spatial (3-D) mechanisms. Note that if you do not allow the two links in Figure 2-3c connected by a roll-slide joint to slide, perhaps by providing a high friction coefficient between them, you can "lock out" the translating (Δx) freedom and make it behave as a full joint. This is then called a **pure rolling joint** and has rotational freedom ($\Delta\theta$) only. A common example of this type of joint is your automobile tire rolling against the road, as shown in Figure 2-3e. In normal use there is pure rolling and no sliding at this joint, unless, of course, you encounter an icy road or become too enthusiastic about accelerating or cornering. If you lock your brakes on ice, this joint converts to a pure sliding one like the slider block in Figure 2-3b. Friction determines the actual number of freedoms at this kind of joint. It can be **pure roll**, **pure slide**, or **roll-slide**.

To visualize the degree of freedom of a joint in a mechanism, it is helpful to "mentally disconnect" the two links which create the joint from the rest of the mechanism. You can then more easily see how many freedoms the two joined links have with respect to one another.

Figure 2-3c also shows examples of both **form-closed** and **force-closed** joints. A **form-closed** joint is kept together or *closed by its geometry*. A pin in a hole or a slider in a two-sided slot are form closed. In contrast, a **force-closed** joint, such as a pin in a half-bearing or a slider on a surface, *requires some external force to keep it together or closed*. This force could be supplied by gravity, a spring, or any external means. There can be substantial differences in the behavior of a mechanism due to the choice of force or form closure, as we shall see. The choice should be carefully considered. In linkages, form closure is usually preferred, and it is easy to accomplish. But for cam-follower systems, force closure is often preferred. This topic will be explored further in later chapters.

Figure 2-3d shows examples of joints of various orders, where **order** is defined as *the number of links joined minus one*. It takes two links to make a single joint; thus the simplest joint combination of two links has order one. As additional links are placed on the same joint, the order is increased on a one for one basis. Joint order has significance in the proper determination of overall degree of freedom for the assembly. We gave definitions for a **mechanism** and a **machine** in Chapter 1. With the kinematic elements of links and joints now defined, we can define those devices more carefully based on Reuleaux's classifications of the kinematic chain, mechanism, and machine. [1]

A kinematic chain is defined as:

An assemblage of links and joints, interconnected in a way to provide a controlled output motion in response to a supplied input motion.

A mechanism is defined as:

A kinematic chain in which at least one link has been "grounded," or attached, to the frame of reference (which itself may be in motion).

A machine is defined as:

A combination of resistant bodies arranged to compel the mechanical forces of nature to do work accompanied by determinate motions.

By Reuleaux's definition [1] a machine is *a collection of mechanisms arranged to transmit forces and do work.* He viewed all energy or force transmitting devices as machines which utilize mechanisms as their building blocks to provide the necessary motion constraints.

We will now define a **crank** as *a link which makes a complete revolution and is pivoted to ground*, a **rocker** as *a link which has oscillatory (back and forth) rotation and is pivoted to ground*, and a **coupler** (or connecting rod) *which has complex motion and is not pivoted to ground*. **Ground** is defined as *any link or links that are fixed* (nonmoving) with respect to the reference frame. Note that the reference frame may in fact itself be in motion.

2.4   DETERMINING DEGREE OF FREEDOM

The concept of **degree of freedom** (*DOF*) is fundamental to both the synthesis and analysis of mechanisms. We need to be able to quickly determine the *DOF* of any collection of links and joints which may be suggested as a solution to a problem. Degree of freedom (also called the **mobility** *M)* of a system can be defined as:

Degree of Freedom

the number of inputs which need to be provided in order to create a predictable output;

also:

the number of independent coordinates required to define its position.

At the outset of the design process, some general definition of the desired output motion is usually available. The number of inputs needed to obtain that output may or may not be specified. Cost is the principal constraint here. Each required input will need some type of actuator, either a human operator or a "slave" in the form of a motor, solenoid, air cylinder, or other energy conversion device. (These devices are discussed in Section 2.15.) These multiple input devices will have to have their actions coordinated by a "controller," which must have some intelligence. This control is now often provided by a computer but can also be mechanically programmed into the mechanism design. There is no requirement that a mechanism have only one *DOF*, although that is often desirable for simplicity. Some machines have many *DOF*. For example, picture the number of control levers or actuating cylinders on a bulldozer or crane. See Figure 1-1b (p. 7).

Kinematic chains or mechanisms may be either **open** or **closed**. Figure 2-4 shows both open and closed mechanisms. A closed mechanism will have no open attachment points or **nodes** and may have one or more degrees of freedom. An open mechanism of more than one link will always have more than one degree of freedom, thus requiring as many actuators (motors) as it has *DOF*. A common example of an open mechanism is an industrial robot. *An open kinematic chain of two binary links and one joint* is called a **dyad**. The sets of links shown in Figure 2-3a and b are **dyads**.

Reuleaux limited his definitions to closed kinematic chains and to mechanisms having only one *DOF*, which he called *constrained*. [1] The somewhat broader definitions above are perhaps better suited to current-day applications. A multi-*DOF* mechanism, such as a robot, will be constrained in its motions as long as the necessary number of inputs are supplied to control all its *DOF*.

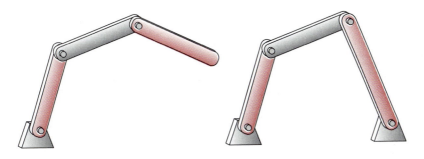

(a) Open mechanism chain   (b) Closed mechanism chain

FIGURE 2-4

Mechanism chains

Degree of Freedom in Planar Mechanisms

To determine the overall *DOF* of any mechanism, we must account for the number of links and joints, and for the interactions among them. The *DOF* of any assembly of links can be predicted from an investigation of the **Gruebler condition**. [2] Any link in a plane has 3 *DOF*. Therefore, a system of *L* unconnected links in the same plane will have $3L$ *DOF*, as shown in Figure 2-5a where the two unconnected links have a total of six *DOF*. When these links are connected by a **full joint** in Figure 2-5b, Δy_1 and Δy_2 are combined as Δy, and Δx_1 and Δx_2 are combined as Δx. This removes two *DOF*, leaving four *DOF*. In Figure 2-5c the half joint removes only one *DOF* from the system (because a half joint has two *DOF*), leaving the system of two links connected by a half joint with a total of five *DOF*. In addition, when any link is grounded or attached to the reference frame, all three of its *DOF* will be removed. This reasoning leads to **Gruebler's equation**:

$$M = 3L - 2J - 3G \qquad (2.1a)$$

where: M = *degree of freedom or mobility*
  L = *number of links*
  J = *number of joints*
  G = *number of grounded links*

Note that in any real mechanism, even if more than one link of the kinematic chain is grounded, the net effect will be to create one larger, higher-order ground link, as there can be only one ground plane. Thus *G* is always one, and Gruebler's equation becomes:

$$M = 3(L-1) - 2J \qquad (2.1b)$$

The value of *J* in equations 2.1a and 2.1b must reflect the value of all joints in the mechanism. That is, half joints count as 1/2 because they only remove one *DOF*. It is less confusing if we use **Kutzbach's** modification of Gruebler's equation in this form:

$$M = 3(L-1) - 2J_1 - J_2 \qquad (2.1c)$$

where: M = *degree of freedom or mobility*
  L = *number of links*
  $J1$ = *number of 1 DOF (full) joints*
  $J2$ = *number of 2 DOF (half) joints*

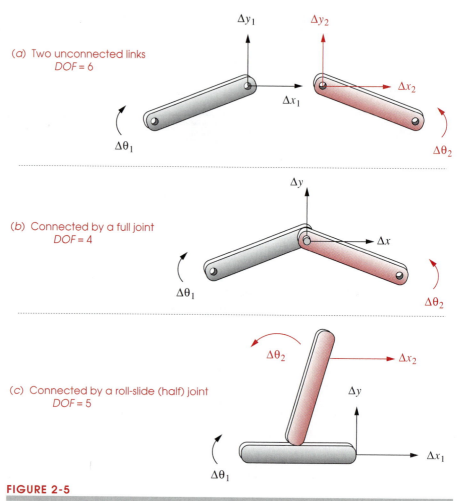

(a) Two unconnected links
DOF = 6

(b) Connected by a full joint
DOF = 4

(c) Connected by a roll-slide (half) joint
DOF = 5

FIGURE 2-5

Joints remove degrees of freedom

The value of J_1 and J_2 in these equations must still be carefully determined to account for all full, half, and multiple joints in any linkage. Multiple joints count as one less than the number of links joined at that joint and add to the "full" (J_1) category. The *DOF* of any proposed mechanism can be quickly ascertained from this expression before investing any time in more detailed design. It is interesting to note that this equation has no information in it about link sizes or shapes, only their quantity. Figure 2-6a shows a mechanism with one *DOF* and only full joints in it.

Figure 2-6b shows a structure with zero *DOF* and which contains both half and multiple joints. Note the schematic notation used to show the ground link. The ground link need not be drawn in outline as long as all the grounded joints are identified. Note also the joints labeled "**multiple**" and "**half**" in Figure 2-6a and b. As an exercise, compute the *DOF* of these examples with **Kutzbach's** equation.

Note:
There are no
roll-slide
(half) joints
in this
linkage

$L = 8, \quad J = 10$

$DOF = 1$

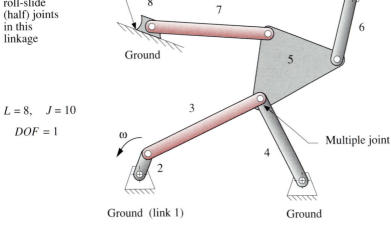

(a) Linkage with full and multiple joints

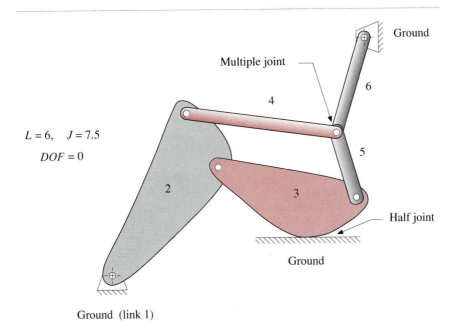

$L = 6, \quad J = 7.5$

$DOF = 0$

(b) Linkage with full, half, and multiple joints

FIGURE 2-6

Linkages containing joints of various types

Degree of Freedom in Spatial Mechanisms

The approach used to determine the mobility of a planar mechanism can be easily extended to three dimensions. Each unconnected link in three-space has 6 *DOF*, and any one of the six lower pairs can be used to connect them, as can higher pairs with more freedom. A one-freedom joint removes 5 *DOF*, a two-freedom joint removes 4 *DOF*, etc. Grounding a link removes 6 *DOF*. This leads to the Kutzbach mobility equation for spatial linkages:

$$M = 6(L-1) - 5J_1 - 4J_2 - 3J_3 - 2J_4 - J_5 \qquad (2.2)$$

where the subscript refers to the number of freedoms of the joint. We will limit our study to 2-D mechanisms in this text.

2.5   MECHANISMS AND STRUCTURES

The degree of freedom of an assembly of links completely predicts its character. There are only three possibilities. *If the DOF is positive, it will be a* **mechanism**, and the links will have relative motion. *If the DOF is exactly zero, then it will be a* **structure**, and no motion is possible. *If the DOF is negative, then it is a* **preloaded structure**, which means that no motion is possible and some stresses may also be present at the time of assembly. Figure 2-7 shows examples of these three cases. One link is grounded in each case.

Figure 2-7a shows four links joined by four full joints which, from the Gruebler equation, gives one *DOF*. It will move, and only one input is needed to give predictable results.

Figure 2-7b shows three links joined by three full joints. It has zero *DOF* and is thus a **structure**. Note that if the link lengths will allow connection,[*] all three pins can be inserted into their respective pairs of link holes (nodes) without straining the structure, as a position can always be found to allow assembly.

Figure 2-7c shows two links joined by two full joints. It has a *DOF* of minus one, making it a **preloaded structure**. In order to insert the two pins without straining the links, the center distances of the holes in both links must be exactly the same. Practically speaking, it is impossible to make two parts exactly the same. There will always be some manufacturing error, even if very small. Thus you may have to force the second pin into place, creating some stress in the links. The structure will then be preloaded. You have probably met a similar situation in a course in applied mechanics in the form of an indeterminate beam, one in which there were too many supports or constraints for the equations available. An indeterminate beam also has negative *DOF*, while a *simply supported* beam has zero *DOF*.

Both structures and preloaded structures are commonly encountered in engineering. In fact the true structure of zero *DOF* is rare in engineering practice. Most buildings, bridges, and machine frames are preloaded structures, due to the use of welded and riveted joints rather than pin joints. Even simple structures like the chair you are sitting in are often preloaded. Since our concern here is with mechanisms, we will concentrate on devices with positive *DOF* only.

---

[*] If the sum of the lengths of any two links is less than the length of the third link, then their interconnection is impossible.

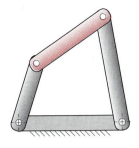

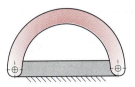

(a) Mechanism—*DOF* = + 1   (b) Structure—*DOF* = 0   (c) Preloaded structure—*DOF* = –1

FIGURE 2-7

Mechanisms, structures and preloaded structures

2.6   NUMBER SYNTHESIS

The term **number synthesis** has been coined to mean *the determination of the number and order of links and joints necessary to produce motion of a particular DOF.* **Order** in this context refers to the number of nodes per link, i.e., **binary**, **ternary**, **quaternary**, etc. The value of number synthesis is to allow the exhaustive determination of all possible combinations of links which will yield any chosen *DOF*. This then equips the designer with a definitive catalog of potential linkages to solve a variety of motion control problems.

As an example we will now derive all the possible link combinations for one *DOF*, including sets of up to eight links, and link orders up to and including hexagonal links. For simplicity we will assume that the links will be connected with only full rotating joints. We can later introduce half joints, multiple joints, and sliding joints through linkage transformation. First let's look at some interesting attributes of linkages as defined by the above assumption regarding full joints.

Hypothesis: If all joints are full joints, an odd number of *DOF* requires an even number of links and vice versa.

Proof: **Given:** All even integers can be denoted by $2m$ or by $2n$, and all odd integers can be denoted by $2m - 1$ or by $2n - 1$, where n and m are any positive integers. The number of joints must be a positive integer.

Let : L = number of links, J = number of joints, and $M = DOF = 2m$ (i.e., all even numbers)

Then: rewriting Gruebler's equation (Equation 2.1b) to solve for J,

$$J = \frac{3}{2}(L-1) - \frac{M}{2} \tag{2.3a}$$

Try: Substituting $M = 2m$, and $L = 2n$ (i.e., both any even numbers):

$$J = 3n - m - \frac{3}{2} \tag{2.3b}$$

This cannot result in J being a positive integer as required.

Try: $M = 2m - 1$ and $L = 2n - 1$ (i.e., both any odd numbers):

$$J = 3n - m - \frac{7}{2}$$

(2.3c)

This also cannot result in J being a positive integer as required.

Try: $M = 2m - 1$, and $L = 2n$ (i.e., odd-even):

$$J = 3n - m - 2$$

(2.3d)

This is a positive integer for $m \geq 1$ and $n \geq 2$.

Try: $M = 2m$ and $L = 2n - 1$ (i.e., even-odd):

$$J = 3n - m - 3$$

(2.3e)

This is a positive integer for $m \geq 1$ and $n \geq 2$.

So, for our example of one-*DOF* mechanisms, we can only consider combinations of 2, 4, 6, 8 . . . links. Letting the order of the links be represented by:

$$B = \text{number of binary links}$$
$$T = \text{number of ternary links}$$
$$Q = \text{number of quaternaries}$$
$$P = \text{number of pentagonals}$$
$$H = \text{number of hexagonals}$$

the total number of links in any mechanism will be:

$$L = B + T + Q + P + H + \cdots$$

(2.4a)

Since *two link nodes* are needed to make *one joint*:

$$J = \frac{nodes}{2}$$

(2.4b)

and

$$nodes = order\ of\ link \times no.\ of\ links\ of\ that\ order$$

(2.4c)

then

$$J = \frac{(2B + 3T + 4Q + 5P + 6H + \cdots)}{2}$$

(2.4d)

Substitute Equations 2.4a and 2.4d into Gruebler's equation (2.1b)

$$M = 3(B + T + Q + P + H - 1) - 2\left(\frac{2B + 3T + 4Q + 5P + 6H}{2}\right)$$

(2.4e)

$$M = B - Q - 2P - 3H - 3$$

Note what is missing from this equation! The ternary links have dropped out. The *DOF* is independent of the number of ternary links in the mechanism. But because each ternary link has three nodes, it can only create or remove 3/2 joints. So we must add or subtract ternary links in pairs to maintain an integer number of joints. *The addition or subtraction of ternary links in pairs will not affect the DOF of the mechanism.*

In order to determine all possible combinations of links for a particular *DOF*, we must combine equations 2.3a and 2.4d:

$$\frac{3}{2}(L-1)-\frac{M}{2}=\frac{(2B+3T+4Q+5P+6H)}{2}$$

(2.5)

$$3L-3-M=2B+3T+4Q+5P+6H$$

Now combine equation 2.5 with equation 2.4a to eliminate *B*:

$$L-3-M=T+2Q+3P+4H \qquad (2.6)$$

We will now solve equations 2.4a and 2.6 simultaneously (by progressive substitution) to determine all compatible combinations of links for *DOF* = 1, up to eight links. The strategy will be to start with the smallest number of links, and the highest-order link possible with that number, eliminating impossible combinations.

(Note: *L must be even for odd DOF*.)

CASE 1.   *L* = 2

$$L-4=T+2Q+3P+4H=-2 \qquad (2.7a)$$

This requires a negative number of links, so *L* = 2 is impossible.

CASE 2.   *L* = 4

$$L-4=T+2Q+3P+4H=0; \qquad so: \ T=Q=P=H=0$$

(2.7b)

$$L=B+0=4; \qquad B=4$$

The simplest one-*DOF* linkage is four binary links—the **fourbar linkage**.

CASE 3.   *L* = 6

$$L-4=T+2Q+3P+4H=2; \qquad so: \quad P=H=0 \qquad (2.7c)$$

T may only be 0, 1, or 2;   *Q* may only be 0 or 1

If *Q* = 0 then *T* must be 2 and:

$$L=B+2T+0Q=6; \qquad B=4, \qquad T=2 \qquad (2.7d)$$

If *Q* = 1, then *T* must be 0 and:

$$L=B+0T+1Q=6; \qquad B=5, \qquad Q=1 \qquad (2.7e)$$

There are then two possibilities for *L* = 6. Note that one of them is in fact the simpler fourbar with two ternaries added as was predicted above.

CASE 4.   *L* = 8

A tabular approach is needed with this many links:

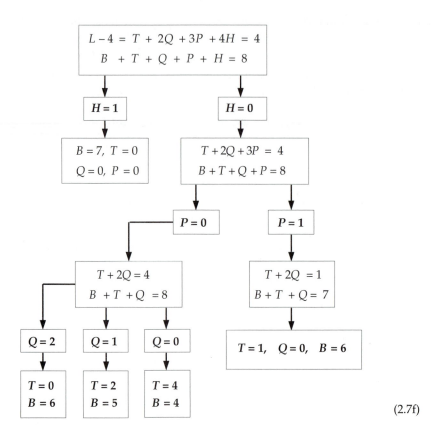

$$L - 4 = T + 2Q + 3P + 4H = 4$$
$$B + T + Q + P + H = 8$$

$H = 1$

$H = 0$

$B = 7, \ T = 0$
$Q = 0, \ P = 0$

$T + 2Q + 3P = 4$
$B + T + Q + P = 8$

$P = 0$

$P = 1$

$T + 2Q = 4$
$B + T + Q = 8$

$T + 2Q = 1$
$B + T + Q = 7$

$Q = 2$

$Q = 1$

$Q = 0$

$T = 1, \quad Q = 0, \quad B = 6$

$T = 0$
$B = 6$

$T = 2$
$B = 5$

$T = 4$
$B = 4$

(2.7f)

From this analysis we can see that, for one *DOF*, there is only one possible four link configuration, two six link configurations and five possibilities for eight links using binary through hexagonal links. Table 2-2 shows the so-called "link sets" for all the possible linkages for one *DOF* up to 8 links and hexagonal order.

TABLE 2-2   1-*DOF* Planar Mechanisms with Revolute Joints and Up to 8 Links

Total Links	Link Sets				
	Binary	Ternary	Quaternary	Pentagonal	Hexagonal
4	4	0	0	0	0
6	4	2	0	0	0
6	5	0	1	0	0
8	7	0	0	0	1
8	4	4	0	0	0
8	5	2	1	0	0
8	6	0	2	0	0
8	6	1	0	1	0

2.7   PARADOXES

Because the Gruebler criterion pays no attention to link sizes or shapes, it *can give mis-leading results* in the face of unique geometric configurations. For example, Figure 2-8a shows a structure (*DOF* = 0) with the ternary links of arbitrary shape. This link arrangement is sometimes called the "**E-quintet**," because of its resemblance to a capital **E** and the fact that it has five links, including the ground.[*] It is the next simplest **structural** building block to the "**delta triplet**."

Figure 2-8b shows the same E-quintet with the ternary links straight and parallel and with equispaced nodes. The three binaries are also equal in length. With this very unique geometry, you can see that it will move despite Gruebler's prediction to the contrary.

Figure 2-8c shows a very common mechanism which also disobeys Gruebler's criterion. The joint between the two wheels can be postulated to allow no slip, provided that sufficient friction is available. If no slip occurs, then this is a one-freedom, or full, joint which allows only relative angular motion ($\Delta\theta$) between the wheels. With that assump-

(*a*) The E-quintet with *DOF* = 0
         —agrees with Gruebler equation

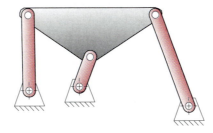

(*b*) The E-quintet with *DOF* = 1
         —disagrees with Gruebler equation
         due to unique geometry

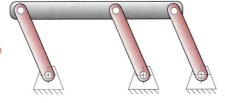

Full joint -
pure rolling
no slip

(*c*) Rolling cylinders with *DOF* = 1
         —disagrees with Gruebler equation
         which predicts *DOF* = 0

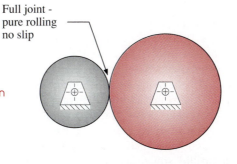

FIGURE 2-8

Gruebler paradoxes—linkages that do not behave as predicted by the Gruebler equation

* It is also called an Assur chain.

tion, there are 3 links and 3 full joints, from which Gruebler's equation predicts zero DOF. However, this linkage does move (actual $DOF = 1$), because the center distance, or length of link 1, is exactly equal to the sum of the radii of the two wheels.

There are other examples of paradoxes which disobey the Gruebler criterion due to their unique geometry. The designer needs to be alert to these possible inconsistencies.

2.8   ISOMERS

The word **isomer** is from the Greek and means *having equal parts*. Isomers in chemistry are compounds that have the same number and type of atoms but which are interconnected differently and thus have different physical properties. Figure 2-9a shows two hydrocarbon isomers, n-butane and isobutane. Note that each has the same number of carbon and hydrogen atoms (C_4H_{10}), but they are differently interconnected and have different properties.

Linkage isomers are analogous to these chemical compounds in that the **links** (like atoms) have various **nodes** (electrons) available to connect to other links' nodes. The assembled linkage is analogous to the chemical compound. Depending on the particular connections of available links, the assembly will have different motion properties. The number of isomers possible from a given collection of links (as in any row of Table 2-2) is far from obvious. In fact the problem of mathematically predicting the number of isomers of all link combinations has been a long-unsolved problem. Many researchers have spent much effort on this problem with some recent success. See references [3] through [7] for more information. Dhararipragada [6] presents a good historical summary of isomer research to 1994. Table 2-3 shows the number of valid isomers found for one-DOF mechanisms with revolute pairs, up to 12 links.

Figure 2-9b shows all the isomers for the simple cases of one DOF with 4 and 6 links. Note that there is only one isomer for the case of 4 links. An isomer is only unique if the interconnections between its types of links are different. That is, all binary links are considered equal, just as all hydrogen atoms are equal in the chemical analog. Link lengths and shapes do not figure into the Gruebler criterion or the condition of isomerism. The 6-link case of 4 binaries and 2 ternaries has only two valid isomers. These are known as the **Watt's chain** and the **Stephenson's chain** after their discoverers. Note the different interconnections of the ternaries to the binaries in these two examples. The Watt's chain has the two ternaries directly connected, but the Stephenson's chain does not.

There is also a third potential isomer for this case of six links, as shown in Figure 2-9c, but it fails the test of **distribution of degree of freedom**, which requires that the overall DOF (here 1) be uniformly distributed throughout the linkage and not concentrated in a subchain. Note that this arrangement (Figure 2-9c) has a **structural subchain** of $DOF = 0$ in the triangular formation of the two ternaries and the single binary connecting them. This creates a truss, or **delta triplet**. The remaining three binaries in series form a fourbar chain ($DOF = 1$) with the structural subchain of the two ternaries and the single binary effectively reduced to a structure which acts like a single link. Thus this arrangement has been reduced to the simpler case of the fourbar linkage despite its six bars. This is an **invalid isomer** and is rejected. It is left as an exercise for the reader to find the 16 valid isomers of the eight bar, one-DOF cases.

TABLE 2-3
Number of Valid Isomers

Links	Valid Isomers
4	1
6	2
8	16
10	230
12	6856 or 6862*

* Researchers disagree.

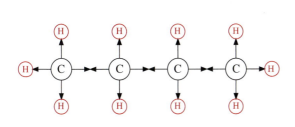

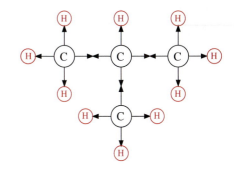

(a) Hydrocarbon isomers n-butane and isobutane

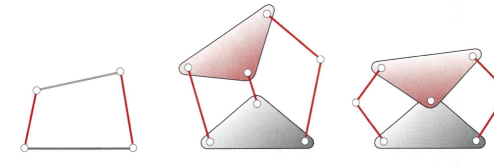

The only fourbar isomer   Stephenson's sixbar isomer   Watt's sixbar isomer

(b) All valid isomers of the fourbar and sixbar linkages

Structural subchain
reduces three links
to a zero *DOF*
"delta triplet" truss

Fourbar subchain
concentrates the
1 *DOF* of the mechanism

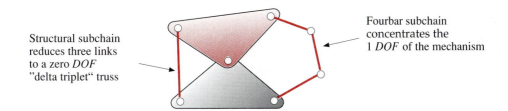

(c) An invalid sixbar isomer which reduces to the simpler fourbar

FIGURE 2-9

Isomers of kinematic chains

2.9   LINKAGE TRANSFORMATION

The number synthesis techniques described above give the designer a toolkit of basic linkages of particular *DOF*. If we now relax the arbitrary constraint which restricted us to only revolute joints, we can transform these basic linkages to a wider variety of mechanisms with even greater usefulness. There are several transformation techniques or rules that we can apply to planar kinematic chains.

1   Revolute joints in any loop can be replaced by prismatic joints with no change in *DOF* of the mechanism, provided that at least two revolute joints remain in the loop.[*]

2   Any full joint can be replaced by a half joint, but this will increase the *DOF* by one.

3   Removal of a link will reduce the *DOF* by one.

4   The combination of rules 2 and 3 above will keep the original *DOF* unchanged.

5   Any ternary or higher-order link can be partially "shrunk" to a lower-order link by coalescing nodes. This will create a multiple joint but will not change the *DOF* of the mechanism.

6   Complete shrinkage of a higher-order link is equivalent to its removal. A multiple joint will be created, and the *DOF* will be reduced.

Figure 2-10a shows a fourbar crank-rocker linkage transformed into the fourbar slider-crank by the application of rule #1. It is still a fourbar linkage. Link 4 has become a sliding block. The Gruebler's equation is unchanged at one *DOF* because the slider block provides a full joint against link 1, as did the pin joint it replaces. Note that this transformation from a rocking output link to a slider output link is equivalent to increasing the length (radius) of rocker link 4 until its arc motion at the joint between links 3 and 4 becomes a straight line. Thus the slider block is equivalent to an infinitely long rocker link 4, which is pivoted at infinity along a line perpendicular to the slider axis as shown in Figure 2-10a.

Figure 2-10b shows a fourbar slider-crank transformed via rule #4 by the substitution of a half joint for the coupler. The first version shown retains the same motion of the slider as the original linkage by use of a curved slot in link 4. The effective coupler is always perpendicular to the tangent of the slot and falls on the line of the original coupler. The second version shown has the slot made straight and perpendicular to the slider axis. The effective coupler now is "pivoted" at infinity. This is called a **Scotch yoke** and gives exact *simple harmonic motion* of the slider in response to a constant speed input to the crank.

Figure 2-10c shows a fourbar linkage transformed into a **cam-follower** linkage by the application of rule #4. Link 3 has been removed and a half joint substituted for a full joint between links 2 and 4. This still has one *DOF*, and the cam-follower is in fact a fourbar linkage in another disguise, in which the coupler (link 3) has become an effective link of *variable length*. We will investigate the fourbar linkage and these variants of it in greater detail in later chapters.

Figure 2-11a shows the **Stephenson's sixbar chain** from Figure 2-9b (p. 39) transformed by *partial shrinkage* of a ternary link (rule #5) to create a multiple joint. It is still a one-*DOF* Stephenson's sixbar. Figure 2-11b shows the **Watt's sixbar chain** from

---

[*] If all revolute joints in a fourbar linkage are replaced by prismatic joints, the result will be a two-*DOF* assembly. Also, if three revolutes in a fourbar loop are replaced with prismatic joints, the one remaining revolute joint will not be able to turn, effectively locking the two pinned links together as one. This effectively reduces the assembly to a threebar linkage which should have zero *DOF*. But, a delta triplet with three prismatic joints has one *DOF*—another Gruebler's paradox.

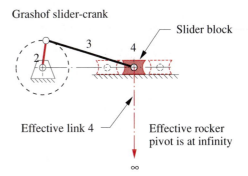

Grashof crank-rocker

3

2

Rocker
pivot

4

Grashof slider-crank

Slider block

3

4

2

Effective link 4

Effective rocker
pivot is at infinity

∞

(*a*) Transforming a fourbar crank-rocker to a slider-crank

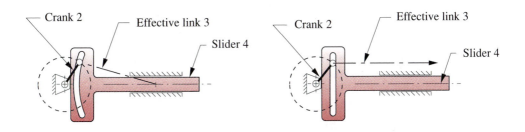

Crank 2

Effective link 3

Slider 4

Crank 2

Effective link 3

Slider 4

(*b*) Transforming the slider-crank to the Scotch yoke

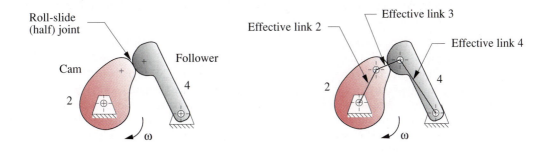

Roll-slide
(half) joint

Cam

Follower

2

4

ω

Effective link 3

Effective link 2

Effective link 4

2

4

ω

(*c*) The cam-follower mechanism has an effective fourbar equivalent

FIGURE 2-10

Linkage transformation

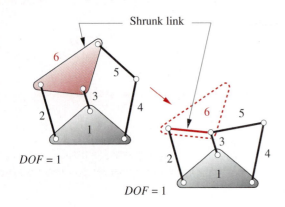

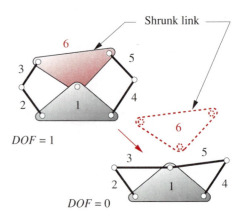

(a) Partial shrinkage of higher link
    retains original *DOF*

(b) Complete shrinkage of higher link
    reduces *DOF* by one

FIGURE 2-11

Link shrinkage

Figure 2-9b with one ternary link *completely shrunk* to create a multiple joint. This is
now a structure with *DOF* = 0. The two triangular subchains are obvious. Just as the
fourbar chain is the basic building block of one-*DOF* mechanisms, this threebar triangle
delta triplet is the *basic building block* of zero-*DOF* structures (trusses).

2.10   INTERMITTENT MOTION

Intermittent motion is *a sequence of motions and dwells*. A **dwell** is *a period in which
the output link remains stationary while the input link continues to move*. There are many
applications in machinery which require intermittent motion. The **cam-follower** variation on the fourbar linkage as shown in Figure 2-10c (p. 41) is often used in these situations. The design of that device for both intermittent and continuous output will be addressed in detail in Chapter 8. Other pure linkage **dwell mechanisms** are discussed in
the next chapter.

GENEVA MECHANISM   A common form of intermittent motion device is the **Geneva mechanism** shown in Figure 2-12a. This is also a transformed fourbar linkage in
which the coupler has been replaced by a half joint. The input crank (link 2) is typically
motor driven at a constant speed. The **Geneva wheel** is fitted with at least three equispaced, radial slots. The crank has a pin that enters a radial slot and causes the Geneva
wheel to turn through a portion of a revolution. When the pin leaves that slot, the Geneva wheel remains stationary until the pin enters the next slot. The result is intermittent
rotation of the Geneva wheel.

The crank is also fitted with an arc segment, which engages a matching cutout on
the periphery of the Geneva wheel when the pin is out of the slot. This keeps the Geneva wheel stationary and in the proper location for the next entry of the pin. The number
of slots determines the number of "stops" of the mechanism, where *stop* is synonymous

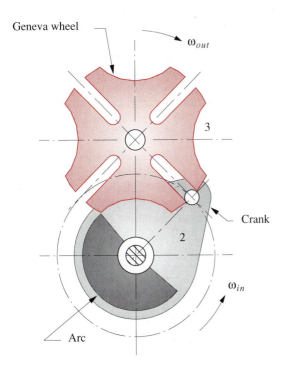

(a) Four-stop Geneva mechanism

(b) Ratchet and pawl mechanism

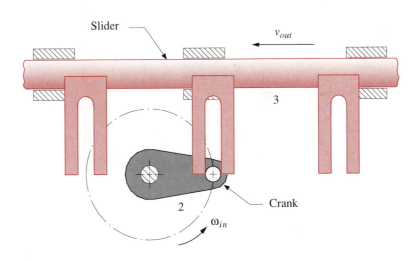

(c) Linear intermittent motion "Geneva" mechanism

FIGURE 2-12

Rotary and linear intermittent motion mechanisms

2

with *dwell*. A Geneva wheel needs a minimum of three stops to work. The maximum number of stops is limited only by the size of the wheel.

RATCHET AND PAWL   Figure 2-12b shows a ratchet and pawl mechanism. The **arm** pivots about the center of the toothed **ratchet wheel** and is moved back and forth to index the wheel. The **driving pawl** rotates the ratchet wheel (or **ratchet**) in the counterclockwise direction and does no work on the return (clockwise) trip. The **locking pawl** prevents the ratchet from reversing direction while the driving pawl returns. Both pawls are usually spring-loaded against the ratchet. This mechanism is widely used in devices such as "ratchet" wrenches, winches, etc.

LINEAR GENEVA MECHANISM   There is also a variation of the Geneva mechanism which has linear translational output, as shown in Figure 2-12c. This mechanism is analogous to an open Scotch yoke device with multiple yokes. It can be used as an intermittent conveyor drive with the slots arranged along the conveyor chain or belt. It also can be used with a reversing motor to get linear, reversing oscillation of a single slotted output slider.

2.11   INVERSION

It should now be apparent that there are many possible linkages for any situation. Even with the limitations imposed in the number synthesis example (one *DOF*, eight links, up to hexagonal order), there are eight linkage combinations shown in Table 2-2 (p. 36), and these together yield 19 valid isomers as shown in Table 2-3 (p. 38). In addition, we can introduce another factor, namely mechanism inversion. An **inversion** is *created by grounding a different link in the kinematic chain*. Thus there are as many inversions of a given linkage as it has links.

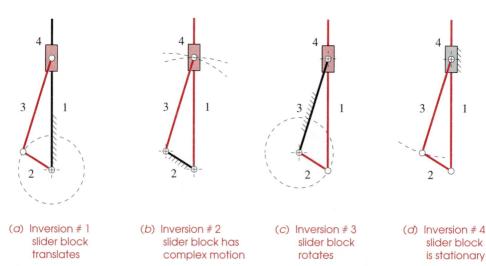

(a) Inversion # 1   (b) Inversion # 2   (c) Inversion # 3   (d) Inversion # 4
    slider block   slider block has   slider block   slider block
    translates   complex motion   rotates   is stationary

FIGURE 2-13

Four distinct inversions of the fourbar slider-crank mechanism (each black link is stationary—all red links move)

The motions resulting from each inversion can be quite different, but some inversions of a linkage may yield motions similar to other inversions of the same linkage. In these cases only some of the inversions may have distinctly different motions. We will denote the *inversions which have distinctly different motions* as **distinct inversions.**

Figure 2-13 (previous page) shows the four inversions of the fourbar slider-crank linkage, all of which have distinct motions. Inversion #1, with link 1 as ground and its slider block in pure translation, is the most commonly seen and is used for **piston engines** and **piston pumps**. Inversion #2 is obtained by grounding link 2 and gives the **Whitworth** or **crank-shaper** quick-return mechanism, in which the slider block has complex motion. (Quick-return mechanisms will be investigated further in the next chapter.) Inversion #3 is obtained by grounding link 3 and gives the slider block pure rotation. Inversion #4 is obtained by grounding the slider link 4 and is used in hand operated, **well pump** mechanisms, in which the handle is link 2 (extended) and link 1 passes down the well pipe to mount a piston on its bottom. (It is upside down in the figure.)

The **Watt's sixbar chain** has two distinct inversions, and the **Stephenson's sixbar** has three distinct inversions, as shown in Figure 2-14. The pin-jointed fourbar has four distinct inversions: the crank-rocker, double-crank, double-rocker, and triple-rocker which are shown in Figures 2-15 and 2-16.

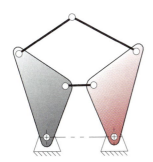
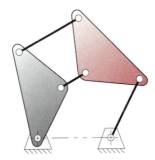
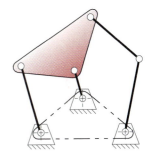

(a) Stephenson's sixbar inversion I   *(b)* Stephenson's sixbar inversion II   *(c)* Stephenson's sixbar inversion III

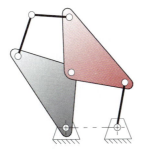
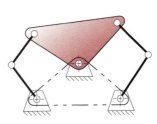

(d) Watt's sixbar inversion I   *(e)* Watt's sixbar inversion II

FIGURE 2-14

All distinct inversions of the sixbar linkage

2

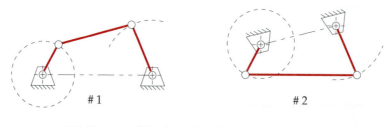

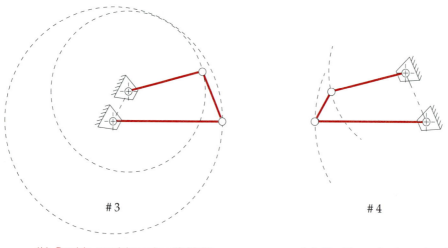

(a) Two non-distinct crank-rocker inversions (GCRR)

(b) Double-crank inversion (GCRC)
(drag link mechanism)

(c) Double-rocker inversion (GRCR)
(coupler rotates)

FIGURE 2-15

All inversions of the Grashof fourbar linkage

2.12   THE GRASHOF CONDITION

The **fourbar linkage** has been shown above to be the *simplest possible pin-jointed mechanism* for single degree of freedom controlled motion. It also appears in various disguises such as the **slider-crank** and the **cam-follower**. It is in fact the most common and ubiquitous device used in machinery. It is also extremely versatile in terms of the types of motion which it can generate.

Simplicity is one mark of good design. The fewest parts that can do the job will usually give the least expensive and most reliable solution. Thus the **fourbar linkage** should be among the first solutions to motion control problems to be investigated. The **Grashof condition** [8] is a very simple relationship which predicts the *rotation behavior* or **rotatability** of a fourbar linkage's inversions based only on the link lengths.

Let: S = *length of shortest link*
              L = *length of longest link*

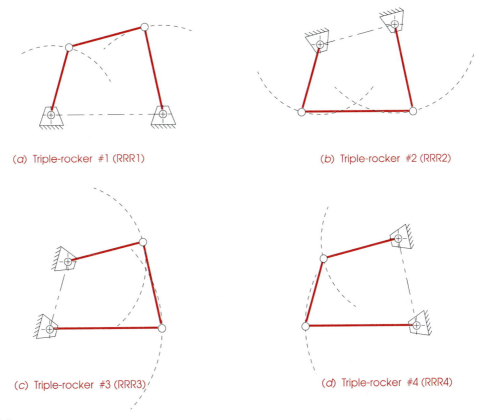

(*a*) Triple-rocker #1 (RRR1)

(*b*) Triple-rocker #2 (RRR2)

(*c*) Triple-rocker #3 (RRR3)

(*d*) Triple-rocker #4 (RRR4)

FIGURE 2-16

All inversions of the non-Grashof fourbar linkage are triple rockers

P = *length of one remaining link*
Q = *length of other remaining link*

Then if : $$S+L \leq P+Q$$ (2.8)

the linkage is **Grashof** and at least one link will be capable of making a full revolution with respect to the ground plane. This is called a **Class I** kinematic chain. If the inequality is not true, then the linkage is **non-Grashof** and *no* link will be capable of a complete revolution relative to any other link. This is a **Class II** kinematic chain.

Note that the above statements apply regardless of the order of assembly of the links. That is, the determination of the Grashof condition can be made on a set of unassembled links. Whether they are later assembled into a kinematic chain in $S, L, P, Q,$ or S, P, L, Q or any other order, will *not* change the Grashof condition.

The motions possible from a fourbar linkage will depend on both the Grashof condition and the **inversion** chosen. The inversions will be defined with respect to the shortest link. The motions are:

For the Class I case, $S + L < P + Q$:

Ground either link adjacent to the shortest and you get a **crank-rocker**, in which the shortest link will fully rotate and the other link pivoted to ground will oscillate.

Ground the shortest link and you will get a **double-crank**, in which both links pivoted to ground make complete revolutions as does the coupler.

Ground the link opposite the shortest and you will get a **Grashof double-rocker**, in which both links pivoted to ground oscillate and only the coupler makes a full revolution.

For the Class II case, $S + L > P + Q$:

All inversions will be **triple-rockers** [9] in which no link can fully rotate.

For the Class III case, $S + L = P + Q$:

Referred to as **special-case Grashof** and also as a **Class III** kinematic chain, all inversions will be either **double-cranks** or **crank-rockers** but will have "**change points**" twice per revolution of the input crank when the links all become colinear. At these change points the output behavior will become indeterminate. The linkage behavior is then unpredictable as it may assume either of two configurations. Its motion must be limited to avoid reaching the change points or an additional, out-of-phase link provided to guarantee a "carry through" of the change points. (See Figure 2-17c.)

Figure 2-15 (p. 46) shows the four possible inversions of the **Grashof case**: two crank-rockers, a double-crank (also called a drag link), and a double-rocker with rotating coupler. The two crank-rockers give similar motions and so are not distinct from one another. Figure 2-16 (p. 47) shows four non-distinct inversions, all triple-rockers, of a **non-Grashof linkage**.

Figure 2-17a and b shows the **parallelogram** and **antiparallelogram** configurations of the **special-case Grashof** linkage. The parallelogram linkage is quite useful as it exactly duplicates the rotary motion of the driver crank at the driven crank. One common use is to couple the two windshield wiper output rockers across the width of the windshield on an automobile. The coupler of the parallelogram linkage is in curvilinear translation, remaining at the same angle while all points on it describe identical circular paths. It is often used for this parallel motion, as in truck tailgate lifts and industrial robots.

The antiparallelogram linkage is also a double-crank, but the output crank has an angular velocity different from the input crank. Note that the change points allow the linkage to switch unpredictably between the parallelogram and antiparallelogram forms every 180 degrees unless some additional links are provided to carry it through those positions. This can be achieved by adding an out-of-phase companion linkage coupled to the same crank, as shown in Figure 2-17c. A common application of this double parallelogram linkage was on steam locomotives, used to connect the drive wheels together. The change points were handled by providing the duplicate linkage, 90 degrees out of phase, on the other side of the locomotive's axle shaft. When one side was at a change point, the other side would drive it through.

The **double-parallelogram** arrangement shown in Figure 2-17c is quite useful as it gives a translating coupler which remains horizontal in all positions. The two parallelo-

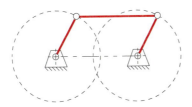

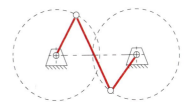

(a) Parallelogram form   *(b)* Antiparallelogram form

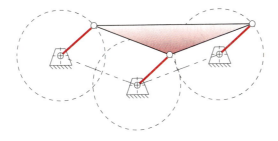

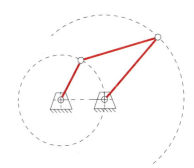

(c) Double-parallelogram linkage gives parallel
motion (pure curvilinear translation) to coupler
and also carries through the change points

(d) Deltoid or kite form

FIGURE 2-17

Some forms of the special-case Grashof linkage

gram stages of the linkage are out of phase so each carries the other through its change points. Figure 2-17d shows the **deltoid** or **kite** configuration which is a crank-rocker.

There is nothing either bad or good about the Grashof condition. Linkages of all three persuasions are equally useful in their place. If, for example, your need is for a motor driven windshield wiper linkage, you may want a non-special-case Grashof crank-rocker linkage in order to have a rotating link for the motor's input, plus a special-case parallelogram stage to couple the two sides together as described above. If your need is to control the wheel motions of a car over bumps, you may want a non-Grashof triple-rocker linkage for short stroke oscillatory motion. If you want to exactly duplicate some input motion at a remote location, you may want a special-case Grashof parallelogram linkage, as used in a drafting machine. In any case, this simply determined condition tells volumes about the behavior to be expected from a proposed fourbar linkage design prior to any construction of models or prototypes.

Classification of the Fourbar Linkage

Barker [10] has developed a classification scheme that allows prediction of the type of motion that can be expected from a fourbar linkage based on the values of its link ratios. A linkage's angular motion characteristics are independent of the absolute values of its link lengths. This allows the link lengths to be normalized by dividing three of them by the fourth to create three dimensionless ratios that define its geometry.

The positive octant of this space, bounded by the λ_1–λ_3, λ_1–λ_4, λ_3–λ_4 planes and the four zero-mobility planes (equation 2.10) contains eight volumes that are separated by the change-point planes (equation 2.11). Each volume contains mechanisms unique to one of the first eight classifications in Table 2-4 (p. 50). These eight volumes are in contact with one another in the solution space, but to show their shapes, they have been "exploded" apart in Figure 2-18 (p. 51). The remaining six change-point mechanisms of Table 2-4 exist only in the change-point planes which are the interfaces between the eight volumes. For more detail on this solution space and Barker's classification system than space permits here, see reference [10].

2.13   LINKAGES OF MORE THAN FOUR BARS

Geared Fivebar Linkages

We have seen that the simplest one-*DOF* linkage is the fourbar mechanism. It is an extremely versatile and useful device. Many quite complex motion control problems can be solved with just four links and four pins. Thus in the interest of simplicity, designers should always first try to solve their problems with a fourbar linkage. However, there will be cases when a more complicated solution is necessary. Adding one link and one joint to form a fivebar (Figure 2-19a) will increase the *DOF* by one, to two. By adding a pair of gears to tie two links together with a new half joint, the *DOF* is reduced again to one, and the **geared fivebar mechanism (GFBM)** of Figure 2-19b is created.

The geared fivebar mechanism provides more complex motions than the fourbar mechanism at the expense of the added link and gearset as can be seen in Appendix E. The reader may also observe the dynamic behavior of the linkage shown in Figure 2-19b by running the program FIVEBAR provided with this text and opening the data file F02-19b.5br. See Appendix A for instructions in running the program. Accept all the default values, and animate the linkage.

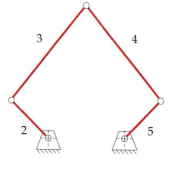
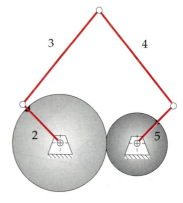

(*a*) Fivebar linkage—2 *DOF*   (*b*) Geared fivebar linkage—1 *DOF*

FIGURE 2-19

Two forms of the fivebar linkage

Sixbar Linkages

We already met the Watt's and Stephenson's sixbar mechanisms. See Figure 2-14 (p. 45). The **Watt's sixbar** can be thought of as *two fourbar linkages connected in series* and sharing two links in common. The **Stephenson's sixbar** can be thought of as *two four-bar linkages connected in parallel* and sharing two links in common. Many linkages can be designed by the technique of combining multiple fourbar chains as *basic building blocks* into more complex assemblages. Many real design problems will require solutions consisting of more than four bars. Some Watt's and Stephenson's linkages are provided as built-in examples to the program SIXBAR supplied with this text. You may run that program to observe these linkages dynamically. Select any example from the menu, accept all default responses, and animate the linkages.

Grashof-Type Rotatability Criteria for Higher-Order Linkages

Rotatability is defined as *the ability of at least one link in a kinematic chain to make a full revolution with respect to the other links* and defines the chain as Class I, II or III. **Revolvability** refers to *a specific link in a chain and indicates that it is one of the links that can rotate*.

ROTATABILITY OF GEARED FIVEBAR LINKAGES Ting [11] has derived an expression for rotatability of the geared fivebar linkage that is similar to the fourbar's Grashof criterion. Let the link lengths be designated L_1 through L_5 in order of increasing length,

$$\text{then if :} \qquad L_1 + L_2 + L_5 < L_3 + L_4 \qquad (2.12)$$

the two shortest links can revolve fully with respect to the others and the linkage is designated a **Class I** kinematic chain. If this inequality is *not* true, then it is a **Class II** chain and may or may not allow any links to fully rotate depending on the gear ratio and phase angle between the gears. If the inequality of equation 2.12 is replaced with an equal sign, the linkage will be a **Class III** chain in which the two shortest links can fully revolve but it will have change points like the special-case Grashof fourbar.

Reference [11] describes the conditions under which a Class II geared fivebar linkage will and will not be rotatable. In practical design terms, it makes sense to obey equation 2.12 in order to guarantee a "Grashof" condition. It also makes sense to avoid the Class III change-point condition. Note that if one of the short links (say L_2) is made zero, equation 2.12 reduces to the Grashof formula of equation 2.8.

In addition to the linkage's rotatability, we would like to know about the kinds of motions that are possible from each of the five inversions of a fivebar chain. Ting [11] describes these in detail. But, if we want to apply a gearset between two links of the five-bar chain (to reduce its *DOF* to 1), we really need it to be a double crank linkage, with the gears attached to the two cranks. A Class I fivebar chain will be a **double-crank** mechanism if *the two shortest links are among the set of three links that comprise the mechanism's ground link and the two cranks pivoted to ground*. [11]

ROTATABILITY OF N-BAR LINKAGES Ting et al. [12], [13] have extended rotatability criteria to all single-loop linkages of *N*-bars connected with revolute joints and have developed general theorems for **linkage rotatability** and the **revolvability** of individual links based on link lengths. Let the links of an *N*-bar linkage be denoted by L_i ($i = 1, 2,$

2

...N), with $L_1 \le L_2 \le \cdots \le L_N$. The links need not be connected in any particular order as rotatability criteria are independent of that factor.

A single-loop, revolute-jointed linkage of N links will have $(N - 3)$ *DOF*. The necessary and sufficient condition for the **assemblability** of an N-Bar linkage is:

$$L_N \le \sum_{k=1}^{N-1} L_k \tag{2.13}$$

A link K will be a so-called *short* link if

$$\{K\}_{k=1}^{N-3} \tag{2.14a}$$

and a so-called *long* link if

$$\{K\}_{k=N-2}^{N} \tag{2.14b}$$

There will be three long links and $(N - 3)$ short links in any linkage of this type.

A single-loop N-bar kinematic chain containing only first-order revolute joints will be a Class I, Class II, or Class III linkage depending on whether the sum of the lengths of its longest link and its $(N - 3)$ shortest links is, respectively, less than, greater than, or equal to the sum of the lengths of the remaining two long links:

Class I : $L_N + (L_1 + L_2 + \cdots + L_{N-3}) < L_{N-2} + L_{N-1}$

Class II : $L_N + (L_1 + L_2 + \cdots + L_{N-3}) > L_{N-2} + L_{N-1}$ (2.15)

Class III : $L_N + (L_1 + L_2 + \cdots + L_{N-3}) = L_{N-2} + L_{N-1}$

and, for a Class I linkage, there must be one and only one long link between two non-input angles. These conditions are necessary and sufficient to define the rotatability.

The **revolvability** of any link L_i is defined as its ability to rotate fully with respect to the other links in the chain and can be determined from:

$$L_i + L_N \le \sum_{k=1, \, k \ne i}^{N-1} L_k \tag{2.16}$$

Also, if L_i is a revolvable link, any link that is not longer than L_i will also be revolvable.

Additional theorems and corollaries regarding limits on link motions can be found in references [12] and [13]. Space does not permit their complete exposition here. Note that the rules regarding the behavior of geared fivebar linkages and fourbar linkages (the Grashof law) stated above are consistent with, and contained within, these general rotatability theorems.

2.14   SPRINGS AS LINKS

We have so far been dealing only with rigid links. In many mechanisms and machines, it is necessary to counterbalance the static loads applied to the device. A common example is the hood hinge mechanism on your automobile. Unless you have the (cheap) model with the strut that you place in a hole to hold up the hood, it will probably have either a fourbar or sixbar linkage connecting the hood to the body on each side. The hood may be the coupler of a non-Grashof linkage whose two rockers are pivoted to the body. A spring is fitted between two of the links to provide a force to hold the hood in the open position. The spring in this case is an additional link of variable length. As long as it can provide the right amount of force, it acts to reduce the *DOF* of the mechanism to zero, and holds the system in static equilibrium. However, you can force it to again be a one-*DOF* system by overcoming the spring force when you pull the hood shut.

Another example, which may now be right next to you, is the ubiquitous adjustable arm desk lamp, shown in Figure 2-20. This device has two springs that counterbalance the weight of the links and lamp head. If well designed and made, it will remain stable over a fairly wide range of positions despite variation in the overturning moment due to the lamp head's changing moment arm. This is accomplished by careful design of the geometry of the spring-link relationships so that, as the spring force changes with increasing length, its moment arm also changes in a way that continually balances the changing moment of the lamp head.

A linear spring can be characterized by its spring constant, $k = F / x$, where F is force and x is spring displacement. Doubling its deflection will double the force. Most coil springs of the type used in these examples are linear. The design of spring-loaded linkages will be addressed in a later chapter.

FIGURE 2-20

A spring-balanced linkage mechanism

2.15 PRACTICAL CONSIDERATIONS

There are many factors that need to be considered to create good-quality designs. Not all of them are contained within the applicable theories. A great deal of art based on experience is involved in design as well. This section attempts to describe a few such practical considerations in machine design.

PIN JOINTS VERSUS SLIDERS AND HALF JOINTS

Proper material selection and good lubrication are the key to long life in any situation, such as a joint, where two materials rub together. Such an interface is called a **bearing**. Assuming the proper materials have been chosen, the choice of joint type can have a significant effect on the ability to provide good, clean lubrication over the lifetime of the machine.

REVOLUTE (PIN) JOINTS The simple revolute or pin joint (Figure 2-21a) is the clear winner here for several reasons. It is relatively easy and inexpensive to design and build a good quality pin joint. In its pure form—a so-called **sleeve** or **journal** bearing—the geometry of pin-in-hole traps a lubricant film within its annular interface by capillary action and promotes a condition called *hydrodynamic lubrication* in which the parts are separated by a thin film of lubricant as shown in Figure 2-22. Seals can easily be provided at the ends of the hole, wrapped around the pin, to prevent loss of the lubricant. Replacement lubricant can be introduced through radial holes into the bearing interface, either continuously or periodically, without disassembly.

(a) Pin joint

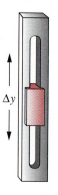

(b) Slider joint

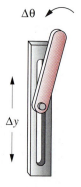

(c) Half joint

FIGURE 2-21

Joints of various types

A convenient form of bearing for linkage pivots is the commercially available **spherical rod end** shown in Figure 2-23. This has a spherical, sleeve-type bearing which *self-aligns* to a shaft that may be out of parallel. Its body threads onto the link, allowing links to be conveniently made from round stock with threaded ends that allow adjustment of link length.

Relatively inexpensive **ball** and **roller bearings** are commercially available in a large variety of sizes for revolute joints as shown in Figure 2-24. Some of these bearings (principally ball type) can be obtained prelubricated and with end seals. Their rolling elements provide low-friction operation and good dimensional control. Note that *rolling-element bearings* actually contain higher-joint interfaces (half joints) at each ball or roller, which is potentially a problem as noted below. However, the ability to trap lubricant within the roll cage (by end seals) combined with the relatively high rolling speed of the balls or rollers promotes hydrodynamic lubrication and long life. For more detailed information on bearings and lubrication, see reference [15].

For revolute joints pivoted to ground, several commercially available bearing types make the packaging easier. **Pillow blocks** and **flange-mount bearings** (Figure 2-25) are available fitted with either rolling-element (ball, roller) bearings or sleeve-type journal bearings. The pillow block allows convenient mounting to a surface parallel to the pin axis, and flange mounts fasten to surfaces perpendicular to the pin axis.

PRISMATIC (SLIDER) JOINTS require a carefully machined and straight slot or rod (Figure 2-21b). The bearings often must be custom made, though linear ball bearings (Figure 2-26) are commercially available but must be run over hardened and ground shafts. Lubrication is difficult to maintain in any sliding joint. The lubricant is not geometrically captured, and it must be resupplied either by running the joint in an oil bath or by periodic manual regreasing. An open slot or shaft tends to accumulate airborne dirt particles which can act as a grinding compound when trapped in the lubricant. This will accelerate wear.

HIGHER (HALF) JOINTS such as a round pin in a slot (Figure 2-21c) or a cam-follower joint (Figure 2-10c, p. 41) suffer even more acutely from the slider's lubrication problems, because they typically have two oppositely curved surfaces in line contact, which tend to squeeze any lubricant out of the joint. This type of joint needs to be run in an oil bath for long life. This requires that the assembly be housed in an expensive, oil-tight box with seals on all protruding shafts.

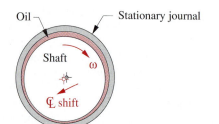

FIGURE 2-22

Hydrodynamic lubrication in a sleeve bearing—clearance and motions exaggerated

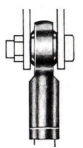

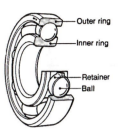

(a) Ball bearing

FIGURE 2-23

Spherical rod end *Courtesy of Emerson Power Transmission, Ithaca, NY.*

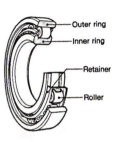

(b) Roller bearing

These joint types are all used extensively in machinery with great success. As long as the proper *attention to engineering detail* is paid, the design can be successful. Some common examples of all three joint types can be found in an automobile. The windshield wiper mechanism is a pure pin-jointed linkage. The pistons in the engine cylinders are true sliders and are bathed in engine oil. The valves in the engine are opened and closed by cam-follower (half) joints which are drowned in engine oil. You probably change your engine oil fairly frequently. When was the last time you lubricated your windshield wiper linkage? Has this linkage (not the motor) ever failed?

Cantilever or Straddle Mount?

Any joint must be supported against the joint loads. Two basic approaches are possible as shown in Figure 2-27. A cantilevered joint has the pin (journal) supported only, as a cantilever beam. This is sometimes necessary as with a crank that must pass over the coupler and cannot have anything on the other side of the coupler. However, a cantile

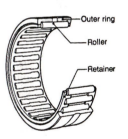

(c) Needle bearing

FIGURE 2-24

Ball, roller and needle bearings for revolute joints. *Courtesy of NTN Corporation, Japan.*

(a) Pillow-block bearing   (b) Flange-mount bearing

FIGURE 2-25

Pillow block and flange-mount bearing units. *Courtesy of Emerson Power Transmission, Ithaca, NY.*

FIGURE 2-26

Linear ball bushing *Courtesy of Thomson Industries, Port Washington, NY*

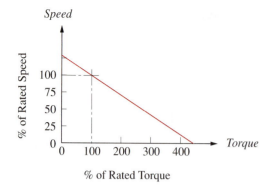

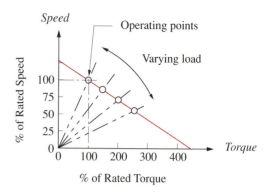

(*a*) Speed–torque characteristic of a PM electric motor   (*b*) Load lines superposed on speed–torque curve

FIGURE 2-32

DC permanent magnet (PM) electric motor's typical speed-torque characteristic

variations in most motors, regardless of their design.[*] If constant speed is required, this may be unacceptable. Other types of DC motors have either more or less speed sensitivity to load than the PM motor. *A motor is typically selected based on its torque-speed curve.*

SHUNT-WOUND DC MOTORS have a torque speed curve like that shown in Figure 2-33a. Note the flatter slope around the rated torque point (at 100%) compared to Figure 2-32. The shunt-wound motor is less speed-sensitive to load variation in its operating range, but stalls very quickly when the load exceeds its maximum overload capacity of about 250% of rated torque. Shunt-wound motors are typically used on fans and blowers.

SERIES-WOUND DC MOTORS have a torque-speed characteristic like that shown in Figure 2-33b. This type is more speed-sensitive than the shunt or PM configurations. However, its starting torque can be as high as 800% of full-load rated torque. It also does not have any theoretical maximum no-load speed which makes it tend to run away if the load is removed. Actually, friction and windage losses will limit its maximum speed which can be as high as 20,000 to 30,000 revolutions per minute (rpm). Overspeed detectors are sometimes fitted to limit its unloaded speed. Series-wound motors are used in sewing machines and portable electric drills where their speed variability can be an advantage as it can be controlled, to a degree, with voltage variation. They are also used in heavy-duty applications such as vehicle traction drives where their high starting torque is an advantage. Also their speed sensitivity (large slope) is advantageous in high-load applications as it gives a "soft-start" when moving high-inertia loads. The motor's tendency to slow down when the load is applied cushions the shock that would be felt if a large step in torque were suddenly applied to the mechanical elements.

COMPOUND-WOUND DC MOTORS have their field and armature coils connected in a combination of series and parallel. As a result their torque-speed characteristic has aspects of both the shunt-wound and series-wound motors as shown in Figure 2-33c.

---

[*] The synchronous AC motor and the speed-controlled DC motor are exceptions.

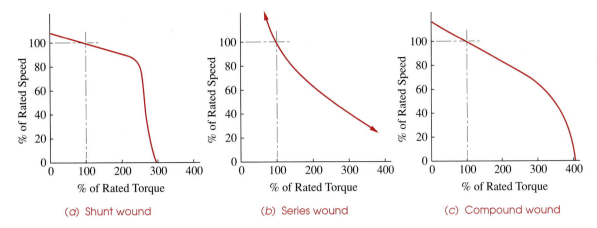

(a) Shunt wound   (b) Series wound   (c) Compound wound

FIGURE 2-33

Torque-speed curves for three types of DC motor

Their speed sensitivity is greater than a shunt-wound but less than a series-wound motor and it will not run away when unloaded. This feature plus its high starting torque and soft-start capability make it a good choice for cranes and hoists which experience high inertial loads and can suddenly lose the load due to cable failure, creating a potential run-away problem if the motor does not have a self-limited no-load speed.

SPEED-CONTROLLED DC MOTORS   If precise speed control is needed, as is often the case in production machinery, another solution is to use a speed-controlled DC motor which operates from a controller that increases and decreases the current to the motor in the face of changing load to try to maintain constant speed. These speed-controlled (typically PM) DC motors will run from an AC source since the controller also converts AC to DC. The cost of this solution is high, however. Another possible solution is to provide a **flywheel** on the input shaft, which will store kinetic energy and help smooth out the speed variations introduced by load variations. Flywheels will be investigated in Chapter 11.

AC MOTORS   are the least expensive way to get continuous rotary motion, and they can be had with a variety of *torque-speed* curves to suit various load applications. They are limited to a few standard speeds that are a function of the AC line frequency (60 Hz in North America, 50 Hz elsewhere). The synchronous motor speed n_s is a function of line frequency f and the number of magnetic poles p present in the rotor.

$$n_s = \frac{120f}{p} \tag{2.17}$$

Synchronous motors "lock on" to the AC line frequency and run exactly at synchronous speed. These motors are used for clocks and timers. Nonsynchronous AC motors have a small amount of slip which makes them lag the line frequency by about 3 to 10%.

Table 2-6 shows the synchronous and nonsynchronous speeds for various AC motor-pole configurations. The most common AC motors have 4 poles, giving nonsynchro-

2

TABLE 2-6
AC Motor Speeds

Poles	Sync rpm	Async rpm
2	3600	3450
4	1800	1725
6	1200	1140
8	900	850
10	720	690
12	600	575

nous *no-load speeds* of about 1725 rpm, which reflects slippage from the 60-Hz synchronous speed of 1800 rpm.

Figure 2-34 shows typical torque-speed curves for single-phase (1φ) and 3-phase (3φ) AC motors of various designs. The single-phase shaded pole and permanent split capacitor designs have a starting torque lower than their full-load torque. To boost the start torque, the split-phase and capacitor-start designs employ a separate starting circuit that is cut off by a centrifugal switch as the motor approaches operating speed. The broken curves indicate that the motor has switched from its starting circuit to its running circuit. The NEMA* three-phase motor designs B, C, and D in Figure 2-34 differ mainly in their starting torque and in speed sensitivity (slope) near the full-load point.

GEARMOTORS   If different single (as opposed to variable) output speeds than the standard ones of Table 2-6 are needed, a gearbox speed reducer can be attached to the motor's output shaft, or a gearmotor can be purchased that has an integral gearbox. Gearmotors are commercially available in a large variety of output speeds and power ratings. The kinematics of gearbox design are covered in Chapter 9.

SERVOMOTORS   are fast-response, closed-loop-controlled motors capable of providing a programmed function of acceleration or velocity, as well as of holding a fixed position against a load. **Closed loop** means that *sensors on the output device being moved feed back information on its position, velocity, and acceleration.* Circuitry in the motor controller responds to the fed back information by reducing or increasing (or reversing) the current flow to the motor. Precise positioning of the output device is then possible, as is control of the speed and shape of the motor's response to changes in load or input commands. These are very expensive devices which are commonly used in applications such as moving the flight control surfaces in aircraft and guided missiles, and in controlling robots, for example. Servomotors have lower power and torque capacity than is available from nonservo AC or DC motors.

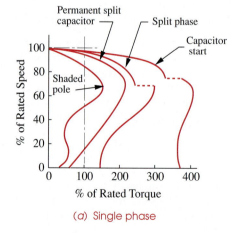

(a) Single phase

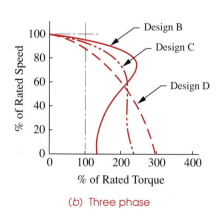

(b) Three phase

* National Electrical
Manufacturers Association.

FIGURE 2-34

Torque-speed curves for single- and three-phase AC motors

2

STEPPER MOTORS   are designed to position an output device.   Unlike servomotors, these run **open loop**, meaning they *receive no feedback as to whether the output device has responded as requested*. Thus they can get out of phase with the desired program. They will, however, happily sit energized for an indefinite period, holding the output in one position. Their internal construction consists of a number of magnetic strips arranged around the circumference of both the rotor and stator. When energized, the rotor will move one step, to the next magnet, for each pulse received. Thus, these are **intermittent motion** devices and do not provide continuous rotary motion like other motors. The number of magnetic strips determines their resolution (typically a few degrees per step). They are relatively small compared to AC/DC motors and have low torque capacity. They are moderately expensive and require special controllers.

Air and Hydraulic Motors

These have more limited application than electric motors, simply because they require the availability of a compressed air or hydraulic source. Both of these devices are less energy efficient than the direct electrical to mechanical conversion of electric motors, because of the losses associated with the conversion of the energy first from chemical or electrical to fluid pressure and then to mechanical form. Every energy conversion involves some losses. Air motors find widest application in factories and shops, where high-pressure compressed air is available for other reasons. A common example is the air impact wrench used in automotive repair shops. Although individual air motors and air cylinders are relatively inexpensive, these pneumatic systems are quite expensive when the cost of all the ancillary equipment is included. Hydraulic motors are most often found within machines or systems such as construction equipment (cranes), aircraft, and ships, where high-pressure hydraulic fluid is provided for many purposes. Hydraulic systems are very expensive when the cost of all the ancillary equipment is included.

Air and Hydraulic Cylinders

These are linear actuators (piston in cylinder) which provide a limited stroke, straight-line output from a pressurized fluid flow input of either compressed air or hydraulic fluid (usually oil). They are the method of choice if you need a linear motion as the input. However, they share the same high cost, low efficiency, and complication factors as listed under their air and hydraulic motor equivalents above.

Another problem is that of control. Most motors, left to their own devices, will tend to run at a constant speed. A linear actuator, when subjected to a constant pressure fluid source, typical of most compressors, will respond with more nearly constant acceleration, which means its velocity will increase linearly with time. This can result in severe impact loads on the driven mechanism when the actuator comes to the end of its stroke at maximum velocity. Servovalve control of the fluid flow, to slow the actuator at the end of its stroke, is possible but is quite expensive.

The most common application of fluid power cylinders is in farm and construction equipment such as tractors and bulldozers, where open loop (nonservo) hydraulic cylinders actuate the bucket or blade through linkages. The cylinder and its piston become two of the links (slider and track) in a slider-crank mechanism. See Figure 1-1b (p. 7).

Solenoids

These are electromechanical (AC or DC) linear actuators which share some of the limitations of air cylinders, and they possess a few more of their own. They are *energy inefficient*, are limited to very short strokes (about 2 to 3 cm), develop a force which varies exponentially over the stroke, and deliver high impact loads. They are, however, inexpensive, reliable, and have very rapid response times. They cannot handle much power, and they are typically used as control or switching devices rather than as devices which do large amounts of work on a system.

A common application of solenoids is in camera shutters, where a small solenoid is used to pull the latch and trip the shutter action when you push the button to take the picture. Its nearly instantaneous response is an asset in this application, and very little work is being done in tripping a latch. Another application is in electric door or trunk locking systems in automobiles, where the click of their impact can be clearly heard when you turn the key (or press the button) to lock or unlock the mechanism.

2.17 REFERENCES

1   **Reuleaux, F.** (1963). *The Kinematics of Machinery*. A. B. W. Kennedy, translator. Dover Publications: New York, pp. 29-55.

2   **Gruebler, M.** (1917). *Getriebelehre*. Springer Verlag: Berlin.

3   **Fang, W. E., and F. Freudenstein**. (1990). "The Stratified Representation of Mechanisms." *Journal of Mechanical Design*. **112**(4), p. 514.

4   **Kim, J. T., and B. M. Kwak.** (1992). "An Algorithm of Topological Ordering for Unique Representation of Graphs." *Journal of Mechanical Design*, **114**(1), p. 103.

5   **Tang, C. S., and T. Liu**. (1993). "The Degree Code—A New Mechanism Identifier." *Journal of Mechanical Design*, **115**(3), p. 627.

6   **Dhararipragada, V. R., et al.** (1994). "A More Direct Method for Structural Synthesis of Simple-Jointed Planar Kinematic Chains." *Proc. of 23rd Biennial Mechanisms Conference*, Minneapolis, MI, p. 507.

7   **Yadav, J. N., et al.** (1995). "Detection of Isomorphism Among Kinematic Chains Using the Distance Concept." *Journal of Mechanical Design*, **117**(4).

8   **Grashof, F.** (1883). *Theoretische Maschinenlehre*. Vol. 2. Voss: Hamburg.

9   **Paul, B.** (1979). "A Reassessment of Grashof's Criterion." *Journal of Mechanical Design*, **101**(3), pp. 515-518.

10 **Barker, C.** (1985). "A Complete Classification of Planar Fourbar Linkages." *Mechanism and Machine Theory*, **20**(6), pp. 535-554.

11 **Ting, K. L.** (1993). "Fully Rotatable Geared Fivebar Linkages." *Proc. of 3rd Applied Mechanisms and Robotics Conference*, Cincinnati, pp. 67-1.

12 **Ting, K. L., and Y. W. Liu**. (1991). "Rotatability Laws for N-Bar Kinematic Chains and Their Proof." *Journal of Mechanical Design*, **113**(1), pp. 32-39.

13 **Shyu, J. H., and K. L. Ting**. (1994). "Invariant Link Rotatability of N-Bar Kinematic Chains." *Journal of Mechanical Design*, **116**(1), p. 343.

14   **Miller, W. S.**, ed. *Machine Design Electrical and Electronics Reference Issue*. Penton
      Publishing: Cleveland, Ohio.

15   **Norton, R. L.** (1998). *Machine Design: An Integrated Approach*. Prentice-Hall:
      Upper Saddle River, NJ.

2.18   PROBLEMS

*2-1   Find three (or other number as assigned) of the following common devices. Sketch
       careful kinematic diagrams and find their total degrees of freedom.

   a.   An automobile hood hinge mechanism
   b.   An automobile hatchback lift mechanism
   c.   An electric can opener
   d.   A folding ironing board
   e.   A folding card table
   f.   A folding beach chair
   g.   A baby swing
   h.   A folding baby walker
   i.   A drafting machine
   j.   A fancy corkscrew
   k.   A windshield wiper mechanism
   l.   A dump truck dump mechanism
   m.   A trash truck dumpster mechanism
   n.   A station wagon tailgate mechanism
   o.   An automobile jack
   p.   A collapsible auto radio antenna
   q.   A record turntable and tone arm

2-2   How many *DOF* do you have in your wrist and hand combined?

*2-3   How many *DOF* do the following joints have?

   a.   Your knee
   b.   Your ankle
   c.   Your shoulder
   d.   Your hip
   e.   Your knuckle

*2-4   How many *DOF* do the following have in their normal environment?

   a.   A submerged submarine   b. An earth-orbiting satellite
   c.   A surface ship   d. A motorcycle
   e.   The print head in a 9-pin dot matrix computer printer
   f.   The pen in an XY plotter

*2-5   Are the joints in Problem 2-3 force closed or form closed?

*2-6   Describe the motion of the following items as pure rotation, pure translation, or
       complex planar motion.

   a.   A windmill
   b.   A bicycle (in the vertical plane, not turning)
   c.   A conventional "double-hung" window
   d.   The keys on a computer keyboard
   e.   The hand of a clock
   f.   A hockey puck on the ice

_____
* Answers in Appendix F

g. The pen in an XY plotter
h. The print head in a computer printer
i. A "casement" window

*2-7 Calculate the *DOF* of the linkages shown in Figure P2-1.

*2-8 Identify the items in Figure P2-1 as mechanisms, structures, or preloaded structures.

2-9 Use linkage transformation on the linkage of Figure P2-1a to make it a 1-*DOF* mechanism.

*2-10 Use linkage transformation on the linkage of Figure P2-1d to make it a 2-*DOF* mechanism.

2-11 Use number synthesis to find all the possible link combinations for 2-*DOF*, up to 9 links, to hexagonal order, using only revolute joints.

2-12 Find all the valid isomers of the eightbar 1-*DOF* link combinations in Table 2-2 (p. 36) having:

a. Four binary and four ternary links
b. Five binaries, two ternaries, and one quaternary link
c. Six binaries and two quaternary links
d. Six binaries, one ternary, and one pentagonal link

2-13 Use linkage transformation to create a 1-*DOF* mechanism with two sliding full joints from a Stephenson's sixbar linkage in Figure 2-14a (p. 45).

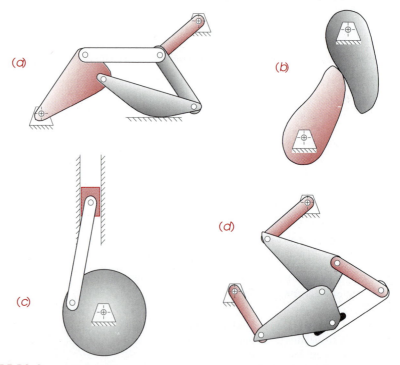

(a)

(b)

(d)

(c)

FIGURE P2-1

Linkages for Problems 2-7 to 2-10

2-14   Use linkage transformation to create a 1-*DOF* mechanism with one sliding full joint and a half joint from a Stephenson's sixbar linkage in Figure 2-14b.

*2-15   Calculate the Grashof condition of the fourbar mechanisms defined below. Build cardboard models of the linkages and describe the motions of each inversion. Link lengths are in inches (or double given numbers for centimeters).

a.	2	4.5	7	9
b.	2	3.5	7	9
c.	2	4.0	6	8

2-16   What type(s) of electric motor would you specify

    a.   To drive a load with large inertia.
    b.   To minimize variation of speed with load variation.
    c.   To maintain accurate constant speed regardless of load variations.

2-17   Describe the difference between a cam-follower (half) joint and a pin joint.

2-18   Examine an automobile hood hinge mechanism of the type described in Section 2.14. Sketch it carefully. Calculate its *DOF* and Grashof condition. Make a cardboard model. Analyze it with a free-body diagram. Describe how it keeps the hood up.

2-19   Find an adjustable arm desk lamp of the type shown in Figure P2-2. Measure it and sketch it to scale. Calculate its *DOF* and Grashof condition. Make a cardboard model. Analyze it with a free-body diagram. Describe how it keeps itself stable. Are there any positions in which it loses stability? Why?

2-20   Make kinematic sketches, define the types of all the links and joints, and determine the *DOF* of the mechanisms shown in Figure P2-3.

*2-21   Find the *DOF* of the mechanisms in Figure P2-4.

2-22   Find the Grashof condition and Barker classifications of the mechanisms in Figure P2-4a, b, and d. Scale the diagrams for dimensions.

FIGURE P2-2

Problem 2-19

2

* Answers in Appendix F

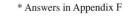

FIGURE P2-3

Problem 2-20 Backhoe and front-end loader *Courtesy of John Deere Co.*

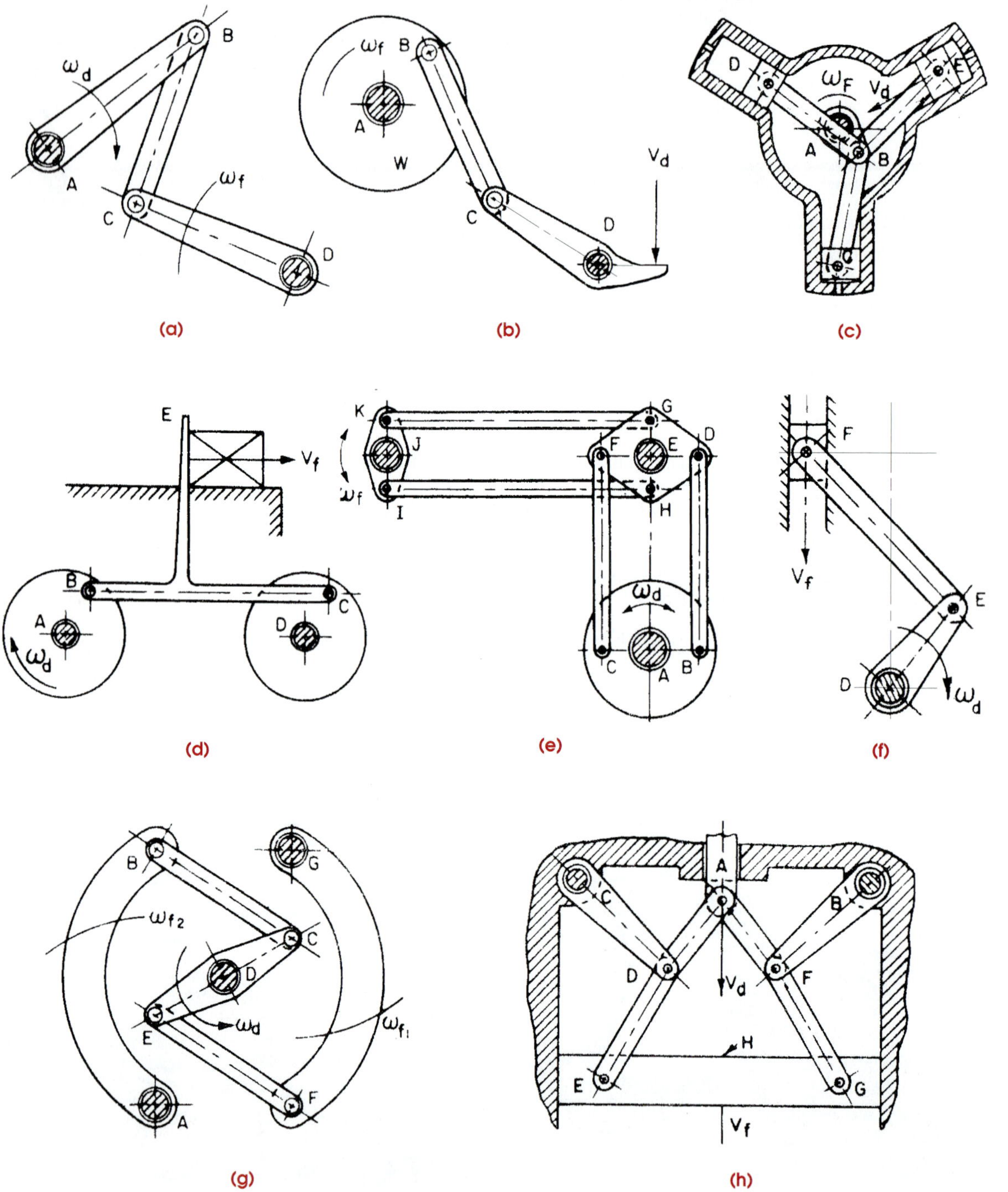

2

(a)

(b)

(c)

(d)

(e)

(f)

(g)

(h)

FIGURE P2-4

Problems 2-21 to 2-23   *Adapted from P. H. Hill and W. P. Rule. (1960) Mechanisms: Analysis and Design, with permission*

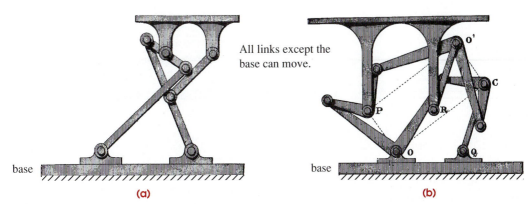

All links except the base can move.

base   base

(a)   (b)

FIGURE P2-5

Problem 2-24 Chebyschev (a) and Sylvester-Kempe (b) straight line mechanisms *Adapted from Kempe, How to Draw a Straight Line, Macmillan: London, 1877*

2-23 Find the rotatability of each loop of the mechanisms in Figure P2-4e, f, and g. Scale the diagrams for dimensions.

*2-24 Find the *DOF* of the mechanisms in Figure P2-5. Scale the diagrams for dimensions.

2-25 Find the *DOF* of the ice tongs in Figure P2-6.

     a.   When operating them to grab the ice block.
     b.   When clamped to the ice block but before it is picked up (ice grounded).
     c.   When the person is carrying the ice block with the tongs.

*2-26 Find the *DOF* of the automotive throttle mechanism in Figure P2-7.

2-27 Sketch a kinematic diagram of the scissors jack shown in Figure P2-8 and determine its *DOF*. Describe how it works.

2-28 Find the *DOF* of the corkscrew in Figure P2-9.

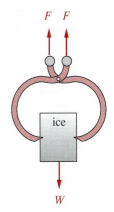

FIGURE P2-6

Problem 2-25

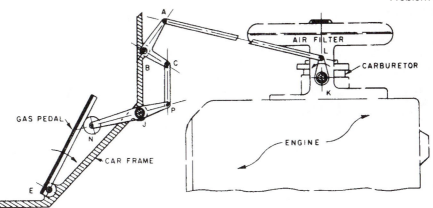

FIGURE P2-7

Problem 2-26. *Adapted from P. H. Hill and W. P Rule. (1960) Mechanisms: Analysis and Design, with permission.*

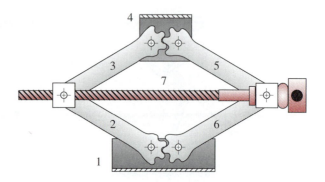

FIGURE P2-8

Problem 2-27

2-29   Figure P2-10 shows Watt's sun and planet drive that he used in his steam engine. The beam 2 is driven in oscillation by the piston of the engine. The planet gear is fixed rigidly to link 3 and its center is guided in the fixed track 1. The output rotation is taken from the sun gear 4. Sketch a kinematic diagram of this mechanism and determine its *DOF*. Can it be classified by the Barker scheme? If so, what Barker class and subclass is it?

2-30   Figure P2-11 shows a bicycle handbrake lever assembly. Sketch a kinematic diagram of this device and draw its equivalent linkage. Determine its *DOF*. Hint: Consider the flexible cable to be a link.

FIGURE 2-9

Problem 2-28

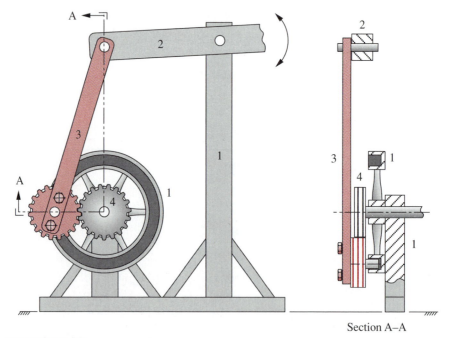

Section A–A

FIGURE P2-10

Problem 2-29   James Watt's sun and planet drive

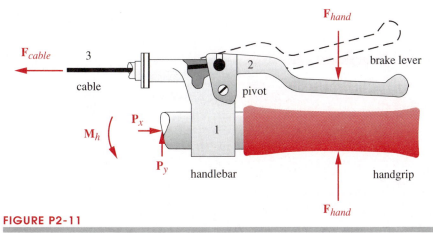

FIGURE P2-11

Problem 2-30   Bicycle hand brake lever assembly

2-31   Figure P2-12 shows a bicycle brake caliper assembly. Sketch a kinematic diagram of this device and draw its equivalent linkage. Determine its *DOF* under two conditions.

a.   Brake pads not contacting the wheel rim.
b.   Brake pads contacting the wheel rim.

Hint: Consider the flexible cables to be replaced by forces in this case.

2-32   Find the *DOF*, the Grashof condition, and the Barker classification of the mechanism in Figure P2-13.

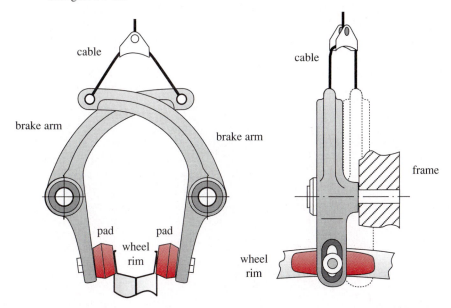

FIGURE P2-12

Problem 2-31   Bicycle brake caliper assembly

2

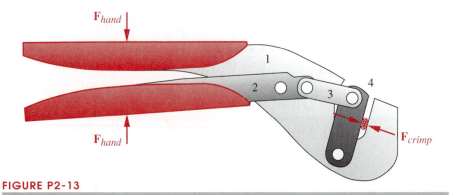

FIGURE P2-13

Problem 2-32   Crimping Tool

2-33   Figure P2-14 shows a "pick-and-place" mechanism in combination with a "walking beam." Sketch its kinematic diagram, determine its *DOF* and its type (i.e., is it a fourbar, a Watts sixbar, a Stephenson's sixbar, an eightbar, or what?) Make a cardboard model of all but the conveyor portion and examine its motions. Describe what it does. (It will help if you xerox the page and enlarge it. Then paste the copies on cardboard and cut out the links.)

2-34   Figure P2-15 shows a power hacksaw, used to cut metal. Sketch its kinematic diagram, determine its *DOF* and its type (i.e., is it a fourbar, a Watts sixbar, a Stephenson's sixbar, an eightbar, or what?) Use reverse linkage transformation to determine its pure revolute-jointed equivalent linkage.

2-35   Figure P2-16 shows a manual press used to compact powdered materials. Sketch its kinematic diagram, determine its *DOF* and its type (i.e., is it a fourbar, a Watts sixbar, a Stephenson's sixbar, an eightbar, or what?) Use reverse linkage transformation to determine its pure revolute-jointed equivalent linkage.

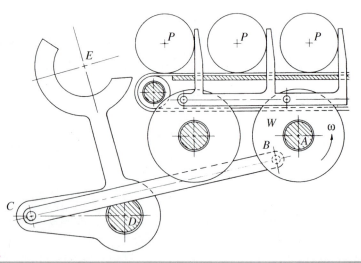

FIGURE P2-14

Problem 2-33   Pusher and pick-and-place mechanism *Adapted from P. H. Hill and W. P. Rule. (1960). Mechanisms: Analysis and Design, with permission*

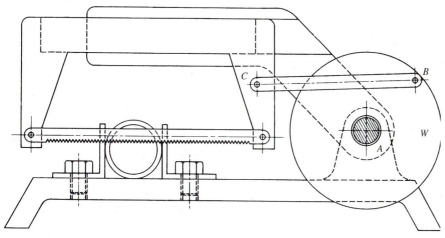

FIGURE P2-15

Problem 2-34   Power hacksaw. *Adapted from P. H. Hill and W. P Rule. (1960). Mechanisms: Analysis and Design, with permission.*

2-36   Sketch the equivalent linkage for the cam and follower mechanism in Figure P2-17 in the position shown. Show that it has the same *DOF* as the original mechanism.

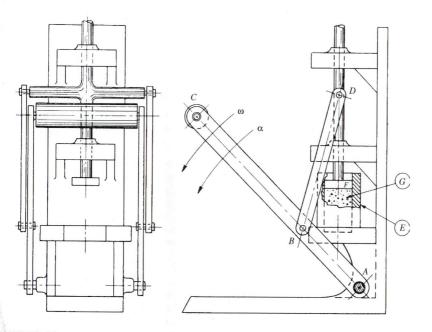

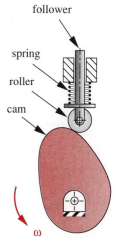

FIGURE P2-16

Problem 2-35   Powder compacting press *Adapted from P. H. Hill and W. P. Rule. (1960). Mechanisms: Analysis and Design, with permission*

FIGURE P2-17

Problem 2-36

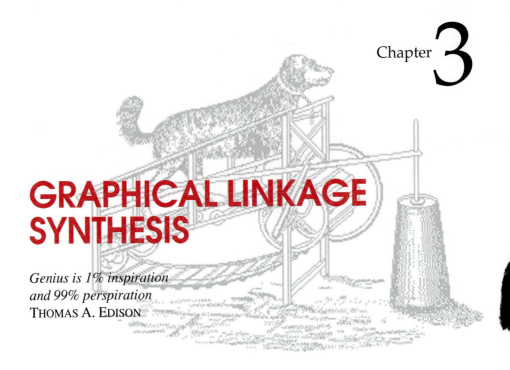

Chapter 3

GRAPHICAL LINKAGE SYNTHESIS

*Genius is 1% inspiration
and 99% perspiration*
THOMAS A. EDISON

3.0   INTRODUCTION

Most engineering design practice involves a combination of synthesis and analysis. Most engineering courses deal primarily with analysis techniques for various situations. However, one cannot analyze anything until it has been synthesized into existence. Many machine design problems require the creation of a device with particular motion characteristics. Perhaps you need to move a tool from position *A* to position *B* in a particular time interval. Perhaps you need to trace out a particular path in space to insert a part into an assembly. The possibilities are endless, but a common denominator is often the need for a linkage to generate the desired motions. So, we will now explore some simple synthesis techniques to enable you to create potential linkage design solutions for some typical kinematic applications.

3.1   SYNTHESIS

QUALITATIVE SYNTHESIS   means *the creation of potential solutions in the absence of a well-defined algorithm which configures or predicts the solution.* Since most real design problems will have many more unknown variables than you will have equations to describe the system's behavior, you cannot simply solve the equations to get a solution. Nevertheless you must work in this fuzzy context to create a potential solution and to also judge its **quality**. You can then analyze the proposed solution to determine its viability, and iterate between synthesis and analysis, as outlined in the **design process**, until you are satisfied with the result. Several tools and techniques exist to assist you in this process. The traditional tool is the drafting board, on which you lay out, to scale, multiple orthographic views of the design, and investigate its motions by drawing arcs, showing multiple positions, and using transparent, movable overlays. *Computer-aided drafting (CAD)* systems can speed this process to some degree, but you will probably find that the quickest way to get a sense of the quality of your linkage design is to model it, to scale, in cardboard or drafting *Mylar®* and see the motions directly.

Other tools are available in the form of computer programs such as FOURBAR, FIVE-BAR, SIXBAR, SLIDER, DYNACAM, ENGINE, and MATRIX (all included with this text), some of which do synthesis, but these are mainly analysis tools. They can analyze a trial mechanism solution so rapidly that their dynamic graphical output gives almost instantaneous visual feedback on the quality of the design. Commercially available programs such as *Working Model** also allow rapid analysis of a proposed mechanical design. The process then becomes one of **qualitative design by successive analysis** which is really *an iteration between synthesis and analysis.* Very many trial solutions can be examined in a short time using these *Computer-aided engineering* (CAE) tools. We will develop the mathematical solutions used in these programs in subsequent chapters in order to provide the proper foundation for understanding their operation. But, if you want to try these programs to reinforce some of the concepts in these early chapters, you may do so. Appendix A is a manual for the use of these programs, and it can be read at any time. Reference will be made to program features which are germane to topics in each chapter, as they are introduced. Data files for input to these computer programs are also provided on disk for example problems and figures in these chapters. The data file names are noted near the figure or example. The student is encouraged to input these sample files to the programs in order to observe more dynamic examples than the printed page can provide. These examples can be run by merely accepting the defaults provided for all inputs.

TYPE SYNTHESIS refers to *the definition of the proper type of mechanism best suited to the problem* and is a form of qualitative synthesis.[†] This is perhaps the most difficult task for the student as it requires some experience and knowledge of the various types of mechanisms which exist and which also may be feasible from a performance and manufacturing standpoint. As an example, assume that the task is to design a device to track the straight-line motion of a part on a conveyor belt and spray it with a chemical coating as it passes by. This has to be done at high, constant speed, with good accuracy and repeatability, and it must be reliable. Moreover, the solution must be inexpensive. Unless you have had the opportunity to see a wide variety of mechanical equipment, you might not be aware that this task could conceivably be accomplished by any of the following devices:

- *A straight-line linkage*
- *A cam and follower*
- *An air cylinder*
- *A hydraulic cylinder*
- *A robot*
- *A solenoid*

Each of these solutions, while possible, may not be optimal or even practical. More detail needs to be known about the problem to make that judgment, and that detail will come from the research phase of the design process. The straight-line linkage may prove to be too large and to have undesirable accelerations; the cam and follower will be expensive, though accurate and repeatable. The air cylinder itself is inexpensive but is noisy and unreliable. The hydraulic cylinder is more expensive, as is the robot. The so-

---

* The student version of *Working Model* is included on CD-ROM with this book. The professional version is available from Knowledge Revolution Inc., San Mateo CA 94402, (800) 766-6615

---

† A good discussion of type synthesis and an extensive bibliography on the topic can be found in Olson, D. G., et al. (1985). "A Systematic Procedure for Type Synthesis of Mechanisms with Literature Review." *Mechanism and Machine Theory*, **20**(4), pp. 285-295.

lenoid, while cheap, has high impact loads and high impact velocity. So, you can see that the choice of device type can have a large effect on the quality of the design. A poor choice at the type synthesis stage can create insoluble problems later on. The design might have to be scrapped after completion, at great expense. **Design is essentially an exercise in trade-offs**. Each proposed type of solution in this example has good and bad points. Seldom will there be a clear-cut, obvious solution to a real engineering design problem. It will be your job as a design engineer to balance these conflicting features and find a solution which gives the best trade-off of functionality against cost, reliability, and all other factors of interest. Remember, *an engineer can do, with one dollar, what any fool can do for ten dollars*. Cost is always an important constraint in engineering design.

QUANTITATIVE SYNTHESIS, OR ANALYTICAL SYNTHESIS   means the generation of one or more solutions of a particular type which you know to be suitable to the problem, and more importantly, one for which there is a synthesis algorithm defined. As the name suggests, this type of solution can be quantified, as some set of equations exists which will give a numerical answer. Whether that answer is a good or suitable one is still a matter for the judgment of the designer and requires analysis and iteration to optimize the design. Often the available equations are fewer than the number of potential variables, in which case you must assume some reasonable values for enough unknowns to reduce the remaining set to the number of available equations. Thus some qualitative judgment enters into the synthesis in this case as well. Except for very simple cases, a CAE tool is needed to do quantitative synthesis. One example of such a tool is the program LINCAGES,* by A. Erdman et al., of the University of Minnesota [1] which solves the three-position and four-position linkage synthesis problems. The computer programs provided with this text also allow you to do three-position **analytical synthesis** and general linkage **design by successive analysis**. The fast computation of these programs allows one to analyze the performance of many trial mechanism designs in a short time and promotes rapid iteration to a better solution.

DIMENSIONAL SYNTHESIS   of a linkage *is the determination of the proportions (lengths) of the links necessary to accomplish the desired motions* and can be a form of quantitative synthesis if an algorithm is defined for the particular problem, but can also be a form of qualitative synthesis if there are more variables than equations. The latter situation is more common for linkages. (Dimensional synthesis of cams is quantitative.) Dimensional synthesis assumes that, through *type synthesis*, you have already determined that a linkage (or a cam) is the most appropriate solution to the problem. This chapter discusses **graphical dimensional synthesis** of linkages in detail. Chapter 5 presents methods of **analytical linkage synthesis**, and Chapter 8 presents **cam synthesis**.

3.2   FUNCTION, PATH, AND MOTION GENERATION

FUNCTION GENERATION   is defined as *the correlation of an input motion with an output motion in a mechanism*. A function generator is conceptually a "black box" which delivers some predictable output in response to a known input. Historically, before the advent of electronic computers, mechanical function generators found wide application in artillery rangefinders and shipboard gun aiming systems, and many other tasks. They are, in fact, **mechanical analog computers**. The development of inexpensive digital electronic microcomputers for control systems coupled with the availability of compact

* Available from
Knowledge Revolution Inc.,
San Mateo CA 94402,
(800) 766-6615.

servo and stepper motors has reduced the demand for these mechanical function generator linkage devices. Many such applications can now be served more economically and
efficiently with electromechanical devices.* Moreover, the computer-controlled electromechanical function generator is programmable, allowing rapid modification of the function generated as demands change. For this reason, while presenting some simple examples in this chapter and a general, analytical design method in Chapter 5, we will not
emphasize mechanical linkage function generators in this text. Note however that the
cam-follower system, discussed extensively in Chapter 8, is in fact a form of mechanical function generator, and it is typically capable of higher force and power levels per
dollar than electromechanical systems.

PATH GENERATION   is defined as *the control of a **point** in the plane such that it
follows some prescribed path.* This is typically accomplished with at least four bars,
wherein a point on the coupler traces the desired path. Specific examples are presented
in the section on coupler curves below. Note that no attempt is made in path generation
to control the orientation of the link which contains the point of interest. However, it is
common for the timing of the arrival of the point at particular locations along the path to
be defined. This case is called *path generation with prescribed timing* and is analogous
to function generation in that a particular output function is specified. Analytical path
and function generation will be dealt with in Chapter 5.

MOTION GENERATION   is defined as *the control of a **line** in the plane such that it
assumes some prescribed set of sequential positions.* Here orientation of the link containing the line is important. This is a more general problem than path generation, and
in fact, path generation is a subset of motion generation. An example of a motion generation problem is the control of the "bucket" on a bulldozer. The bucket must assume a
set of positions to dig, pick up, and dump the excavated earth. Conceptually, the motion
of a line, painted on the side of the bucket, must be made to assume the desired positions.
A linkage is the usual solution.

PLANAR MECHANISMS VERSUS SPATIAL MECHANISMS   The above discussion of
controlled movement has assumed that the motions desired are planar (2-D). We live in
a three-dimensional world, however, and our mechanisms must function in that world.
Spatial mechanisms *are 3-D devices*. Their design and analysis is much more complex
than that of **planar mechanisms**, which *are 2-D devices*. The study of spatial mechanisms is beyond the scope of this introductory text. Some references for further study
are in the bibliography to this chapter. However, the study of planar mechanisms is not
as practically limiting as it might first appear since many devices in three dimensions are
constructed of multiple sets of 2-D devices coupled together. An example is any folding
chair. It will have some sort of linkage in the left side plane which allows folding. There
will be an identical linkage on the right side of the chair. These two *XY* planar linkages
will be connected by some structure along the *Z* direction, which ties the two planar linkages into a 3-D assembly. Many real mechanisms are arranged in this way, as **duplicate
planar linkages**, displaced in the *Z* direction in parallel planes and rigidly connected.
When you open the hood of a car, take note of the hood hinge mechanism. It will be duplicated on each side of the car. The hood and the car body tie the two planar linkages
together into a 3-D assembly. Look and you will see many other such examples of assemblies of planar linkages into 3-D configurations. So, the 2-D techniques of synthesis
and analysis presented here will prove to be of practical value in designing in 3-D as well.

---

* It is worth noting that
the day is long past when a
mechanical engineer can
be content to remain
ignorant of electronics and
electromechanics.
Virtually all modern
machines are controlled by
electronic devices.
Mechanical engineers must
understand their operation.

3.3　LIMITING CONDITIONS

The manual, graphical, dimensional synthesis techniques presented in this chapter and the computerizable, analytical synthesis techniques presented in Chapter 5 are reasonably rapid means to obtain a trial solution to a motion control problem. Once a potential solution is found, it must be evaluated for its quality. There are many criteria which may be applied. In later chapters, we will explore the analysis of these mechanisms in detail. However, one does not want to expend a great deal of time analyzing, in great detail, a design which can be shown to be inadequate by some simple and quick evaluations.

TOGGLE　One important test can be applied within the synthesis procedures described below. You need to check that the linkage can in fact reach all of the specified design positions without encountering a limit or **toggle** position, also called a **stationary configuration**. Linkage synthesis procedures often only provide that the particular positions specified will be obtained. They say nothing about the linkage's behavior between those positions. Figure 3-1a shows a non-Grashof fourbar linkage in an arbitrary position CD (dashed lines), and also in its two **toggle positions**, C_1D_1 (solid black lines) and C_2D_2 (solid red lines). *The toggle positions are determined by the **colinearity** of two of the moving links*. A fourbar double- or triple-rocker mechanism will have at least two of these toggle positions in which the linkage assumes a triangular configuration. When in a triangular (toggle) position, it will not allow further input motion in one direction from one of its rocker links (either of link 2 from position C_1D_1 or link 4 from position C_2D_2). The other rocker will then have to be driven to get the linkage out of toggle. A Grashof fourbar crank-rocker linkage will also assume two toggle positions as shown in Figure 3-1b, when the shortest link (crank O_2C) is colinear with the coupler CD (link 3), either *extended colinear* ($O_2C_2D_2$) or *overlapping colinear* ($O_2C_1D_1$). It cannot be *back driven* from the rocker O_4D (link 4) through these colinear positions, but when the crank O_2C (link 2) is driven, it will carry through both toggles because it is Grashof. Note that these toggle positions also define the limits of motion of the driven rocker (link 4), at which its angular velocity will go through zero. Use program FOURBAR to read the data files F03-01A.4BR and F03-1b.4br and animate these examples.

After synthesizing a **double- or triple-rocker** solution to a multiposition (motion generation) problem, you **must** check for the presence of toggle positions *between* your

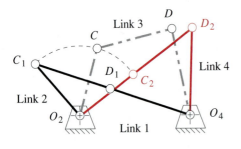

(*a*) Double-rocker toggle positions

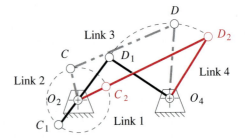

(*b*) Crank-rocker toggle positions

FIGURE 3-1

Linkages in toggle

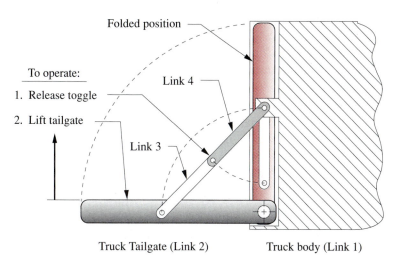

To operate:

1. Release toggle

2. Lift tailgate

Folded position

Link 4

Link 3

Truck Tailgate (Link 2)   Truck body (Link 1)

FIGURE 3-2

Deltoid toggle linkage used to control truck tailgate motion

design positions. *An easy way to do this is with a cardboard model of the linkage design.* A CAE tool such as FOURBAR or *Working Model* will also check for this problem. It is important to realize that a toggle condition is only undesirable if it is preventing your linkage from getting from one desired position to the other. In other circumstances the toggle is very useful. It can provide a self-locking feature when a linkage is moved slightly beyond the toggle position and against a fixed stop. Any attempt to reverse the motion of the linkage then causes it merely to jam harder against the stop. It must be manually pulled "over center," out of toggle, before the linkage will move. You have encountered many examples of this application, as in card table or ironing board leg linkages and also pickup truck or station wagon tailgate linkages. An example of such a toggle linkage is shown in Figure 3-2. It happens to be a special-case Grashof linkage in the deltoid configuration (see also Figure 2-17d, p. 49), which provides a locking toggle position when open, and folds on top of itself when closed, to save space. We will analyze the toggle condition in more detail in a later chapter.

TRANSMISSION ANGLE   Another useful test that can be very quickly applied to a linkage design to judge its quality is the measurement of its transmission angle. This can be done analytically, graphically on the drawing board, or with the cardboard model for a rough approximation. (Extend the links beyond the pivot to measure the angle.) The **transmission angle** μ is shown in Figure 3-3a and is defined as *the angle between the output link and the coupler.*[*] It is usually taken as the *absolute value of the acute angle of the pair of angles at the intersection of the two links* and *varies continuously from some minimum to some maximum value as the linkage goes through its range of motion.* It is a measure of the quality of force and velocity transmission at the joint.[†] Note in Figure 3-2 that the linkage cannot be moved from the open position shown by any force applied to the tailgate, link 2, since the transmission angle is then between links 3 and 4 and is zero at that position. But a force applied to link 4 as the input link will move it. The transmission angle is now between links 3 and 2 and is 45 degrees.

* As defined by Alt. [2]

† The transmission angle has limited application. It only predicts the quality of force or torque transmission if the input and output links are pivoted to ground. If the output force is taken from a floating link (coupler), then the transmission angle is of no value. A different index of merit called the joint force index (JFI) is presented in Chapter 11 which discusses force analysis in linkages. (See Section 11.12 on p. 554.) The JFI is useful for situations in which the output link is floating as well as for giving the same kind of information when the output is taken from a link rotating against the ground. However, the JFI requires that a complete force analysis of the linkage be done, whereas the transmission angle is determined from linkage geometry alone.

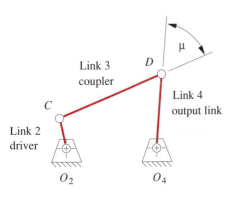

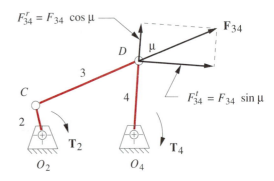

(a) Linkage transmission angle μ   (b) Static forces at a linkage joint

FIGURE 3-3

Transmission angle in the fourbar linkage

Figure 3-3b shows a torque T_2 applied to link 2. Even before any motion occurs, this causes a static, colinear force F_{34} to be applied by link 3 to link 4 at point D. Its radial and tangential components F_{34}^r and F_{34}^t are resolved parallel and perpendicular to link 4, respectively. Ideally, we would like all of the force F_{34} to go into producing output torque T_4 on link 4. However, only the tangential component creates torque on link 4. The radial component F_{34}^r provides only tension or compression in link 4. This radial component only increases pivot friction and does not contribute to the output torque. Therefore, the optimum value for the **transmission angle** is 90°. When μ is less than 45° the radial component will be larger than the tangential component. Most machine designers try to keep the **minimum transmission angle above about 40°** to promote smooth running and good force transmission. However, if in your particular design there will be little or no external force or torque applied to link 4, you may be able to get away with even lower values of μ.* The transmission angle provides one means to judge the quality of a newly synthesized linkage. If it is unsatisfactory, you can iterate through the synthesis procedure to improve the design. We will investigate the transmission angle in more detail in later chapters.

3.4   DIMENSIONAL SYNTHESIS

Dimensional synthesis of a linkage is *the determination of the proportions (lengths) of the links necessary to accomplish the desired motions.* This section assumes that, through *type synthesis*, you have determined that a linkage is the most appropriate solution to the problem. Many techniques exist to accomplish this task of **dimensional synthesis of a fourbar linkage**. The simplest and quickest methods are graphical. These work well for up to three design positions. Beyond that number, a numerical, analytical synthesis approach as described in Chapter 5, using a computer, is usually necessary.

Note that the principles used in these graphical synthesis techniques are simply those of **euclidean geometry**. The rules for bisection of lines and angles, properties of parallel and perpendicular lines, and definitions of arcs, etc., are all that are needed to generate these linkages. **Compass**, **protractor**, and **rule** are the only tools needed for graphical linkage synthesis. Refer to any introductory (high school) text on geometry if your geometric theorems are rusty.

Two-Position Synthesis

Two-position synthesis subdivides into two categories: **rocker output** (pure rotation) and **coupler output** (complex motion). Rocker output is most suitable for situations in which a Grashof crank-rocker is desired and is, in fact, a trivial case of *function generation* in which the output function is defined as two discrete angular positions of the rocker. Coupler output is more general and is a simple case of *motion generation* in which two positio   ⁀f a line in the plane are defined as the output. This solution will frequently lead to a t.   ocker. However, the fourbar triple-rocker can be motor driven by the addition of a d̲y̲ad (twobar chain), which makes the final result a **Watt's sixbar** containing a **Grashof fourbar subchain**. We will now explore the synthesis of each of these types of solution for the two-position problem.

✎ EXAMPLE 3-1

Rocker Output - Two Positions with Angular Displacement. (Function Generation)

Problem: Design a fourbar Grashof crank-rocker to give 45° of rocker rotation with equal time forward and back, from a constant speed motor input.

Solution: (see Figure 3-4)

1 Draw the output link O_4B in both extreme positions, B_1 and B_2 in any convenient location, such that the desired angle of motion θ_4 is subtended.

2 Draw the chord B_1B_2 and extend it in any convenient direction.

3 Select a convenient point O_2 on line B_1B_2 extended.

4 Bisect line segment B_1B_2, and draw a circle of that radius about O_2.

5 Label the two intersections of the circle and B_1B_2 extended, A_1 and A_2.

6 Measure the length of the coupler as A_1 to B_1 or A_2 to B_2.

7 Measure ground length 1, crank length 2, and rocker length 4.

8 Find the Grashof condition. If non-Grashof, redo steps 3 to 8 with O_2 further from O_4.

9 Make a cardboard model of the linkage and articulate it to check its function and its transmission angles.

10 You can input the file F03-04.4br to program FOURBAR to see this example come alive.

Note several things about this synthesis process. We started with the output end of the system, as it was the only aspect defined in the problem statement. We had to make

3

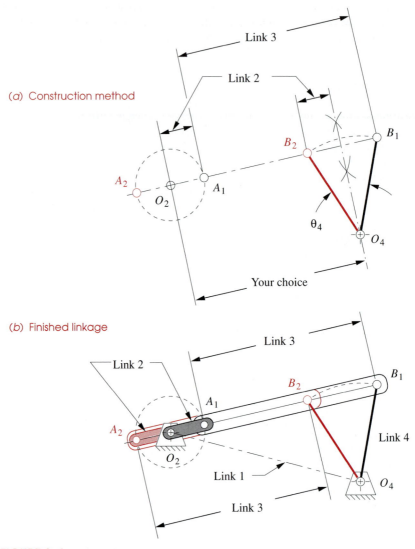

(a) Construction method

Your choice

(b) Finished linkage

FIGURE 3-4

Two-position function synthesis with rocker output (non-quick-return)

many quite arbitrary decisions and assumptions to proceed because there were many more variables than we could have provided "equations" for. We are frequently forced to make "free choices" of "a convenient angle or length." These free choices are actually definitions of design parameters. A poor choice will lead to a poor design. Thus these are **qualitative synthesis** approaches and require an iterative process, even for this simple an example. The first solution you reach will probably not be satisfactory, and several attempts (iterations) should be expected to be necessary. As you gain more experience in designing kinematic solutions you will be able to make better choices for these design parameters with fewer iterations. **The value of making a simple model of your design cannot be overstressed**! You will get the *most insight* into your design's quality for the *least effort* by making, articulating, and studying the model. These general observations will hold for most of the linkage synthesis examples presented.

✎ EXAMPLE 3-2

Rocker Output - Two Positions with Complex Displacement. (Motion Generation)

Problem: Design a fourbar linkage to move link CD from position C_1D_1 to C_2D_2.

Solution: (see Figure 3-5)

1   Draw the link CD in its two desired positions, C_1D_1 and C_2D_2, in the plane as shown.

2   Draw construction lines from point C_1 to C_2 and from point D_1 to D_2.

3   Bisect line C_1C_2 and line D_1D_2 and extend their perpendicular bisectors to intersect at O_4. Their intersection is the **rotopole**.

4   Select a convenient radius and draw an arc about the rotopole to intersect both lines O_4C_1 and O_4C_2. Label the intersections B_1 and B_2.

5   Do steps 2 to 8 of Example 3-1 above to complete the linkage.

6   Make a cardboard model of the linkage and articulate it to check its function and its transmission angles.

Note that Example 3-2 reduces to the method of Example 3-1 once the **rotopole** is found. Thus a link represented by a line in complex motion can be reduced to the simpler problem of pure rotation and moved to any two positions in the plane as the rocker on a fourbar linkage. The next example moves the same link through the same two positions as the coupler of a fourbar linkage.

✎ EXAMPLE 3-3

Coupler Output - Two Positions with Complex Displacement. (Motion Generation)

Problem: Design a fourbar linkage to move the link CD shown from position C_1D_1 to C_2D_2 (with moving pivots at C and D).

Solution: (see Figure 3-6)

1   Draw the link CD in its two desired positions, C_1D_1 and C_2D_2, in the plane as shown.

2   Draw construction lines from point C_1 to C_2 and from point D_1 to D_2.

3   Bisect line C_1C_2 and line D_1D_2 and extend the perpendicular bisectors in convenient directions. The rotopole will **not** be used in this solution.

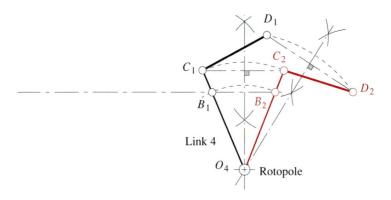

(a) Finding the rotopole for Example 3-2

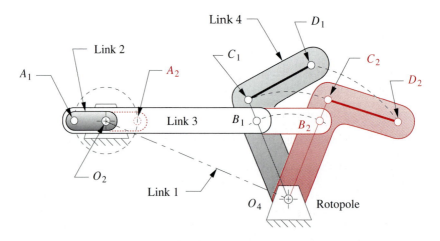

(b) Constructing the linkage by the method in Example 3-1

FIGURE 3-5

Two-position motion synthesis with rocker output (non-quick-return)

4   Select any convenient point on each bisector as the fixed pivots O_2 and O_4, respectively.

5   Connect O_2 with C_1 and call it link 2. Connect O_4 with D_1 and call it link 4.

6   Line C_1D_1 is link 3. Line O_2O_4 is link 1.

7   Check the Grashof condition, and repeat steps 4 to 7 if unsatisfied. Note that any Grashof condition is potentially acceptable in this case.

8   Construct a cardboard model and check its function to be sure it can get from the initial to final position without encountering any limit (toggle) positions.

9   Check transmission angles.

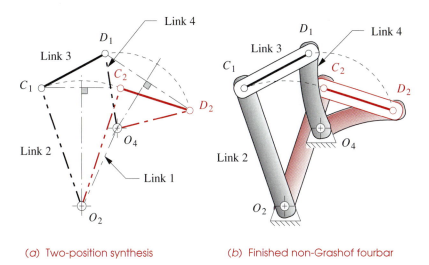

(a) Two-position synthesis   (b) Finished non-Grashof fourbar

FIGURE 3-6

Two-position motion synthesis with coupler output

Input file F03-06.4br to program FOURBAR to see Example 3-3. Note that this example had nearly the same problem statement as Example 3-2, but the solution is quite different. Thus a link can also be moved to any two positions in the plane as the coupler of a fourbar linkage, rather than as the rocker. However, to limit its motions to those two coupler positions as extrema, two additional links are necessary. These additional links can be designed by the method shown in Example 3-4 and Figure 3-7.

✎EXAMPLE 3-4

Adding a Dyad (Twobar Chain) to Control Motion in Example 3-3.

Problem: Design a **dyad** to control and limit the extremes of motion of the linkage in Example 3-3 to its two design positions.

Solution: (see Figure 3-7a)

1 Select a convenient point on link 2 of the linkage designed in Example 3-3. Note that it need not be on the line O_2C_1. Label this point B_1.

2 Draw an arc about center O_2 through B_1 to intersect the corresponding line O_2B_2 in the second position of link 2. Label this point B_2. The chord B_1B_2 provides us with the same problem as in Example 3-1.

3 Do steps 2 to 9 of Example 3-1 to complete the linkage, except add links 5 and 6 and center O_6 rather than links 2 and 3 and center O_2. Link 6 will be the driver crank. The fourbar subchain of links O_6, A_1, B_1, O_2 must be a Grashof crank-rocker.

3

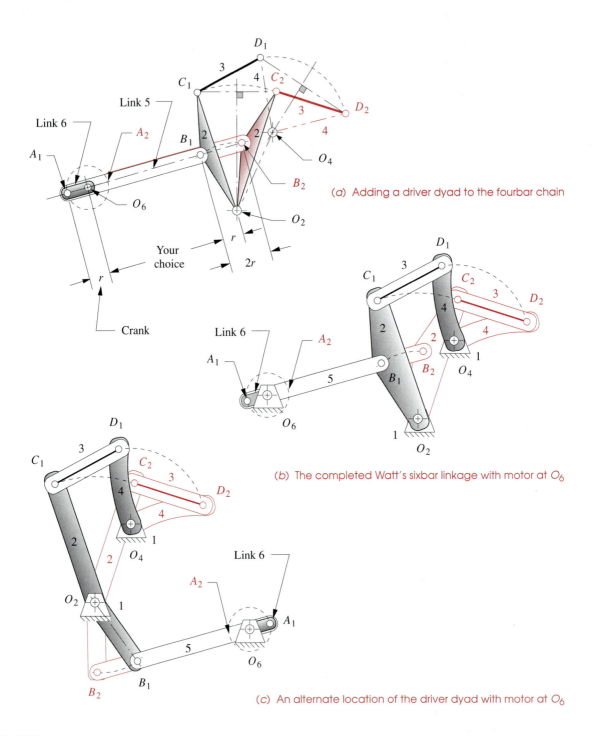

(a) Adding a driver dyad to the fourbar chain

(b) The completed Watt's sixbar linkage with motor at O_6

(c) An alternate location of the driver dyad with motor at O_6

FIGURE 3-7

Driving a non-Grashof linkage with a dyad (non-quick-return)

Note that we have used the approach of Example 3-1 to add a **dyad** to serve as a *driver stage* for our existing fourbar. This results in a **sixbar Watt's mechanism** whose first stage is Grashof as shown in Figure 3-7b. Thus we can drive this with a motor on link 6. Note also that we can locate the motor center O_6 anywhere in the plane by judicious choice of point B_1 on link 2. If we had put B_1 below center O_2, the motor would be to the right of links 2, 3, and 4 as shown in Figure 3-7c. There is *an infinity of driver dyads* possible which will drive any double-rocker assemblage of links. Input the files F03-07b.6br and F03-07c.6br to program SIXBAR to see Example 3-4 in motion for these two solutions.

Three-Position Synthesis with Specified Moving Pivots

Three-position synthesis allows the definition of three positions of a line in the plane and will create a fourbar linkage configuration to move it to each of those positions. This is a **motion generation** problem. The synthesis technique is a logical extension of the method used in Example 3-3 for two-position synthesis with coupler output. The resulting linkage may be of any Grashof condition and will usually require the addition of a dyad to control and limit its motion to the positions of interest. Compass, protractor, and rule are the only tools needed in this graphical method.

✎ EXAMPLE 3-5

Coupler Output - 3 Positions with Complex Displacement. (Motion Generation)

Problem: Design a fourbar linkage to move the link CD shown from position C_1D_1 to C_2D_2 and then to position C_3D_3. Moving pivots are at C and D. Find the fixed pivot locations.

Solution: (see Figure 3-8)

1   Draw link CD in its three design positions C_1D_1, C_2D_2, C_3D_3 in the plane as shown.

2   Draw construction lines from point C_1 to C_2 and from point C_2 to C_3.

3   Bisect line C_1C_2 and line C_2C_3 and extend their perpendicular bisectors until they intersect. Label their intersection O_2.

4   Repeat steps 2 and 3 for lines D_1D_2 and D_2D_3. Label the intersection O_4.

5   Connect O_2 with C_1 and call it link 2. Connect O_4 with D_1 and call it link 4.

6   Line C_1D_1 is link 3. Line O_2O_4 is link 1.

7   Check the Grashof condition. Note that any Grashof condition is potentially acceptable in this case.

8   Construct a cardboard model and check its function to be sure it can get from initial to final position without encountering any limit (toggle) positions.

9   Construct a driver dyad according to the method in Example 3-4 using an extension of link 3 to attach the dyad.

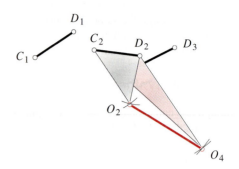

(a) Original coupler three-position problem with specified pivots

(b) Position of the ground plane relative to the second coupler position

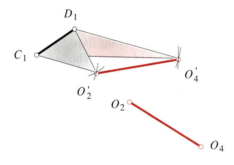

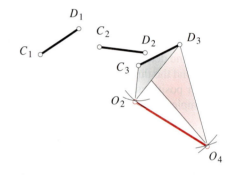

(c) Transferring second ground plane position to reference location at first position

(d) Position of the ground plane relative to the third coupler position

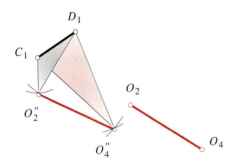

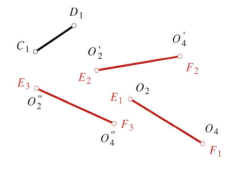

(e) Transferring third ground plane position to reference location at first position

(f) The three inverted positions of the ground plane corresponding to the original coupler position

FIGURE 3-10

Inverting the three-position motion synthesis problem

7 We must now reinvert the linkage to ret
the ground link O_2O_4, and GH is really
the linkage in which points G and H are
resumed its real identity as ground link

8 Figure 3-11d reintroduces the original
the initial position as shown in the orig
the required coupler plane and defines :

9 The angular motions required to reach
Figure 3-11e are the same as those def
angle F_1HF_2 in Figure 3-11b is the sam
same as angle $H_2O_4H_3$. The angular
tween Figure 3-11b and e as well. The a
inversions as the link excursions are re

10 Check the Grashof condition. Note tha
this case provided that the linkage has r
non-Grashof linkage.

11 Construct a cardboard model and check
position without encountering any limi
toggle position between points H_1 and
from link 2 as it will hang up at that to

By inverting the original problem, w
allows a direct solution by the general r
ples 3-5 and 3-6.

Position Synthesis for More Tho

It should be obvious that the more cons
the more complicated the task becomes
three positions of the output link, the di

FOUR-POSITION SYNTHESIS doe
lutions, though Hall [3] does present on
used by Sandor and Erdman [4] and othe
requires a computer to execute it. Briefl
ten to represent the desired four positio
after some free choices of variable valu
gram LINCAGES [1] by Erdman et al., and
vide a convenient and user-friendly con
sary design choices to solve the four-posit

3.5   QUICK-RETURN MECHA

Many machine design applications hav
tween their "forward" and "return" strol
by the linkage on the forward stroke, ar

sition C_2D_2. By doing this, we have pretended that the ground plane moved from O_2O_4 to $O_2'O_4'$ instead of the coupler moving from C_1D_1 to C_2D_2. That is, we have *inverted* the problem.

6 Repeat the process for the third coupler position as shown in Figure 3-10d and transfer the third relative ground link position to the first, or reference, position as shown in Figure 3-10e.

7 The three inverted positions of the ground plane which correspond to the three desired coupler positions are labeled $O_2O_4, O_2'O_4'$, and $O_2''O_4''$ and have also been renamed E_1F_1, E_2F_2, and E_3F_3 as shown in Figure 3-10f. These correspond to the three coupler positions shown in Figure 3-10a. Note that the original three lines C_1D_1, C_2D_2, and C_3D_3 are not now needed for the linkage synthesis.

We can use these three new lines E_1F_1, E_2F_2, and E_3F_3 to find the attachment points GH (moving pivots) on link 3 which will allow the desired fixed pivots O_2 and O_4 to be used for the three specified output positions. In effect we will now consider the ground link O_2O_4 to be a coupler moving through the inverse of the original three positions, find the "ground pivots" GH needed for that inverted motion, and put them on the real coupler instead. The inversion process done in Example 3-7 and Figure 3-10 has swapped the roles of coupler and ground plane. The remaining task is identical to that done in Example 3-5 and Figure 3-8. The result of the synthesis then must be reinverted to obtain the solution.

✎ EXAMPLE 3-8

Finding the Moving Pivots for Three Positions and Specified Fixed Pivots.

Problem: Design a fourbar linkage to move the link CD shown from position C_1D_1 to C_2D_2 and then to position C_3D_3. Use specified fixed pivots O_2 and O_4. Find the required moving pivot locations on the coupler by inversion.

Solution: Using the inverted ground link positions E_1F_1, E_2F_2, and E_3F_3 found in Example 3-7, find the fixed pivots for that inverted motion, then reinvert the resulting linkage to create the moving pivots for the three positions of coupler CD which use the selected fixed pivots O_2 and O_4 as shown in Figure 3-10a (see also Figure 3-11).

1 Start with the inverted three positions in the plane as shown in Figures 3-10f and 3-11a. Lines E_1F_1, E_2F_2, and E_3F_3 define the three positions of the inverted link to be moved.

2 Draw construction lines from point E_1 to E_2 and from point E_2 to E_3.

3 Bisect line E_1E_2 and line E_2E_3 and extend the perpendicular bisectors until they intersect. Label the intersection G.

4 Repeat steps 2 and 3 for lines F_1F_2 and F_2F_3. Label the intersection H.

5 Connect G with E_1 and call it link 2. Connect H with F_1 and call it link 4. See Figure 3-11b.

6 In this inverted linkage, line E_1F_1 is the coupler, link 3. Line GH is the "ground" link 1.

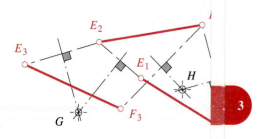

E_2

E_3

E_1

H

G F_3

(a) Construction to find rotopoles G o

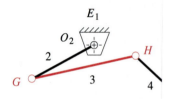

E_1

O_2

2

G 3 H

4

(c) Reinvert to obtain the result

C_1

3

(e) The three positions (link 4 driving (

FIGURE 3-11

Constructing the linkage for three positions

---

9   Make a cardboard model of the linkage and articulate it to check its function.

10   Check the transmission angles.

This method works well for time ratios down to about 1:1.5. Beyond that value the transmission angles become poor, and a more complex linkage is needed. Input the file F03-12.4br to program FOURBAR to see Example 3-9.

Sixbar Quick-Return

Larger time ratios, up to about 1:2, can be obtained by designing a sixbar linkage. The strategy here is to first design a fourbar drag link mechanism which has the desired time ratio between its driver crank and its driven or "dragged" crank, and then add a dyad (twobar) output stage, driven by the dragged crank. This dyad can be arranged to have either a rocker or a translating slider as the output link. First the drag link fourbar will be synthesized; then the dyad will be added.

EXAMPLE 3-10

Sixbar Drag Link Quick-Return Linkage for Specified Time Ratio.

Problem: Provide a time ratio of 1:1.4 with 90° rocker motion.

Solution: (see Figure 3-13)

1   Calculate α and β using equations 3.1.   For this example, $\alpha = 150°$ and $\beta = 210°$.

2   Draw a line of centers *XX* at any convenient location.

3   Choose a crank pivot location O_2 on line *XX* and draw an axis *YY* perpendicular to *XX* through O_2.

4   Draw a circle of convenient radius O_2A about center O_2.

5   Lay out angle α with vertex at O_2, symmetrical about quadrant one.

6   Label points A_1 and A_2 at the intersections of the lines subtending angle α and the circle of radius O_2A .

7   Set the compass to a convenient radius *AC* long enough to cut *XX* in two places on either side of O_2 when swung from both A_1 and A_2. Label the intersections C_1 and C_2.

8   The line O_2A_1 is the driver crank, link 2, and line A_1C_1 is the coupler, link 3.

9   The distance C_1C_2 is twice the driven (dragged) crank length. Bisect it to locate the fixed pivot O_4.

10   The line O_2O_4 now defines the ground link. Line O_4C_1 is the driven crank, link 4.

11   Calculate the Grashof condition. If non-Grashof, repeat steps 7 to 11 with a shorter radius in step 7.

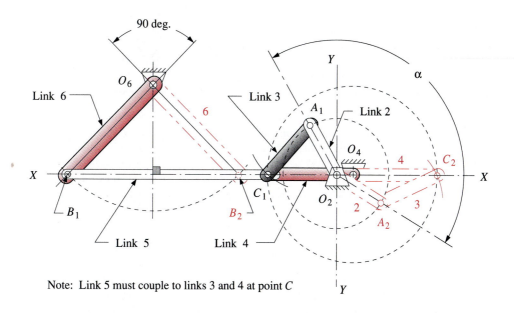

Note: Link 5 must couple to links 3 and 4 at point C

(a) Rocker output sixbar drag link quick-return mechanism

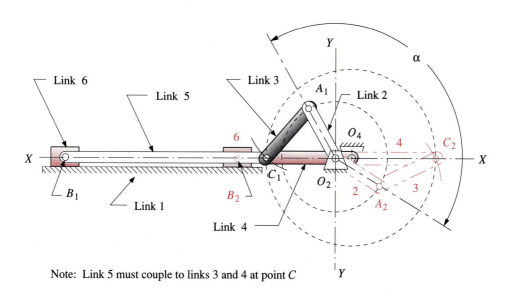

Note: Link 5 must couple to links 3 and 4 at point C

(b) Slider output sixbar drag link quick-return mechanism

FIGURE 3-13

Synthesizing a sixbar drag link quick-return mechanism

12   Invert the method of Example 3-1 to create the output dyad using XX as the chord and O_4C_1 as the driving crank. The points B_1 and B_2 will lie on line XX and be spaced apart a distance $2O_4C_1$. The pivot O_6 will lie on the perpendicular bisector of B_1B_2, at a distance from line XX which subtends the specified output rocker angle.

13   Check the transmission angles.

This linkage provides a quick-return when a constant-speed motor is attached to link 2. Link 2 will go through angle α while link 4 (which is dragging the output dyad along) goes through the first 180 degrees, from position C_1 to C_2. Then, while link 2 completes its cycle through β degrees, the output stage will complete another 180 degrees from C_2 to C_1. Since angle β is greater than α, the forward stroke takes longer. Note that the chordal stroke of the output dyad is twice the crank length O_4C_1. This is independent of the angular displacement of the output link which can be tailored by moving the pivot O_6 closer to or further from the line XX.

The transmission angle at the joint between link 5 and link 6 will be optimized if the fixed pivot O_6 is placed on the perpendicular bisector of the chord B_1B_2 as shown in Figure 3-13a (p. 101). If a translating output is desired, the slider (link 6) will be located on line XX and will oscillate between B_1 and B_2 as shown in Figure 3-13b. The arbitrarily chosen size of this or any other linkage can be scaled up or down, simply by multiplying all link lengths by the same scale factor. Thus a design made to arbitrary size can be fit to any package. Input the file F03-13a.6br to program SIXBAR to see Example 3-10 in action.

CRANK-SLIDER QUICK-RETURN   A commonly used mechanism, capable of large time ratios is shown in Figure 3-14. It is often used in metal shaper machines to provide

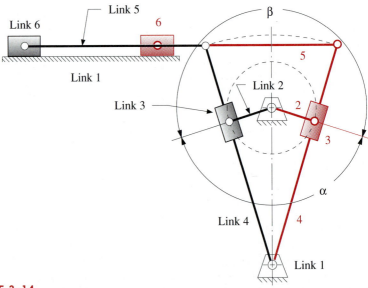

FIGURE 3-14

Crank-shaper quick-return mechanism

a slow cutting stroke and a quick-return stroke when the tool is doing no work. It is the inversion #2 of the slider-crank mechanism as was shown in Figure 2-13b (p. 44). It is very easy to synthesize this linkage by simply moving the rocker pivot O_4 along the vertical centerline O_2O_4 while keeping the two extreme positions of link 4 tangent to the circle of the crank, until the desired time ratio (α / β) is achieved. Note that the angular displacement of link 4 is then defined as well. Link 2 is the input and link 6 is the output.

Depending on the relative lengths of the links this mechanism is known as a **Whitworth** or **crank-shaper** mechanism. If the ground link is the shortest, then it will behave as a double-crank linkage, or *Whitworth mechanism*, with both pivoted links making full revolutions as shown in Figure 2-13b (p. 44). If the driving crank is the shortest link, then it will behave as a crank-rocker linkage, or *crank-shaper mechanism*, as shown in Figure 3-14. They are the same inversion as the slider block is in complex motion in each case.

3.6   COUPLER CURVES

A **coupler** is the most interesting link in any linkage. It is in complex motion, and thus points on the coupler can have path motions of high degree.[*] In general, the more links, the higher the degree of curve generated, where degree here means *the highest power of any term in its equation*. A curve (function) can have *up to* as many intersections (roots) with any straight line as the degree of the function. The *fourbar slider-crank* has, in general, fourth-degree coupler curves; the *pin-jointed fourbar*, up to sixth degree.[†] The geared fivebar, the sixbar, and more complicated assemblies will have still higher degree curves. Wunderlich[7b] derived an expression for the highest degree m possible for a coupler curve of a mechanism of n links connected with only revolute joints.

$$m = 2 \cdot 3^{(n/2-1)} \tag{3.3}$$

This gives, respectively, degrees of 6, 18, and 54 for the fourbar, sixbar, and eightbar linkage coupler curves. Specific points on their couplers may have degenerate curves of lower degree as, for example, the pin joints between any crank or rocker and the coupler that describes second-degree curves (circles). The parallelogram fourbar linkage has degenerate coupler curves, all of which are circles.

All linkages that possess one or more "floating" coupler links will generate coupler curves. It is interesting to note that these will be closed curves even for non-Grashof linkages. The coupler (or any link) can be extended infinitely in the plane. Figure 3-15 shows a fourbar linkage with its coupler extended to include a large number of points, each of which describes a different coupler curve. Note that these points may be anywhere on the coupler, including along line *AB*. There is, of course, an infinity of points on the coupler, each of which generates a different curve.

Coupler curves can be used to generate quite useful path motions for machine design problems. They are capable of *approximating straight lines* and *large circle arcs* with remote centers. Recognize that the coupler curve is a solution to the path generation problem described in Section 3.2 (p. 78). It is not by itself a solution to the motion generation problem, since the attitude or orientation of a line on the coupler is not predicted by the information contained in the path. Nevertheless it is a very useful device, and it can be converted to a parallel motion generator by adding two links as described

[*] In 1876, Kempe[7a] proved his theory that a linkage with only revolute (pin) and prismatic (slider) joints can be found that will trace any algebraic curve of any order or complexity. But the linkage for a particular curve may be excessively complex, may be unable to traverse the curve without encountering limit (toggle) positions, and may even need to be disassembled and reassembled to reach all points on the curve. See the discussion of circuit and branch defects in Section 4.12 (p. 173). Nevertheless this theory points to the potential for interesting motions from the coupler curve.

[†] The algebraic equation of the coupler curve is sometimes referred to as a "tricircular sextic" referring respectively to its circularity of 3 and its degree of 6.

(a) Pseudo ellipse

(b) Kidney bean

(c) Banana

(d) Crescent

(e) Single straight

(f) Double straight

FIGURE 3-16 Part 1

A "Cursory Catalog" of coupler curve shapes

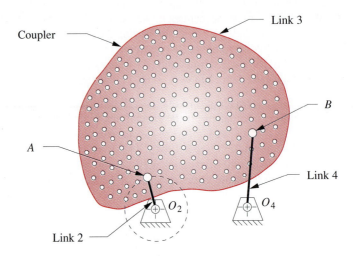

FIGURE 3-15

The fourbar coupler extended to include a large number of coupler points

in the next section. As we shall see, approximate straight-line motions, dwell motions, and more complicated symphonies of timed motions are available from even the simple fourbar linkage and its infinite variety of often surprising coupler curve motions.

FOURBAR COUPLER CURVES come in a variety of shapes which can be crudely categorized as shown in Figure 3-16. There is an infinite range of variation between these generalized shapes. Some features of interest are the curve's **double points**, ones that have two tangents. They occur in two types, the **cusp** and the **crunode**. A **cusp** is *a sharp point on the curve which has the useful property of instantaneous zero velocity*. The simplest example of a curve with a cusp is the cycloid curve which is generated by a point on the rim of a wheel rotating on a flat surface. When the point touches the surface, it has the same (zero) velocity as all points on the stationary surface, provided there is pure rolling and no slip between the elements. Anything attached to a cusp point will come smoothly to a stop along one path and then accelerate smoothly away from that point on a different path. The cusp's feature of zero velocity has value in such applications as transporting, stamping and feeding processes. Note that *the acceleration at the cusp is not zero*. A **crunode** *creates a multiloop curve which has double points at the crossovers*. The two slopes (tangents) at a crunode give the point two different velocities, neither of which is zero in contrast to the cusp. In general, a fourbar coupler curve can have up to three real double points[*] which may be a combination of cusps and crunodes as can be seen in Figure 3-16.

The *Hrones and Nelson* (H&N) atlas of fourbar coupler curves [8a] is a useful reference which can provide the designer with a starting point for further design and

---

[*] Actually, the fourbar coupler curve has 9 double points of which 6 are usually imaginary. However, Fichter and Hunt [8b] point out that some unique configurations of the fourbar linkage (i.e., rhombus parallelograms and those close to this configuration) can have up to 6 real double points which they denote as comprising 3 "proper" and 3 "improper" real double points. For non-special-case Grashof fourbar linkages with minimum transmission angles suitable for engineering applications, only the 3 "proper" double points will appear.

analysis. It contains about 7000 coupler curves and defines the linkage geometry for each of its Grashof crank-rocker linkages. Figure 3-17a reproduces a page from this book. The H&N atlas is logically arranged, with all linkages defined by their link ratios, based on a unit length crank. The coupler is shown as a matrix of fifty coupler points for each linkage geometry, arranged ten to a page. Thus each linkage geometry occupies five pages. Each page contains a schematic "key" in the upper right corner which defines the link ratios.

Figure 3-17b shows a "fleshed out" linkage drawn on top of the H&N atlas page to illustrate its relationship to the atlas information. The double circles in Figure 3-17a define the fixed pivots. The crank is always of unit length. The ratios of the other link lengths to the crank are given on each page. The actual link lengths can be scaled up or down to suit your package constraints and this will affect the size but not the shape of the coupler curve. Any one of the ten coupler points shown can be used by incorporating it into a triangular coupler link. The location of the chosen coupler point can be scaled from the atlas and is defined within the coupler by the position vector **R** whose constant angle ϕ is measured with respect to the line of centers of the coupler. The H&N coupler curves are shown as dashed lines. Each dash station represents **five degrees** of crank rotation. So, for an assumed constant crank velocity, the dash spacing is proportional to path velocity. The changes in velocity and the quick-return nature of the coupler path motion can be clearly seen from the dash spacing.

One can peruse this linkage atlas resource and find an approximate solution to virtually any path generation problem. Then one can take the tentative solution from the atlas to a CAE resource such as the FOURBAR program or other package such as *Working Model* * and further refine the design, based on the complete analysis of positions, velocities, and accelerations provided by the program. The only data needed for the FOURBAR program are the four link lengths and the location of the chosen coupler point with respect to the line of centers of the coupler link as shown in Figure 3-17. These parameters can be changed within program FOURBAR to alter and refine the design. Input the file F03-17b.4br to program FOURBAR to animate the linkage shown in that figure.

An example of an application of a fourbar linkage to a practical problem is shown in Figure 3-18 which is a movie camera (or projector) film advance mechanism. Point O_2 is the crank pivot which is motor driven at constant speed. Point O_4 is the rocker pivot, and points A and B are the moving pivots. Points A, B, and C define the coupler where C is the coupler point of interest. A movie is really a series of still pictures, each "frame" of which is projected for a small fraction of a second on the screen. Between each picture, the film must be moved very quickly from one frame to the next while the shutter is closed to blank the screen. The whole cycle takes only 1/24 of a second. The human eye's response time is too slow to notice the flicker associated with this discontinuous stream of still pictures, so it appears to us to be a continuum of changing images.

The linkage shown in Figure 3-18 is cleverly designed to provide the required motion. A hook is cut into the coupler of this fourbar Grashof crank-rocker at point C which generates the coupler curve shown. The hook will enter one of the sprocket holes in the film as it passes point F_1. Notice that the direction of motion of the hook at that point is nearly perpendicular to the film, so it enters the sprocket hole cleanly. It then turns abruptly downward and follows a crudely approximate straight line as it rapidly pulls the film downward to the next frame. The film is separately guided in a straight track called

(g) Teardrop

(h) Scimitar

(i) Umbrella

(j) Triple cusp

(k) Figure eight

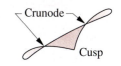

(l) Triple loop

FIGURE 3-16 Part 2

A "Cursory Catalog" of coupler curve shapes

* Included on CD-ROM with this book.

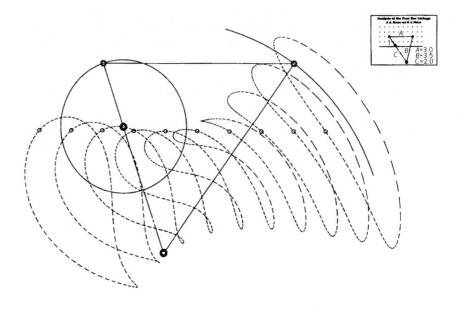

(a) A page from the Hrones and Nelson atlas of fourbar coupler curves *
Hrones, J. A., and G. L. Nelson (1951). Analysis of the Fourbar Linkage.
MIT Technology Press, Cambridge, MA. Reprinted with permission.

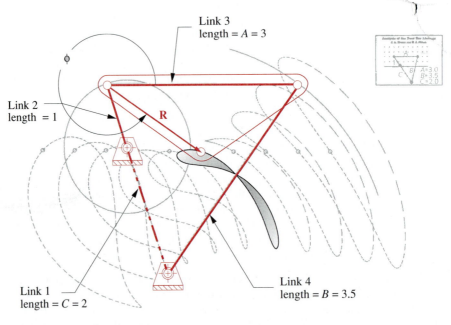

* The Hrones and Nelson
atlas is long out of print but
may be available from
University Microfilms, Ann
Arbor, MI. Also, the *Atlas
of Linkage Design and
Analysis Vol 1: The Four Bar
Linkage* similar to the H&N
atlas, has been recently
published and is available
from Saltire Software, 9725
SW Gemini Drive,
Beaverton, OR 97005, (800)
659-1874.

(b) Creating the linkage from the information in the atlas

FIGURE 3-17

Selecting a coupler curve and constructing the linkage from the Hrones and Nelson atlas

the "gate." The shutter (driven by another linkage from the same driveshaft at O_2) is closed during this interval of film motion, blanking the screen. At point F_2 there is a cusp on the coupler curve which causes the hook to decelerate smoothly to zero velocity in the vertical direction, and then as smoothly accelerate up and out of the sprocket hole. The abrupt transition of direction at the cusp allows the hook to back out of the hole without jarring the film, which would make the image jump on the screen as the shutter opens. The rest of the coupler curve motion is essentially "wasting time" as it proceeds up the back side, to be ready to enter the film again to repeat the process. Input the file F03-18.4br to program FOURBAR to animate the linkage shown in that figure.

Some advantages of using this type of device for this application are that it is very simple and inexpensive (only four links, one of which is the frame of the camera), is extremely reliable, has low friction if good bearings are used at the pivots, and can be reliably timed with the other events in the overall camera mechanism through common shafting from a single motor. There are a myriad of other examples of fourbar coupler curves used in machines and mechanisms of all kinds.

One other example of a very different application is that of the automobile suspension (Figure 3-19). Typically, the up and down motions of the car's wheels are controlled by some combination of planar fourbar linkages, arranged in duplicate to provide three-dimensional control as described in Section 3.2. Only a few manufacturers currently use a true spatial linkage in which the links are not arranged in parallel planes. In all cases the wheel assembly is attached to the coupler of the linkage assembly, and its motion is along a set of coupler curves. The orientation of the wheel is also of concern in this case, so this is not strictly a path generation problem. By designing the linkage to control the paths of multiple points on the wheel (tire contact patch, wheel center, etc.—all of which are points on the same coupler link extended), motion generation is achieved as the coupler has complex motion. Figure 3-19a and b shows parallel planar fourbar linkages suspending the wheels. The coupler curve of the wheel center is nearly a straight line over the small vertical displacement required. This is desirable as the idea is to keep the tire perpendicular to the ground for best traction under all cornering and attitude changes of the car body. This is an application in which a non-Grashof linkage is perfectly acceptable, as full rotation of the wheel in this plane might have some undesirable results and surprise the driver. Limit stops are of course provided to prevent such behavior, so even a Grashof linkage could be used. The springs support the weight of the vehicle and provide a fifth, variable-length "force link" which stabilizes the mechanism as was described in Section 2.14 (p. 54). The function of the fourbar linkage is solely to guide and control the wheel motions. Figure 3-19c shows a true spatial linkage of seven links (including frame and wheel) and nine joints (some of which are ball-and-socket joints) used to control the motion of the rear wheel. These links do not move in parallel planes but rather control the three-dimensional motion of the coupler which carries the wheel assembly.

SYMMETRICAL FOURBAR COUPLER CURVES When a fourbar linkage's geometry is such that the coupler and rocker are the same length pin-to-pin, all coupler points that lie on a circle centered on the coupler-rocker joint with radius equal to the coupler length will generate symmetrical coupler curves. Figure 3-20 shows such a linkage, its symmetrical coupler curve, and the locus of all points that will give symmetrical curves. Using the notation of that figure, the criterion for coupler curve symmetry can be stated as:

$$AB = O_4 B = BP \qquad (3.4)$$

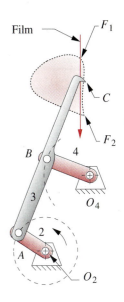

FIGURE 3-18

Movie camera film-advance mechanism. *From Die Wissenschaftliche und Angenwandte Photographie, Michel, Kurt,(ed.). (1955). Vol. 3, Harold Weise, Die Kinematographische Kamera, p. 202, Springer Verlag, OHG, Vienna.* (Input the file F03-18.4br to program FOURBAR to animate this linkage.)

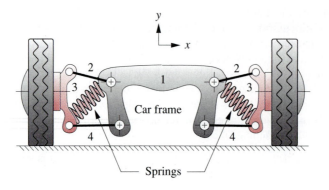

(*a*) Fourbar planar linkages are duplicated in parallel planes,
displaced in the *z* direction, behind the links shown

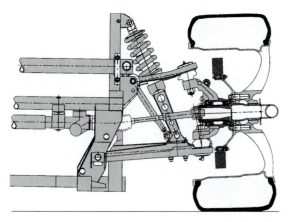

(*b*) Parallel-planar linkage used to control Viper wheel motion
(*Courtesy of Chrysler Corporation*)

(*c*) Multi-link true spatial linkage used to
control rear wheel motion
(*Courtesy of Mercedes-Benz of North
America Inc.*)

FIGURE 3-19

Linkages used in automotive chassis suspensions

A linkage for which equation 3.4 is true is referred to as a **symmetrical fourbar
linkage.** The axis of symmetry of the coupler curve is the line O_4P drawn when the crank
O_2A and the ground link O_2O_4 are colinear-extended (i.e., $\theta_2 = 180°$). Symmetrical coupler curves prove to be quite useful as we shall see in the next several sections. Some
give good approximations to circular arcs and others give very good approximations to
straight lines (over a portion of the coupler curve).

In the general case, nine parameters are needed to define the geometry of a **nonsymmetrical fourbar linkage** with one coupler point.[*] We can reduce this to five as follows.
Three parameters can be eliminated by fixing the location and orientation of the ground
link. The four link lengths can be reduced to three parameters by normalizing three link
lengths to the fourth. The shortest link (the crank if a Grashof linkage) is usually taken
as the reference link, and three link ratios are formed as $L_1 / L_2, L_3 / L_2, L_4 / L_2$, where

* The nine independent
parameters of a fourbar
linkage are: four link
lengths, two coordinates of
the coupler point with
respect to the coupler link,
and three parameters that
define the location and
orientation of the fixed link
in the global coordinate
system.

L_1 = ground, L_2 = crank, L_3 = coupler, and L_4 = rocker length as shown in Figure 3-20. Two parameters are needed to locate the coupler point: the distance from a convenient reference point on the coupler (either B or A in Figure 3-20) to the coupler point P, and the angle that the line BP (or AP) makes with the line of centers of the coupler AB (either δ or γ). Thus, with a defined ground link, five parameters that will define the geometry of a nonsymmetrical fourbar linkage (using point B as the reference in link 3 and the labels of Figure 3-20) are: L_1 / L_2, L_3 / L_2, L_4 / L_2, BP / L_2, and γ. Note that multiplying these parameters by a scaling factor will change the size of the linkage and its coupler curve but will not change the coupler curve's shape.

A **symmetrical fourbar linkage** with a defined ground link needs only *three parameters* to define its geometry because three of the five nonsymmetrical parameters are now equal per equation 3.4: $L_3 / L_2 = L_4 / L_2 = BP / L_2$. Three possible parameters to define the geometry of a symmetrical fourbar linkage in combination with equation 3.4 are then: L_1 / L_2, L_3 / L_2, and γ. Having only three parameters to deal with rather than five greatly simplifies an analysis of the behavior of the coupler curve shape when the linkage geometry is varied. Other relationships for the isosceles-triangle coupler are shown in Figure 3-20. Length AP and angle δ are needed for input of the linkage geometry to program FOURBAR.

Kota [9] did an extensive study of the characteristics of coupler curves of symmetrical fourbar linkages and mapped coupler curve shape as a function of the three linkage parameters defined above. He defined a three-dimensional design space to map the coupler curve shape. Figure 3-21 shows two orthogonal plane sections taken through this design space for particular values of link ratios,* and Figure 3-22 shows a schematic of

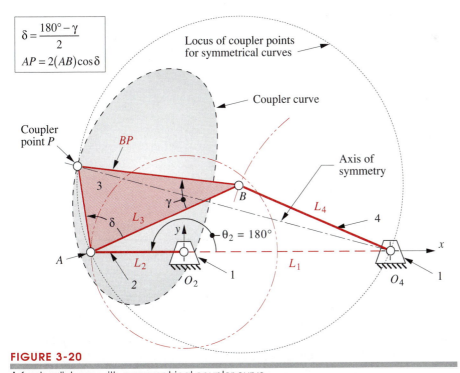

FIGURE 3-20

A fourbar linkage with a symmetrical coupler curve

* Adapted from materials provided by Professor Sridhar Kota, University of Michigan.

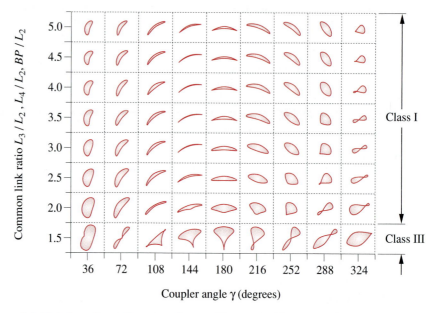

Common link ratio L_3 / L_2, L_4 / L_2, BP / L_2

Coupler angle γ (degrees)

Class I

Class III

(*a*) Variation of coupler curve shape with common link ratio and coupler angle
        for a ground link ratio $L_1 / L_2 = 2.0$

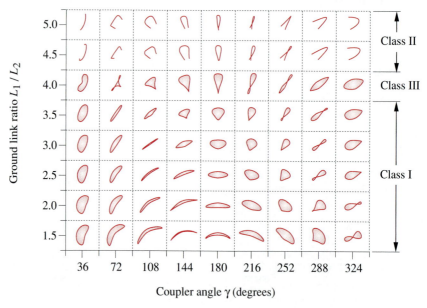

Ground link ratio L_1 / L_2

Coupler angle γ (degrees)

Class II

Class III

Class I

(*b*) Variation of coupler curve shape with ground link ratio and coupler angle
        for a common link ratio $L_3 / L_2 = L_4 / L_2 = BP / L_2 = 2.5$

FIGURE 3-21

Coupler curve shapes of symmetrical fourbar linkages   *Adapted from reference (9)*

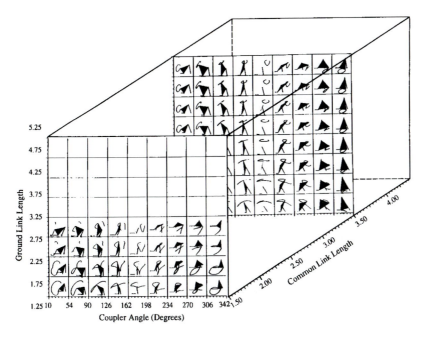

FIGURE 3-22

A three-dimensional map of coupler curve shapes of symmetrical fourbar linkages (9)

the design space. Though the two cross sections of Figure 3-21 show only a small fraction of the information in the 3-D design space of Figure 3-22, they nevertheless give a sense of the way that variation of the three linkage parameters affects the coupler curve shape. Used in combination with a linkage design tool such as program FOURBAR, these design charts can help guide the designer in choosing suitable values for the linkage parameters to achieve a desired path motion.

GEARED FIVEBAR COUPLER CURVES (Figure 3-23) are more complex than the fourbar variety. Because there are three additional, independent design variables in the geared fivebar compared to the fourbar (an additional link ratio, the gear ratio, and the phase angle between the gears), the coupler curves can be of higher degree than those of the fourbar. This means that the curves can be more convoluted, having more cusps and crunodes (loops). In fact, if the gear ratio used is noninteger, the input link will have to make a number of revolutions equal to the factor necessary to make the ratio an integer before the coupler curve pattern will repeat. The Zhang, Norton, Hammond (ZNH) *Atlas of Geared FiveBar Mechanisms* (GFBM) [10] shows typical coupler curves for these linkages limited to symmetrical geometry (e.g., link 2 = link 5 and link 3 = link 4) and gear ratios of ±1 and ±2. A page from the ZNH atlas is reproduced in Figure 3-23. Additional pages are in Appendix E. Each page shows the family of coupler curves obtained by variation of the phase angle for a particular set of link ratios and gear ratio. A key in the upper right corner of each page defines the ratios: α = link 3 / link 2, β = link 1 / link 2, λ = gear 5 / gear 2. Symmetry defines links 4 and 5 as noted above. The phase angle ϕ is noted on the axes drawn at each coupler curve and can be seen to have a significant effect on the resulting coupler curve shape.

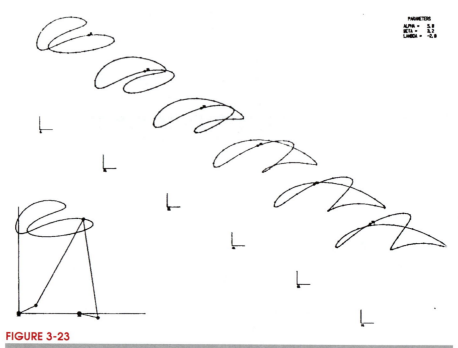

PARAMETERS
ALPHA =   5.8
BETA =   3.2
LAMBDA = -2.8

FIGURE 3-23

A page from the Zhang-Norton-Hammond atlas of geared fivebar coupler curves [10]

This reference atlas is intended to be used as a starting point for a geared fivebar linkage design. The link ratios, gear ratio, and phase angle can be input to the program FIVEBAR and then varied to observe the effects on coupler curve shape, velocities, and accelerations. Asymmetry of links can be introduced, and a coupler point location other than the pin joint between links 3 and 4 defined within the FIVEBAR program as well. Note that program FIVEBAR expects the gear ratio to be in the form gear 2 / gear 5 which is the inverse of the ratio λ in the ZNH atlas.

3.7   COGNATES

It sometimes happens that a good solution to a linkage synthesis problem will be found that satisfies path generation constraints but which has the fixed pivots in inappropriate locations for attachment to the available ground plane or frame. In such cases, the use of a **cognate** to the linkage may be helpful. The term **cognate** was used by Hartenberg and Denavit [11] to describe *a linkage, of different geometry, which generates the same coupler curve*. Samuel Roberts (1875) and Chebyschev (1878) independently discovered the theorem which now bears their names:

Roberts-Chebyschev Theorem

Three different planar, pin-jointed fourbar linkages will trace identical coupler curves.

Hartenberg and Denavit [11] presented extensions of this theorem to the slider-crank and the sixbar linkages:

Two different planar slider-crank linkages will trace identical coupler curves.

The coupler-point curve of a planar fourbar linkage is also described by the joint of a dyad of an appropriate sixbar linkage.

Figure 3-24a shows a fourbar linkage for which we want to find the two cognates. The first step is to release the fixed pivots O_A and O_B. While holding the coupler stationary, rotate links 2 and 4 into colinearity with the line of centers (A_1B_1) of link 3 as shown in Figure 3-24b. We can now construct lines parallel to all sides of the links in the original linkage to create the **Cayley diagram** in Figure 3-24c. This schematic arrangement defines the lengths and shapes of links 5 through 10 which belong to the cognates. All three fourbars share the original coupler point P and will thus generate the same path motion on their coupler curves.

In order to find the correct location of the fixed pivot O_C from the Cayley diagram, the ends of links 2 and 4 are returned to the original locations of the fixed pivots O_A and O_B as shown in Figure 3-25a. The other links will follow this motion, maintaining the parallelogram relationships between links, and fixed pivot O_C will then be in its proper location on the ground plane. This configuration is called a **Roberts diagram**—three fourbar linkage cognates which share the same coupler curve.

The Roberts diagram can be drawn directly from the original linkage without resort to the Cayley diagram by noting that the parallelograms which form the other cognates are also present in the Roberts diagram and the three couplers are similar triangles. It is also possible to locate fixed pivot O_C directly from the original linkage as shown in Figure 3-25a. Construct a similar triangle to that of the coupler, placing its base (*AB*) between O_A and O_B. Its vertex will be at O_C.

The ten-link Roberts configuration (Cayley's nine plus the ground) can now be articulated up to any toggle positions, and point P will describe the original coupler path which is the same for all three cognates. Point O_C will not move when the Roberts linkage is articulated, proving that it is a ground pivot. The cognates can be separated as shown in Figure 3-25b and any one of the three linkages used to generate the same coupler curve. Corresponding links in the cognates will have the same angular velocity as the original mechanism as defined in Figure 3-25.

Nolle [12] reports on work by Luck [13] (in German) that defines the character of all fourbar cognates and their transmission angles. If the original linkage is a Grashof crank-rocker, then one cognate will be also, and the other will be a Grashof double rocker. The minimum transmission angle of the crank-rocker cognate will be the same as that of the original crank-rocker. If the original linkage is a Grashof double-crank (drag link), then both cognates will be also and their minimum transmission angles will be the same in pairs that are driven from the same fixed pivot. If the original linkage is a non-Grashof triple-rocker, then both cognates are also triple-rockers.

These findings indicate that cognates of Grashof linkages do not offer improved transmission angles over the original linkage. Their main advantages are the different fixed pivot location and different velocities and accelerations of other points in the linkage. While the coupler path is the same for all cognates, its velocities and accelerations will not generally be the same since each cognate's overall geometry is different.

When the coupler point lies on the line of centers of link 3, the Cayley diagram degenerates to a group of colinear lines. A different approach is needed to determine the

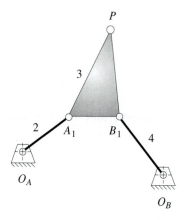

(a) Original fourbar linkage
(cognate #1)

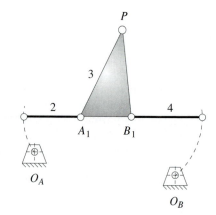

(b) Align links 2 and 4 with coupler

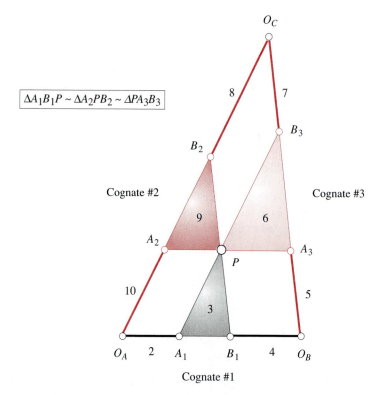

$$\Delta A_1B_1P \sim \Delta A_2PB_2 \sim \Delta PA_3B_3$$

(c) Construct lines parallel to all sides of the original fourbar linkage to create cognates

FIGURE 3-24

Cayley diagram to find cognates of a fourbar linkage

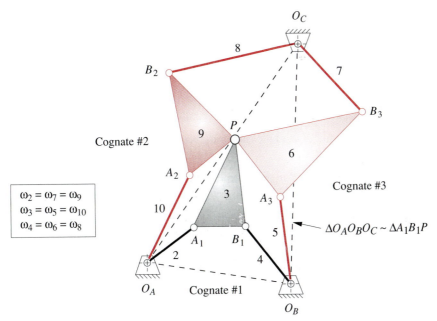

$\omega_2 = \omega_7 = \omega_9$
$\omega_3 = \omega_5 = \omega_{10}$
$\omega_4 = \omega_6 = \omega_8$

$\Delta O_A O_B O_C \sim \Delta A_1 B_1 P$

(a) Return links 2 and 4 to their fixed pivots O_A and O_B.
     Point O_C will assume its proper location

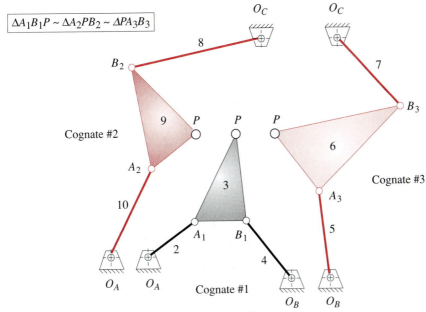

$\Delta A_1 B_1 P \sim \Delta A_2 P B_2 \sim \Delta P A_3 B_3$

(b) Separate the three cognates.
     Point P has the same path motion in each cognate.

FIGURE 3-25

Roberts diagram of three fourbar cognates

geometry of the cognates. Hartenberg and Denavit [11] give the following set of steps to find the cognates in this case. The notation refers to Figure 3-26.

1　Fixed pivot O_C lies on the line of centers $O_A O_B$ extended and divides it in the same ratio as point P divides AB (i.e., $O_C / O_A = PA / AB$).

2　Line $O_A A_2$ is parallel to $A_1 P$ and $A_2 P$ is parallel to $O_A A_1$, locating A_2.

3　Line $O_B A_3$ is parallel to $B_1 P$ and $A_3 P$ is parallel to $O_B B_1$, locating A_3.

4　Joint B_2 divides line $A_2 P$ in the same ratio as point P divides AB. This defines the first cognate $O_A A_2 B_2 O_C$.

5　Joint B_3 divides line $A_3 P$ in the same ratio as point P divides AB. This defines the second cognate $O_B A_3 B_3 O_C$.

The three linkages can then be separated and each will independently generate the same coupler curve. The example chosen for Figure 3-26 is unusual in that the two cognates of the original linkage are identical, mirror-image twins. These are special linkages and will be discussed further in the next section.

Program FOURBAR will automatically calculate the two cognates for any linkage configuration input to it. The velocities and accelerations of each cognate can then be calculated and compared. The program also draws the Cayley diagram for the set of cognates. Input the file F03-24.4br to program FOURBAR to display the Cayley diagram of Figure 3-24 (p.114). Input the files COGNATE1.4br, COGNATE2.4br, and COGNATE3.4br to animate and view the motion of each cognate shown in Figure 3-25 (p. 115). Their coupler curves (at least those portions that each cognate can reach) will be seen to be identical.

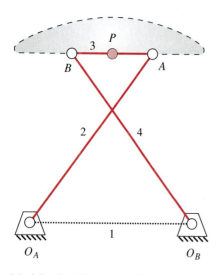

(a) A fourbar linkage and its coupler curve

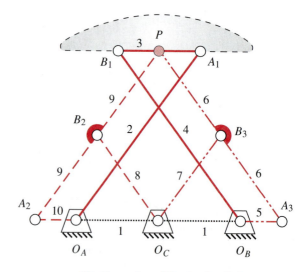

(b) Cognates of the fourbar linkage

FIGURE 3-26

Finding cognates of a fourbar linkage when its coupler point lies on the line of centers of the coupler

Parallel Motion

It is quite common to want the output link of a mechanism to follow a particular path without any rotation of the link as it moves along the path. Once an appropriate path motion in the form of a coupler curve and its fourbar linkage have been found, a cognate of that linkage provides a convenient means to replicate the coupler path motion and provide curvilinear translation (i.e., no rotation) of a new, output link that follows the coupler path. This is referred to as **parallel motion.** Its design is best described with an example, the result of which will be a Watt's I sixbar linkage[*] that incorporates the original fourbar and parts of one of its cognates. The method shown is as described in Soni.[14]

✎ EXAMPLE 3-11

Parallel Motion from a Fourbar Linkage Coupler Curve.

Problem: Design a sixbar linkage for parallel motion over a fourbar linkage coupler path.

Solution: (see Figure 3-27)

1   Figure 3-27a shows the chosen Grashof crank-rocker fourbar linkage and its coupler curve. The first step is to create the Roberts diagram and find its cognates as shown in Figure 3-27b. The Roberts linkage can be found directly, without resort to the Cayley diagram, as described above. The fixed center O_C is found by drawing a triangle similar to the coupler triangle A_1B_1P with base O_AO_B.

2   One of a crank-rocker linkage's cognates will also be a crank-rocker (here cognate #3) and the other is a Grashof double-rocker (here cognate #2). Discard the double-rocker, keeping the links numbered 2, 3, 4, 5, 6, and 7 in Figure 3-27b. Note that links 2 and 7 are the two cranks, and both have the same angular velocity. The strategy is to coalesce these two cranks on a common center (O_A) and then combine them into a single link.

3   Draw the line qq parallel to line O_AO_C and through point O_B as shown in Figure 3-27c.

4   Without allowing links 5, 6, and 7 to rotate, slide them as an assembly along lines O_AO_C and qq until the free end of link 7 is at point O_A. The free end of link 5 will then be at point O'_B and point P on link 6 will be at P'.

5   Add a new link of length O_AO_C between P and P'. This is the *new output link* 8 and all points on it describe the original coupler curve as depicted at points P, P', and P'' in Figure 3-27c.

6   The mechanism in Figure 3-27c has 8 links, 10 revolute joints, and one DOF. When driven by **either** crank 2 or 7, all points on link 8 will duplicate the coupler curve of point P.

7   This is an *overclosed linkage* with redundant links. Because links 2 and 7 have the same angular velocity, they can be joined into one link as shown in Figure 3-27d. Then link 5 can be removed and link 6 reduced to a binary link supported and constrained as part of the loop 2, 6, 8, 3. The resulting mechanism is a Watt-I sixbar (see Figure 2-14, p. 45) with the links numbered 1, 2, 3, 4, 6, and 8. Link 8 is in *curvilinear translation* and follows the coupler path of the original point P.

* Another common method used to obtain parallel motion is to duplicate the same linkage (i.e., the identical cognate), connect them with a parallelogram loop, and remove two redundant links. This results in an eight-link mechanism. See Figure P3-7 on p. 135 for an example of such a mechanism. The method shown here using a different cognate results in a simpler linkage, but either approach will accomplish the desired goal.

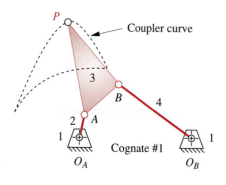

$$\omega_2 = \omega_7 = \omega_9$$
$$\omega_3 = \omega_5 = \omega_{10}$$
$$\omega_4 = \omega_6 = \omega_8$$

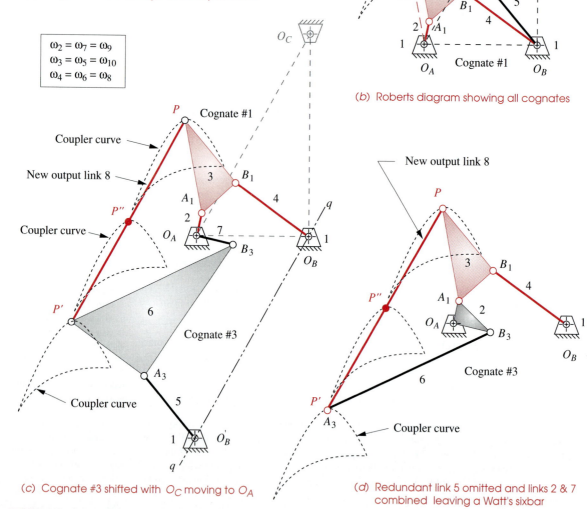

(a) Original fourbar linkage with coupler curve

(b) Roberts diagram showing all cognates

$\Delta O_A O_B O_C \sim \Delta A_1 B_1 P$

(c) Cognate #3 shifted with O_C moving to O_A

(d) Redundant link 5 omitted and links 2 & 7 combined leaving a Watt's sixbar

FIGURE 3-27

Method to construct a Watt-I sixbar that replicates a coupler path with curvilinear translation (parallel motion) (14)

Geared Fivebar Cognates of the Fourbar

Chebyschev also discovered that any fourbar coupler curve can be duplicated with a **geared fivebar mechanism whose gear ratio is plus one,** meaning that the gears turn with the same speed and direction. The geared fivebar's link lengths will be different from those of the fourbar but can be determined directly from the fourbar. Figure 3-28a shows the construction method, as described by Hall [15], to obtain the geared fivebar which will give the same coupler curve as a fourbar. The original fourbar is $O_A A_1 B_1 O_B$ (links 1, 2, 3, 4). The fivebar is $O_A A_2 P B_2 O_B$ (links 1, 5, 6, 7, 8). The two linkages share only the coupler point P and fixed pivots O_A and O_B. The fivebar is constructed by simply drawing link 6 parallel to link 2, link 7 parallel to link 4, link 5 parallel to $A_1 P$, and link 8 parallel to $B_1 P$.

A three-gear set is needed to couple links 5 and 8 with a ratio of plus one (gear 5 and gear 8 are the same diameter and have the same direction of rotation, due to the idler gear), as shown in Figure 3-28b. Link 5 is attached to gear 5, as is link 8 to gear 8. This construction technique may be applied to each of the three fourbar cognates, yielding three geared fivebars (which may or may not be Grashof). The three fivebar cognates can actually be seen in the Roberts diagram. Note that in the example shown, a non-Grashof triple-rocker fourbar yields a Grashof fivebar, which can be motor driven. This conversion to a GFBM linkage could be an advantage when the "right" coupler curve has been found on a non-Grashof fourbar linkage, but continuous output through the fourbar's toggle positions is needed. Thus we can see that there are at least seven linkages which will generate the same coupler curve, three fourbars, three GFBMs and one or more sixbars.

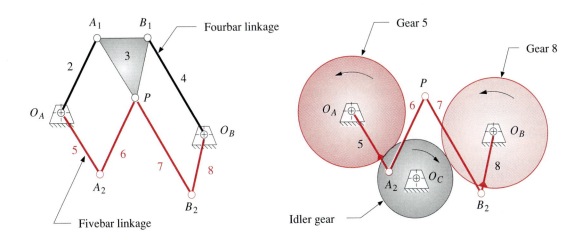

(a) Construction of equivalent fivebar linkage   (b) Resulting geared fivebar linkage

FIGURE 3-28

A geared fivebar linkage cognate of a fourbar linkage

Program FOURBAR calculates the equivalent geared fivebar configuration for any fourbar linkage and will export its data to a disk file that can be opened in program FIVE-BAR for analysis. The file F03-28a.4br can be opened in FOURBAR to animate the linkage shown in Figure 3-28a. Then also open the file F03-28b.5br in program FIVEBAR to see the motion of the equivalent geared fivebar linkage. Note that the original fourbar linkage is a triple-rocker, so cannot reach all portions of the coupler curve when driven from one rocker. But, its geared fivebar equivalent linkage can make a full revolution and traverses the entire coupler path. To export a FIVEBAR disk file for the equivalent GFBM of any fourbar linkage from program FOURBAR, use the *Export* selection under the *File* pull-down menu.

3.8   STRAIGHT-LINE MECHANISMS

A very common application of coupler curves is the generation of approximate straight lines. Straight-line linkages have been known and used since the time of James Watt in the 18th century. Many kinematicians such as Watt, Chebyschev, Peaucellier, Kempe, Evans, and Hoeken (as well as others) over a century ago, developed or discovered either approximate or exact straight-line linkages, and their names are associated with those devices to this day.

The first recorded application of a coupler curve to a motion problem is that of **Watt's straight-line linkage,** patented in 1784, and shown in Figure 3-29a. Watt devised his straight-line linkage to guide the long-stroke piston of his steam engine at a time when metal-cutting machinery that could create a long, straight guideway did not yet exist.* This triple-rocker linkage is still used in automobile suspension systems to guide the rear axle up and down in a straight line as well as in many other applications.

Richard Roberts (1789-1864) (not to be confused with Samuel Roberts of the cognates) discovered the **Roberts' straight-line linkage** shown in Figure 3-29b. This is a triple-rocker. **Chebyschev** (1821-1894) also devised a **straight-line linkage**—a Grashof double-rocker—shown in Figure 3-29c.

The **Hoeken linkage** [16] in Figure 3-29d is a Grashof crank-rocker, which is a significant practical advantage. In addition, the Hoeken linkage has the feature of very *nearly constant velocity along the center portion of its straight-line motion.* It is interesting to note that the **Hoecken** and **Chebyschev** linkages are cognates of one another.[†] The cognates shown in Figure 3-26 (p. 116) are the Chebyschev and Hoeken linkages.

These straight-line linkages are provided as built-in examples in program FOURBAR. A quick look in the Hrones and Nelson atlas of coupler curves will reveal a large number of coupler curves with **approximate straight-line** segments. They are quite common.

To generate an **exact straight line** with only pin joints requires more than four links. At least six links and seven pin joints are needed to generate an exact straight line with a pure revolute-jointed linkage, i.e., a Watt's or Stephenson's sixbar. A geared fivebar mechanism, with a gear ratio of −1 and a phase angle of π radians, will generate an exact straight line at the joint between links 3 and 4. But this linkage is merely a transformed Watt's sixbar obtained by replacing one binary link with a higher joint in the form of a gear pair. This geared fivebar's straight-line motion can be seen by reading the file STRAIGHT.5BR into program FIVEBAR, calculating and animating the linkage.

---

* In Watt's time, straight-line motion was dubbed "parallel motion" though we use that term somewhat differently now. James Watt is reported to have told his son *"Though I am not over anxious after fame, yet I am more proud of the parallel motion than of any other mechanical invention I have made."* Quoted in Muirhead, J. P. (1854). *The Origin and Progress of the Mechanical Inventions of James Watt*, Vol. 3, London, p. 89.

---

† Hain [17] (1967) cites the Hoeken reference [16] (1926) for this linkage. Nolle [18] (1974) shows the Hoeken mechanism but refers to it as a Chebyschev crank-rocker without noting its cognate relationship to the Chebyschev double-rocker, which he also shows. It is certainly conceivable that Chebyschev, as one of the creators of the theorem of cognate linkages, would have discovered the "Hoeken" cognate of his own double-rocker. However, this author has been unable to find any mention of its genesis in the English literature other than the ones cited here.

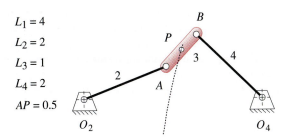

$L_1 = 4$
$L_2 = 2$
$L_3 = 1$
$L_4 = 2$
$AP = 0.5$

(a) Watt straight-line linkage

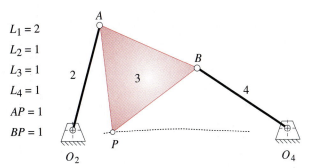

$L_1 = 2$
$L_2 = 1$
$L_3 = 1$
$L_4 = 1$
$AP = 1$
$BP = 1$

(b) Roberts straight-line linkage

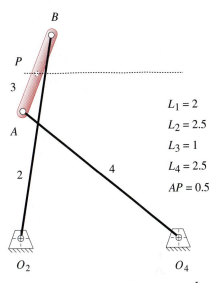

$L_1 = 2$
$L_2 = 2.5$
$L_3 = 1$
$L_4 = 2.5$
$AP = 0.5$

(c) Chebyschev straight-line linkage[*]

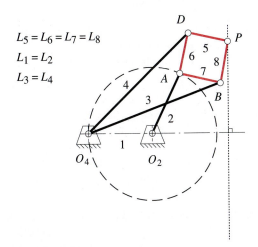

$L_5 = L_6 = L_7 = L_8$
$L_1 = L_2$
$L_3 = L_4$

(e) Peaucellier exact straight-line linkage

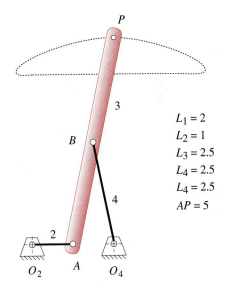

$L_1 = 2$
$L_2 = 1$
$L_3 = 2.5$
$L_4 = 2.5$
$L_4 = 2.5$
$AP = 5$

(d) Hoeken straight-line linkage

FIGURE 3-29

Some common and classic approximate, and one exact, straight-line linkages

* The link ratios of the Chebyschev straight-line linkage have been reported differently by various authors. The ratios used here are those first reported (in English) by Kempe (1877). But Kennedy (1893) describes the same linkage, reportedly "as Chebyschev demonstrated it at the Vienna Exhibition of 1893" as having the link ratios 1, 3.25, 2.5, 3.25. We will assume the earliest reference by Kempe to be correct as listed in the figure.

Peaucellier[*] (1864) discovered an **exact straight-line** mechanism of eight bars and six pins, shown in Figure 3-29e. Links 5, 6, 7, 8 form a rhombus of convenient size. Links 3 and 4 can be any convenient but equal lengths. When O_2O_4 exactly equals O_2A, point C generates an *arc of infinite radius*, i.e., **an exact straight line**. By moving the pivot O_2 left or right from the position shown, changing only the length of link 1, this mechanism *will generate true circle arcs with radii much larger than the link lengths*.

Designing Optimum Straight-Line Fourbar Linkages

Given the fact that an exact straight line can be generated with six or more links using only revolute joints, why use a fourbar approximate straight-line linkage at all? One reason is the desire for simplicity in machine design. The pin-jointed fourbar is the simplest possible one-DOF mechanism. Another reason is that a very good approximation to a true straight line can be obtained with just four links, and this is often "good enough" for the needs of the machine being designed. Manufacturing tolerances will, after all, cause any mechanism's performance to be less than ideal. As the number of links and joints increases, the probability that an exact-straight-line mechanism will deliver its theoretical performance in practice is obviously reduced.

There is a real need for straight-line motions in machinery of all kinds, especially in automated production machinery. Many consumer products such as cameras, film, toiletries, razors, and bottles are manufactured, decorated, or assembled on sophisticated and complicated machines that contain a myriad of linkages and cam-follower systems. Traditionally, most of this kind of production equipment has been of the intermittent-motion variety. This means that the product is carried through the machine on a linear or rotary conveyor that stops for any operation to be done on the product, and then indexes the product to the next work station where it again stops for another operation to be performed. The forces and power required to accelerate and decelerate the large mass of the conveyor (which is independent of, and typically larger than, the mass of the product) severely limit the speeds at which these machines can be run.

Economic considerations continually demand higher production rates, requiring higher speeds or additional, expensive machines. This economic pressure has caused many manufacturers to redesign their assembly equipment for continuous conveyor motion. When the product is in continuous motion in a straight line and at constant velocity, every workhead that operates on the product must be articulated to chase the product and match both its straight-line path and its constant velocity while performing the task. These factors have increased the need for straight-line mechanisms, including ones capable of near-constant velocity over the straight-line path.

A (near) perfect straight-line motion is easily obtained with a fourbar slider-crank mechanism. Ball-bushings (Figure 2-26, p. 57) and hardened ways are available commercially at moderate cost and make this a reasonable, low-friction solution to the straight-line path guidance problem. But, the cost and lubrication problems of a properly guided slider-crank mechanism are still greater than those of a pin-jointed fourbar linkage. Moreover, a crank-slider-block has a velocity profile that is nearly sinusoidal (with some harmonic content) and is far from having constant velocity over any of its motion.

The Hoeken-type linkage offers an optimum combination of straightness and near constant velocity and is a crank-rocker, so it can be motor driven. Its geometry, dimen-

[*] Peaucellier was a French army captain and military engineer who first proposed his "compas compose" or *compound compass* in 1864 but received no immediate recognition therefor. The British-American mathematician, James Sylvester, reported on it to the *Atheneum Club* in London in 1874. He observed that *"The perfect parallel motion of Peaucellier looks so simple and moves so easily that people who see it at work almost universally express astonishment that it waited so long to be discovered."* A model of the Peaucellier linkage was passed around the table. The famous physicist, Sir William Thomson (later Lord Kelvin), refused to relinquish it, declaring *"No. I have not had nearly enough of it—it is the most beautiful thing I have ever seen in my life."* Source: Strandh, S. (1979). *A History of the Machine.* A&W Publishers: New York, p. 67.

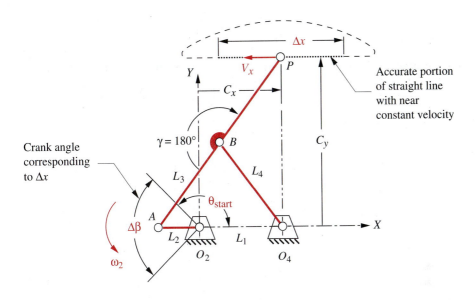

FIGURE 3-30

Hoekens linkage geometry. Linkage shown with P at center of straight-line portion of path

sions, and coupler path are shown in Figure 3-30. This is a symmetrical fourbar linkage. Since the angle γ of line BP is specified and $L_3 = L_4 = BP$, only two link ratios are needed to define its geometry, say L_1 / L_2 and L_3 / L_2. If the crank L_2 is driven at constant angular velocity ω_2, the linear velocity V_x along the straight line portion Δx of the coupler path will be very close to constant over a significant portion of crank rotation $\Delta \beta$.

A study was done to determine the errors in straightness and constant velocity of the Hoeken-type linkage over various fractions $\Delta \beta$ of the crank cycle as a function of the link ratios. [19] The structural error in position (i.e., straightness) ε_S and the structural error in velocity ε_V are defined using notation from Figure 3-30 as:

$$\varepsilon_S = \frac{MAX_{i=1}^{n}\left(C_{y_i}\right) - MIN_{i=1}^{n}\left(C_{y_i}\right)}{\Delta x}$$

$$\varepsilon_V = \frac{MAX_{i=1}^{n}\left(V_{x_i}\right) - MIN_{i=1}^{n}\left(V_{x_i}\right)}{\overline{V}_x}$$

(3.5)*

The structural errors were computed separately for each of nine crank angle ranges $\Delta \beta$ from 20° to 180°. Table 3-1 shows the link ratios that give the smallest possible structural error in either position or velocity over values of $\Delta \beta$ from 20° to 180°. Note that one cannot attain optimum straightness and minimum velocity error in the same linkage. However, reasonable compromises between the two criteria can be achieved, especially for smaller ranges of crank angle. The errors in both straightness and velocity increase as longer portions of the curve are used (larger $\Delta \beta$). The use of Table 3-1 to design a straight-line linkage will be shown with an example.

* See reference [19] for the derivation of equations 3.5.

TABLE 3-1   Link Ratios for Smallest Attainable Errors in Straightness and Velocity for Various
                  Crank-Angle Ranges of a Hoeken-Type Fourbar Approximate Straight-Line Linkage (19)

Range of Motion			Optimized for Straightness							Optimized for Constant Velocity						
$\Delta\beta$ (deg)	θ_{start} (deg)	% of cycle	Maximum ΔC_y %	ΔV %	$\dfrac{V_x}{(L_2\,\omega_2)}$	L_1/L_2	L_3/L_2	$\Delta x/L_2$	Maximum ΔV_x %	ΔC_y %	$\dfrac{V_x}{(L_2\,\omega_2)}$	L_1/L_2	L_3/L_2	$\Delta x/L_2$
20	170	5.6%	0.00001%	0.38%	1.436	2.975	3.963	0.601	0.006%	0.137%	1.045	2.075	2.613	0.480
40	160	11.1%	0.00004%	1.53%	1.504	2.950	3.925	1.193	0.038%	0.274%	1.124	2.050	2.575	0.950
60	150	16.7%	0.00027%	3.48%	1.565	2.900	3.850	1.763	0.106%	0.387%	1.178	2.025	2.538	1.411
80	140	22.2%	0.001%	6.27%	1.611	2.825	3.738	2.299	0.340%	0.503%	1.229	1.975	2.463	1.845
100	130	27.8%	0.004%	9.90%	1.646	2.725	3.588	2.790	0.910%	0.640%	1.275	1.900	2.350	2.237
120	120	33.3%	0.010%	14.68%	1.679	2.625	3.438	3.238	1.885%	0.752%	1.319	1.825	2.238	2.600
140	110	38.9%	0.023%	20.48%	1.702	2.500	3.250	3.623	3.327%	0.888%	1.347	1.750	2.125	2.932
160	100	44.4%	0.047%	27.15%	1.717	2.350	3.025	3.933	5.878%	1.067%	1.361	1.675	2.013	3.232
180	90	50.0%	0.096%	35.31%	1.725	2.200	2.800	4.181	9.299%	1.446%	1.374	1.575	1.863	3.456

EXAMPLE 3-12

Designing a Hoeken-Type Straight-Line Linkage.

Problem: A 100-mm-long straight line motion is needed over 1/3 of the total cycle (120° of
               crank rotation). Determine the dimensions of a Hoeken-type linkage that will

               (a) Provide minimum deviation from a straight line. Determine its maximum
                   deviation from constant velocity.
               (b) Provide minimum deviation from constant velocity. Determine its maximum
                   deviation from a straight line.

Solution: (see Figure 3-30, p. 123, and Table 3-1)

1 Part (a) requires the most accurate straight line. Enter Table 3-1 at the 6th row which is for
   a crank angle duration $\Delta\beta$ of the required 120°. The 4th column shows the minimum possible deviation from straight to be 0.01% of the length of the straight line portion used. For a
   100-mm length the absolute deviation will then be 0.01 mm (0.0004 in). The 5th column
   shows that its velocity error will be 14.68% of the average velocity over the 100-mm length.
   The absolute value of this velocity error of course depends on the speed of the crank.

2 The linkage dimensions for part (a) are found from the ratios in columns 7, 8, and 9. The
   crank length required to obtain the 100-mm length of straight line Δx is:

   from Table 3-1: $$\frac{\Delta x}{L_2} = 3.238$$

                                                                                    (a)

   $$L_2 = \frac{\Delta x}{3.238} = \frac{100 \text{ mm}}{3.23} = 30.88 \text{ mm}$$

The other link lengths are then:

from Table 3-1:
$$\frac{L_1}{L_2} = 2.625$$

(b)

$$L_1 = 2.625L_2 = 2.625(30.88 \text{ mm}) = 81.07 \text{ mm}$$

from Table 3-1:
$$\frac{L_3}{L_2} = 3.438$$

(c)

$$L_3 = 3.438L_2 = 3.438(30.88 \text{ mm}) = 106.18 \text{ mm}$$

The complete linkage is then: $L_1 = 81.07, L_2 = 30.88, L_3 = L_4 = BP = 106.18$ mm. The nominal velocity V_x of the coupler point at the center of the straight line ($\theta_2 = 180°$) can be found from the factor in the 6th column which must be multiplied by the crank length L_2 and the crank angular velocity ω_2 in radians per second (rad/sec).

3   Part (b) requires the most accurate velocity. Again enter Table 3-1 at the 6th row which is for a crank angle duration $\Delta\beta$ of the required 120°. The 10th column shows the minimum possible deviation from constant velocity to be 1.885% of the average velocity V_x over the length of the straight line portion used. The 11th column shows the deviation from straight to be 0.752% of the length of the straight line portion used. For a 100-mm length the absolute deviation in straightness for this optimum constant velocity linkage will then be 0.75 mm (0.030 in).

The link lengths for this mechanism are found in the same way as was done in step 2 above except that the link ratios 1.825, 2.238, and 2.600 from columns 13, 14, and 15 are used. The result is: $L_1 = 70.19, L_2 = 38.46, L_3 = L_4 = BP = 86.08$ mm. The nominal velocity V_x of the coupler point at the center of the straight line ($\theta_2 = 180°$) can be found from the factor in the 12th column which must be multiplied by the crank length L_2 and the crank angular velocity ω_2 in rad/sec.

4   The first solution (step 2) gives an extremely accurate straight line over a significant part of the cycle but its 15% deviation in velocity would probably be unacceptable if that factor was considered important. The second solution (step 3) gives less than 2% deviation from constant velocity, which may be viable for a design application. Its 3/4% deviation from straightness, while much greater than the first design, may be acceptable in some situations.

3.9   DWELL MECHANISMS

A common requirement in machine design problems is the need for a dwell in the output motion. A **dwell** is defined as *zero output motion for some nonzero input motion.* In other words, the motor keeps going, but the output link stops moving. Many production machines perform a series of operations which involve feeding a part or tool into a workspace, and then holding it there (in a dwell) while some task is performed. Then the part must be removed from the workspace, and perhaps held in a second dwell while the rest of the machine "catches up" by indexing or performing some other tasks. Cams and followers (Chapter 8) are often used for these tasks because it is trivially easy to create a

7   Locate fixed pivot O_6 on the bisector of DE such that lines O_6D and O_6E subtend the desired output angle, in this example, 90°.

8   Draw link 6 from D (or E) through O_6 and extend to any convenient length. This is the output link which will dwell for the specified portion of the crank cycle.

9   Check the transmission angles.

10   Make a cardboard model of the linkage and articulate it to check its function.

This linkage dwells because, during the time that the coupler point P is traversing the pseudo-arc portion of the coupler curve, the other end of link 5, attached to P and the same length as the arc radius, is essentially stationary at its other end, which is the arc center. However the dwell at point D will have some "jitter" or oscillation, due to the fact that D is only an approximate center of the pseudo-arc on the sixth-degree coupler curve. When point P leaves the arc portion, it will smoothly drive link 5 from point D to point E, which will in turn rotate the output link 6 through its arc as shown in Figure 3-31c (p. 127). Note that we can have any angular displacement of link 6 we desire with the same links 2 to 5, as they alone completely define the dwell aspect. Moving pivot O_6 left and right along the bisector of line DE will change the angular displacement of link 6 but not its timing. In fact, a slider block could be substituted for link 6 as shown in Figure 3-31d, and linear translation along line DE with the same timing and dwell at D will result. Input the file F03-31c.6br to program SIXBAR and animate to see the linkage of Example 3-13 in motion. The dwell in the motion of link 6 can be clearly seen in the animation, including the jitter due to its approximate nature.

Double-Dwell Linkages

It is also possible, using a fourbar coupler curve, to create a double-dwell output motion. One approach is the same as that used in the single-dwell of Example 3-11. Now a coupler curve is needed which has *two* approximate circle arcs of the same radius but with different centers, both convex or both concave. A link 5 of length equal to the radius of the two arcs will be added such that it and link 6 will remain nearly stationary at the center of each of the arcs, while the coupler point traverses the circular parts of its path. Motion of the output link 6 will occur only when the coupler point is between those arc portions. Higher-order linkages, such as the geared fivebar, can be used to create multiple-dwell outputs by a similar technique since they possess coupler curves with multiple, approximate circle arcs. See the built-in example double-dwell linkage in program SIXBAR for a demonstration of this approach.

A second approach uses a coupler curve with two approximate straight-line segments of appropriate duration. If a pivoted slider block (link 5) is attached to the coupler at this point, and link 6 is allowed to slide in link 5, it only remains to choose a pivot O_6 at the intersection of the straight-line segments extended. The result is shown in Figure 3-32. While block 5 is traversing the "straight-line" segments of the curve, it will not impart any angular motion to link 6. The approximate nature of the fourbar straight line causes some jitter in these dwells also.

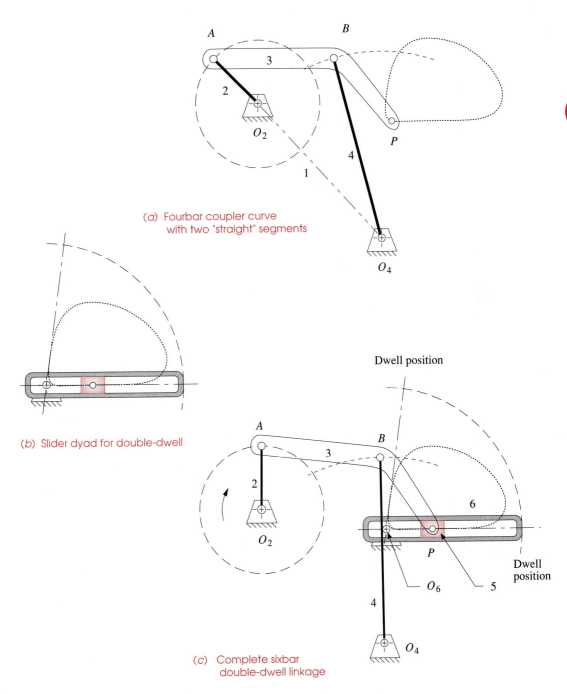

(a) Fourbar coupler curve with two "straight" segments

(b) Slider dyad for double-dwell

(c) Complete sixbar double-dwell linkage

FIGURE 3-32

Double-dwell sixbar linkage

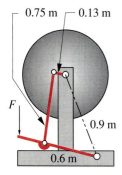

FIGURE P3-3

Problem 3-14. Treadle operated grinding wheel

3-11 Find the three equivalent geared fivebar linkages for the three fourbar cognates in Figure 3-25a (p. 115). Check your results by comparing the coupler curves with programs FOURBAR and FIVEBAR.

3-12 Design a sixbar single-dwell linkage for a dwell of 90° of crank motion, with an output rocker motion of 45°.

3-13 Design a sixbar double-dwell linkage for a dwell of 90° of crank motion, with an output rocker motion of 60°, followed by a second dwell of about 60° of crank motion.

3-14 Figure P3-3 shows a treadle operated grinding wheel driven by a fourbar linkage. Make a cardboard model of the linkage to any convenient scale. Determine its minimum transmission angles. Comment on its operation. Will it work? If so, explain how it does.

3-15 Figure P3-4 shows a non-Grashof fourbar linkage that is driven from link O_2A. All dimensions are in centimeters (cm).

   (a) Find the transmission angle at the position shown.
   (b) Find the toggle positions in terms of angle AO_2O_4.
   (c) Find the maximum and minimum transmission angles over its range of motion.
   (d) Draw the coupler curve of point P over its range of motion.

3-16 Draw the Roberts diagram for the linkage in Figure P3-4 and find its two cognates. Are they Grashof or non-Grashof?

3-17 Design a Watt-I sixbar to give parallel motion that follows the coupler path of point P of the linkage in Figure P3-4.

3-18 Add a driver dyad to the solution of Problem 3-17 to drive it over its possible range of motion with no quick return. (The result will be an 8-bar linkage.)

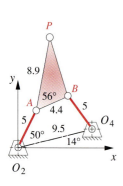

FIGURE P3-4

Problems 3-15 to 3-18

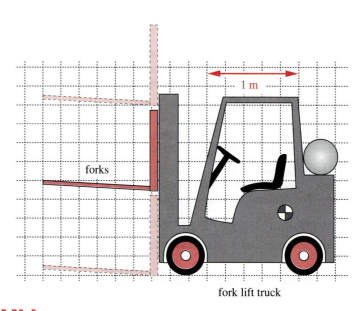

fork lift truck

FIGURE P3-5

Problem 3-19

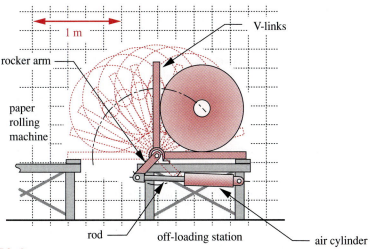

FIGURE P3-6

Problem 3-20

3-19 Design a pin-jointed linkage that will guide the forks of the fork lift truck in Figure P3-5 up and down in an approximate straight line over the range of motion shown. Arrange the fixed pivots so they are close to some part of the existing frame or body of the truck.

3-20 Figure P3-6 shows a "V-link" off-loading mechanism for a paper roll conveyor. Design a pin-jointed linkage to replace the air cylinder driver that will rotate the rocker arm and V-link through the 90° motion shown. Keep the fixed pivots as close to the existing frame as possible. Your fourbar linkage should be Grashof and be in toggle at each extreme position of the rocker arm.

3-21 Figure P3-7 shows a walking-beam transport mechanism that uses a fourbar coupler curve, replicated with a parallelogram linkage for parallel motion. Note the duplicate crank and coupler shown ghosted in the right half of the mechanism—they are redundant

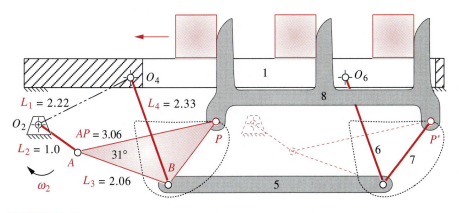

FIGURE P3-7

Problems 3-21 to 3-22. Straight-line walking beam eightbar transport mechanism

* Answers in Appendix F.

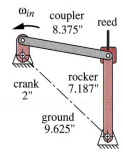

ω_{in} coupler   reed
          8.375"

crank   rocker
2"   7.187"

ground
9.625"

FIGURE P3-8

Problem 3-23.
Loom laybar drive

and have been removed from the duplicate fourbar linkage. Using the same fourbar driving stage (links L_1, L_2, L_3, L_4 with coupler point P), design a Watt-I sixbar linkage that will drive link 8 in the same parallel motion using two fewer links.

*3-22   Find the maximum and minimum transmission angles of the fourbar driving stage (links L_1, L_2, L_3, L_4) in Figure P3-7 (to graphical accuracy).

*3-23   Figure P3-8 shows a fourbar linkage used in a power loom to drive a comblike reed against the thread, "beating it up" into the cloth. Determine its Grashof condition and its minimum and maximum transmission angles to graphical accuracy.

3-24   Draw the Roberts diagram and find the cognates of the linkage in Figure P3-9.

3-25   Find the equivalent geared fivebar mechanism cognate of the linkage in Figure P3-9.

3-26   Use the linkage in Figure P3-9 to design an eightbar double-dwell mechanism that has a rocker output through 45°.

3-27   Use the linkage in Figure P3-9 to design an eightbar double-dwell mechanism that has a slider output stroke of 5 crank units.

3-28   Use two of the cognates in Figure 3-26 (p. 116) to design a Watt-I sixbar parallel motion mechanism that carries a link through the same coupler curve at all points. Comment on its similarities to the original Roberts diagram.

3-29   Find the cognates of the Watt straight-line mechanism in Figure 3-29a (p. 121).

3-30   Find the cognates of the Roberts straight-line mechanism in Figure 3-29b (p. 121).

3-31   Design a Hoeken straight-line linkage to give minimum error in velocity over 22% of the cycle for a 15-cm-long straight-line motion. Specify all linkage parameters.

3-32   Design a Hoeken straight-line linkage to give minimum error in straightness over 39% of the cycle for a 20-cm-long straight-line motion. Specify all linkage parameters.

3-33   Design a linkage that will give a symmetrical "kidney bean" shaped coupler curve as shown in Figure 3-16 (p. 104 and 105). Use the data in Figure 3-21 (p. 110) to determine the required link ratios and generate the coupler curve with program FOURBAR.

3-34   Repeat Problem 3-33 for a "double straight" coupler curve.

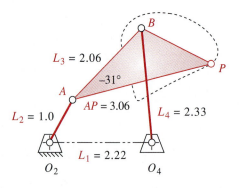

$L_3 = 2.06$

$-31°$

A

$AP = 3.06$

B

P

$L_4 = 2.33$

$L_2 = 1.0$

$L_1 = 2.22$

O_2 O_4

FIGURE P3-9

Problems 3-24 to 3-27

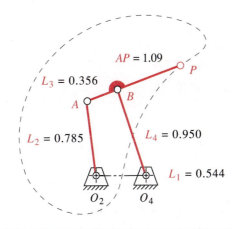

FIGURE P3-10

Problems 3-36 to 3-38

3-35 Repeat problem 3-33 for a "scimitar" coupler curve with two distinct cusps. Show that there are (or are not) true cusps on the curve by using program FOURBAR. (Hint: Think about the definition of a cusp and how you can use the program's data to show it.)

*3-36 Find the Grashof condition, inversion, any limit positions, and the extreme values of the transmission angle (to graphical accuracy) of the linkage in Figure P3-10.

3-37 Draw the Roberts diagram and find the cognates of the linkage in Figure P3-10.

3-38 Find the three geared fivebar cognates of the linkage in Figure P3-10.

3-39 Find the Grashof condition, any limit positions, and the extreme values of the transmission angle (to graphical accuracy) of the linkage in Figure P3-11.

3-40 Draw the Roberts diagram and find the cognates of the linkage in Figure P3-11.

3-41 Find the three geared fivebar cognates of the linkage in Figure P3-11.

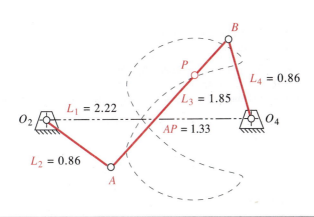

FIGURE P3-11

Problems 3-39 to 3-41

* Answers in Appendix F.

3

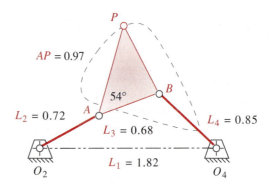

FIGURE P3-12

Problems 3-42 to 3-44

[*]3-42   Find the Grashof condition, any limit positions, and the extreme values of the transmission angle (to graphical accuracy) of the linkage in Figure P3-12.

3-43   Draw the Roberts diagram and find the cognates of the linkage in Figure P3-12.

3-44   Find the three geared fivebar cognates of the linkage in Figure P3-12.

3-45   Prove that the relationships between the angular velocities of various links in the Roberts diagram as shown in Figure 3-25 (p. 115) are true.

3-46   Design a fourbar linkage to move the object in Figure P3-13 from position 1 to 2 using points A and B for attachment. Add a driver dyad to limit its motion to the range of positions shown making it a sixbar. All fixed pivots should be on the base.

3-47   Design a fourbar linkage to move the object in Figure P3-13 from position 2 to 3 using points A and B for attachment. Add a driver dyad to limit its motion to the range of positions shown making it a sixbar. All fixed pivots should be on the base.

3-48   Design a fourbar linkage to move the object in Figure P3-13 through the three positions shown using points A and B for attachment. Add a driver dyad to limit its motion to the range of positions shown making it a sixbar. All fixed pivots should be on the base.

3-49   Design a fourbar linkage to move the object in Figure P3-14 from position 1 to 2 using points A and B for attachment. Add a driver dyad to limit its motion to the range of positions shown making it a sixbar. All fixed pivots should be on the base.

3-50   Design a fourbar linkage to move the object in Figure P3-14 from position 2 to 3 using points A and B for attachment. Add a driver dyad to limit its motion to the range of positions shown making it a sixbar. All fixed pivots should be on the base.

3-51   Design a fourbar linkage to move the object in Figure P3-14 through the three positions shown using points A and B for attachment. Add a driver dyad to limit its motion to the range of positions shown making it a sixbar. All fixed pivots should be on the base.

3-52   Design a fourbar linkage to move the object in Figure P3-15 from position 1 to 2 using points A and B for attachment. Add a driver dyad to limit its motion to the range of positions shown making it a sixbar. All fixed pivots should be on the base.

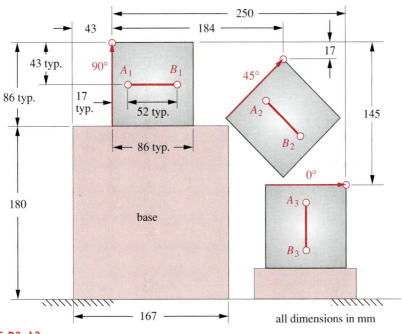

FIGURE P3-13

Problems 3-46 to 3-48

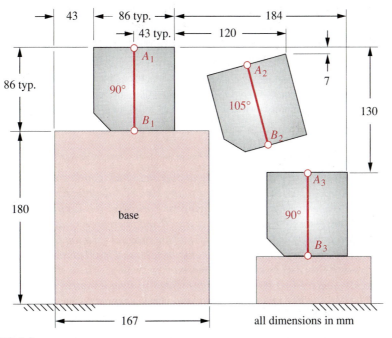

FIGURE P3-14

Problems 3-49 to 3-51

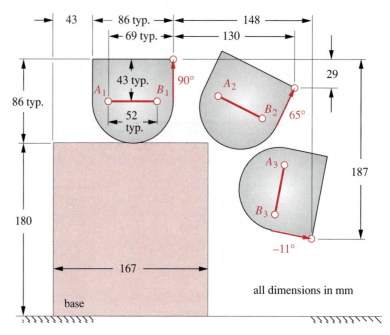

FIGURE P3-15

Problems 3-52 to 3-54

3-53   Design a fourbar linkage to move the object in Figure P3-15 from position 2 to 3 using
        points *A* and *B* for attachment. Add a driver dyad to limit its motion to the range of
        positions shown making it a sixbar. All fixed pivots should be on the base.

3-54   Design a fourbar linkage to move the object in Figure P3-15 through the three positions
        shown using points *A* and *B* for attachment. Add a driver dyad to limit its motion to the
        range of positions shown making it a sixbar. All fixed pivots should be on the base.

3.13   PROJECTS

*These larger-scale project statements deliberately lack detail and structure and are
loosely defined. Thus, they are similar to the kind of "identification of need" problem
statement commonly encountered in engineering practice. It is left to the student to structure the problem through background research and to create a clear goal statement and
set of performance specifications before attempting to design a solution. This design process is spelled out in Chapter 1 and should be followed in all of these examples. These
projects can be done as an exercise in mechanism synthesis alone or can be revisited and
thoroughly analyzed by the methods presented in later chapters as well. All results
should be documented in a professional engineering report.*

P3-1   The tennis coach needs a better tennis ball server for practice. This device must fire a
        sequence of standard tennis balls from one side of a standard tennis court over the net
        such that they land and bounce within each of the three court areas defined by the court's
        white lines. The order and frequency of a ball's landing in any one of the three court areas

P3-23 Click and Clack, the tappet brothers, need a better transmission jack for their Good News Garage. This device should position a transmission under a car (on a lift) and allow it to be maneuvered into place safely and quickly.

P3-24 A paraplegic who was an avid golfer before his injury, wants a mechanism to allow him to stand up in his wheelchair in order to once again play golf. It must not interfere with normal wheelchair use, though it could be removed from the chair when he is not golfing.

P3-25 A wheelchair lift is needed to raise the wheelchair and person 3 ft from the garage floor to the level of the first floor of the house. Safety and reliability are of major concern, as is cost.

P3-26 A paraplegic needs a mechanism that can be installed on a full-size 3-door pickup truck that will lift the wheelchair into the area behind the driver's seat. This person has excellent upper body strength and, with the aid of specially installed handles on the truck, can get into the cab from the chair. The truck can be modified as necessary to accommodate this task. For example, attachment points can be added to its structure and the back seat of the truck can be removed if necessary.

P3-27 There is demand for a better *Baby Transport Device.* Many such devices are on the market. Some are called carriages, some strollers. Some are convertible to multiple uses. Our marketing survey data so far seems to indicate that the customers want portability (i.e., foldability), light weight, one-handed operation, and large wheels. Some of these features are present in existing devices. We need a better design that more completely meets the needs of the customer. The device must be stable, effective, and safe for the baby and the operator. Full joints are preferred to half joints and simplicity is the mark of good design. A linkage solution with manual input is desired.

P3-28 There is a need for a *dining-table leaf insertion device.* The device must be simple to use, preferably using the action of opening the table-halves as the actuating motion. That is, as you pull the table open, the stored leaf should be carried by the mechanism of your design into its proper place in order to extend the dining surface.

P3-29 A boat owner has requested that we design her a lift mechanism to automatically move a 1000-lb, 15-ft boat from a cradle on land to the water. A seawall protects the owner's yard, and the boat cradle sits above the seawall. The tidal variation is 4 ft. Your mechanism will be attached to land and move the boat from its stored position on the cradle to the water and return it to the cradle. The device must be safe and easy to use and not overly expensive.

P3-30 The landfills are full! We're about to be up to our ears in trash! The world needs a better trash compactor. It should be simple, inexpensive, quiet, compact, and safe. It can either be manually powered or motorized, but manual operation is preferred to keep the cost down. The device must be stable, effective, and safe for the operator.

P3-31 A small contractor needs a mini-dumpster attachment for his pickup truck. He has made several trash containers which are 4 ft x 4 ft x 3.5 ft high. The empty container weighs 150 lb. He needs a mechanism which he can attach to his fleet of standard, full-size pickup trucks (Chevrolet, Ford, or Dodge). This mechanism should be able to pick up the full trash container from the ground, lift it over the closed tailgate of the truck, dump its contents into the truck bed, and then return it empty to the ground. He would like not to tip his truck over in the process. The mechanism should store permanently on the truck in such a manner as to allow the normal use of the pickup truck at all other times. You may specify any means of attachment of your mechanism to the container and to the truck.

Chapter 4

POSITION ANALYSIS

Theory is the distilled essence of practice
RANKINE

4.0   INTRODUCTION

Once a tentative mechanism design has been **synthesized**, it must then be **analyzed**. A principal goal of kinematic analysis is to determine the accelerations of all the moving parts in the assembly. **Dynamic forces** are proportional to acceleration, from Newton's second law. We need to know the dynamic forces in order to calculate the **stresses** in the components. The design engineer must ensure that the proposed mechanism or machine will not fail under its operating conditions. Thus the stresses in the materials must be kept well below allowable levels. To calculate the stresses, we need to know the static and dynamic forces on the parts. To calculate the dynamic forces, we need to know the **accelerations**. In order to calculate the accelerations, we must first find the **positions** of all the links or elements in the mechanism for each increment of input motion, and then differentiate the position equations versus time to find **velocities**, and then differentiate again to obtain the expressions for acceleration. For example, in a simple Grashof four-bar linkage, we would probably want to calculate the positions, velocities, and accelerations of the output links (coupler and rocker) for perhaps every two degrees (180 positions) of input crank position for one revolution of the crank.

This can be done by any of several methods. We could use a **graphical approach** to determine the position, velocity, and acceleration of the output links for all 180 positions of interest, or we could **derive the general equations** of motion for any position, differentiate for velocity and acceleration, and then solve these **analytical expressions** for our 180 (or more) crank locations. A computer will make this latter task much more palatable. If we choose to use the graphical approach to analysis, we will have to do an independent graphical solution for each of the positions of interest. None of the information obtained graphically for the first position will be applicable to the second position or to any others. In contrast, once the analytical solution is derived for a particular mechanism, it can be quickly solved (with a computer) for all positions. If you want information for more than 180 positions, it only means you will have to wait longer for the computer to generate those data. The derived equations are the same. So, have another cup of coffee while the computer crunches the numbers! In this chapter, we will present and derive analytical solutions to the position analysis problem for various planar mechanisms. We will also discuss graphical solutions which are useful for checking your analytical results. In Chapters 6 and 7 we will do the same for velocity and acceleration analysis of planar mechanisms.

It is interesting to note that **graphical position analysis** of linkages is a truly trivial exercise, while the algebraic approach to position analysis is much more complicated. If you can draw the linkage to scale, you have then solved the position analysis problem graphically. It only remains to measure the link angles on the scale drawing to protractor accuracy. But, the converse is true for velocity and especially for acceleration analysis. Analytical solutions for these are less complicated to derive than is the analytical position solution. However, graphical velocity and acceleration analysis becomes quite complex and difficult. Moreover, the graphical vector diagrams must be redone *de novo* (meaning literally *from new*) for each of the linkage positions of interest. This is a very tedious exercise and was the only practical method available in the days *B.C.* (*Before Computer*), not so long ago. The proliferation of inexpensive microcomputers in recent years has truly revolutionized the practice of engineering. As a graduate engineer, you will never be far from a computer of sufficient power to solve this type of problem and may even have one in your pocket. Thus, in this text we will emphasize analytical solutions which are easily solved with a microcomputer. The computer programs provided with this text use the same analytical techniques as derived in the text.

Geez Joe, - now I wish I took that programming course!

4.1   COORDINATE SYSTEMS

Coordinate systems and reference frames exist for the pleasure and convenience of the engineer who defines them. In the next chapters we will freely adorn our systems with multiple coordinate systems as we see fit, to aid in understanding and solving the problem. We will denote one of these as the *global* or *absolute* coordinate system, and the others will be *local* coordinate systems within the global framework. The global system is often taken to be attached to Mother Earth, though it could as well be attached to another ground plane such as the frame of an automobile. If our goal is to analyze the motion of a windshield wiper blade, we may not care to include the gross motion of the automobile in the analysis. In that case a global coordinate system (GCS—denoted as *X,Y*) attached to the car would be useful, and we could consider it to be an **absolute** coordinate system. Even if we use the earth as an absolute reference frame, we must realize that it is not stationary either, and as such is not very useful as a reference frame for a space probe. Though we will speak of absolute positions, velocities, and accelerations, keep in mind that ultimately, until we discover some stationary point in the universe, all motions are really relative. The term **inertial reference frame** is used to denote *a system which itself has no acceleration*. All angles in this text will be measured according to the *right-hand rule*. That is, **counterclockwise angles**, angular velocities, and angular accelerations *are positive in sign*.

Local coordinate systems are typically attached to a link at some point of interest. This might be a pin joint, a center of gravity, or a line of centers of a link. These local coordinate systems may be either rotating or nonrotating as we desire. If we want to measure the angle of a link as it rotates in the global system, we probably will want to attach a local nonrotating coordinate system (LNCS—denoted as *x, y*) to some point on the link (say a pin joint). This nonrotating system will move with its origin on the link but remains always parallel to the global system. If we want to measure some parameters within a link, independent of its rotation, then we will want to construct a local rotating coordinate system (LRCS—denoted as *x', y'*) along some line on the link. This system will both move and rotate with the link in the global system. Most often we will need to have both types of local coordinate systems (LNCS and LRCS) on our moving links to do a complete analysis. Obviously we must define the angles and/or positions of these moving, local coordinate systems in the global system at all positions of interest.

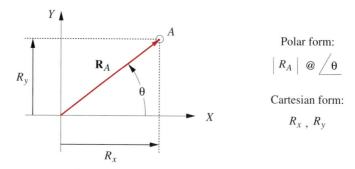

FIGURE 4-1

A position vector in the plane

4.2 POSITION AND DISPLACEMENT

Position

The **position** of a point in the plane can be defined by the use of a **position vector** as shown in Figure 4-1. The choice of **reference axes** is arbitrary and is selected to suit the observer. A two-dimensional vector has two attributes, which can be expressed in either *polar* or *cartesian* coordinates. The **polar form** provides the magnitude and the angle of the vector. The **cartesian form** provides the X and Y components of the vector. Each form is directly convertible into the other by:[*]

the Pythagorean theorem:

$$R_A = \sqrt{R_x^2 + R_y^2}$$

and trigonometry:

$$\theta = \arctan\left(\frac{R_y}{R_x}\right) \tag{4.0}$$

Displacement

Displacement of a point is the change in its position and can be defined as *the straight-line distance between the initial and final position of a point which has moved in the reference frame.* Note that displacement is not necessarily the same as the path length which the point may have traveled to get from its initial to final position. Figure 4-2a shows a point in two positions, A and B. The curved line depicts the path along which the point traveled. The position vector $\mathbf{R}_{BA}$ defines the displacement of the point B with respect to point A. Figure 4-2b defines this situation more rigorously and with respect to a reference frame XY. The notation $\mathbf{R}$ will be used to denote a position vector. The vectors $\mathbf{R}_A$ and $\mathbf{R}_B$ define, respectively, the absolute positions of points A and B with respect to this *global XY* reference frame. The vector $\mathbf{R}_{BA}$ denotes the difference in position, or the *displacement*, between A and B. This can be expressed as the *position difference equation:*

$$\mathbf{R}_{BA} = \mathbf{R}_B - \mathbf{R}_A \tag{4.1a}$$

This expression is read: *The position of B with respect to A is equal to the (absolute) position of B minus the (absolute) position of A*, where *absolute* means with respect to the origin of the *global* reference frame. This expression could also be written as:

$$\mathbf{R}_{BA} = \mathbf{R}_{BO} - \mathbf{R}_{AO} \tag{4.1b}$$

with the second subscript O denoting the origin of the XY reference frame. When a position vector is rooted at the origin of the reference frame, it is customary to omit the second subscript. It is understood, in its absence, to be the origin. Also, a vector referred to the origin, such as $\mathbf{R}_A$, is often called an absolute vector. This means that it is taken with respect to a reference frame which is assumed to be stationary, e.g. *the ground*. It is important to realize, however, that the ground is usually also in motion in some larger frame of reference. Figure 4-2c shows a graphical solution to equation 4.1.

[*] Note that a two-argument arctangent function must be used to obtain angles in all four quadrants. The single argument arctangent function found in most calculators and computer programming languages returns angle values in only the first and fourth quadrants. You can calculate your own two-argument arctangent function very easily by testing the sign of the x component of the arguments and, if x is minus, adding π radians or 180° to the result obtained from the available single-argument arctangent function.

For example (in Fortran):

```
FUNCTION Atan2( x, y )
IF x <> 0 THEN Q = y / x
Temp = ATAN(Q)
IF x < 0 THEN
    Atan2 = Temp + 3.14159
ELSE
    Atan2 = Temp
END IF
RETURN
END
```

The above code assumes that the language used has a built-in single-argument arctangent function called ATAN(x) which returns an angle between ± π/2 radians when given a signed argument representing the value of the tangent of that angle.

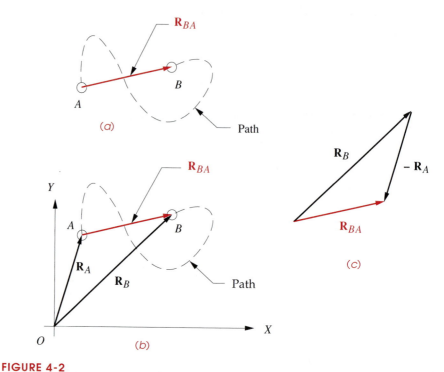

FIGURE 4-2

Position difference and relative position

In our example of Figure 4-2, we have tacitly assumed so far that this point, which is first located at A and later at B, is, in fact, the same particle, moving within the reference frame. It could be, for example, one automobile moving along the road from A to B. With that assumption, it is conventional to refer to the vector $\mathbf{R}_{BA}$ as a **position difference**. There is, however, another situation which leads to the same diagram and equation but needs a different name. Assume now that points A and B in Figure 4-2b represent not the same particle but two independent particles moving in the same reference frame, as perhaps two automobiles traveling on the same road. The vector equations 4.1 and the diagram in Figure 4-2b still are valid, but we now refer to $\mathbf{R}_{BA}$ as a **relative position**, or **apparent position**. We will use the relative position term here. A more formal way to distinguish between these two cases is as follows:

CASE 1:   *One body in two successive positions* => ***position difference***

CASE 2:   *Two bodies simultaneously in separate positions* => ***relative position***

This may seem a rather fine point to distinguish, but the distinction will prove useful, and the reasons for it more clear, when we analyze velocities and accelerations, especially when we encounter (CASE 2 type) situations in which the two bodies occupy the same position at the same time but have different motions.

4.3   TRANSLATION, ROTATION, AND COMPLEX MOTION

So far we have been dealing with a particle, or point, in plane motion. It is more interesting to consider the motion of a **rigid body**, or link. Figure 4-3a shows a link AB denoted by a position vector $\mathbf{R}_{BA}$. An axis system has been set up at the root of the vector, at point A, for convenience.

Translation

Figure 4-3b shows link AB moved to a new position $A'B'$ by translation through the displacement AA' or BB' which are equal, i.e., $\mathbf{R}_{A'A} = \mathbf{R}_{B'B}$.

A definition of translation is:

All points on the body have the same displacement.

As a result the link retains its angular orientation. Note that the translation need not be along a straight path. The curved lines from A to A' and B to B' are the **curvilinear translation** path of the link. There is no rotation of the link if these paths are parallel. If the path happens to be straight, then it will be the special case of **rectilinear translation**, and the path and the displacement will be the same.

Rotation

Figure 4-3c shows the same link AB moved from its original position at the origin by rotation through an angle. Point A remains at the origin, but B moves through the position difference vector $\mathbf{R}_{B'B} = \mathbf{R}_{B'A} - \mathbf{R}_{BA}$.

A definition of rotation is:

Different points in the body undergo different displacements and thus there is a displacement difference between any two points chosen.

The link now changes its angular orientation in the reference frame, and all points have different displacements.

Complex Motion

The general case of **complex motion** is the sum of the translation and rotation components. Figure 4-3d shows the same link moved through both the translation and the rotation applied above. Note that the order in which these two components are added is immaterial. The resulting complex displacement will be the same whether you first rotate and then translate or vice versa. This is because the two factors are independent. The total complex displacement of point B is defined by the following expression:

$$\text{Total displacement} = \text{translation component} + \text{rotation component}$$

$$\mathbf{R}_{B''B} = \mathbf{R}_{B'B} + \mathbf{R}_{B''B'} \qquad (4.1c)$$

The new absolute position of point B referred to the origin at A is:

$$\mathbf{R}_{B''A} = \mathbf{R}_{A'A} + \mathbf{R}_{B''A'} \qquad (4.1d)$$

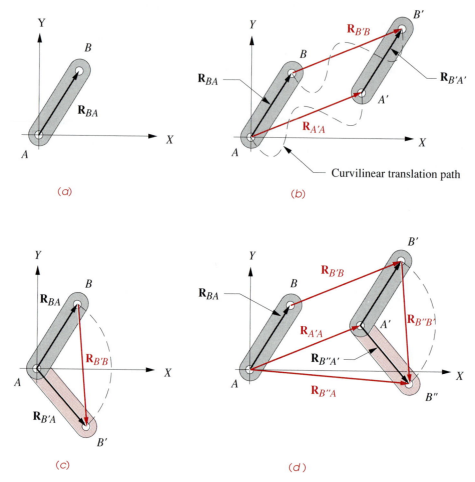

FIGURE 4-3

Translation, rotation, and complex motion

Note that the above two formulas are merely applications of the position difference equation 4.1a. See also Section 2.2 (p. 23) for definitions and discussion of *rotation, translation,* and *complex motion.* These motion states can be expressed as the following theorems.

Theorems

Euler's theorem:

The general displacement of a rigid body with one point fixed is a rotation about some axis.

This applies to pure rotation as defined above and in Section 2.2 (p. 23). Chasles (1793-1880) provided a corollary to Euler's theorem now known as:

Chasles' theorem:

Any displacement of a rigid body is equivalent to the sum of a translation of any one point on that body and a rotation of the body about an axis through that point.

This describes complex motion as defined above and in Section 2.2 (p. 23). Note that equation 4.1c is an expression of Chasles' theorem.

4.4   GRAPHICAL POSITION ANALYSIS OF LINKAGES

For any one-*DOF* linkage, such as a fourbar, only one parameter is needed to complete-ly define the positions of all the links. The parameter usually chosen is the angle of the input link. This is shown as θ_2 in Figure 4-4. We want to find θ_3 and θ_4. The link lengths are known. Note that we will consistently number the ground link as 1 and the driver link as 2 in these examples.

   The graphical analysis of this problem is trivial and can be done using only high-school geometry. If we draw the linkage carefully to scale with rule, compass, and pro-tractor in a particular position (given θ_2), then it is only necessary to measure the angles of links 3 and 4 with the protractor. Note that all link angles are measured from a posi-tive *X* axis. In Figure 4-4, a *local xy* axis system, parallel to the *global XY* system, has been created at point *A* to measure θ_3. The accuracy of this graphical solution will be limited by our care and drafting ability and by the crudity of the protractor used. Never-theless, a very rapid approximate solution can be found for any one position.

   Figure 4-5 shows the construction of the graphical position solution. The four link lengths *a, b, c, d* and the angle θ_2 of the input link are given. First, the ground link (1) and the input link (2) are drawn to a convenient scale such that they intersect at the ori-gin O_2 of the global *XY* coordinate system with link 2 placed at the input angle θ_2. Link 1 is drawn along the X axis for convenience. The compass is set to the scaled length of link 3, and an arc of that radius swung about the end of link 2 (point *A*). Then the com-pass is set to the scaled length of link 4, and a second arc swung about the end of link 1

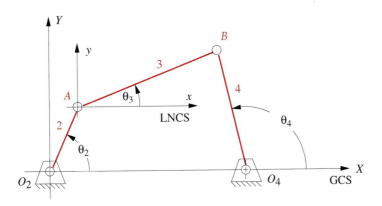

FIGURE 4-4

Measurement of angles in the fourbar linkage

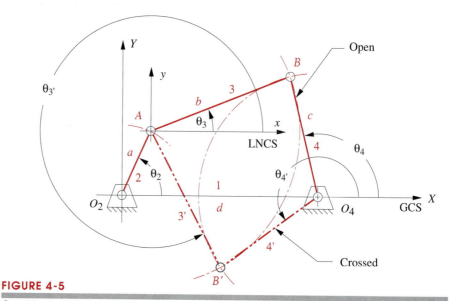

FIGURE 4-5

Graphical position solution to the open and crossed configurations of the fourbar linkage

(point O_4). These two arcs will have two intersections at B and B' that define the two solutions to the position problem for a fourbar linkage which can be assembled in two configurations, called circuits, labeled open and crossed in Figure 4-5. Circuits in linkages will be discussed in a later section.

The angles of links 3 and 4 can be measured with a protractor. One circuit has angles θ_3 and θ_4, the other $\theta_{3'}$ and $\theta_{4'}$. A graphical solution is only valid for the particular value of input angle used. For each additional position analysis we must completely redraw the linkage. This can become burdensome if we need a complete analysis at every 1- or 2-degree increment of θ_2. In that case we will be better off to derive an analytical solution for θ_3 and θ_4 which can be solved by computer.

4.5   ALGEBRAIC POSITION ANALYSIS OF LINKAGES

The same procedure that was used in Figure 4-5 to solve geometrically for the intersections B and B' and angles of links 3 and 4 can be encoded into an algebraic algorithm. The coordinates of point A are found from

$$A_x = a\cos\theta_2$$
$$A_y = a\sin\theta_2$$

(4.2a)

The coordinates of point B are found using the equations of circles about A and O_4.

$$b^2 = \left(B_x - A_x\right)^2 + \left(B_y - A_y\right)^2$$

(4.2b)

$$c^2 = \left(B_x - d\right)^2 + B_y^2$$

(4.2c)

which provide a pair of simultaneous equations in B_x and B_y.

Subtracting equation 4.2c from 4.2b gives an expression for B_x.

$$B_x = \frac{a^2 - b^2 + c^2 - d^2}{2(A_x - d)} - \frac{2A_y B_y}{2(A_x - d)} = S - \frac{2A_y B_y}{2(A_x - d)} \qquad (4.2d)$$

Substituting equation 4.2d into 4.2c gives a quadratic equation in B_y which has two solutions corresponding to those in Figure 4-5.

$$B_y^2 + \left(S - \frac{A_y B_y}{A_x - d} - d\right)^2 - c^2 = 0 \qquad (4.2e)$$

This can be solved with the familiar expression for the roots of a quadratic equation,

$$B_y = \frac{-Q \pm \sqrt{Q^2 - 4PR}}{2P} \qquad (4.2f)$$

where :

$$P = \frac{A_y^2}{(A_x - d)^2} + 1 \qquad\qquad Q = \frac{2A_y(d - S)}{A_x - d}$$

$$R = (d - S)^2 - c^2 \qquad\qquad S = \frac{a^2 - b^2 + c^2 - d^2}{2(A_x - d)}$$

Note that the solutions to this equation set can be real or imaginary. If the latter, it indicates that the links cannot connect at the given input angle or at all. Once the two values of B_y are found (if real), they can be substituted into equation 4.2d to find their corresponding x components. The link angles for this position can then be found from

$$\theta_3 = \tan^{-1}\left(\frac{B_y - A_y}{B_x - A_x}\right)$$

$$\qquad\qquad (4.2g)$$

$$\theta_4 = \tan^{-1}\left(\frac{B_y}{B_x - d}\right)$$

A two-argument arctangent function must be used to solve equations 4.2g since the angles can be in any quadrant. Equations 4.2 can be encoded in any computer language or equation solver, and the value of θ_2 varied over the linkage's usable range to find all corresponding values of the other two link angles.

Vector Loop Representation of Linkages

An alternate approach to linkage position analysis creates a vector loop (or loops) around the linkage.[*] This approach offers some advantages in the synthesis of linkages which will be addressed in Chapter 5. The links are represented as **position vectors**. Figure 4-6 shows the same fourbar linkage as in Figure 4-4 (p. 151) , but the links are now drawn as position vectors which form a vector loop. This loop closes on itself making the sum of the vectors around the loop zero. The lengths of the vectors are the link lengths which

* This method was originated by Prof. F. H. Raven in "Velocity and Acceleration Analysis of Plane and Space Mechanisms by Means of Independent Position Equations," *Trans.* ASME, Vol. 25, 1958, pp. 1-6.

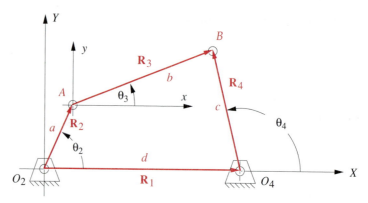

FIGURE 4-6

Position vector loop for a fourbar linkage

are known. The current linkage position is defined by the input angle θ_2 as it is a one-DOF mechanism. We want to solve for the unknown angles θ_3 and θ_4. To do so we need a convenient notation to represent the vectors.

Complex Numbers as Vectors

There are many ways to represent vectors. They may be defined in **polar coordinates**, by their *magnitude* and *angle*, or in **cartesian coordinates** as x and y components. These forms are of course easily convertible from one to the other using equations 4.0. The position vectors in Figure 4-6 can be represented as any of these expressions:

Polar form	Cartesian form	
$R @ \angle \theta$	$r\cos\theta\,\hat{\mathbf{i}} + r\sin\theta\,\hat{\mathbf{j}}$	(4.3a)
$r e^{j\theta}$	$r\cos\theta + jr\sin\theta$	(4.3b)

Equation 4.3a uses **unit vectors** to represent the x and y vector component directions in the cartesian form. Figure 4-7 shows the unit vector notation for a position vector. Equation 4.3b uses **complex number notation** wherein the X direction component is called the *real portion* and the Y direction component is called the *imaginary portion*. This unfortunate term *imaginary* comes about because of the use of the notation j to represent the square root of minus one, which of course cannot be evaluated numerically. However, this *imaginary* number is used in a **complex number** as an **operator**, *not as a value*. Figure 4-8a shows the **complex plane** in which the *real* axis represents the X-directed component of the vector in the plane, and the *imaginary* axis represents the Y-directed component of the same vector. So, any term in a complex number which has no j operator is an x component, and a j indicates a y component.

Note in Figure 4-8b that each multiplication of the vector $\mathbf{R}_A$ by the operator j results in a *counterclockwise rotation* of the vector through 90 degrees. The vector $\mathbf{R}_B = j\mathbf{R}_A$ is directed along the *positive imaginary* or j axis. The vector $\mathbf{R}_C = j^2 \mathbf{R}_A$ is directed along the *negative real* axis because $j^2 = -1$ and thus $\mathbf{R}_C = -\mathbf{R}_A$. In similar fashion, $\mathbf{R}_D = j^3 \mathbf{R}_A = -j\mathbf{R}_A$ and this component is directed along the *negative j axis*.

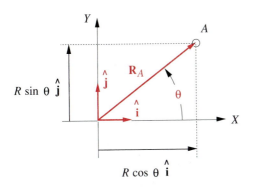

Polar form:

$$\left| \mathbf{R}_A \right| @ \underline{/\theta}$$

Cartesian form:

$$R \cos \theta \, \hat{\mathbf{i}}, \; R \sin \theta \, \hat{\mathbf{j}}$$

FIGURE 4-7

Unit vector notation for position vectors

One advantage of using this complex number notation to represent planar vectors comes from the **Euler identity**:

$$e^{\pm j\theta} = \cos\theta \pm j\sin\theta \qquad (4.4a)$$

Any two-dimensional vector can be represented by the compact polar notation on the left side of equation 4.4a. There is no easier function to differentiate or integrate, since it is its own derivative:

Polar form: $R\,e^{\,j\theta}$

Cartesian form: $R \cos \theta + j R \sin \theta$

$$R = \left| \mathbf{R}_A \right|$$

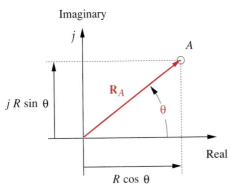

(a) Complex number representation of a position vector

(b) Vector rotations in the complex plane

FIGURE 4-8

Complex number representation of vectors in the plane

$$\frac{de^{j\theta}}{d\theta} = je^{j\theta} \qquad\qquad (4.4b)$$

We will use this **complex number notation** for vectors to develop and derive the equations for position, velocity, and acceleration of linkages.

The Vector Loop Equation for a Fourbar Linkage

The directions of the position vectors in Figure 4-6 are chosen so as to define their angles where we desire them to be measured. By definition, *the angle of a vector is always measured at its root, **not at its head***. We would like angle θ_4 to be measured at the fixed pivot O_4, so vector $\mathbf{R}_4$ is arranged to have its root at that point. We would like to measure angle θ_3 at the point where links 2 and 3 join, so vector $\mathbf{R}_3$ is rooted there. A similar logic dictates the arrangement of vectors $\mathbf{R}_1$ and $\mathbf{R}_2$. Note that the X (*real*) axis is taken for convenience along link 1 and the origin of the global coordinate system is taken at point O_2, the root of the input link vector, $\mathbf{R}_2$. These choices of vector directions and senses, as indicated by their arrowheads, lead to this vector loop equation:

$$\mathbf{R}_2 + \mathbf{R}_3 - \mathbf{R}_4 - \mathbf{R}_1 = 0 \qquad\qquad (4.5a)$$

An **alternate notation** for these position vectors is to use the labels of the points at the vector **tips** and **roots** (*in that order*) as subscripts. The second subscript is conventionally omitted if it is the origin of the global coordinate system (point O_2):

$$\mathbf{R}_A + \mathbf{R}_{BA} - \mathbf{R}_{BO_4} - \mathbf{R}_{O_4} = 0 \qquad\qquad (4.5b)$$

Next, we substitute the complex number notation for each position vector. To simplify the notation and minimize the use of subscripts, we will denote the scalar lengths of the four links as a, b, c, and d. These are so labeled in Figure 4-6. The equation then becomes:

$$ae^{j\theta_2} + be^{j\theta_3} - ce^{j\theta_4} - de^{j\theta_1} = 0 \qquad\qquad (4.5c)$$

These are three forms of the same vector equation, and as such can be solved for two unknowns. There are four variables in this equation, namely the four link angles. The link lengths are all constant in this particular linkage. Also, the value of the angle of link 1 is fixed (at zero) since this is the ground link. The *independent variable* is θ_2 which we will control with a motor or other driver device. That leaves the angles of link 3 and 4 to be found. We need algebraic expressions which define θ_3 and θ_4 as functions only of the constant link lengths and the one input angle, θ_2. These expressions will be of the form:

$$\theta_3 = f\{a, b, c, d, \theta_2\}$$
$$\theta_4 = g\{a, b, c, d, \theta_2\} \qquad\qquad (4.5d)$$

To solve the polar form, vector equation 4.5c, we must substitute the *Euler equivalents* (equation 4.4a) for the $e^{j\theta}$ terms, and then separate the resulting cartesian form vector equation into two scalar equations which can be solved simultaneously for θ_3 and θ_4. Substituting equation 4.4a into equation 4.5c:

$$a(\cos\theta_2 + j\sin\theta_2) + b(\cos\theta_3 + j\sin\theta_3) - c(\cos\theta_4 + j\sin\theta_4) - d(\cos\theta_1 + j\sin\theta_1) = 0 \qquad (4.5e)$$

This equation can now be separated into its real and imaginary parts and each set to zero.

real part (x component):

$$a\cos\theta_2 + b\cos\theta_3 - c\cos\theta_4 - d\cos\theta_1 = 0$$

but: $\theta_1 = 0$, so: $\qquad\qquad$ (4.6a)

$$a\cos\theta_2 + b\cos\theta_3 - c\cos\theta_4 - d = 0$$

imaginary part (y component):

$$ja\sin\theta_2 + jb\sin\theta_3 - jc\sin\theta_4 - jd\sin\theta_1 = 0$$

but: $\theta_1 = 0$, and the j's divide out, so: $\qquad\qquad$ (4.6b)

$$a\sin\theta_2 + b\sin\theta_3 - c\sin\theta_4 = 0$$

The scalar equations 4.6a and 4.6b can now be solved simultaneously for θ_3 and θ_4. To solve this set of two simultaneous trigonometric equations is straightforward but tedious. Some substitution of trigonometric identities will simplify the expressions. The first step is to rewrite equations 4.6a and 4.6b so as to isolate one of the two unknowns on the left side. We will isolate θ_3 and solve for θ_4 in this example.

$$b\cos\theta_3 = -a\cos\theta_2 + c\cos\theta_4 + d \qquad\qquad (4.6c)$$

$$b\sin\theta_3 = -a\sin\theta_2 + c\sin\theta_4 \qquad\qquad (4.6d)$$

Now square both sides of equations 4.6c and 4.6d and add them:

$$b^2\left(\sin^2\theta_3 + \cos^2\theta_3\right) = \left(-a\sin\theta_2 + c\sin\theta_4\right)^2 + \left(-a\cos\theta_2 + c\cos\theta_4 + d\right)^2 \qquad (4.7a)$$

Note that the quantity in parentheses on the left side is equal to 1, eliminating θ_3 from the equation, leaving only θ_4 which can now be solved for.

$$b^2 = \left(-a\sin\theta_2 + c\sin\theta_4\right)^2 + \left(-a\cos\theta_2 + c\cos\theta_4 + d\right)^2 \qquad (4.7b)$$

The right side of this expression must now be expanded and terms collected.

$$b^2 = a^2 + c^2 + d^2 - 2ad\cos\theta_2 + 2cd\cos\theta_4 - 2ac\left(\sin\theta_2\sin\theta_4 + \cos\theta_2\cos\theta_4\right) \qquad (4.7c)$$

To further simplify this expression, the constants K_1, K_2, and K_3 are defined in terms of the constant link lengths in equation 4.7c:

$$K_1 = \frac{d}{a} \qquad K_2 = \frac{d}{c} \qquad K_3 = \frac{a^2 - b^2 + c^2 + d^2}{2ac} \qquad (4.8a)$$

and:

$$K_1\cos\theta_4 - K_2\cos\theta_2 + K_3 = \cos\theta_2\cos\theta_4 + \sin\theta_2\sin\theta_4 \qquad (4.8b)$$

If we substitute the identity $\cos(\theta_2 - \theta_4) = \cos\theta_2\cos\theta_4 + \sin\theta_2\sin\theta_4$, we get the form known as Freudenstein's equation.

$$K_1\cos\theta_4 - K_2\cos\theta_2 + K_3 = \cos(\theta_2 - \theta_4) \qquad (4.8c)$$

In order to reduce equation 4.8b to a more tractable form for solution, it will be useful to substitute the *half angle identities* which will convert the $\sin \theta_4$ and $\cos \theta_4$ terms to $\tan \theta_4$ terms:

$$\sin \theta_4 = \frac{2\tan\left(\dfrac{\theta_4}{2}\right)}{1+\tan^2\left(\dfrac{\theta_4}{2}\right)}; \qquad \cos \theta_4 = \frac{1-\tan^2\left(\dfrac{\theta_4}{2}\right)}{1+\tan^2\left(\dfrac{\theta_4}{2}\right)} \qquad (4.9)$$

This results in the following simplified form, where the link lengths and known input value (θ_2) terms have been collected as constants A, B, and C.

$$A\, \tan^2\left(\frac{\theta_4}{2}\right) + B\, \tan\left(\frac{\theta_4}{2}\right) + C = 0$$

$$(4.10a)$$

where:
$$A = \cos\theta_2 - K_1 - K_2 \cos\theta_2 + K_3$$
$$B = -2\sin\theta_2$$
$$C = K_1 - (K_2 + 1)\cos\theta_2 + K_3$$

Note that equation 4.10a is quadratic in form, and the solution is:

$$\tan\left(\frac{\theta_4}{2}\right) = \frac{-B \pm \sqrt{B^2 - 4AC}}{2A}$$

$$(4.10b)$$

$$\theta_{4_{1,2}} = 2\arctan\left(\frac{-B \pm \sqrt{B^2 - 4AC}}{2A}\right)$$

Equation 4.10b has two solutions, obtained from the $\pm$ conditions on the radical. These two solutions, as with any quadratic equation, may be of three types: *real and equal*, *real and unequal*, *complex conjugate*. If the discriminant under the radical is negative, then the solution is complex conjugate, which simply means that the link lengths chosen are not capable of connection for the chosen value of the input angle θ_2. This can occur either when the link lengths are completely incapable of connection in any position or, in a non-Grashof linkage, when the input angle is beyond a toggle limit position. There is then no real solution for that value of input angle θ_2. Excepting this situation, the solution will usually be real and unequal, meaning there are two values of θ_4 corresponding to any one value of θ_2. These are referred to as the **crossed** and **open** configurations of the linkage and also as the two **circuits** of the linkage. In the fourbar linkage, the minus solution gives θ_4 for the open configuration and the positive solution gives θ_4 for the crossed configuration.

Figure 4-5 (p. 152) shows both crossed and open solutions for a Grashof crank-rocker linkage. The terms crossed and open are based on the assumption that the input link 2, for which θ_2 is defined, is placed in the first quadrant (i.e., $0 < \theta_2 < \pi/2$). A Grashof linkage is then defined as **crossed** if the two links adjacent to the shortest link cross one another, and as **open** if they do not cross one another in this position. Note that the configuration of the linkage, either crossed or open, is solely dependent upon the way that the links are assembled. You cannot predict, based on link lengths alone, which of the solutions will be the desired one. In other words, you can obtain either solution with the same linkage by simply taking apart the pin which connects links 3 and 4 in Figure 4-5, and moving those links to the only other positions at which the pin will again connect them. In so doing, you will have switched from one position solution, or **circuit**, to the other.

The solution for angle θ_3 is essentially similar to that for θ_4. Returning to equations 4.6, we can rearrange them to isolate θ_4 on the left side.

$$c\cos\theta_4 = a\cos\theta_2 + b\cos\theta_3 - d \qquad (4.6e)$$

$$c\sin\theta_4 = a\sin\theta_2 + b\sin\theta_3 \qquad (4.6f)$$

Squaring and adding these equations will eliminate θ_4. The resulting equation can be solved for θ_3 as was done above for θ_4, yielding this expression:

$$K_1\cos\theta_3 + K_4\cos\theta_2 + K_5 = \cos\theta_2\cos\theta_3 + \sin\theta_2\sin\theta_3 \qquad (4.11a)$$

The constant K_1 is the same as defined in equation 4.8b. K_4 and K_5 are:

$$K_4 = \frac{d}{b}; \qquad\qquad K_5 = \frac{c^2 - d^2 - a^2 - b^2}{2ab} \qquad (4.11b)$$

This also reduces to a quadratic form:

$$D\tan^2\left(\frac{\theta_3}{2}\right) + E\tan\left(\frac{\theta_3}{2}\right) + F = 0$$

$$(4.12)$$

where :

$$D = \cos\theta_2 - K_1 + K_4\cos\theta_2 + K_5$$

$$E = -2\sin\theta_2$$

$$F = K_1 + (K_4 - 1)\cos\theta_2 + K_5$$

and the solution is:

$$\theta_{3_{1,2}} = 2\arctan\left(\frac{-E \pm \sqrt{E^2 - 4DF}}{2D}\right) \qquad (4.13)$$

As with the angle θ_4, this also has two solutions, corresponding to the crossed and open circuits of the linkage, as shown in Figure 4-5.

4.6   THE FOURBAR SLIDER-CRANK POSITION SOLUTION

The same vector loop approach as used above can be applied to a linkage containing sliders. Figure 4-9 shows an offset fourbar slider-crank linkage, inversion #1. The term **off-set** means that *the slider axis extended does not pass through the crank pivot*. This is the general case. (The nonoffset slider-crank linkages shown in Figure 2-13 (p. 44) are the special cases.) This linkage could be represented by only three position vectors, $\mathbf{R}_2$, $\mathbf{R}_3$, and $\mathbf{R}_s$, but one of them ($\mathbf{R}_s$) will be a vector of varying magnitude and angle. It will be easier to use four vectors, $\mathbf{R}_1$, $\mathbf{R}_2$, $\mathbf{R}_3$, and $\mathbf{R}_4$ with $\mathbf{R}_1$ arranged parallel to the axis of slid-

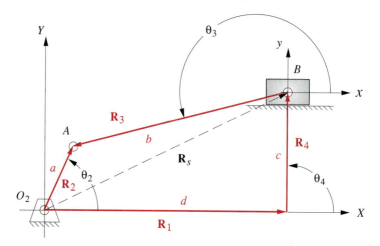

FIGURE 4-9

Position vector loop for a fourbar slider-crank linkage

ing and $\mathbf{R}_4$ perpendicular. In effect the pair of vectors $\mathbf{R}_1$ and $\mathbf{R}_4$ are orthogonal components of the position vector $\mathbf{R}_s$ from the origin to the slider.

It simplifies the analysis to arrange one coordinate axis parallel to the axis of sliding. The variable-length, constant-direction vector $\mathbf{R}_1$ then represents the slider position with magnitude d. The vector $\mathbf{R}_4$ is orthogonal to $\mathbf{R}_1$ and defines the constant magnitude **offset** of the linkage. Note that for the special-case, nonoffset version, the vector $\mathbf{R}_4$ will be zero and $\mathbf{R}_1 = \mathbf{R}_s$. The vectors $\mathbf{R}_2$ and $\mathbf{R}_3$ complete the vector loop. The coupler's position vector $\mathbf{R}_3$ is placed with its root at the slider which then defines its angle θ_3 at point B. This particular arrangement of position vectors leads to a vector loop equation similar to the pin-jointed fourbar example:

$$\mathbf{R}_2 - \mathbf{R}_3 - \mathbf{R}_4 - \mathbf{R}_1 = 0 \tag{4.14a}$$

Compare equation 4.14a to equation 4.5a (p. 156) and note that the only difference is the sign of $\mathbf{R}_3$. This is due solely to the somewhat arbitrary choice of the sense of the position vector $\mathbf{R}_3$ in each case. The angle θ_3 must always be measured at the root of vector $\mathbf{R}_3$, and in this example it will be convenient to have that angle θ_3 at the joint labeled B. Once these arbitrary choices are made it is crucial that the resulting algebraic signs be carefully observed in the equations, or the results will be completely erroneous. Letting the vector magnitudes (link lengths) be represented by a, b, c, d as shown, we can substitute the complex number equivalents for the position vectors.

$$ae^{j\theta_2} - be^{j\theta_3} - ce^{j\theta_4} - de^{j\theta_1} = 0 \tag{4.14b}$$

Substitute the Euler equivalents:

$$a(\cos\theta_2 + j\sin\theta_2) - b(\cos\theta_3 + j\sin\theta_3)$$
$$- c(\cos\theta_4 + j\sin\theta_4) - d(\cos\theta_1 + j\sin\theta_1) = 0 \tag{4.14c}$$

Separate the real and imaginary components:

real part (x component):

$$a\cos\theta_2 - b\cos\theta_3 - c\cos\theta_4 - d\cos\theta_1 = 0$$

but: $\theta_1 = 0$, so: (4.15a)

$$a\cos\theta_2 - b\cos\theta_3 - c\cos\theta_4 - d = 0$$

imaginary part (y component):

$$ja\sin\theta_2 - jb\sin\theta_3 - jc\sin\theta_4 - jd\sin\theta_1 = 0$$

but: $\theta_1 = 0$, and the j's divide out, so: (4.15b)

$$a\sin\theta_2 - b\sin\theta_3 - c\sin\theta_4 = 0$$

We want to solve equations 4.15 simultaneously for the two unknowns, link length d and link angle θ_3. The independent variable is crank angle θ_2. Link lengths a and b, the offset c, and angle θ_4 are known. But note that since we set up the coordinate system to be parallel and perpendicular to the axis of the slider block, the angle θ_1 is zero and θ_4 is 90°. Equation 4.15b can be solved for θ_3 and the result substituted into equation 4.15a to solve for d. The solution is:

$$\theta_{3_1} = \arcsin\left(\frac{a\sin\theta_2 - c}{b}\right)$$ (4.16a)

$$d = a\cos\theta_2 - b\cos\theta_3$$ (4.16b)

Note that there are again two valid solutions corresponding to the two circuits of the linkage. The arcsine function is multivalued. Its evaluation will give a value between ±90° representing only one circuit of the linkage. The value of d is dependent on the calculated value of θ_3. The value of θ_3 for the second circuit of the linkage can be found from:

$$\theta_{3_2} = \arcsin\left(-\frac{a\sin\theta_2 - c}{b}\right) + \pi$$ (4.17)

4.7 AN INVERTED SLIDER-CRANK POSITION SOLUTION

Figure 4-10a shows inversion #3 of the common fourbar slider-crank linkage in which the sliding joint is between links 3 and 4 at point B. This is shown as an **offset** slider-crank mechanism. The slider block has pure rotation with its center offset from the slide axis. (Figure 2-13c, p. 44, shows the nonoffset version of this linkage in which the vector $\mathbf{R}_4$ is zero.)

The global coordinate system is again taken with its origin at input crank pivot O_2 and the positive X axis along link 1, the ground link. A local axis system has been placed at point B in order to define θ_3. Note that there is a fixed angle γ within link 4 which defines the slot angle with respect to that link.

In Figure 4-10b the links have been represented as position vectors having senses consistent with the coordinate systems that were chosen for convenience in defining the

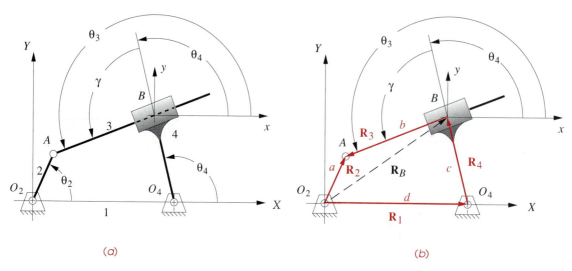

FIGURE 4-10

Inversion #3 of the slider-crank fourbar linkage

link angles. This particular arrangement of position vectors leads to the same vector loop equation as the previous slider-crank example. Equations 4.14 and 4.15 apply to this inversion as well. Note that the absolute position of point B is defined by vector $\mathbf{R}_B$ which varies in both magnitude and direction as the linkage moves. We choose to represent $\mathbf{R}_B$ as the vector difference $\mathbf{R}_2 - \mathbf{R}_3$ in order to use the actual links as the position vectors in the loop equation.

All slider linkages will have at least one link whose effective length between joints will vary as the linkage moves. In this example the length of link 3 between points A and B, designated as b, will change as it passes through the slider block on link 4. Thus the value of b will be one of the variables to be solved for in this inversion. Another variable will be θ_4, the angle of link 4. Note however, that we also have an unknown in θ_3, the angle of link 3. This is a total of three unknowns. Equations 4.15 can only be solved for two unknowns. Thus we require another equation to solve the system. There is a fixed relationship between angles θ_3 and θ_4, shown as γ in Figure 4-10, which gives the equation:

$$\theta_3 = \theta_4 \pm \gamma \tag{4.18}$$

where the + sign is used for the open and the – sign for the crossed configuration.

Substituting equation 4.18 into equations 4.15 yields:

$$a\cos\theta_2 - b\cos\theta_3 - c\cos\theta_4 - d = 0 \tag{4.19a}$$
$$a\sin\theta_2 - b\sin\theta_3 - c\sin\theta_4 = 0 \tag{4.19b}$$

These have only two unknowns and can be solved simultaneously for θ_4 and b. Equation 4.19b can be solved for link length b and substituted into equation 4.19a.

$$b = \frac{a\sin\theta_2 - c\sin\theta_4}{\sin\theta_3} \qquad (4.20a)$$

$$a\cos\theta_2 - \frac{a\sin\theta_2 - c\sin\theta_4}{\sin\theta_3}\cos\theta_3 - c\cos\theta_4 - d = 0 \qquad (4.20b)$$

Substitute equation 4.18 and after some algebraic manipulation, equation 4.20 can be reduced to:

$$P\sin\theta_4 + Q\cos\theta_4 + R = 0$$

where :   (4.21)

$$P = a\sin\theta_2 \sin\gamma + (a\cos\theta_2 - d)\cos\gamma$$
$$Q = -a\sin\theta_2 \cos\gamma + (a\cos\theta_2 - d)\sin\gamma$$
$$R = -c\sin\gamma$$

Note that the factors P, Q, R are constant for any input value of θ_2. To solve this for θ_4, it is convenient to substitute the tangent half angle identities (equation 4.9, p.158) for the $\sin\theta_4$ and $\cos\theta_4$ terms. This will result in a quadratic equation in $\tan(\theta_4/2)$ which can be solved for the two values of θ_4.

$$P\frac{2\tan\left(\dfrac{\theta_4}{2}\right)}{1+\tan^2\left(\dfrac{\theta_4}{2}\right)} + Q\frac{1-\tan^2\left(\dfrac{\theta_4}{2}\right)}{1+\tan^2\left(\dfrac{\theta_4}{2}\right)} + R = 0 \qquad (4.22a)$$

This reduces to:

$$(R-Q)\tan^2\left(\frac{\theta_4}{2}\right) + 2P\tan\left(\frac{\theta_4}{2}\right) + (Q+R) = 0$$

let :

$$S = R - Q; \qquad T = 2P; \qquad U = Q + R$$

then :

$$S\tan^2\left(\frac{\theta_4}{2}\right) + T\tan\left(\frac{\theta_4}{2}\right) + U = 0 \qquad (4.22b)$$

and the solution is:

$$\theta_{4_{1,2}} = 2\arctan\left(\frac{-T \pm \sqrt{T^2 - 4SU}}{2S}\right) \qquad (4.22c)$$

As was the case with the previous examples, this also has a crossed and an open solution represented by the plus and minus signs on the radical. Note that we must also calculate the values of link length b for each θ_4 by using equation 4.20a. The coupler angle θ_3 is found from equation 4.18.

4.8   LINKAGES OF MORE THAN FOUR BARS

With some exceptions,* the same approach as shown here for the fourbar linkage can be used for any number of links in a closed-loop configuration. More complicated linkages may have multiple loops which will lead to more equations to be solved simultaneously and may require an iterative solution.

The Geared Fivebar Linkage

Another example, which can be reduced to two equations in two unknowns, is the **geared fivebar linkage**, which was introduced in Section 2.13 (p. 52) and is shown in Figure 4-11a and program FIVEBAR disk file F04-11.5br. The vector loop for this linkage is shown in Figure 4-11b. It obviously has one more position vector than the fourbar. Its vector loop equation is:

$$\mathbf{R}_2 + \mathbf{R}_3 - \mathbf{R}_4 - \mathbf{R}_5 - \mathbf{R}_1 = 0 \tag{4.23a}$$

Note that the vector senses are again chosen to suit the analyst's desires to have the vector angles defined at a convenient end of the respective link. Equation 4.23b substitutes the complex polar notation for the position vectors in equation 4-23a, using a, b, c, d, f to represent the scalar lengths of the links as shown in Figure 4-11.

$$a e^{j\theta_2} + b e^{j\theta_3} - c e^{j\theta_4} - d e^{j\theta_5} - f e^{j\theta_1} = 0 \tag{4.23b}$$

Note also that this vector loop equation has three unknown variables in it, namely the angles of links 3, 4, and 5. (The angle of link 2 is the input, or independent, variable, and link 1 is fixed with constant angle). Since a two-dimensional vector equation can only be solved for two unknowns, we will need another equation to solve this system. Because this is a geared fivebar linkage, there exists a relationship between the two geared links, here links 2 and 5. Two factors determine how link 5 behaves with respect to link 2, namely, the **gear ratio** λ and the **phase angle** ϕ. The relationship is:

$$\theta_5 = \lambda \theta_2 + \phi \tag{4.23c}$$

This allows us to express θ_5 in terms of θ_2 in equation 4.23b and reduce the unknowns to two by substituting equation 4.23c into equation 4.23b.

$$a e^{j\theta_2} + b e^{j\theta_3} - c e^{j\theta_4} - d e^{j(\lambda\theta_2 + \phi)} - f e^{j\theta_1} = 0 \tag{4.24a}$$

Note that the gear ratio λ is the ratio of the diameters of the gears connecting the two links ($\lambda = dia_2 / dia_5$), and the phase angle ϕ is the *initial angle* of link 5 with respect to link 2. When link 2 is at zero degrees, link 5 is at the **phase angle** ϕ. Equation 4.23c defines the relationship between θ_2 and θ_5. Both λ and ϕ are design parameters selected by the design engineer along with the link lengths. With these parameters defined, the only unknowns left in equation 4.24 are θ_3 and θ_4.

The behavior of the geared fivebar linkage can be modified by changing the link lengths, the gear ratio, or the phase angle. The phase angle can be changed simply by lifting the gears out of engagement, rotating one gear with respect to the other, and reengaging them. Since links 2 and 5 are rigidly attached to gears 2 and 5, respectively, their relative angular rotations will be changed also. It is this fact that results in different po-

---

* Waldron and Sreenivasan[1] report that the common solution methods for position analysis are not general, i.e., are not extendable to *n*-link mechanisms. Conventional position analysis methods, such as those used here, rely on the presence of a 4-bar loop in the mechanism that can be solved first, followed by a decomposition of the remaining links into a series of dyads. Not all mechanisms contain fourbar loops. (One 8-bar, 1-*DOF* linkage contains no 4-bar loops.) Even if they do, the 4-bar loop's pivots may not be grounded, requiring that the linkage be inverted to start the solution. Also, if the driving joint is not in the fourbar loop, then interpolation is needed to solve for link positions.

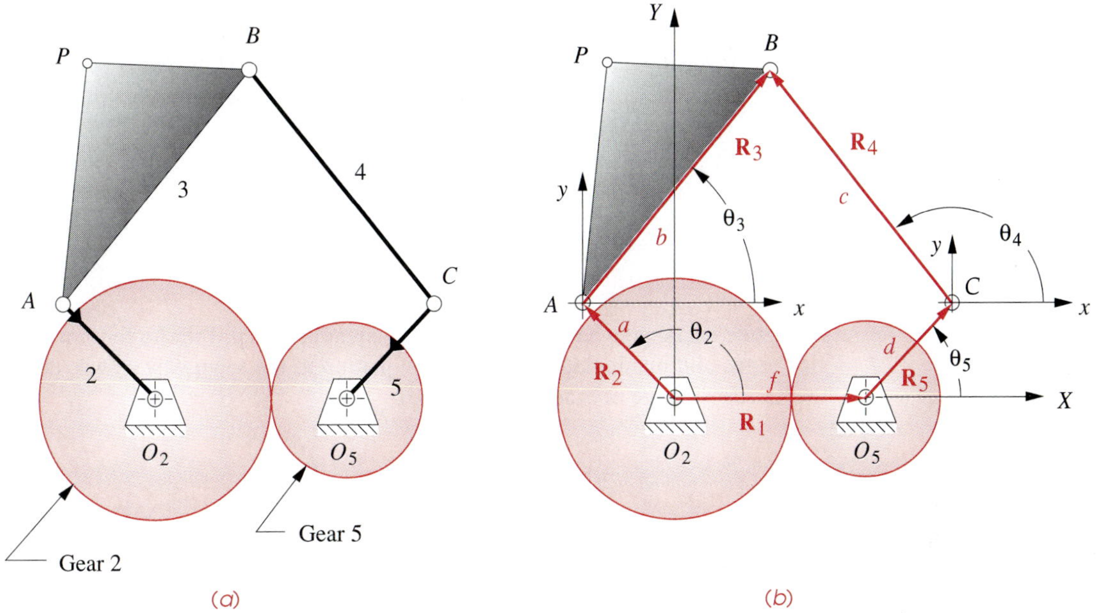

B
P
3
4
A
2
5
O_2
O_5

Gear 5

Gear 2

(a)

Y
P
B
R_3
R_4
y
θ_3
b
c
y
θ_4
A
a
θ_2
x
C
x
R_2
f
d
θ_5
O_2
O_5
R_1
R_5
X

(b)

FIGURE 4-11

The geared fivebar linkage and its vector loop

sitions of links 3 and 4 with any change in phase angle. The coupler curve's shapes will also change with variation in any of these parameters as can be seen in Figure 3-23 (p. 112) and in Appendix E.

The procedure for solution of this vector loop equation is the same as that used above for the fourbar linkage:

1   Substitute the Euler equivalent (equation 4.4a, p. 155) into each term in the vector loop equation 4.24a.

$$a(\cos\theta_2 + j\sin\theta_2) + b(\cos\theta_3 + j\sin\theta_3) - c(\cos\theta_4 + j\sin\theta_4)$$
$$-d\left[\cos(\lambda\theta_2 + \phi) + j\sin(\lambda\theta_2 + \phi)\right] - f(\cos\theta_1 + j\sin\theta_1) = 0 \qquad (4.24b)$$

2   Separate the real and imaginary parts of the cartesian form of the vector loop equation.

$$a\cos\theta_2 + b\cos\theta_3 - c\cos\theta_4 - d\cos(\lambda\theta_2 + \phi) - f\cos\theta_1 = 0 \qquad (4.24c)$$
$$a\sin\theta_2 + b\sin\theta_3 - c\sin\theta_4 - d\sin(\lambda\theta_2 + \phi) - f\sin\theta_1 = 0 \qquad (4.24d)$$

3   Rearrange to isolate one unknown (either θ_3 or θ_4) in each scalar equation. Note that θ_1 is zero.

$$b\cos\theta_3 = -a\cos\theta_2 + c\cos\theta_4 + d\cos(\lambda\theta_2 + \phi) + f \qquad (4.24e)$$
$$b\sin\theta_3 = -a\sin\theta_2 + c\sin\theta_4 + d\sin(\lambda\theta_2 + \phi) \qquad (4.24f)$$

4 Square both equations and add them to eliminate one unknown, say θ_3.

$$b^2 = 2c\big[d\cos(\lambda\theta_2 + \phi) - a\cos\theta_2 + f\big]\cos\theta_4$$
$$+ 2c\big[d\sin(\lambda\theta_2 + \phi) - a\sin\theta_2\big]\sin\theta_4$$
$$+ a^2 + c^2 + d^2 + f^2 - 2af\cos\theta_2$$
$$- 2d(a\cos\theta_2 - f)\cos(\lambda\theta_2 + \phi)$$
$$- 2ad\sin\theta_2\sin(\lambda\theta_2 + \phi) \qquad (4.24g)$$

5 Substitute the tangent half-angle identities (equation 4.9, p. 158) for the sine and co-sine terms and manipulate the resulting equation in the same way as was done for the fourbar linkage in order to solve for θ_4.

$$A = 2c\big[d\cos(\lambda\theta_2 + \phi) - a\cos\theta_2 + f\big]$$
$$B = 2c\big[d\sin(\lambda\theta_2 + \phi) - a\sin\theta_2\big]$$
$$C = a^2 - b^2 + c^2 + d^2 + f^2 - 2af\cos\theta_2$$
$$- 2d(a\cos\theta_2 - f)\cos(\lambda\theta_2 + \phi)$$
$$- 2ad\sin\theta_2\sin(\lambda\theta_2 + \phi)$$
$$D = C - A; \qquad E = 2B; \qquad F = A + C$$

$$\theta_{4_{1,2}} = 2\arctan\left(\frac{-E \pm \sqrt{E^2 - 4DF}}{2D}\right) \qquad (4.24h)$$

6 Repeat steps 3 to 5 for the other unknown angle θ_3.

$$G = 2b\big[a\cos\theta_2 - d\cos(\lambda\theta_2 + \phi) - f\big]$$
$$H = 2b\big[a\sin\theta_2 - d\sin(\lambda\theta_2 + \phi)\big]$$
$$K = a^2 + b^2 - c^2 + d^2 + f^2 - 2af\cos\theta_2$$
$$- 2d(a\cos\theta_2 - f)\cos(\lambda\theta_2 + \phi)$$
$$- 2ad\sin\theta_2\sin(\lambda\theta_2 + \phi)$$
$$L = K - G; \qquad M = 2H; \quad N = G + K$$

$$\theta_{3_{1,2}} = 2\arctan\left(\frac{-M \pm \sqrt{M^2 - 4LN}}{2L}\right) \qquad (4.24i)$$

Note that these derivation steps are essentially identical to those for the pin-jointed fourbar linkage once θ_2 is substituted for θ_5 using equation 4.23c.

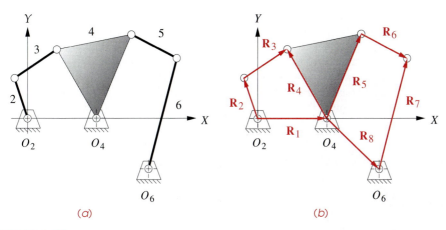

FIGURE 4-12

Watt's sixbar linkage and vector loop

Sixbar Linkages

WATT'S SIXBAR   is essentially two fourbar linkages in series, as shown in Figure 4-12a, and can be analyzed as such. Two vector loops are drawn as shown in Figure 4-12b. These vector loop equations can be solved in succession with the results of the first loop applied as input to the second loop. Note that there is a constant angular relationship between vectors $\mathbf{R}_4$ and $\mathbf{R}_5$ within link 4. The solution for the fourbar linkage (equations 4.10 and 4.13, pp. 158 and 159, respectively) is simply applied twice in this case. Depending on the inversion of the Watts linkage being analyzed, there may be two four-link loops or one four-link and one five-link loop. (See Figure 2-14, p. 45.) In either case, if the four-link loop is analyzed first, there will not be more than two unknown link angles to be found at one time.

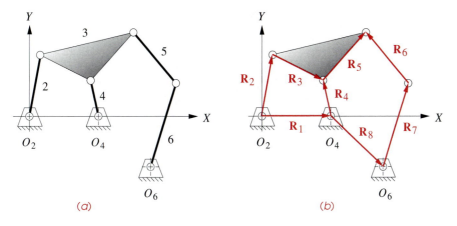

FIGURE 4-13

Stephenson's sixbar linkage and vector loop

STEPHENSON'S SIXBAR   is a more complicated mechanism to analyze. Two vector loops can be drawn, but depending on the inversion being analyzed, either one or both loops will have five links[*] and thus three unknown angles as shown in Figure 4-13a and b (p. 167). However, the two loops will have at least one nonground link in common and so a solution can be found. In the other cases an iterative solution such as a Newton-Raphson method (see Section 4.13) must be used to find the roots of the equations. Program SIXBAR is limited to the inversions which allow a closed-form solution, one of which is shown in Figure 4-13. Program SIXBAR does not do the iterative solution.

4.9   POSITION OF ANY POINT ON A LINKAGE

Once the angles of all the links are found, it is simple and straightforward to define and calculate the position of any point on any link for any input position of the linkage. Figure 4-14 shows a fourbar linkage whose coupler, link 3, is enlarged to contain a coupler point P. The crank and rocker have also been enlarged to show points S and U which might represent the centers of gravity of those links. We want to develop algebraic expressions for the positions of these (or any) points on the links.

To find the position of point S, draw a position vector from the fixed pivot O_2 to point S. This vector $\mathbf{R}_{SO2}$ makes an angle δ_2 with the vector $\mathbf{R}_{AO2}$. This angle δ_2 is completely defined by the geometry of link 2 and is constant. The position vector for point S is then:

$$\mathbf{R}_{SO_2} = \mathbf{R}_S = se^{j(\theta_2 + \delta_2)} = s\left[\cos(\theta_2 + \delta_2) + j\sin(\theta_2 + \delta_2)\right]$$ (4.25)

The position of point U on link 4 is found in the same way, using the angle δ_4 which is a constant angular offset within the link. The expression is:

$$\mathbf{R}_{UO_4} = ue^{j(\theta_4 + \delta_4)} = u\left[\cos(\theta_4 + \delta_4) + j\sin(\theta_4 + \delta_4)\right]$$ (4.26)

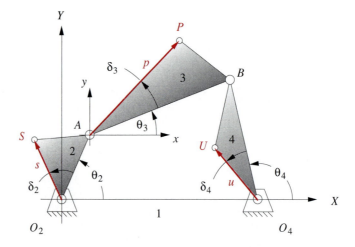

FIGURE 4-14

Positions of points on the links

The position of point P on link 3 can be found from the addition of two position vectors $\mathbf{R}_A$ and $\mathbf{R}_{PA}$. $\mathbf{R}_A$ is already defined from our analysis of the link angles in equation 4.5. $\mathbf{R}_{PA}$ is the relative position of point P with respect to point A. $\mathbf{R}_{PA}$ is defined in the same way as $\mathbf{R}_S$ or $\mathbf{R}_U$, using the internal link offset angle δ_3 and the position angle of link 3, θ_3.

$$\mathbf{R}_{PA} = pe^{j(\theta_3 + \delta_3)} = p\left[\cos(\theta_3 + \delta_3) + j\sin(\theta_3 + \delta_3)\right] \tag{4.27a}$$

$$\mathbf{R}_P = \mathbf{R}_A + \mathbf{R}_{PA} \tag{4.27b}$$

Compare equation 4.27 with equation 4.1. This is also the position difference equation.

4.10 TRANSMISSION ANGLES

The transmission angle was defined in Section 3.3 (p. 81) for a fourbar linkage. That definition is repeated here for your convenience.

The **transmission angle** μ is shown in Figure 3-3a (p. 82) and is defined as *the angle between the output link and the coupler*. It is usually taken as the absolute value of the acute angle of the pair of angles at the intersection of the two links and varies continuously from some minimum to some maximum value as the linkage goes through its range of motion. It is a measure of the quality of force transmission at the joint.[*]

We will expand that definition here to represent the angle between any two links in a linkage, as a linkage can have many transmission angles. The angle between any output link and the coupler which drives it is a transmission angle. Now that we have developed the analytic expressions for the angles of all the links in a mechanism, it is easy to define the transmission angle algebraically. It is merely the difference between the angles of the two joined links through which we wish to pass some force or velocity. For our fourbar linkage example it will be the difference between θ_3 and θ_4. By convention we take the absolute value of the difference and force it to be an acute angle.

$$\theta_{trans} = |\theta_3 - \theta_4|$$

$$\text{if} \qquad \theta_{trans} > \frac{\pi}{2} \qquad \text{then} \qquad \theta_{trans} = \pi - \theta_{trans} \tag{4.28}$$

This computation can be done for any joint in a linkage by using the appropriate link angles.

Extreme Values of the Transmission Angle

For a Grashof crank-rocker foufbar linkage the extreme values of the transmission angle will occur when the crank is colinear with the ground link as shown in Figure 4-15. The values of the transmission angle in these positions are easily calculated from the law of cosines since the linkage is then in a triangular configuration. The sides of the two triangles are link 3, link 4, and either the sum or difference of links 1 and 2. One extreme value of the transmission angle occurs when links 1 and 2 are *colinear and nonoverlapping* as shown in Figure 4-15a. The other extreme transmission angle occurs when links 1 and 2 are *colinear and overlapping* as shown in Figure 4-15b. Using notation consistent with Section 4.5 and Figure 4-7 (p. 155) we will label the links:

[*] The transmission angle has limited application. It only predicts the quality of force or torque transmission if the input and output links are pivoted to ground. If the output force is taken from a floating link (coupler), then the transmission angle is of no value. A different index of merit called the joint force index (JFI) is presented in Chapter 11 which discusses force analysis in linkages. (See Section 11.12 on p. 554.) The JFI is useful for situations in which the output link is floating as well as giving the same kind of information when the output is taken from a link rotating against the ground. However, the JFI requires a complete force analysis of the linkage be done whereas the transmission angle is determined from linkage geometry alone.

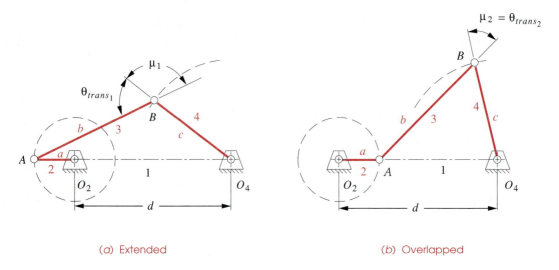

(a) Extended   (b) Overlapped

FIGURE 4-15

Extreme transmission angles in the Grashof fourbar linkage

a = link 2; b = link 3; c = link 4; d = link 1

For one extreme transmission angle the cosine law gives

$$\mu_1 = \arccos\left[\frac{b^2 + c^2 - (d+a)^2}{2bc}\right] \tag{4.29a}$$

and for the other extreme transmission angle

$$\mu_2 = \arccos\left[\frac{b^2 + c^2 - (d-a)^2}{2bc}\right] \tag{4.29b}$$

For a **Grashof double-rocker** linkage the transmission angle can vary from 0 to 90 degrees because the coupler can make a full revolution with respect to the other links. For a **non-Grashof triple-rocker** linkage the transmission angle will be zero degrees in the toggle positions which occur when the output rocker c and the coupler b are colinear as shown in Figure 4-16a. In the other toggle positions when input rocker a and coupler b are colinear (Figure 4-16b), the transmission angle can be calculated from the cosine law as:

when $v = 0$,

$$\mu = \arccos\left[\frac{(a+b)^2 + c^2 - d^2}{2c(a+b)}\right] \tag{4.30}$$

This is not the smallest value that the transmission angle μ can have in a triple-rocker, as that will obviously be zero. Of course, when analyzing any linkage, the transmis-

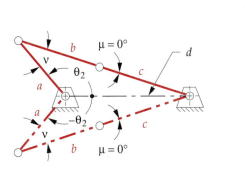

(a) Toggle positions for links b and c

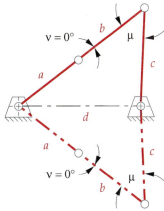

(b) Toggle positions for links a and b

FIGURE 4-16

Non-Grashof triple-rocker linkages in toggle

sion angles can easily be computed and plotted for all positions using equation 4.28. Programs FOURBAR, FIVEBAR, and SIXBAR do this. The student should investigate the variation in transmission angle for the example linkages in those programs. Disk file F04-15.4br can be opened in program FOURBAR to observe that linkage in motion.

4.11 TOGGLE POSITIONS

The input link angles which correspond to the toggle positions (stationary configurations) of the **non-Grashof triple-rocker** can be calculated by the following method, using trigonometry. Figure 4-17 shows a non-Grashof fourbar linkage in a general position. A construction line h has been drawn between points A and O_4. This divides the quadrilateral loop into two triangles, O_2AO_4 and ABO_4. Equation 4.31 uses the cosine law to express the transmission angle μ in terms of link lengths and the input link angle θ_2.

$$h^2 = a^2 + d^2 - 2ad\cos\theta_2$$

also:

$$h^2 = b^2 + c^2 - 2bc\cos\mu$$

so:

$$a^2 + d^2 - 2ad\cos\theta_2 = b^2 + c^2 - 2bc\cos\mu$$

and:

$$\cos\mu = \frac{b^2 + c^2 - a^2 - d^2}{2bc} + \frac{ad}{bc}\cos\theta_2 \qquad (4.31)$$

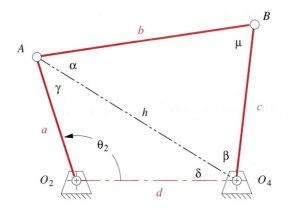

FIGURE 4-17

Finding the crank angle corresponding to the toggle positions

To find the maximum and minimum values of input angle θ_2, we can differentiate equation 4.31, form the derivative of θ_2 with respect to μ, and set it equal to zero.

$$\frac{d\theta_2}{d\mu} = \frac{bc}{ad}\frac{\sin\mu}{\sin\theta_2} = 0 \qquad (4.32)$$

The link lengths a, b, c, d are never zero, so this expression can only be zero when $\sin\mu$ is zero. This will be true when angle μ in Figure 4-17 is either zero or 180°. This is consistent with the definition of toggle given in Section 3.3 (p. 80). If μ is zero or 180° then $\cos\mu$ will be ±1. Substituting these two values for $\cos\mu$ into equation 4.31 will give a solution for the value of θ_2 between zero and 180° which corresponds to the toggle position of a triple-rocker linkage when driven from one rocker.

$$\cos\mu = \frac{b^2 + c^2 - a^2 - d^2}{2bc} + \frac{ad}{bc}\cos\theta_2 = \pm 1$$

or :

$$\cos\theta_2 = \frac{a^2 + d^2 - b^2 - c^2}{2ad} \pm \frac{bc}{ad} \qquad (4.33)$$

and :

$$\theta_{2_{toggle}} = \arccos\left(\frac{a^2 + d^2 - b^2 - c^2}{2ad} \pm \frac{bc}{ad}\right); \qquad 0 \le \theta_{2_{toggle}} \le \pi$$

One of these ± cases will produce an argument for the arccosine function which lies between ±1. The toggle angle which is in the first or second quadrant can be found from this value. The other toggle angle will then be the negative of the one found, due to the mirror symmetry of the two toggle positions about the ground link as shown in Figure 4-16 (p. 171). Program FOURBAR computes the values of these toggle angles for any non-Grashof linkage.

4.12   CIRCUITS AND BRANCHES IN LINKAGES

In Section 4.5 it was noted that the fourbar linkage position problem has two solutions which correspond to the two circuits of the linkage. This section will explore the topics of circuits and branches in linkages in more detail.

Chase and Mirth[2] define a **circuit** in a linkage as "*all possible orientations of the links that can be realized without disconnecting any of the joints*" and a **branch** as "*a continuous series of positions of the mechanism on a circuit between two stationary configurations. . . . The stationary configurations divide a circuit into a series of branches.*" A linkage may have one or more circuits each of which may contain one or more branches. The number of circuits corresponds to the number of solutions possible from the position equations for the linkage.

Circuit defects are fatal to linkage operation, but branch defects are not. A mechanism that must change circuits to move from one desired position to the other (referred to as a **circuit defect**) is not useful as it cannot do so without disassembly and reassembly. A mechanism that changes branch when moving from one circuit to another (referred to as a **branch defect**) may or may not be usable depending on the designer's intent.

The tailgate linkage shown in Figure 3-2 (p. 81) is an example of a linkage with a deliberate branch defect in its range of motion (actually at the limit of its range of motion). The toggle position (stationary configuration) that it reaches with the tailgate fully open serves to hold it open. But the user can move it out of this stationary configuration by rotating one of the links out of toggle. Folding chairs and tables often use a similar scheme as do fold-down seats in automobiles and station wagons (shooting brakes).

Another example of a common linkage with a branch defect is the slider-crank linkage (crankshaft, connecting rod, piston) used in every piston engine and shown in Figure 13-3 (p. 601). This linkage has two toggle positions (top and bottom dead center) giving it two branches within one revolution of its crank. It works nevertheless because it is carried through these stationary configurations by the angular momentum of the rotating crank and its attached flywheel. One penalty is that the engine must be spun to start it in order to build sufficient momentum to carry it through these toggle positions.

The Watt sixbar linkage can have four circuits, and the Stephenson sixbar can have either four or six circuits depending on which link is driving. Eightbar linkages can have as many as 16 or 18 circuits, not all of which may be real, however.[2]

The number of circuits and branches in the fourbar linkage depends on its Grashof condition and the inversion used. A non-Grashof, triple-rocker fourbar linkage has only one circuit but has two branches. All Grashof fourbar linkages have two circuits, but the number of branches per circuit differs with the inversion. The crank-rocker and double-crank have only one branch within each circuit. The double-rocker and rocker-crank have two branches within each circuit. Table 4-1 summarizes these relationships.[2]

Any solution for the position of a linkage must take into account the number of possible circuits that it contains. A closed-form solution, if available, will contain all the circuits. An iterative solution such as is described in the next section will only yield the position data for one circuit, and it may not be the one you expect.

TABLE 4-1
Circuits & Branches
In the Fourbar Linkage

Fourbar Linkage Type	Number of Circuits	Branch per Circuit
Non-Grashof triple-rocker	1	2
Grashof* crank-rocker	2	1
Grashof* double-crank	2	1
Grashof* double-rocker	2	2
Grashof* rocker-crank	2	2

* Valid only for non-special case Grashof linkages.

4

4.13   NEWTON-RAPHSON SOLUTION METHOD

The solution methods for position analysis shown so far in this chapter are all of "closed form," meaning that they provide the solution with a direct, noniterative approach.* In some situations, particularly with multiloop mechanisms, a closed-form solution may not be attainable. Then an alternative approach is needed, and the Newton-Raphson method (sometimes just called Newton's method) provides one that can solve sets of simultaneous nonlinear equations. Any iterative solution method requires that one or more guess values be provided to start the computation. It then uses the guess values to obtain a new solution that may be closer to the correct one. This process is repeated until it converges to a solution close enough to the correct one for practical purposes. However, there is no guarantee that an iterative method will converge at all. It may diverge, taking successive solutions further from the correct one, especially if the initial guess is not sufficiently close to the real solution.

Though we will need to use the multidimensional (Newton-Raphson version) of Newton's method for these linkage problems, it is easier to understand how the algorithm works by first discussing the one-dimensional Newton's method for finding the roots of a single nonlinear function in one independent variable. Then we will discuss the multidimensional Newton-Raphson method.

One-Dimensional Root-Finding (Newton's Method)

A nonlinear function may have multiple roots, where a root is defined as the intersection of the function with any straight line. Typically the zero axis of the independent variable is the straight line for which we desire the roots. Take, for example, a cubic polynomial which will have three roots, with either one or all three being real.

$$y = f(x) = -x^3 - 2x^2 + 50x + 60 \tag{4.34}$$

There is a closed-form solution for the roots of a cubic function† which allows us to calculate in advance that the roots of this particular cubic are all real and are $x = -7.562$, -1.177, and 6.740.

Figure 4-18 shows this function plotted over a range of x. In Figure 4-18a, an initial guess value of $x_1 = 1.8$ is chosen. Newton's algorithm evaluates the function for this guess value, finding y_1. The value of y_1 is compared to a user-selected tolerance (say 0.001) to see if it is close enough to zero to call x_1 the root. If not, then the slope (m) of the function at x_1, y_1 is calculated either by using an analytic expression for the derivative of the function or by doing a numerical differentiation (less desirable). The equation of the tangent line is then evaluated to find its intercept at x_2 which is used as a new guess value. The above process is repeated, finding y_2; testing it against the user selected tolerance; and, if it is too large, calculating another tangent line whose x intercept is used as a new guess value. This process is repeated until the value of the function y_i at the latest x_i is close enough to zero to satisfy the user.

The Newton's algorithm described above can be expressed algebraically (in pseudocode) as shown in equation 4.35. The function for which the roots are sought is $f(x)$, and its derivative is $f'(x)$. The slope m of the tangent line is equal to $f'(x)$ at the current point x_i y_i.

* Kramer [3] states that: "In theory, any nonlinear algebraic system of equations can be manipulated into the form of a single polynomial in one unknown. The roots of this polynomial can then be used to determine all unknowns in the system. However, if the derived polynomial is greater than degree four, factoring and/or some form of iteration are necessary to obtain the roots. In general,. systems that have more than a fourth degree polynomial associated with the eliminant of all but one variable must be solved by iteration. However, if factoring of the polynomial into terms of degree four or less is possible, all roots may be found without iteration. Therefore the only truly symbolic solutions are those that can be factored into terms of fourth degree or less. This is the formal definition of a closed form solution."

† Viete's method from "De Emendatione" by Francois Viete (1615) as described in reference [4].

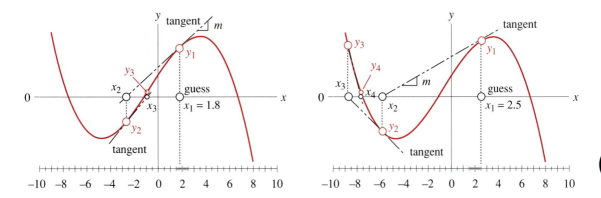

(a) A guess of $x = 1.8$ converges to the root at $x = -1.177$ (b) A guess of $x = 2.5$ converges to the root at $x = -7.562$

FIGURE 4-18

Newton-Raphson method of solution for roots of nonlinear functions

step 1 $y_i = f(x_i)$

step 2   IF $y_i \le tolerance$ THEN STOP

step 3 $m = f'(x_i)$

step 4 $x_{i+1} = x_i - \dfrac{y_i}{m}$

step 5 $y_{i+1} = f(x_{i+1})$

step 6   IF $y_{i+1} \le tolerance$ THEN STOP

            ELSE $x_i = x_{i+1}$: $y_i = y_{i+1}$:   GOTO step 1   (4.35)

If the initial guess value is close to a root, this algorithm will converge rapidly to the solution. However, it is quite sensitive to the initial guess value. Figure 4-18b shows the result of a slight change in the initial guess from $x_1 = 1.8$ to $x_1 = 2.5$. With this slightly different guess it converges to another root. Note also that if we choose an initial guess of $x_1 = 3.579$ which corresponds to a local maximum of this function, the tangent line will be horizontal and will not intersect the x axis at all. The method fails in this situation. Can you suggest a value of x_1 that would cause it to converge to the root at $x = 6.74$?

So this method has its drawbacks. It may fail to converge. It may behave chaotically.* It is sensitive to the guess value. It also is incapable of distinguishing between multiple circuits in a linkage. The circuit solution it finds is dependent on the initial guess. It requires that the function be differentiable, and the derivative as well as the function must be evaluated at every step. Nevertheless, it is the method of choice for functions whose derivatives can be efficiently evaluated and which are continuous in the region of the root. Furthermore, it is about the only choice for systems of nonlinear equations.

---

* Kramer[3] points out that "the Newton Raphson algorithm can exhibit chaotic behavior when there are multiple solutions to kinematic constraint equations. . . . Newton Raphson has no mechanism for distinguishing between the two solutions" (circuits). He does an experiment with just two links, exactly analogous to finding the angles of the coupler and rocker in the fourbar linkage position problem, and finds that the initial guess values need to be quite close to the desired solution (one of the two possible circuits) to avoid divergence or chaotic oscillation between the two solutions.

Multidimensional Root-Finding (Newton-Raphson Method)

The one-dimensional Newton's method is easily extended to multiple, simultaneous, nonlinear equation sets and is then called the Newton-Raphson method. First, let's generalize the expression developed for the one-dimensional case in step 4 of equation 4.35. Refer also to Figure 4-18 (p. 175).

$$x_{i+1} = x_i - \frac{y_i}{m} \qquad \text{or} \qquad m(x_{i+1} - x_i) = -y_i$$

but: $\qquad y_i = f(x_i) \qquad m = f'(x_i) \qquad x_{i+1} - x_i = \Delta x$

substituting: $\qquad f'(x_i) \cdot \Delta x = -f(x_i)$ $\qquad\qquad\qquad$ (4.36)

Here a Δx term is introduced which will approach zero as the solution converges. The Δx term rather than y_i will be tested against a selected tolerance in this case. Note that this form of the equation avoids the division operation which is acceptable in a scalar equation but impossible with a matrix equation.

A multidimensional problem will have a set of equations of the form

$$\begin{bmatrix} f_1(x_1, x_2, x_3, \ldots, x_n) \\ f_2(x_1, x_2, x_3, \ldots, x_n) \\ \vdots \qquad\qquad \vdots \\ f_n(x_1, x_2, x_3, \ldots, x_n) \end{bmatrix} = \mathbf{B} \qquad\qquad (4.37)$$

where the set of equations constitutes a vector, here called $\mathbf{B}$.

Partial derivatives are required to obtain the slope terms

$$\begin{bmatrix} \dfrac{\partial f_1}{\partial x_1} & \dfrac{\partial f_1}{\partial x_2} & \cdots & \dfrac{\partial f_1}{\partial x_n} \\ \vdots & \vdots & & \vdots \\ \dfrac{\partial f_n}{\partial x_1} & \dfrac{\partial f_n}{\partial x_2} & \cdots & \dfrac{\partial f_n}{\partial x_n} \end{bmatrix} = \mathbf{A} \qquad\qquad (4.38)$$

which form the *jacobian matrix* of the system, here called $\mathbf{A}$.

The error terms are also a vector, here called $\mathbf{X}$.

$$\begin{bmatrix} \Delta x_1 \\ \Delta x_2 \\ \vdots \\ \Delta x_n \end{bmatrix} = \mathbf{X} \qquad\qquad (4.39)$$

Equation 4.36 then becomes a matrix equation for the multidimensional case.

$$\mathbf{A}\mathbf{X} = -\mathbf{B} \qquad\qquad (4.40)$$

Equation 4.40 can be solved for **X** either by matrix inversion or by gaussian elimination. The values of the elements of **A** and **B** are calculable for any assumed (guess) values of the variables. A criterion for convergence can be taken as the sum of the error vector **X** at each iteration where the sum approaches zero at a root.

Let's set up this Newton-Raphson solution for the fourbar linkage.

Newton-Raphson Solution for the Fourbar Linkage

The vector loop equation of the fourbar linkage, separated into its real (equation 4.6a) and imaginary (equation 4.6b) parts provides the set of functions that define the two unknown link angles θ_3 and θ_4. The link lengths, a, b, c, d, and the input angle θ_2 are given.

$$f_1 = a\cos\theta_2 + b\cos\theta_3 - c\cos\theta_4 - d = 0$$
$$f_2 = a\sin\theta_2 + b\sin\theta_3 - c\sin\theta_4 = 0$$
(4.41a)

$$\mathbf{B} = \begin{bmatrix} a\cos\theta_2 + b\cos\theta_3 - c\cos\theta_4 - d \\ a\sin\theta_2 + b\sin\theta_3 - c\sin\theta_4 \end{bmatrix}$$
(4.41b)

The error vector is:

$$\mathbf{X} = \begin{bmatrix} \Delta\theta_3 \\ \Delta\theta_4 \end{bmatrix}$$
(4.42)

The partial derivatives are:

$$\mathbf{A} = \begin{bmatrix} \dfrac{\partial f_1}{\partial \theta_3} & \dfrac{\partial f_1}{\partial \theta_4} \\ \dfrac{\partial f_2}{\partial \theta_3} & \dfrac{\partial f_2}{\partial \theta_4} \end{bmatrix} = \begin{bmatrix} -b\sin\theta_3 & c\sin\theta_4 \\ b\cos\theta_3 & -c\cos\theta_4 \end{bmatrix}$$
(4.43)

This matrix is known as the **jacobian** of the system, and, in addition to its usefulness in this solution method, it also tells something about the solvability of the system. The system of equations for position, velocity, and acceleration (in all of which the jacobian appears) can only be solved if the value of the determinant of the jacobian is nonzero.

Substituting equations 4.41b, 4.42, and 4.43 into equation 4.40 gives:

$$\begin{bmatrix} -b\sin\theta_3 & c\sin\theta_4 \\ b\cos\theta_3 & -c\cos\theta_4 \end{bmatrix} \begin{bmatrix} \Delta\theta_3 \\ \Delta\theta_4 \end{bmatrix} = -\begin{bmatrix} a\cos\theta_2 + b\cos\theta_3 - c\cos\theta_4 - d \\ a\sin\theta_2 + b\sin\theta_3 - c\sin\theta_4 \end{bmatrix}$$
(4.44)

To solve this matrix equation, guess values will have to be provided for θ_3 and θ_4 and the two equations then solved simultaneously for $\Delta\theta_3$ and $\Delta\theta_4$. For a larger system of equations, a matrix reduction algorithm will need to be used. For this simple system in two unknowns, the two equations can be solved by combination and reduction. The test described above which compares the sum of the values of $\Delta\theta_3$ and $\Delta\theta_4$ to a selected tolerance must be applied after each iteration to determine if a root has been found.

Equation Solvers

Some commercially available equation solver software packages include the ability to do a Newton-Raphson iterative solution on sets of nonlinear simultaneous equations. *TKSolver*[*] and *Mathcad*[†] are examples. *TKSolver* automatically invokes its Newton-Raphson solver when it cannot directly solve the presented equation set, provided that enough guess values have been supplied for the unknowns. These equation solver tools are quite convenient in that the user need only supply the equations for the system in "raw" form such as equation 4.41a. It is not necessary to arrange them into the Newton-Raphson algorithm as shown in the previous section. Lacking such a commercial equation solver, you will have to write your own computer code to program the solution as described above. Reference [5] is a useful aid in this regard. The CD-ROM included with this text contains example *TKSolver* files for the solution of this fourbar position problem as well as others.

4.14   REFERENCES

1   **Waldron, K. J., and S. V. Sreenivasan**. (1996). "A Study of the Solvability of the Position Problem for Multi-Circuit Mechanisms by Way of Example of the Double Butterfly Linkage." *Journal of Mechanical Design*, **118**(3), p. 390.

2   **Chase, T. R., and J. A. Mirth.** (1993). "Circuits and Branches of Single-Degree-of-Freedom Planar Linkages." *Journal of Mechanical Design*, **115**, p. 223.

3   **Kramer, G.** (1992). *Solving Geometric Constraint Systems: A Case Study in Kinematics.* MIT Press: Cambridge, pp. 155-158.

4   **Press, W. H., et al.** (1986). *Numerical Recipes: The Art of Scientific Computing.* Cambridge University Press: Cambridge, pp. 145-146.

5   Ibid, pp. 254-273.

4.15   PROBLEMS

4-1   A position vector is defined as having a length equal to your height in inches (or centimeters). The tangent of its angle is defined as your weight in pounds (or kilograms) divided by your age in years. Calculate the data for this vector and:

    a.   Draw the position vector to scale on cartesian axes.
    b.   Write an expression for the position vector using unit vector notation.
    c.   Write an expression for the position vector using complex number notation, in both polar and cartesian forms.

4-2   A particle is traveling along an arc of 6.5-in radius. The arc center is at the origin of a coordinate system. When the particle is at position *A*, its position vector makes a 45° angle with the *X* axis. At position *B*, its vector makes a 75° angle with the *X* axis. Draw this system to some convenient scale and:

    a.   Write an expression for the particle's position vector in position *A* using complex number notation, in both polar and cartesian forms.
    b.   Write an expression for the particle's position vector in position *B* using complex number notation, in both polar and cartesian forms.
    c.   Write a vector equation for the position difference between points *B* and *A*. Substitute the complex number notation for the vectors in this equation and solve for the position difference numerically.
    d.   Check the result of part c with a graphical method.

[*] Universal Technical Systems, 1220 Rock St. Rockford, IL 61101, USA. (800) 435-7887

[†] Mathsoft, 201 Broadway, Cambridge, MA 02139 (800) 628-4223

4-3 Repeat problem 4-2 considering points A and B to represent separate particles, and find their relative position.

4-4 Repeat Problem 4-2 with the particle's path defined as being along the line $y = -2x + 10$.

4-5 Repeat Problem 4-3 with the path of the particle defined as being along the curve $y = -2x^2 - 2x + 10$.

*4-6 The link lengths and the value of θ_2 for some fourbar linkages are defined in Table P4-1. The linkage configuration and terminology are shown in Figure P4-1. For the rows assigned, draw the linkage to scale and graphically find all possible solutions (both open and crossed) for angles θ_3 and θ_4. Determine the Grashof condition.

*†4-7 Repeat Problem 4-6 except solve by the vector loop method.

4-8 Expand equation 4.7b and prove that it reduces to equation 4.7c (p. 157).

*4-9 The link lengths and the value of θ_2 and offset for some fourbar slider-crank linkages are defined in Table P4-2. The linkage configuration and terminology are shown in Figure P4-2. For the rows assigned, draw the linkage to scale and graphically find all possible solutions (both open and crossed) for angle θ_3 and slider position d.

*†4-10 Repeat Problem 4-9 except solve by the vector loop method.

*4-11 The link lengths and the value of θ_2 and γ for some inverted fourbar slider-crank linkages are defined in Table P4-3. The linkage configuration and terminology are shown in Figure P4-3. For the rows assigned, draw the linkage to scale and graphically find both open and crossed solutions for angles θ_3 and θ_4 and vector $\mathbf{R}_B$.

*†4-12 Repeat Problem 4-11 except solve by the vector loop method.

*†4-13 Find the transmission angles of the linkages in the assigned rows in Table P4-1.

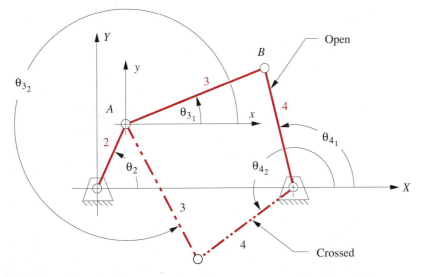

FIGURE P4-1

Problems 4-6 to 4-7. General configuration and terminology for the fourbar linkage

TABLE P4-1   Data for Problems 4-6, 4-7 and 4-13 to 4-15

Row	Link 1	Link 2	Link 3	Link 4	θ_2
a	6	2	7	9	30
b	7	9	3	8	85
c	3	10	6	8	45
d	8	5	7	6	25
e	8	5	8	6	75
f	5	8	8	9	15
g	6	8	8	9	25
h	20	10	10	10	50
i	4	5	2	5	80
j	20	10	5	10	33
k	4	6	10	7	88
l	9	7	10	7	60
m	9	7	11	8	50
n	9	7	11	6	120

*†4-14 Find the minimum and maximum values of the transmission angle for all the Grashof crank-rocker linkages in Table P4-1.

*†4-15 Find the input angles corresponding to the toggle positions of the non-Grashof linkages in Table P4-1. (For this problem, ignore the values of θ_2 given in the table.)

*4-16 The link lengths, gear ratio (λ), phase angle (ϕ), and the value of θ_2 for some geared fivebar linkages are defined in Table P4-4. The linkage configuration and terminology are shown in Figure P4-4. For the rows assigned, draw the linkage to scale and graphically find all possible solutions for angles θ_3 and θ_4.

*†4-17 Repeat Problem 4-16 except solve by the vector loop method.

---

* Answers in Appendix F.

---

† These problems are suited to solution using *Mathcad, Matlab,* or *TKSolver* equation solver programs. In most cases, your solution can be checked with program FOURBAR, SLIDER, or SIXBAR.

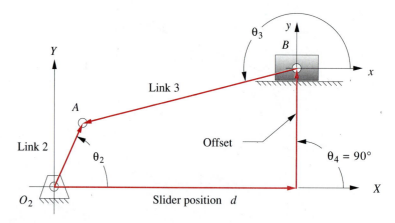

FIGURE P4-2

Problems 4-9 to 4-10 Open configuration and terminology for a fourbar slider-crank linkage

TABLE P4-2 Data for Problems 4-9 to 4-10

Row	Link 2	Link 3	Offset	θ_2
a	1.4	4	1	45
b	2	6	-3	60
c	3	8	2	-30
d	3.5	10	1	120
e	5	20	-5	225
f	3	13	0	100
g	7	25	10	330

TABLE P4-3 Data for Problems 4-11 to 4-12

Row	Link 1	Link 2	Link 4	γ	θ_2
a	6	2	4	90	30
b	7	9	3	75	85
c	3	10	6	45	45
d	8	5	3	60	25
e	8	4	2	30	75
f	5	8	8	90	150

4-18 Scale Figure P4-5 to obtain dimensions for the following problems each of which refers to the part of the figure with the same letter. Reference all angular measurements to a line of centers.

   a. Find the angular displacement of link *CD* when link *AB* rotates clockwise from the position shown to horizontal. What is the smallest transmission angle between

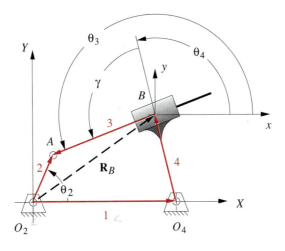

FIGURE P4-3

Problems 4-11 to 4-12. Terminology for inversion #3 of the fourbar slider-crank linkage

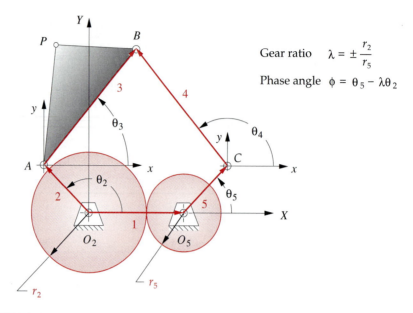

Gear ratio $\quad \lambda = \pm \dfrac{r_2}{r_5}$

Phase angle $\quad \phi = \theta_5 - \lambda\theta_2$

FIGURE P4-4

Problems 4-16 to 4-17. Open configuration and terminology for the geared fivebar linkage

those two positions? Find the toggle positions of this linkage in terms of the angle of link *AB*.

b. Find and plot the angular position of links *BC* and *CD* and the transmission angle as a function of the angle of wheel *W* as it rotates through one revolution.

c. Find and plot the position of any one piston as a function of the angle of crank *AB* as it rotates through one revolution. Once one piston's motion is defined, find the motions of the other two pistons and their phase relationship to the first piston.

d. Find the total angular displacement of link *BC* as link *AB* makes a complete revolution. Determine the total stroke of the box at point *E* as link *AB* makes a complete revolution.

TABLE P4-4   Data for Problems 4-16 to 4-17

Row	Link 1	Link 2	Link 3	Link 4	Link 5	λ	ϕ	θ_2
a	6	1	7	9	4	2	30	60
b	6	5	7	8	4	-2.5	60	30
c	3	5	7	8	4	-0.5	0	45
d	4	5	7	8	4	-1	120	75
e	5	9	11	8	8	3.2	-50	-39
f	10	2	7	5	3	1.5	30	120
g	15	7	9	11	4	2.5	-90	75
h	12	8	7	9	4	-2.5	60	55
i	9	7	8	9	4	-4	120	100

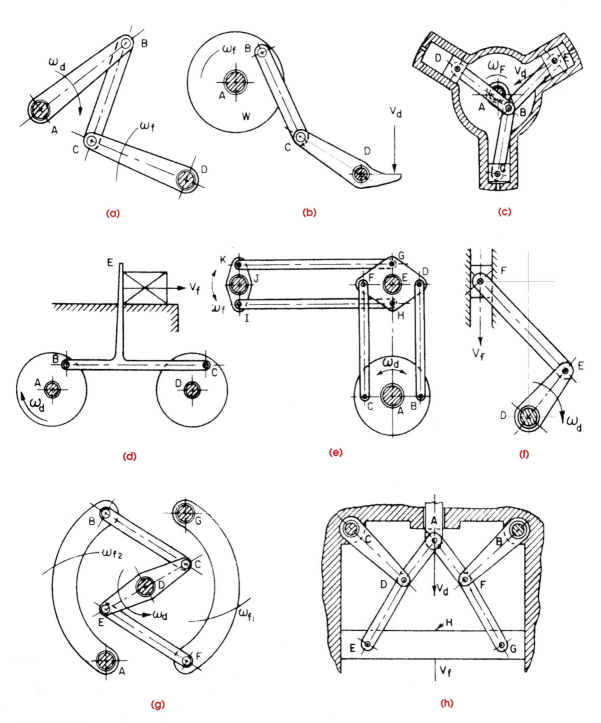

FIGURE P4-5

Mechanisms for Problem 4-18. *Adapted from P. H. Hill and W. P. Rule. (1960). Mechanisms: Analysis and Design, with permission.*

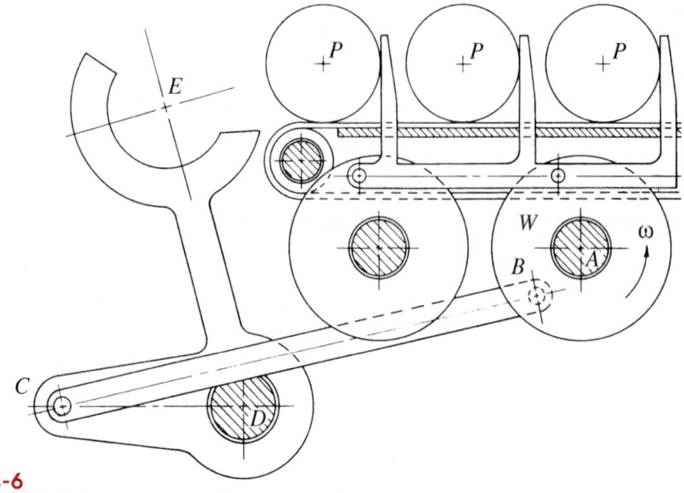

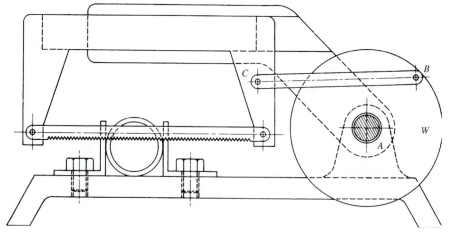

FIGURE P4-6

Pusher and pick-and-place mechanism for Problem 4-19. *Adapted from P. H. Hill and W. P. Rule. (1960). Mechanisms: Analysis and Design, with permission.*

e. Determine the ratio of angular displacement between link *AB* and *JK* as a function of angular displacement of input crank *JK*. Plot the transmission angle at point *G* for one revolution of crank *JK*. Comment on the behavior of this linkage.

f. Find and plot the displacement of piston *F* and the angular displacement of link *EF* as a function of the angular displacement of crank *DE*.

g. Find and plot the angular displacement of link *AB* versus the angle of input link *DC* as it is rotated from the position shown to a vertical position. Find the toggle positions of this linkage in terms of the angle of link *DC*.

h. Find point *G's* maximum displacement vertically downward from the position shown. What will the angle of input link *BF* be at that position?

FIGURE P4-7

Power hacksaw for Problem 4-20. *Adapted from P. H. Hill and W. P. Rule. (1960). Mechanisms: Analysis and Design, with permission.*

†4-19 For one revolution of driving wheel W of the pusher and pick-and-place mechanism in Figure P4-6, find the horizontal stroke of the pushers for the portion of their motion where their tips are above the top belt. Express the stroke as a percentage of the crank length AB. What portion of a revolution of the driver crank AB does this stroke correspond to? Also find the total angular displacement of link DE over one revolution of wheel W. Scale the drawing for all necessary dimensions.

†4-20 For one revolution of driving wheel W of the hacksaw mechanism on the cutting stroke depicted in Figure P4-7, find and plot the horizontal stroke of the saw blade as a function of the angle of input crank AB.

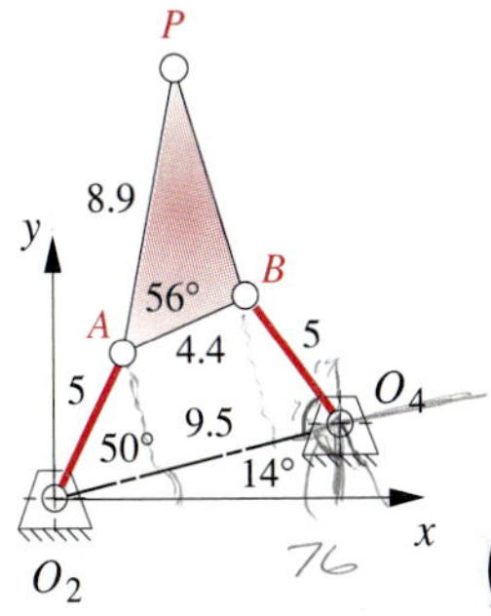

*†4-21 For the linkage in Figure P4-8, find its limit (toggle) positions in terms of the angle of link O_2A referenced to the line of centers O_2O_4 when driven from link O_2A. Then calculate and plot the xy coordinates of coupler point P between those limits, referenced to the line of centers O_2O_4.

FIGURE P4-8

Problem 4-21

†4-22 For the walking beam mechanism of Figure P4-9, calculate and plot the x and y components of the position of the coupler point P for one complete revolution of the crank O_2A. Hint: Calculate them first with respect to the ground link O_2O_4 and then transform them into the global XY coordinate system (i.e., horizontal and vertical in the figure). Scale the figure for any additional information needed.

*†4-23 For the linkage in Figure P4-10, calculate and plot the angular displacement of links 3 and 4 and the path coordinates of point P with respect to the angle of the input crank O_2A for one revolution.

†4-24 For the linkage in Figure P4-11, calculate and plot the angular displacement of links 3 and 4 with respect to the angle of the input crank O_2A for one revolution.

*†4-25 For the linkage in Figure P4-12, find its limit (toggle) positions in terms of the angle of link O_2A referenced to the line of centers O_2O_4 when driven from link O_2A. Then calculate and plot the angular displacement of links 3 and 4 and the path coordinates of point P with respect to the angle of the input crank O_2A over its possible range of motion referenced to the line of centers O_2O_4.

*†4-26 For the linkage in Figure P4-13, find its limit (toggle) positions in terms of the angle of link O_2A referenced to the line of centers O_2O_4 when driven from link O_2A. Then calculate and plot the angular displacement of links 3 and 4 and the path coordinates

* Answers in Appendix F.

† These problems are suited to solution using *Mathcad, Matlab,* or *TKSolver* equation solver programs. In most cases, your solution can be checked with program FOURBAR, SLIDER, or SIXBAR.

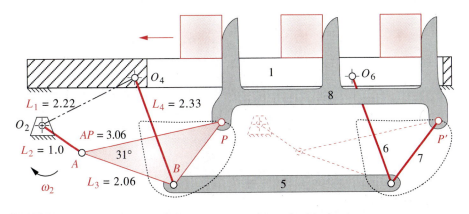

FIGURE P4-9

Problem 4-22. Straight-line walking beam eightbar transport mechanism

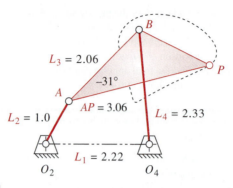

FIGURE P4-10

Problem 4-23

of point P between those limits, with respect to the angle of the input crank O_2A over its possible range of motion referenced to the line of centers O_2O_4.

†4-27   For the linkage in Figure P4-13, find its limit (toggle) positions in terms of the angle of link O_4B referenced to the line of centers O_4O_2 when driven from link O_4B. Then calculate and plot the angular displacement of links 2 and 3 and the path coordinates of point P between those limits, with respect to the angle of the input crank O_4B over its possible range of motion referenced to the line of centers O_4O_2.

†4-28   For the rocker-crank linkage in Figure P4-14, find the maximum angular displacement possible for the treadle link (to which force F is applied). Determine the toggle positions. How does this work? Explain why the grinding wheel is able to fully rotate despite the presence of toggle positions when driven from the treadle. How would you get it started if it was in a toggle position?

*†4-29   For the linkage in Figure P4-15, find its limit (toggle) positions in terms of the angle of link O_2A referenced to the line of centers O_2O_4 when driven from link O_2A. Then calculate and plot the angular displacement of links 3 and 4 and the path coordinates of point P between those limits, with respect to the angle of the input crank O_2A over its possible range of motion referenced to the line of centers O_2O_4.

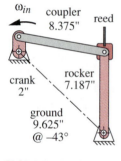

FIGURE P4-11

Problem 4-24

---

* Answers in Appendix F.

---

† These problems are suited to solution using *Mathcad, Matlab,* or *TKSolver* equation solver programs. In most cases, your solution can be checked with program FOURBAR, SLIDER, or SIXBAR.

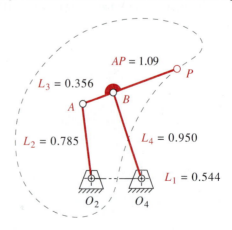

FIGURE P4-12

Problem 4-25

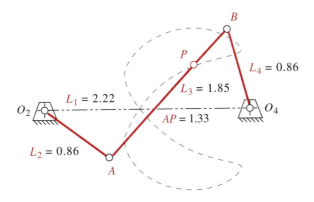

FIGURE P4-13

Problems 4-26 to 4-27

*†4-30 For the linkage in Figure P4-15, find its limit (toggle) positions in terms of the angle of link O_4B referenced to the line of centers O_4O_2 when driven from link O_4B. Then calculate and plot the angular displacement of links 2 and 3 and the path coordinates of point P between those limits, with respect to the angle of the input crank O_4B over its possible range of motion referenced to the line of centers O_4O_2.

†4-31 Write a computer program (or use an equation solver such as *Mathcad*, *Matlab*, or *TKSolver*) to find the roots of $y = 9x^2 + 50x - 40$. Hint: Plot the function to determine good guess values.

†4-32 Write a computer program (or use an equation solver such as *Mathcad*, *Matlab*, or *TKSolver*) to find the roots of $y = -x^3 - 4x^2 + 80x - 40$. Hint: Plot the function to determine good guess values.

†4-33 Figure 4-18 (p.175) plots the cubic function from equation 4.34. Write a computer program (or use an equation solver such as *Mathcad*, *Matlab*, or *TKSolver* to solve the matrix equation) to investigate the behavior of the Newton-Raphson algorithm as the initial guess value is varied from $x = 1.8$ to 2.5 in steps of 0.1. Determine the guess value at which the convergence switches roots. Explain this root-switching phenomenon based on your observations from this exercise.

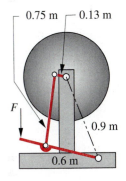

FIGURE P4-14

Problem 4-28

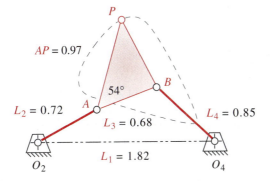

FIGURE P4-15

Problems 4-29 to 4-30

† These problems are suited to solution using *Mathcad*, *Matlab*, or *TKSolver* equation solver programs. In most cases, your solution can be checked with program FOURBAR, SLIDER, or SIXBAR.

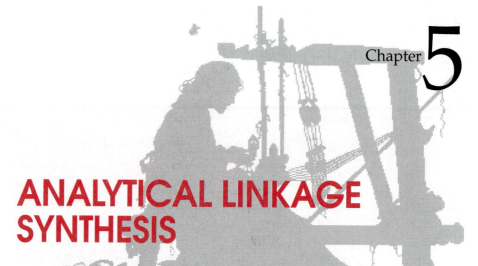

Chapter **5**

ANALYTICAL LINKAGE SYNTHESIS

Imagination is more important than knowledge
ALBERT EINSTEIN

5.0   INTRODUCTION

With the fundamentals of position analysis established, we can now use these techniques to **synthesize linkages** for specified output positions **analytically**. The synthesis techniques presented in Chapter 3 were strictly graphical and somewhat intuitive. The **analytical synthesis** procedure is algebraic rather than graphical and is less intuitive. However, its algebraic nature makes it quite suitable for computerization. These analytical synthesis methods were originated by Sandor[1] and further developed by his students, Erdman,[2] Kaufman,[3] and Loerch et al.[4,5]

5.1   TYPES OF KINEMATIC SYNTHESIS

Erdman and Sandor[6] define three types of kinematic synthesis, **function**, **path**, and **motion generation**, which were discussed in Section 3.2. Brief definitions are repeated here for your convenience.

   FUNCTION GENERATION   is defined as *the correlation of an **input function** with an **output function** in a mechanism*. Typically, a double-rocker or crank-rocker is the result, with pure rotation input and pure rotation output. A slider-crank linkage can be a function generator as well, driven from either end, i.e., rotation in and translation out or vice versa.

   PATH GENERATION   is defined as *the control of a **point** in the plane such that it follows some prescribed path*. This is typically accomplished with a fourbar crank-rocker or double-rocker, wherein a point on the coupler traces the desired output path. No attempt is made in path generation to control the orientation of the link which contains the point of interest. The coupler curve is made to pass through a set of desired output points. However, it is common for the timing of the arrival of the coupler point at particular locations along the path to be defined. This case is called *path generation with prescribed timing* and is analogous to function generation in that a particular output function is specified.

MOTION GENERATION   is defined as *the control of a **line** in the plane such that it assumes some sequential set of prescribed positions.* Here orientation of the link containing the line is important. This is typically accomplished with a fourbar crank-rocker or double-rocker, wherein a point on the coupler traces the desired output path and the linkage also controls the angular orientation of the coupler link containing the output line of interest.

5.2   PRECISION POINTS

The *points, or positions, prescribed for successive locations of the output (coupler or rocker) link in the plane* are generally referred to as **precision points** or **precision positions**. The number of precision points which can be synthesized is limited by the number of equations available for solution. The fourbar linkage can be synthesized by closed-form methods for up to five precision points for motion or path generation with prescribed timing (coupler output) and up to seven points for function generation (rocker output). Synthesis for two or three precision points is relatively straightforward, and each of these cases can be reduced to a system of linear simultaneous equations easily solved on a calculator. The four or more position synthesis problems involve the solution of nonlinear, simultaneous equation systems, and so are more complicated to solve, requiring a computer.

Note that these analytical synthesis procedures provide a solution which will be able to "be at" the specified precision points, but no guarantee is provided regarding the linkage's behavior between those precision points. It is possible that the resulting linkage will be incapable of moving from one precision point to another due to the presence of a toggle position or other constraint. This situation is actually no different than that of the graphical synthesis cases in Chapter 3, wherein there was also the possibility of a toggle position between design points. In fact, these analytical synthesis methods are just an alternate way to solve the same multiposition synthesis problems. One should still build a simple cardboard model of the synthesized linkage to observe its behavior and check for the presence of problems, even if the synthesis was performed by an esoteric analytical method.

5.3   TWO-POSITION MOTION GENERATION BY ANALYTICAL SYNTHESIS

Figure 5-1 shows a fourbar linkage in one position with a coupler point located at a first precision position P_1. It also indicates a second precision position (point P_2) to be achieved by the rotation of the input rocker, link 2, through an as yet unspecified angle β_2. Note also that the angle of the coupler link 3 at each of the precision positions is defined by the angles of the position vectors $\mathbf{Z}_1$ and $\mathbf{Z}_2$. The angle ϕ corresponds to the

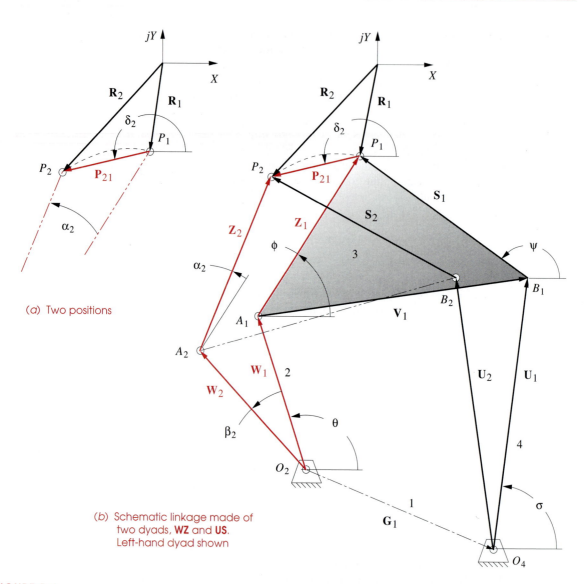

(a) Two positions

(b) Schematic linkage made of
two dyads, **WZ** and **US**.
Left-hand dyad shown

FIGURE 5-1

Two-position analytical synthesis

angle θ_3 of link 3 in its first position. This angle is unknown at the start of the synthesis and will be found. The angle α_2 represents the angular change of link 3 from position one to position two. This angle is defined in the problem statement.

It is important to realize that the linkage as shown in the figure is schematic. Its dimensions are unknown at the outset and are to be found by this synthesis technique. Thus, for example, the length of the position vector Z_1 as shown is not indicative of the final length of that edge of link 3, nor are the lengths (W, Z, U, V) or angles ($\theta, \phi, \sigma, \psi$) of any of the links as shown predictive of the final result.

The problem statement is:

Design a fourbar linkage which will move a line on its coupler link such that a point P on that line will be first at P_1 and later at P_2 and will also rotate the line through an angle α_2 between those two precision positions. Find the lengths and angles of the four links and the coupler link dimensions A_1P_1 and B_1P_1 as shown in Figure 5-1.

The two-position analytical motion synthesis procedure is as follows:

Define the two desired precision positions in the plane with respect to an arbitrarily chosen global coordinate system XY using position vectors $\mathbf{R}_1$ and $\mathbf{R}_2$ as shown in Figure 5-1a. The change in angle α_2 of vector $\mathbf{Z}$ is the rotation required of the coupler link. Note that the position difference vector $\mathbf{P}_{21}$ defines the displacement of the output motion of point P and is defined as:

$$\mathbf{P}_{21} = \mathbf{R}_2 - \mathbf{R}_1 \tag{5.1}$$

The dyad $\mathbf{W}_1\mathbf{Z}_1$ defines the left half of the linkage. The dyad $\mathbf{U}_1\mathbf{S}_1$ defines the right half of the linkage. Note that $\mathbf{Z}_1$ and $\mathbf{S}_1$ are both embedded in the rigid coupler (link 3), and both of these vectors will undergo the same rotation through angle α_2 from position 1 to position 2. The pin-to-pin length and angle of link 3 (vector $\mathbf{V}_1$) is defined in terms of vectors $\mathbf{Z}_1$ and $\mathbf{S}_1$.

$$\mathbf{V}_1 = \mathbf{Z}_1 - \mathbf{S}_1 \tag{5.2a}$$

The ground link 1 is also definable in terms of the two dyads.

$$\mathbf{G}_1 = \mathbf{W}_1 + \mathbf{V}_1 - \mathbf{U}_1 \tag{5.2b}$$

Thus if we can define the two dyads $\mathbf{W}_1$, $\mathbf{Z}_1$, and $\mathbf{U}_1$, $\mathbf{S}_1$, we will have defined a linkage that meets the problem specifications.

We will first solve for the left side of the linkage (vectors $\mathbf{W}_1$ and $\mathbf{Z}_1$) and later use the same procedure to solve for the right side (vectors $\mathbf{U}_1$ and $\mathbf{S}_1$). To solve for $\mathbf{W}_1$ and $\mathbf{Z}_1$ we need only write a vector loop equation around the loop which includes both positions P_1 and P_2 for the left-side dyad. We will go clockwise around the loop, starting with $\mathbf{W}_2$.

$$\mathbf{W}_2 + \mathbf{Z}_2 - \mathbf{P}_{21} - \mathbf{Z}_1 - \mathbf{W}_1 = 0 \tag{5.3}$$

Now substitute the complex number equivalents for the vectors.

$$we^{j(\theta+\beta_2)} + ze^{j(\phi+\alpha_2)} - p_{21}e^{j\delta_2} - ze^{j\phi} - we^{j\theta} = 0 \tag{5.4}$$

The sums of angles in the exponents can be rewritten as products of terms.

$$we^{j\theta}e^{j\beta_2} + ze^{j\phi}e^{j\alpha_2} - p_{21}e^{j\delta_2} - ze^{j\phi} - we^{j\theta} = 0 \tag{5.5a}$$

Simplifying and rearranging:

$$we^{j\theta}\left(e^{j\beta_2} - 1\right) + ze^{j\phi}\left(e^{j\alpha_2} - 1\right) = p_{21}e^{j\delta_2} \tag{5.5b}$$

Note that the lengths of vectors $\mathbf{W}_1$ and $\mathbf{W}_2$ are the same magnitude w because they represent the same rigid link in two different positions. The same can be said about vectors $\mathbf{Z}_1$ and $\mathbf{Z}_2$ whose common magnitude is z.

Equations 5.5 are vector equations, each of which contains two scalar equations and so can be solved for two unknowns. The two scalar equations can be revealed by substituting Euler's identity (equation 4.4a, p. 155) and separating the real and imaginary terms as was done in Section 4.5 (p. 152).

real part:

$$[w\cos\theta](\cos\beta_2 - 1) - [w\sin\theta]\sin\beta_2$$
$$+ [z\cos\phi](\cos\alpha_2 - 1) - [z\sin\phi]\sin\alpha_2 = p_{21}\cos\delta_2 \qquad (5.6a)$$

imaginary part (with complex operator j divided out):

$$[w\sin\theta](\cos\beta_2 - 1) + [w\cos\theta]\sin\beta_2$$
$$+ [z\sin\phi](\cos\alpha_2 - 1) + [z\cos\phi]\sin\alpha_2 = p_{21}\sin\delta_2 \qquad (5.6b)$$

There are eight variables in these two equations: w, θ, β_2, z, ϕ, α_2, p_{21}, and δ_2. We can only solve for two. Three of the eight are defined in the problem statement, namely α_2, p_{21}, and δ_2. Of the remaining five, w, θ, β_2, z, ϕ, we are forced to choose three as "free choices" (assumed values) in order to solve for the other two.

One strategy is to assume values for the three angles, θ, β_2, ϕ, on the premise that we may want to specify the orientation θ, ϕ of the two link vectors $\mathbf{W}_1$ and $\mathbf{Z}_1$ to suit packaging constraints, and also specify the angular excursion β_2 of link 2 to suit some driving constraint. This choice also has the advantage of leading to a set of equations which are linear in the unknowns and are thus easy to solve. For this solution, the equations can be simplified by setting the assumed and specified terms to be equal to some constants.

In equations 5.6a, let:

$$A = \cos\theta(\cos\beta_2 - 1) - \sin\theta\sin\beta_2$$
$$B = \cos\phi(\cos\alpha_2 - 1) - \sin\phi\sin\alpha_2 \qquad (5.7a)$$
$$C = p_{21}\cos\delta_2$$

and in equations 5.6b let:

$$D = \sin\theta(\cos\beta_2 - 1) + \cos\theta\sin\beta_2$$
$$E = \sin\phi(\cos\alpha_2 - 1) + \cos\phi\sin\alpha_2 \qquad (5.7b)$$
$$F = p_{21}\sin\delta_2$$

then:

$$Aw + Bz = C$$
$$Dw + Ez = F \qquad (5.7c)$$

and solving simultaneously,

$$w = \frac{CE - BF}{AE - BD}; \qquad z = \frac{AF - CD}{AE - BD} \qquad (5.7d)$$

A second strategy is to assume a length z and angle ϕ for vector $\mathbf{Z}_1$ and the angular excursion β_2 of link 2 and then solve for the vector $\mathbf{W}_1$. This is a commonly used approach. Note that the terms in square brackets in each of equations 5.6 are respectively the x and y components of the vectors $\mathbf{W}_1$ and $\mathbf{Z}_1$.

$$W_{1_x} = w\cos\theta; \qquad\qquad Z_{1_x} = z\cos\phi$$

$$W_{1_y} = w\sin\theta; \qquad\qquad Z_{1_y} = z\sin\phi \tag{5.8a}$$

Substituting in equation 5.6,

$$W_{1_x}(\cos\beta_2 - 1) - W_{1_y}\sin\beta_2$$
$$+ Z_{1_x}(\cos\alpha_2 - 1) - Z_{1_y}\sin\alpha_2 = p_{21}\cos\delta_2$$

$$W_{1_y}(\cos\beta_2 - 1) + W_{1_x}\sin\beta_2$$
$$+ Z_{1_y}(\cos\alpha_2 - 1) + Z_{1_x}\sin\alpha_2 = p_{21}\sin\delta_2 \tag{5.8b}$$

Z_{1x} and Z_{1y} are known from equation 5.8a with z and ϕ assumed as free choices. To further simplify the expression, combine other known terms as:

$$A = \cos\beta_2 - 1; \qquad B = \sin\beta_2; \qquad C = \cos\alpha_2 - 1$$

$$D = \sin\alpha_2; \qquad E = p_{21}\cos\delta_2; \qquad F = p_{21}\sin\delta_2 \tag{5.8c}$$

substituting,

$$AW_{1_x} - BW_{1_y} + CZ_{1_x} - DZ_{1_y} = E$$

$$AW_{1_y} + BW_{1_x} + CZ_{1_y} + DZ_{1_x} = F \tag{5.8d}$$

and the solution is:

$$W_{1_x} = \frac{A\left(-CZ_{1_x} + DZ_{1_y} + E\right) + B\left(-CZ_{1_y} - DZ_{1_x} + F\right)}{-2A}$$

$$W_{1_y} = \frac{A\left(-CZ_{1_y} - DZ_{1_x} + F\right) + B\left(CZ_{1_x} - DZ_{1_y} - E\right)}{-2A} \tag{5.8e}$$

Either of these strategies results in the definition of a left dyad $\mathbf{W}_1\mathbf{Z}_1$ and its pivot locations which will provide the motion generation specified.

We must repeat the process for the right-hand dyad, $\mathbf{U}_1\mathbf{S}_1$. Figure 5-2 highlights the two positions $\mathbf{U}_1\mathbf{S}_1$ and $\mathbf{U}_2\mathbf{S}_2$ of the right dyad. Vector $\mathbf{U}_1$ is initially at angle σ and moves through angle γ_2 from position 1 to 2. Vector $\mathbf{S}_1$ is initially at angle ψ. Note that the rotation of vector $\mathbf{S}$ from $\mathbf{S}_1$ to $\mathbf{S}_2$ is through the same angle α_2 as vector $\mathbf{Z}$, since they are in the same link. A vector loop equation similar to equation 5.3 can be written for this dyad.

$$\mathbf{U}_2 + \mathbf{S}_2 - \mathbf{P}_{21} - \mathbf{S}_1 - \mathbf{U}_1 = 0 \tag{5.9a}$$

Rewrite in complex variable form and collect terms.

$$ue^{j\sigma}\left(e^{j\gamma_2}-1\right)+se^{j\psi}\left(e^{j\alpha_2}-1\right)=p_{21}e^{j\delta_2} \tag{5.9b}$$

When this is expanded and the proper angles substituted, the x and y component equations become:

real part:

$$u\cos\sigma(\cos\gamma_2-1)-u\sin\sigma\sin\gamma_2$$
$$+\,s\cos\psi(\cos\alpha_2-1)-s\sin\psi\sin\alpha_2=p_{21}\cos\delta_2 \tag{5.10a}$$

imaginary part (with complex operator j divided out):

$$u\sin\sigma(\cos\gamma_2-1)+u\cos\sigma\sin\gamma_2$$
$$+\,s\sin\psi(\cos\alpha_2-1)+s\cos\psi\sin\alpha_2=p_{21}\sin\delta_2 \tag{5.10b}$$

Compare equations 5.10 to equations 5.6.

The same first strategy can be applied to equations 5.10 as was used for equations 5.6 to solve for the magnitudes of vectors $\mathbf{U}$ and $\mathbf{S}$, assuming values for angles σ, ψ, and γ_2. The quantities p_{21}, δ_2, and α_2 are defined from the problem statement as before.

In equations 5.10a let:

$$A=\cos\sigma(\cos\gamma_2-1)-\sin\sigma\sin\gamma_2$$
$$B=\cos\psi(\cos\alpha_2-1)-\sin\psi\sin\alpha_2 \tag{5.11a}$$
$$C=p_{21}\cos\delta_2$$

and in equations 5.10b let:

$$D=\sin\sigma(\cos\gamma_2-1)+\cos\sigma\sin\gamma_2$$
$$E=\sin\psi(\cos\alpha_2-1)+\cos\psi\sin\alpha_2 \tag{5.11b}$$
$$F=p_{21}\sin\delta_2$$

then:

$$Au+Bs=C$$
$$Du+Es=F \tag{5.11c}$$

and solving simultaneously,

$$u=\frac{CE-BF}{AE-BD}; \qquad\qquad s=\frac{AF-CD}{AE-BD} \tag{5.11d}$$

If the second strategy is used, assuming angle γ_2 and the magnitude and direction of vector $\mathbf{S}_1$ (which will define link 3), the result will be:

$$U_{1_x}=u\cos\sigma; \qquad\qquad S_{1_x}=s\cos\psi$$
$$U_{1_y}=u\sin\sigma; \qquad\qquad S_{1_y}=s\sin\psi \tag{5.12a}$$

Substitute in equation 5.10:

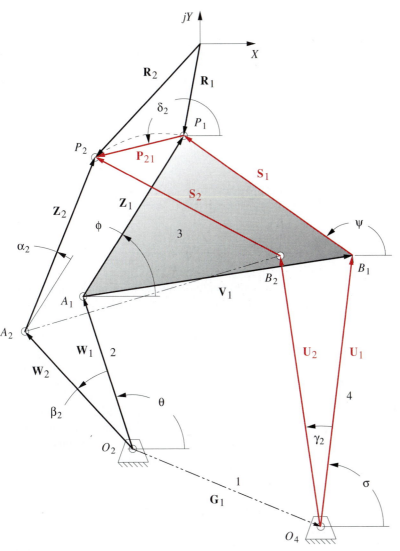

FIGURE 5-2

Right-side dyad shown in two positions

$$U_{1_x}(\cos\gamma_2 - 1) - U_{1_y}\sin\gamma_2$$
$$+ S_{1_x}(\cos\alpha_2 - 1) - S_{1_y}\sin\alpha_2 = p_{21}\cos\delta_2$$

$$(5.12b)$$

$$U_{1_y}(\cos\gamma_2 - 1) + U_{1_x}\sin\gamma_2$$
$$+ S_{1_y}(\cos\alpha_2 - 1) + S_{1_x}\sin\alpha_2 = p_{21}\sin\delta_2$$

Let : $A = \cos\gamma_2 - 1;$ $B = \sin\gamma_2;$ $C = \cos\alpha_2 - 1$

$D = \sin\alpha_2;$ $E = p_{21}\cos\delta_2;$ $F = p_{21}\sin\delta_2$ (5.12c)

Substitute in equation 5.12b,

$$AU_{1_x} - BU_{1_y} + CS_{1_x} - DS_{1_y} = E$$

$$AU_{1_y} + BU_{1_x} + CS_{1_y} + DS_{1_x} = F$$ (5.12d)

and the solution is:

$$U_{1_x} = \frac{A\left(-CS_{1_x} + DS_{1_y} + E\right) + B\left(-CS_{1_y} - DS_{1_x} + F\right)}{-2A}$$

(5.12e)

$$U_{1_y} = \frac{A\left(-CS_{1_y} - DS_{1_x} + F\right) + B\left(CS_{1_x} - DS_{1_y} - E\right)}{-2A}$$

Note that there are infinities of possible solutions to this problem because we may choose any set of values for the three free choices of variables in this two-position case. Technically there is an infinity of solutions for each free choice. Three choices then give infinity cubed solutions! But since infinity is defined as a number larger than the largest number you can think of, infinity cubed is not any more impressively large than just plain infinity. While not strictly correct mathematically, we will, for simplicity, refer to all of these cases as having "an infinity of solutions," regardless of the power to which infinity may be raised as a result of the derivation. There are plenty of solutions to pick from, at any rate. *Unfortunately, not all will work.* Some will have circuit, branch, or order (CBO) defects such as toggle positions between the precision points. Others will have poor transmission angles or poor pivot locations or overlarge links. Design judgment is still most important in selecting the assumed values for your free choices. Despite their name you must pay for those "free choices" later. Make a model!

5.4   COMPARISON OF ANALYTICAL AND GRAPHICAL TWO-POSITION SYNTHESIS

Note that in the **graphical solution** to this two-position synthesis problem (in Example 3-3 and Figure 3-6, p. 87), we also had to make *three free choices* to solve the problem. The identical two-position synthesis problem from Figure 3-6 is reproduced in Figure 5-3. The approach taken in Example 3-3 used the two points A and B as the attachments for the moving pivots. Figure 5-3a shows the graphical construction used to find the fixed pivots O_2 and O_4. For the analytical solution we will use those points A and B as the joints of the two dyads **WZ** and **US**. These dyads meet at point P, which is the precision point. The relative position vector $\mathbf{P}_{21}$ defines the displacement of the precision point.

Note that in the graphical solution, we implicitly defined the left dyad vector **Z** by locating attachment points A and B on link 3 as shown in Figure 5-3a. This defined the two variables, z and ϕ. We also implicitly chose the value of w by selecting an arbitrary location for pivot O_2 on the perpendicular bisector. When that third choice was made, the remaining two unknowns, angles β_2 and θ, were solved for graphically at the same time, because the geometric construction was in fact a graphical "computation" for the solution of the two simultaneous equations 5.8.

The graphical and analytical methods represent two alternate solutions to the same problem. All of these problems can be solved both analytically and graphically. One method can provide a good check for the other. We will now solve this problem analytically and correlate the results with the graphical solution from Chapter 3.

✍️ EXAMPLE 5-1

Two-Position Analytical Motion Synthesis.

Problem: Design a fourbar linkage to move the link APB shown from position $A_1P_1B_1$ to $A_2P_2.B_2$.

Solution: (see Figure 5-3)

1   Draw the link APB in its two desired positions, $A_1P_1B_1$ and $A_2P_2B_2$, to scale in the plane as shown.

2   Measure or calculate the values of the magnitude and angle of vector $\mathbf{P}_{21}$, namely, p_{21} and δ_2. In this example they are:

$$p_{21} = 2.416; \qquad \delta_2 = 165.2°$$

3   Measure or calculate the value of the change in angle, α_2, of vector $\mathbf{Z}$ from position 1 to position 2. In this example it is:

$$\alpha_2 = 43.3°$$

4   The three values in steps 2 and 3 are the only ones defined in the problem statement. We must assume three additional "free choices" to solve the problem. Method two (see equations 5.8) chooses the length z and angle ϕ of vector $\mathbf{Z}$ and β_2, the change in angle of vector $\mathbf{W}$. In order to obtain the same solution as the graphical method produced in Figure 5-3a (from the infinities of solutions available), we will choose those values consistent with the graphical solution.

$$z = 1.298; \qquad \phi = 26.5°; \qquad \beta_2 = 38.4°$$

5   Substitute these six values in equations 5.8 and obtain:

$$w = 2.467 \qquad \theta = 71.6°$$

6   Compare these to the graphical solution;

$$w = 2.48 \qquad \theta = 71°$$

which is a reasonable match given the graphical accuracy. This vector $\mathbf{W}_1$ is link 2 of the fourbar.

7   Repeat the procedure for the link-4 side of the linkage. The free choices will now be:

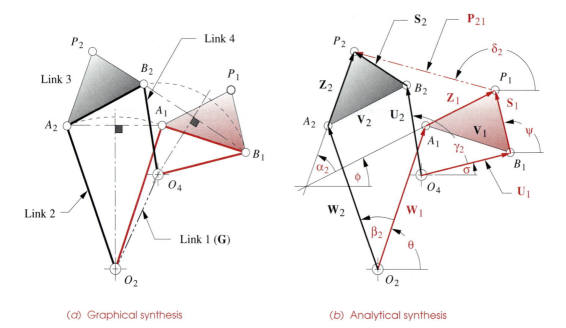

(a) Graphical synthesis   (b) Analytical synthesis

FIGURE 5-3

Two-position motion synthesis with coupler output

$$s = 1.035; \qquad \psi = 104.1°; \qquad \gamma_2 = 85.6°$$

8   Substitute these three values along with the original three values from steps 2 and 3 in equations 5.8 and obtain:

$$u = 1.486 \qquad \sigma = 15.4°$$

9   Compare these to the graphical solution:

$$u = 1.53 \qquad \sigma = 14°$$

These are a reasonable match for graphical accuracy. Vector U_1 is link 4 of the fourbar.

10 Line A_1B_1 is link 3 and can be found from equation 5.2a. Line O_2O_4 is link 1 and can be found from equation 5.2b.

11 Check the Grashof condition, and repeat steps 4 to 7 if unsatisfied. Note that any Grashof condition is potentially acceptable in this case.

12 Construct a cardboard model and check its function to be sure it can get from initial to final position without encountering any limit (toggle) positions.

13 Check transmission angles.

Input the file E05-01.4br to program FOURBAR to see Example 5-1.

5.5 SIMULTANEOUS EQUATION SOLUTION

These methods of analytical synthesis lead to sets of linear simultaneous equations. The two-position synthesis problem results in two simultaneous equations which can be solved by direct substitution. The three-position synthesis problem will lead to a system of four simultaneous linear equations and will require a more complicated method of solution. A convenient approach to the solution of sets of linear simultaneous equations is to put them in a standard matrix form and use a numerical matrix solver to obtain the answers. Matrix solvers are built into most engineering and scientific pocket calculators. Some spreadsheet packages and equation solvers will also do a matrix solution.

As an example of this general approach, consider the following set of simultaneous equations:

$$
\begin{aligned}
-2x_1 \ -x_2 \ +x_3 &= -1 \\
x_1 \ +x_2 \ +x_3 &= 6 \\
3x_1 \ +x_2 \ -x_3 &= 2
\end{aligned}
\tag{5.13a}
$$

A system this small can be solved longhand by the elimination method, but we will put it in matrix form to show the general approach which will work regardless of the number of equations. The equations 5.13a can be written as the product of two matrices set equal to a third matrix.

$$
\begin{bmatrix} -2 & -1 & 1 \\ 1 & 1 & 1 \\ 3 & 1 & -1 \end{bmatrix} \times \begin{bmatrix} x_1 \\ x_2 \\ x_3 \end{bmatrix} = \begin{bmatrix} -1 \\ 6 \\ 2 \end{bmatrix}
\tag{5.13b}
$$

We will refer to these matrices as $\mathbf{A}$, $\mathbf{B}$, and $\mathbf{C}$,

$$
[\mathbf{A}] \quad \times \quad [\mathbf{B}] \quad = \quad [\mathbf{C}]
\tag{5.13c}
$$

where $\mathbf{A}$ is the matrix of coefficients of the unknowns, $\mathbf{B}$ is a column vector of the unknown terms and $\mathbf{C}$ is a column vector of the constant terms. When matrix $\mathbf{A}$ is multiplied by $\mathbf{B}$, the result will be the same as the left sides of equation 5.13a. See any text on linear algebra such as reference 7 for a discussion of the procedure for matrix multiplication.

If equation 5.13c were a scalar equation,

$$
ab = c
\tag{5.14a}
$$

rather than a vector (matrix) equation, it would be very easy to solve it for the unknown b when a and c are known. We would simply divide c by a to find b.

$$
b = \frac{c}{a}
\tag{5.14b}
$$

Unfortunately, division is not defined for matrices, so another approach must be used. Note that we could also express the division in equation 5.14b as:

$$
b = a^{-1}c
\tag{5.14c}
$$

If the equations to be solved are linearly independent, then we can find the inverse of matrix **A** and multiply it by matrix **C** to find **B**. The inverse of a matrix is defined as that matrix which when multiplied by the original matrix yields the identity matrix. The **identity matrix** is a square matrix with ones on the main diagonal and zeros everywhere else. The inverse of a matrix is denoted by adding a superscript of negative one to the symbol for the original matrix.

$$[A]^{-1} \times [A] = [I] = \begin{bmatrix} 1 & 0 & 0 \\ 0 & 1 & 0 \\ 0 & 0 & 1 \end{bmatrix} \qquad (5.15)$$

Not all matrices will possess an inverse. The determinant of the matrix must be nonzero for an inverse to exist. The class of problems dealt with here will yield matrices which have inverses provided that all data are correctly calculated for input to the matrix and represent a real physical system. The calculation of the terms of the inverse for a matrix is a complicated numerical process which requires a computer or preprogrammed pocket calculator to invert any matrix of significant size. A Gauss-Jordan-elimination numerical method is usually used to find an inverse. For our simple example in equation 5.13 the inverse of matrix **A** is found to be:

$$\begin{bmatrix} -2 & -1 & 1 \\ 1 & 1 & 1 \\ 3 & 1 & -1 \end{bmatrix}^{-1} = \begin{bmatrix} 1.0 & 0.0 & 1.0 \\ -2.0 & 0.5 & -1.5 \\ 1.0 & 0.5 & 0.5 \end{bmatrix} \qquad (5.16)$$

If the inverse of matrix **A** can be found, we can solve equation 5.13 for the unknowns **B** by multiplying both sides of the equation by the inverse of **A**. Note that unlike scalar multiplication, matrix multiplication is not commutative; i.e., **A** x **B** is not equal to **B** x **A**. We will premultiply each side of the equation by the inverse.

$$[A]^{-1} \times [A] \times [B] = [A]^{-1} \times [C]$$

but:

$$[A]^{-1} \times [A] = [I] \qquad (5.17)$$

so:

$$[B] = [A]^{-1} \times [C]$$

The product of **A** and its inverse on the left side of the equation is equal to the identity matrix **I**. Multiplying by the identity matrix is equivalent, in scalar terms, to multiplying by one, so it has no effect on the result. Thus the unknowns can be found by premultiplying the inverse of the coefficient matrix **A** times the matrix of constant terms **C**.

This method of solution works no matter how many equations are present as long as the inverse of **A** can be found and enough computer memory and/or time is available to do the computation. Note that it is not actually necessary to find the inverse of matrix **A** to solve the set of equations. The Gauss-Jordan algorithm which finds the inverse can also be used to directly solve for the unknowns **B** by assembling the **A** and **C** matrices into an **augmented matrix** of n rows and $n + 1$ columns. The added column is the **C** vector. This approach requires fewer calculations, so it is faster and more accurate. The augmented matrix for this example is:

$$\begin{bmatrix} -2 & -1 & 1 & \vdots & -1 \\ 1 & 1 & 1 & \vdots & 6 \\ 3 & 1 & -1 & \vdots & 2 \end{bmatrix} \qquad (5.18a)$$

The Gauss-Jordan algorithm manipulates this augmented matrix until it is in the form shown below, in which the left, square portion has been reduced to the identity matrix and the rightmost column contains the values of the column vector of unknowns. In this case the results are $x_1 = 1$, $x_2 = 2$, and $x_3 = 3$ which are the correct solution to the original equations 5.13.

$$\begin{bmatrix} 1 & 0 & 0 & \vdots & 1 \\ 0 & 1 & 0 & \vdots & 2 \\ 0 & 0 & 1 & \vdots & 3 \end{bmatrix} \qquad (5.18b)$$

The program MATRIX, supplied with this text, solves these problems with this Gauss-Jordan elimination method and operates on the augmented matrix without actually finding the inverse of $\mathbf{A}$ in explicit form. See Appendix A for instructions on running program MATRIX. For a review of matrix algebra see reference 7.

5.6 THREE-POSITION MOTION GENERATION BY ANALYTICAL SYNTHESIS

The same approach of defining two dyads, one at each end of the fourbar linkage, as used for two-position motion synthesis can be extended to three, four, and five positions in the plane. The three-position motion synthesis problem will now be addressed. Figure 5-4 shows a fourbar linkage in one general position with a coupler point located at its first precision position P_1. Second and third precision positions (points P_2 and P_3) are also shown. These are to be achieved by the rotation of the input rocker, link 2, through as yet unspecified angles β_2 and β_3. Note also that the angles of the coupler link 3 at each of the precision positions are defined by the angles of the position vectors $\mathbf{Z}_1$, $\mathbf{Z}_2$, and $\mathbf{Z}_3$. The linkage shown in the figure is schematic. Its dimensions are unknown at the outset and are to be found by this synthesis technique. Thus, for example, the length of the position vector $\mathbf{Z}_1$ as shown is not indicative of the final length of that edge of link 3 nor are the lengths or angles of any of the links shown predictive of the final result.

The problem statement is:

Design a fourbar linkage which will move a line on its coupler link such that a point P on that line will be first at P_1, later at P_2, and still later at P_3, and also will rotate the line through an angle α_2 between the first two precision positions and through an angle α_3 between the first and third precision positions. Find the lengths and angles of the four links and the coupler link dimensions A_1P_1 and B_1P_1 as shown in Figure 5-4.

The three-position analytical motion synthesis procedure is as follows:

For convenience, we will place the global coordinate system XY at the first precision point P_1. We define the other two desired precision positions in the plane with respect to this global system as shown in Figure 5-4. The position difference vectors $\mathbf{P}_{21}$, drawn from P_1 to P_2, and $\mathbf{P}_{31}$, drawn from P_1 to P_3, have angles δ_2 and δ_3, respectively.

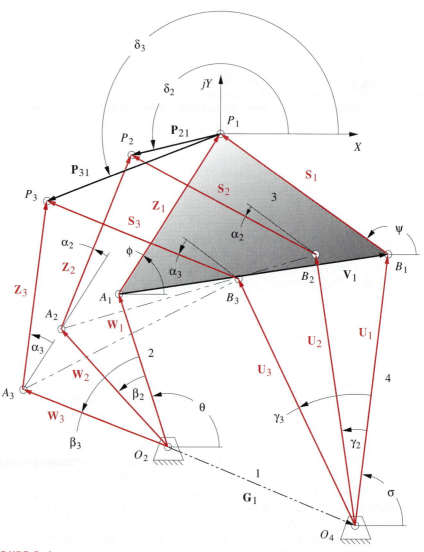

FIGURE 5-4

Three-position analytical synthesis

The position difference vectors $\mathbf{P}_{21}$ and $\mathbf{P}_{31}$ define the displacements of the output motion of point P from point 1 to 2 and from 1 to 3, respectively.

The dyad $\mathbf{W}_1\,\mathbf{Z}_1$ defines the left half of the linkage. The dyad $\mathbf{U}_1\,\mathbf{S}_1$ defines the right half of the linkage. Vectors $\mathbf{Z}_1$ and $\mathbf{S}_1$ are both embedded in the rigid coupler (link 3), and both will undergo the same rotations, through angle α_2 from position 1 to position 2 and through angle α_3 from position 1 to position 3. The pin-to-pin length and angle of link 3 (vector $\mathbf{V}_1$) is defined in terms of vectors $\mathbf{Z}_1$ and $\mathbf{S}_1$ as in equation 5.2a (p. 191). The ground link is defined by equation 5.2b as before.

As we did in the two-position case, we will first solve for the left side of the linkage (vectors $\mathbf{W}_1$ and $\mathbf{Z}_1$) and later use the same procedure to solve for the right side (vectors $\mathbf{U}_1$ and $\mathbf{S}_1$). To solve for $\mathbf{W}_1$ and $\mathbf{Z}_1$ we need to now write **two vector loop equations**, one around the loop which includes positions P_1 and P_2 and the second one around the loop which includes positions P_1 and P_3 (see Figure 5-4). We will go clockwise around the first loop for motion from position 1 to 2, starting with $\mathbf{W}_2$, and then write the second loop equation for motion from position 1 to 3 starting with $\mathbf{W}_3$.

$$\mathbf{W}_2 + \mathbf{Z}_2 - \mathbf{P}_{21} - \mathbf{Z}_1 - \mathbf{W}_1 = 0$$
$$\mathbf{W}_3 + \mathbf{Z}_3 - \mathbf{P}_{31} - \mathbf{Z}_1 - \mathbf{W}_1 = 0$$

(5.19)

Substituting the complex number equivalents for the vectors.

$$we^{j(\theta+\beta_2)} + ze^{j(\phi+\alpha_2)} - p_{21}e^{j\delta_2} - ze^{j\phi} - we^{j\theta} = 0$$
$$we^{j(\theta+\beta_3)} + ze^{j(\phi+\alpha_3)} - p_{31}e^{j\delta_3} - ze^{j\phi} - we^{j\theta} = 0$$

(5.20)

Rewriting the sums of angles in the exponents as products of terms.

$$we^{j\theta}e^{j\beta_2} + ze^{j\phi}e^{j\alpha_2} - p_{21}e^{j\delta_2} - ze^{j\phi} - we^{j\theta} = 0$$
$$we^{j\theta}e^{j\beta_3} + ze^{j\phi}e^{j\alpha_3} - p_{31}e^{j\delta_3} - ze^{j\phi} - we^{j\theta} = 0$$

(5.21a)

Simplifying and rearranging:

$$we^{j\theta}\left(e^{j\beta_2} - 1\right) + ze^{j\phi}\left(e^{j\alpha_2} - 1\right) = p_{21}e^{j\delta_2}$$
$$we^{j\theta}\left(e^{j\beta_3} - 1\right) + ze^{j\phi}\left(e^{j\alpha_3} - 1\right) = p_{31}e^{j\delta_3}$$

(5.21b)

The magnitude w of vectors $\mathbf{W}_1$, $\mathbf{W}_2$, and $\mathbf{W}_3$ is the same in all three positions because it represents the same line in a rigid link. The same can be said about vectors $\mathbf{Z}_1$, $\mathbf{Z}_2$, and $\mathbf{Z}_3$ whose common magnitude is z.

Equations 5.21 are a set of two vector equations, each of which contains two scalar equations. This set of four equations can be solved for four unknowns. The scalar equations can be revealed by substituting Euler's identity (equation 4.4a, p. 155) and separating the real and imaginary terms as was done in the two-position example above.

real part:

$$w\cos\theta(\cos\beta_2 - 1) - w\sin\theta\sin\beta_2$$
$$+z\cos\phi(\cos\alpha_2 - 1) - z\sin\phi\sin\alpha_2 = p_{21}\cos\delta_2$$

(5.22a)

$$w\cos\theta(\cos\beta_3 - 1) - w\sin\theta\sin\beta_3$$
$$+z\cos\phi(\cos\alpha_3 - 1) - z\sin\phi\sin\alpha_3 = p_{31}\cos\delta_3$$

(5.22b)

imaginary part (with complex operator j divided out):

$$w\sin\theta(\cos\beta_2 - 1) + w\cos\theta\sin\beta_2$$

$$+ z\sin\phi(\cos\alpha_2 - 1) + z\cos\phi\sin\alpha_2 = p_{21}\sin\delta_2 \qquad (5.22c)$$

$$w\sin\theta(\cos\beta_3 - 1) + w\cos\theta\sin\beta_3$$

$$+ z\sin\phi(\cos\alpha_3 - 1) + z\cos\phi\sin\alpha_3 = p_{31}\sin\delta_3 \qquad (5.22d)$$

There are **twelve variables** in these four equations 5.22: w, θ, β_2, β_3, z, ϕ, α_2, α_3, p_{21}, p_{31}, δ_2, and δ_3. **We can solve for only four.** Six of them are defined in the problem statement, namely α_2, α_3, p_{21}, p_{31}, δ_2, and δ_3. Of the remaining six, w, θ, β_2, β_3, z, ϕ, **we must choose two as free choices** (assumed values) in order to solve for the other four. One strategy is to assume values for the two angles, β_2 and β_3, on the premise that we may want to specify the angular excursions of link 2 to suit some driving constraint. (This choice also has the benefit of leading to a set of linear equations for simultaneous solution.)

This leaves the magnitudes and angles of vectors **W** and **Z** to be found (w, θ, z, ϕ). To simplify the solution, we can substitute the following relationships to obtain the x and y components of the two unknown vectors **W** and **Z**, rather than their polar coordinates.

$$W_{1_x} = w\cos\theta; \qquad\qquad Z_{1_x} = z\cos\phi$$
$$\qquad\qquad\qquad\qquad\qquad\qquad\qquad\qquad\qquad (5.23)$$
$$W_{1_y} = w\sin\theta; \qquad\qquad Z_{1_y} = z\sin\phi$$

Substituting equations 5.23 into 5.22 we obtain:

$$W_{1_x}(\cos\beta_2 - 1) - W_{1_y}\sin\beta_2$$

$$+ Z_{1_x}(\cos\alpha_2 - 1) - Z_{1_y}\sin\alpha_2 = p_{21}\cos\delta_2 \qquad (5.24a)$$

$$W_{1_x}(\cos\beta_3 - 1) - W_{1_y}\sin\beta_3$$

$$+ Z_{1_x}(\cos\alpha_3 - 1) - Z_{1_y}\sin\alpha_3 = p_{31}\cos\delta_3 \qquad (5.24b)$$

$$W_{1_y}(\cos\beta_2 - 1) + W_{1_x}\sin\beta_2$$

$$+ Z_{1_y}(\cos\alpha_2 - 1) + Z_{1_x}\sin\alpha_2 = p_{21}\sin\delta_2 \qquad (5.24c)$$

$$W_{1_y}(\cos\beta_3 - 1) + W_{1_x}\sin\beta_3$$

$$+ Z_{1_y}(\cos\alpha_3 - 1) + Z_{1_x}\sin\alpha_3 = p_{31}\sin\delta_3 \qquad (5.24d)$$

These are four equations in the four unknowns W_{1x}, W_{1y}, Z_{1x}, and Z_{1y}. By setting the coefficients which contain the assumed and specified terms equal to some constants, we can simplify the notation and obtain the following solutions.

$$
\begin{array}{lll}
A = \cos\beta_2 - 1; & B = \sin\beta_2; & C = \cos\alpha_2 - 1 \\
D = \sin\alpha_2; & E = p_{21}\cos\delta_2; & F = \cos\beta_3 - 1 \\
G = \sin\beta_3; & H = \cos\alpha_3 - 1; & K = \sin\alpha_3 \\
L = p_{31}\cos\delta_3; & M = p_{21}\sin\delta_2; & N = p_{31}\sin\delta_3
\end{array}
\qquad (5.25)
$$

Substituting equations 5.25 in 5.24 to simplify:

$$AW_{1_x} - BW_{1_y} + CZ_{1_x} - DZ_{1_y} = E \qquad (5.26a)$$

$$FW_{1_x} - GW_{1_y} + HZ_{1_x} - KZ_{1_y} = L \qquad (5.26b)$$

$$BW_{1_x} + AW_{1_y} + DZ_{1_x} + CZ_{1_y} = M \qquad (5.26c)$$

$$GW_{1_x} + FW_{1_y} + KZ_{1_x} + HZ_{1_y} = N \qquad (5.26d)$$

This system can be put into standard matrix form:

$$\begin{bmatrix} A & -B & C & -D \\ F & -G & H & -K \\ B & A & D & C \\ G & F & K & H \end{bmatrix} \times \begin{bmatrix} W_{1_x} \\ W_{1_y} \\ Z_{1_x} \\ Z_{1_y} \end{bmatrix} = \begin{bmatrix} E \\ L \\ M \\ N \end{bmatrix} \qquad (5.27)$$

This is of the general form of equation 5.13c. The vector of unknowns **B** can be solved for by premultiplying the inverse of the coefficient matrix **A** by the constant vector **C** or by forming the augmented matrix as in equation 5.18. For any numerical problem, the inverse of a 4 x 4 matrix can be found with many pocket calculators. The computer program MATRIX, supplied with this text, will also solve the augmented matrix equation.

Equations 5.25 and 5.26 solve the three-position synthesis problem for the left-hand side of the linkage using any pair of assumed values for β_2 and β_3. We must repeat the above process for the right-hand side of the linkage to find vectors **U** and **S**. Figure 5-4 (p. 202) also shows the three positions of the **US** dyad, and the angles σ, γ_2, γ_3, ψ, α_2, and α_3, which define those vector rotations for all three positions. The solution derivation for the right-side dyad, **US**, is identical to that just done for the left dyad **WZ**. The angles and vector labels are the only difference. The vector loop equations are:

$$\mathbf{U}_2 + \mathbf{S}_2 - \mathbf{P}_{21} - \mathbf{S}_1 - \mathbf{U}_1 = 0$$
$$\mathbf{U}_3 + \mathbf{S}_3 - \mathbf{P}_{31} - \mathbf{S}_1 - \mathbf{U}_1 = 0 \qquad (5.28)$$

Substituting, simplifying, and rearranging,

$$ue^{j\sigma}\left(e^{j\gamma_2} - 1\right) + se^{j\psi}\left(e^{j\alpha_2} - 1\right) = p_{21}e^{j\delta_2}$$
$$ue^{j\sigma}\left(e^{j\gamma_3} - 1\right) + se^{j\psi}\left(e^{j\alpha_3} - 1\right) = p_{31}e^{j\delta_3} \qquad (5.29)$$

The solution requires that two free choices be made. We will assume values for the angles γ_2 and γ_3. Note that α_2 and α_3 are the same as for dyad **WZ**. We will, in effect, solve for angles σ and ψ by finding the x and y components of the vectors **U** and **S**. The solution is:

$$A = \cos\gamma_2 - 1; \qquad B = \sin\gamma_2; \qquad C = \cos\alpha_2 - 1$$
$$D = \sin\alpha_2; \qquad E = p_{21}\cos\delta_2; \qquad F = \cos\gamma_3 - 1$$
$$G = \sin\gamma_3; \qquad H = \cos\alpha_3 - 1; \qquad K = \sin\alpha_3$$
$$L = p_{31}\cos\delta_3; \qquad M = p_{21}\sin\delta_2; \qquad N = p_{31}\sin\delta_3$$

(5.30)

$$AU_{1_x} - BU_{1_y} + CS_{1_x} - DS_{1_y} = E \tag{5.31a}$$

$$FU_{1_x} - GU_{1_y} + HS_{1_x} - KS_{1_y} = L \tag{5.31b}$$

$$BU_{1_x} + AU_{1_y} + DS_{1_x} + CS_{1_y} = M \tag{5.31c}$$

$$GU_{1_x} + FU_{1_y} + KS_{1_x} + HS_{1_y} = N \tag{5.31d}$$

Equations 5.31 can be solved using the approach of equations 5.27 and 5.18, by changing W to U and Z to S and using the definitions of the constants given in equation 5.30 in equation 5.27.

It should be apparent that there are infinities of solutions to this three-position synthesis problem as well. An inappropriate selection of the two free choices could lead to a solution which has circuit, branch, or order problems in moving among all specified positions. Thus we must check the function of the solution synthesized by this or any other method. A simple model is the quickest check.

5.7   COMPARISON OF ANALYTICAL AND GRAPHICAL THREE-POSITION SYNTHESIS

Figure 5-5 shows the same three-position synthesis problem as was done graphically in Example 3-6 in Section 3.4. Compare this figure to Figure 3-9. The labeling has been changed to be consistent with the notation in this chapter. The points P_1, P_2, and P_3 correspond to the three points labeled D in the earlier figure. Points A_1, A_2, and A_3 correspond to points E; points B_1, B_2, and B_3 correspond to points F. The old line AP becomes the present $\mathbf{Z}$ vector. Point P is the coupler point which will go through the specified precision points, P_1, P_2, and P_3. Points A and B are the attachment points for the rockers (links 2 and 4, respectively) on the coupler (link 3). We wish to solve for the coordinates of vectors $\mathbf{W}$, $\mathbf{Z}$, $\mathbf{U}$, and $\mathbf{S}$, which define not only the lengths of those links but also the locations of the fixed pivots O_2 and O_4 in the plane and the lengths of links 3 and 1. Link 1 is defined as vector $\mathbf{G}$ in Figure 5-4 (p. 202) and can be found from equation 5.2b (p. 191). Link 3 is vector $\mathbf{V}$ found from equation 5.2a.

Four free choices must be made to constrain the problem to a particular solution out of the infinities of solutions available. In this case the values of link angles β_2, β_3, γ_2, and γ_3 have been chosen to be the same values as those which were found in the graphical solution to Example 3-6 in order to obtain the same solution as a check and comparison. Recall that in doing the graphical three-position synthesis solution to this same problem we in fact also had to make four free choices. These were the x,y coordinates of the moving pivot locations E and F in Figure 3-9 which correspond in concept to our four free choices of link angles here.

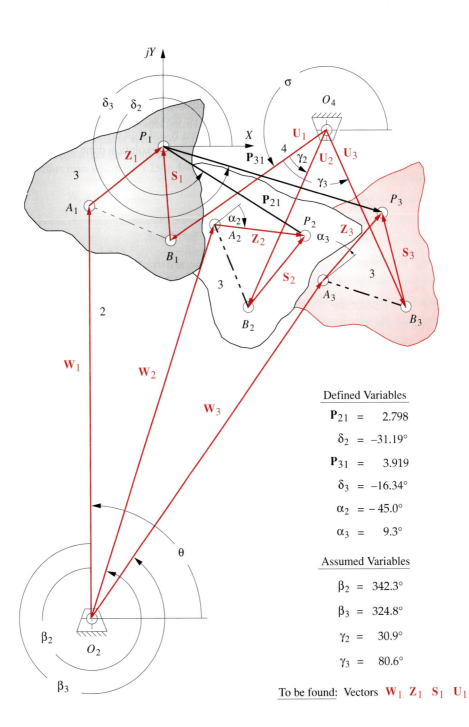

Defined Variables

P_{21} = 2.798

δ_2 = −31.19°

P_{31} = 3.919

δ_3 = −16.34°

α_2 = − 45.0°

α_3 = 9.3°

Assumed Variables

β_2 = 342.3°

β_3 = 324.8°

γ_2 = 30.9°

γ_3 = 80.6°

To be found: Vectors W_1 Z_1 S_1 U_1

FIGURE 5-5

Data needed for three-position analytical synthesis

where :

$$K_1 = A_2 A_4 + A_3 A_6$$
$$K_2 = A_3 A_4 + A_5 A_6$$
$$K_3 = \frac{A_1^2 - A_2^2 - A_3^2 - A_4^2 - A_6^2}{2}$$

(5.34b)

and :

$$A_1 = -C_3{}^2 - C_4{}^2; \qquad\qquad A_2 = C_3 C_6 - C_4 C_5$$
$$A_3 = -C_4 C_6 - C_3 C_5; \qquad\qquad A_4 = C_2 C_3 + C_1 C_4$$
$$A_5 = C_4 C_5 - C_3 C_6; \qquad\qquad A_6 = C_1 C_3 - C_2 C_4$$

(5.34c)

$$C_1 = R_3 \cos(\alpha_2 + \zeta_3) - R_2 \cos(\alpha_3 + \zeta_2)$$
$$C_2 = R_3 \sin(\alpha_2 + \zeta_3) - R_2 \sin(\alpha_3 + \zeta_2)$$
$$C_3 = R_1 \cos(\alpha_3 + \zeta_1) - R_3 \cos\zeta_3$$
$$C_4 = -R_1 \sin(\alpha_3 + \zeta_1) + R_3 \sin\zeta_3$$
$$C_5 = R_1 \cos(\alpha_2 + \zeta_1) - R_2 \cos\zeta_2$$
$$C_6 = -R_1 \sin(\alpha_2 + \zeta_1) + R_2 \sin\zeta_2$$

(5.34d)

The ten variables in these equations are: α_2, α_3, β_2, β_3, ζ_1, ζ_2, ζ_3, R_1, R_2, and R_3. The constants C_1 to C_6 are defined in terms of the eight known variables, $R_1, R_2, R_3, \zeta_1,$ ζ_2, and ζ_3 (which are the magnitudes and angles of position vectors $\mathbf{R}_1$, $\mathbf{R}_2$, and $\mathbf{R}_3$) and the angles α_2 and α_3 which define the change in angle of the coupler. See Figure 5-6 (p. 212) for depictions of these variables.

Note in equation 5.34a that there are two solutions for each angle (just as there were to the position analysis of the fourbar linkage in Section 4.5 and Figure 4-8, p. 155). One solution in this case will be a trivial one wherein $\beta_2 = \alpha_2$ and $\beta_3 = \alpha_3$. The nontrivial solution is the one desired.

This procedure is then repeated, solving equations 5.34 for the right-hand end of the linkage using the desired location of fixed pivot O_4 to calculate the necessary angles γ_2 and γ_3 for link 4.

We have now reduced the problem to that of three-position synthesis without specified pivots as described in Section 5.6 and Example 5-2. In effect we have found the particular values of β_2, β_3, γ_2, and γ_3 which correspond to the solution that uses the desired fixed pivots. The remaining task is to solve for the values of W_x, W_y, Z_x, Z_y using equations 5.25 through 5.31.

✏️ EXAMPLE 5-3

Three-Position Analytical Synthesis with Specified Fixed Pivots.

Problem: Design a fourbar linkage to move the line AP shown from position $A_1 P_1$ to $A_2 P_2$ and then to position $A_3 P_3$ using fixed pivots O_2 and O_4 in the locations specified.

Solution: (see Figure 5-7)

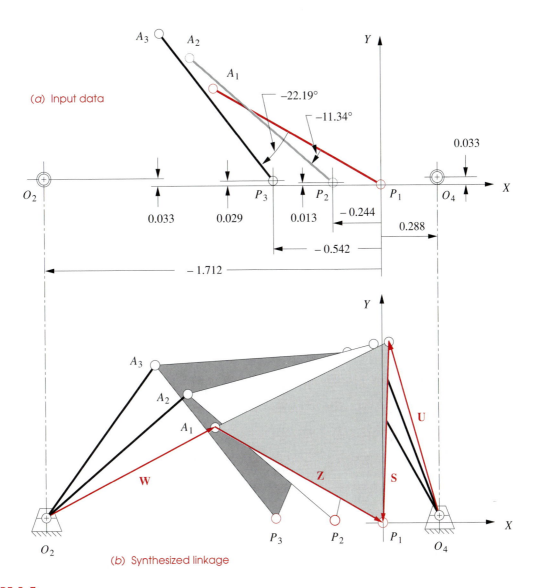

FIGURE 5-7

Three-position synthesis example for specified fixed pivots

1 Draw the link AP in its three desired positions, A_1P_1, A_2P_2, and A_3P_3 to scale in the plane as shown in Figure 5-7. The three positions are defined with respect to a global origin positioned at the first precision point P_1. The given data are specified in parts 2 to 4 below.

2 The position difference vectors between precision points are:

$P_{21x} = -0.244$ $\qquad$ $P_{21y} = 0.013$ $\qquad\qquad$ $P_{31x} = -0.542$ $\qquad$ $P_{31y} = 0.029$

3   The angle changes of the coupler between precision points are:

$$\alpha_2 = -11.34° \qquad\qquad \alpha_3 = -22.19°$$

4   The assumed free choices are the desired fixed pivot locations.

$$O_{2x} = -1.712 \qquad O_{2y} = 0.033 \qquad\qquad O_{4x} = 0.288 \qquad O_{4y} = 0.033$$

5   Solve equations 5.34 twice, once using the O_2 pivot location coordinates and again using the O_4 pivot location coordinates.

For pivot O_2:

$$C_1 = -0.205 \qquad C_2 = 0.3390 \qquad C_3 = 0.4028$$
$$C_4 = 0.6731 \qquad C_5 = 0.2041 \qquad C_6 = 0.3490$$
$$A_1 = -0.6152 \qquad A_2 = 0.0032 \qquad A_3 = -0.3171$$
$$A_4 = -0.0017 \qquad A_5 = -0.0032 \qquad A_6 = -0.3108$$
$$K_1 = 0.0986 \qquad K_2 = 0.0015 \qquad K_3 = 0.0907$$

The values found for the link angles to match this choice of fixed pivot location O_2 are:

$$\beta_2 = 11.96° \qquad\qquad \beta_3 = 23.96°$$

For pivot O_4:

$$C_1 = -0.3144 \qquad C_2 = -0.0231 \qquad C_3 = 0.5508$$
$$C_4 = -0.0822 \qquad C_5 = 0.2431 \qquad C_6 = -0.0443$$
$$A_1 = -0.3102 \qquad A_2 = -0.0044 \qquad A_3 = -0.1376$$
$$A_4 = 0.0131 \qquad A_5 = 0.0044 \qquad A_6 = -0.1751$$
$$K_1 = 0.0240 \qquad K_2 = -0.0026 \qquad K_3 = 0.0232$$

The values found for the link angles to match this choice of fixed pivot location O_4 are:

$$\gamma_2 = 2.78° \qquad\qquad \gamma_3 = 9.96°$$

6   At this stage, the problem has been reduced to the same one as in the previous section; i.e., find the linkage given the free choices of the above angles β_2, β_3, γ_2, γ_3, using equations 5.25 through 5.31. The data needed for the remaining calculations are those given in steps 2, 3, and 5 of this example, namely:

for dyad 1:

$$P_{21x} \qquad P_{21y} \qquad P_{31x} \qquad P_{31y} \qquad a_2 \qquad a_3 \qquad b_2 \qquad b_3$$

for dyad 2:

$$P_{21x} \qquad P_{21y} \qquad P_{31x} \qquad P_{31y} \qquad a_2 \qquad a_3 \qquad g_2 \qquad g_3$$

See Example 5-2 and Section 5.6 for the procedure. A matrix solving calculator, Mathcad, TKSolver, Matlab, program MATRIX, or program FOURBAR will solve this and compute the coordinates of the link vectors:

$W_x = 0.866$	$W_y = 0.500$	$Z_x = 0.846$	$Z_y = -0.533$
$U_x = -0.253$	$U_y = 0.973$	$S_x = -0.035$	$S_y = -1.006$

7   The link lengths are computed as was done in Example 5-2 and are shown in Table 5-2.

TABLE 5-2
Example 5-3 Results

Link 1 = 2.00 in
Link 2 = 1.00 in
Link 3 = 1.00 in
Link 4 = 1.01 in
Coupler Pt. =1.0 in
@ –60.73°
Circuit = Open
Start Theta2 = 30°
Final Theta2 = 54°
Delta Theta2 = 12°

This example can be read into program FOURBAR from diskfile E05-03.4br and animated.

5.9   CENTER-POINT AND CIRCLE-POINT CIRCLES

It would be quite convenient if we could find the loci of all possible solutions to the three-position synthesis problem, as we would then have an overview of the potential locations of the ends of the vectors **W**, **Z**, **U**, and **S**. Loerch[5] shows that by holding one of the free choices (say β_2) at an arbitrary value, and then solving equations 5.25 and 5.26 while iterating the other free choice (β_3) through all possible values from 0 to 2π, a circle will be generated. This circle is the locus of all possible locations of the root of vector **W** (for the particular value of β_2 used). The root of the vector **W** is the location of the fixed pivot or *center O_2*. Thus, this circle is called a ***center-point* circle**. The vector **N** in Figure 5-8 defines points on the *center-point* circle with respect to the global coordinate system which is placed at precision point P_1 for convenience.

If the same thing is done for vector **Z**, holding α_2 constant at some arbitrary value and iterating α_3 from 0 to 2π, another circle will be generated. This circle is the locus of all possible locations of the root of vector **Z** for the chosen value of α_2. Because the root of vector **Z** is joined to the tip of vector **W** and **W**'s tip describes a circle about pivot O_2 in the finished linkage, this locus is called the ***circle-point* circle**. Vector (**–Z**) defines points on the *circle-point* circle with respect to the global coordinate system.

The x,y components of vectors **W** and **Z** are defined by equations 5.25 and 5.26. Negating the x,y components of **Z** will give the coordinates of points on the circle-point circle for any assumed value of α_2 as angle α_3 is iterated from 0 to 2π. The x,y components of **N** = **–Z** – **W**, define points on the O_2 center-point circle for any assumed value of β_2 as β_3 is iterated through 0 to 2π. Vector **W** is calculated using angles β_2 and β_3, and vector **Z** using angles α_2 and α_3, both from equations 5.25 and 5.26.

For the right-hand dyad, there will also be separate center-point circles and circle-point circles. The x,y components of **M** = **–S** – **U**, define points on the O_4 center-point circle for any assumed value of γ_2 as γ_3 is iterated through 0 to 2π. (See Figure 5-8 and also Figure 5-4. p. 202.) Negating the x,y components of **S** will give the coordinates of points on the circle-point circle for any assumed value of α_2 as α_3 is iterated through 0 to 2π. Vector **U** is calculated using angles γ_2 and γ_3, and vector **S** using angles α_2 and α_3, both from equations 5.30 and 5.31.

Note that there is still an infinity of solutions because we are choosing the value of one angle arbitrarily. Thus there will be an **infinite number of sets of** *center-point* **and** *circle-point* **circles**. A computer program can be of help in choosing a linkage design which has pivots in convenient locations. Program FOURBAR, provided with this text, will calculate the solutions to the analytical synthesis equations derived in this section, for user-selected values of all the free choices needed for three-position synthesis, both

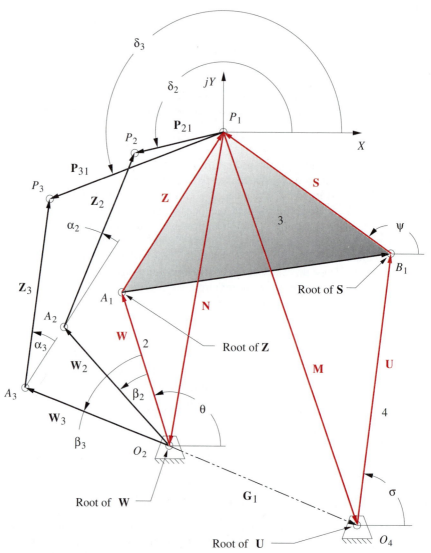

FIGURE 5-8

Definition of vectors to define center-point and circle-point circles

with and without specification of fixed pivot locations. The computer programs FOUR-BAR, FIVEBAR, and SIXBAR and their use are discussed in detail in Appendix A.

Figure 5-9 shows the circle-point and center-point circles for the Chebyschev straight line linkage for choices of $\beta_2 = 26°$, $\alpha_2 = 97.41°$, $\alpha_3 = 158.18°$ for the left dyad and $\gamma_2 = 36°$, $\alpha_2 = 97.41°$, $\alpha_3 = 158.18°$ for the right dyad. In this example the two larger circles are the center-point circles which define the loci of possible fixed pivot locations O_2 and O_4. The smaller two circles define the loci of possible moving pivot loca-

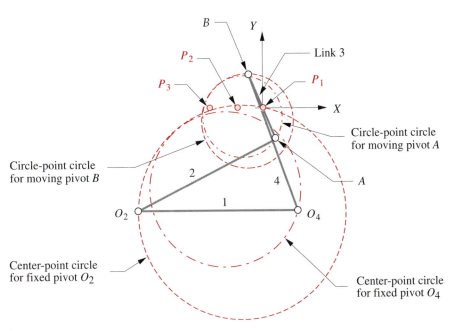

FIGURE 5-9

Circle-point and center-point circles and a linkage that reaches the precision points

tions I_{23} and I_{34}. Note that the coordinate system has its origin at the reference precision point, in this case P_1, from which we measured all parameters used in the analysis. These circles define the pivot loci of all possible linkages which will reach the three precision points P_1, P_2, and P_3 that were specified for particular choices of angles β_2, γ_2, and α_2. An example linkage is drawn on the diagram to illustrate one possible solution.

5.10   FOUR- AND FIVE-POSITION ANALYTICAL SYNTHESIS

The same techniques derived above for two- and three-position synthesis can be extended to four and five positions by writing more vector loop equations, one for each precision point. To facilitate this we will now put the vector loop equations in a more general form, applicable to any number of precision positions. Figure 5-4 (p. 202) will still serve to illustrate the notation for the general solution. The angles α_2, α_3, β_2, β_3, γ_2, and γ_3 will now be designated as α_k, β_k, and γ_k, $k = 2$ to n, where k represents the precision position and $n = 2$, 3, 4, or 5 represents the total number of positions to be solved for. The vector loop general equation set then becomes:

$$\mathbf{W}_k + \mathbf{Z}_k - \mathbf{P}_{k1} - \mathbf{Z}_1 - \mathbf{W}_1 = 0, \qquad k = 2 \text{ to } n \qquad (5.35a)$$

Which, after substituting the complex number forms and simplifying becomes:

$$we^{j\theta}\left(e^{j\beta_k} - 1\right) + ze^{j\phi}\left(e^{j\alpha_k} - 1\right) = p_{k1}e^{j\delta_k}, \qquad k = 2 \text{ to } n \qquad (5.35b)$$

This can be put in a more compact form by substituting vector notation for those terms to which it applies, let:

$$\mathbf{W} = we^{j\theta}; \qquad\qquad \mathbf{Z} = ze^{j\phi}; \qquad\qquad \mathbf{P}_{k1} = p_{k1}e^{j\delta_k} \qquad\qquad (5.35c)$$

then:

$$\mathbf{W}\left(e^{j\beta_k} - 1\right) + \mathbf{Z}\left(e^{j\alpha_k} - 1\right) = \mathbf{P}_{k1}e^{j\delta_k}, \qquad k = 2 \text{ to } n \qquad (5.35d)$$

Equation 5.35d is called the *standard form equation* by Erdman and Sandor.[6] By substituting the values of α_k, β_k, and δ_k, in equation 5.35d for all the precision positions desired, the requisite set of simultaneous equations can be written for the left dyad of the linkage. The standard form equation applies to the right-hand dyad **US** as well, with appropriate changes to variable names as required.

$$\mathbf{U}\left(e^{j\beta_k} - 1\right) + \mathbf{S}\left(e^{j\alpha_k} - 1\right) = \mathbf{P}_{k1}e^{j\delta_k}, \qquad k = 2 \text{ to } n \qquad (5.35e)$$

The number of resulting equations, variables, and free choices for each value of n is shown in Table 5-3 (after Erdman and Sandor). They provide solutions for the four- and five-position problems in reference 6. The circle-point and center-point circles of the three-position problem become cubic curves, called **Burmester curves**, in the four-position problem. Erdman's commercially available computer program LINCAGES[8] solves the **four-position problem** in an interactive way, allowing the user to select center and circle pivot locations on their Burmester curve loci, which are drawn on the graphics screen of the computer.

5.11   ANALYTICAL SYNTHESIS OF A PATH GENERATOR WITH PRESCRIBED TIMING

The approach derived above for motion generation synthesis is also applicable to the case of **path generation with prescribed timing**. In path generation, the precision points are to be reached, but the angle of a line on the coupler is not of concern. Instead, the timing at which the coupler reaches the precision point is specified in terms of input rocker angle β_2. In the three-position motion generation problem we specified the angles α_2 and α_3 of vector **Z** in order to control the angle of the coupler. Here we instead want to specify angles β_2 and β_3 of the input rocker, to define the timing. Before, the free choices were β_2 and β_3. Now they will be α_2 and α_3. In either case, all four angles are either specified or assumed as free choices and the solution is identical. Figure 5-4 (p. 202) and equations 5.25, 5.26, 5.30, and 5.31 apply to this case as well. This case can be extended to as many as five precision points as shown in Table 5-3.

5.12   ANALYTICAL SYNTHESIS OF A FOURBAR FUNCTION GENERATOR

A similar process to that used for the synthesis of path generation with prescribed timing can be applied to the problem of function generation. In this case we do not care about motion of the coupler at all. In a fourbar function generator, the coupler exists only to **couple** the input link to the output link. Figure 5-10 shows a fourbar linkage in three

TABLE 5-3   Number of Variables and Free Choices for Analytical Precision
Point Motion and Timed Path Synthesis.[7]

No. of Positions (n)	No. of Scalar Variables	No. of Scalar Equations	No. of Prescribed Variables	No. of Free Choices	No. of Available Solutions
2	8	2	3	3	∞^3
3	12	4	6	2	∞^2
4	16	6	9	1	∞^1
5	20	8	12	0	Finite

positions. Note that the coupler, link 3, is merely a line from point A to point P. Point P can be thought of as a coupler point which happens to coincide with the pin joint between links 3 and 4. As such it will have simple arc motion, pivoting about O_4, rather than, for example, the higher-order path motion of the coupler point P_1 in Figure 5-4.

Our **function generator** uses *link 2 as the input link and takes the output from link 4.* The "**function**" generated is the **relationship between the angles of link 2 and link 4** for the specified three-position positions, P_1, P_2, and P_3. These are located in the plane with respect to an arbitrary global coordinate system by position vectors $\mathbf{R}_1$, $\mathbf{R}_2$, and $\mathbf{R}_3$. The function is:

$$\gamma_k = f(\beta_k), \qquad k = 1, 2, \ldots, n; \qquad n \leq 7 \qquad (5.36)$$

This is *not* a **continuous function.** The relationship holds only for the discrete points (k) specified.

To synthesize the lengths of the links needed to satisfy equation 5.36, we will write vector loop equations around the linkage in pairs of positions, as was done for the previous examples. However, we now wish to include both link 2 and link 4 in the loop, since link 4 is the output. See Figure 5-10.

$$\mathbf{W}_2 + \mathbf{Z}_2 - \mathbf{U}_2 + \mathbf{U}_1 - \mathbf{Z}_1 - \mathbf{W}_1 = 0$$
$$\mathbf{W}_3 + \mathbf{Z}_3 - \mathbf{U}_3 + \mathbf{U}_1 - \mathbf{Z}_1 - \mathbf{W}_1 = 0 \qquad (5.37a)$$

rearranging:

$$\mathbf{W}_2 + \mathbf{Z}_2 - \mathbf{Z}_1 - \mathbf{W}_1 = \mathbf{U}_2 - \mathbf{U}_1$$
$$\mathbf{W}_3 + \mathbf{Z}_3 - \mathbf{Z}_1 - \mathbf{W}_1 = \mathbf{U}_3 - \mathbf{U}_1 \qquad (5.37b)$$

but,

$$\mathbf{P}_{21} = \mathbf{U}_2 - \mathbf{U}_1$$
$$\mathbf{P}_{31} = \mathbf{U}_3 - \mathbf{U}_1 \qquad (5.37c)$$

substituting:

$$\mathbf{W}_2 + \mathbf{Z}_2 - \mathbf{Z}_1 - \mathbf{W}_1 = \mathbf{P}_{21}$$
$$\mathbf{W}_3 + \mathbf{Z}_3 - \mathbf{Z}_1 - \mathbf{W}_1 = \mathbf{P}_{31} \qquad (5.37d)$$

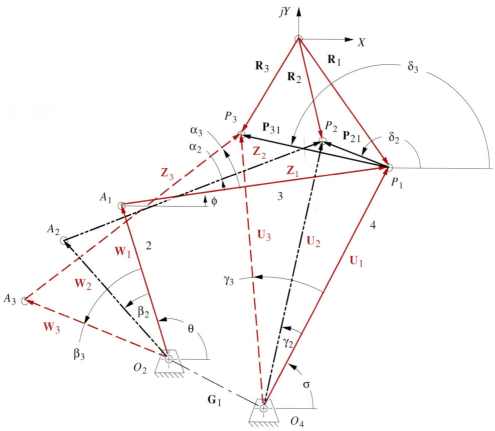

FIGURE 5-10

Analytical synthesis of a fourbar function generator

$$we^{j(\theta+\beta_2)} + ze^{j(\phi+\alpha_2)} - ze^{j\phi} - we^{j\theta} = p_{21}e^{j\delta_2}$$

$$we^{j(\theta+\beta_3)} + ze^{j(\phi+\alpha_3)} - ze^{j\phi} - we^{j\theta} = p_{31}e^{j\delta_3}$$

(5.37e)

Note that equations 5.37d and 5.37e are identical to equations 5.19 and 5.20 derived for the three-position motion generation case and can also be put into Erdman's *standard form*[6] of equation 5.35 for the *n*-position case. The twelve variables in equation 5.37e are the same as those in equation 5.20: w, θ, β_2, β_3, z, ϕ, α_2, α_3, p_{21}, p_{31}, δ_2, and δ_3.

For the three-position function generation case the solution procedure then can be the same as that described by equations 5.20 through 5.27 for the motion synthesis problem. In other words, the solution equations are the same for **all three types** of kinematic synthesis, *function generation, motion generation,* and *path generation with prescribed timing.* This is why Erdman and Sandor called equation 5.35 the *standard form equation.* To develop the data for the function generation solution, expand equation 5.37b:

$$we^{j(\theta+\beta_2)} + ze^{j(\phi+\alpha_2)} - ze^{j\phi} - we^{j\theta} = ue^{j(\sigma+\gamma_2)} - ue^{j\sigma}$$

$$we^{j(\theta+\beta_3)} + ze^{j(\phi+\alpha_3)} - ze^{j\phi} - we^{j\theta} = ue^{j(\sigma+\gamma_3)} - ue^{j\sigma}$$

$$(5.37f)$$

There are also **twelve variables** in equation 5.37f: w, θ, z, ϕ, α_2, α_3, β_2, β_3, u, σ, γ_2, and γ_3. We can solve for any four. Four angles, β_2, β_3, γ_2, and γ_3 are specified from the function to be generated in equation 5.36. This leaves **four free choices**. In the function generation problem it is often convenient to define the length of the output rocker, u, and its initial angle σ to suit the package constraints. Thus, selecting the components u and σ of vector $\mathbf{U}_1$ can provide two convenient free choices of the four required.

With u, σ, γ_2, and γ_3 known, $\mathbf{U}_2$ and $\mathbf{U}_3$ can be found. Vectors $\mathbf{P}_{21}$ and $\mathbf{P}_{31}$ can then be found from equation 5.37c. Six of the unknowns in equation 5.37e are then defined, namely, β_2, β_3, p_{21}, p_{31}, δ_2, and δ_3. Of the remaining six (w, θ, z, ϕ, α_2, α_3), we must assume values for two more as free choices in order to solve for the remaining four. We will assume values (free choices) for the two angles, α_2 and α_3 (as was done for path generation with prescribed timing) and solve equations 5.37e for the components of $\mathbf{W}$ and $\mathbf{Z}$ (w, θ, z, ϕ). We have now reduced the problem to that of Section 5.6 and Example 5-2. See equations 5.20 through 5.27 for the solution.

Having chosen vector $\mathbf{U}_1$ (u, σ) as a free choice in this case, we only have to solve for one dyad, $\mathbf{WZ}$. Though we arbitrarily choose the length of vector $\mathbf{U}_1$, the resulting function generator linkage can be scaled up or down to suit packaging constraints without affecting the input/output relation defined in equation 5.36, because it is a function of angles only. This fact is not true for the motion or path generation cases, as scaling them will change the absolute coordinates of the path or motion output precision points which were specified in the problem statement.

Table 5-4 shows the relationships between number of positions, variables, free choices, and solutions for the function generation case. Note that up to seven angular output positions can be solved for with this method.

TABLE 5-4   Number of Variables and Free Choices for Function Generation Synthesis.[7]

No. of Positions (n)	No. of Scalar Variables	No. of Scalar Equations	No. of Prescribed Variables	No. of Free Choices	No. of Available Solutions
2	8	2	1	5	∞^5
3	12	4	4	4	∞^4
4	16	6	7	3	∞^3
5	20	8	10	2	∞^2
6	24	10	13	1	∞^1
7	28	12	16	0	Finite

5.13   OTHER LINKAGE SYNTHESIS METHODS

Many other techniques for the synthesis of linkages to provide a prescribed motion have been created or discovered in recent years. Most of these approaches are somewhat involved and many are mathematically complicated. Only a few allow a closed-form solution; most require an iterative numerical solution. Most address the path synthesis problem with or without concern for prescribed timing. As Erdman and Sandor point out, the path, motion, and function generation problems are closely related.[6]

Space does not permit a complete exposition of even one of these approaches in this text. We choose instead to present brief synopses of a number of synthesis methods along with complete references to their full descriptions in the engineering and scientific literature. The reader interested in a detailed account of any method listed may consult the referenced papers which can be obtained through any university library or large public library. Also, some of the authors of these methods may make copies of their computer code available to interested parties.

Table 5-5 summarizes some of the existing fourbar linkage synthesis methods and for each one lists the method type, the maximum number of positions synthesized, the approach, special features, and a bibliographic reference (see the end of this chapter for the complete reference). The list in Table 5-5 is not exhaustive; other methods than these also exist.

The listed methods are divided into three types labeled **precision**, **equation**, and **optimized** (first column of Table 5-5). By **precision** (from precision point) is meant a method, such as the ones described in previous sections of this chapter, that attempts to find a solution which will pass exactly through the desired (precision) points but may deviate from the desired path between these points. Precision point methods are limited to matching a number of points equal to the number of independently adjustable parameters that define the mechanism. For a fourbar linkage, this is nine.[*] (Higher-order linkages with more links and joints will have a larger number of possible precision points.)

For up to 5 precision points in the fourbar linkage, the equations can be solved in closed form without iteration. (The four-point solution is used as a tool to solve for 5 positions in closed form, but for 6 points or more the nonlinear equations are difficult to handle.) For 6 to 9 precision points an iterative method is needed to solve the equation set. There can be problems of nonconvergence, or convergence to singular or imaginary solutions, when iterating nonlinear equations. Regardless of the number of points solved for, the solution found may be unusable due to circuit, branch, or order (CBO) defects. A circuit defect means that the linkage must be disassembled and reassembled to reach some positions, and a branch defect means that a toggle position is encountered between successive positions (see Section 4.12, p. 173). An order defect means that the points are all reachable on the same branch but are encountered in the wrong order.

The type labeled **equation** in Table 5-5 refers to methods that solve the tricircular, trinodal sextic coupler curve to find a linkage that will generate an entire coupler curve that closely approximates a set of desired points on the curve.

The type labeled **optimized** in Table 5-5 refers to an iterative optimization procedure that attempts to minimize an **objective function** that can be defined in many ways, such as the least-squares deviation between the calculated and desired coupler point po-

---

* The nine independent parameters of a fourbar linkage are: four link lengths, two coordinates of the coupler point with respect to the coupler link, and three parameters that define the location and orientation of the fixed link in the global coordinate system.

TABLE 5-5   Some Methods for the Analytic Synthesis of Linkages

Type	Max Pos.	Approach	Special Features	Bibliography	References
Precision	4	Loop equations—closed form	Linear equations extendable to five positions	Freudenstein (1959) Sandor (1959) Erdman (1981)	1, 2, 4, 5, 6, 8, 10
Precision	5	Loop equations—Newton Raphson	Uses displacement matrix	Suh (1967)	11
Precision	5	Loop equations—continuation	Specified fixed pivots, specified moving pivots	Morgan (1990) Subbian (1991)	14, 15, 16, 17
Precision	7	Closed form 5 pt. — iterative to 7 pt.	Extendable to Watt I sixbar	Tylaska (1994)	19, 20
Precision	9	loop equations - Newton-Raphson	Exhaustive solution	Morgan (1987) Wampler (1992)	12, 13, 18
Equation	10	Coupler curve eqn.	Iterative solution	Blechschmidt (1986)	21
Equation	15	Coupler curve eqn.	Builds on Blechschmidt	Ananthasuresh (1993)	22
Optimized	N	Loop equations—least squares	Specified fixed pivots, control force and torque	Fox (1966)	24
Optimized	N	Loop equations—various criteria	Path or function generation	Youssef (1975)	25
Optimized	N	Least squares on linear equations	Prescribed timing, rapid convergence	Nolle (1971)	9
Optimized	N	Selective precision synthesis (SPS)	Relaxes precision requirements	Kramer (1975)	26, 27
Optimized	N	SPS + fuzzy logic	Extends Kramer's SPS	Krishnamurthi (1993)	28
Optimized	N	Quasi-precision pos.	Builds on Kramer	Mirth (1994)	29
Optimized	3 or 4	Loop equations and dynamic criteria	Kinematics and dynamic forces and torques	Conte (1975) Kakatsios (1987)	30, 31, 32
Optimized	N	Loop equations—least squares	Avoids branch problems, rapid convergence	Angeles (1988)	33
Optimized	N	Energy method	FEA approach	Aviles (1994)	34
Optimized	N	Genetic algorithm	Whole curve synthesis	Fang (1994)	35
Optimized	N	Fourier descriptors	Whole curve synthesis	Ullah (1996)	36, 37
Optimized	2, 3, or 4	Loop equations—various criteria	Automatic generation CBO defect free	Bawab (1997)	38

sitions, for example. The calculated points are found by solving a set of equations that define the behavior of the linkage geometry, using assumed initial values for the linkage parameters. A set of inequality constraints that limit the range of variation of parameters such as link length ratios, Grashof condition, or transmission angle may also be included in the calculation. New values of linkage parameters are generated with each iteration step according to the particular optimization scheme used. The closest achievable fit between the calculated solution points and the desired points is sought, defined as minimization of the chosen objective function. None of the desired points will be exactly matched by these methods, but for most engineering tasks this is an acceptable result.

Optimization methods allow larger numbers of points to be specified than do the precision methods, limited only by available computer time and numerical roundoff error. Table 5-5 shows a variety of optimization schemes ranging from the mundane (least squares) to the esoteric (fuzzy logic, genetic algorithms). All require a computer-programmed solution. Most can be run on current desktop computers in reasonably short times. Each different optimization approach has advantages and disadvantages in respect to convergence, accuracy, reliability, complexity, speed, and computational burden. Convergence often depends on a good choice of initial assumptions (guess values) for the linkage parameters. Some methods, if they converge at all, do so to a local minimum (only one of many possible solutions), and it may not be the best one for the task.

Precision Point Methods

Table 5-5 shows several precision point synthesis methods. Some of these are based on original work by Freudenstein and Sandor.[10] Sandor [1] and Erdman [2], [6] developed this approach into the "standard form" which is described in detail in this chapter. This method yields closed-form solutions for 2, 3, and 4 precision positions and is extendable to 5 positions. It suffers from the possible circuit, branch, and order (CBO) defects common to all precision point methods.

The method of Suh and Radcliffe [11] is similar to that of Freudenstein and others [1], [2], [6], [10] but leads to a set of simultaneous nonlinear equations which are solved for up to 5 positions using the Newton-Raphson numerical method (see Section 4.13, p. 174). This approach adds to the usual CBO problems the possibilities of nonconvergence, or convergence to singular or imaginary solutions.

Recent developments in the mathematical theory of polynomials have created new methods of solution called **continuation methods** (also called **homotopy methods**) which do not suffer from the same convergence problems as other methods and can also determine all the solutions of the equations starting from any set of assumed values. [12], [13] Continuation methods are a general solution to this class of problem and are reliable and fast enough to allow multiple designs to be investigated in a reasonable time (typically measured in CPU *hours* on a powerful computer).

Several researchers have developed solutions for the 5- to 9-precision point problem using this technique. Morgan and Wampler [14] solved the fourbar linkage 5-point problem with specified fixed pivots completely and found a maximum of 36 real solutions. Subbian and Flugrad [15] used specified moving pivots for the 5-point problem, extended the 5-point method to sixbar linkages,[16] and also synthesized eightbar and geared fivebar mechanisms for 6 and 7 precision points using continuation methods. [17]

Only the continuation method has yet been able to completely solve the fourbar linkage 9-precision-point problem and yield all its possible solutions. Wampler, Morgan, and Sommese [18] used a combination of analytical equation reduction and numerical continuation methods to exhaustively compute all possible nondegenerate, generic solutions to the 9-point problem.* They proved that there is a maximum of 4326 distinct, nondegenerate linkages (occurring in 1442 sets of cognate triples) that will potentially solve a generic 9-precision-point fourbar problem. Their method does not eliminate physically impossible (complex link) linkages or those with CBO defects. These still have to be removed by examination of the various solutions. They also solved four examples and

---

* The authors report that this calculation took 332 CPU hours on an IBM 3081 computer.

found the maximum number of linkages with real link lengths that generated these particular 9-point paths to be, respectively, 21, 45, 64, and 120 cognate triples. Computation times ranged from 69 to 321 CPU minutes on an IBM 3090 for these four examples.

Tylaska and Kazerounian [19], [20] took a different approach and devised a method that synthesizes a fourbar linkage for up to 7 precision points and also synthesized a Watt I sixbar linkage for up to six body guidance (motion specification) positions with control over locations of some ground and moving pivots. Their method yields the entire set of solutions for any set of design data and is an improvement over iterative methods that are sensitive to initial guesses. It is less computationally intensive than the continuation methods.

Coupler Curve Equation Methods

Blechschmidt and Uicker [21] and Ananthasuresh and Kota [22] used the algebraic coupler curve equation rather than a vector loop approach to calculate the coupler point path. The equation of the coupler curve is a tricircular, trinodal sextic of 15 terms. Nolle [23] states that:

> *The coupler curve equation itself is very complex and as far as is known in the study of mechanics (or for that matter elsewhere) no other mathematical result has been found having algebraic characteristics matching those of the coupler curve.*

Its solution is quite involved and requires iteration. Blechschmidt and Uicker's approach [21] chose coordinates for 10 points on the desired curve. Ananthasuresh used 15 points with some trial and error required in their selection. The advantage of these coupler curve equation approaches is that they define the entire curve which can be plotted and examined for suitability and defects prior to calculating the link dimensions which requires significant additional computing time.

Optimization Methods

The methods listed as **optimized** in Table 5-5 are a diverse group and some have little in common except the goal of finding a linkage that will generate a desired path. All allow a theoretically unlimited number of design points to be specified, but making N too large will increase the computation time and may not improve the result. One inherent limitation of optimization methods is that they may converge to a local minimum near the starting conditions. The result may not be as good as other minima located elsewhere in the N-space of the variables. Finding the global optimum is possible but more difficult and time consuming.

Perhaps the earliest application (1966) of optimization techniques to this fourbar linkage path synthesis problem is that of Fox and Willmert [24] in which they minimized the area between the desired and calculated curves subject to a number of equality and inequality constraints. They controlled link lengths to be positive and less than some maximum, controlled for Grashof condition, limited forces and torques, and restricted the locations of the fixed pivots. They used Powell's method to find the minimum of the objective function.

Youssef et al.,[25] used sum of squares, sum of absolute values, or area error criteria to minimize the objective function. They accommodated path and function generation

desired curve ·············
actual curve ———

(a) Coupler curve

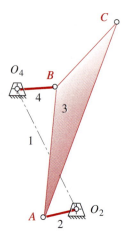

(b) Synthesized
        Linkage

FIGURE 5-11

Linkage synthesized to
generate a desired
coupler curve by an
optimization method
*Reproduced from
"Optimal Kinematic
Synthesis of Planar
Linkage Mechanisms"* (25)
*with the kind permission of
Professional Engineering
Publishing, Bury St.
Edmunds, UK.*

for single-loop (fourbar) or multiloop (more than four bar) linkages with both pin and slider joints. They allowed constraints to be imposed on the allowable ranges of link lengths and angles, any of which also may be held constant during the iteration. An example of an optimization done with this method for 19 evenly spaced points around a desired fourbar coupler path is shown in Figure 5-11.[25] Another example of this method is the 10-bar crank-slider linkage in Figure 5-12 [25] which also shows the desired and actual coupler curve generated by point P for 24 points corresponding to equal increments of input crank angle.

Nolle and Hunt [9] derived analytical expressions that lead to a set of ten linear simultaneous nonhomogeneous equations whose solution gives values for all the independent variables. They used a least squares approach to the optimization and also allowed specified timing of the input crank to each position on the coupler. Because their equations are linear, convergence is rapid requiring only about one second per iteration.

Kramer and Sandor [26], [27] described a variant on the precision point technique which they call *selective precision synthesis* (SPS). It relaxes the requirement that the curve pass exactly through the precision points by defining "accuracy neighborhoods" around each point. The size of these tolerance zones can be different for each point, and more than nine points can be used. They point out that exact correspondence to a set of points is often not necessary in engineering applications and even if achieved theoretically would be compromised by manufacturing tolerances.

The SPS approach is suitable to any linkage constructible from dyads or triads and so can accommodate sixbar and geared fivebar linkages as well as fourbars. Fourbar function, motion, or path generation (with prescribed timing) can all be synthesized, using the standard form approach which considers all three forms equivalent in terms of equation formulation. Spatial mechanisms can also be accommodated. The solutions are stable and less sensitive to small changes in the data than precision point methods. Krishnamurthi et al.,[28] extended the SPS approach by using fuzzy set theory which gives a mechanism path as close to the specified points as is possible for a given start point; but it is sensitive to start point selection and may find local optima rather than global.

Mirth [29] provided a variation on Kramer's SPS technique called quasi-precision position synthesis which uses three precision positions and *N* quasi positions which are defined as tolerance zones. This approach retains the computational advantages of the Burmester (precision point) approach while also allowing the specification of a larger number of points to improve and refine the design.

Conte et al., [30] and Kakatsios and Tricamo [31], [32] described methods to satisfy a small number of precision points and simultaneously optimize the linkage's dynamic characteristics. The link lengths are controlled to reasonable size, the Grashof condition constrained, and the input torque, dynamic bearing and reaction forces, and shaking moments simultaneously minimized.

Many of the optimization methods listed above use some form of inequality constraints to limit the allowable values of design parameters such as link lengths and transmission angles. These constraints often cause problems that lead to nonconvergence, or to CBO defects. Angeles et al., [33] described an unconstrained nonlinear least-square-method that avoids these problems. Continuation methods are employed, and good convergence is claimed with no branch defects.

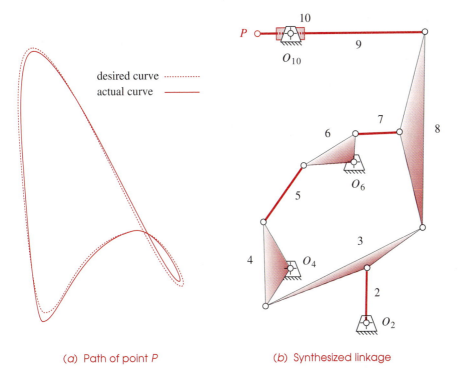

(a) Path of point *P*   *(b)* Synthesized linkage

FIGURE 5-12

Example of synthesis of a 10 link mechanism to generate a coupler path. *Reproduced from "Optimal Kinematic Synthesis of Planar Linkage Mechanisms"* [25] *with the kind permission of Professional Engineering Publishing, Bury St. Edmunds, UK.*

Aviles et al., [34] proposed a novel approach to the linkage synthesis problem that uses the elastic energy that would be stored in the links if they were allowed to deform elastically such that the coupler point reaches the desired location. The objective function is defined as the minimum energy condition in the set of deformed links which of course will occur when their rigid body positions most closely approach the desired path. This is essentially a finite element method approach that considers each link to be a bar element. Newton's method is used for the iteration and, in this case, converges to a minimum even when the initial guess is far from a solution.

Fang [35] described an unusual approach to linkage synthesis using genetic algorithms. Genetic algorithms emulate the way that living organisms adapt to nature. Initially, a population of random "organisms" are generated that represent the system to be optimized. This takes the form of a bit string, analogous to a cell's chromosomes which is called the first generation. Two operations are performed on a given population, called crossover and mutation. Crossover combines part of the "genetic code" of a "father" organism with part of the code of a "mother" organism. Mutation changes values of the genetic code at random points in the bit string. An objective function is created that expresses the "fitness" of the organism for the desired task. Each successive generation is produced by selecting the organisms that best fit the task. The population "evolves" through generations until a termination criterion is met based on the objective function.

Some advantages of this approach are that it searches from population to population rather than point to point, and this makes it less likely to be trapped at local optima. The population also preserves a number of valid solutions rather than converging to only one. The disadvantages are long computation times due to the large number of objective function evaluations required. Nevertheless it is more efficient than random walk or exhaustive search algorithms. All other optimization approaches listed here deal only with dimensional synthesis, but genetic algorithms can also deal with type synthesis.

Ullah and Kota [36], [37] separated the linkage synthesis problem into two steps. The first step seeks an acceptable match for the shape of the desired curve without regard to the size, orientation, or location of the curve in space. Once a curve of suitable shape and its associated linkage are found, the result can be translated, rotated, and scaled as desired. This approach simplifies the optimization task compared to the algorithms that seek a structural optimization that includes size, orientation, and location of the coupler curve all at once in the objective function. Fourier descriptors are used to characterize the shape of the curve as is done in many pattern matching applications such as for automated robotic assembly tasks. A stochastic global optimization algorithm is used which avoids unwanted convergence to suboptimal local minima.

Bawab et al., [38] described an approach that will automatically (within the software program) synthesize a fourbar linkage for two, three, or four positions using Burmester theory and eliminate all solutions having CBO defects. Limits on link length ratios and transmission angle are specified and the objective function is based on these criteria with weighting factors applied. Regions in the plane within which the fixed or moving pivots must be located may also be specified.

5.14   REFERENCES

1   **Sandor, G. N.** (1959). "A General Complex Number Method for Plane Kinematic Synthesis with Applications." Ph.D. Thesis, Columbia University, University Microfilms, Ann Arbor, MI.

2   **Erdman, A. G.** (1981). "Three and Four Precision Point Kinematic Synthesis of Planar Linkages." *Mechanism and Machine Theory*, **16**, pp. 227-245.

3   **Kaufman, R. E.** (1978). "Mechanism Design by Computer." *Machine Design*, October 26, 1978, pp. 94-100.

4   **Loerch, R. J., et al.** (1975). "Synthesis of Fourbar Linkages with Specified Ground Pivots." *Proc. of 4th Applied Mechanisms Conference*, Chicago, IL, pp. 10.1-10.6.

5   **Loerch, R. J., et al.** (1979). "On the Existence of Circle-Point and Center-Point Circles for Three Position Dyad Synthesis." *Journal of Mechanical Design*, **101**(3), pp. 554-562.

6   **Erdman, A. G., and G. N. Sandor**. (1997). *Mechanism Design: Analysis and Synthesis*. Vol. 1, 3d ed., and *Advanced Mechanism Design, Analysis and Synthesis*, Vol. 2 (1984). Prentice-Hall: Upper Saddle River, NJ.

7   **Jennings, A.** (1977). *Matrix Computation for Engineers and Scientists*. John Wiley & Sons: New York.

8 **Erdman, A. G.,** and **J. E. Gustafson**. (1977). "LINCAGES: Linkage INteractive Computer Analysis and Graphically Enhanced Synthesis." ASME Paper: 77-DTC-5.

9 **Nolle, H., and K. H. Hunt**. (1971). "Optimum Synthesis of Planar Linkages to Generate Coupler Curves." *Journal of Mechanisms*, **6**, pp. 267-287.

10 **Freudenstein, F., and G. N. Sandor.** (1959). "Synthesis of Path Generating Mechanisms by Means of a Programmed Digital Computer." *ASME Journal for Engineering in Industry*, 81, p. 2.

11 **Suh, C. H., and C. W. Radcliffe**. (1966). "Synthesis of Planar Linkages With Use of the Displacement Matrix." ASME Paper: 66-MECH-19, 9 pp.

12 **Morgan, A. P., and A. J. Sommese**. (1987). "Computing all Solutions to Polynomial Systems Using Homotopy Continuation." *Applied Mathematics and Computation*, **24**, pp. 115-138.

13 **Morgan, A. P.** (1987). *Solving Polynomial Systems Using Continuation for Scientific and Engineering Problems.* Prentice-Hall: Upper Saddle River, NJ.

14 **Morgan, A. P., and C. W. Wampler**. (1990). "Solving a Planar Fourbar Design Problem Using Continuation." *Journal of Mechanical Design*, **112**(4), p. 544.

15 **Subbian, T., and J. D. R. Flugrad**. (1991). "Fourbar Path Generation Synthesis by a Continuation Method." *Journal of Mechanical Design*, **113**(1), p. 63.

16 **Subbian, T., and J. D. R. Flugrad**. (1993). "Five Position Triad Synthesis with Applications to Four and Sixbar Mechanisms." *Journal of Mechanical Design*, **115**(2), p. 262.

17 **Subbian, T., and J. D. R. Flugrad**. (1994). "Six and Seven Position Triad Synthesis Using Continuation Methods." *Journal of Mechanical Design*, **116**(2), p. 660.

18 **Wampler, C. W., et al.** (1992). "Complete Solution of the Nine-Point Path Synthesis Problem for Fourbar Linkages." *Journal of Mechanical Design*, **114**(1), p. 153.

19 **Tylaska, T., and K. Kazerounian**. (1994). "Synthesis of Defect-Free Sixbar Linkages for Body Guidance Through Up to Six Finitely Separated positions." *Proc. of 23rd Biennial Mechanisms Conference*, Minneapolis, MI, p. 369.

20 **Tylaska, T., and K. Kazerounian**. (1993). "Design of a Six Position Body Guidance Watt I Sixbar Linkage and Related Concepts." *Proc. of 3rd Applied Mechanisms and Robotics Conference*, Cincinnati, pp. 93-1.

21 **Blechschmidt, J. L., and J. J. Uicker**. (1986). "Linkage Synthesis Using Algebraic Curves." *J. Mechanisms, Transmissions, and Automation in Design*, **108** (December 1986), pp. 543-548.

22 **Ananthasuresh, G. K., and S. Kota**. (1993). "A Renewed Approach to the Synthesis of Fourbar Linkages for Path Generation via the Coupler Curve Equation." *Proc. of 3rd Applied Mechanisms and Robotics Conference*, Cincinnati, pp. 83-1.

23 **Nolle, H.** (1975). "Linkage Coupler Curve Synthesis: A Historical Review - III. Spatial Synthesis and Optimization." *Mechanism and Machine Theory*, **10**, 1975, pp. 41-55.

24 **Fox, R. L., and K. D. Willmert**. (1967). "Optimum Design of Curve-Generating Linkages with Inequality Constraints." *Journal of Engineering for Industry (Feb 1967)*, pp. 144-152.

25   **Youssef, A. H., et al.** (1975). "Optimal Kinematic Synthesis of Planar Linkage Mechanisms." *I. Mech. E.*, pp. 393-398.

26   **Kramer, S. N., and G. N. Sandor**. (1975). "Selective Precision Synthesis—A General Method of Optimization for Planar Mechanisms." *Trans ASME J. Eng. for Industry*, **97B**(2), pp. 689-701.

27   **Kramer, S. N.** (1987). "Selective Precision Synthesis—A General Design Method for Planar and Spatial Mechanisms." *Proc. of 7th World Congress on Theory of Machines and Mechanisms*, Seville Spain.

28   **Krishnamurthi, S., et al.** (1993). "Fuzzy Synthesis of Mechanisms." *Proc. of 3rd Applied Mechanisms and Robotics Conference*, Cincinnati, pp. 94-1.

29   **Mirth, J. A.** (1994). "Quasi-Precision Position Synthesis of Fourbar Linkages." *Proc. of 23rd Biennial Mechanisms Conference*, Minneapolis, MI, p. 215.

30   **Conte, F. L., et al.** (1975). "Optimum Mechanism Design Combining Kinematic and Dynamic-Force Considerations." *Journal of Engineering for Industry (*May 1975), pp. 662-670.

31   **Kakatsios, A. J., and S. J. Tricamo**. (1987). "Precision Point Synthesis of Mechanisms with Optimal Dynamic Characteristics." *Proc. of 7th World Congress on the Theory of Machines and Mechanisms*, Seville, Spain, pp. 1041-1046.

32   **Kakatsios, A. J., and S. J. Tricamo**. (1986). "Design of Planar Rigid Body Guidance Mechanisms with Simultaneously Optimized Kinematic and Dynamic Characteristics." ASME Paper: 86-DET-142.

33   **Angeles, J., et al.** (1988). "An Unconstrained Nonlinear Least-Square Method of Optimization of RRRR Planar Path Generators." *Mechanism and Machine Theory*, **23**(5), pp. 343-353.

34   **Aviles, R., et al.** (1994). "An Energy-Based General Method for the Optimum Synthesis of Mechanisms." *Journal of Mechanical Design*, **116**(1), p. 127.

35   **Fang, W. E.** (1994). "Simultaneous Type and Dimensional Synthesis of Mechanisms by Genetic Algorithms." *Proc. of 23rd Biennial Mechanisms Conference*, Minneapolis, MI, p. 36.

36   **Ullah, I., and S. Kota**. (1994). "A More Effective Formulation of the Path Generation Mechanism Synthesis Problem." *Proc. of 23rd Biennial Mechanisms Conference*, Minneapolis, Minn, p. 239.

37   **Ullah, I., and S. Kota**. (1996). "Globally-Optimal Synthesis of Mechanisms for Path Generation using Simulated Annealing and Powell's Method." *Proc. of ASME Design Engineering Conference*, Irvine, CA, pp. 1-8.

38   **Bawab, S., et al.** (1997). "Automatic Synthesis of Crank Driven Fourbar Mechanisms for Two, Three, or Four Position Motion Generation." *Journal of Mechanical Design*, **119**(June), pp. 225-231.

5.15   PROBLEMS

Note that all three-position synthesis problems below may be done using a matrix solving calculator, equation solver such as Mathcad, Matlab, or TKSolver, program Matrix, or program Fourbar. Two-position synthesis problems can be done with a four-function calculator.

5-1 Redo Problem 3-3 using the analytical methods of this chapter.

5-2 Redo Problem 3-4 using the analytical methods of this chapter.

5-3 Redo Problem 3-5 using the analytical methods of this chapter.

5-4 Redo Problem 3-6 using the analytical methods of this chapter.

5-5 See Project P3-8. Define three positions of the boat and analytically synthesize a linkage to move through them.

5-6 See Project P3-20. Define three positions of the dumpster and analytically synthesize a linkage to move through them. The fixed pivots must be located on the existing truck.

5-7 See Project P3-7. Define three positions of the computer monitor and analytically synthesize a linkage to move through them. The fixed pivots must be located on the floor or wall.

*†5-8 Design a linkage to carry the body in Figure P5-1 through the two positions P_1 and P_2 at the angles shown in the figure. Use analytical synthesis without regard for the fixed pivots shown. Hint: Try the free choice values $z = 1.075$, $\phi = 204.4°$, $\beta_2 = -27°$; $s = 1.24$, $\psi = 74°$, $\gamma_2 = -40°$.

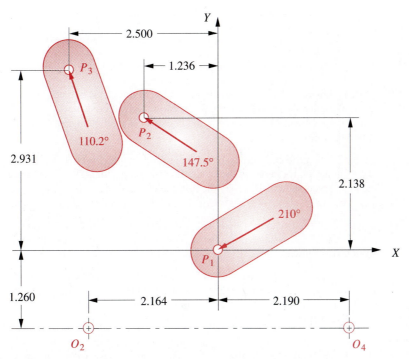

FIGURE P5-1

Data for Problems 5-8 to 5-11

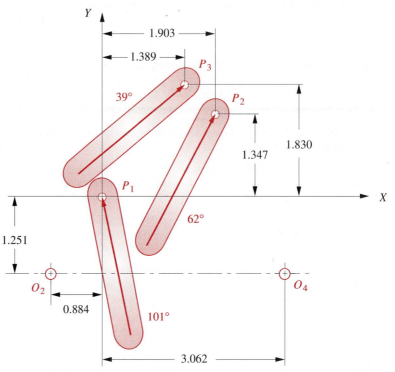

FIGURE P5-2

Data for Problems 5-12 to 5-15

†5-9　Design a linkage to carry the body in Figure P5-1 through the two positions P_2 and P_3 at the angles shown in the figure. Use analytical synthesis without regard for the fixed pivots shown. Hint: First try a rough graphical solution to create realistic values for free choices.

†5-10　Design a linkage to carry the body in Figure P5-1 through the three positions P_1, P_2, and P_3 at the angles shown in the figure. Use analytical synthesis without regard for the fixed pivots shown. Hint: Try the free choice values $\beta_2 = 30°$, $\beta_3 = 60°$; $\gamma_2 = -10°$, $\gamma_3 = 25°$.

*†5-11　Design a linkage to carry the body in Figure P5-1 through the three positions P_1, P_2, and P_3 at the angles shown in the figure. Use analytical synthesis and design it for the fixed pivots shown.

* Answers in Appendix F.

† These problems are suited to solution using *Mathcad,* or *TKSolver* equation solver programs. In most cases, your solution can be checked with program FOURBAR.

†5-12　Design a linkage to carry the body in Figure P5-2 through the two positions P_1 and P_2 at the angles shown in the figure. Use analytical synthesis without regard for the fixed pivots shown. Hint: Try the free choice values $z = 2$, $\phi = 150°$, $\beta_2 = 30°$; $s = 3$, $\psi = -50°$, $\gamma_2 = 40°$.

†5-13　Design a linkage to carry the body in Figure P5-2 through the two positions P_2 and P_3 at the angles shown in the figure. Use analytical synthesis without regard for the fixed pivots shown. Hint: First try a rough graphical solution to create realistic values for free choices.

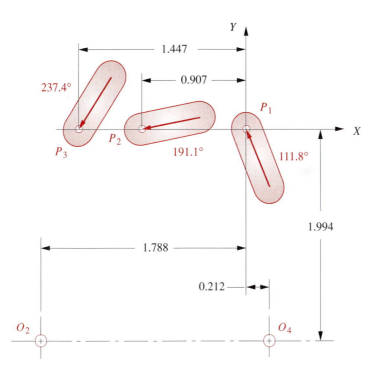

FIGURE P5-3

Data for Problems 5-16 to 5-20

†5-14 Design a linkage to carry the body in Figure P5-2 through the three positions P_1, P_2, and P_3 at the angles shown in the figure. Use analytical synthesis without regard for the fixed pivots shown.

*†5-15 Design a linkage to carry the body in Figure P5-2 through the three positions P_1, P_2, and P_3 at the angles shown in the figure. Use analytical synthesis and design it for the fixed pivots shown.

†5-16 Design a linkage to carry the body in Figure P5-3 through the two positions P_1 and P_2 at the angles shown in the figure. Use analytical synthesis without regard for the fixed pivots shown.

†5-17 Design a linkage to carry the body in Figure P5-3 through the two positions P_2 and P_3 at the angles shown in the figure. Use analytical synthesis without regard for the fixed pivots shown.

†5-18 Design a linkage to carry the body in Figure P5-3 through the three positions P_1, P_2, and P_3 at the angles shown in the figure. Use analytical synthesis without regard for the fixed pivots shown.

*†5-19 Design a linkage to carry the body in Figure P5-3 through the three positions P_1, P_2, and P_3 at the angles shown in the figure. Use analytical synthesis and design it for the fixed pivots shown.

* Answers in Appendix F.

† These problems are suited to solution using *Mathcad*, or *TKSolver* equation solver programs. In most cases, your solution can be checked with program FOURBAR.

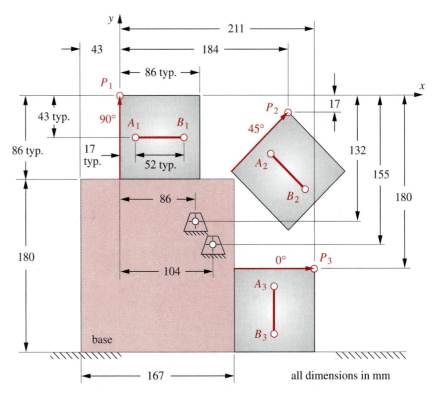

FIGURE P5-4

Data for Problems 5-21 to 5-26

† These problems are suited to solution using *Mathcad,* or *TKSolver* equation solver programs. In most cases, your solution can be checked with program FOURBAR.

†5-20   Write a program to generate and plot the circle-point and center-point circles for Problem 5-19 using an equation solver or any programming language.

†5-21   Design a fourbar linkage to carry the box in Figure P5-4 from position 1 to 2 without regard for the fixed pivots shown. Use points *A* and *B* for your attachment points. Determine the range of the transmission angle. The fixed pivots should be on the base.

†5-22   Design a fourbar linkage to carry the box in Figure P5-4 from position 1 to 3 without regard for the fixed pivots shown. Use points *A* and *B* for your attachment points. Determine the range of the transmission angle. The fixed pivots should be on the base.

†5-23   Design a fourbar linkage to carry the box in Figure P5-4 from position 2 to 3 without regard for the fixed pivots shown. Use points *A* and *B* for your attachment points. Determine the range of the transmission angle. The fixed pivots should be on the base.

†5-24   Design a fourbar linkage to carry the box in Figure P5-4 through the three positions shown in their numbered order without regard for the fixed pivots shown. Determine the range of the transmission angle. Use any points on the object as your attachment points. The fixed pivots should be on the base.

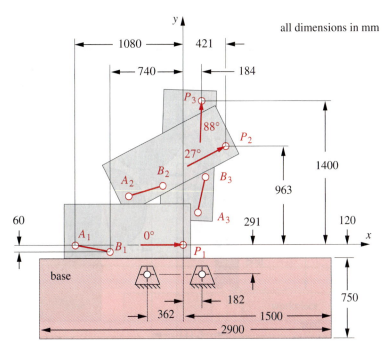

FIGURE P5-5

Data for Problems 5-27 to 5-30

†5-25 Design a fourbar linkage to carry the box in Figure P5-4 through the three positions shown in their numbered order without regard for the fixed pivots shown. Use points A and B for your attachment points. Determine the range of the transmission angle. Add a driver dyad with a crank to control the motion of your fourbar so that it cannot move beyond positions one and three.

*†5-26 Design a fourbar linkage to carry the box in Figure P5-4 through the three positions shown in their numbered order using the fixed pivots shown. Determine the range of the transmission angle. Add a driver dyad with a crank to control the motion of your fourbar so that it cannot move beyond positions one and three.

†5-27 Design a fourbar linkage to carry the object in Figure P5-5 through the three positions shown in their numbered order without regard for the fixed pivots shown. Use any points on the object as attachment points. The fixed pivots should be on the base. Determine the range of the transmission angle.

†5-28 Design a fourbar linkage to carry the object in Figure P5-5 through the three positions shown in their numbered order without regard for the fixed pivots shown. Use points A and B for your attachment points. Determine the range of the transmission angle.

†5-29 Design a fourbar linkage to carry the object in Figure P5-5 through the three positions shown in their numbered order using the fixed pivots shown. Determine the range of the transmission angle.

* Answers in Appendix F.

† These problems are suited to solution using *Mathcad,* or *TKSolver* equation solver programs. In most cases, your solution can be checked with program FOURBAR.

✏️ EXAMPLE 6-2

Finding All Instant Centers for a Fourbar Linkage.

Problem: Given a fourbar linkage in one position, find all *ICs* by graphical methods.

Solution: (see Figure 6-5)

1   Draw a circle with all links numbered around the circumference as shown in Figure 6-5a.

2   Locate as many *ICs* as possible by inspection. All pin joints will be permanent *ICs*. Connect the link numbers on the circle to create a linear graph and record those *ICs* found, as shown in Figure 6-5a.

3   Identify a link combination on the linear graph for which the *IC* has not been found, and draw a dotted line connecting those two link numbers. Identify two triangles on the graph which each contain the dotted line and whose other two sides are solid lines representing *ICs* already found. On the graph in Figure 6-5b, link numbers 1 and 3 have been connected with a dotted line. This line forms one triangle with sides 13, 34, 14 and another with sides 13, 23, 12. These triangles define trios of *ICs* which obey **Kennedy's rule**. Thus *ICs* 13, 34, and 14 **must lie on the same straight line**. Also *ICs* 13, 23 and 12 will **lie on a different straight line**.

4   On the linkage diagram draw a line through the two known *ICs* which form a trio with the unknown *IC*. Repeat for the other trio. In Figure 6-5b, a line has been drawn through $I_{1,2}$ and $I_{2,3}$ and extended. $I_{1,3}$ must lie on this line. Another line has been drawn through $I_{1,4}$ and $I_{3,4}$ and extended to intersect the first line. By Kennedy's rule, instant center $I_{1,3}$ must also lie on this line, so their intersection is $I_{1,3}$.

5   Connect link numbers 2 and 4 with a dotted line on the linear graph as shown in Figure 6-5c. This line forms one triangle with sides 24, 23, 34 and another with sides 24, 12, 14. These sides represent trios of *ICs* which obey Kennedy's rule. Thus *ICs* 24, 23, and 34 must lie on the same straight line. Also *ICs* 24, 12, and 14 lie on a different straight line.

6   On the linkage diagram draw a line through the two known *ICs* which form a trio with the unknown *IC*. Repeat for the other trio. In Figure 6-5c, a line has been drawn through $I_{1,2}$ and $I_{1,4}$ and extended. $I_{2,4}$ must lie on this line. Another line has been drawn through $I_{2,3}$ and $I_{3,4}$ and extended to intersect the first line. By Kennedy's rule, instant center $I_{2,4}$ must also lie on this line, so their intersection is $I_{2,4}$.

7   If there were more links, this procedure would be repeated until all *ICs* were found.

The presence of slider joints makes finding the instant centers a little more subtle as is shown in the next example. Figure 6-6a shows a **fourbar slider-crank linkage**. Note that there are only three pin joints in this linkage. All pin joints are *permanent instant centers*. But the joint between links 1 and 4 is a rectilinear, sliding full joint. A sliding joint is kinematically equivalent to an infinitely long link, "pivoted" at infinity. Figure 6-6b shows a nearly equivalent pin-jointed version of the slider-crank in which link 4 is a very long rocker. Point *B* now swings through a shallow arc which is nearly a straight

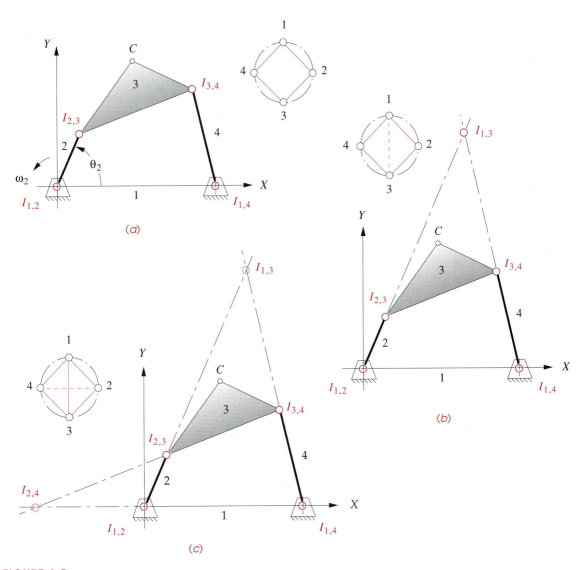

FIGURE 6-5

Locating instant centers in the pin-jointed linkage

line. It is clear in Figure 6-6b that, in this linkage, $I_{1,4}$ is at pivot O_4. Now imagine increasing the length of this long, link 4 rocker even more. In the limit, link 4 approaches infinite length, the pivot O_4 approaches infinity along the line which was originally the long rocker, and the arc motion of point B approaches a straight line. Thus, *a slider joint will have its instant center at infinity along a line perpendicular to the direction of sliding* as shown in Figure 6-6a.

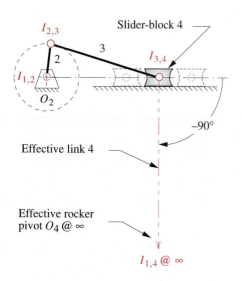

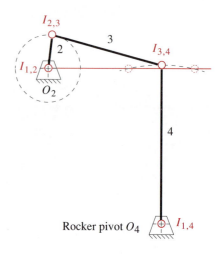

(a) Slider-crank linkage   (b) Crank-rocker linkage

FIGURE 6-6

A rectilinear slider's instant center is at infinity

EXAMPLE 6-3

Finding All Instant Centers for a Slider-Crank Linkage.

Problem: Given a slider-crank linkage in one position, find all *ICs* by graphical methods.

Solution: (see Figure 6-7)

1   Draw a circle with all links numbered around the circumference as shown in Figure 6-7a.

2   Locate all *ICs* possible by inspection. All pin joints will be permanent *ICs*. The slider joint's instant center will be at infinity along a line perpendicular to the axis of sliding. Connect the link numbers on the circle to create a linear graph and record those *ICs* found, as shown in Figure 6-7a.

3   Identify a link combination on the linear graph for which the *IC* has not been found, and draw a dotted line connecting those two link numbers. Identify two triangles on the graph which each contain the dotted line and whose other two sides are solid lines representing *ICs* already found. In the graph on Figure 6-7b, link numbers 1 and 3 have been connected with a dotted line. This line forms one triangle with sides 13, 34, 14 and another with sides 13, 23, 12. These sides represent trios of *ICs* which obey Kennedy's rule. Thus *ICs* 13, 34, and 14 must lie on the same straight line. Also *ICs* 13, 23, and 12 lie on a different straight line.

4   On the linkage diagram draw a line through the two known *ICs* which form a trio with the unknown *IC*. Repeat for the other trio. In Figure 6-7b, a line has been drawn from $I_{1,2}$ through $I_{2,3}$ and extended. $I_{1,3}$ must lie on this line. Another line has been drawn from $I_{1,4}$

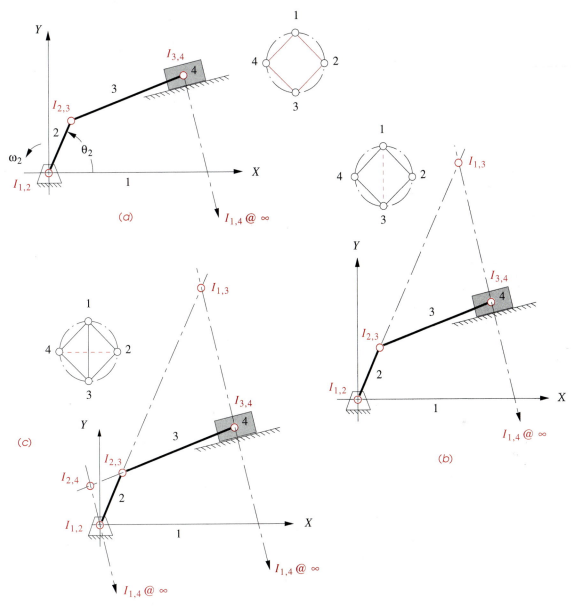

FIGURE 6-7

Locating instant centers in the slider-crank linkage

(at infinity) through $I_{3,4}$ and extended to intersect the first line. By Kennedy's rule, instant center $I_{1,3}$ must also lie on this line, so their intersection is $I_{1,3}$.

5   Connect link numbers 2 and 4 with a dotted line on the graph as shown in Figure 6-7c. This line forms one triangle with sides 24, 23, 34 and another with sides 24, 12, 14. These sides also represent trios of *ICs* which obey Kennedy's rule. Thus *ICs* 24, 23, and 34 must lie on the same straight line. Also *ICs* 24, 12, and 14 lie on a different straight line.

6   On the linkage diagram draw a line through the two known *ICs* which form a trio with the unknown *IC*. Repeat for the other trio. In Figure 6-7c, a line has been drawn from $I_{1,2}$ to intersect $I_{1,4,}$ and extended. Note that the only way to "intersect" $I_{1,4}$ at infinity is to draw a line parallel to the line $I_{3,4}$ $I_{1,4}$ since all parallel lines intersect at infinity. Instant center $I_{2,4}$ must lie on this parallel line. Another line has been drawn through $I_{2,3}$ and $I_{3,4}$ and extended to intersect the first line. By Kennedy's rule, instant center $I_{2,4}$ must also lie on this line, so their intersection is $I_{2,4}$.

7   If there were more links, this procedure would be repeated until all *ICs* were found.

The procedure in this slider example is identical to that used in the pin-jointed fourbar, except that it is complicated by the presence of instant centers located at infinity.

In Section 2.9 and Figure 2-10c (p. 41) we showed that a cam-follower mechanism is really a fourbar linkage in disguise. As such it will also possess instant centers. The presence of the half joint in this, or any linkage, makes the location of the instant centers a little more complicated. We have to recognize that the instant center between any two links will be along a line that is perpendicular to the *relative velocity* vector between the links at the half joint, as shown in the following example. Figure 6-8 shows the same cam-follower mechanism as in Figure 2-14 (p. 45). The effective links 2, 3, and 4 are also shown.

EXAMPLE 6-4

Finding All Instant Centers for a Cam-Follower Mechanism.

Problem: Given a cam and follower in one position, find all *ICs* by graphical methods.

Solution: (see Figure 6-8)

1   Draw a circle with all links numbered around the circumference as shown in Figure 6-8b. In this case there are only three links and thus only three *ICs* to be found as shown by equation 6.8. Note that the links are numbered 1, 2, and 4. The missing link 3 is the variable-length effective coupler.

2   Locate all *ICs* possible by inspection. All pin joints will be permanent *ICs*. The two fixed pivots $I_{1,2}$ and $I_{1,4}$ are the only pin joints here. Connect the link numbers on the circle to create a linear graph and record those *ICs* found, as shown in Figure 6-8b. The only link combination on the linear graph for which the *IC* has not been found is $I_{2,4}$, so draw a dotted line connecting those two link numbers.

3   Kennedy's rule says that all three *ICs* must lie on the same straight line; thus the remaining instant center $I_{2,4}$ must lie on the line $I_{1,2}$ $I_{1,4}$ extended. Unfortunately in this example, we have too few links to find a second line on which $I_{2,4}$ must lie.

4   On the linkage diagram draw a line through the two known *ICs* which form a trio with the unknown *IC*. In Figure 6-8c, a line has been drawn from $I_{1,2}$ through $I_{1,4}$ and extended. This is, of course, link 1. By Kennedy's rule, $I_{2,4}$ must lie on this line.

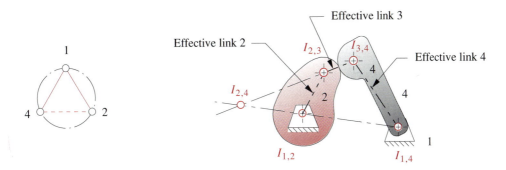

(b) The linkage graph

(c) The instantaneously equivalent "effective linkage"

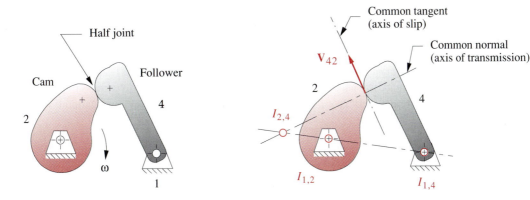

(a) The cam and follower

(d) Finding $I_{2,4}$ without using the effective linkage

FIGURE 6-8

Locating instant centers in the cam-follower mechanism

5 Looking at Figure 6-8c which shows the effective links of the equivalent fourbar linkage for this position, we can extend effective link 3 until it intersects link 1 extended. Just as in the "pure" fourbar linkage, instant center 2,4 lies on the intersection of links 1 and 3 extended (see Example 6-2, p. 250).

6 Figure 6-8d shows that it is not necessary to construct the effective fourbar linkage to find $I_{2,4}$. Note that the **common tangent** to links 2 and 4 at their contact point (the half joint) has been drawn. This line is also called the **axis of slip** because it is the line along which all relative (slip) velocity will occur between the two links. Thus the velocity of link 4 versus 2, $\mathbf{V}_{42}$, is directed along the axis of slip. Instant center $I_{2,4}$ must therefore lie along a line perpendicular to the common tangent, called the **common normal.** Note that this line is the same as the effective link 3 line in Figure 6-8c.

6.4   VELOCITY ANALYSIS WITH INSTANT CENTERS

Once the *ICs* have been found, they can be used to do a very rapid graphical velocity analysis of the linkage. Note that, depending on the particular position of the linkage being analyzed, some of the *ICs* may be very far removed from the links. For example, if links 2 and 4 are nearly parallel, their extended lines will intersect at a point far away and not be practically available for velocity analysis. Figure 6-9 shows the same linkage as Figure 6-5 (p. 251) with $I_{1,3}$ located and labeled. From the definition of the instant center, both links sharing the instant center will have identical velocity at that point. Instant center $I_{1,3}$ involves the coupler (link 3) which is in complex motion, and the ground link 1, which is stationary. All points on link 1 have zero velocity in the global coordinate system, which is embedded in link 1. Therefore, $I_{1,3}$ must have zero velocity at this instant. If $I_{1,3}$ has zero velocity, then it can be considered to be an instantaneous "fixed pivot" about which link 3 is in pure rotation with respect to link 1. A moment later, $I_{1,3}$ will move to a new location and link 3 will be "pivoting" about a new instant center.

The velocity of point A is shown on Figure 6-9. The magnitude of $\mathbf{V}_A$ can be computed from equation 6.7 (p. 244). Its direction and sense can be determined by inspection as was done in Example 6-1 (p. 246). Note that point A is also instant center $I_{2,3}$. It has the same velocity as part of link 2 and as part of link 3. Since link 3 is effectively pivoting about $I_{1,3}$ at this instant, the angular velocity ω_3 can be found by rearranging equation 6.7:

$$\omega_3 = \frac{v_A}{\left(AI_{1,3}\right)} \tag{6.9a}$$

Once w_3 is known, the magnitude of $\mathbf{V}_B$ can also be found from equation 6.7:

$$v_B = \left(BI_{1,3}\right)\omega_3 \tag{6.9b}$$

Once $\mathbf{V}_B$ is known, w_4 can also be found from equation 6.7:

$$\omega_4 = \frac{v_B}{\left(BO_4\right)} \tag{6.9c}$$

Finally, the magnitude of $\mathbf{V}_C$ (or the velocity of any other point on the coupler) can be found from equation 6.7:

$$v_C = \left(CI_{1,3}\right)\omega_3 \tag{6.9d}$$

Note that equations 6.7 and 6.9 provide only the **scalar magnitude** of these velocity vectors. We have to determine their **direction** from the information in the scale diagram (Figure 6-9). Since we know the location of $I_{1,3}$, which is an instantaneous "fixed" pivot for link 3, all of that link's absolute velocity vectors for this instant will be **perpendicular to their radii from $I_{1,3}$ to the point in question.** $\mathbf{V}_B$ and $\mathbf{V}_C$ can be seen to be perpendicular to their radii from $I_{1,3}$. Note that $\mathbf{V}_B$ is also perpendicular to the radius from O_4 because B is also pivoting about that point as part of link 4.

A rapid graphical solution to equations 6.9 is shown in the figure. Arcs centered at $I_{1,3}$ are swung from points B and C to intersect line $AI_{1,3}$. The magnitudes of velocities

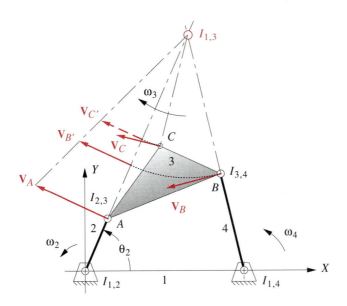

FIGURE 6-9

Velocity analysis using instant centers

$\mathbf{V}_{B'}$ and $\mathbf{V}_{C'}$ are found from the vectors drawn perpendicular to that line at the intersections of the arcs and line $AI_{1,3}$. The lengths of the vectors are defined by the line from the tip of $\mathbf{V}_A$ to the instant center $I_{1,3}$. These vectors can then be slid along their arcs back to points B and C, maintaining their tangency to the arcs.

Thus, we have in only a few steps found all the same velocities that were found in the more tedious method of Example 6-1. The instant center method is a quick graphical method to analyze velocities, but it will only work if the instant centers are in reachable locations for the particular linkage position analyzed. However, the graphical method using the velocity difference equation shown in Example 6-1 will always work, regardless of linkage position.

Angular Velocity Ratio

The **angular velocity ratio** mV is defined as *the output angular velocity divided by the input angular velocity.* For a fourbar mechanism this is expressed as:

$$m_V = \frac{\omega_4}{\omega_2} \qquad (6.10)$$

We can derive this ratio for any linkage by constructing a **pair of effective links** as shown in Figure 6-10a. The definition of **effective link pairs** is *two lines, mutually parallel, drawn through the fixed pivots and intersecting the coupler extended.* These are shown as O_2A' and O_4B' in Figure 6-10a. Note that there is an infinity of possible effective link pairs. They must be parallel to one another but may make any angle with link 3. In the figure they are shown perpendicular to link 3 for convenience in the derivation to follow. The angle between links 2 and 3 is shown as ν. The transmission angle between links 3 and 4 is μ. We will now derive an expression for the angular velocity ratio using these effective links, the actual link lengths, and angles ν and μ.

From geometry:

$$O_2A' = (O_2A)\sin\nu \qquad\qquad O_4B' = (O_4B)\sin\mu \qquad\qquad (6.11a)$$

From equation 6.7

$$V_{A'} = (O_2A')\omega_2 \qquad\qquad (6.11b)$$

The component of velocity $V_{A'}$ lies along the link AB. Just as with a two-force member in which a force applied at one end transmits only its component that lies along the link to the other end, this velocity component can be transmitted along the link to point B. This is sometimes called the **principle of transmissibility**. We can then equate these components at either end of the link.

$$V_{A'} = V_{B'} \qquad\qquad (6.11c)$$

Then:

$$O_2A'\omega_2 = O_4B'\omega_4 \qquad\qquad (6.11d)$$

rearranging:

$$\frac{\omega_4}{\omega_2} = \frac{O_2A'}{O_4B'} \qquad\qquad (6.11e)$$

and substituting:

$$\frac{\omega_4}{\omega_2} = \frac{O_2A\sin\nu}{O_4B\sin\mu} = m_V \qquad\qquad (6.11f)$$

Note in equation 6.11f that as angle ν goes through zero, the angular velocity ratio will be zero regardless of the values of ω_2 or the link lengths, and thus ω_4 will be zero. When angle ν is zero, links 2 and 3 will be colinear and thus be in their toggle positions. We learned in Section 3.3 (p. 80) that the limiting positions of link 4 are defined by these toggle conditions. We should expect that the velocity of link 4 will be zero when it has come to the end of its travel. An even more interesting situation obtains if we allow angle μ to go to zero. Equation 6.11f shows that ω_4 **will go to infinity** when $\mu = 0$, regardless of the values of ω_2 or the link lengths. We clearly cannot allow μ to reach zero. In fact, we learned in Section 3.3 that we should keep this transmission angle μ above about 40 degrees to maintain good quality of motion and force transmission.[*]

Figure 6-10b shows the same linkage as in Figure 6-10a, but the effective links have now been drawn so that they are not only parallel but are colinear, and thus lie on top of one another. Both intersect the extended coupler at the same point, which is instant center $I_{2,4}$. So, A' and B' of Figure 6-10a are now coincident at $I_{2,4}$. This allows us to write an equation for the **angular velocity ratio** in terms of the distances from the fixed pivots to instant center $I_{2,4}$.

$$m_V = \frac{\omega_4}{\omega_2} = \frac{O_2I_{2,4}}{O_4I_{2,4}} \qquad\qquad (6.11g)$$

Thus, the instant center $I_{2,4}$ can be used to determine the **angular velocity ratio**.

[*] This limitation on transmission angle is only critical if the output load is applied to a link that is pivoted to ground (i.e., to link 4 in the case of a fourbar linkage). If the load is applied to a floating link (e.g., a coupler), then other measures of the quality of force transmission than the transmission angle are more appropriate, as discussed in Chapter 11, Section 11.12, p. 554, where the joint force index (JFI) is defined.

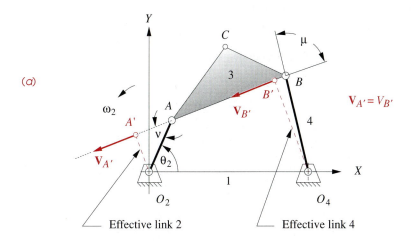

(a)

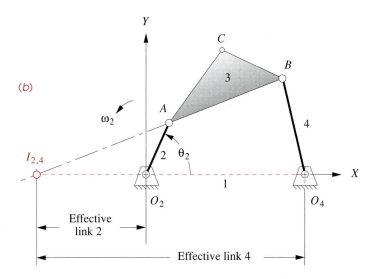

(b)

FIGURE 6-10

Effective links and the angular velocity ratio

Mechanical Advantage

The power P in a mechanical system can be defined as the dot or scalar product of the force vector $\mathbf{F}$ and the velocity vector $\mathbf{V}$ at any point :

$$P = \mathbf{F} \cdot \mathbf{V} = F_x V_x + F_y V_y \qquad (6.12a)$$

For a rotating system, power P becomes the product of torque T and angular velocity ω which, in two dimensions, have the same (z) direction:

$$P = T\omega \qquad (6.12b)$$

The power flows through a passive system and:

$$P_{in} = P_{out} + losses \tag{6.12c}$$

Mechanical efficiency can be defined as:

$$\varepsilon = \frac{P_{out}}{P_{in}} \tag{6.12d}$$

Linkage systems can be very efficient if they are well made with low friction bearings on all pivots. Losses are often less than 10%. For simplicity in the following analysis we will assume that the losses are zero (i.e., a conservative system). Then, letting T_{in} and ω_{in} represent input torque and angular velocity, and T_{out} and ω_{out} represent output torque and angular velocity, then:

$$P_{in} = T_{in}\omega_{in}$$
$$P_{out} = T_{out}\omega_{out} \tag{6.12e}$$

and :
$$P_{out} = P_{in}$$
$$T_{out}\omega_{out} = T_{in}\omega_{in}$$
$$\frac{T_{out}}{T_{in}} = \frac{\omega_{in}}{\omega_{out}} \tag{6.12f}$$

Note that the **torque ratio** ($m_T = T_{out}/T_{in}$) is the inverse of the angular velocity ratio.

Mechanical advantage (m_A) can be defined as:

$$m_A = \frac{F_{out}}{F_{in}} \tag{6.13a}$$

Assuming that the input and output forces are applied at some radii r_{in} and r_{out}, perpendicular to their respective force vectors,

$$F_{out} = \frac{T_{out}}{r_{out}}$$
$$F_{in} = \frac{T_{in}}{r_{in}} \tag{6.13b}$$

substituting equations 6.13b in 6.13a gives an expression in terms of torque.

$$m_A = \left(\frac{T_{out}}{T_{in}}\right)\left(\frac{r_{in}}{r_{out}}\right) \tag{6.13c}$$

Substituting equation 6.12f in 6.13c gives

$$m_A = \left(\frac{\omega_{in}}{\omega_{out}}\right)\left(\frac{r_{in}}{r_{out}}\right) \tag{6.13d}$$

and substituting equation 6.11f gives

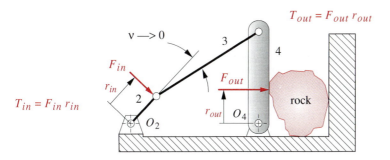

FIGURE 6-11

"Rock-crusher" toggle mechanism

$$m_A = \left(\frac{O_4 B \sin \mu}{O_2 A \sin \nu} \right) \left(\frac{r_{in}}{r_{out}} \right) \tag{6.13e}$$

See Figure 6-11 and compare equation 6.13e to equation 6.11f and its discussion under **angular velocity ratio** above. Equation 6.13e shows that for any choice of r_{in} and r_{out}, the mechanical advantage responds to changes in angles ν and μ in opposite fashion to that of the angular velocity ratio. If the transmission angle μ goes to zero (which we don't want it to do), the mechanical advantage also goes to zero regardless of the amount of input force or torque applied. But, when angle ν goes to zero (which it can and does, twice per cycle in a Grashof linkage), the mechanical advantage becomes infinite! This is the principle of a rock-crusher mechanism as shown in Figure 6-11. A quite moderate force applied to link 2 can generate a huge force on link 4 to crush the rock. Of course, we cannot expect to achieve the theoretical output of infinite force or torque magnitude, as the strengths of the links and joints will limit the maximum forces and torques obtainable. Another common example of a linkage which takes advantage of this theoretically infinite mechanical advantage at the toggle position is a ViseGrip locking pliers (see Figure P6-21, p. 296).

These two ratios, **angular velocity ratio** and **mechanical advantage**, provide useful, dimensionless **indices of merit** by which we can judge the relative quality of various linkage designs which may be proposed as solutions.

Using Instant Centers in Linkage Design

In addition to providing a quick numerical velocity analysis, instant center analysis more importantly gives the designer a remarkable overview of the linkage's global behavior. It is quite difficult to mentally visualize the complex motion of a "floating" coupler link even in a simple fourbar linkage, unless you build a model or run a computer simulation. Because this complex coupler motion in fact reduces to an instantaneous pure rotation about the instant center $I_{1,3}$, finding that center allows the designer to visualize the motion of the coupler as a pure rotation. One can literally *see* the motion and the directions of velocities of any points of interest by relating them to the instant center. It is only necessary to draw the linkage in a few positions of interest, showing the instant center locations for each position.

Figure 6-12 shows a practical example of how this visual, qualitative analysis technique could be applied to the design of an automobile rear suspension system. Most automobile suspension mechanisms are either fourbar linkages or fourbar slider-cranks, with the wheel assembly carried on the coupler (as was also shown in Figure 3-19, p. 108). Figure 6-12a shows a rear suspension design from a domestic car of 1970's vintage which was later redesigned because of a disturbing tendency to "bump steer," i.e., turn the rear axle when hitting a bump on one side of the car. The figure is a view looking from the center of the car outward, showing the fourbar linkage which controls the up and down motion of one side of the rear axle and one wheel. Links 2 and 4 are pivoted to the frame of the car which is link 1. The wheel and axle assembly is rigidly attached to the coupler, link 3. Thus the wheel assembly has complex motion in the vertical plane. Ideally, one would like the wheel to move up and down in a straight vertical line when hitting a bump. Figure 6-12b shows the motion of the wheel and the new instant center ($I_{1,3}$) location for the situation when one wheel has hit a bump. The velocity vector for the center of the wheel in each position is drawn perpendicular to its radius

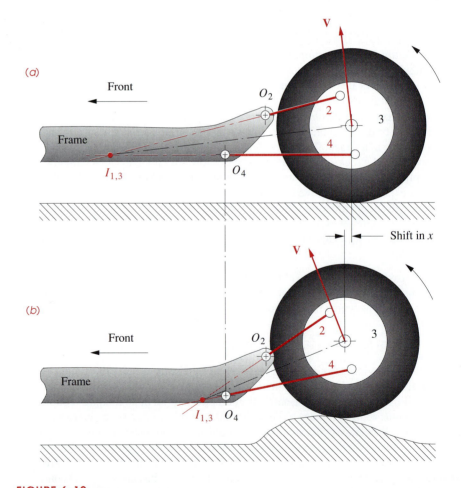

FIGURE 6-12

"Bump steer" due to shift in instant center location

from $I_{1,3}$. You can see that the wheel center has a significant horizontal component of motion as it moves up over the bump. This horizontal component causes the wheel center on that side of the car to move forward while it moves upward, thus turning the axle (about a vertical axis) and steering the car with the rear wheels in the same way that you steer a toy wagon. Viewing the path of the instant center over some range of motion gives a clear picture of the behavior of the coupler link. The undesirable behavior of this suspension linkage system could have been predicted from this simple instant center analysis before ever building the mechanism.

Another practical example of the effective use of instant centers in linkage design is shown in Figure 6-13, which is an optical adjusting mechanism used to position a mirror and allow a small amount of rotational adjustment. [1] A more detailed account of this design case study [2] is provided in Chapter 16. The designer, K. Towfigh, recognized that $I_{1,3}$ at point E is an instantaneous "fixed pivot" and will allow very small pure rotations about that point with very small translational error. He then designed a one-piece, plastic fourbar linkage whose "pin joints" are thin webs of plastic which flex to allow slight rotation. This is termed a **compliant linkage**, one that uses elastic deformations of the links as hinges instead of pin joints. He then placed the mirror on the coupler at $I_{1,3}$. Even the fixed link 1 is the same piece as the "movable links" and has a small set screw to provide the adjustment. A simple and elegant design.

6.5   CENTRODES

Figure 6-14 illustrates the fact that the successive positions of an instant center (or **centro**) form a path of their own. *This path, or locus, of the instant center is called the **centrode**.* Since there are two links needed to create an instant center, there will be two centrodes associated with any one instant center. These are formed by projecting the path of the instant center first on one link and then on the other. Figure 6-14a shows the locus of instant center $I_{1,3}$ as projected onto link 1. Because link 1 is stationary, or fixed, this is called the **fixed centrode**. By temporarily inverting the mechanism and fixing link 3

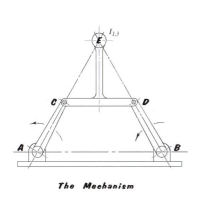

The Mechanism

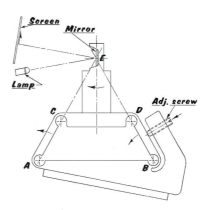

The Final Product of Keivan Towfigh

FIGURE 6-13

An optical adjustment linkage *(Reproduced from reference (2) with permission)*

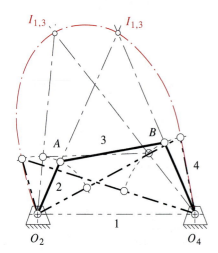

(a) The fixed centrode

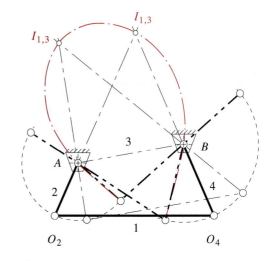

(b) The moving centrode

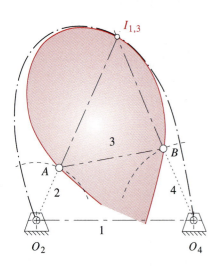

(c) The centrodes in contact

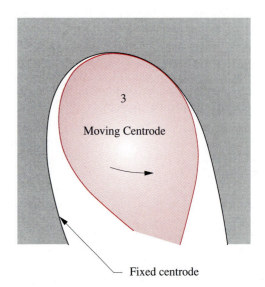

(d) Roll the moving centrode against the
fixed centrode to produce the same
coupler motion as the original linkage

FIGURE 6-14

Open-loop fixed and moving centrodes (or polodes) of a fourbar linkage

as the ground link, as shown in Figure 6-14b, we can move link 1 as the coupler and project the locus of $I_{1,3}$ onto link 3. In the original linkage, link 3 was the moving coupler, so this is called the **moving centrode**. Figure 6-14c shows the original linkage with both fixed and moving centrodes superposed.

The definition of the instant center says that both links have the same velocity at that point, at that instant. Link 1 has zero velocity everywhere, as does the fixed centrode. So, as the linkage moves, the moving centrode must roll against the fixed centrode without slipping. If you cut the fixed and moving centrodes out of metal, as shown in Figure 6-14d, and roll the moving centrode (which is link 3) against the fixed centrode (which is link 1), the complex motion of link 3 will be identical to that of the original linkage. *All of the coupler curves of points on link 3 will have the same path shapes as in the original linkage.* We now have, in effect, a "linkless" fourbar linkage, really one composed of two bodies which have these centrode shapes rolling against one another. Links 2 and 4 have been eliminated. Note that the example shown in Figure 6-14 is a non-Grashof fourbar. The lengths of its centrodes are limited by the double-rocker toggle positions.

All instant centers of a linkage will have centrodes. If the links are directly connected by a joint, such as $I_{2,3}$, $I_{3,4}$, $I_{1,2}$, and $I_{1,4}$, their fixed and moving centrodes will degenerate to a point at that location on each link. The most interesting centrodes are those involving links not directly connected to one another such as $I_{1,3}$ and $I_{2,4}$. If we look at the double-crank linkage in Figure 6-15a in which links 2 and 4 both revolve fully, we see that the centrodes of $I_{1,3}$ form closed curves. The motion of link 3 with respect to link 1 could be duplicated by causing these two centrodes to roll against one another without slipping. Note that there are two loops to the moving centrode. Both must roll on the single-loop fixed centrode to complete the motion of the equivalent double-crank linkage.

We have so far dealt largely with the instant center $I_{1,3}$. Instant center $I_{2,4}$ involves two links which are each in pure rotation and not directly connected to one another. If we use a special-case Grashof linkage with the links crossed (sometimes called an **antiparallelogram** linkage), the centrodes of $I_{2,4}$ become ellipses as shown in Figure 6-15b. To guarantee no slip, it will probably be necessary to put meshing teeth on each centrode. We then will have a pair of elliptical, **noncircular gears**, or *gearset*, which gives the *same output motion as the original double-crank linkage* and will have the *same variations in the angular velocity ratio and mechanical advantage as the linkage* had. Thus we can see that *gearsets are also just fourbar linkages in disguise.* Noncircular gears find much use in machinery, such as printing presses, where rollers must be speeded and slowed with some pattern during each cycle or revolution. More complicated shapes of noncircular gears are analogous to cams and followers in that the equivalent fourbar linkage must have variable-length links. **Circular gears** are just a special case of noncircular gears which give a **constant angular velocity ratio** and are widely used in all machines. Gears and gearsets will be dealt with in more detail in Chapter 10.

In general, centrodes of crank-rockers and double- or triple-rockers will be open curves with asymptotes. Centrodes of double-crank linkages will be closed curves. Program FOURBAR will calculate and draw the fixed and moving centrodes for any linkage input to it. Input the datafiles F06-14.4br, F06-15a.4br, and F06-15b.4br into program FOURBAR to see the centrodes of these linkage drawn as the linkages rotate.

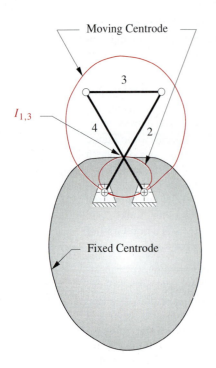

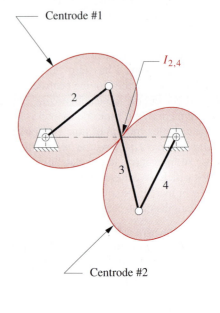

(a) Closed-loop centrodes of $I_{1,3}$
     for a Grashof double-crank linkage

(b) Ellipsoidal centrodes of $I_{2,4}$
     for a special-case Grashof
     anti-parallelogram linkage

FIGURE 6-15

Closed-loop fixed and moving centrodes

A "Linkless" Linkage

A common example of a mechanism made of centrodes is shown in Figure 6-16a. You have probably rocked in a *Boston* or *Hitchcock* rocking chair and experienced the soothing motions that it delivers to your body. You may have also rocked in a *platform* rocker as shown in Figure 6-16b and noticed that its motion did not feel as soothing.

There are good kinematic reasons for the difference. The platform rocker has a fixed pin joint between the seat and the base (floor). Thus all parts of your body are in pure rotation along concentric arcs. You are in effect riding on the rocker of a linkage.

The Boston rocker has a shaped (curved) base, or "runners," which rolls against the floor. These runners are usually *not* circular arcs. They have a higher-order curve contour. They are, in fact, **moving centrodes**. The floor is the **fixed centrode**. When one is rolled against the other, the chair and its occupant experience coupler curve motion. Every part of your body travels along a different sixth-order coupler curve which provides smooth accelerations and velocities and feels better than the cruder second-order (circular) motion of the platform rocker. Our ancestors, who carved these rocking chairs, probably had never heard of fourbar linkages and centrodes, but they knew intuitively how to create comfortable motions.

Cusps

Another example of a centrode which you probably use frequently is the path of the tire on your car or bicycle. As your tire rolls against the road without slipping, the road becomes a fixed centrode and the circumference of the tire is the moving centrode. The tire is, in effect, the coupler of a linkless fourbar linkage. All points on the contact surface of the tire move along cycloidal coupler curves and pass through a cusp of zero velocity when they reach the fixed centrode at the road surface as shown in Figure 6-17a. All other points on the tire and wheel assembly travel along coupler curves which do not have cusps. This last fact is a clue to a means to identify coupler points which will have cusps in their coupler curve. *If a coupler point is chosen to be on the moving centrode at one extreme of its path motion (i.e., at one of the positions of $I_{1,3}$), then it will have a cusp in its coupler curve.* Figure 6-17b shows a coupler curve of such a point, drawn with program FOURBAR. The right end of the coupler path touches the moving centrode and as a result has a cusp at that point. So, if you desire a cusp in your coupler motion, many are available. Simply choose a coupler point on the moving centrode of link 3. Read the diskfile F06-17b.4br into program FOURBAR to animate that linkage with its coupler curve or centrodes. Note in Figure 6-14 (p. 264) that choosing any location of instant center $I_{1,3}$ on the coupler as the coupler point will provide a cusp at that point.

6.6 VELOCITY OF SLIP

When there is a sliding joint between two links and neither one is the ground link, the velocity analysis is more complicated. Figure 6-18 shows an inversion of the fourbar slider-crank mechanism in which the sliding joint is floating, i.e., not grounded. To solve for the velocity at the sliding joint A, we have to recognize that there is more than one point A at that joint. There is a point A as part of link 2 (A_2), a point A as part of link 3 (A_3), and a point A as part of link 4 (A_4). This is a CASE 2 situation in which we have at least two points belonging to different links but occupying the same location at a given instant. Thus, the **relative velocity** equation 6.6 (p. 243) will apply. We can usually solve for the velocity of at least one of these points directly from the known input information using equation 6.7 (p. 244). It and equation 6.6 are all that are needed to solve for everything else. In this example link 2 is the driver, and θ_2 and ω_2 are given for the "freeze frame" position shown. We wish to solve for ω_4, the angular velocity of link 4, and also for the velocity of slip at the joint labeled A.

   In Figure 6-18 the **axis of slip** is shown to be tangent to the slider motion and is the line along which all sliding occurs between links 3 and 4. The **axis of transmission** is defined to be perpendicular to the axis of slip and pass through the slider joint at A. This *axis of transmission is the* **only line** *along which we can transmit motion or force across the slider joint, except for friction.* We will assume friction to be negligible in this example. Any force or velocity vector applied to point A can be resolved into two components along these two axes which provide a *translating and rotating, local coordinate system* for analysis at the joint. The component along the axis of transmission will do useful work at the joint. But, the component along the axis of slip does no work, except *friction work.*

Coupler
motion

Moving
centrode

Fixed
centrode

(*a*) Boston rocker

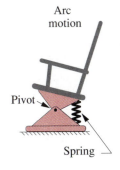

Arc
motion

Pivot

Spring

(*b*) Platform rocker

FIGURE 6-16

Some rocking chairs use centrodes of a fourbar linkage

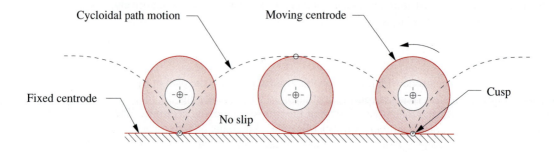

Cycloidal path motion

Moving centrode

Fixed centrode

No slip

Cusp

(a) Cycloidal motion of a circular, moving centrode rolling on a straight, fixed centrode

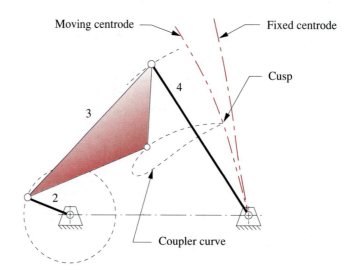

Moving centrode

Fixed centrode

Cusp

4

3

2

Coupler curve

(b) Coupler curve cusps exist only on the moving centrode

FIGURE 6-17

Examples of centrodes

 **EXAMPLE 6-5**

Graphical Velocity Analysis at a Sliding Joint.

Problem: Given θ_2, θ_3, θ_4, ω_2, find ω_3, ω_4, V_A, by graphical methods.

Solution: (see Figure 6-18)

1   Start at the end of the linkage for which you have the most information. Calculate the mag-
    nitude of the velocity of **point A as part of link 2** (A_2) using scalar equation 6.7 (p. 244).

$$v_{A_2} = (AO_2)\omega_2$$

                                                                                                (a)

2   Draw the velocity vector $\mathbf{V}_{A_2}$ with its length equal to its magnitude v_{A_2} at some convenient scale and with its root at point A and its direction perpendicular to the radius AO_2. Its sense is the same as that of ω_2 as is shown in Figure 6-18.

3   Draw the **axis of slip** and **axis of transmission** through point A.

4   Project $\mathbf{V}_{A_2}$ onto the axis of slip and onto the axis of transmission to create the components $\mathbf{V}_{A_2slip}$ and $\mathbf{V}_{trans}$ of $\mathbf{V}_{A_2}$ on the axes of slip and transmission, respectively. Note that the **transmission component** is shared by all true velocity vectors at this point, as it is the only component which can transmit across the joint.

5   Note that link 3 is pin-jointed to link 2, so $\mathbf{V}_{A_3} = \mathbf{V}_{A_2}$.

6   Note that the direction of the velocity of point $\mathbf{V}_{A_4}$ is predictable since all points on link 4 are pivoting in pure rotation about point O_4. Draw the line pp through point A and perpendicular to the effective link 4, AO_4. Line pp is the direction of velocity $\mathbf{V}_{A_4}$.

7   Construct the true magnitude of velocity vector $\mathbf{V}_{A_4}$ by extending the projection of the **transmission component** $\mathbf{V}_{trans}$ until it intersects line pp.

8   Project $\mathbf{V}_{A_4}$ onto the axis of slip to create the **slip component** $\mathbf{V}_{A_4slip}$.

9   Write the relative velocity vector equation 6.6 (p. 243) for the **slip components** of point A_2 versus point A_4.

$$V_{slip42} = V_{A4slip} - V_{A2slip} \qquad\qquad (b)$$

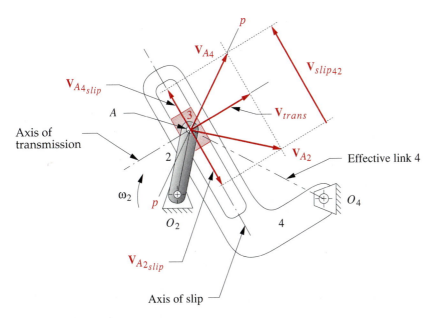

FIGURE 6-18

Velocity of slip and velocity of transmission (note that the applied ω is negative as shown)

10   The angular velocities of links 3 and 4 are identical because they share the slider joint and must rotate together. They can be calculated from equation 6.7:

$$\omega_4 = \omega_3 = \frac{V_{A_4}}{AO_4} \qquad (c)$$

Instant center analysis can also be used to solve sliding joint velocity problems graphically.

✎ EXAMPLE 6-6

Graphical Velocity Analysis in the Fourbar Inverted Slider-Crank Mechanism Using Instant Centers.

Problem: Given θ_2, θ_3, θ_4, ω_2, find ω_3, ω_4, V_A, by graphical methods.

Solution: (see Figure 6-19)

1   Start at the end of the linkage about which you have the most information. Calculate the magnitude of the velocity of point A as part of link 2 (A_2) using scalar equation 6.7 (p. 244).

$$v_{A_2} = (AO_2)\omega_2 \qquad (a)$$

2   Draw the velocity vector $\mathbf{V}_{A_2}$ with its length equal to its magnitude v_{A_2} at some convenient scale and with its root at point A and its direction perpendicular to the radius AO_2. Its sense is the same as that of ω_2 as is shown in Figure 6-19. Note that link 3 is pin-jointed to link 2, so $\mathbf{V}_{A_3} = \mathbf{V}_{A_2}$.

3   Find the instant centers of the linkage as shown in Figure 6-19.

4   Define a point (B) on the slider block for analysis. Draw the **axis of slip** and **axis of transmission** through point B. Note that point B is a multiple point, belonging to both link 3 and link 4, and has different linear velocities in each.

5   Project $\mathbf{V}_{A_2}$ onto the axis of slip to create the orthogonal component $\mathbf{V}_{A_{3slip}}$ along link 3. Translate this slip component along link 3 and place it at point B. Rename it $\mathbf{V}_{B_{3slip}}$.

6   The direction of the true velocity of point B as part of link 3 ($\mathbf{V}_{B_3}$) is along a line perpendicular to the radius from $I_{1,3}$ to B. Construct a perpendicular to $\mathbf{V}_{B_{3slip}}$ at its tip and create $\mathbf{V}_{B_3}$.

7   Project $\mathbf{V}_{B_3}$ onto the axis of transmission to create the component $\mathbf{V}_{trans}$. Note that the **transmission component** is shared by all true velocity vectors at this point, as it is the only component which can transmit across the joint.

8   Note that the direction of the velocity of point $\mathbf{V}_{B_4}$ is predictable since all points on link 4 are pivoting in pure rotation about point O_4. Construct a line in the direction of $\mathbf{V}_{B_4}$ perpendicular to the effective link 4. Construct the true magnitude of velocity vector $\mathbf{V}_{B_4}$ by extending the projection of the transmission component $\mathbf{V}_{trans}$ until it intersects the line of $\mathbf{V}_{B_4}$.

9   Project $\mathbf{V}_{B_4}$ onto the axis of slip to create the **slip component** $\mathbf{V}_{B_{4slip}}$.

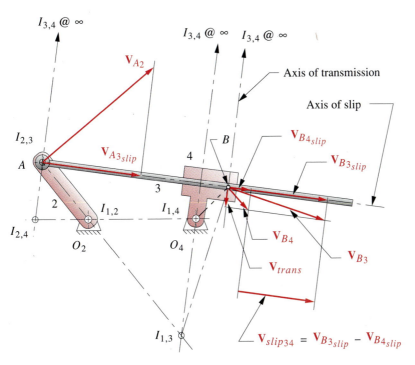

FIGURE 6-19

Graphical velocity analysis of an inverted slider-crank linkage

10   The total slip velocity at B is the difference between the two slip components. Write the relative velocity vector equation 6.6 (p. 243) for the slip components of point B_3 versus point B_4.

$$V_{slip34} = V_{B_3slip} - V_{B_4slip} \qquad (b)$$

11   The angular velocities of links 3 and 4 are identical because they share the slider joint and must rotate together. They can be calculated from equation 6.7:

$$\omega_4 = \omega_3 = \frac{V_{B_4}}{AO_4} \qquad (c)$$

The above examples show how a sliding joint linkage can be solved graphically for velocities at one position. In the next section, we will develop the general solution using algebraic equations to solve the same type of problem.

6.7   ANALYTICAL SOLUTIONS FOR VELOCITY ANALYSIS

The Fourbar Pin-Jointed Linkage

The position equations for the fourbar pin-jointed linkage were derived in Section 4.5 (p. 152). The linkage was shown in Figure 4-7 (p. 155) and is shown again in Figure

6-20 on which we also show an input angular velocity ω_2 applied to link 2. This ω_2 can be a time-varying input velocity. The vector loop equation is shown in equations 4.5a and 4.5c, repeated here for your convenience.

$$\mathbf{R}_2 + \mathbf{R}_3 - \mathbf{R}_4 - \mathbf{R}_1 = 0 \tag{4.5a}$$

As before, we substitute the complex number notation for the vectors, denoting their scalar lengths as a, b, c, d as shown in Figure 6-20a.

$$ae^{j\theta_2} + be^{j\theta_3} - ce^{j\theta_4} - de^{j\theta_1} = 0 \tag{4.5c}$$

To get an expression for velocity, differentiate equation 4.5c with respect to time.

$$jae^{j\theta_2}\frac{d\theta_2}{dt} + jbe^{j\theta_3}\frac{d\theta_3}{dt} - jce^{j\theta_4}\frac{d\theta_4}{dt} = 0 \tag{6.14a}$$

But,

$$\frac{d\theta_2}{dt} = \omega_2; \qquad \frac{d\theta_3}{dt} = \omega_3; \qquad \frac{d\theta_4}{dt} = \omega_4 \tag{6.14b}$$

and:

$$ja\omega_2 e^{j\theta_2} + jb\omega_3 e^{j\theta_3} - jc\omega_4 e^{j\theta_4} = 0 \tag{6.14c}$$

Note that the θ_1 term has dropped out because that angle is a constant, and thus its derivative is zero. Note also that equation 6.14 is, in fact the **relative velocity** or **velocity difference equation**.

$$\mathbf{V}_A + \mathbf{V}_{BA} - \mathbf{V}_B = 0 \tag{6.15a}$$

where:

$$\begin{aligned} \mathbf{V}_A &= ja\omega_2 e^{j\theta_2} \\ \mathbf{V}_{BA} &= jb\omega_3 e^{j\theta_3} \\ \mathbf{V}_B &= jc\omega_4 e^{j\theta_4} \end{aligned} \tag{6.15b}$$

Please compare equations 6.15 to equations 6.3, 6.5, and 6.6 (pp. 242 and 243). This equation is solved graphically in the vector diagram of Figure 6-20b.

We now need to solve equation 6.14 for ω_3 and ω_4, knowing the input velocity ω_2, the link lengths, and all link angles. Thus the position analysis derived in Section 4.5 must be done first to determine the link angles before this velocity analysis can be completed. We wish to solve equation 6.14 to get expressions in this form:

$$\omega_3 = f(a, b, c, d, \theta_2, \theta_3, \theta_4, \omega_2) \qquad \omega_4 = g(a, b, c, d, \theta_2, \theta_3, \theta_4, \omega_2) \tag{6.16}$$

The strategy of solution will be the same as was done for the position analysis. First, substitute the Euler identity from equation 4.4a (p. 155) in each term of equation 6.14c:

$$ja\omega_2(\cos\theta_2 + j\sin\theta_2) + jb\omega_3(\cos\theta_3 + j\sin\theta_3)$$
$$- jc\omega_4(\cos\theta_4 + j\sin\theta_4) = 0 \tag{6.17a}$$

Multiply through by the operator j:

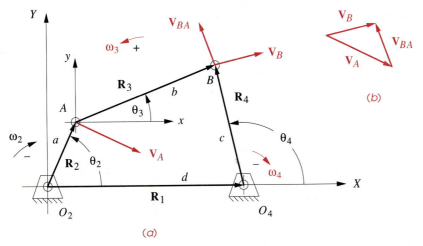

FIGURE 6-20

Position vector loop for a fourbar linkage showing velocity vectors for a negative (cw) ω_2

$$a\omega_2\left(j\cos\theta_2 + j^2\sin\theta_2\right) + b\omega_3\left(j\cos\theta_3 + j^2\sin\theta_3\right)$$
$$-c\omega_4\left(j\cos\theta_4 + j^2\sin\theta_4\right) = 0 \tag{6.17b}$$

The cosine terms have become the imaginary, or y-directed terms, and because $j^2 = -1$, the sine terms have become real or x-directed.

$$a\omega_2\left(-\sin\theta_2 + j\cos\theta_2\right) + b\omega_3\left(-\sin\theta_3 + j\cos\theta_3\right)$$
$$-c\omega_4\left(-\sin\theta_4 + j\cos\theta_4\right) = 0 \tag{6.17c}$$

We can now separate this vector equation into its two components by collecting all real and all imaginary terms separately:

real part (x component):

$$-a\omega_2\sin\theta_2 - b\omega_3\sin\theta_3 + c\omega_4\sin\theta_4 = 0 \tag{6.17d}$$

imaginary part (y component):

$$a\omega_2\cos\theta_2 + b\omega_3\cos\theta_3 - c\omega_4\cos\theta_4 = 0 \tag{6.17e}$$

Note that the j's have cancelled in equation 6.17e. We can solve these two equations, 6.17d and 6.17e, simultaneously by direct substitution to get:

$$\omega_3 = \frac{a\omega_2}{b}\frac{\sin(\theta_4 - \theta_2)}{\sin(\theta_3 - \theta_4)} \tag{6.18a}$$

$$\omega_4 = \frac{a\omega_2}{c}\frac{\sin(\theta_2 - \theta_3)}{\sin(\theta_4 - \theta_3)} \tag{6.18b}$$

Once we have solved for ω_3 and ω_4, we can then solve for the linear velocities by substituting the Euler identity into equations 6.15,

$$V_A = ja\omega_2\left(\cos\theta_2 + j\sin\theta_2\right) = a\omega_2\left(-\sin\theta_2 + j\cos\theta_2\right) \tag{6.19a}$$

$$V_{BA} = jb\omega_3\left(\cos\theta_3 + j\sin\theta_3\right) = b\omega_3\left(-\sin\theta_3 + j\cos\theta_3\right) \tag{6.19b}$$

$$V_B = jc\omega_4\left(\cos\theta_4 + j\sin\theta_4\right) = c\omega_4\left(-\sin\theta_4 + j\cos\theta_4\right) \tag{6.19c}$$

where the real and imaginary terms are the x and y components, respectively. Equations 6.18 and 6.19 provide a complete solution for the angular velocities of the links and the linear velocities of the joints in the pin-jointed fourbar linkage. Note that there are also two solutions to this velocity problem, corresponding to the open and crossed branches of the linkage. They are found by the substitution of the open or crossed branch values of θ_3 and θ_4 obtained from equations 4.10 (p. 158) and 4.13 (p. 159) into equations 6.18 and 6.19. Figure 6-20a (p. 273) shows the open branch.

The Fourbar Slider-Crank

The position equations for the fourbar offset slider-crank linkage (inversion #1) were derived in Section 4.6 (p. 159). The linkage was shown in Figure 4-9 (p. 160) and is shown again in Figure 6-21a on which we also show an input angular velocity ω_2 applied to link 2. This ω_2 can be a time-varying input velocity. The vector loop equation 4.14 is repeated here for your convenience.

$$\mathbf{R}_2 - \mathbf{R}_3 - \mathbf{R}_4 - \mathbf{R}_1 = 0 \tag{4.14a}$$

$$ae^{j\theta_2} - be^{j\theta_3} - ce^{j\theta_4} - de^{j\theta_1} = 0 \tag{4.14b}$$

Differentiate equation 4.14b with respect to time noting that a, b, c, θ_1, and θ_4 are constant but the length of link d varies with time in this inversion.

$$ja\omega_2 e^{j\theta_2} - jb\omega_3 e^{j\theta_3} - \dot{d} = 0 \tag{6.20a}$$

The term d *dot* is the linear velocity of the slider block. Equation 6.20a is the velocity difference equation 6.5 (p. 243) and can be written in that form.

$$V_A - V_{AB} - V_B = 0$$

or :

$$V_A = V_B + V_{AB}$$

but :
$$\hspace{4cm} \tag{6.20b}$$

$$V_{AB} = -V_{BA}$$

then :

$$V_B = V_A + V_{BA}$$

Equation 6.20 is identical in form to equations 6.5 and 6.15a. Note that because we arranged the position vector $\mathbf{R}_3$ in Figures 4-9 and 6-21 with its root at point B, directed from B to A, its derivative represents the velocity difference of point A with respect to point B, the opposite of that in the previous fourbar example. Compare this also to equa-

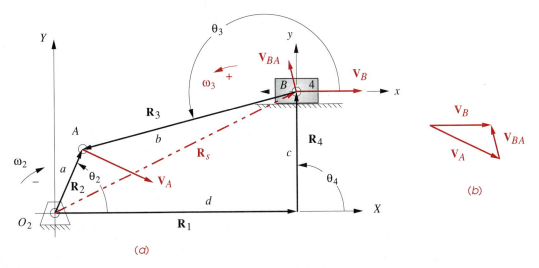

FIGURE 6-21

Position vector loop for a fourbar slider-crank linkage showing velocity vectors for a negative (cw) ω_2

tion 6.15b noting that its vector $\mathbf{R}_3$ is directed from A to B. Figure 6-21b shows the vector diagram of the graphical solution to equation 6.20b.

Substitute the Euler equivalent, equation 4.4a (p. 155), in equation 6.20a,

$$ja\omega_2(\cos\theta_2 + j\sin\theta_2) - jb\omega_3(\cos\theta_3 + j\sin\theta_3) - \dot{d} = 0 \qquad (6.21a)$$

simplify,

$$a\omega_2(-\sin\theta_2 + j\cos\theta_2) - b\omega_3(-\sin\theta_3 + j\cos\theta_3) - \dot{d} = 0 \qquad (6.21b)$$

and separate into real and imaginary components.

real part (x component):

$$-a\omega_2\sin\theta_2 + b\omega_3\sin\theta_3 - \dot{d} = 0 \qquad (6.21c)$$

imaginary part (y component):

$$a\omega_2\cos\theta_2 - b\omega_3\cos\theta_3 = 0 \qquad (6.21d)$$

These are two simultaneous equations in the two unknowns, d dot and ω_3. Equation 6.21d can be solved for ω_3 and substituted into 6.21c to find d dot.

$$\omega_3 = \frac{a\cos\theta_2}{b\cos\theta_3}\omega_2 \qquad (6.22a)$$

$$\dot{d} = -a\omega_2\sin\theta_2 + b\omega_3\sin\theta_3 \qquad (6.22b)$$

The absolute velocity of point A and the velocity difference of point A versus point B are found from equation 6.20:

$$\mathbf{V}_A = a\omega_2\left(-\sin\theta_2 + j\cos\theta_2\right) \tag{6.23a}$$

$$\mathbf{V}_{AB} = b\omega_3\left(-\sin\theta_3 + j\cos\theta_3\right) \tag{6.23b}$$

$$\mathbf{V}_{BA} = -\mathbf{V}_{AB} \tag{6.23c}$$

The Fourbar Inverted Slider-Crank

The position equations for the fourbar inverted slider-crank linkage were derived in Section 4.7 (p. 161). The linkage was shown in Figure 4-10 (p. 162) and is shown again in Figure 6-22 on which we also show an input angular velocity ω_2 applied to link 2. This ω_2 can vary with time. The vector loop equations 4.14 shown on p. 274 are valid for this linkage as well.

All slider linkages will have at least one link whose effective length between joints varies as the linkage moves. In this inversion the length of link 3 between points A and B, designated as b, will change as it passes through the slider block on link 4. To get an expression for velocity, differentiate equation 4.14b with respect to time noting that a, c, d, and θ_1 are constant and b varies with time.

$$ja\omega_2 e^{j\theta_2} - jb\omega_3 e^{j\theta_3} - \dot{b}e^{j\theta_3} - jc\omega_4 e^{j\theta_4} = 0 \tag{6.24}$$

The value of db/dt will be one of the variables to be solved for in this case and is the b *dot* term in the equation. Another variable will be ω_4, the angular velocity of link 4. Note, however, that we also have an unknown in ω_3, the angular velocity of link 3. This is a total of three unknowns. Equation 6.24 can only be solved for two unknowns. Thus we require another equation to solve the system. There is a fixed relationship between angles θ_3 and θ_4, shown as γ in Figure 6-22 and defined in equation 4.18, repeated here:

$$\theta_3 = \theta_4 \pm \gamma \tag{4.18}$$

Differentiate it with respect to time to obtain:

$$\omega_3 = \omega_4 \tag{6.25}$$

We wish to solve equation 6.24 to get expressions in this form:

$$\omega_3 = \omega_4 = f\left(a, b, c, d, \theta_2, \theta_3, \theta_4, \omega_2\right)$$

$$\frac{db}{dt} = \dot{b} = g\left(a, b, c, d, \theta_2, \theta_3, \theta_4, \omega_2\right) \tag{6.26}$$

Substitution of the Euler identity (equation 4.4a, p. 155) into equation 6.24 yields:

$$ja\omega_2\left(\cos\theta_2 + j\sin\theta_2\right) - jb\omega_3\left(\cos\theta_3 + j\sin\theta_3\right)$$
$$-\dot{b}\left(\cos\theta_3 + j\sin\theta_3\right) - jc\omega_4\left(\cos\theta_4 + j\sin\theta_4\right) = 0 \tag{6.27a}$$

Multiply by the operator j and substitute ω_4 for ω_3 from equation 6.25:

$$a\omega_2\left(-\sin\theta_2 + j\cos\theta_2\right) - b\omega_4\left(-\sin\theta_3 + j\cos\theta_3\right)$$
$$-\dot{b}\left(\cos\theta_3 + j\sin\theta_3\right) - c\omega_4\left(-\sin\theta_4 + j\cos\theta_4\right) = 0 \tag{6.27b}$$

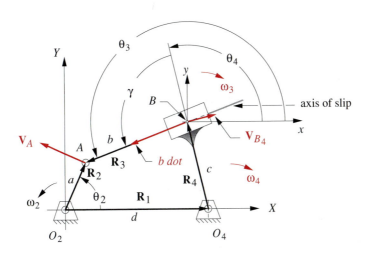

FIGURE 6-22

Velocity analysis of inversion #3 of the slider-crank fourbar linkage

We can now separate this vector equation into its two components by collecting all real and all imaginary terms separately:

real part (x component):

$$-a\omega_2 \sin\theta_2 + b\omega_4 \sin\theta_3 - \dot{b}\cos\theta_3 + c\omega_4 \sin\theta_4 = 0 \qquad (6.28a)$$

imaginary part (y component):

$$a\omega_2 \cos\theta_2 - b\omega_4 \cos\theta_3 - \dot{b}\sin\theta_3 - c\omega_4 \cos\theta_4 = 0 \qquad (6.28b)$$

Collect terms and rearrange equations 6.28 to isolate one unknown on the left side.

$$\dot{b}\cos\theta_3 = -a\omega_2 \sin\theta_2 + \omega_4(b\sin\theta_3 + c\sin\theta_4) \qquad (6.29a)$$

$$\dot{b}\sin\theta_3 = a\omega_2 \cos\theta_2 - \omega_4(b\cos\theta_3 + c\cos\theta_4) \qquad (6.29b)$$

Either equation can be solved for b *dot* and the result substituted in the other. Solving equation 6.29a:

$$\dot{b} = \frac{-a\omega_2 \sin\theta_2 + \omega_4(b\sin\theta_3 + c\sin\theta_4)}{\cos\theta_3} \qquad (6.30a)$$

Substitute in equation 6.29b and simplify:

$$\omega_4 = \frac{a\omega_2 \cos(\theta_2 - \theta_3)}{b + c\cos(\theta_4 - \theta_3)} \qquad (6.30b)$$

Equation 6.30a provides the **velocity of slip** at point B. Equation 6.30b gives the **angular velocity** of link 4. Note that we can substitute $-\gamma = \theta_4 - \theta_3$ from equation 4.18 (for an open linkage) into equation 6.30b to further simplify it. Note that $\cos(-\gamma) = \cos(\gamma)$.

$$\omega_4 = \frac{a\omega_2 \cos(\theta_2 - \theta_3)}{b + c \cos\gamma} \tag{6.30c}$$

The **velocity of slip** from equation 6.30a is always directed along the **axis of slip** as shown in Figures 6-19 (p. 271) and 6-22 (p. 277). There is also a component orthogonal to the axis of slip called the **velocity of transmission**. This lies along the **axis of transmission** which is the only line along which any useful work can be transmitted across the sliding joint. All energy associated with motion along the slip axis is converted to heat and lost.

The absolute linear velocity of point A is found from equation 6.23a. We can find the absolute velocity of point B on link 4 since ω_4 is now known. From equation 6.15b:

$$\mathbf{V}_{B_4} = jc\omega_4 e^{j\theta_4} = c\omega_4\left(-\sin\theta_4 + j\cos\theta_4\right) \tag{6.31}$$

6.8   VELOCITY ANALYSIS OF THE GEARED FIVEBAR LINKAGE

The position loop equation for the geared fivebar mechanism was derived in Section 4.8 and is repeated here. See Figure P6-4 for notation.

$$ae^{j\theta_2} + be^{j\theta_3} - ce^{j\theta_4} - de^{j\theta_5} - fe^{j\theta_1} = 0 \tag{4.23b}$$

Differentiate this with respect to time to get an expression for velocity.

$$a\omega_2 je^{j\theta_2} + b\omega_3 je^{j\theta_3} - c\omega_4 je^{j\theta_4} - d\omega_5 je^{j\theta_5} = 0 \tag{6.32a}$$

Substitute the Euler equivalents:

$$a\omega_2 j(\cos\theta_2 + j\sin\theta_2) + b\omega_3 j(\cos\theta_3 + j\sin\theta_3)$$
$$-c\omega_4 j(\cos\theta_4 + j\sin\theta_4) - d\omega_5 j(\cos\theta_5 + j\sin\theta_5) = 0 \tag{6.32b}$$

Note that the angle θ_5 is defined in terms of θ_2, the gear ratio λ, and the phase angle ϕ.

$$\theta_5 = \lambda\theta_2 + \phi \tag{4.23c}$$

Differentiate with respect to time:

$$\omega_5 = \lambda\omega_2 \tag{6.32c}$$

Since a complete position analysis must be done before a velocity analysis, we will assume that the values of θ_5 and ω_5 have been found and will leave these equations in terms of θ_5 and ω_5.

Separating the real and imaginary terms in equation 6.32b:

real : $-a\omega_2 \sin\theta_2 - b\omega_3 \sin\theta_3 + c\omega_4 \sin\theta_4 + d\omega_5 \sin\theta_5 = 0 \tag{6.32d}$

imaginary : $a\omega_2 \cos\theta_2 + b\omega_3 \cos\theta_3 - c\omega_4 \cos\theta_4 - d\omega_5 \cos\theta_5 = 0 \tag{6.32e}$

The only two unknowns are ω_3 and ω_4. Either equation 6.32d or 6.32e can be solved for one unknown and the result substituted in the other. The solution for ω_3 is:

$$\omega_3 = -\frac{2\sin\theta_4\left[a\omega_2\sin(\theta_2-\theta_4)+d\omega_5\sin(\theta_4-\theta_5)\right]}{b\left[\cos(\theta_3-2\theta_4)-\cos\theta_3\right]} \tag{6.33a}$$

The angular velocity ω_4 can be found from equation 6.32d using ω_3.

$$\omega_4 = \frac{a\omega_2\sin\theta_2 + b\omega_3\sin\theta_3 - d\omega_5\sin\theta_5}{c\sin\theta_4} \tag{6.33b}$$

With all link angles and angular velocities known, the linear velocities of the pin joints can be found from:

$$\mathbf{V}_A = a\omega_2\left(-\sin\theta_2 + j\cos\theta_2\right) \tag{6.33c}$$

$$\mathbf{V}_{BA} = b\omega_3\left(-\sin\theta_3 + j\cos\theta_3\right) \tag{6.33d}$$

$$\mathbf{V}_C = d\omega_5\left(-\sin\theta_5 + j\cos\theta_5\right) \tag{6.33e}$$

$$\mathbf{V}_B = \mathbf{V}_A + \mathbf{V}_{BA} \tag{6.33f}$$

6.9 VELOCITY OF ANY POINT ON A LINKAGE

Once the angular velocities of all the links are found it is easy to define and calculate the velocity of *any point on any link* for any input position of the linkage. Figure 6-23 shows the fourbar linkage with its coupler, link 3, enlarged to contain a coupler point P. The crank and rocker have also been enlarged to show points S and U which might represent the centers of gravity of those links. We want to develop algebraic expressions for the velocities of these (or any) points on the links.

To find the velocity of point S, draw the position vector from the fixed pivot O_2 to point S. This vector, $\mathbf{R}_{SO_2}$ makes an angle δ_2 with the vector $\mathbf{R}_{AO_2}$. The angle δ_2 is completely defined by the geometry of link 2 and is constant. The position vector for point S is then:

$$\mathbf{R}_{SO_2} = \mathbf{R}_S = se^{j(\theta_2+\delta_2)} = s\left[\cos(\theta_2+\delta_2)+j\sin(\theta_2+\delta_2)\right] \tag{4.25}$$

Differentiate this position vector to find the velocity of that point.

$$\mathbf{V}_S = jse^{j(\theta_2+\delta_2)}\omega_2 = s\omega_2\left[-\sin(\theta_2+\delta_2)+j\cos(\theta_2+\delta_2)\right] \tag{6.34}$$

The position of point U on link 4 is found in the same way, using the angle δ_4 which is a constant angular offset within the link. The expression is:

$$\mathbf{R}_{UO_4} = ue^{j(\theta_4+\delta_4)} = u\left[\cos(\theta_4+\delta_4)+j\sin(\theta_4+\delta_4)\right] \tag{4.26}$$

Differentiate this position vector to find the velocity of that point.

$$\mathbf{V}_U = jue^{j(\theta_4+\delta_4)}\omega_4 = u\omega_4\left[-\sin(\theta_4+\delta_4)+j\cos(\theta_4+\delta_4)\right] \tag{6.35}$$

The velocity of point P on link 3 can be found from the addition of two velocity vectors, such as $\mathbf{V}_A$ and $\mathbf{V}_{PA}$. $\mathbf{V}_A$ is already defined from our analysis of the link velocities.

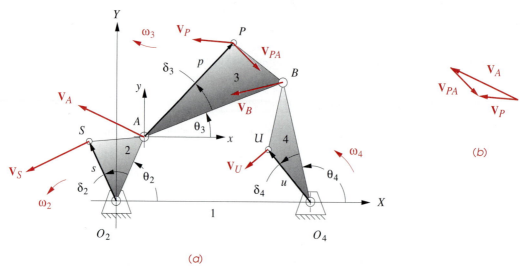

FIGURE 6-23

Finding the velocities of points on the links

$\mathbf{V}_{PA}$ is the velocity difference of point P with respect to point A. Point A is chosen as the reference point because angle θ_3 is defined in a local coordinate system whose origin is at A. Position vector $\mathbf{R}_{PA}$ is defined in the same way as $\mathbf{R}_S$ or $\mathbf{R}_U$ using the internal link offset angle δ_3 and the angle of link 3, θ_3. This was done in equation 4.27.

$$\mathbf{R}_{PA} = pe^{j(\theta_3 + \delta_3)} = p\left[\cos(\theta_3 + \delta_3) + j\sin(\theta_3 + \delta_3)\right] \tag{4.27a}$$

$$\mathbf{R}_P = \mathbf{R}_A + \mathbf{R}_{PA} \tag{4.27b}$$

Differentiate this position vector to find the velocity of that point.

$$\mathbf{V}_{PA} = jpe^{j(\theta_3 + \delta_3)}\omega_3 = p\omega_3\left[-\sin(\theta_3 + \delta_3) + j\cos(\theta_3 + \delta_3)\right] \tag{6.36a}$$

$$\mathbf{V}_P = \mathbf{V}_A + \mathbf{V}_{PA} \tag{6.36b}$$

Please compare equation 6.36 with equations 6.5 (p. 243) and 6.15 (p. 272). It is, again, the velocity difference equation.

6.10   REFERENCES

1   **Towfigh, K.** (1969). "The Fourbar Linkage as an Adjustment Mechanism." *Proc. of Applied Mechanism Conference*, Tulsa, OK, pp. 27-1 to 27-4.

2   **Wood, G. A.** (1977). "Educating for Creativity in Engineering." *Proc. of ASEE 85th Annual Conference*, University of North Dakota, pp. 1-13.

6.11   PROBLEMS

6-1   Use the relative velocity equation and solve graphically or analytically.

   a.   A ship is steaming due north at 20 knots (nautical miles per hour). A submarine is laying in wait 1/2 mile due west of the ship. The sub fires a torpedo on a course of 85 degrees. The torpedo travels at a constant speed of 30 knots. Will it strike the ship? If not, by how many nautical miles will it miss?

   b.   A plane is flying due south at 500 mph at 35,000 ft altitude, straight and level. A second plane is initially 40 miles due east of the first plane, also at 35,000 feet altitude, flying straight and level and traveling at 550 mph. Determine the compass angle at which the second plane would be on a collision course with the first. How long will it take for the second plane to catch the first?

6-2   A point is at a 6.5 in radius on a body in pure rotation with $\omega = 100$ rad/sec. The rotation center is at the origin of a coordinate system. When the point is at position A, its position vector makes a $45°$ angle with the X axis. At position B, its position vector makes a $75°$ angle with the X axis. Draw this system to some convenient scale and:

   a.   Write an expression for the particle's velocity vector in position A using complex number notation, in both polar and cartesian forms.

   b.   Write an expression for the particle's velocity vector in position B using complex number notation, in both polar and cartesian forms.

   c.   Write a vector equation for the velocity difference between points B and A. Substitute the complex number notation for the vectors in this equation and solve for the position difference numerically.

   d.   Check the result of part c with a graphical method.

6-3   Repeat problem 6-2 considering points A and B to be on separate bodies rotating about the origin with ω's of -50 (A) and $+75$ rad/sec (B). Find their relative velocity.

*6-4   A general fourbar linkage configuration and its notation are shown in Figure P6-1. The link lengths, coupler point location, and the values of θ_2 and ω_2 for the same fourbar linkages as used for position analysis in Chapter 4 are redefined in Table

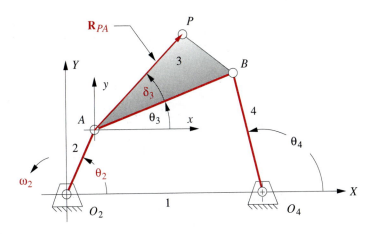

FIGURE P6-1

Configuration and terminology for the pin-jointed fourbar linkage of Problems 6-4 to 6-5

* Answers in Appendix F.

TABLE P6-1   Data for Problems 6-4 to 6-5

Row	Link 1	Link 2	Link 3	Link 4	θ_2	ω_2	R_{pa}	δ_3
a	6	2	7	9	30	10	6	30
b	7	9	3	8	85	−12	9	25
c	3	10	6	8	45	−15	10	80
d	8	5	7	6	25	24	5	45
e	8	5	8	6	75	−50	9	300
f	5	8	8	9	15	−45	10	120
g	6	8	8	9	25	100	4	300
h	20	10	10	10	50	−65	6	20
i	4	5	2	5	80	25	9	80
j	20	10	5	10	33	25	1	0
k	4	6	10	7	88	−80	10	330
l	9	7	10	7	60	−90	5	180
m	9	7	11	8	50	75	10	90
n	9	7	11	6	120	15	15	60

P6-1, which is the same as Table P4-1. *For the row(s) assigned,* draw the linkage to scale and find the velocities of the pin joints A and B and of instant centers $I_{1,3}$ and $I_{2,4}$ using a graphical method. Then calculate ω_3 and ω_4 and find the velocity of point P.

*†6-5   Repeat Problem 6-4 using an analytical method. Draw the linkage to scale and label it before setting up the equations.

*6-6   The general linkage configuration and terminology for an offset fourbar slider-crank linkage are shown in Figure P6-2. The link lengths and the values of θ_2 and ω_2 are defined in Table P6-2. *For the row(s) assigned,* draw the linkage to scale and find the velocities of the pin joints A and B and the velocity of slip at the sliding joint using a graphical method.

* Answers in Appendix F.

† These problems are suited to solution using *Mathcad, Matlab,* or *TKSolver* equation solver programs.

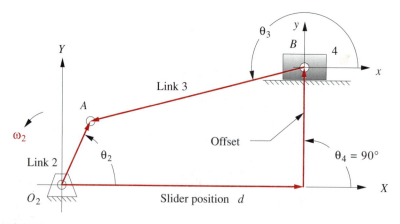

FIGURE P6-2

Configuration and terminology for Problems 6-6 to 6-7

Row	Link 2	Link 3	Offset	θ_2	ω_2
a	1.4	4	1	45	10
b	2	6	-3	60	-12
c	3	8	2	-30	-15
d	3.5	10	1	120	24
e	5	20	-5	225	-50
f	3	13	0	100	-45
g	7	25	10	330	100

*†6-7 Repeat Problem 6-6 using an analytical method. Draw the linkage to scale and label it before setting up the equations.

*6-8 The general linkage configuration and terminology for an inverted fourbar slider-crank linkage are shown in Figure P6-3. The link lengths and the values of θ_2, ω_2, and γ are defined in Table P6-3. *For the row(s) assigned,* draw the linkage to scale

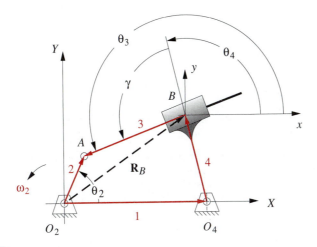

FIGURE P6-3

Configuration and terminology for Problems 6-8 to 6-9

Row	Link 1	Link 2	Link 4	γ	θ_2	ω_2
a	6	2	4	90	30	10
b	7	9	3	75	85	-15
c	3	10	6	45	45	24
d	8	5	3	60	25	-50
e	8	4	2	30	75	-45
f	5	8	8	90	150	100

* Answers in Appendix F.

† These problems are suited to solution using *Mathcad, Matlab,* or *TKSolver* equation solver programs.

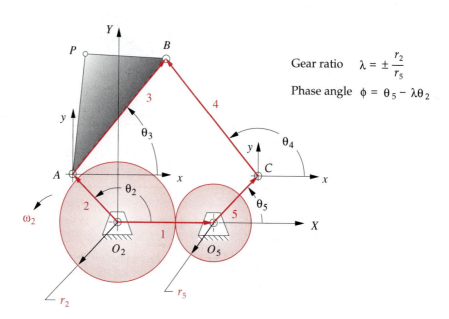

FIGURE P6-4

Configuration and terminology for problems 6-10 to 6-11

and find the velocities of the pin joints A and B and velocity of slip at the sliding joint using a graphical method.

*†6-9   Repeat Problem 6-8 using an analytical method. Draw the linkage to scale and label it before setting up the equations.

*6-10   The general linkage configuration and terminology for a geared fivebar linkage are shown in Figure P6-4. The link lengths, gear ratio (λ), phase angle (ϕ), and the values of θ_2 and ω_2 are defined in Table P6-4. *For the row(s) assigned,* draw the linkage to scale and find ω_3 and ω_4 using a graphical method.

TABLE P6-4   Data for Problems 6-10 to 6-11

Row	Link 1	Link 2	Link 3	Link 4	Link 5	λ	ϕ	ω_2	θ_2
a	6	1	7	9	4	2	30	10	60
b	6	5	7	8	4	-2.5	60	-12	30
c	3	5	7	8	4	-0.5	0	-15	45
d	4	5	7	8	4	-1	120	24	75
e	5	9	11	8	8	3.2	-50	-50	-39
f	10	2	7	5	3	1.5	30	-45	120
g	15	7	9	11	4	2.5	-90	100	75
h	12	8	7	9	4	-2.5	60	-65	55
i	9	7	8	9	4	-4	120	25	100

* Answers in Appendix F.

† These problems are suited to solution using *Mathcad, Matlab,* or *TKSolver* equation solver programs.

*†6-11 Repeat Problem 6-10 using an analytical method. Draw the linkage to scale and label it before setting up the equations.

6-12 Find all the instant centers of the linkages shown in Figure P6-5.

6-13 Find all the instant centers of the linkages shown in Figure P6-6.

6-14 Find all the instant centers of the linkages shown in Figure P6-7.

6-15 Find all the instant centers of the linkages shown in Figure P6-8.

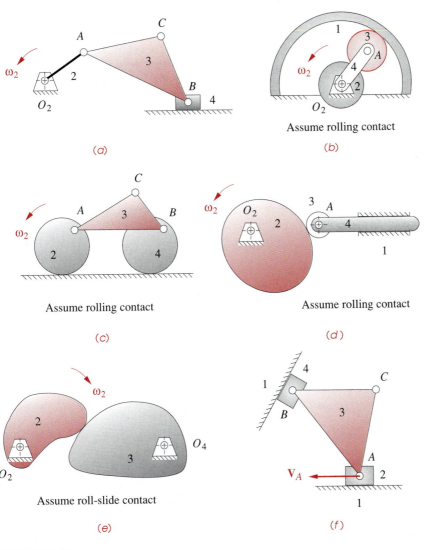

Assume rolling contact

(a)

(b)

Assume rolling contact

Assume rolling contact

(c)

(d)

Assume roll-slide contact

(e)

(f)

* Answers in Appendix F.

† These problems are suited to solution using *Mathcad, Matlab,* or *TKSolver* equation solver programs.

FIGURE P6-5

Velocity analysis and instant center problems. Problems 6-12 and 6-16 to 6-20

6

6-16   The linkage in Figure P6-5a has $O_2A = 0.8$, $AB = 1.93$, $AC = 1.33$, and offset $= 0.38$ in. The crank angle in the position shown is 34.3° and angle $BAC = 38.6°$. Find ω_3, $\mathbf{V}_A$, $\mathbf{V}_B$, and $\mathbf{V}_C$ for the position shown for $\omega_2 = 15$ rad/sec in the direction shown.

   a.   Using the velocity difference graphical method.
   b.   Using the instant center graphical method.
   †c.   Using an analytical method.

6-17   The linkage in Figure P6-5c has $I_{12}A = 0.75$, $AB = 1.5$, and $AC = 1.2$ in. The effective crank angle in the position shown is 77° and angle $BAC = 30°$. Find ω_3, ω_4, $\mathbf{V}_A$, $\mathbf{V}_B$, and $\mathbf{V}_C$ for the position shown for $\omega_2 = 15$ rad/sec in direction shown.

   a.   Using the velocity difference graphical method.
   b.   Using the instant center graphical method.
   †c.   Using an analytical method. (Hint: Create an effective linkage for the position shown and analyze as a pin-jointed fourbar.)

6-18   The linkage in Figure P6-5f has $AB = 1.8$ and $AC = 1.44$ in. The angle of AB in the position shown is 128° and angle $BAC = 49°$. The slider at B is at an angle of 59°. Find ω_3, $\mathbf{V}_A$, $\mathbf{V}_B$, and $\mathbf{V}_C$ for the position shown for $\mathbf{V}_A = 10$ in/sec in the direction shown.

   a.   Using the velocity difference graphical method.
   b.   Using the instant center graphical method.
   †c.   Using an analytical method.

6-19   Measure the link geometry in Figure P6-5d and find $\mathbf{V}_4$, $\mathbf{V}_{trans}$, and $\mathbf{V}_{slip}$ for the position shown with $\omega_2 = 20$ rad/sec in the direction shown.

6-20   Measure the link geometry in Figure P6-5e and find ω_3, $\mathbf{V}_{trans}$, and $\mathbf{V}_{slip}$ for the position shown for $\omega_2 = 10$ rad/sec in the direction shown.

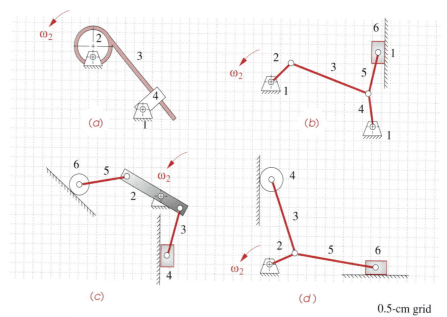

(a)   (b)

(c)   (d)

0.5-cm grid

† These problems are
suited to solution using
Mathcad, Matlab, or
TKSolver equation solver
programs.

FIGURE P6-6

Problems 6-13, 6-21, and 6-22

6-21 Measure the link geometry in Figure P6-6b and find, for the position shown, the velocity ratio $V_{I5,6}/V_{I2,3}$ and the mechanical advantage from link 2 to link 6.

   a. Using the velocity difference graphical method.
   b. Using the instant center graphical method.

6-22 Repeat Problem 6-21 for the mechanism in Figure P6-6d.

†6-23 Generate and draw the fixed and moving centrodes of links 1 and 3 for the linkage in Figure P6-7a.

6-24 The linkage in Figure P6-8a has $AB = 116$, $BC = 108$, $CD = 110$, and $AD = 174$ mm. AD is at $-25°$ and AB is at $37°$ in the global XY coordinate system. Find ω_f, V_B, and V_C in the global coordinate system for the position shown if $\omega_d = 15$ rad/sec clockwise (CW). Use the velocity difference graphical method. (Hint: Make an enlarged copy of the figure and draw on it.)

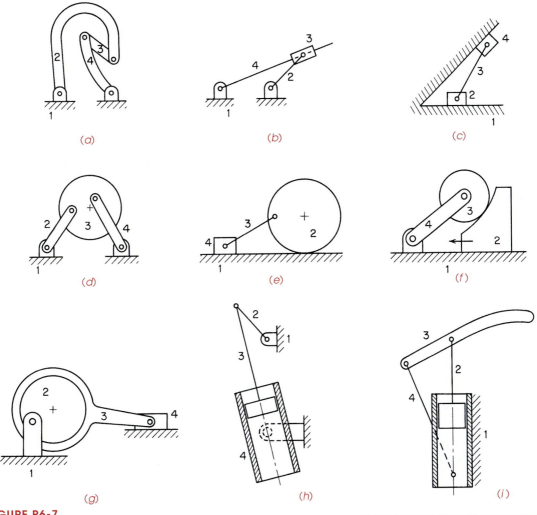

(a)   (b)   (c)

(d)   (e)   (f)

(g)   (h)   (i)

FIGURE P6-7

Problems 6-14 and 6-23. *From R. T. Hinkle, Problems in Kinematics, Prentice-Hall. Englewood Cliffs, NJ, 1954.*

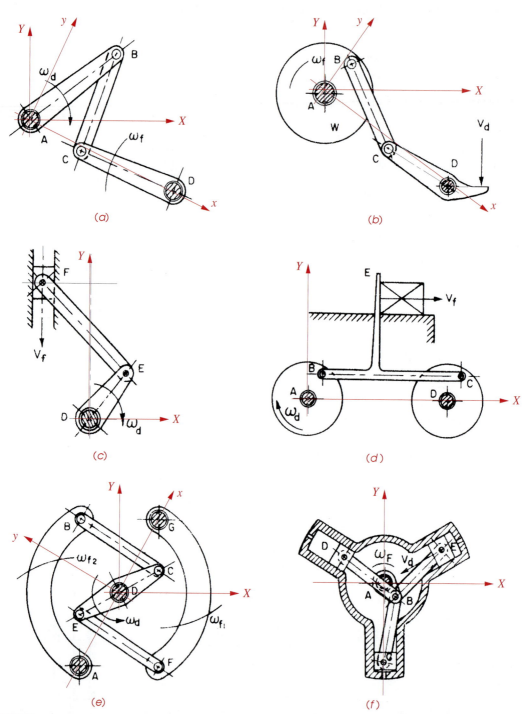

FIGURE P6-8

Problems 6-15 and 6-24 to 6-45.   *Adapted from P. H. Hill and W. P. Rule. (1960). Mechanisms: Analysis and Design, with permission.*

6-25   The linkage in Figure P6-8a has $AB = 116$, $BC = 108$, $CD = 110$, and $AD = 174$ mm. AD is at $-25°$ and AB is at $37°$ in the global XY coordinate system. Find ω_f, $\mathbf{V}_B$, and $\mathbf{V}_C$ in the global coordinate system for the position shown if $\omega_d = 15$ rad/sec CW. Use the instant center graphical method. (Hint: Make an enlarged copy of the figure and draw on it.)

†6-26   The linkage in Figure P6-8a has $AB = 116$, $BC = 108$, $CD = 110$, and $AD = 174$ mm. AB is at $62°$ in the local $x'y'$ coordinate system. Find ω_f, $\mathbf{V}_B$, and $\mathbf{V}_C$ in the local coordinate system for the position shown if $\omega_d = 15$ rad/sec CW. Use an analytical method.

†6-27   The linkage in Figure P6-8a has $AB = 116$, $BC = 108$, $CD = 110$, and $AD = 174$ mm. Write a computer program or use an equation solver to find and plot ω_f, $\mathbf{V}_B$, and $\mathbf{V}_C$ in the local coordinate system for the maximum range of motion that this linkage allows if $\omega_d = 15$ rad/sec CW.

6-28   The linkage in Figure P6-8b has $AB = 40$, $BC = 96$, $CD = 75$, and $AD = 162$ mm. The perpendicular distance from D to $\mathbf{V}_D$ is 36 mm. AD is at $-36°$ and AB is at $47°$ in the global XY coordinate system. Find ω_d, $\mathbf{V}_B$, $\mathbf{V}_C$, and $\mathbf{V}_d$ in the global coordinate system for the position shown if $\omega_f = 20$ rad/sec counterclockwise (CCW). Use the velocity difference graphical method. (Hint: Make an enlarged copy of the figure and draw on it.)

6-29   The linkage in Figure P6-8b has $AB = 40$, $BC = 96$, $CD = 75$, and $AD = 162$ mm. The perpendicular distance from D to $\mathbf{V}_D$ is 36 mm. AD is at $-36°$ and AB is at $47°$ in the global XY coordinate system. Find ω_d, $\mathbf{V}_B$, $\mathbf{V}_C$, and $\mathbf{V}_d$ in the global coordinate system for the position shown if $\omega_f = 20$ rad/sec CCW. Use the instant center graphical method. (Hint: Make an enlarged copy of the figure and draw on it.)

†6-30   The linkage in Figure P6-8b has $AB = 40$, $BC = 96$, $CD = 75$, and $AD = 162$ mm. The perpendicular distance from D to $\mathbf{V}_D$ is 36 mm. AB is at $83°$ in the local $x'y'$ coordinate system. Find ω_d, $\mathbf{V}_B$, $\mathbf{V}_C$, and $\mathbf{V}_d$ in the local coordinate system for the position shown if $\omega_f = 20$ rad/sec CCW. Use an analytical method.

†6-31   The linkage in Figure P6-8b has $AB = 40$, $BC = 96$, $CD = 75$, and $AD = 162$ mm. The perpendicular distance from D to $\mathbf{V}_D$ is 36 mm. Write a computer program or use an equation solver to find and plot ω_d, $\mathbf{V}_B$, $\mathbf{V}_C$, and $\mathbf{V}_d$ in the local coordinate system for the maximum range of motion that this linkage allows if $\omega_f = 20$ rad/sec CCW.

6-32   The offset slider-crank linkage in Figure P6-8c has $DE = 63$, $EF = 130$, and offset = 52 mm. DE is at $51°$ in the global XY coordinate system. Find $\mathbf{V}_E$ and $\mathbf{V}_f$ in the global coordinate system for the position shown if $\omega_d = 25$ rad/sec CW. Use the velocity difference graphical method. (Hint: Make an enlarged copy of the figure and draw on it.)

6-33   The offset slider-crank linkage in Figure P6-8c has $DE = 63$, $EF = 130$, and offset = 52 mm. DE is at $51°$ in the global XY coordinate system. Find $\mathbf{V}_E$ and $\mathbf{V}_f$ in the global coordinate system for the position shown if $\omega_d = 25$ rad/sec CW. Use the instant center graphical method. (Hint: Make an enlarged copy of the figure and draw on it.)

†6-34   The offset slider-crank linkage in Figure P6-8c has $DE = 63$, $EF = 130$, and offset = 52 mm. DE is at $51°$ in the global XY coordinate system. Find $\mathbf{V}_E$ and $\mathbf{V}_f$ in the global coordinate system for the position shown if $\omega_d = 25$ rad/sec CW. Use an analytical method.

† These problems are suited to solution using *Mathcad, Matlab,* or *TKSolver* equation solver programs.

†6-35   The offset slider-crank linkage in Figure P6-8c has $DE = 63$, $EF = 130$, and offset = 52 mm. Write a computer program or use an equation solver to find and plot $\mathbf{V}_E$ and $\mathbf{V}_f$ in the global coordinate system for the maximum range of motion that this linkage allows if $\omega_d = 25$ rad/sec CW.

6-36   The linkage in Figure P6-8d has $AB = 30$, $BC = 150$, $CD = 30$, and $AD = 150$ mm. AB is at 58° in the global XY coordinate system. Find $\mathbf{V}_B$, $\mathbf{V}_C$, and $\mathbf{V}_f$ in the global coordinate system for the position shown if $\omega_d = 30$ rad/sec CW. Use the velocity difference graphical method. (Make an enlarged copy of the figure and draw on it.)

†6-37   The linkage in Figure P6-8d has $AB = 30$, $BC = 150$, $CD = 30$, and $AD = 150$ mm. AB is at 58° in the global coordinate system. Find $\mathbf{V}_B$, $\mathbf{V}_C$, and $\mathbf{V}_f$ in the global coordinate system for the position shown if $\omega_d = 30$ rad/sec CW. Use an analytical method.

†6-38   The linkage in Figure P6-8d has $AB = 30$, $BC = 150$, $CD = 30$, and $AD = 150$ mm. Write a computer program or use an equation solver to find and plot $\mathbf{V}_B$, $\mathbf{V}_C$, and $\mathbf{V}_f$ in the global coordinate system for the maximum range of motion that this linkage allows if $\omega_d = 30$ rad/sec CW.

6-39   The linkage in Figure P6-8e has $AB = GF = 153$, $BC = EF = 100$, $CD = DE = 49$, and $AD = DG = 87$ mm. DG is at 61° and DC is at 29° in the global XY coordinate system. Find ω_f, $\mathbf{V}_B$, and $\mathbf{V}_C$ in the global coordinate system for the position shown if $\omega_d = 15$ rad/sec CW. Use the velocity difference graphical method. (Make an enlarged copy of the figure and draw on it.)

6-40   The linkage in Figure P6-8e has $AB = GF = 153$, $BC = EF = 100$, $CD = DE = 49$, and $AD = DG = 87$ mm. DG is at 61° and DC is at 29° in the global XY coordinate system. Find ω_f, $\mathbf{V}_B$, and $\mathbf{V}_C$ in the global coordinate system for the position shown if $\omega_d = 15$ rad/sec CW. Use the instant center graphical method. (Make an enlarged copy of the figure and draw on it.)

†6-41   The linkage in Figure P6-8e has $AB = GF = 153$, $BC = EF = 100$, $CD = DE = 49$, and $AD = DG = 87$ mm. DC is at –38° in the local $x'y'$ coordinate system. Find ω_f, $\mathbf{V}_B$, and $\mathbf{V}_C$ in the local coordinate system for the position shown if $\omega_d = 15$ rad/sec CW. Use an analytical method.

†6-42   The linkage in Figure P6-8e has $AB = GF = 153$, $BC = EF = 100$, $CD = DE = 49$, and $AD = DG = 87$ mm. Write a computer program or use an equation solver to find and plot ω_f, $\mathbf{V}_B$, and $\mathbf{V}_C$ in the local coordinate system for the maximum range of motion that this linkage allows if $\omega_d = 15$ rad/sec CW.

6-43   The 3-cylinder radial compressor in Figure P6-8f has a crank length $AB = 19$ mm and connecting rods $BC = BD = BE = 70$ mm. The cylinders are equispaced at 120°. Find the piston velocities $\mathbf{V}_C$, $\mathbf{V}_D$, $\mathbf{V}_E$ with the crank at –53° using a graphical method if $\omega = 15$ rad/sec CW. (Make an enlarged copy of the figure and draw on it.)

†6-44   The 3-cylinder radial compressor in Figure P6-8f has a crank length $AB = 19$ mm and connecting rods $BC = BD = BE = 70$ mm. The cylinders are equispaced at 120°. Find the piston velocities $\mathbf{V}_C$, $\mathbf{V}_D$, $\mathbf{V}_E$ with the crank at –53° using an analytical method if $\omega = 15$ rad/sec CW.

†6-45   The 3-cylinder radial compressor in Figure P6-8f has a crank length $AB = 19$ mm and connecting rods $BC = BD = BE = 70$ mm. The cylinders are equispaced at 120°. Write a program or use an equation solver to find and plot the piston velocities $\mathbf{V}_C$, $\mathbf{V}_D$, $\mathbf{V}_E$ for one revolution of the crank if $\omega = 15$ rad/sec CW.

† These problems are suited to solution using *Mathcad, Matlab,* or *TKSolver* equation solver programs.

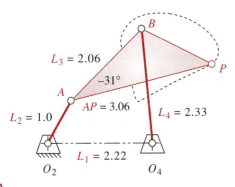

FIGURE P6-10

Problem 6-47 A fourbar linkage with a double straight-line coupler curve

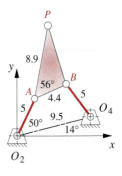

FIGURE P6-9

Problem 6-46

6-46   Figure P6-9 shows a linkage in one position. Find the instantaneous velocities of points A, B, and P if link O_2A is rotating CW at 40 rad/sec.

*†6-47   Figure P6-10 shows a linkage and its coupler curve. Write a computer program or use an equation solver to calculate and plot the magnitude and direction of the velocity of the coupler point P at 2° increments of crank angle for $\omega_2 = 100$ rpm. Check your result with program FOURBAR.

*†6-48   Figure P6-11 shows a linkage that operates at 500 crank rpm. Write a computer program or use an equation solver to calculate and plot the magnitude and direction of the velocity of point B at 2° increments of crank angle. Check your result with program FOURBAR.

*†6-49   Figure P6-12 shows a linkage and its coupler curve. Write a computer program or use an equation solver to calculate and plot the magnitude and direction of the velocity of the coupler point P at 2° increments of crank angle for $\omega_2 = 20$ rpm over the maximum range of motion possible. Check your result with program FOURBAR.

†6-50   Figure P6-13 shows a linkage and its coupler curve. Write a computer program or use an equation solver to calculate and plot the magnitude and direction of the

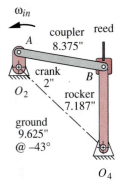

FIGURE P6-11

Problem 6-48 Loom laybar drive

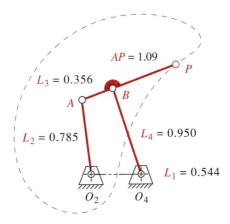

FIGURE P6-12

Problem 6-49

* Answers in Appendix F.

† These problems are suited to solution using *Mathcad, Matlab,* or *TKSolver* equation solver programs.

6

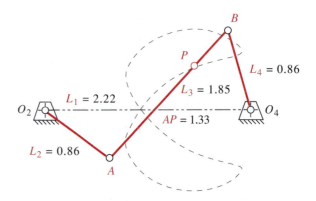

FIGURE P6-13

Problem 6-50

velocity of the coupler point P at $2°$ increments of crank angle for $\omega_2 = 80$ rpm over the maximum range of motion possible. Check your result with program FOURBAR.

*†6-51   Figure P6-14 shows a linkage and its coupler curve. Write a computer program or use an equation solver to calculate and plot the magnitude and direction of the velocity of the coupler point P at $2°$ increments of crank angle for $\omega_2 = 80$ rpm over the maximum range of motion possible. Check your result with program FOURBAR.

†‡6-52   Figure P6-15 shows a power hacksaw which is an offset slider-crank mechanism. The crank is 75 mm, the connecting rod is 170 mm, and the offset is 45 mm. Draw an equivalent linkage diagram; then calculate and plot the velocity of the sawblade with respect to the piece being cut over one revolution of the crank at 50 rpm.

†‡6-53   Figure P6-16 shows a pick-and-place mechanism which can be analyzed as two fourbar linkages driven by a common crank. The parallelogram stage has 40-mm cranks and a 108-mm coupler. Crank $AB = 32$ mm. $BC = 260$, $CD = 96$, $DE = 160$, and $AD = 200$ mm. Angle $CDE = 75°$. AD is at $205°$. The phase angle between the two crankpins on wheel W is $120°$. The cylinders P being pushed have 60-mm diameters. The point of contact between the vertical finger and the left-most cylinder in the position shown is 58 mm at $80°$ versus the left end of the parallelogram's

* Answers in Appendix F.

† These problems are suited to solution using *Mathcad, Matlab,* or *TKSolver* equation solver programs.

‡ These problems are suited to solution using the *Working Model* program,.which is on the attached CD-ROM.

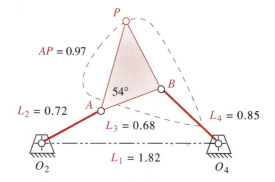

FIGURE P6-14

Problem 6-51

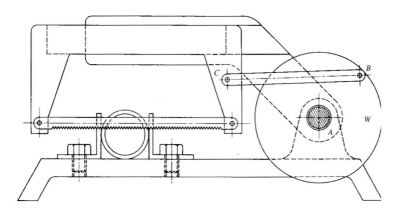

FIGURE P6-15

Problem 6-52. *From P. H. Hill and W. P. Rule. (1960). Mechanisms: Analysis and Design, with permission.*

6

coupler. Calculate and plot the relative velocity between point E and the center of the left-most cylinder P.

[†‡]6-54 Figure P6-17 shows a paper roll off-loading mechanism driven by an air cylinder. In the position shown, $AO_2 = 1.1$ m at $178°$ and O_4A is 0.3 m at $226°$. $O_2O_4 = 0.93$ m at $163°$. The V-links are rigidly attached to O_4A. The air cylinder is retracted at a constant velocity of 0.2 m/sec. Draw a kinematic diagram of the mechanism, write the necessary equations, and calculate and plot the angular velocity of the paper roll and the linear velocity of its center as it rotates through $90°$ CCW from the position shown.

[†‡]6-55 Figure P6-18 shows a powder compaction mechanism.

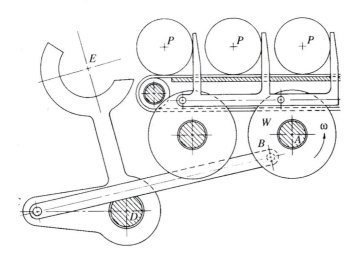

FIGURE P6-16

Problem 6-53. *From P. H. Hill and W. P. Rule. (1960). Mechanisms: Analysis and Design, with permission.*

[†] These problems are suited to solution using *Mathcad, Matlab,* or *TKSolver* equation solver programs.

[‡] These problems are suited to solution using the *Working Model* program,.which is on the attached CD-ROM.

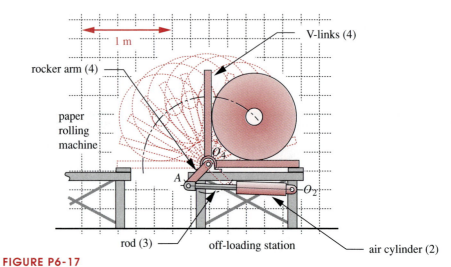

FIGURE P6-17

Problem 6-54

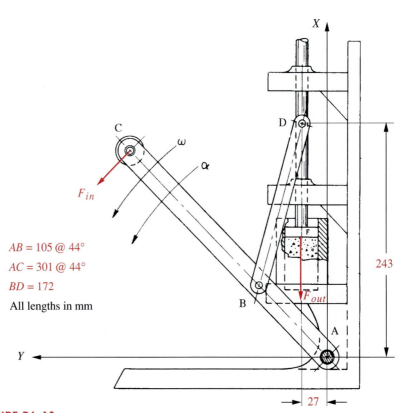

$AB = 105 \ @ \ 44°$

$AC = 301 \ @ \ 44°$

$BD = 172$

All lengths in mm

FIGURE P6-18

Problem 6-55. *From P. H. Hill and W. P. Rule. (1960). Mechanisms: Analysis and Design, with permission.*

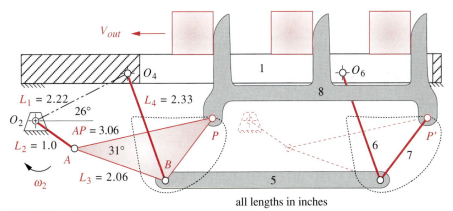

all lengths in inches

FIGURE P6-19

Problem 6-56. Straight-line walking beam eightbar transport mechanism

  a. Calculate its mechanical advantage for the position shown.
  b. Calculate and plot its mechanical advantage as a function of the angle of link *AC* as it rotates from 15 to 60°.

‡6-56 Figure P6-19 shows a walking beam mechanism. Calculate and plot the velocity V_{out} for one revolution of the input crank 2 rotating at 100 rpm.

†6-57 Figure P6-20 shows a crimping tool.

  a. Calculate its mechanical advantage for the position shown.
  b. Calculate and plot its mechanical advantage as a function of the angle of link *AB* as it rotates from 60 to 45°.

†6-58 Figure P6-21 shows a locking pliers. Calculate its mechanical advantage for the position shown. Scale any dimensions needed from the diagram.

†6-59 Figure P6-22 shows a fourbar toggle clamp used to hold a workpiece in place by clamping it at *D*. $O_2A = 70$, $O_2C = 138$, $AB = 35$, $O_4B = 34$, $O_4D = 82$, and $O_2O_4 =$

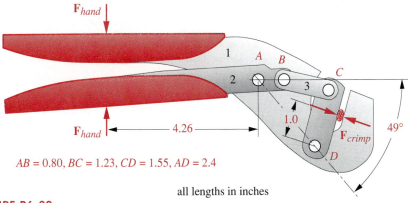

$AB = 0.80$, $BC = 1.23$, $CD = 1.55$, $AD = 2.4$

all lengths in inches

FIGURE P6-20

Problem 6-57. A crimping tool

† These problems are suited to solution using *Mathcad, Matlab,* or *TKSolver* equation solver programs.

‡ These problems are suited to solution using the *Working Model* program,.which is on the attached CD-ROM.

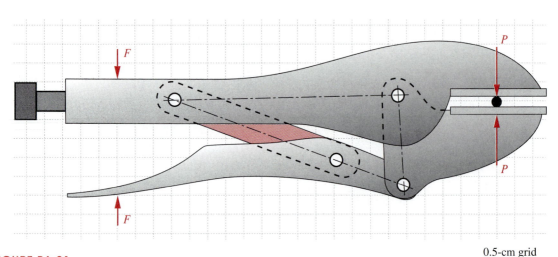

0.5-cm grid

FIGURE P6-21

Problem 6-58

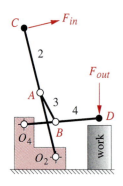

FIGURE P6-22

Problem 6-59

48 mm. At the position shown, link 2 is at 104°. The linkage will toggle when link 2 reaches 90°.

a. Calculate its mechanical advantage for the position shown.

b. Calculate and plot its mechanical advantage as a function of the angle of link AB as link 2 rotates from 120 to 90°.

†‡6-60 Figure P6-23 shows a surface grinder. The workpiece is oscillated under the spinning 90-mm-diameter grinding wheel by the slider-crank linkage which has a 22-mm crank, a 157-mm connecting rod, and a 40-mm offset. The crank turns at 120 rpm, and the grinding wheel at 3450 rpm. Calculate and plot the velocity of the grinding wheel contact point relative to the workpiece over one revolution of the crank.

6-61 Figure P6-24 shows an inverted slider-crank mechanism. Link 2 is 2.5 in long. The distance O_4A is 4.1 in and O_2O_4 is 3.9 in. Find $\omega_2, \omega_3, \omega_4, V_{A4}, V_{trans}$, and V_{slip} for the position shown with $V_{A2} = 20$ in/sec in the direction shown.

**†6-62 Figure P6-25 shows a drag link mechanism. Scale the diagram for dimensions, write the necessary equations, and solve them to calculate the angular velocity of link 4 for an input of $\omega_2 = 1$ rad/sec. Comment on uses for this mechanism.

†6-63 Figure P6-25 shows a drag link mechanism. Scale the diagram for dimensions, write the necessary equations, and solve them to calculate and plot the centrodes of instant center $I_{2,4}$.

‡6-64 Figure P6-26 shows a mechanism. Scale the diagram for dimensions and use a graphical method to calculate the velocities of points B, D, and E and the velocity of slip for the position shown. $\omega_{AB} = 20$ rad/sec.

6-65 Figure P6-27 shows a cam and follower. Find the velocities of points A and B, the velocity of transmission, velocity of slip, and ω_3 if $\omega_2 = 50$ rad/sec. Use a graphical method. Scale the diagram for dimensions.

‡6-66 Figure P6-28 shows a quick-return mechanism. Scale the diagram for dimensions and use a graphical method to calculate the velocities of points B, C, and E and the velocity of slip for the position shown. $\omega_{CD} = 10$ rad/sec.

---

† These problems are suited to solution using *Mathcad, Matlab,* or *TKSolver* equation solver programs.

---

‡ These problems are suited to solution using the *Working Model* program,.which is on the attached CD-ROM.

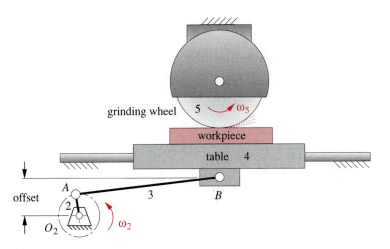

FIGURE P6-23

Problem 6-60. A surface grinder

†‡6-67 Figure P6-29 shows a drum pedal mechanism. $O_2A = 100$ mm at $162°$ and rotates to $171°$ at A'. $O_2O_4 = 56$ mm, $AB = 28$ mm, $AP = 124$ mm, and $O_4B = 64$ mm. The distance from O_4 to F_{in} is 48 mm. Find and plot the mechanical advantage and the velocity ratio of the linkage over its range of motion. If the input velocity V_{in} is a constant magnitude of 3 m/sec, and F_{in} is constant at 50 N, find the output velocity and output force over the range of motion and the power in.

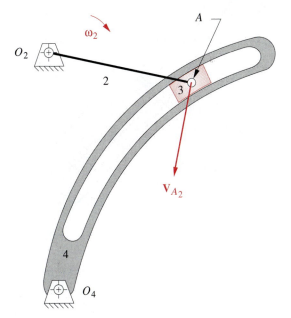

† These problems are suited to solution using *Mathcad, Matlab,* or *TKSolver* equation solver programs.

‡ These problems are suited to solution using the *Working Model* program,.which is on the attached CD-ROM.

FIGURE P6-24

Problem 6-61. *From P. H. Hill and W. P. Rule. (1960). Mechanisms: Analysis and Design, with permission.*

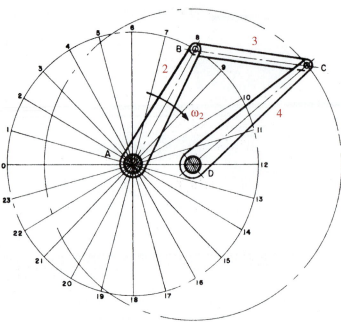

FIGURE P6-25

Problems 6-62 and 6-63. *From P. H. Hill and W. P. Rule. (1960). Mechanisms: Analysis and Design, with permission.*

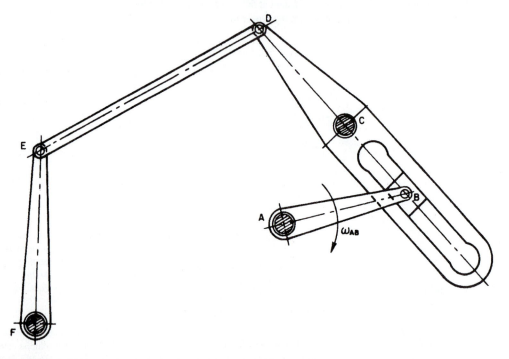

FIGURE P6-26

Problem 6-64. *From P. H. Hill and W. P. Rule. (1960). Mechanisms: Analysis and Design, with permission.*

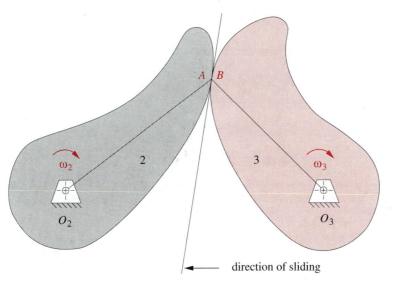

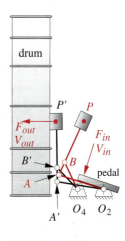

direction of sliding

FIGURE P6-29

Problem 6-67.

FIGURE P6-27

Problem 6-65. *From P. H. Hill and W. P. Rule. (1960). Mechanisms: Analysis and Design, with permission.*

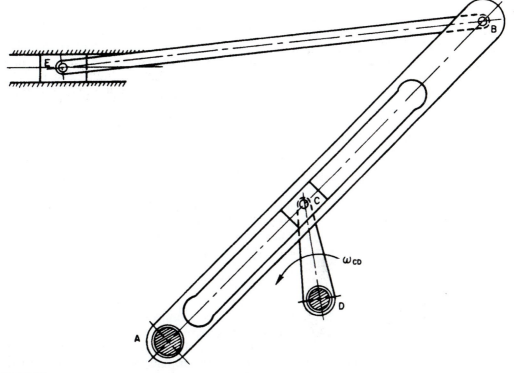

FIGURE P6-28

Problem 6-66. *From P. H. Hill and W. P. Rule. (1960). Mechanisms: Analysis and Design, with permission.*

Chapter 7

ACCELERATION ANALYSIS

Take it to warp five, Mr. Sulu
CAPTAIN KIRK

7.0   INTRODUCTION

Once a velocity analysis is done, the next step is to determine the accelerations of all links and points of interest in the mechanism or machine. We need to know the accelerations to calculate the dynamic forces from $\mathbf{F} = m\mathbf{a}$. The dynamic forces will contribute to the stresses in the links and other components. Many methods and approaches exist to find accelerations in mechanisms. We will examine only a few of these methods here. We will first develop a manual graphical method, which is often useful as a check on the more complete and accurate analytical solution. Then we will derive the analytical solution for accelerations in the fourbar and inverted slider-crank linkages as examples of the general vector loop equation solution to acceleration analysis problems.

7.1   DEFINITION OF ACCELERATION

Acceleration is defined as *the rate of change of velocity with respect to time*. Velocity ($\mathbf{V}$, ω) is a vector quantity and so is acceleration. Accelerations can be **angular** or **linear**. **Angular acceleration** will be denoted as α and **linear acceleration** as $\mathbf{A}$.

$$\alpha = \frac{d\omega}{dt}; \qquad \mathbf{A} = \frac{d\mathbf{V}}{dt} \qquad (7.1)$$

Figure 7-1 shows a link PA in pure rotation, pivoted at point A in the xy plane. We are interested in the acceleration of point P when the link is subjected to an angular velocity ω and an angular acceleration α, which need not have the same sense. The link's position is defined by the position vector $\mathbf{R}$, and the velocity of point P is $\mathbf{V}_{PA}$. These vectors were defined in equations 6.2 and 6.3 which are repeated here for convenience. (See also Figure 6-1, p. 242.)

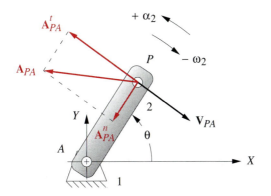

FIGURE 7-1

Acceleration of a link in pure rotation with a positive (CCW) α_2 and a negative (CW) ω_2

$$\mathbf{R}_{PA} = pe^{j\theta} \tag{6.2}$$

$$\mathbf{V}_{PA} = \frac{d\mathbf{R}_{PA}}{dt} = p\,je^{j\theta}\,\frac{d\theta}{dt} = p\omega\,je^{j\theta} \tag{6.3}$$

where p is the scalar length of the vector $\mathbf{R}_{PA}$. We can easily differentiate equation 6.3 to obtain an expression for the acceleration of point P:

$$\mathbf{A}_{PA} = \frac{d\mathbf{V}_{PA}}{dt} = \frac{d\left(p\omega\,je^{j\theta}\right)}{dt}$$

$$\mathbf{A}_{PA} = j\,p\left(e^{j\theta}\,\frac{d\omega}{dt} + \omega\,je^{j\theta}\,\frac{d\theta}{dt}\right) \tag{7.2}$$

$$\mathbf{A}_{PA} = p\alpha\,je^{j\theta} - p\omega^2\,e^{j\theta}$$

$$\mathbf{A}_{PA} = \mathbf{A}_{PA}^t + \mathbf{A}_{PA}^n$$

Note that there are two functions of time in equation 6.3, θ and ω. Thus there are two terms in the expression for acceleration, the tangential component of acceleration $\mathbf{A}_{PA}^t$ involving α, and the normal (or centripetal) component $\mathbf{A}_{PA}^n$ involving ω^2. As a result of the differentiation, the tangential component is multiplied by the (constant) complex operator j. This causes a rotation of this acceleration vector through 90 ° with respect to the original position vector. (See also Figure 4-5b, p. 152.) This 90° rotation is nominally positive, or counterclockwise (CCW). However, the tangential component is also multiplied by α, which may be either positive or negative. As a result, the tangential component of acceleration will be **rotated 90°** from the angle θ of the *position vector* **in a direction dictated by the sign of** α. This is just mathematical verification of what you already knew, namely that *tangential acceleration is always in a direction perpendicular to the radius of rotation and is thus tangent to the path of motion* as shown in Figure 7-1. The normal, or centripetal, acceleration component is multiplied by j^2, or -1. This directs *the centripetal component at 180° to the angle θ of the original position vector,* i.e., toward the center (centripetal means *toward the center*). The total acceleration $\mathbf{A}_{PA}$ of point P is the vector sum of the tangential $\mathbf{A}_{PA}^{t}$ and normal $\mathbf{A}_{PA}^{n}$ components as shown in Figure 7-1 and equation 7.2.

Substituting the Euler identity (equation 4.4a, p. 155) into equation 7.2 gives us the real and imaginary (or x and y) components of the acceleration vector.

$$\mathbf{A}_{PA} = p\alpha\left(-\sin\theta + j\cos\theta\right) - p\omega^{2}\left(\cos\theta + j\sin\theta\right) \tag{7.3}$$

The acceleration $\mathbf{A}_{PA}$ in Figure 7-1 can be referred to as an **absolute acceleration** since it is referenced to A, which is the origin of the global coordinate axes in that system. As such, we could have referred to it as $\mathbf{A}_{P}$, with the absence of the second subscript implying reference to the global coordinate system.

Figure 7-2a shows a different and slightly more complicated system in which the pivot A is no longer stationary. It has a known linear acceleration $\mathbf{A}_{A}$ as part of the translating carriage, link 3. If α is unchanged, the acceleration of point P versus A will be the same as before, but $\mathbf{A}_{PA}$ can no longer be considered an absolute acceleration. It is now an **acceleration difference** and **must** carry the second subscript as $\mathbf{A}_{PA}$. The absolute acceleration $\mathbf{A}_{P}$ must now be found from the **acceleration difference** equation whose graphical solution is shown in Figure 7-2b:

$$\mathbf{A}_{P} = \mathbf{A}_{A} + \mathbf{A}_{PA}$$

$$\tag{7.4}$$

$$\left(\mathbf{A}_{P}^{t} + \mathbf{A}_{P}^{n}\right) = \left(\mathbf{A}_{A}^{t} + \mathbf{A}_{A}^{n}\right) + \left(\mathbf{A}_{PA}^{t} + \mathbf{A}_{PA}^{n}\right)$$

Note the similarity of equation 7.4 to the **velocity difference equation** (equation 6.5, p. 243). Note also that the solution for $\mathbf{A}_{P}$ in equation 7.4 can be found either by adding the resultant vector $\mathbf{A}_{PA}$ or its normal and tangential components, $\mathbf{A}_{PA}^{n}$ and $\mathbf{A}_{PA}^{t}$ to the vector $\mathbf{A}_{A}$ in Figure 7-2b. The vector $\mathbf{A}_{A}$ has a zero normal component in this example because link 3 is in pure translation.

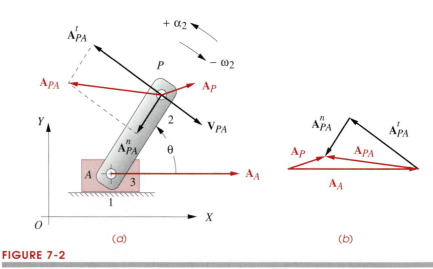

(a)   (b)

FIGURE 7-2

Acceleration difference in a system with a positive (CCW) α_2 and a negative (CW) ω_2

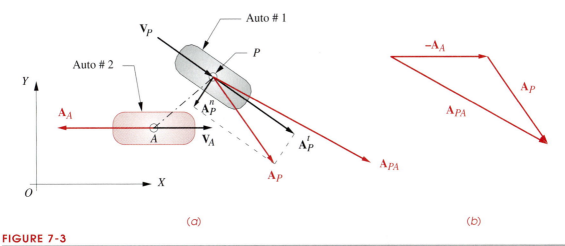

FIGURE 7-3

Relative acceleration

Figure 7-3 shows two independent bodies P and A, which could be two automobiles, moving in the same plane. Auto #1 is turning and accelerating into the path of auto #2, which is decelerating to avoid a crash. If their independent accelerations $\mathbf{A}_P$ and $\mathbf{A}_A$ are known, their **relative acceleration** $\mathbf{A}_{PA}$ can be found from equation 7.4 arranged algebraically as:

$$\mathbf{A}_{PA} = \mathbf{A}_P - \mathbf{A}_A \qquad (7.5)$$

The graphical solution to this equation is shown in Figure 7-3b.

As we did for velocity analysis, we give these two cases different names despite the fact that the same equation applies. Repeating the definition from Section 6.1 (p. 241), modified to refer to acceleration:

CASE 1:   Two points in the same body => *acceleration difference*

CASE 2:   Two points in different bodies => *relative acceleration*

7.2   GRAPHICAL ACCELERATION ANALYSIS

The comments made in regard to graphical velocity analysis in Section 6.2 (p. 244) apply as well to graphical acceleration analysis. Historically, graphical methods were the only practical way to solve these acceleration analysis problems. With some practice, and with proper tools such as a drafting machine or CAD package, one can fairly rapidly solve for the accelerations of particular points in a mechanism for any one input position by drawing vector diagrams. However, if accelerations for many positions of the mechanism are to be found, each new position requires a completely new set of vector diagrams be drawn. Very little of the work done to solve for the accelerations at position 1 carries over to position 2, etc. This is an even more tedious process than that for graphical velocity analysis because there are more components to draw. Nevertheless, this method still has more than historical value as it can provide a quick check on the results from a computer program solution. Such a check only needs to be done for a few positions to prove the validity of the program.

To solve any acceleration analysis problem graphically, we need only three equations, equation 7.4 and equations 7.6 (which are merely the scalar magnitudes of the terms in equation 7.2, p. 301):

$$\left|\mathbf{A}^t\right| = A^t = r\alpha$$

$$\left|\mathbf{A}^n\right| = A^n = r\omega^2$$

(7.6)

Note that the scalar equations 7.6 define only the **magnitudes** (A^t, A^n) of the components of acceleration of any point in rotation. In a CASE 1 graphical analysis, the **directions** of the vectors due to the centripetal and tangential components of the acceleration difference must be understood from equation 7.2 to be perpendicular to and along the radius of rotation, respectively. Thus, if the center of rotation is known or assumed, the directions of the acceleration difference components due to that rotation are known and their senses will be consistent with the angular velocity ω and angular acceleration α of the body.

Figure 7-4 shows a fourbar linkage in one particular position. We wish to solve for the angular accelerations of links 3 and 4 (α_3, α_4) and the linear accelerations of points A, B, and C ($\mathbf{A}_A$, $\mathbf{A}_B$, $\mathbf{A}_C$). Point C represents any general point of interest such as a coupler point. The solution method is valid for any point on any link. To solve this problem we need to know the *lengths of all the links*, the *angular positions of all the links*, the *angular velocities of all the links*, and the *instantaneous input acceleration of any one driving link or driving point*. Assuming that we have designed this linkage, we will know or can measure the link lengths. We must also first do a **complete position and velocity analysis** to find the link angles θ_3 and θ_4 and angular velocities ω_3 and ω_4 given the input link's position θ_2, input angular velocity ω_2, and input acceleration α_2. This can be done by any of the methods in Chapters 4 and 6. In general we must solve these problems in stages, first for link positions, then for velocities, and finally for accelerations. For the following example, we will assume that a complete position and velocity analysis has been done and that the input is to link 2 with known θ_2, ω_2, and α_2 for this one "freeze-frame" position of the moving linkage.

✎ EXAMPLE 7-1

Graphical Acceleration Analysis for One Position of a Fourbar Linkage.

Problem: Given θ_2, θ_3, θ_4, ω_2, ω_3, ω_4, α_2, find α_3, α_4, $\mathbf{A}_A$, $\mathbf{A}_B$, $\mathbf{A}_P$ by graphical methods.

Solution: (see Figure 7-4)

1   Start at the end of the linkage about which you have the most information. Calculate the magnitudes of the centripetal and tangential components of acceleration of point A using scalar equations 7.6.

$$A_A^n = (AO_2)\omega_2^2 ; \qquad\qquad A_A^t = (AO_2)\alpha_2 \qquad\qquad (a)$$

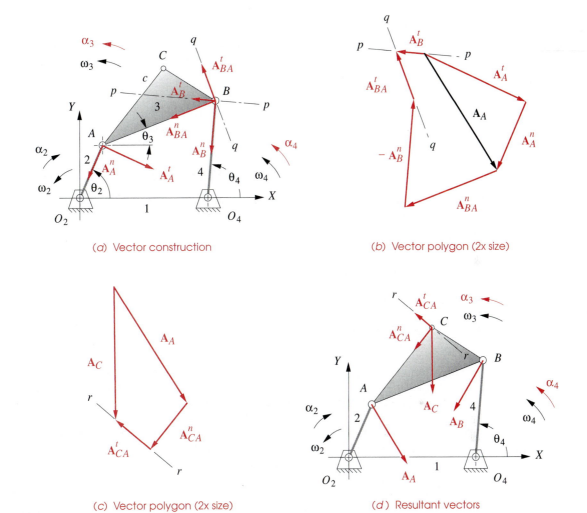

(a) Vector construction

(b) Vector polygon (2x size)

(c) Vector polygon (2x size)

(d) Resultant vectors

FIGURE 7-4

Graphical solution for acceleration in a pin-jointed linkage with a negative (CW) α_2 and a positive (CCW) ω_2

2 On the linkage diagram, Figure 7-4a, draw the acceleration component vectors A_A^n and A_A^t with their lengths equal to their magnitudes at some convenient scale. Place their roots at point A with their directions respectively along and perpendicular to the radius AO_2. The sense of A_A^t is defined by that of α_2 (according to the right-hand rule), and the sense of A_A^n is the opposite of that of the position vector R_A as shown in Figure 7-4a.

3 Move next to a point about which you have some information, such as B on link 4. Note that the directions of the tangential and normal components of acceleration of point B are predictable since this link is in pure rotation about point O_4. Draw the construction line pp through point B perpendicular to BO_4, to represent the direction of A_B^t as shown in Figure 7-4a.

4 Write the acceleration difference vector equation 7.4 for point B versus point A.

$$\mathbf{A}_B = \mathbf{A}_A + \mathbf{A}_{BA} \qquad (b)$$

Substitute the normal and tangential components for each term:

$$\left(\mathbf{A}_B^t + \mathbf{A}_B^n\right) = \left(\mathbf{A}_A^t + \mathbf{A}_A^n\right) + \left(\mathbf{A}_{BA}^t + \mathbf{A}_{BA}^n\right) \qquad (c)$$

We will use point A as the reference point to find $\mathbf{A}_B$ because A is in the same link as B and we have already solved for $\mathbf{A}_A^t$ and $\mathbf{A}_A^n$. Any two-dimensional vector equation can be solved for two unknowns. Each term has two parameters, namely magnitude and direction. There are then potentially twelve unknowns in this equation, two per term. We must know ten of them to solve it. We know both the magnitudes and directions of $\mathbf{A}_A^t$ and $\mathbf{A}_A^n$ and the directions of $\mathbf{A}_B^t$ and $\mathbf{A}_B^n$ which are along line pp and line BO_4, respectively. We can also calculate the magnitude of $\mathbf{A}_B^n$ from equation 7.6 since we know ω_4. This provides seven known values. We need to know three more parameters to solve the equation.

5 The term $\mathbf{A}_{BA}$ represents the acceleration difference of B with respect to A. This has two components. The normal component $\mathbf{A}_{BA}^n$ is directed along the line BA because we are using point A as the reference center of rotation for the free vector ω_3, and its magnitude can be calculated from equation 7.6. The direction of $\mathbf{A}_{BA}^t$ must then be perpendicular to the line BA. Draw construction line qq through point B and perpendicular to BA to represent the direction of $\mathbf{A}_{BA}^t$ as shown in Figure 7-4a (p. 305). The calculated magnitude and direction of component $\mathbf{A}_{BA}^n$ and the known direction of $\mathbf{A}_{BA}^t$ provide the needed additional three parameters.

6 Now the vector equation can be solved graphically by drawing a vector diagram as shown in Figure 7-4b. Either drafting tools or a CAD package is necessary for this step. The strategy is to first draw all vectors for which we know both magnitude and direction, being careful to arrange their senses according to equation 7.4 (p. 302).

First draw acceleration vectors $\mathbf{A}_A^t$ and $\mathbf{A}_A^n$ tip to tail, carefully to some scale, maintaining their directions. (They are drawn twice size in the figure.) Note that the sum of these two components is the vector $\mathbf{A}_A$. The equation in step 4 says to add $\mathbf{A}_{BA}$ to $\mathbf{A}_A$. We know $\mathbf{A}_{BA}^n$, so we can draw that component at the end of $\mathbf{A}_A$. We also know $\mathbf{A}_B^n$, but this component is on the left side of equation 7.4, so we must subtract it. Draw the negative (opposite sense) of $\mathbf{A}_B^n$ at the end of $\mathbf{A}_{BA}^n$.

This exhausts our supply of components for which we know both magnitude and direction. Our two remaining knowns are the directions of $\mathbf{A}_B^t$ and $\mathbf{A}_{BA}^t$ which lie along the lines pp and qq, respectively. Draw a line parallel to line qq across the tip of the vector representing $minus$ $\mathbf{A}_B^n$. The resultant, or left side of the equation, must close the vector diagram, from the tail of the first vector drawn ($\mathbf{A}_A$) to the tip of the last, so draw a line parallel to pp across the tail of $\mathbf{A}_A$. The intersection of these lines parallel to pp and qq defines the lengths of $\mathbf{A}_B^t$ and $\mathbf{A}_{BA}^t$. The senses of these vectors are determined from reference to equation 7.4. Vector $\mathbf{A}_A$ was added to $\mathbf{A}_{BA}$, so their components must be arranged tip to tail. Vector $\mathbf{A}_B$ is the resultant, so its component $\mathbf{A}_B^t$ must be from the tail of the first to the tip of the last. The resultant vectors are shown in Figure 7-4b and d.

7 The angular accelerations of links 3 and 4 can be calculated from equation 7.6:

$$\alpha_4 = \frac{A_B^t}{BO_4} \qquad\qquad \alpha_3 = \frac{A_{BA}^t}{BA} \qquad (d)$$

Note that the acceleration difference term $\mathbf{A}_{BA}^t$ represents the rotational component of acceleration of link 3 due to α_3. The rotational acceleration α of any body is a "**free vector**" which has no particular point of application to the body. It exists everywhere on the body.

8   Finally we can solve for $\mathbf{A}_C$ using equation 7.4 again. We select any point in link 3 for which we know the absolute velocity to use as the reference, such as point A.

$$\mathbf{A}_C = \mathbf{A}_A + \mathbf{A}_{CA} \qquad\qquad (e)$$

In this case, we can calculate the magnitude of $\mathbf{A}_{CA}^t$ from equation 7.6 (p. 304) as we have already found α_3,

$$A_{CA}^t = c\alpha_3 \qquad\qquad (f)$$

The magnitude of the component $\mathbf{A}_{CA}^n$ can be found from equation 7.6 using ω_3.

$$A_{CA}^n = c\omega_3^2 \qquad\qquad (g)$$

Since both $\mathbf{A}_A$ and $\mathbf{A}_{CA}$ are known, the vector diagram can be directly drawn as shown in Figure 7-4c. Vector $\mathbf{A}_C$ is the resultant which closes the vector diagram. Figure 7-4d shows the calculated acceleration vectors on the linkage diagram.

The above example contains some interesting and significant principles which deserve further emphasis. Equation 7.4 is repeated here for discussion.

$$\mathbf{A}_P = \mathbf{A}_A + \mathbf{A}_{PA}$$

$$\qquad\qquad (7.4)$$

$$\left(\mathbf{A}_P^t + \mathbf{A}_P^n \right) = \left(\mathbf{A}_A^t + \mathbf{A}_A^n \right) + \left(\mathbf{A}_{PA}^t + \mathbf{A}_{PA}^n \right)$$

This equation represents the *absolute* acceleration of some general point P referenced to the origin of the global coordinate system. The right side defines it as the sum of the absolute acceleration of some other reference point A in the same system and the acceleration difference (or relative acceleration) of point P versus point A. These terms are then further broken down into their normal (centripetal) and tangential components which have definitions as shown in equation 7.2 (p. 301).

Let us review what was done in Example 7-1 in order to extract the general strategy for solution of this class of problem. We started at the input side of the mechanism, as that is where the driving angular acceleration α_2 was defined. We first looked for a point (A) for which the motion was pure rotation. We then solved for the absolute acceleration of that point ($\mathbf{A}_A$) using equations 7.4 and 7.6 by breaking $\mathbf{A}_A$ into its normal and tangential components. *(Steps 1 and 2)*

We then used the point (A) just solved for as a reference point to define the translation component in equation 7.4 written for a new point (B). Note that we needed to choose a second point (B) which was in the same rigid body as the reference point (A) which we had already solved, and about which we could predict some aspect of the new point's (B's) acceleration components. In this example, we knew the direction of the component $\mathbf{A}_B^t$, though we did not yet know its magnitude. We could also calculate both magnitude and direction of the centripetal component, $\mathbf{A}_B^n$, since we knew ω_3 and the link length. In general this situation will obtain for any point on a link which is jointed to ground (as is link 4). In this example, we could not have solved for point C until we solved for B, because point C is on a floating link for which we do not yet know the angular acceleration or absolute acceleration direction. *(Steps 3 and 4)*

To solve the equation for the second point (B), we also needed to recognize that the tangential component of the acceleration difference $\mathbf{A}_{BA}^t$ is always directed perpendicular to the line connecting the two related points in the link (B and A in the example). In addition, you will always know the magnitude and direction of the centripetal acceleration components in equation 7.4 *if* **it represents an acceleration difference** (CASE 1) **situation**. *If the two points are in the* **same rigid** *body, then that acceleration difference centripetal component has a magnitude of $r\omega^2$ and is always directed along the line connecting the two points, pointing toward the reference point as the center* (see Figure 7-2, p. 302). These observations will be true regardless of the two points selected. But, *note this is not true in a* CASE 2 *situation* as shown in Figure 7-3a (p. 303) where the normal component of acceleration of auto #2 is **not** directed along the line connecting points A and P. *(Steps 5 and 6)*

Once we found the absolute acceleration ($\mathbf{A}_B$) of a second point on the same link (CASE 1) we could solve for the angular acceleration of that link. (Note that points A and B are both on link 3 and the acceleration of point O_4 is zero.) Once the angular accelerations of all the links were known, we could solve for the linear acceleration of any point (such as C) in any link using equation 7.4. To do so, we had to understand the concept of angular acceleration as a **free vector**, which means that it exists everywhere on the link at any given instant. It has no particular center. *It has an infinity of potential centers.* The link simply *has an angular acceleration.* It is this property that allows us to solve equation 7.4 for literally **any point** on a rigid body in complex motion **referenced to any other point** on that body. *(Steps 7 and 8)*

7.3   ANALYTICAL SOLUTIONS FOR ACCELERATION ANALYSIS

The Fourbar Pin-Jointed Linkage

The position equations for the fourbar pin-jointed linkage were derived in Section 4.5 (p. 152). The linkage was shown in Figure 4-7 and is shown again in Figure 7-5a on which we also show an input angular acceleration α_2 applied to link 2. This input angular acceleration α_2 may vary with time. The vector loop equation was shown in equations 4.5a and c, repeated here for your convenience.

$$\mathbf{R}_2 + \mathbf{R}_3 - \mathbf{R}_4 - \mathbf{R}_1 = 0 \tag{4.5a}$$

As before, we substitute the complex number notation for the vectors, denoting their scalar lengths as a, b, c, d as shown in Figure 7-5.

$$ae^{j\theta_2} + be^{j\theta_3} - ce^{j\theta_4} - de^{j\theta_1} = 0 \tag{4.5c}$$

In Section 6.7 (p. 271), we differentiated equation 4.5c versus time to get an expression for velocity which is repeated here.

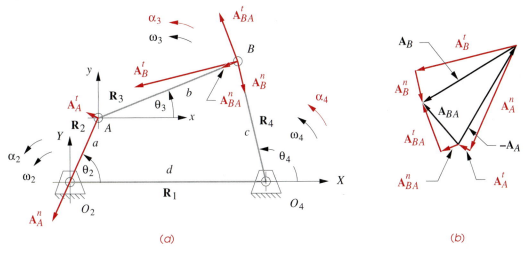

FIGURE 7-5

Position vector loop for a fourbar linkage showing acceleration vectors

$$ja\omega_2 e^{j\theta_2} + jb\omega_3 e^{j\theta_3} - jc\omega_4 e^{j\theta_4} = 0 \qquad (6.14c)$$

We will now differentiate equation 6.14c versus time to obtain an expression for accelerations in the linkage. Each term in equation 6.14c contains two functions of time, θ and ω. Differentiating with the chain rule in this example will result in two terms in the acceleration expression for each term in the velocity equation.

$$\left(j^2 a\omega_2^2 e^{j\theta_2} + ja\alpha_2 e^{j\theta_2} \right) + \left(j^2 b\omega_3^2 e^{j\theta_3} + jb\alpha_3 e^{j\theta_3} \right) - \left(j^2 c\omega_4^2 e^{j\theta_4} + jc\alpha_4 e^{j\theta_4} \right) = 0 \quad (7.7a)$$

Simplifying and grouping terms:

$$\left(a\alpha_2 \, je^{j\theta_2} - a\omega_2^2 e^{j\theta_2} \right) + \left(b\alpha_3 \, je^{j\theta_3} - b\omega_3^2 e^{j\theta_3} \right) - \left(c\alpha_4 \, je^{j\theta_4} - c\omega_4^2 e^{j\theta_4} \right) = 0 \quad (7.7b)$$

Compare the terms grouped in parentheses with equations 7.2 (p. 301). Equation 7.7 contains the tangential and normal components of the accelerations of points A and B and of the acceleration difference of B to A. Note that these are the same relationships which we used to solve this problem graphically in Section 7.2 (p. 303). Equation 7.7 is, in fact, the **acceleration difference equation** 7.4 which, with the labels used here, is:

$$\mathbf{A}_A + \mathbf{A}_{BA} - \mathbf{A}_B = 0 \qquad (7.8a)$$

where:

$$\mathbf{A}_A = \left(\mathbf{A}_A^t + \mathbf{A}_A^n \right) = \left(a\alpha_2 \, je^{j\theta_2} - a\omega_2^2 e^{j\theta_2} \right)$$

$$\mathbf{A}_{BA} = \left(\mathbf{A}_{BA}^t + \mathbf{A}_{BA}^n \right) = \left(b\alpha_3 \, je^{j\theta_3} - b\omega_3^2 e^{j\theta_3} \right) \qquad (7.8b)$$

$$\mathbf{A}_B = \left(\mathbf{A}_B^t + \mathbf{A}_B^n \right) = \left(c\alpha_4 \, je^{j\theta_4} - c\omega_4^2 e^{j\theta_4} \right)$$

The vector diagram in Figure 7-5b (p. 309) shows these components and is a graphical solution to equation 7.8a. The vector components are also shown acting at their respective points on Figure 7-5a.

We now need to solve equation 7.7 for α_3 and α_4, knowing the input angular acceleration α_2, the link lengths, all link angles, and angular velocities. Thus, the position analysis derived in Section 4.5 (p. 152) and the velocity analysis from Section 6.7 (p. 271) must be done first to determine the link angles and angular velocities before this acceleration analysis can be completed. We wish to solve equation 7.8 to get expressions in this form:

$$\alpha_3 = f(a, b, c, d, \theta_2, \theta_3, \theta_4, \omega_2, \omega_3, \omega_4, \alpha_2) \tag{7.9a}$$

$$\alpha_4 = g(a, b, c, d, \theta_2, \theta_3, \theta_4, \omega_2, \omega_3, \omega_4, \alpha_2) \tag{7.9b}$$

The strategy of solution will be the same as was done for the position and velocity analysis. First, substitute the Euler identity from equation 4.4a in each term of equation 7.7:

$$\begin{aligned}
& \left[a\alpha_2 \, j(\cos\theta_2 + j\sin\theta_2) - a\omega_2^2(\cos\theta_2 + j\sin\theta_2) \right] \\
& + \left[b\alpha_3 \, j(\cos\theta_3 + j\sin\theta_3) - b\omega_3^2(\cos\theta_3 + j\sin\theta_3) \right] \\
& - \left[c\alpha_4 \, j(\cos\theta_4 + j\sin\theta_4) - c\omega_4^2(\cos\theta_4 + j\sin\theta_4) \right] = 0
\end{aligned} \tag{7.10a}$$

Multiply by the operator j and rearrange:

$$\begin{aligned}
& \left[a\alpha_2 (-\sin\theta_2 + j\cos\theta_2) - a\omega_2^2(\cos\theta_2 + j\sin\theta_2) \right] \\
& + \left[b\alpha_3 (-\sin\theta_3 + j\cos\theta_3) - b\omega_3^2(\cos\theta_3 + j\sin\theta_3) \right] \\
& - \left[c\alpha_4 (-\sin\theta_4 + j\cos\theta_4) - c\omega_4^2(\cos\theta_4 + j\sin\theta_4) \right] = 0
\end{aligned} \tag{7.10b}$$

We can now separate this vector equation into its two components by collecting all real and all imaginary terms separately:

real part (x component):

$$-a\alpha_2 \sin\theta_2 - a\omega_2^2 \cos\theta_2 - b\alpha_3 \sin\theta_3 - b\omega_3^2 \cos\theta_3 + c\alpha_4 \sin\theta_4 + c\omega_4^2 \cos\theta_4 = 0 \tag{7.11a}$$

imaginary part (y component):

$$a\alpha_2 \cos\theta_2 - a\omega_2^2 \sin\theta_2 + b\alpha_3 \cos\theta_3 - b\omega_3^2 \sin\theta_3 - c\alpha_4 \cos\theta_4 + c\omega_4^2 \sin\theta_4 = 0 \tag{7.11b}$$

Note that the j's have cancelled in equation 7.11b. We can solve equations 7.11a and 7.11b simultaneously to get:

$$\alpha_3 = \frac{CD - AF}{AE - BD} \tag{7.12a}$$

$$\alpha_4 = \frac{CE - BF}{AE - BD} \tag{7.12b}$$

where:

$$A = c\sin\theta_4$$

$$B = b\sin\theta_3$$

$$C = a\alpha_2\sin\theta_2 + a\omega_2^2\cos\theta_2 + b\omega_3^2\cos\theta_3 - c\omega_4^2\cos\theta_4$$

$$D = c\cos\theta_4 \tag{7.12c}$$

$$E = b\cos\theta_3$$

$$F = a\alpha_2\cos\theta_2 - a\omega_2^2\sin\theta_2 - b\omega_3^2\sin\theta_3 + c\omega_4^2\sin\theta_4$$

Once we have solved for α_3 and α_4, we can then solve for the linear accelerations by substituting the Euler identity into equations 7.8b,

$$\mathbf{A}_A = a\alpha_2\left(-\sin\theta_2 + j\cos\theta_2\right) - a\omega_2^2\left(\cos\theta_2 + j\sin\theta_2\right) \tag{7.13a}$$

$$\mathbf{A}_{BA} = b\alpha_3\left(-\sin\theta_3 + j\cos\theta_3\right) - b\omega_3^2\left(\cos\theta_3 + j\sin\theta_3\right) \tag{7.13b}$$

$$\mathbf{A}_B = c\alpha_4\left(-\sin\theta_4 + j\cos\theta_4\right) - c\omega_4^2\left(\cos\theta_4 + j\sin\theta_4\right) \tag{7.13c}$$

where the real and imaginary terms are the x and y components, respectively. Equations 7.12 and 7.13 provide a complete solution for the angular accelerations of the links and the linear accelerations of the joints in the pin-jointed fourbar linkage.

The Fourbar Slider-Crank

The first inversion of the offset slider-crank has its slider block sliding against the ground plane as shown in Figure 7-6a. Its accelerations can be solved for in similar manner as was done for the pin-jointed fourbar.

The position equations for the fourbar offset slider-crank linkage (inversion #1) were derived in Section 4.6 (p. 159). The linkage was shown in Figures 4-9 (p. 160) and 6-21 (p. 275) and is shown again in Figure 7-6a on which we also show an input angular acceleration α_2 applied to link 2. This α_2 can be a time-varying input acceleration. The vector loop equation 4.14 is repeated here for your convenience.

$$\mathbf{R}_2 - \mathbf{R}_3 - \mathbf{R}_4 - \mathbf{R}_1 = 0 \tag{4.14a}$$

$$ae^{j\theta_2} - be^{j\theta_3} - ce^{j\theta_4} - de^{j\theta_1} = 0 \tag{4.14b}$$

In Section 6.7 (p. 267) we differentiated equation 4.14b with respect to time noting that a, b, c, θ_1, and θ_4 are constant but the length of link d varies with time in this inversion.

$$ja\omega_2 e^{j\theta_2} - jb\omega_3 e^{j\theta_3} - \dot{d} = 0 \tag{6.20a}$$

The term d dot is the linear velocity of the slider block. Equation 6.20a is the velocity difference equation.

We now will differentiate equation 6.20a with respect to time to get an expression for acceleration in this inversion of the slider-crank mechanism.

$$\left(ja\alpha_2 e^{j\theta_2} + j^2 a\omega_2^2 e^{j\theta_2}\right) - \left(jb\alpha_3 e^{j\theta_3} + j^2 b\omega_3^2 e^{j\theta_3}\right) - \ddot{d} = 0 \tag{7.14a}$$

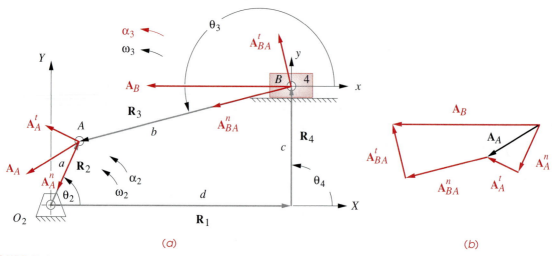

FIGURE 7-6

Position vector loop for a fourbar slider-crank linkage showing acceleration vectors

Simplifying:

$$\left(a\alpha_2\, je^{j\theta_2} - a\omega_2^2 e^{j\theta_2}\right) - \left(b\alpha_3\, je^{j\theta_3} - b\omega_3^2\, e^{j\theta_3}\right) - \ddot{d} = 0 \qquad (7.14b)$$

Note that equation 7.14 is again the acceleration difference equation:

$$\mathbf{A}_A - \mathbf{A}_{AB} - \mathbf{A}_B = 0$$
$$\mathbf{A}_{BA} = -\mathbf{A}_{AB} \qquad (7.15a)$$
$$\mathbf{A}_B = \mathbf{A}_A + \mathbf{A}_{BA}$$

$$\mathbf{A}_A = \left(\mathbf{A}_A^t + \mathbf{A}_A^n\right) = \left(a\alpha_2\, je^{j\theta_2} - a\omega_2^2\, e^{j\theta_2}\right)$$
$$\mathbf{A}_{BA} = \left(\mathbf{A}_{BA}^t + \mathbf{A}_{BA}^n\right) = \left(b\alpha_3\, je^{j\theta_3} - b\omega_3^2\, e^{j\theta_3}\right) \qquad (7.15b)$$
$$\mathbf{A}_B = \mathbf{A}_B^t = \ddot{d}$$

Note that in this mechanism, link 4 is in pure translation and so has zero ω_4 and zero α_4. The acceleration of link 4 has only a "tangential" component of acceleration along its path.

The two unknowns in the vector equation 7.14 are the angular acceleration of link 3, α_3, and the linear acceleration of link 4, *d double dot*. To solve for them, substitute the Euler identity,

$$a\alpha_2\left(-\sin\theta_2 + j\cos\theta_2\right) - a\omega_2^2\left(\cos\theta_2 + j\sin\theta_2\right)$$
$$- b\alpha_3\left(-\sin\theta_3 + j\cos\theta_3\right) + b\omega_3^2\left(\cos\theta_3 + j\sin\theta_3\right) - \ddot{d} = 0 \qquad (7.16a)$$

and separate the real (x) and imaginary (y) components:

real part (x component):

$$-a\alpha_2\sin\theta_2 - a\omega_2^2\cos\theta_2 + b\alpha_3\sin\theta_3 + b\omega_3^2\cos\theta_3 - \ddot{d} = 0 \qquad (7.16b)$$

imaginary part (y component):

$$a\alpha_2\cos\theta_2 - a\omega_2^2\sin\theta_2 - b\alpha_3\cos\theta_3 + b\omega_3^2\sin\theta_3 = 0 \qquad (7.16c)$$

Equation 7.16c can be solved directly for α_3 and the result substituted in equation 7.16b to find d *double dot*.

$$\alpha_3 = \frac{a\alpha_2\cos\theta_2 - a\omega_2^2\sin\theta_2 + b\omega_3^2\sin\theta_3}{b\cos\theta_3} \qquad (7.16d)$$

$$\ddot{d} = -a\alpha_2\sin\theta_2 - a\omega_2^2\cos\theta_2 + b\alpha_3\sin\theta_3 + b\omega_3^2\cos\theta_3 \qquad (7.16e)$$

The other linear accelerations can be found from equation 7.15b and are shown in the vector diagram of Figure 7-6b.

Coriolis Acceleration

The examples used for acceleration analysis above have involved only pin-jointed linkages or the inversion of the slider-crank in which the slider block has no rotation. When a sliding joint is present on a rotating link, an additional component of acceleration will be present, called the **Coriolis component**, after its discoverer. Figure 7-7a shows a simple, two-link system consisting of a link with a radial slot, and a slider block free to slip within that slot.

The instantaneous location of the block is defined by a position vector ($\mathbf{R}_P$) referenced to the global origin at the link center. *This vector is both rotating and changing length as the system moves.* As shown this is a two-degree-of-freedom system. The **two inputs to the system** are the angular acceleration (α) of the link and the relative linear slip velocity ($\mathbf{V}_{P_{slip}}$) of the block versus the disk. The angular velocity ω is a result of the time history of the angular acceleration. The situation shown, with a counterclockwise α and a clockwise ω, implies that earlier in time the link had been accelerated up to a clockwise angular velocity and is now being slowed down. The transmission component of velocity ($\mathbf{V}_{P_{trans}}$) is a result of the ω of the link acting at the radius R_P whose magnitude is p.

We show the situation in Figure 7-7 at one instant of time. However, the equations to be derived will be valid for all time. We want to determine the acceleration at the center of the block (P) under this combined motion of rotation and sliding. To do so we first write the expression for the position vector $\mathbf{R}_P$ which locates point P.

$$\mathbf{R}_P = pe^{j\theta_2} \qquad (7.17)$$

Note that there are two functions of time in equation 7.17, p and θ. When we differentiate versus time we get two terms in the velocity expression:

$$\mathbf{V}_P = p\omega_2 je^{j\theta_2} + \dot{p}e^{j\theta_2} \qquad (7.18a)$$

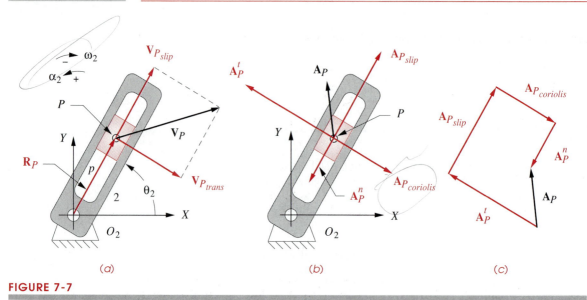

FIGURE 7-7

The Coriolis component of acceleration shown in a system with a positive (CCW) α_2 and a negative (CW) ω_2

These are the transmission component and the slip component of velocity.

$$\mathbf{V}_P = \mathbf{V}_{P_{trans}} + \mathbf{V}_{P_{slip}} \tag{7.18b}$$

The $p\omega$ term is the transmission component and is directed at 90 degrees to the axis of slip which, in this example, is coincident with the position vector $\mathbf{R}_P$. The p *dot* term is the **slip component** and is directed along the **axis of slip** in the same direction as the position vector in this example. Their vector sum is $\mathbf{V}_P$ as shown in Figure 7-7a.

To get an expression for acceleration, we must differentiate equation 7.18 versus time. Note that the transmission component has **three** functions of time in it, p, ω, and θ. The chain rule will yield three terms for this one term. The slip component of velocity contains two functions of time, p and θ, yielding two terms in the derivative for a total of five terms, two of which turn out to be the same.

$$\mathbf{A}_P = \left(p\alpha_2 je^{j\theta_2} + p\omega_2^2 j^2 e^{j\theta_2} + \dot{p}\omega_2 je^{j\theta_2} \right) + \left(\dot{p}\omega_2 je^{j\theta_2} + \ddot{p}e^{j\theta_2} \right) \tag{7.19a}$$

Simplifying and collecting terms:

$$\mathbf{A}_P = p\alpha_2 je^{j\theta_2} - p\omega_2^2 e^{j\theta_2} + 2\dot{p}\omega_2 je^{j\theta_2} + \ddot{p}e^{j\theta_2} \tag{7.19b}$$

These terms represent the following components:

$$\mathbf{A}_P = \mathbf{A}_{P_{tangential}} + \mathbf{A}_{P_{normal}} + \mathbf{A}_{P_{coriolis}} + \mathbf{A}_{P_{slip}} \tag{7.19c}$$

Note that the Coriolis term has appeared in the acceleration expression as a result of the differentiation simply because the length of the vector p is a function of time. The Coriolis component magnitude is twice the product of the velocity of slip (equation 7.18) and the angular velocity of the link containing the slider slot. Its direction is rotated 90 degrees from that of the original position vector $\mathbf{R}_P$ either clockwise or counterclockwise depending on the sense of ω.* (Note that we chose to align the position vector $\mathbf{R}_P$ with the axis of slip in Figure 7-7 which can always be done regardless of the location of the center of rotation—also see Figure 7-6 (p. 312) where $\mathbf{R}_1$ is aligned with the axis of slip.) All four components from equation 7.19 are shown acting on point P in Figure 7-7b. The total acceleration $\mathbf{A}_P$ is the vector sum of the four terms as shown in Figure 7-7c. Note that the normal acceleration term in equation 7.19b is negative in sign, so it becomes a subtraction when substituted in equation 7.19c.

This Coriolis component of acceleration will always be present when there is a velocity of slip associated with any member which also has an angular velocity. In the absence of either of those two factors the Coriolis component will be zero. You have probably experienced Coriolis acceleration if you have ever ridden on a carousel or merry-go-round. If you attempted to walk radially from the outside to the inside (or vice versa) while the carousel was turning, you were thrown sideways by the inertial force due to the Coriolis acceleration. You were the *slider block* in Figure 7-7, and your *slip velocity* combined with the rotation of the carousel created the Coriolis component. As you walked from a large radius to a smaller one, your tangential velocity had to change to match that of the new location of your foot on the spinning carousel. Any change in velocity requires an acceleration to accomplish. It was the "*ghost of Coriolis*" that pushed you sideways on that carousel.

Another example of the Coriolis component is its effect on weather systems. Large objects which exist in the earth's lower atmosphere, such as hurricanes, span enough area to be subject to significantly different velocities at their northern and southern extremities. The atmosphere turns with the earth. The earth's surface tangential velocity due to its angular velocity varies from zero at the poles to a maximum of about 1000 mph at the equator. The winds of a storm system are attracted toward the low pressure at its center. These winds have a slip velocity with respect to the surface, which in combination with the earth's ω, creates a Coriolis component of acceleration on the moving air masses. This Coriolis acceleration causes the inrushing air to rotate about the center, or "eye" of the storm system. This rotation will be counterclockwise in the northern hemisphere and clockwise in the southern hemisphere. The movement of the entire storm system from south to north also creates a Coriolis component which will tend to deviate the storm's track eastward, though this effect is often overridden by the forces due to other large air masses such as high-pressure systems which can deflect a storm. These complicated factors make it difficult to predict a large storm's true track.

Note that in the analytical solution presented here, the Coriolis component will be accounted for automatically as long as the differentiations are correctly done. However, when doing a graphical acceleration analysis one must be on the alert to recognize the presence of this component, calculate it, and include it in the vector diagrams when its two constituents $\mathbf{V}_{slip}$ and ω are both nonzero.

The Fourbar Inverted Slider-Crank

The position equations for the fourbar inverted slider-crank linkage were derived in Section 4.7 (p. 159). The linkage was shown in Figures 4-10 (p. 162) and 6-22 (p. 277) and is shown again in Figure 7-8a on which we also show an input angular acceleration α_2 applied to link 2. This α_2 can vary with time. The vector loop equations 4.14 (p. 311) are valid for this linkage as well.

7

* This approach works in the 2-D case. Coriolis acceleration is the cross product of 2ω and the velocity of slip. The cross product operation will define its magnitude, sign, and direction in the 3-D case.

nate system whose origin is at A. Position vector $\mathbf{R}_{PA}$ is defined in the same way as $\mathbf{R}_U$ or $\mathbf{R}_S$, using the internal link offset angle δ_3 and the angle of link 3, θ_3. We previously analyzed this position vector and differentiated it in Section 6.9 to find the velocity difference of that point with respect to point A. Those equations are repeated here for your convenience.

$$\mathbf{R}_{PA} = pe^{j(\theta_3 + \delta_3)} = p\left[\cos(\theta_3 + \delta_3) + j\sin(\theta_3 + \delta_3)\right] \tag{4.27a}$$

$$\mathbf{R}_P = \mathbf{R}_A + \mathbf{R}_{PA} \tag{4.27b}$$

$$\mathbf{V}_{PA} = jpe^{j(\theta_3 + \delta_3)}\omega_3 = p\omega_3\left[-\sin(\theta_3 + \delta_3) + j\cos(\theta_3 + \delta_3)\right] \tag{6.36a}$$

$$\mathbf{V}_P = \mathbf{V}_A + \mathbf{V}_{PA} \tag{6.36b}$$

We can differentiate equation 6.36 again versus time to find $\mathbf{A}_{PA}$, the acceleration of point P versus A. This vector can then be added to the vector $\mathbf{A}_A$ already found to define the absolute acceleration $\mathbf{A}_P$ of point P.

$$\mathbf{A}_P = \mathbf{A}_A + \mathbf{A}_{PA} \tag{7.32a}$$

where:

$$
\begin{aligned}
\mathbf{A}_{PA} &= p\alpha_3\, je^{j(\theta_3 + \delta_3)} - p\omega_3^2\, e^{j(\theta_3 + \delta_3)} \\
&= p\alpha_3\left[-\sin(\theta_3 + \delta_3) + j\cos(\theta_3 + \delta_3)\right] \\
&\quad - p\omega_3^2\left[\cos(\theta_3 + \delta_3) + j\sin(\theta_3 + \delta_3)\right]
\end{aligned}
\tag{7.32b}
$$

Please compare equation 7.32 with equation 7.4 (p. 302). It is again the acceleration difference equation. Note that this equation applies to **any point** on **any link** at any position for which the positions and velocities are defined. It is a general solution for any rigid body.

7.6   HUMAN TOLERANCE OF ACCELERATION

It is interesting to note that the human body does not sense velocity, except with the eyes, but is very sensitive to acceleration. Riding in an automobile, in the daylight, one can see the scenery passing by and have a sense of motion. But, traveling at night in a commercial airliner at a 500 mph constant velocity, we have no sensation of motion as long as the flight is smooth. What we will sense in this situation is any change in velocity due to atmospheric turbulence, takeoffs, or landings. The semicircular canals in the inner ear are sensitive accelerometers which report to us on any accelerations which we experience. You have no doubt also experienced the sensation of acceleration when riding in an elevator and starting, stopping, or turning in an automobile. Accelerations produce dynamic forces on physical systems, as expressed in Newton's second law, $\mathbf{F}=m\mathbf{a}$. Force is proportional to acceleration, for a constant mass. The dynamic forces produced within the human body in response to acceleration can be harmful if excessive.

The human body is, after all, not rigid. It is a loosely coupled bag of water and tissue, most of which is quite internally mobile. Accelerations in the headward or footward directions will tend to either starve or flood the brain with blood as this liquid responds to Newton's law and effectively moves within the body in a direction opposite to the imposed acceleration as it lags the motion of the skeleton. Lack of blood supply to the brain causes blackout; excess blood supply causes redout. Either results in death if sustained for a long enough period.

A great deal of research has been done, largely by the military and NASA, to determine the limits of human tolerance to sustained accelerations in various directions. Figure 7-10 shows data developed from such tests. [1] The units of linear acceleration were defined in Table 1-4 (p. 19) as in/sec^2, ft/sec^2, or m/sec^2. Another common unit for acceleration is the g, defined as the acceleration due to gravity, which on Earth at sea level is approximately 386 in/sec^2, 32.2 ft/sec^2, or 9.8 m/sec^2. The g is a very convenient unit to use for accelerations involving the human as we live in a 1 g environment. Our weight, felt on our feet or buttocks, is defined by our mass times the acceleration due to gravity or mg. Thus an imposed acceleration of 1 g above the baseline of our gravity, or 2 g's, will be felt as a doubling of our weight. At 6 g's we would feel six times as heavy as normal and would have great difficulty even moving our arms against that acceleration. Figure 7-10 shows that the body's tolerance of acceleration is a function of its direction versus the body, its magnitude, and its duration. Note also that the data used for this chart were developed from tests on young, healthy military personnel in prime physical condition. The general population, children and elderly in particular, should not be expected to be able to withstand such high levels of acceleration. Since much machinery is designed for human use, these acceleration tolerance data should be of great interest and value to the machine designer. Several references dealing with these human factors data are provided in the bibliography to Chapter 1 (p. 20).

Another useful benchmark when designing machinery for human occupation is to attempt to relate the magnitudes of accelerations which you commonly experience to the calculated values for your potential design. Table 7-1 lists some approximate levels of acceleration, in g's, which humans can experience in everyday life. Your own experience of these will help you develop a "feel" for the values of acceleration which you encounter in designing machinery intended for human occupation.

Note that machinery which does not carry humans is limited in its acceleration levels only by considerations of the stresses in its parts. These stresses are often generated in large part by the dynamic forces due to accelerations. The range of acceleration values in such machinery is so wide that it is not possible to comprehensively define any guidelines for the designer as to acceptable or unacceptable levels of acceleration. If the moving mass is small, then very large numerical values of acceleration are reasonable. If the mass is large, the dynamic stresses which the materials can sustain may limit the allowable accelerations to low values. Unfortunately, the designer does not usually know how much acceleration is too much in a design until completing it to the point of calculating stresses in the parts. This usually requires a fairly complete and detailed design. If the stresses turn out to be too high and are due to dynamic forces, then the only recourse is to iterate back through the design process and reduce the accelerations and or masses in the design. This is one reason that the design process is a circular and not a linear one.

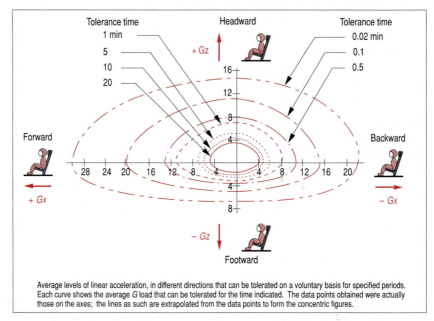

Average levels of linear acceleration, in different directions that can be tolerated on a voluntary basis for specified periods. Each curve shows the average *G* load that can be tolerated for the time indicated. The data points obtained were actually those on the axes; the lines as such are extrapolated from the data points to form the concentric figures.

(Adapted from reference [1], Fig. 17-17, p. 505, reprinted with permission)

FIGURE 7-10

Human tolerance of acceleration

As one point of reference, the acceleration of the piston in a small, four-cylinder economy car engine (about 1.5L displacement) at idle speed is about 40 *g*'s. At highway speeds the piston acceleration may be as high as 700 *g*'s. At the engine's top speed of 6000 rpm the peak piston acceleration is 2000 *g*'s! As long as you're not riding on the piston, this is acceptable. These engines last a long time in spite of the high accelerations they experience. One key factor is the choice of low-mass, high-strength materials for the moving parts to both keep the dynamic forces down at these high accelerations and to enable them to tolerate high stresses.

7.7   JERK

No, not you! The **time derivative of acceleration** is called *jerk, pulse,* or *shock.* The name is apt, as it conjures the proper image of this phenomenon. **Jerk** is *the time rate of change of acceleration.* Force is proportional to acceleration. Rapidly changing acceleration means a rapidly changing force. Rapidly changing forces tend to "jerk" the object about! You have probably experienced this phenomenon when riding in an automobile. If the driver is inclined to "jackrabbit" starts and accelerates violently away from the traffic light, you will suffer from large jerk because your acceleration will go from zero to a large value quite suddenly. But, when Jeeves, the chauffeur, is driving the *Rolls,* he always attempts to minimize jerk by accelerating gently and smoothly, so that *Madame* is entirely unaware of the change.

TABLE 7-1   Common Values of Acceleration in Human Activities

Gentle acceleration in an automobile	0.1 g
Jet aircraft on takeoff	0.3 g
Hard acceleration in an automobile	0.5 g
Panic stop in an automobile	0.7 g
Fast cornering in an automobile	0.8 g
Roller coaster	3.5 g
F-16 Air Force jet	9.0 g

Controlling and minimizing jerk in machine design is often of interest, especially if low vibration is desired. Large magnitudes of jerk will tend to excite the natural frequencies of vibration of the machine or structure to which it is attached and cause increased vibration and noise levels. Jerk control is of greater interest in the design of cams than of linkages, and we will investigate it in more detail in Chapter 8 on cam design.

The procedure for calculating the jerk in a linkage is a straightforward extension of the methods shown for acceleration analysis. Let angular jerk be represented by:

$$\varphi = \frac{d\alpha}{dt} \tag{7.33a}$$

and linear jerk by:

$$\mathbf{J} = \frac{d\mathbf{A}}{dt} \tag{7.33b}$$

To solve for jerk in a fourbar linkage, for example, the vector loop equation for acceleration (equation 7.7) is differentiated versus time. Refer to Figure 7-5 (p. 309) for notation.

$$-a\omega_2^3 je^{j\theta_2} - 2a\omega_2\alpha_2 e^{j\theta_2} + a\alpha_2\omega_2 j^2 e^{j\theta_2} + a\varphi_2 je^{j\theta_2}$$
$$- b\omega_3^3 je^{j\theta_3} - 2b\omega_3\alpha_3 e^{j\theta_3} + b\alpha_3\omega_3 j^2 e^{j\theta_3} + b\varphi_3 je^{j\theta_3} \tag{7.34a}$$
$$+ c\omega_4^3 je^{j\theta_4} + 2c\omega_4\alpha_4 e^{j\theta_4} - c\alpha_4\omega_4 j^2 e^{j\theta_4} - c\varphi_4 je^{j\theta_4} = 0$$

Collect terms and simplify:

$$-a\omega_2^3 je^{j\theta_2} - 3a\omega_2\alpha_2 e^{j\theta_2} + a\varphi_2 je^{j\theta_2}$$
$$- b\omega_3^3 je^{j\theta_3} - 3b\omega_3\alpha_3 e^{j\theta_3} + b\varphi_3 je^{j\theta_3} \tag{7.34b}$$
$$+ c\omega_4^3 je^{j\theta_4} + 3c\omega_4\alpha_4 e^{j\theta_4} - c\varphi_4 je^{j\theta_4} = 0$$

Substitute the Euler identity and separate into x and y components:

real part (x component):

$$a\omega_2^3 \sin\theta_2 - 3a\omega_2\alpha_2 \cos\theta_2 - a\varphi_2 \sin\theta_2$$
$$+ b\omega_3^3 \sin\theta_3 - 3b\omega_3\alpha_3 \cos\theta_3 - b\varphi_3 \sin\theta_3$$
$$- c\omega_4^3 \sin\theta_4 + 3c\omega_4\alpha_4 \cos\theta_4 + c\varphi_4 \sin\theta_4 = 0 \tag{7.35a}$$

imaginary part (y component):

$$-a\omega_2^3 \cos\theta_2 - 3a\omega_2\alpha_2 \sin\theta_2 + a\varphi_2 \cos\theta_2$$
$$- b\omega_3^3 \cos\theta_3 - 3b\omega_3\alpha_3 \sin\theta_3 + b\varphi_3 \cos\theta_3$$
$$+ c\omega_4^3 \cos\theta_4 + 3c\omega_4\alpha_4 \sin\theta_4 - c\varphi_4 \cos\theta_4 = 0 \qquad (7.35b)$$

These can be solved simultaneously for φ_3 and φ_4, which are the only unknowns. The driving angular jerk, φ_2, if nonzero, must be known in order to solve the system. All the other factors in equations 7.35 are defined or have been calculated from the position, velocity, and acceleration analyses. To simplify these expressions we will set the known terms to temporary constants.

In equation 7.35a, let:

$A = a\omega_2^3 \sin\theta_2$	$D = b\omega_3^3 \sin\theta_3$	$G = 3c\omega_4\alpha_4 \cos\theta_4$
$B = 3a\omega_2\alpha_2 \cos\theta_2$	$E = 3b\omega_3\alpha_3 \cos\theta_3$	$H = c\sin\theta_4$
$C = a\varphi_2 \sin\theta_2$	$F = c\omega_4^3 \sin\theta_4$	$K = b\sin\theta_3$

$$(7.36a)$$

Equation 7.35a then reduces to:

$$\varphi_3 = \frac{A - B - C + D - E - F + G + H\varphi_4}{K} \qquad (7.36b)$$

Note that equation 7.36b defines angle φ_3 in terms of angle φ_4. We will now simplify equation 7.35b and substitute equation 7.36b into it.

In equation 7.35b, let:

$L = a\omega_2^3 \cos\theta_2$	$P = b\omega_3^3 \cos\theta_3$	$S = c\omega_4^3 \cos\theta_4$
$M = 3a\omega_2\alpha_2 \sin\theta_2$	$Q = 3b\omega_3\alpha_3 \sin\theta_3$	$T = 3c\omega_4\alpha_4 \sin\theta_4$
$N = a\varphi_2 \cos\theta_2$	$R = b\cos\theta_3$	$U = c\cos\theta_4$

$$(7.37a)$$

Equation 7.35b then reduces to:

$$R\varphi_3 - U\varphi_4 - L - M + N - P - Q + S + T = 0 \qquad (7.37b)$$

Substituting equation 7.36b in equation 7.35b:

$$R\left(\frac{A - B - C + D - E - F + G + H\varphi_4}{K}\right) - U\varphi_4 - L - M + N - P - Q + S + T = 0 \qquad (7.38)$$

The solution is:

$$\varphi_4 = \frac{KN - KL - KM - KP - KQ + AR - BR - CR + DR - ER - FR + GR + KS + KT}{KU - HR} \qquad (7.39)$$

The result from equation 7.39 can be substituted into equation 7.36b to find φ_3. Once the angular jerk values are found, the linear jerk at the pin joints can be found from:

$$\mathbf{J}_A = -a\omega_2^3 je^{j\theta_2} - 3a\omega_2\alpha_2 e^{j\theta_2} + a\varphi_2 je^{j\theta_2}$$
$$\mathbf{J}_{BA} = -b\omega_3^3 je^{j\theta_3} - 3b\omega_3\alpha_3 e^{j\theta_3} + b\varphi_3 je^{j\theta_3} \qquad (7.40)$$
$$\mathbf{J}_B = -c\omega_4^3 je^{j\theta_4} - 3c\omega_4\alpha_4 e^{j\theta_4} + c\varphi_4 je^{j\theta_4} = 0$$

The same approach as used in Section 7.4 (p. 319) to find the acceleration of any point on any link can be used to find the linear jerk at any point.

$$\mathbf{J}_P = \mathbf{J}_A + \mathbf{J}_{PA} \qquad (7.41)$$

The jerk difference equation 7.41 can be applied to any point on any link if we let P represent any arbitrary point on any link and A represent any reference point on the same link for which we know the value of the jerk vector. Note that if you substitute equations 7.40 into 7.41, you will get equation 7.34.

7.8   LINKAGES OF *N* BARS

The same analysis techniques presented here for position, velocity, acceleration, and jerk, using the fourbar and fivebar linkage as the examples, can be extended to more complex assemblies of links. Multiple vector loop equations can be written around a linkage of arbitrary complexity. The resulting vector equations can be differentiated and solved simultaneously for the variables of interest. In some cases, the solution will require simultaneous solution of a set of nonlinear equations. A root-finding algorithm such as the Newton-Raphson method will be needed to solve these more complicated cases. A computer is necessary. An equation solver software package such as *TKSolver* or *Mathcad* that will do an iterative root-finding solution will be a useful aid to the solution of any of these analysis problems, including the examples shown here.

7.9   REFERENCES

1   Sanders, M. S., and E. J. McCormick, *Human Factors in Engineering and Design*, 6th ed., McGraw-Hill Co., New York, 1987, p. 505.

7.10   PROBLEMS

7-1   A point at a 6.5-in radius is on a body which is in pure rotation with $\omega = 100$ rad/sec and a constant $\alpha = -500$ rad/sec^2 at point A. The rotation center is at the origin of a coordinate system. When the point is at position A, its position vector makes a 45° angle with the X axis. It takes 0.01 sec to reach point B. Draw this system to some convenient scale, calculate the θ and ω of position B, and:

   a.   Write an expression for the particle's acceleration vector in position A using complex number notation, in both polar and cartesian forms.

   b.   Write an expression for the particle's acceleration vector in position B using complex number notation, in both polar and cartesian forms.

   c.   Write a vector equation for the acceleration difference between points B and A. Substitute the complex number notation for the vectors in this equation and solve for the acceleration difference numerically.

   d.   Check the result of part c with a graphical method.

7-2   In problem 7-1 let A and B represent points on separate, rotating bodies both having the given ω and α at $t = 0$, $\theta_A = 45°$, and $\theta_B = 120°$. Find their relative acceleration.

*7-3   The link lengths, coupler point location, and the values of θ_2, ω_2, and α_2 for the same fourbar linkages as used for position and velocity analysis in Chapters 4 and 6 are redefined in Table P7-1, which is the same as Table P6-1. The general linkage configuration and terminology are shown in Figure P7-1. *For the row(s) assigned,* draw the linkage to scale and graphically find the accelerations of points A and B. Then calculate α_3 and α_4 and the acceleration of point P.

*†7-4   Repeat problem 7-3 except solve by the analytical vector loop method of Section 7.3 (p. 308).

*7-5   The link lengths and offset and the values of θ_2, ω_2, and α_2 for some noninverted, offset fourbar slider-crank linkages are defined in Table P7-2. The general linkage configuration and terminology are shown in Figure P7-2. *For the row(s) assigned,* draw the linkage to scale and graphically find the accelerations of the pin joints A and B and the acceleration of slip at the sliding joint.

*†7-6   Repeat problem 7-5 using an analytical method.

*†7-7   The link lengths and the values of θ_2, ω_2, and γ for some inverted fourbar slider-crank linkages are defined in Table P7-3. The general linkage configuration and terminology are shown in Figure P7-3. *For the row(s) assigned*, find the accelerations of the pin joints A and the acceleration of slip at the sliding joint.   Solve by the analytical vector loop method of Section 7.3 for the open configuration of the linkage.

*†7-8   Repeat problem 7-7 for the crossed configuration of the linkage.

*7-9   The link lengths, gear ratio (λ), phase angle (ϕ), and the values of θ_2, ω_2, and α_2 for some geared fivebar linkages are defined in Table P7-4. The general linkage configuration and terminology are shown in Figure P7-4. *For the row(s) assigned*, find α_3 and α_4 and the linear acceleration of point P.

†7-10 An automobile driver took a curve too fast. The car spun out of control about its center of gravity (CG) and slid off the road in a northeasterly direction. The friction of the

* Answers in Appendix F.

† These problems are suited to solution using *Mathcad, Matlab,* or *TKSolver* equation solver programs.

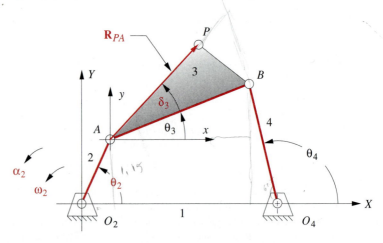

FIGURE P7-1

Configuration and terminology for Problems 7-3, 7-4 and 7-11

TABLE P7-1 Data for Problems 7-3 and 7-4

Row	Link 1	Link 2	Link 3	Link 4	θ_2	ω_2	α_2	R_{pa}	δ_3
a	6	2	7	9	30	10	0	6	30
b	7	9	3	8	85	− 12	5	9	25
c	3	10	6	8	45	− 15	− 10	10	80
d	8	5	7	6	25	24	− 4	5	45
e	8	5	8	6	75	− 50	10	9	300
f	5	8	8	9	15	− 45	50	10	120
g	6	8	8	9	25	100	18	4	300
h	20	10	10	10	50	− 65	25	6	20
i	4	5	2	5	80	25	− 25	9	80
j	20	10	5	10	33	25	− 40	1	0
k	4	6	10	7	88	− 80	30	10	330
l	9	7	10	7	60	− 90	20	5	180
m	9	7	11	8	50	75	− 5	10	90
n	9	7	11	6	120	15	− 65	15	60

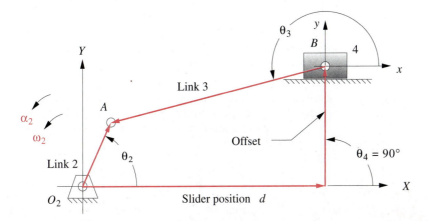

FIGURE P7-2

Configuration and terminology for problems 7-5 to 7-6

TABLE P7-2 Data for Problems 7-5 and 7-6

Row	Link 2	Link 3	Offset	θ_2	ω_2	α_2
a	1.4	4	1	45	10	0
b	2	6	− 3	60	− 12	5
c	3	8	2	− 30	− 15	− 10
d	3.5	10	1	120	24	− 4
e	5	20	− 5	225	− 50	10
f	3	13	0	100	− 45	50
g	7	25	10	330	100	18

7

TABLE P7-3   Data for Problems 7-7 to 7-8

Row	Link 1	Link 2	Link 4	γ	θ_2	ω_2	α_2
a	6	2	4	90	30	10	− 25
b	7	9	3	75	85	−15	− 40
c	3	10	6	45	45	24	30
d	8	5	3	60	25	− 50	20
e	8	4	2	30	75	− 45	− 5
f	5	8	8	90	150	100	− 65

skidding tires provided a 0.25 g linear deceleration. The car rotated at 100 rpm. When the car hit the tree head-on at 30 mph, it took 0.1 sec to come to rest.

a. What was the acceleration experienced by the child seated on the middle of the rear seat, 2 ft behind the car's *CG*, just prior to impact?

b. What force did the 100-lb child exert on her seatbelt harness as a result of the acceleration, just prior to impact?

c. Assuming a constant deceleration during the 0.1 sec of impact, what was the magnitude of the average deceleration felt by the passengers in that interval?

†7-11 For the row(s) assigned in Table P7-1, find the angular jerk of links 3 and 4 and the linear jerk of the pin joint between links 3 and 4 (point *B*). Assume an angular jerk of zero on link 2. The linkage configuration and terminology are shown in Figure P7-1.

†7-12 You are riding on a carousel which is rotating at a constant 15 rpm. It has an inside radius of 3 ft and an outside radius of 10 ft. You begin to run from the inside to the outside along a radius. Your peak velocity with respect to the carousel is 5 mph and occurs at a radius of 7 ft. What is your maximum Coriolis acceleration magnitude and its direction with respect to the carousel?

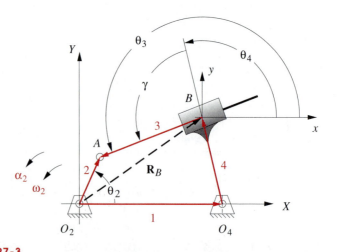

† These problems are suited to solution using *Mathcad, Matlab,* or *TKSolver* equation solver programs.

FIGURE P7-3

Configuration and terminology for problems 7-7 to 7-8

TABLE P7-4   Data for Problem 7-9

Row	Link 1	Link 2	Link 3	Link 4	Link 5	λ	ϕ	θ_2	ω_2	α_2	R_{pa}	δ_3
a	6	1	7	9	4	2.0	30	60	10	0	6	30
b	6	5	7	8	4	−2.5	60	30	−12	5	9	25
c	3	5	7	8	4	−0.5	0	45	−15	−10	10	80
d	4	5	7	8	4	−1.0	120	75	24	−4	5	45
e	5	9	11	8	8	3.2	−50	−39	−50	10	9	300
f	10	2	7	5	3	1.5	30	120	−45	50	10	120
g	15	7	9	11	4	2.5	−90	75	100	18	4	300
h	12	8	7	9	4	−2.5	60	55	−65	25	6	20
i	9	7	8	9	4	−4.0	120	100	25	−25	9	80

7-13   The linkage in Figure P7-5a has $O_2A = 0.8$, $AB = 1.93$, $AC = 1.33$, and offset = 0.38 in. The crank angle in the position shown is 34.3° and angle $BAC = 38.6°$. Find α_3, $\mathbf{A}_A$, $\mathbf{A}_B$, and $\mathbf{A}_C$ for the position shown for $\omega_2 = 15$ rad/sec and $\alpha_2 = 10$ rad/sec^2 in directions shown,

    a.   Using the acceleration difference graphical method.
    †b.   Using an analytical method.

7-14   The linkage in Figure P7-5b has $I_{12}A = 0.75$, $AB = 1.5$, and $AC = 1.2$ in. The effective crank angle in the position shown is 77° and angle $BAC = 30°$. Find α_3, $\mathbf{A}_A$, $\mathbf{A}_B$, and $\mathbf{A}_C$ for the position shown for $\omega_2 = 15$ rad/sec and $\alpha_2 = 10$ rad/sec^2 in the directions shown,

    a.   Using the acceleration difference graphical method.

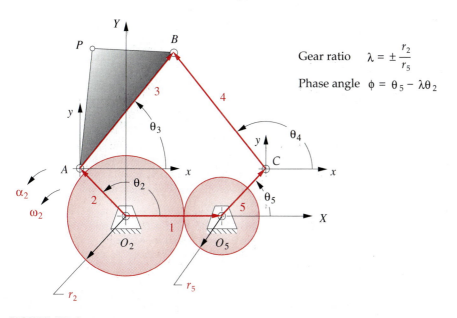

Gear ratio $\lambda = \pm \dfrac{r_2}{r_5}$

Phase angle $\phi = \theta_5 - \lambda\theta_2$

FIGURE P7-4

Configuration and terminology for Problem 7-9

† These problems are suited to solution using *Mathcad, Matlab,* or *TKSolver* equation solver programs.

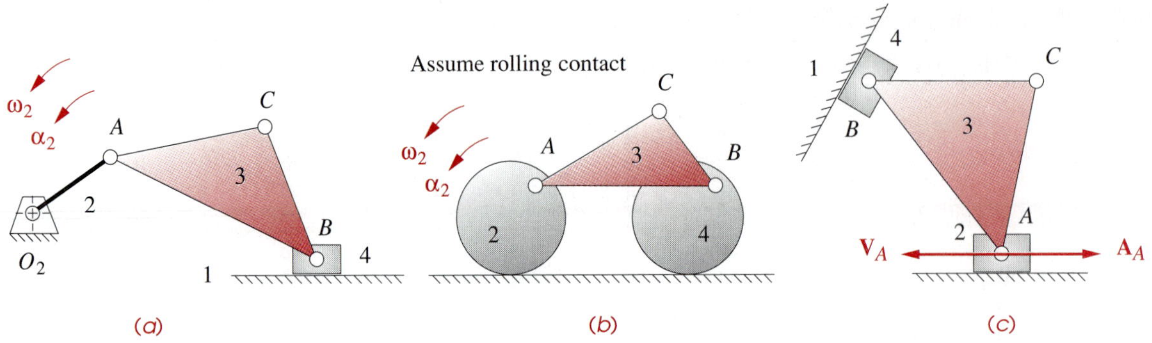

FIGURE P7-5

Problems 7-13 to 7-15

†b. Using an analytical method. (Hint: Create an effective linkage for the position shown and analyze it as a pin-jointed fourbar.)

7-15   The linkage in Figure P7-5c has $AB = 1.8$ and $AC = 1.44$ in. The angle of AB in the position shown is $128°$ and angle $BAC = 49°$. The slider at B is at an angle of $59°$. Find α_3, $\mathbf{A}_B$, and $\mathbf{A}_C$ for the position shown for $\mathbf{V}_A = 10$ in/sec and $\mathbf{A}_A = 15$ in/sec² in the directions shown.

    a. Using the acceleration difference graphical method.
    †b. Using an analytical method.

†7-16   For the linkage shown in Figure P7-6a, write the vector loop equations; differentiate them, and do a complete position, velocity, and acceleration analysis of the linkage. Measure the linkage geometry from the figure. Assume $\omega_2 = 10$ rad/sec and $\alpha_2 = 20$ rad/sec².

†7-17   Repeat Problem 7-16 for the linkage shown in Figure P7-6b.

†7-18   Repeat Problem 7-16 for the linkage shown in Figure P7-6c.

†7-19   Repeat Problem 7-16 for the linkage shown in Figure P7-6d.

†7-20   Figure P7-7 shows a sixbar linkage with $O_2B = 1$, $BD = 1.5$, $DC = 3.5$, $DO_6 = 3$, and $h = 1.3$ in. Find the angular acceleration of link 6 if ω_2 is a constant 1 rad/sec.

7-21   The linkage in Figure P7-8a has $AB = 116$, $BC = 108$, $CD = 110$, and $AD = 174$ mm. AD is at $-25°$ and AB is at $37°$ in the global XY coordinate system. Find $\mathbf{A}_B$, and $\mathbf{A}_C$ in the global coordinate system for the position shown if $\omega_d = 15$ rad/sec CW and $\alpha_d = 25$ rad/sec² CCW. Use the acceleration difference graphical method. (Hint: Make an enlarged copy of the figure and draw on it.)

†7-22   The linkage in Figure P7-8a has $AB = 116$, $BC = 108$, $CD = 110$, and $AD = 174$ mm. AB is at $62°$ in the local $x'y'$ coordinate system. Find α_f, $\mathbf{A}_B$, and $\mathbf{A}_C$ in the local coordinate system for the position shown if $\omega_d = 15$ rad/sec CW and $\alpha_d = 25$ rad/sec² CCW. Use an analytical method.

† These problems are suited to solution using *Mathcad, Matlab,* or *TKSolver* equation solver programs.

†7-23   The linkage in Figure P7-8a has $AB = 116$, $BC = 108$, $CD = 110$, and $AD = 174$ mm. Write a computer program or use an equation solver to find and plot α_f, $\mathbf{A}_B$, and $\mathbf{A}_C$ in the local coordinate system for the maximum range of motion that this linkage allows if $\omega_d = 15$ rad/sec CW.

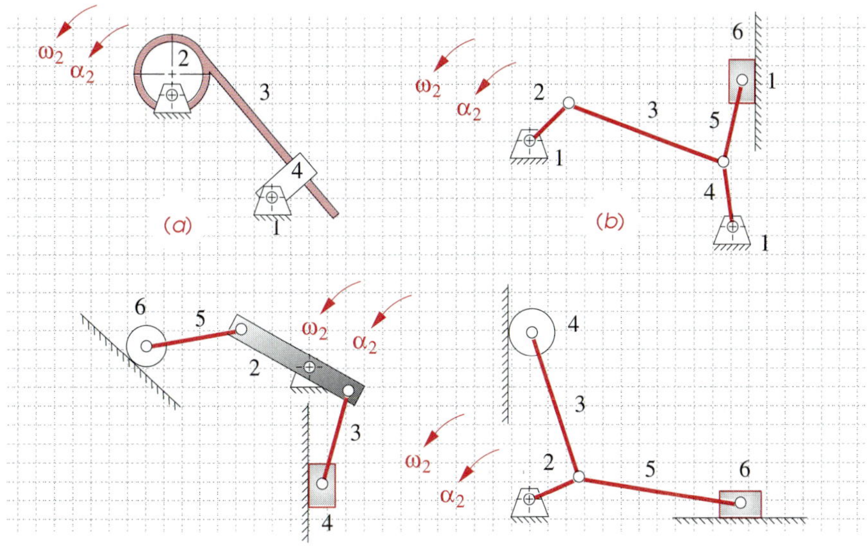

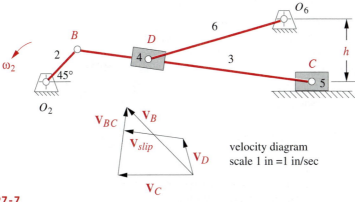

FIGURE P7-6

Problems 7-16 to 7-19

7-24   The linkage in Figure P7-8b has $AB = 40$, $BC = 96$, $CD = 75$, and $AD = 162$ mm. The perpendicular distance from D to $\mathbf{V}_D$ is 36 mm. AD is at $-36°$ and AB is at $47°$ in the global XY coordinate system. Find α_d, $\mathbf{A}_B$, and $\mathbf{A}_C$ in the global coordinate system for the position shown if $\omega_f = 20$ rad/sec CCW. Use the acceleration difference graphical method. (Hint: Make an enlarged copy of the figure and draw on it.)

†7-25   The linkage in Figure P7-8b has $AB = 40$, $BC = 96$, $CD = 75$, and $AD = 162$ mm. The perpendicular distance from D to $\mathbf{V}_D$ is 36 mm. AB is at $83°$ in the local $x'y'$

FIGURE P7-7

Problem 7-20. *Courtesy of Prof. J. M. Vance, Iowa State University*

7

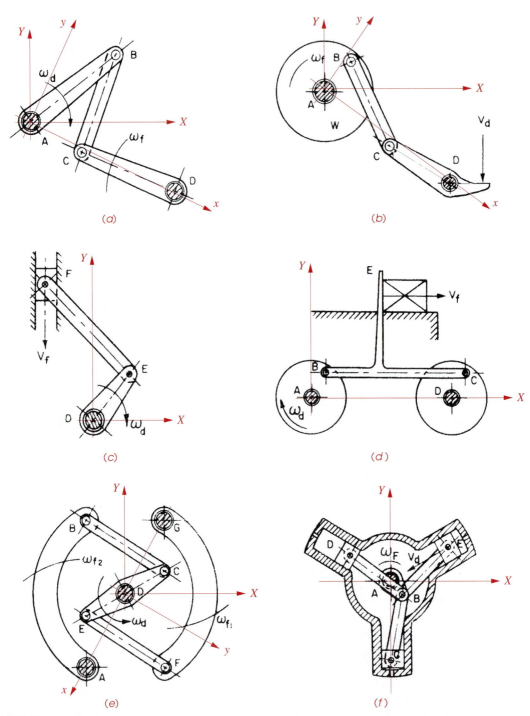

(a)

(b)

(c)

(d)

(e)

(f)

FIGURE P7-8

Problems 7-21 to 7-38. *Adapted from P. H. Hill and W. P. Rule. (1960). Mechanisms: Analysis and Design, with permission.*

coordinate system. Find α_d, $\mathbf{A}_B$, and $\mathbf{A}_C$ in the global coordinate system for the position shown if $\omega_f = 20$ rad/sec CCW. Use an analytical method.

†7-26   The linkage in Figure P7-8b has $AB = 40$, $BC = 96$, $CD = 75$, and $AD = 162$ mm. The perpendicular distance from D to $\mathbf{V}_D$ is 36 mm. Write a computer program or use an equation solver to find and plot α_d, $\mathbf{A}_B$, and $\mathbf{A}_C$ in the local coordinate system for the maximum range of motion that this linkage allows if $\omega_f = 20$ rad/sec CCW.

7-27   The offset slider-crank linkage in Figure P7-8c has $DE = 63$, $EF = 130$, and offset = 52 mm. DE is at $51°$ in the global XY coordinate system. Find $\mathbf{A}_E$ and $\mathbf{A}_f$ in the global coordinate system for the position shown if $\omega_d = 25$ rad/sec CW. Use the acceleration difference graphical method. (Hint: Make an enlarged copy of the figure and draw on it.)

†7-28   The offset slider-crank linkage in Figure P7-8c has $DE = 63$, $EF = 130$, and offset = 52 mm. DE is at $51°$ in the global XY coordinate system. Find $\mathbf{A}_E$ and $\mathbf{A}_f$ in the global coordinate system for the position shown if $\omega_d = 25$ rad/sec CW. Use an analytical method.

†7-29   The offset slider-crank linkage in Figure P7-8c has $DE = 63$, $EF = 130$, and offset = 52 mm. Write a computer program or use an equation solver to find and plot $\mathbf{A}_E$ and $\mathbf{A}_f$ in the global coordinate system for the maximum range of motion that this linkage allows if $\omega_d = 25$ rad/sec CW.

7-30   The linkage in Figure P7-8d has $AB = 30$, $BC = 150$, $CD = 30$, and $AD = 150$ mm. AB is at $58°$ in the global XY coordinate system. Find $\mathbf{A}_B$, $\mathbf{A}_C$, and $\mathbf{A}_f$ (the acceleration of the box) in the global coordinate system for the position shown if $\omega_d = 30$ rad/sec CW. Use the acceleration difference graphical method. (Make an enlarged copy of the figure and draw on it.)

†7-31   The linkage in Figure P7-8d has $AB = 30$, $BC = 150$, $CD = 30$, and $AD = 150$ mm. AB is at $58°$ in the global coordinate system. Find $\mathbf{A}_B$, $\mathbf{A}_C$, and $\mathbf{A}_f$ (the acceleration of the box) in the global coordinate system for the position shown if $\omega_d = 30$ rad/sec CW. Use an analytical method.

†7-32   The linkage in Figure P7-8d has $AB = 30$, $BC = 150$, $CD = 30$, and $AD = 150$ mm. Write a computer program or use an equation solver to find and plot $\mathbf{A}_B$, $\mathbf{A}_C$, and $\mathbf{A}_f$ (the acceleration of the box) in the global coordinate system for the maximum range of motion that this linkage allows if $\omega_d = 30$ rad/sec CW.

7-33   The linkage in Figure P7-8e has $AB = GF = 153$, $BC = EF = 100$, $CD = DE = 49$, and $AD = DG = 87$ mm. DG is at $61°$ and DC is at $29°$ in the global XY coordinate system. Find α_f, $\mathbf{A}_B$, and $\mathbf{A}_C$ in the global coordinate system for the position shown if $\omega_d = 15$ rad/sec CW. Use the acceleration difference graphical method. (Make an enlarged copy of the figure and draw on it.)

†7-34   The linkage in Figure P7-8e has $AB = GF = 153$, $BC = EF = 100$, $CD = DE = 49$, and $AD = DG = 87$ mm. DC is at $148°$ in the local xy coordinate system. Find α_f, $\mathbf{A}_B$, and $\mathbf{A}_C$ in the local coordinate system for the position shown if $\omega_d = 15$ rad/sec CW. Use an analytical method.

†7-35   The linkage in Figure P7-8e has $AB = GF = 153$, $BC = EF = 100$, $CD = DE = 49$, and $AD = DG = 87$ mm. Write a computer program or use an equation solver to find and plot α_f, $\mathbf{A}_B$, and $\mathbf{A}_C$ in the local coordinate system for the maximum range of motion that this linkage allows if $\omega_d = 15$ rad/sec CW.

7

† These problems are suited to solution using *Mathcad, Matlab,* or *TKSolver* equation solver programs.

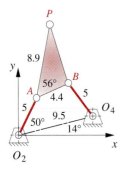

FIGURE P7-9

Problem 7-39

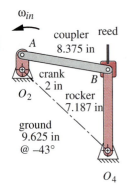

FIGURE P7-11

Problem 7-41. Loom
laybar drive

7-36   The 3-cylinder radial compressor in Figure P7-8f has a crank length $AB = 19$ mm and connecting rods $BC = BD = BE = 70$ mm. The crank is at $-53°$ in the position shown and $\omega = 15$ rad/sec CW. The cylinders are equispaced at $120°$. Find the piston accelerations A_C, A_D, A_E for the crank position shown using a graphical method. (Hint: Make an enlarged copy of the figure and draw on it.)

†7-37   The 3-cylinder radial compressor in Figure P7-8f has a crank length $AB = 19$ mm and connecting rods $BC = BD = BE = 70$ mm. The crank is at $-53°$ in the position shown and $\omega = 15$ rad/sec CW. The cylinders are equispaced at $120°$. Find the piston accelerations A_C, A_D, A_E for the crank position shown using an analytical method.

†7-38   The 3-cylinder radial compressor in Figure P7-8f has a crank length $AB = 19$ mm and connecting rods $BC = BD = BE = 70$ mm. The crank is at $-53°$ in the position shown and $\omega = 15$ rad/sec CW. The cylinders are equispaced at $120°$. Write a program or use an equation solver to find and plot the piston accelerations A_C, A_D, A_E for one revolution of the crank.

*†7-39   Figure P7-9 shows a linkage in one position. Find the instantaneous accelerations of points A, B, and P if link O_2A is rotating CW at 40 rad/sec.

*†7-40   Figure P7-10 shows a linkage and its coupler curve. Write a computer program or use an equation solver to calculate and plot the magnitude and direction of the acceleration of the coupler point P at $2°$ increments of crank angle for $\omega_2 = 100$ rpm. Check your result with program FOURBAR.

*†7-41   Figure P7-11 shows a linkage that operates at 500 crank rpm. Write a computer program or use an equation solver to calculate and plot the magnitude and direction of the acceleration of point B at $2°$ increments of crank angle. Check your result with program FOURBAR.

*†7-42   Figure P7-12 shows a linkage and its coupler curve. Write a computer program or use an equation solver to calculate and plot the magnitude and direction of the acceleration of the coupler point P at $2°$ increments of crank angle for $\omega_2 = 20$ rpm over the maximum range of motion possible. Check your result with program FOURBAR.

†7-43   Figure P7-13 shows a linkage and its coupler curve. Write a computer program or use an equation solver to calculate and plot the magnitude and direction of the acceleration of the coupler point P at $2°$ increments of crank angle for $\omega_2 = 80$ rpm

---

* Answers in Appendix F.

---

† These problems are suited to solution using *Mathcad, Matlab,* or *TKSolver* equation solver programs.

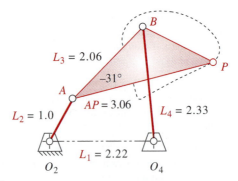

FIGURE P7-10

Problem 7-40

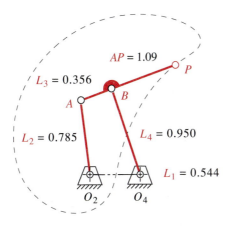

FIGURE P7-12

Problem 7-42

over the maximum range of motion possible. Check your result with program
FOURBAR.

*†7-44 Figure P7-14 shows a linkage and its coupler curve. Write a computer program or
use an equation solver to calculate and plot the magnitude and direction of the
acceleration of the coupler point P at 2° increments of crank angle for $\omega_2 = 80$ rpm
over the maximum range of motion possible. Check your result with program
FOURBAR.

†‡7-45 Figure P7-15 shows a power hacksaw which is an offset slider-crank mechanism.
The crank is 75 mm, the connecting rod is 170 mm, and the offset is 45 mm. Draw
an equivalent linkage diagram, and then calculate and plot the acceleration of the saw
blade with respect to the piece being cut over one revolution of the crank at 50 rpm.

†‡7-46 Figure P7-16 shows a pick-and-place mechanism which can be analyzed as two
fourbar linkages driven by a common crank. The parallelogram stage has 40 mm
cranks and a 108-mm coupler. Crank $AB = 32$ mm. $BC = 260$, $CD = 96$, $DE = 160$,
and $AD = 200$ mm. Angle $CDE = 75°$. AD is at 205°. The phase angle between the

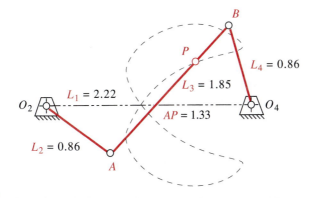

FIGURE P7-13

Problem 7-43

---

* Answers in Appendix F.

---

† These problems are
suited to solution using
Mathcad, Matlab, or
TKsolver equation solver
programs.

---

‡ These problems are
suited to solution using the
Working Model
program,.which is on the
attached CD-ROM.

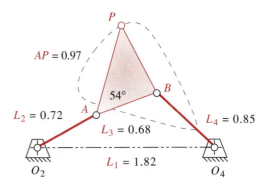

FIGURE P7-14

Problem 7-44

two crankpins on wheel W is 120°. The cylinders P being pushed have 60-mm diameters. The point of contact between the vertical finger and the leftmost cylinder in the position shown is 58 mm at 80° versus the left end of the parallelogram's coupler. Calculate and plot the relative acceleration between point E and the center of the leftmost cylinder P.

†‡7-47   Figure P7-17 shows a paper roll off-loading mechanism driven by an air cylinder. In the position shown, $AO_2 = 1.1$ m at 178° and O_4A is 0.3 m at 226°. $O_2AO_4A = 0.93$ m at 163°. The V-links are rigidly attached to O_4A. The air cylinder is retracted at a constant acceleration of 0.1 m/sec². Draw a kinematic diagram of the mechanism, write the necessary equations, and calculate and plot the angular acceleration of the paper roll and the linear acceleration of its center as it rotates through 90° CCW from the position shown.

† These problems are suited to solution using *Mathcad, Matlab,* or *TKsolver* equation solver programs.

‡ These problems are suited to solution using the *Working Model* program,.which is on the attached CD-ROM.

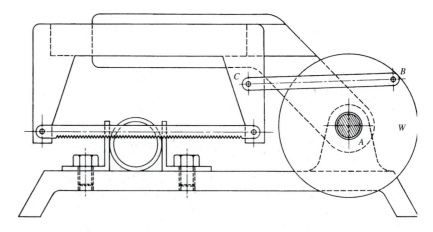

FIGURE P7-15

Problem 7-45   *From P. H. Hill and W. P. Rule. (1960). Mechanisms: Analysis and Design, with permission*

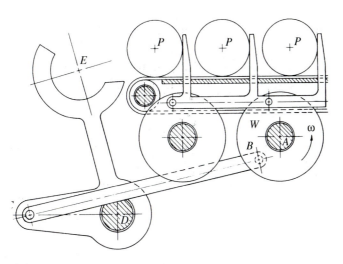

FIGURE P7-16

Problem 7-46. *From P. H. Hill and W. P. Rule. (1960). Mechanisms: Analysis and Design, with permission*

†7-48 Figure P7-18 shows a mechanism. Scale the diagram for dimensions and find the accelerations of points B, C, E, and F for the position shown.

†‡7-49 Figure P7-19 shows a walking beam mechanism. Calculate and plot the acceleration A_{out} for one revolution of the input crank 2 rotating at 100 rpm.

†‡7-50 Figure P7-20 shows a surface grinder. The workpiece is oscillated under the spinning 90-mm diameter grinding wheel by the slider-crank linkage which has a 22-mm crank, a 157-mm connecting rod, and a 40-mm offset. The crank turns at 30 rpm, and

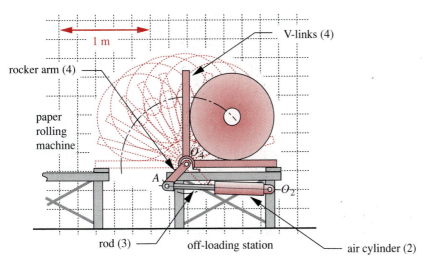

FIGURE P7-17

Problem 7-47

† These problems are suited to solution using *Mathcad, Matlab,* or *TKsolver* equation solver programs.

‡ These problems are suited to solution using the *Working Model* program.which is on the attached CD-ROM

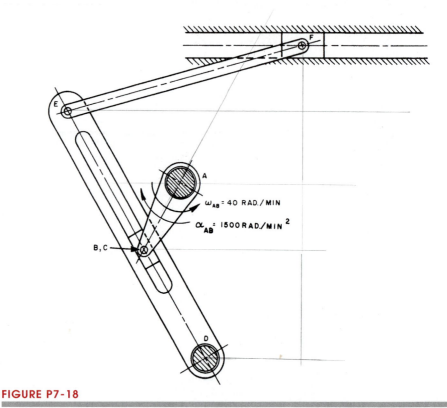

FIGURE P7-18

Problem 7-48. *From P. H. Hill and W. P. Rule. (1960). Mechanisms: Analysis and Design, with permission*

the grinding wheel at 3450 rpm. Calculate and plot the acceleration of the grinding wheel contact point relative to the workpiece over one revolution of the crank.

† These problems are suited to solution using *Mathcad, Matlab,* or *TKsolver* equation solver programs.

‡ These problems are suited to solution using the *Working Model* program.which is on the attached CD-ROM

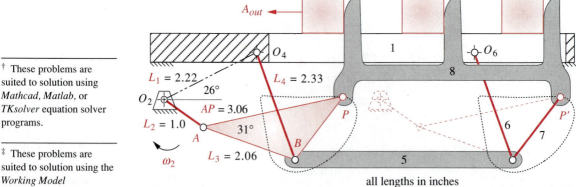

FIGURE P7-19

Problem 7-49: Straight-line walking beam eightbar transport mechanism

†‡7-51   Figure P7-21 shows a drag link mechanism. Scale the diagram for dimensions, write the necessary equations, and solve them to calculate the angular acceleration of link 4 for an input of $\omega_2 = 1$ rad/sec. Comment on uses for this mechanism.

7-52   Figure P7-22 shows a mechanism. Scale the diagram for dimensions and use a graphical method to calculate the accelerations of points B, D, and E for the position shown. $\omega_{AB} = 20$ rad/sec.

7-53   Figure P7-23 shows a quick-return mechanism. Scale the diagram for dimensions and use a graphical method to calculate the accelerations of points B, C, and E for the position shown. $\omega_{CD} = 10$ rad/sec.

†‡7-54   Figure P7-23 shows a quick-return mechanism. Scale the diagram for dimensions and use an analytical method to calculate the accelerations of points B, C, and E for one revolution of the input link. $\omega_{CD} = 10$ rad/sec.

†‡7-55   Figure P7-24 shows a drum-pedal mechanism. $O_2A = 100$ mm at 162° and rotates to 171° at A'. $O_2O_4 = 56$ mm, $AB = 28$ mm, $AP = 124$ mm, and $O_4B = 64$ mm. The distance from O_4 to F_{in} is 48 mm. If the input velocity V_{in} is a constant magnitude of 3 m/sec, find the output acceleration over the range of motion.

*†‡7-56   A tractor-trailer tipped over while negotiating an on-ramp to the New York Thruway. The road has a 50-ft radius at that point and tilts 3° toward the outside of the curve. The 45-ft-long by 8-ft-wide by 8.5-ft-high trailer box (13 ft from ground to top) was loaded with 44 415 lb of paper rolls in two rows by two layers as shown in Figure P7-25. The rolls are 40 in diameter by 38 in long, and weigh about 900 lb each. They are wedged against backward rolling but not against sidewards sliding. The empty trailer weighed 14 000 lb. The driver claims that he was traveling at less than 15 mph and that the load of paper shifted inside the trailer, struck the trailer sidewall, and tipped the truck. The paper company that loaded the truck claims the load was properly stowed and would not shift at that speed. Independent tests of the coefficient of friction between similar paper rolls and a similar trailer floor give a value of

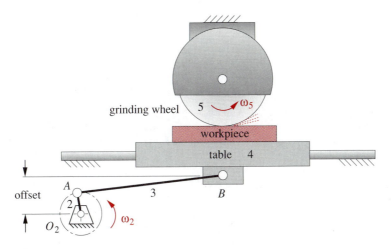

grinding wheel

offset

FIGURE P7-20

Problem 7-50. A surface grinder

* Answers in Appendix F.

† These problems are suited to solution using *Mathcad, Matlab,* or *TKsolver* equation solver programs.

‡ These problems are suited to solution using the *Working Model* program, which is on the attached CD-ROM.

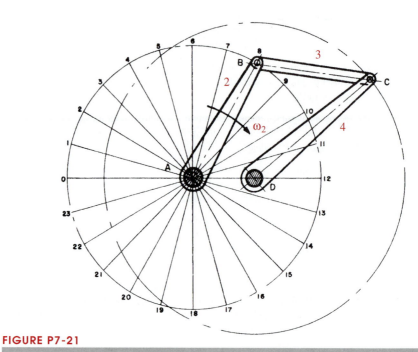

FIGURE P7-21

Problem 7-51. *From P. H. Hill and W. P. Rule. (1960). Mechanisms: Analysis and Design with permission*

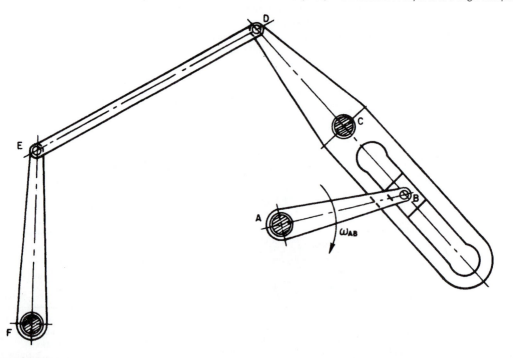

FIGURE P7-22

Problem 7-52. *From P. H. Hill and W. P. Rule. (1960). Mechanisms: Analysis and Design, with permission*

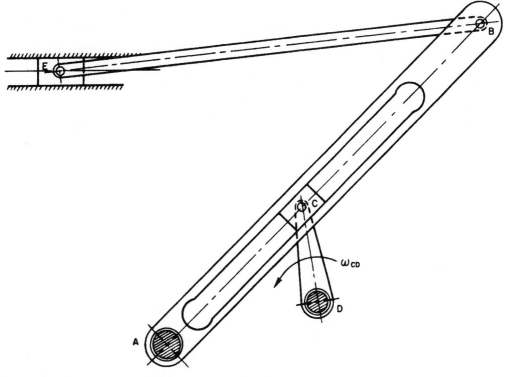

FIGURE P7-23

Problems 7-53 to 7-54. *From P. H. Hill and W. P. Rule. (1960). Mechanisms: Analysis and Design, with permission*

0.43 ± 0.08. The composite center of gravity of the loaded trailer is estimated to be 7.5 ft above the road. Determine the truck speed that would cause the truck to just begin to tip and the speed at which the rolls will just begin to slide sideways. What do you think caused the accident?

[†]7-57   Figure P7-26 shows a V-belt drive. The sheaves (pulleys) have pitch diameters of 150 and 300 mm, respectively. The smaller sheave is driven at a constant 1750 rpm. For a cross-sectional differential element of the belt, write the equations of its acceleration for one complete trip around both pulleys including its travel between the pulleys. Compute and plot the acceleration of the differential element versus time for one circuit around the belt path. What does your analysis tell about the dynamic behavior of the belt? Relate your findings to your personal observation of a belt of this type in operation. (Look in your school's machine shop or under the hood of an automobile—but mind your fingers!)

[†]7-58   Write a program using an equation solver or any computer language to solve for the displacements, velocities, and accelerations in an offset slider-crank linkage as shown in Figure P7-2 (p. 329). Plot the variation in all link's angular and all pin's linear positions, velocities, and accelerations with a constant angular velocity input to the crank over one revolution for both open and crossed configurations of the linkage. To test the program, use data from row *a* of Table P7-2. Check your results with program SLIDER.

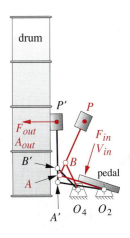

FIGURE P7-24

Problem 7-55.

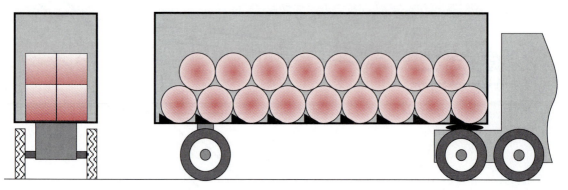

FIGURE P7-25

Problem 7-56

†7-59   Write a program using an equation solver or any computer language to solve for the displacements, velocities, and accelerations in an inverted slider-crank linkage as shown in Figure P7-3 (p. 330). Plot the variation in all link's angular and all pin's linear positions, velocities, and accelerations with a constant angular velocity input to the crank over one revolution for both open and crossed configurations of the linkage. To test the program, use data from row e of Table P7-3 except for the value of α_2 which will be set to zero for this exercise.

†7-60   Write a program using an equation solver or any computer language to solve for the displacements, velocities, and accelerations in a geared-fivebar linkage as shown in Figure P7-4 (p. 331). Plot the variation in all link's angular and all pin's linear positions, velocities, and accelerations with a constant angular velocity input to the crank over one revolution for both open and crossed configurations of the linkage. To test the program, use data from row a of Table P7-4. Check your results with program FIVEBAR.

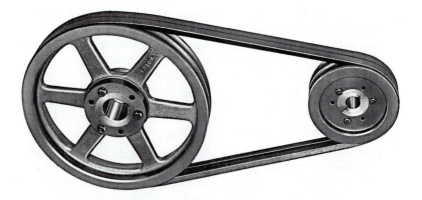

† These problems are suited to solution using *Mathcad, Matlab,* or *TKSolver* equation solver programs.

FIGURE P7-26

Problem 7-57. A two-groove V-belt drive. *Courtesy of T. B. Wood's Sons Co., Chambersburg, PA*

Chapter **8**

CAM DESIGN

It is much easier to design than to perform
SAMUEL JOHNSON

8.0   INTRODUCTION

Cam-follower systems are frequently used in all kinds of machines. The valves in your automobile engine are opened by cams. Machines used in the manufacture of many consumer goods are full of cams. Compared to linkages, cams are easier to design to give a specific output function, but they are much more difficult and expensive to make than a linkage. Cams are a form of degenerate fourbar linkage in which the coupler link has been replaced by a half joint as shown in Figure 8-1. This topic was discussed in Section 2.9 (p. 40) on linkage transformation (see also Figure 2-10, p. 41). For any one instantaneous position of cam and follower, we can substitute an effective linkage which will, for that instantaneous position, have the same motion as the original. In effect, the cam-follower is a fourbar linkage with variable-length (effective) links. It is this conceptual difference that makes the cam-follower such a flexible and useful **function generator**. We can specify virtually any output function we desire and quite likely create a curved surface on the cam to generate that function in the motion of the follower. We are not limited to fixed-length links as we were in linkage synthesis. The cam-follower is an extremely useful mechanical device, without which the machine designer's tasks would be more difficult to accomplish. But, as with everything else in engineering, there are trade-offs. These will be discussed in later sections. A list of the variables used in this chapter is provided in Table 8-1.

This chapter will present the proper approach to designing a cam-follower system, and in the process also present some less than proper designs as examples of the problems which inexperienced cam designers often get into. Theoretical considerations of the mathematical functions commonly used for cam curves will be discussed. Methods for the derivation of custom polynomial functions, to suit any set of boundary conditions, will be presented. The task of sizing the cam with considerations of pressure angle and radius of curvature will be addressed, and manufacturing processes and their limitations

---

TABLE 8-1   Notation Used in This Chapter

t = time, seconds

θ = camshaft angle, degrees or radians (rad)

ω = camshaft angular velocity, rad/sec

β = total angle of any segment, rise, fall, or dwell, degrees or rad

h = total lift (rise or fall) of any one segment, length units

s or S = follower displacement, length units

$v = ds/d\theta$ = follower velocity, length/rad

$V = dS/dt$ = follower velocity, length/sec

$a = dv/d\theta$ = follower acceleration, length/rad^2

$A = dV/dt$ = follower acceleration, length/sec^2

$j = da/d\theta$ = follower jerk, length/rad^3

$J = dA/dt$ = follower jerk, length/sec^3

$s\ v\ a\ j$ refer to the group of diagrams, length units versus radians

$S\ V\ A\ J$ refer to the group of diagrams, length units versus time

R_b = base circle radius, length units

R_p = prime circle radius, length units

R_f = roller follower radius, length units

ε = eccentricity of cam-follower, length units

ϕ = pressure angle, degrees or radians

ρ = radius of curvature of cam surface, length units

ρ_{pitch} = radius of curvature of pitch curve, length units

ρ_{min} = minimum radius of curvature of pitch curve or cam surface, length units

---

discussed. The computer program DYNACAM will be used throughout the chapter as a tool to present and illustrate design concepts and solutions. A user manual for this program is in Appendix A. The reader can refer to that section at any time without loss of continuity in order to become familiar with the program's operation.

8.1   CAM TERMINOLOGY

Cam-follower systems can be classified in several ways: by *type of follower motion*, either **translating** or **rotating** (oscillating); by type of cam, radial, cylindrical, three-dimensional; by *type of joint closure*, either **force-** or **form**-closed; by *type of follower*, **curved** or **flat**, **rolling** or **sliding**; by *type of motion constraints*, **critical extreme position** (CEP), **critical path motion** (CPM); by *type of motion program*, **rise-fall** (RF), **rise-fall-dwell** (RFD), **rise-dwell-fall-dwell** (RDFD). We will now discuss each of these classification schemes in more detail.

Type of Follower Motion

Figure 8-1a shows a system with an oscillating, or **rotating, follower**. Figure 8-1b shows a **translating follower**. These are analogous to the crank-rocker fourbar and the slider-crank fourbar linkages, respectively. An effective fourbar linkage can be substituted for the cam-follower system for any instantaneous position. The lengths of the effective links are determined by the instantaneous locations of the centers of curvature of cam and follower as shown in Figure 8-1. The velocities and accelerations of the cam-follower system can be found by analyzing the behavior of the effective linkage for any position. A proof of this can be found in reference [1]. Of course, the effective links change length as the cam-follower moves giving it an advantage over a pure linkage as this allows more flexibility in meeting the desired motion constraints.

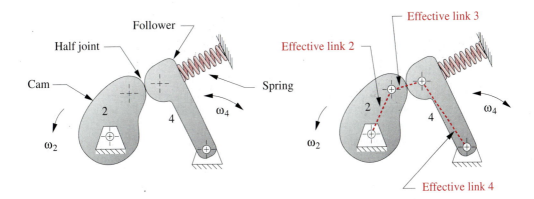

(a) An oscillating cam-follower has an effective pin-jointed fourbar equivalent

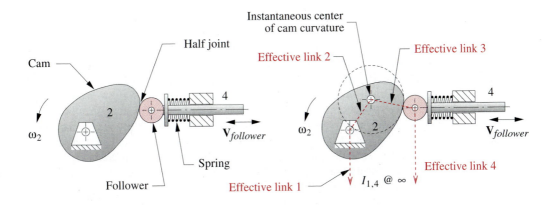

(b) A translating cam-follower has an effective fourbar slider-crank equivalent

FIGURE 8-1

Effective linkages in the cam-follower mechanism

The choice between these two forms of the cam-follower is usually dictated by the type of output motion desired. If true rectilinear translation is required, then the translating follower is dictated. If pure rotation output is needed, then the oscillator is the obvious choice. There are advantages to each of these approaches, separate from their motion characteristics, depending on the type of follower chosen. These will be discussed in a later section.

Type of Joint Closure

Force and form closure were discussed in Section 2.3 (p. 24) on the subject of joints and have the same meaning here. **Force closure**, as shown in Figure 8-1 (p. 347), *requires an external force be applied to the joint* in order to keep the two links, cam and follower, physically in contact. This force is usually provided by a spring. This force, defined as positive in a direction which closes the joint, cannot be allowed to become negative. If it does, the links have lost contact because a *force-closed joint can only push, not pull*. **Form closure**, as shown in Figure 8-2, *closes the joint by geometry*. No external force is required. There are really two cam surfaces in this arrangement, one surface on each side of the follower. Each surface pushes, in its turn, to drive the follower in both directions.

Figure 8-2a and b shows track or groove cams which capture a single follower in the groove and both push and pull on the follower. Figure 8-2c shows another variety of form-closed cam-follower arrangement, called **conjugate cams**. There are two cams fixed on a common shaft which are mathematical conjugates of one another. Two roller followers, attached to a common arm, are each pushed in opposite directions by the conjugate cams. When form-closed cams are used in automobile or motorcycle engine valve trains, they are called **desmodromic** cams. There are advantages and disadvantages to both force- and form-closed arrangements which will be discussed in a later section.

Type of Follower

Follower, in this context, refers only to that part of the follower link which contacts the cam. Figure 8-3 shows three common arrangements, **flat-faced**, **mushroom** (curved), and **roller**. The roller follower has the advantage of lower (rolling) friction than the sliding contact of the other two but can be more expensive. **Flat-faced followers** can package smaller than roller followers for some cam designs and are often favored for that reason as well as cost for automotive valve trains. **Roller followers** are more frequently used in production machinery where their ease of replacement and availability from bearing manufacturers' stock in any quantities are advantages. Grooved or track cams require roller followers. Roller followers are essentially ball or roller bearings with customized mounting details. Figure 8-5a shows two common types of commercial roller followers. Flat-faced or **mushroom followers** are usually custom designed and manufactured for each application. For high-volume applications such as automobile engines, the quantities are high enough to warrant a custom-designed follower.

Type of Cam

The direction of the follower's motion relative to the axis of rotation of the cam determines whether it is a **radial** or **axial** cam. All cams shown in Figures 8-1 to 8-3 are ra-

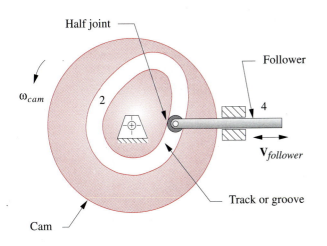

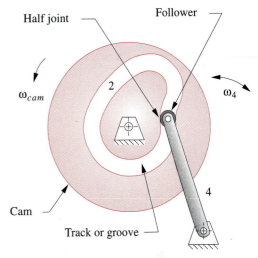

(a) Form-closed cam with translating follower

(b) Form-closed cam with oscillating follower

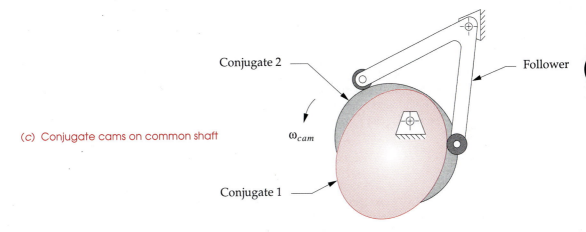

(c) Conjugate cams on common shaft

FIGURE 8-2

Form-closed cam-follower systems

dial cams because the follower motion is generally in a radial direction. Open **radial cams** are also called **plate cams**.

Figure 8-4 shows an **axial cam** whose follower moves parallel to the axis of cam rotation. This arrangement is also called a **face** cam if open (force-closed) and a **cylindrical** or **barrel** cam if grooved or ribbed (form-closed).

Figure 8-5b shows a selection of cams of various types. Clockwise from the lower left, they are: an open (force-closed) axial or face cam; an axial grooved (track) cam (form-closed) with external gear; an open radial, or plate cam (force-closed); a ribbed axial cam (form-closed); an axial grooved (barrel) cam.

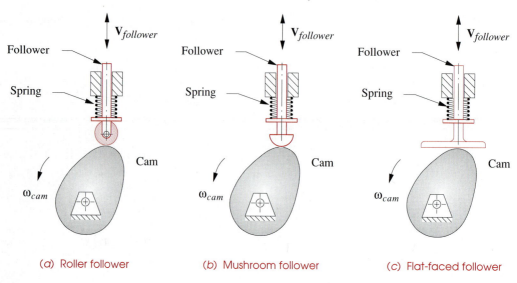

(a) Roller follower   (b) Mushroom follower   (c) Flat-faced follower

FIGURE 8-3

Three common types of cam followers

A **three-dimensional cam** or **camoid** (not shown) is a combination of radial and axial cams. It is a two-degree-of-freedom system. The two inputs are rotation of the cam about its axis and translation of the cam along its axis. The follower motion is a function of both inputs. The follower tracks along a different portion of the cam depending on the axial input.

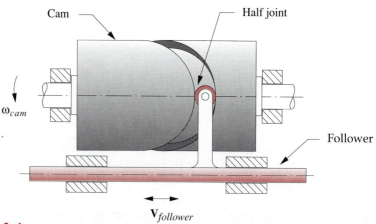

FIGURE 8-4

Axial, cylindrical, or barrel cam with form-closed, translating follower

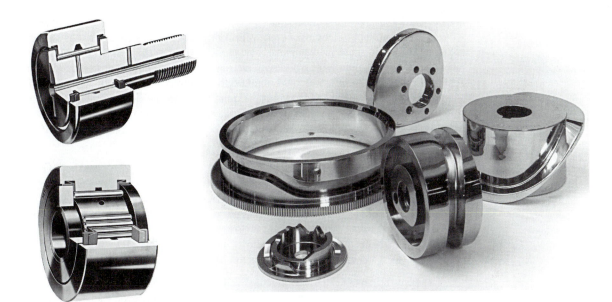

(*a*) Commercial roller followers
Courtesy of McGill Manufacturing Co.
South Bend, IN

(*b*) Commercial cams of various types
Courtesy of The Ferguson Co.
St. Louis, MS

FIGURE 8-5

Cams and roller followers

Type of Motion Constraints

There are two general categories of motion constraint, **critical extreme position** (CEP; also called endpoint specification) and **critical path motion** (CPM). **Critical extreme position** refers to the case in which the design specifications define the start and finish positions of the follower (i.e., extreme positions) but do not specify any constraints on the path motion between the extreme positions. This case is discussed in Sections 8.3 and 8.4 and is the easier of the two to design as the designer has great freedom to choose the cam functions which control the motion between extremes. **Critical path motion** is a more constrained problem than CEP because the path motion, and/or one or more of its derivatives are defined over all or part of the interval of motion. This is analogous to **function generation** in the linkage design case except that with a cam we can achieve a continuous output function for the follower. Section 8.6 discusses this CPM case. It may only be possible to create an approximation of the specified function and still maintain suitable dynamic behavior.

Type of Motion Program

The motion programs **rise-fall** (RF), **rise-fall-dwell** (RFD), and **rise-dwell-fall-dwell** (RDFD) all refer mainly to the CEP case of motion constraint and in effect define how many dwells are present in the full cycle of motion, either none (RF), one (RFD), or more than one (RDFD). **Dwells,** defined as *no output motion for a specified period of input motion*, are an important feature of cam-follower systems because it is very easy to create exact dwells in these mechanisms. The cam-follower is the design type of choice whenever a dwell is required. We saw in Section 3.9 (p. 125) how to design dwell linkages and found that at best we could obtain only an approximate dwell. The resulting single- or double-dwell linkages tend to be quite large for their output motion and are somewhat difficult to design. (See program SIXBAR for some built-in examples of these dwell linkages.) Cam-follower systems tend to be more compact than linkages for the same output motion.

If your need is for a **rise-fall** (RF) CEP motion, with no dwell, then you should really be considering a crank-rocker linkage rather than a cam-follower to obtain all the linkage's advantages over cams of reliability, ease of construction, and lower cost which were discussed in Section 2.15 (p. 55). If your needs for compactness outweigh those considerations, then the choice of a cam-follower in the RF case may be justified. Also, if you have a CPM design specification, and the motion or its derivatives are defined over the interval, then a cam-follower system is the logical choice in the RF case.

The **rise-fall-dwell** (RFD) and **rise-dwell-fall-dwell** (RDFD) cases are obvious choices for cam-followers for the reasons discussed above. However, each of these two cases has its own set of constraints on the behavior of the cam functions at the interfaces between the segments which control the rise, the fall, and the dwells. In general, we must match the **boundary conditions** (BCs) of the functions and their derivatives at all interfaces between the segments of the cam. This topic will be thoroughly discussed in the following sections.

8.2   *S V A J* DIAGRAMS

The first task faced by the cam designer is to select the mathematical functions to be used to define the motion of the follower. The easiest approach to this process is to "linearize" the cam, i.e., "unwrap it" from its circular shape and consider it as a function plotted on cartesian axes. We plot the displacement function s, its first derivative velocity v, its second derivative acceleration a, and its third derivative jerk j, all on aligned axes as a function of camshaft angle θ as shown in Figure 8-6. Note that we can consider the independent variable in these plots to be either time t or shaft angle θ, as we know the constant angular velocity ω of the camshaft and can easily convert from angle to time and vice versa.

$$\theta = \omega t \qquad\qquad\qquad (8.1)$$

Figure 8-6a shows the specifications for a four-dwell cam that has eight segments, RDFDRDFD. Figure 8-6b shows the $s\ v\ a\ j$ curves for the whole cam over 360 degrees of camshaft rotation. A cam design begins with a definition of the required cam functions and their $s\ v\ a\ j$ diagrams. Functions for the nondwell cam segments should be chosen based on their velocity, acceleration, and jerk characteristics and the relationships at the interfaces between adjacent segments including the dwells. These function characteristics can be conveniently and quickly investigated with program DYNACAM which generated the data and plots shown in Figure 8-6.

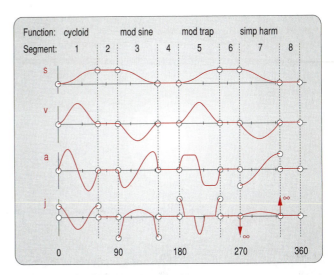

Segment Number	Function Used	Start Angle	End Angle	Delta Angle
1	Cycloid rise	0	60	60
2	Dwell	60	90	30
3	ModSine fall	90	150	60
4	Dwell	150	180	30
5	ModTrap rise	180	240	60
6	Dwell	240	270	30
7	SimpHarm fall	270	330	60
8	Dwell	330	360	30

(a) Cam program specifications

(b) Plots of cam-follower's s v a j diagrams

FIGURE 8-6

Cycloidal, modified sine, modified trapezoid, and simple harmonic motion functions on a four-dwell cam

8.3 DOUBLE-DWELL CAM DESIGN—CHOOSING *S V A J* FUNCTIONS

Many cam design applications require multiple dwells. The double-dwell case is quite common. Perhaps a **double-dwell** cam is driving a part feeding station on a production machine that makes toothpaste. This hypothetical cam's follower is fed an empty toothpaste tube (during the low dwell), then moves the empty tube into a loading station (during the rise), holds the tube absolutely still in a **critical extreme position** (CEP) while toothpaste is squirted into the open bottom of the tube (during the high dwell), and then retracts the filled tube back to the starting (zero) position and holds it in this other critical extreme position. At this point, another mechanism (during the low dwell) picks the tube up and carries it to the next operation, which might be to seal the bottom of the tube. A similar cam could be used to feed, align, and retract the tube at the bottom-sealing station as well.

Cam specifications such as this are often depicted on a timing diagram as shown in Figure 8-7 which is a graphical representation of the specified events in the machine cycle. A **machine's cycle** is defined as *one revolution of its master driveshaft*. In a complicated machine, such as our toothpaste maker, there will be a **timing diagram** for each subassembly in the machine. The time relationships among all subassemblies are defined by their timing diagrams which are all drawn on a common time axis. Obviously all these operations must be kept in precise synchrony and time phase for the machine to work.

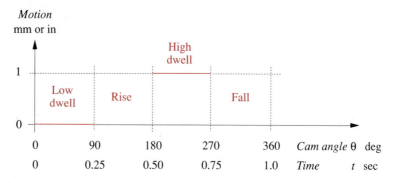

FIGURE 8-7

Cam timing diagram

This simple example in Figure 8-7 is a critical extreme position (CEP) case, because nothing is specified about the functions to be used to get from the low dwell position (one extreme) to the high dwell position (other extreme). The designer is free to choose any function that will do the job. Note that these specifications contain only information about the displacement function. The higher derivatives are not specifically constrained in this example. We will now use this problem to investigate several different ways to meet the specifications.

EXAMPLE 8-1

Naive Cam Design—A Bad Cam.

Problem: Consider the following cam design CEP specification:

dwell	at zero displacement for 90 degrees (low dwell)
rise	1 in (25 mm) in 90 degrees
dwell	at 1 in (25 mm) for 90 degrees (high dwell)
fall	1 in (25 mm) in 90 degrees.
cam ω	2π rad/sec = 1 rev/sec

Solution:

1   The naive or inexperienced cam designer might proceed with a design as shown in Figure 8-8a. Taking the given specifications literally, it is tempting to merely "connect the dots" on the timing diagram to create the displacement (s) diagram. (After all, when we wrap this s diagram around a circle to create the actual cam, it will look quite smooth despite the sharp corners on the s diagram.) The mistake our beginning designer is making here is to ignore the effect on the higher derivatives of the displacement function which results from this simplistic approach.

2   Figure 8-8b, c, and d shows the problem. Note that we have to treat each segment of the cam (rise, fall, dwell) as a separate entity in developing mathematical functions for the cam. Taking the rise segment (#2) first, the displacement function in Figure 8-8a during this portion is a straight line, or first-degree polynomial. The general equation for a straight line is:

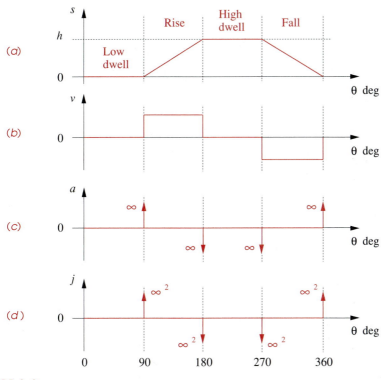

FIGURE 8-8

The *s v a j* diagrams of a "bad" cam design

$$y = mx + b \tag{8.2}$$

where m is the slope of the line and b is the y intercept. Substituting variables appropriate to this example in equation 8.2, angle θ replaces the independent variable x, and the displacement s replaces the dependent variable y. By definition, the constant slope m of the displacement is the velocity constant K_v.

3   For the rise segment, the y intercept b is zero because the low dwell position typically is taken as zero displacement by convention. Equation 8.2 then becomes:

$$s = K_v\theta \tag{8.3}$$

4   Differentiating with respect to θ gives a function for velocity during the rise.

$$v = K_v = \text{constant} \tag{8.4}$$

5   Differentiating again with respect to θ gives a function for acceleration during the rise.

$$a = 0 \tag{8.5}$$

This seems too good to be true (and it is). Zero acceleration means zero dynamic force. This cam appears to have no dynamic forces or stresses in it!

Figure 8-8 shows what is really happening here. If we return to the displacement function and graphically differentiate it twice, we will observe that, from the definition of the derivative as the instantaneous slope of the function, the acceleration is in fact zero **during the interval**. *But, at the boundaries of the interval*, where rise meets low dwell on one side and high dwell on the other, note that *the velocity function is multivalued. There are discontinuities at these boundaries*. The effect of these discontinuities is to create a portion of the velocity curve which has **infinite slope** and zero duration. This results in the *infinite spikes of acceleration* shown at those points.

These spikes are more properly called **Dirac delta functions**. Infinite acceleration cannot really be obtained, as it requires infinite force. Clearly the dynamic forces will be very large at these boundaries and will create high stresses and rapid wear. In fact, if this cam were built and run at any significant speeds, the sharp corners on the displacement diagram which are creating these theoretical infinite accelerations would be quickly worn to a smoother contour by the unsustainable stresses generated in the materials. *This is an unacceptable design.*

The unacceptability of this design is reinforced by the **jerk** diagram which shows theoretical values of **infinity squared** at the discontinuities. The problem has been engendered by an inappropriate choice of displacement function. In fact, the cam designer should not be as concerned with the displacement function as with its higher derivatives.

The Fundamental Law of Cam Design

Any cam designed for operation at other than very low speeds must be designed with the following constraints:

The cam function must be continuous through the first and second derivatives of displacement across the entire interval (360 degrees).

corollary:

The jerk function must be finite across the entire interval (360 degrees).

In any but the simplest of cams, the cam motion program cannot be defined by a single mathematical expression, but rather must be defined by several separate functions, each of which defines the follower behavior over one segment, or piece, of the cam. These expressions are sometimes called *piecewise functions*. These functions must have **third-order continuity** (the function plus two derivatives) at all boundaries. **The displacement, velocity and acceleration functions must have no discontinuities in them.**[*]

* This rule is stated by Neklutin[2] but is disputed by some other authors.[3],[4] Despite the minor controversy over this issue, this author believes that it is a good (and simple) rule to follow in order to get acceptable dynamic results with high-speed cams.

If any discontinuities exist in the acceleration function, then there will be infinite spikes, or Dirac delta functions, appearing in the derivative of acceleration, jerk. Thus the corollary merely restates the fundamental law of cam design. Our naive designer failed to recognize that by starting with a low-degree (linear) polynomial as the displacement function, discontinuities would appear in the upper derivatives.

Polynomial functions are one of the best choices for cams as we shall shortly see, but they do have one fault that can lead to trouble in this application. Each time they are differentiated, they reduce by one degree. Eventually, after enough differentiations, polynomials degenerate to zero degree (a constant value) as the velocity function in Figure 8-8b (p. 355) shows. Thus, by starting with a first-degree polynomial as a displacement function, it was inevitable that discontinuities would soon appear in its derivatives.

In order to obey the fundamental law of cam design, one must start with at least a fifth-degree polynomial (quintic) as the displacement function for a double-dwell cam. This will degenerate to a cubic function in the acceleration. The parabolic jerk function will have discontinuities, and the (unnamed) derivative of jerk will have infinite spikes in it. This is acceptable, as the jerk is still finite.

Simple Harmonic Motion (SHM)

Our naive cam designer recognized his mistake in choosing a straight-line function for the displacement. He also remembered a family of functions he had met in a calculus course which have the property of remaining continuous throughout any number of differentiations. These are the harmonic functions. On repeated differentiation, sine becomes cosine, which becomes negative sine, which becomes negative cosine, etc., ad infinitum. One never runs out of derivatives with the harmonic family of curves. In fact, differentiation of a harmonic function really only amounts to a 90° phase shift of the function. It is though, as you differentiated it, you cut out, with a scissors, a different portion of the same continuous sine wave function, which is defined from minus infinity to plus infinity. The equations of simple harmonic motion (SHM) for a rise motion are:

$$s = \frac{h}{2}\left[1 - \cos\left(\pi\frac{\theta}{\beta}\right)\right]$$
(8.6a)

$$v = \frac{\pi}{\beta}\frac{h}{2}\sin\left(\pi\frac{\theta}{\beta}\right)$$
(8.6b)

$$a = \frac{\pi^2}{\beta^2}\frac{h}{2}\cos\left(\pi\frac{\theta}{\beta}\right)$$
(8.6c)

$$j = -\frac{\pi^3}{\beta^3}\frac{h}{2}\sin\left(\pi\frac{\theta}{\beta}\right)$$
(8.6d)

where h is the total rise, or lift, θ is the camshaft angle, and β is the total angle of the rise interval.

We have here introduced a notation to simplify the expressions. The independent variable in our cam functions is θ, the camshaft angle. The period of any one segment is defined as the angle β. Its value can, of course, be different for each segment. We normalize the independent variable θ by dividing it by the period of the segment β. Both θ and β are measured in radians (or both in degrees). The value of θ/β will then vary from 0 to 1 over any segment. It is a dimensionless ratio. Equations 8.6 define simple harmonic motion and its derivatives for this rise segment in terms of θ/β.

This family of harmonic functions appears, at first glance, to be well suited to our cam design problem above. If we define the displacement function to be one of the harmonic functions, we should not "run out of derivatives" before reaching the acceleration.

**EXAMPLE 8-2**

Sophomoric* Cam Design—Simple Harmonic Motion—Still a Bad Cam.

Problem: Consider the same cam design CEP specification as in Example 8-1:

dwell	at zero displacement for 90 degrees (low dwell)
rise	1 in (25 mm) in 90 degrees
dwell	at 1 in (25 mm) for 90 degrees (high dwell)
fall	1 in (25 mm) in 90 degrees
cam ω	2π rad/sec = 1 rev/sec

Solution:

1   Figure 8-9 shows a full-rise simple harmonic function[†] applied to the rise segment of our cam design problem.

2   Note that the velocity function is continuous, as it matches the zero velocity of the dwells at each end. The peak value is 6.28 in/sec (160 mm/sec) at the midpoint of the rise.

3   The acceleration function, however, is **not** continuous. It is a half-period cosine curve and has nonzero values at start and finish which are $\pm$ 78.8 in/sec^2 (2.0 m/sec^2).

4   Unfortunately, the dwell functions, which adjoin this rise on each side, have zero acceleration as can be seen in Figure 8-6 (p. 353). Thus there are **discontinuities in the acceleration at each end of the interval** which uses this simple harmonic displacement function.

5   This violates the fundamental law of cam design and creates **infinite spikes of jerk** at the ends of this fall interval. **This is also an unacceptable design.**

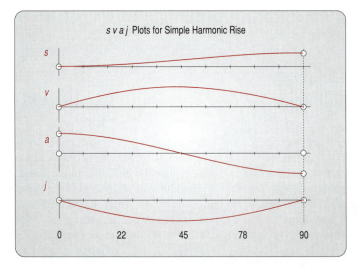

FIGURE 8-9

Simple harmonic motion with dwells has discontinuous acceleration

What went wrong? While it is true that harmonic functions are differentiable ad infinitum, we are not dealing here with single harmonic functions. Our cam function over the entire interval is a **piecewise function**, (Figure 8-6, p. 353) made up of several segments, some of which may be dwell portions or other functions. A dwell will always have zero velocity and zero acceleration. Thus we must match the dwells' zero values at the ends of those derivatives of any nondwell segments that adjoin them. The simple harmonic displacement function, when used with dwells, does **not** satisfy the fundamental law of cam design. Its second derivative, acceleration, is nonzero at its ends and thus does not match the dwells required in this example.

The only case in which the simple harmonic displacement function will satisfy the fundamental law is the non-quick-return RF case, i.e., rise in 180° and fall in 180° with no dwells. Then the cam becomes an eccentric as shown in Figure 8-10. As a single continuous (not piecewise) function, its derivatives are continuous also. Figure 8-11 shows the displacement (in inches) and acceleration functions (in *g*'s) of the eccentric cam in Figure 8-10 as actually measured on the follower. The noise, or "ripple," on the acceleration curve is due to small, unavoidable, manufacturing errors. Manufacturing limitations will be discussed in a later section.

$s = a \cos \omega t$

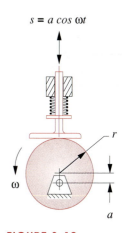

FIGURE 8-10

An eccentric cam has simple harmonic motion

Cycloidal Displacement

The two bad examples of cam design described above should lead the cam designer to the conclusion that consideration only of the displacement function when designing a cam is erroneous. The better approach is to start with consideration of the higher derivatives, especially acceleration. The acceleration function, and to a lesser extent the jerk

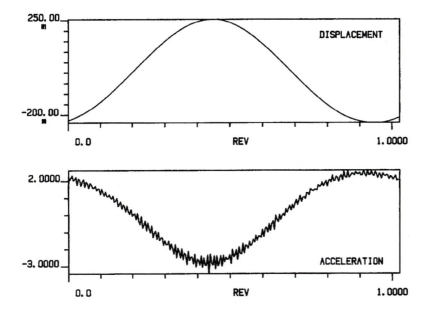

FIGURE 8-11

Displacement and acceleration as measured on the follower of an eccentric cam

function, should be the principal concern of the designer. In some cases, especially when the mass of the follower train is large, or when there is a specification on velocity, that function must be carefully designed as well.

With this in mind, we will redesign the cam for the same example specifications as above. This time we will start with the acceleration function. The harmonic family of functions still have advantages which make them attractive for these applications. Figure 8-12 shows a full-period sinusoid applied as the acceleration function. It meets the constraint of zero magnitude at each end to match the dwell segments which adjoin it. The equation for a sine wave is:

$$a = C \sin\left(2\pi \frac{\theta}{\beta}\right) \tag{8.7}$$

We have again normalized the independent variable θ by dividing it by the period of the segment β with both θ and β measured in radians. The value of θ/β ranges from 0 to 1 over any segment and is a dimensionless ratio. Since we want a full-cycle sine wave, we must multiply the argument by 2π. The argument of the sine function will then vary between 0 and 2π regardless of the value of β. The constant C defines the amplitude of the sine wave.

Integrate to obtain velocity,

$$a = \frac{dv}{d\theta} = C \sin\left(2\pi \frac{\theta}{\beta}\right)$$

$$\int dv = \int C \sin\left(2\pi \frac{\theta}{\beta}\right) d\theta \tag{8.8}$$

$$v = -C \frac{\beta}{2\pi} \cos\left(2\pi \frac{\theta}{\beta}\right) + k_1$$

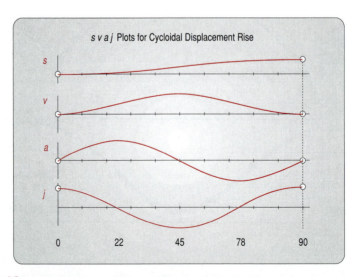

s v a j Plots for Cycloidal Displacement Rise

FIGURE 8-12
Sinusoidal acceleration gives cycloidal displacement

where k_1 is the constant of integration. To evaluate k_1, substitute the boundary condition $v = 0$ at $\theta = 0$, since we must match the zero velocity of the dwell at that point. The constant of integration is then:

$$k_1 = C \frac{\beta}{2\pi}$$

and:

$$v = C \frac{\beta}{2\pi} \left[1 - \cos\left(2\pi \frac{\theta}{\beta} \right) \right]$$

(8.9)

Note that substituting the boundary values at the other end of the interval, $v = 0$, $\theta = \beta$, will give the same result for k_1. Integrate again to obtain displacement:

$$v = \frac{ds}{d\theta} = C \frac{\beta}{2\pi} \left[1 - \cos\left(2\pi \frac{\theta}{\beta} \right) \right]$$

$$\int ds = \int \left\{ C \frac{\beta}{2\pi} \left[1 - \cos\left(2\pi \frac{\theta}{\beta} \right) \right] \right\} d\theta$$

(8.10)

$$s = C \frac{\beta}{2\pi} \theta - C \frac{\beta^2}{4\pi^2} \sin\left(2\pi \frac{\theta}{\beta} \right) + k_2$$

To evaluate k_2, substitute the boundary condition $s = 0$ at $\theta = 0$, since we must match the zero displacement of the dwell at that point. To evaluate the amplitude constant C, substitute the boundary condition $s = h$ at $\theta = \beta$, where h is the maximum follower rise (or lift) required over the interval and is a constant for any one cam specification.

$$k_2 = 0$$

$$C = 2\pi \frac{h}{\beta^2}$$

(8.11)

Substituting the value of the constant C in equation 8.7 for acceleration gives:

$$a = 2\pi \frac{h}{\beta^2} \sin\left(2\pi \frac{\theta}{\beta} \right)$$

(8.12a)

Differentiating with respect to θ gives the expression for jerk.

$$j = 4\pi^2 \frac{h}{\beta^3} \cos\left(2\pi \frac{\theta}{\beta} \right)$$

(8.12b)

Substituting the values of the constants C and k_1 in equation 8.9 for velocity gives:

$$v = \frac{h}{\beta} \left[1 - \cos\left(2\pi \frac{\theta}{\beta} \right) \right]$$

(8.12c)

This velocity function is the sum of a negative cosine term and a constant term. The coefficient of the cosine term is equal to the constant term. This results in a velocity curve which starts and ends at zero and reaches a maximum magnitude at $\beta/2$ as can be

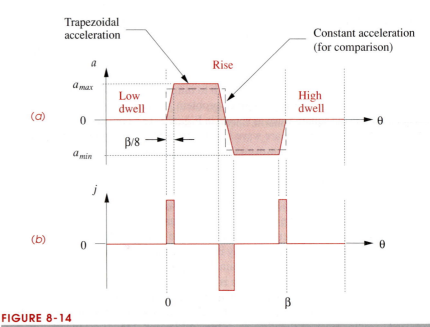

FIGURE 8-14

Trapezoidal acceleration gives finite jerk

MODIFIED TRAPEZOIDAL ACCELERATION An improvement can be made to the trapezoidal acceleration function by substituting pieces of sine waves for the sloped sides of the trapezoids as shown in Figure 8-15. This function is called the **modified trapezoidal acceleration** curve.[*] This function is a marriage of the sine acceleration and constant acceleration curves. Conceptually, a full period sine wave is cut into fourths and "pasted into" the square wave to provide a smooth transition from the zeros at the endpoints to the maximum and minimum peak values, and to make the transition from maximum to minimum in the center of the interval. The portions of the total segment period (β) used for the sinusoidal parts of the function can be varied. The most common arrangement is to cut the square wave at $\beta/8$, $3\beta/8$, $5\beta/8$, and $7\beta/8$ to insert the pieces of sine wave as shown in Figure 8-15. The $s\ v\ a\ j$ formulas for that arrangement of a modified trapezoidal rise are:

for $\quad 0 \leq \theta < \dfrac{1}{8}\beta$

$$s = h\left[0.38898448\frac{\theta}{\beta} - 0.0309544\sin\left(4\pi\frac{\theta}{\beta}\right)\right]$$

$$v = 0.38898448\frac{h}{\beta}\left[1 - \cos\left(4\pi\frac{\theta}{\beta}\right)\right]$$

$$a = 4.888124\frac{h}{\beta^2}\sin\left(4\pi\frac{\theta}{\beta}\right)$$

$$j = 61.425769\frac{h}{\beta^3}\cos\left(4\pi\frac{\theta}{\beta}\right)$$

(8.13a)

[*] Developed by C. N. Neklutin of the Universal Match Corp. See reference [2].

for $\quad \dfrac{1}{8}\beta \le \theta < \dfrac{3}{8}\beta$

$$s = h\left[2.44406184\left(\dfrac{\theta}{\beta}\right)^2 - 0.22203097\left(\dfrac{\theta}{\beta}\right) + 0.00723407\right]$$

$$v = \dfrac{h}{\beta}\left[4.888124\left(\dfrac{\theta}{\beta}\right) - 0.22203097\right] \qquad (8.13b)$$

$$a = 4.888124\dfrac{h}{\beta^2}$$

$$j = 0$$

for $\quad \dfrac{3}{8}\beta \le \theta < \dfrac{5}{8}\beta$

$$s = h\left[1.6110154\dfrac{\theta}{\beta} - 0.0309544\sin\left(4\pi\dfrac{\theta}{\beta} - \pi\right) - 0.3055077\right]$$

$$v = \dfrac{h}{\beta}\left[1.6110154 - 0.38898448\cos\left(4\pi\dfrac{\theta}{\beta} - \pi\right)\right] \qquad (8.13c)$$

$$a = 4.888124\dfrac{h}{\beta^2}\sin\left(4\pi\dfrac{\theta}{\beta} - \pi\right)$$

$$j = 61.425769\dfrac{h}{\beta^3}\cos\left(4\pi\dfrac{\theta}{\beta} - \pi\right)$$

for $\quad \dfrac{5}{8}\beta \le \theta < \dfrac{7}{8}\beta$

$$s = h\left[-2.44406184\left(\dfrac{\theta}{\beta}\right)^2 + 4.6660917\left(\dfrac{\theta}{\beta}\right) - 1.2292648\right]$$

$$v = \dfrac{h}{\beta}\left[-4.888124\left(\dfrac{\theta}{\beta}\right) + 4.6660917\right] \qquad (8.13d)$$

$$a = -4.888124\dfrac{h}{\beta^2}$$

$$j = 0$$

for $\quad \dfrac{7}{8}\beta \le \theta \le \beta$

$$s = h\left[0.6110154 + 0.38898448\dfrac{\theta}{\beta} + 0.0309544\sin\left(4\pi\dfrac{\theta}{\beta} - 3\pi\right)\right]$$

$$v = 0.38898448\dfrac{h}{\beta}\left[1 + \cos\left(4\pi\dfrac{\theta}{\beta} - 3\pi\right)\right] \qquad (8.13e)$$

$$a = -4.888124\dfrac{h}{\beta^2}\sin\left(4\pi\dfrac{\theta}{\beta} - 3\pi\right)$$

$$j = -61.425769\dfrac{h}{\beta^3}\cos\left(4\pi\dfrac{\theta}{\beta} - 3\pi\right)$$

8

(a) Take a sine wave

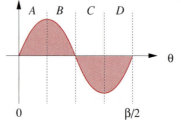

(b) Split the sine
wave apart

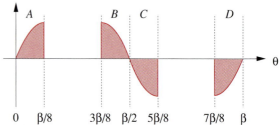

(c) Take a constant
acceleration
square wave

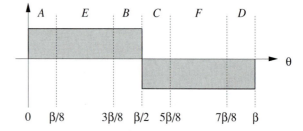

(d) Combine the two

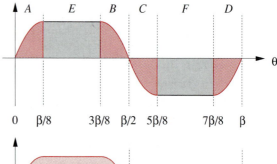

(e) Modified trapezoidal
acceleration

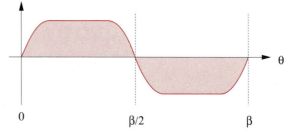

FIGURE 8-15

Creating the modified trapezoidal acceleration function

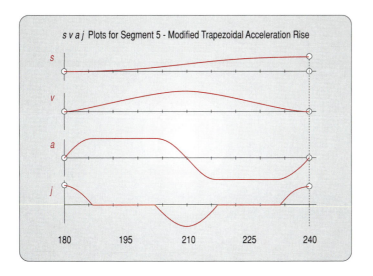

svaj Plots for Segment 5 - Modified Trapezoidal Acceleration Rise

FIGURE 8-16

Modified trapezoidal acceleration

The modified trapezoidal function defined above is one of many combined functions created for cams by piecing together various functions, while being careful to match the values of the s, v, and a curves at all the interfaces between the joined functions. It has the advantage of relatively low theoretical peak acceleration, and reasonably rapid, smooth transitions at the beginning and end of the interval. Note that in the form presented, with θ (in radians) as the independent variable, the units of the expressions in equations 8.13 are length, length/rad, length/rad^2, and length/rad^3 for s, v, a, j, respectively. To convert equations 8.13 to a time base, multiply velocity v by the camshaft angular velocity ω (in rad/sec), acceleration a by ω^2, and jerk j by ω^3. The modified trapezoidal cam function is a popular and often used program for double-dwell cams. Its s v a j curves are shown in Figure 8-16.

MODIFIED SINUSOIDAL ACCELERATION[*]   The sine acceleration curve (cycloidal displacement) has the advantage of smoothness (less ragged jerk curve) compared to the modified trapezoid but has higher theoretical peak acceleration. By combining two harmonic (sinusoid) curves of different frequencies, we can retain some of the smoothness characteristics of the cycloid and also reduce the peak acceleration. As an added bonus we will find that the peak velocity is also lower than in either the cycloidal or modified trapezoid. Figure 8-17 shows how the modified sine acceleration curve is made up of pieces of two sinusoid functions, one of higher frequency than the other. The first and last quarter of the high-frequency (short period, $\beta/2$) sine curve is used for the first and last eighths of the combined function. The center half of the low-frequency (long period, $3\beta/2$) sine wave is used to fill in the center three-fourths of the combined curve. Obviously, the magnitudes of the two curves and their derivatives must be matched at their interfaces in order to avoid discontinuities. The equations for the **modified sine** curve for a rise of height h over a period β, with the functions joined at the $\beta/8$ and $7\beta/8$ points are as follows:

* Developed by E. H. Schmidt of DuPont.

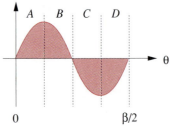

(a) Sine wave #1 of period $\beta/2$

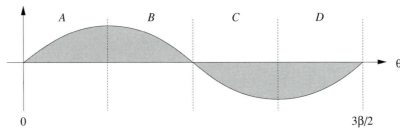

(b) Sine wave #2 of period $3\beta/2$

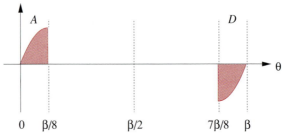

(c) Take 1st and 4th quarters of #1

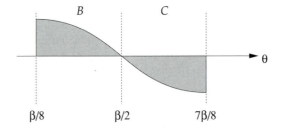

(d) Take 2nd and 3rd quarters of #2

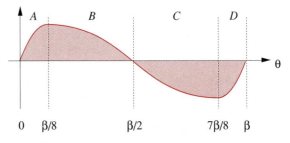

(e) Combine to get modified sine

FIGURE 8-17

Creating the modified sine acceleration function

for $\quad 0 \le \theta < \dfrac{1}{8}\beta$

$$s = h\left[0.43990085\dfrac{\theta}{\beta} - 0.0350062\sin\left(4\pi\dfrac{\theta}{\beta}\right)\right]$$

$$v = 0.43990085\dfrac{h}{\beta}\left[1 - \cos\left(4\pi\dfrac{\theta}{\beta}\right)\right] \qquad (8.14a)$$

$$a = 5.5279571\dfrac{h}{\beta^2}\sin\left(4\pi\dfrac{\theta}{\beta}\right)$$

$$j = 69.4663577\dfrac{h}{\beta^3}\cos\left(4\pi\dfrac{\theta}{\beta}\right)$$

for $\quad \dfrac{1}{8}\beta \le \theta < \dfrac{7}{8}\beta$

$$s = h\left[0.28004957 + 0.43990085\dfrac{\theta}{\beta} - 0.31505577\cos\left(\dfrac{4\pi}{3}\dfrac{\theta}{\beta} - \dfrac{\pi}{6}\right)\right]$$

$$v = 0.43990085\dfrac{h}{\beta}\left[1 + 3\sin\left(\dfrac{4\pi}{3}\dfrac{\theta}{\beta} - \dfrac{\pi}{6}\right)\right] \qquad (8.14b)$$

$$a = 5.5279571\dfrac{h}{\beta^2}\cos\left(\dfrac{4\pi}{3}\dfrac{\theta}{\beta} - \dfrac{\pi}{6}\right)$$

$$j = -23.1553\dfrac{h}{\beta^3}\sin\left(\dfrac{4\pi}{3}\dfrac{\theta}{\beta} - \dfrac{\pi}{6}\right)$$

for $\quad \dfrac{7}{8}\beta \le \theta \le \beta$

$$s = h\left\{0.56009915 + 0.43990085\dfrac{\theta}{\beta} - 0.0350062\sin\left[2\pi\left(2\dfrac{\theta}{\beta} - 1\right)\right]\right\}$$

$$v = 0.43990085\dfrac{h}{\beta}\left\{1 - \cos\left[2\pi\left(2\dfrac{\theta}{\beta} - 1\right)\right]\right\} \qquad (8.14c)$$

$$a = 5.5279571\dfrac{h}{\beta^2}\sin\left[2\pi\left(2\dfrac{\theta}{\beta} - 1\right)\right]$$

$$j = 69.4663577\dfrac{h}{\beta^3}\cos\left[2\pi\left(2\dfrac{\theta}{\beta} - 1\right)\right]$$

Figure 8-18 shows a comparison of the shapes and relative magnitudes of five cam acceleration programs including the cycloidal, modified trapezoid, and modified sine acceleration curves.[*] The cycloidal curve has a theoretical peak acceleration which is approximately 1.3 times that of the modified trapezoid's peak value for the same cam

[*] The 3-4-5 and 4-5-6-7 polynomial functions also shown in the figure will be discussed in a later section.

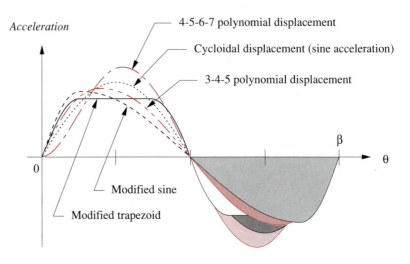

FIGURE 8-18

Comparison of five double-dwell cam acceleration functions

specification. The peak value of acceleration for the modified sine is between the cycloidal and modified trapezoid. Table 8-2 lists the peak values of acceleration, velocity, and jerk for these functions in terms of the total rise h and period β.

Figure 8-19 compares the jerk curves for the same functions. The modified sine jerk is somewhat less ragged than the modified trapezoid jerk but not as smooth as that of the cycloid, which is a full-period cosine. Figure 8-20 compares their velocity curves. The peak velocities of the cycloidal and modified trapezoid functions are the same, so each will store the same peak kinetic energy in the follower train. The peak velocity of the modified sine is the lowest of the five functions shown. This is the principal advantage of the modified sine acceleration curve and the reason it is usually chosen for applications in which the follower mass is very large.

TABLE 8-2   Factors for Peak Velocity and Acceleration of Some Cam Functions

Function	Max. Veloc.	Max. Accel.	Max. Jerk	Comments
Constant accel.	$2.000\ h/\beta$	$4.000\ h/\beta^2$	infinite	∞ jerk - not acceptable.
Harmonic disp.	$1.571\ h/\beta$	$4.945\ h/\beta^2$	infinite	∞ jerk - not acceptable.
Trapezoid accel.	$2.000\ h/\beta$	$5.300\ h/\beta^2$	$44\ h/\beta^3$	Not as good as mod. trap.
Mod. trap. accel.	$2.000\ h/\beta$	$4.888\ h/\beta^2$	$61\ h/\beta^3$	Low accel but rough jerk.
Mod. sine. accel.	$1.760\ h/\beta$	$5.528\ h/\beta^2$	$69\ h/\beta^3$	Low veloc - good accel.
3-4-5 Poly. disp.	$1.875\ h/\beta$	$5.777\ h/\beta^2$	$60\ h/\beta^3$	Good compromise.
Cycloidal disp.	$2.000\ h/\beta$	$6.283\ h/\beta^2$	$40\ h/\beta^3$	Smooth accel. & jerk.
4-5-6-7 Poly. disp.	$2.188\ h/\beta$	$7.526\ h/\beta^2$	$52\ h/\beta^3$	Smooth jerk-high accel.

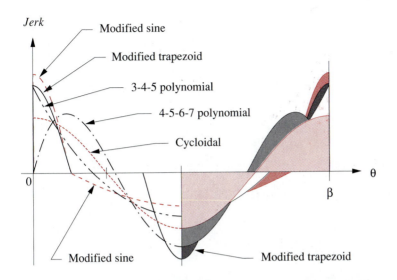

FIGURE 8-19

Comparison of five double-dwell cam jerk functions

An example of such an application is shown in Figure 8-21 which is an indexing table drive used for automated assembly lines. The round indexing table is mounted on

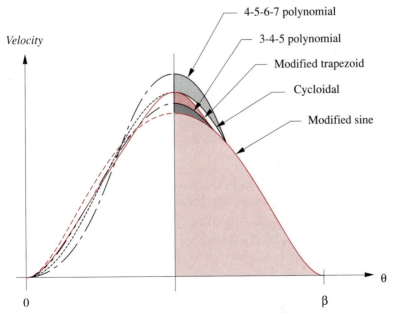

FIGURE 8-20

Comparison of five double-dwell cam velocity functions

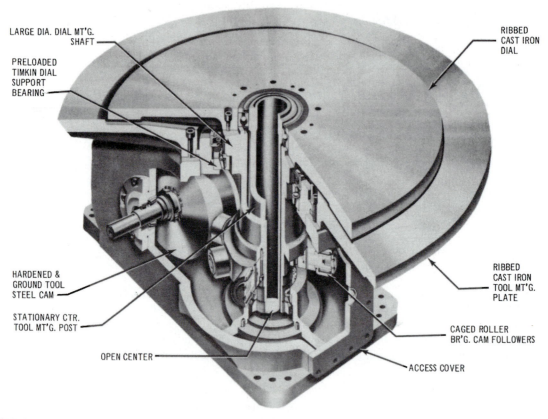

LARGE DIA. DIAL MT'G.
SHAFT

PRELOADED
TIMKIN DIAL
SUPPORT
BEARING

RIBBED
CAST IRON
DIAL

HARDENED &
GROUND TOOL
STEEL CAM

STATIONARY CTR.
TOOL MT'G. POST

RIBBED
CAST IRON
TOOL MT'G.
PLATE

CAGED ROLLER
BR'G. CAM FOLLOWERS

OPEN CENTER

ACCESS COVER

FIGURE 8-21

Multistop rotary indexer. *Courtesy of The Ferguson Co., St. Louis. MO*

a tapered vertical spindle and driven as part of the follower train by a form-closed barrel cam which moves it through some angular displacement, and then holds the table still in a dwell (called a "stop") while an assembly operation is performed on the workpiece carried on the table. These indexers may have three or more stops, each corresponding to an index position. The table (not shown) is solid steel and may be several feet in diameter; thus its mass is large. To minimize the stored kinetic energy, which must be dissipated each time the table is brought to a stop, the manufacturers often use the modified sine program on these multidwell cams, because of its lower peak velocity.

Let us again try to improve the double-dwell cam example with these combined functions of modified trapezoid and modified sine acceleration.

**EXAMPLE 8-4**

Senior Cam Design—Combined Functions—Better Cams.

Problem: Consider the same cam design CEP specification as in Examples 8-1 to 8-3:

dwell	at zero displacement for 90 degrees (low dwell)
rise	1 in (25 mm) in 90 degrees
dwell	at 1 in (25 mm) for 90 degrees (high dwell)
fall	1 in (25 mm) in 90 degrees.
cam ω	2π rad/sec = 1 rev/sec.

Solution:

1   The modified trapezoidal function is an acceptable one for this double-dwell cam specification. Its derivatives are continuous through the acceleration function as shown in Figures 8-16, 8-18, and 8-20 (pp. 367-371). The peak acceleration is 78.1 in/sec^2 (1.98 m/sec^2).

2   The modified trapezoidal jerk curve in Figures 8-16 and 8-19 (p. 371) is discontinuous at its boundaries but has finite magnitude of 3925 in/sec^3 (100 m/sec^3), and this is acceptable.

3   The modified trapezoidal velocity in Figures 8-16 and 8-20 is smooth and matches the zeros of the dwell at each end. Its peak magnitude is 8 in/sec (0.2 m/sec).

4   The advantage of this modified trapezoidal function is that it has smaller peak acceleration than the cycloidal but its peak velocity is identical to that of the cycloidal.

5   The modified sinusoid function is also an acceptable one for this double-dwell cam specification. Its derivatives are also continuous through the acceleration function as shown in Figures 8-18 and 8-20. Its peak acceleration is 88.3 in/sec^2 (2.24 m/sec^2).

6   The modified sine jerk curve in Figure 8-19 is discontinuous at its boundaries but is of finite magnitude and is larger in magnitude at 4439 in/sec^3 (113 m/sec^3) but smoother than that of the modified trapezoid.

7   The modified sine velocity (Figure 8-20) is smooth, matches the zeros of the dwell at each end, and is lower in peak magnitude than either the cycloid or modified trapezoidal at 7 in/sec (0.178 m/sec). This is an advantage for high-mass follower systems as it reduces the kinetic energy. This, coupled with a peak acceleration lower than the cycloidal (but higher than the modified trapezoidal), is its chief advantage.

Figure 8-22 shows the displacement curves for these three cam programs. (Open the diskfile E08-04.cam in program DYNACAM also.) Note how little difference there is between the displacement curves despite the large differences in their acceleration waveforms in Figure 8-18. This is evidence of the smoothing effect of the integration process. Differentiating any two functions will exaggerate their differences. Integration tends to mask their differences. It is nearly impossible to recognize these very differently behaving cam functions by looking only at their displacement curves. This is further evidence of the folly of our earlier naive approach to cam design which dealt exclusively with the displacement function. The cam designer must be concerned with the higher derivatives of displacement. The displacement function is primarily of value to the manufacturer of the cam who needs its coordinate information in order to cut the cam.

FALL FUNCTIONS   We have used only the rise portion of the cam for these examples. The fall is handled similarly. The rise functions presented here are applicable to the fall with slight modification. To convert rise equations to fall equations, it is only necessary to subtract the rise displacement function s from the maximum lift h and to negate the higher derivatives, v, a, and j.

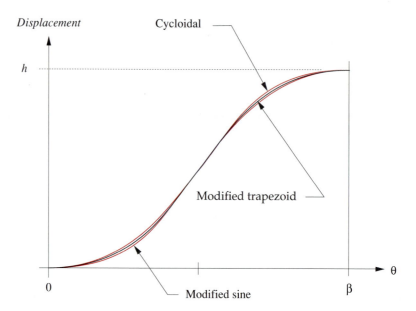

FIGURE 8-22

Comparison of three double-dwell cam displacement functions

SUMMARY   This section has attempted to present an approach to the selection of appropriate double-dwell cam functions, using the common rise-dwell-fall-dwell cam as the example, and to point out some of the pitfalls awaiting the cam designer. The particular functions described are only a few of the ones that have been developed for this double-dwell case over many years, by many designers, but they are probably the most used and most popular among cam designers. Most of them are also included in program DYNACAM. There are many trade-offs to be considered in selecting a cam program for any application, some of which have already been mentioned, such as function continuity, peak values of velocity and acceleration, and smoothness of jerk. There are many other trade-offs still to be discussed in later sections of this chapter, involving the sizing and the manufacturability of the cam.

8.4   SINGLE-DWELL CAM DESIGN—CHOOSING *S V A J* FUNCTIONS

Many applications in machinery require a **single-dwell** cam program, **rise-fall-dwell** (RFD). Perhaps a single-dwell cam is needed to lift and lower a roller which carries a moving paper web on a production machine that makes envelopes. This cam's follower lifts the paper up to one critical extreme position at the right time to contact a roller which applies a layer of glue to the envelope flap. Without dwelling in the up position, it immediately retracts the web back to the starting (zero) position and holds it in this other critical extreme position (low dwell) while the rest of the envelope passes by. It repeats the cycle for the next envelope as it comes by. Another common example of a single-dwell application is the cam which opens the valves in your automobile engine. This lifts the valve open on the rise, immediately closes it on the fall, and then keeps the valve closed in a dwell while the compression and combustion take place.

If we attempt to use the same type of cam programs as were defined for the double-dwell case for a single-dwell application, we will achieve a solution which may work but is not optimal. We will nevertheless do so here as an example in order to point out the problems that result. Then we will redesign the cam to eliminate those problems.

✎ EXAMPLE 8-5

Using Cycloidal Motion for Single-Dwell.

Problem: Consider the following single-dwell cam specification:

rise	1 in (25 mm) in 90 degrees.
fall	1 in (25 mm) in 90 degrees.
dwell	at zero displacement for 180 degrees (low dwell).
cam ω	15 rad/sec.

Solution:

1 Figure 8-23 shows a cycloidal displacement rise and separate cycloidal displacement fall applied to this single-dwell example. Note that the displacement (*s*) diagram looks acceptable in that it moves the follower from the low to the high position and back in the required intervals.

2 The velocity (*v*) also looks acceptable in shape in that it takes the follower from zero velocity at the low dwell to a peak value of 19.1 in/sec (0.49 m/sec) to zero again at the maximum displacement, where the glue is applied.

3 Figure 8-23 shows the acceleration function for this solution. Its maximum absolute value is about 573 in/sec^2.

4 The problem is that this acceleration curve has an **unnecessary return to zero** at the end of the rise. It is unnecessary because the acceleration during the first part of the fall is also negative. It would be better to keep it in the negative region at the end of the rise.

5 This unnecessary oscillation to zero in the acceleration causes the jerk to have more abrupt changes and discontinuities. The only real justification for taking the acceleration to zero is the need to change its sign (as is the case halfway through the rise or fall) or to match an adjacent segment which has zero acceleration.

The reader may input the file E08-05.cam to program DYNACAM to investigate this example in more detail.

For the single-dwell case we would like a function for the rise which does not return its acceleration to zero at the end of the interval. The function for the fall should begin with the same nonzero acceleration value as ended the rise and then be zero at its terminus to match the dwell. One function which meets those criteria is the **double harmonic** which gets its name from its two cosine terms, one of which is a half-period harmonic and the other a full-period wave. The equations for the double harmonic functions are:

8

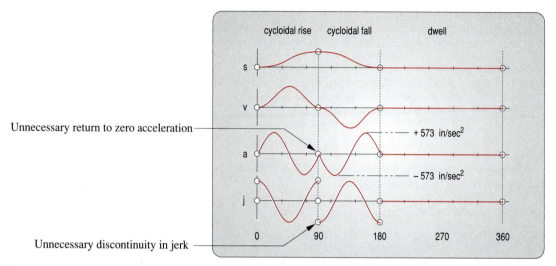

Unnecessary return to zero acceleration

Unnecessary discontinuity in jerk

FIGURE 8-23

Cycloidal motion (or any double-dwell program) is a poor choice for the single-dwell case

for the rise:

$$s = \frac{h}{2}\left\{\left[1-\cos\left(\pi\frac{\theta}{\beta}\right)\right] - \frac{1}{4}\left[1-\cos\left(2\pi\frac{\theta}{\beta}\right)\right]\right\}$$

$$v = \frac{\pi}{\beta}\frac{h}{2}\left[\sin\left(\pi\frac{\theta}{\beta}\right) - \frac{1}{2}\sin\left(2\pi\frac{\theta}{\beta}\right)\right]$$

$$a = \frac{\pi^2}{\beta^2}\frac{h}{2}\left[\cos\left(\pi\frac{\theta}{\beta}\right) - \cos\left(2\pi\frac{\theta}{\beta}\right)\right] \qquad (8.15a)$$

$$j = -\frac{\pi^3}{\beta^3}\frac{h}{2}\left[\sin\left(\pi\frac{\theta}{\beta}\right) - 2\sin\left(2\pi\frac{\theta}{\beta}\right)\right]$$

for the fall:

$$s = \frac{h}{2}\left\{\left[1+\cos\left(\pi\frac{\theta}{\beta}\right)\right] - \frac{1}{4}\left[1-\cos\left(2\pi\frac{\theta}{\beta}\right)\right]\right\}$$

$$v = -\frac{\pi}{\beta}\frac{h}{2}\left[\sin\left(\pi\frac{\theta}{\beta}\right) + \frac{1}{2}\sin\left(2\pi\frac{\theta}{\beta}\right)\right]$$

$$a = -\frac{\pi^2}{\beta^2}\frac{h}{2}\left[\cos\left(\pi\frac{\theta}{\beta}\right) + \cos\left(2\pi\frac{\theta}{\beta}\right)\right] \qquad (8.15b)$$

$$j = \frac{\pi^3}{\beta^3}\frac{h}{2}\left[\sin\left(\pi\frac{\theta}{\beta}\right) + 2\sin\left(2\pi\frac{\theta}{\beta}\right)\right]$$

Note that these double harmonic functions should **never** be used for the double-dwell case because their acceleration is nonzero at one end of the interval.

EXAMPLE 8-6

Double Harmonic Motion for Single-Dwell.

Problem: Consider the same single-dwell cam specification as in example 8-5:

rise	1 in (25 mm) in 90 degrees
fall	1 in (25 mm) in 90 degrees
dwell	at zero displacement for 180 degrees (low dwell)
cam ω	15 rad/sec

Solution:

1   Figure 8-24 shows a double harmonic rise and a double harmonic fall. The peak velocity is 19.5 in/sec (0.50 m/sec) which is similar to that of the cycloidal solution of Example 8-5.

2   Note that the acceleration of this double harmonic function does not return to zero at the end of the rise. This makes it more suitable for a single-dwell case in that respect.

3   The double harmonic jerk function peaks at 36 931 in/sec^3 (938 m/sec^3) and is quite smooth compared to the cycloidal solution.

4   Unfortunately, the peak negative acceleration is 900 in/sec^2, nearly twice that of the cycloidal solution. This is a smoother function but will develop higher dynamic forces. Open the diskfile E08-06.cam in program DYNACAM to see this example in more detail.

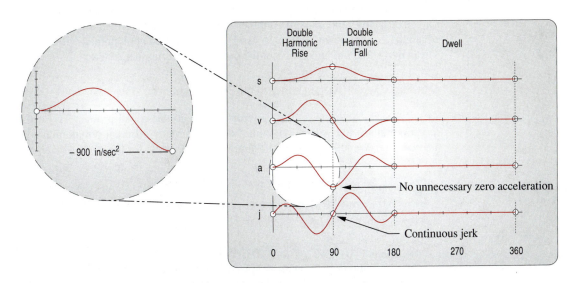

FIGURE 8-24

Double harmonic motion can be used for the single-dwell case if rise and fall durations are equal

Neither of the solutions in Examples 8-5 and 8-6 is optimal. We will revisit this example problem after introducing polynomial functions and redesign it yet again to improve both its smoothness and reduce its peak acceleration.

SUMMARY   This section has presented an approach to the selection of appropriate single-dwell cam functions and pointed out some of their limitations. The particular functions described (cycloidal and double harmonic) are only two of those that have been developed for this single-dwell case. We will visit this single-dwell problem again in the next section and develop a superior solution using other techniques.

8.5   POLYNOMIAL FUNCTIONS

The class of polynomial functions is one of the more versatile types that can be used for cam design. They are not limited to single- or double-dwell applications and can be tailored to many design specifications.   The general form of a polynomial function is:

$$s = C_0 + C_1 x + C_2 x^2 + C_3 x^3 + C_4 x^4 + C_5 x^5 + C_6 x^6 + \cdots + C_n x^n \qquad (8.16)$$

where s is the follower displacement; x is the independent variable, which in our case will be replaced by either θ/β or time t. The constant coefficients C_n are the unknowns to be determined in our development of the particular polynomial equation to suit a design specification. The degree of a polynomial is defined as the highest power present in any term. Note that a polynomial of degree n will have $n + 1$ terms because there is an x^0 or constant term with coefficient C_0, as well as coefficients through and including C_n.

We structure a polynomial cam design problem by deciding how many boundary conditions (BCs) we want to specify on the $s\ v\ a\ j$ diagrams. The number of BCs then determines the degree of the resulting polynomial. We can write an independent equation for each BC by substituting it into equation 8.16 or one of its derivatives. We will then have a system of linear equations which can be solved for the unknown coefficients $C_0, \ldots, C_n$. If k represents the number of chosen boundary conditions, there will be k equations in k unknowns $C_0, \ldots, C_n$ and the **degree** of the polynomial will be $n = k - 1$. The **order** of the n-degree polynomial is equal to the number of terms, k.

Double-Dwell Applications of Polynomials

THE 3-4-5 POLYNOMIAL   Let us return to the double-dwell problem of Section 8.3 (p. 353) and solve it with polynomial functions. Many different polynomial solutions are possible. We will start with the simplest one possible for the double-dwell case.

✎ EXAMPLE 8-7

The 3-4-5 Polynomial for the Double-Dwell Case.

Problem: Consider the same cam design CEP specification as in Examples 8-1 to 8-4:

dwell	at zero displacement for 90 degrees (low dwell)
rise	1 in (25 mm) in 90 degrees
dwell	at 1 in (25 mm) for 90 degrees (high dwell)
fall	1 in (25 mm) in 90 degrees
cam ω	2π rad/sec = 1 rev/sec

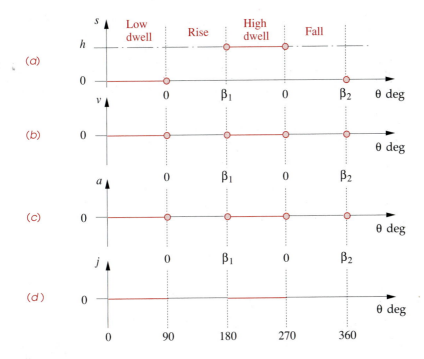

FIGURE 8-25

Minimum boundary conditions for the double-dwell case

8

Solution:

1 To satisfy the fundamental law of cam design the values of the rise (and fall) functions at their boundaries with the dwells must match with no discontinuities in, at a minimum, s, v, and a.

2 Figure 8-25 shows the axes for the s v a j diagrams on which the known data have been drawn. The dwells are the only fully defined segments at this stage. The requirement for continuity through the acceleration defines a minimum of **six boundary conditions** for the rise segment and six more for the fall in this problem. They are shown as filled circles on the plots. For generality, we will let the specified total rise be represented by the variable h. The minimum set of required BCs for this example is then:

for the rise:

$$\text{when} \quad \theta = 0; \quad \text{then} \quad s = 0, \quad v = 0, \quad a = 0$$
$$\text{when} \quad \theta = \beta_1; \quad \text{then} \quad s = h, \quad v = 0, \quad a = 0 \tag{8.17a}$$

for the fall:

$$\text{when} \quad \theta = 0; \quad \text{then} \quad s = h, \quad v = 0, \quad a = 0$$
$$\text{when} \quad \theta = \beta_2; \quad \text{then} \quad s = 0, \quad v = 0, \quad a = 0 \tag{8.17b}$$

3   We will use the rise for an example solution. (The fall is a similar derivation.) We have six BCs on the rise. This requires six terms in the equation. The highest term will be fifth degree. We will use the normalized angle θ/β as our independent variable, as before. Because our boundary conditions involve velocity and acceleration as well as displacement, we need to differentiate equation 8.16 versus θ to obtain expressions into which we can substitute those BCs. Rewriting equation 8.16 to fit these constraints and differentiating twice, we get:

$$s = C_0 + C_1\left(\frac{\theta}{\beta}\right) + C_2\left(\frac{\theta}{\beta}\right)^2 + C_3\left(\frac{\theta}{\beta}\right)^3 + C_4\left(\frac{\theta}{\beta}\right)^4 + C_5\left(\frac{\theta}{\beta}\right)^5 \qquad (8.18a)$$

$$v = \frac{1}{\beta}\left[C_1 + 2C_2\left(\frac{\theta}{\beta}\right) + 3C_3\left(\frac{\theta}{\beta}\right)^2 + 4C_4\left(\frac{\theta}{\beta}\right)^3 + 5C_5\left(\frac{\theta}{\beta}\right)^4\right] \qquad (8.18b)$$

$$a = \frac{1}{\beta^2}\left[2C_2 + 6C_3\left(\frac{\theta}{\beta}\right) + 12C_4\left(\frac{\theta}{\beta}\right)^2 + 20C_5\left(\frac{\theta}{\beta}\right)^3\right] \qquad (8.18c)$$

4   Substitute the boundary conditions $\theta = 0$, $s = 0$ into equation 8.18a:

$$0 = C_0 + 0 + 0 + \cdots$$
$$C_0 = 0 \qquad (8.19a)$$

5   Substitute $\theta = 0$, $v = 0$ into equation 8.18b:

$$0 = \frac{1}{\beta}\left[C_1 + 0 + 0 + \cdots\right]$$
$$C_1 = 0 \qquad (8.19b)$$

6   Substitute $\theta = 0$, $a = 0$ into equation 8.18c:

$$0 = \frac{1}{\beta^2}\left[C_2 + 0 + 0 + \cdots\right]$$
$$C_2 = 0 \qquad (8.19c)$$

7   Substitute $\theta = \beta$, $s = h$ into equation 8.18a:

$$h = C_3 + C_4 + C_5 \qquad (8.19d)$$

8   Substitute $\theta = \beta$, $v = 0$ into equation 8.18b:

$$0 = \frac{1}{\beta}\left[3C_3 + 4C_4 + 5C_5\right] \qquad (8.19e)$$

9   Substitute $\theta = \beta$, $a = 0$ into equation 8.18c:

$$0 = \frac{1}{\beta^2}\left[6C_3 + 12C_4 + 20C_5\right] \qquad (8.19f)$$

10 Three of our unknowns are found to be zero, leaving three unknowns to be solved for, C_3, C_4, C_5. Equations 8.19d, e, and f can be solved simultaneously to get:

$$C_3 = 10h; \qquad C_4 = -15h; \qquad C_5 = 6h \qquad (8.19g)$$

11 The equation for this cam design's displacement is then:

$$s = h\left[10\left(\frac{\theta}{\beta}\right)^3 - 15\left(\frac{\theta}{\beta}\right)^4 + 6\left(\frac{\theta}{\beta}\right)^5\right] \qquad (8.20)$$

12 The expressions for velocity and acceleration can be obtained by substituting the values of C_3, C_4, and C_5 into equations 8.18b and c. This function is referred to as the **3-4-5 polynomial**, after its exponents. Open the file E08-07.cam in program DYNACAM to investigate this example in more detail.

Figure 8-26 shows the resulting *s v a j* diagrams for a **3-4-5 polynomial rise** function from program DYNACAM. Note that the acceleration is continuous but the jerk is not, because we did not place any constraints on the boundary values of the jerk function. It is also interesting to note that the acceleration waveform looks very similar to the sinusoidal acceleration of the cycloidal function in Figure 8-12 (p. 360). Figure 8-18 (p. 370) shows the relative peak accelerations of this 3-4-5 polynomial compared to four other functions with the same *h* and β. Table 8-2 (p. 370) lists factors for the maximum velocity, acceleration and jerk of these functions.

THE 4-5-6-7 POLYNOMIAL   We left the jerk unconstrained in the previous example. We will now redesign the cam for the same specifications but will also constrain the jerk function to be zero at both ends of the rise. It will then match the dwells in the jerk

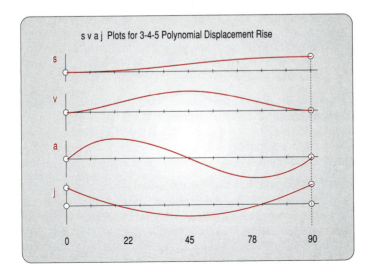

FIGURE 8-26

3-4-5 polynomial rise. Its acceleration is very similar to the sinusoid of cycloidal motion

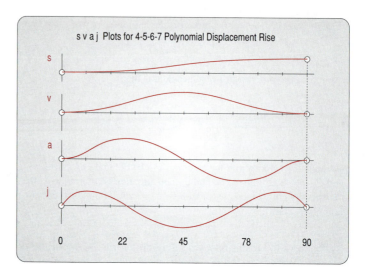

FIGURE 8-27

4-5-6-7 polynomial rise. Its jerk is piecewise continuous with the dwells

function with no discontinuities. This gives eight boundary conditions and yields a seventh-degree polynomial. The solution procedure to find the eight unknown coefficients is identical to that used in the previous example. Write the polynomial with the appropriate number of terms. Differentiate it to get expressions for all orders of boundary conditions. Substitute the boundary conditions and solve the resulting set of simultaneous equations.[*] This problem reduces to four equations in four unknowns, as the coefficients $C_0, C_1, C_2,$ and C_3 turn out to be zero. For this set of boundary conditions the displacement equation for the rise is:

$$s = h\left[35\left(\frac{\theta}{\beta}\right)^4 - 84\left(\frac{\theta}{\beta}\right)^5 + 70\left(\frac{\theta}{\beta}\right)^6 - 20\left(\frac{\theta}{\beta}\right)^7 \right] \tag{8.21}$$

This is known as the **4-5-6-7 polynomial**, after its exponents. Figure 8-27 shows the *s v a j* diagrams for this function. Compare them to the 3-4-5 polynomial functions shown in Figure 8-26 (p. 381). Note that the acceleration of the 4-5-6-7 starts off slowly, with zero slope (as we demanded with our zero jerk BC), and as a result goes to a larger peak value of acceleration in order to replace the missing area in the leading edge.

This **4-5-6-7 polynomial** function has the advantage of smoother jerk for better vibration control, compared to the **3-4-5 polynomial**, the **cycloidal**, and all other functions so far discussed (except the double harmonic), but it pays a stiff price in the form of significantly higher acceleration than all those functions. See also Table 8-2 (p. 370).

Single-Dwell Applications of Polynomials

Let us return to the single-dwell example used in the previous section and attempt to solve it with a polynomial cam function. Restating the original problem for reference:

[*] Any matrix solving calculator, equation solver such as *Matlab, Mathcad,* or *TKSolver,* or programs MATRIX and DYNACAM (supplied with this text) will do the simultaneous equation solution for you. Programs MATRIX and DYNACAM are discussed in Appendix A. You need only to supply the desired boundary conditions to DYNACAM and the coefficients will be computed. The reader is encouraged to do so and examine the example problems presented here with the DYNACAM program.

rise	1 in (25 mm) in 90°
fall	1 in (25 mm) in 90°
dwell	at zero displacement for 18° (low dwell)
cam ω	15 rad/sec

To solve this with a polynomial we must decide on a suitable set of boundary conditions. We must also first decide how many segments to divide the cam cycle into. The problem statement seems to imply three segments, a rise, a fall, and a dwell. We could use those three segments to create the functions as we did in the two previous examples, but a better approach is to use only **two segments**, one for the rise-fall combined and one for the dwell. *As a general rule we would like to minimize the number of segments in our polynomial cam functions.* Any dwell requires its own segment. So, the minimum number possible in this case is two segments.

Another rule of thumb is that *we would like to minimize the number of boundary conditions specified*, because the degree of the polynomial is tied to the number of BCs. As the degree of the function increases, so will the number of its **inflection points** and its number of **minima and maxima**. The polynomial derivation process will guarantee that the function will pass through all specified BCs but says nothing about the function's behavior between the BCs. *A high-degree function may have undesirable oscillations between its BCs.*

With these assumptions we can select a set of boundary conditions for a trial solution. First we will restate the problem to reflect our two-segment configuration.

✎ EXAMPLE 8-8

Designing a Polynomial for the Single-Dwell Case.

Problem: Redefine the CEP specification from Examples 8-5 and 8-6.

rise-fall	1 in (25 mm) in 90° and fall 1 in (25 mm) in 90° for a total of 180°
dwell	at zero displacement for 180° (low dwell)
cam ω	15 rad/sec

Solution:

1   Figure 8-28 shows the minimum set of seven BCs for this problem which will give a sixth-degree polynomial. The dwell on either side of the combined rise-fall segment has zero values of s, v, a, and j. The fundamental law of cam design requires that we match these zero values, through the acceleration function, at each end of the rise-fall segment.

2   These then account for six BCs; s, v, $a = 0$ at each end of the rise-fall segment.

3   We also must specify a value of displacement at the 1-in peak of the rise which occurs at θ = 90°. This is the seventh BC.

4   Figure 8-28 also shows the coefficients of the displacement polynomial which result from the simultaneous solution of the equations for the chosen BCs. For generality we have substituted the variable h for the specified 1-in rise. The function turns out to be a 3-4-5-6 polynomial whose equation is:

Segment number	Function used	Start angle	End angle	Delta angle
1	Poly 6	0	180	180

Boundary Conditions Imposed				Equation Resulting	
Function	Theta	% Beta	Boundary Cond.	Exponent	Coefficient
Displ	0	0	0	0	0
Veloc	0	0	0	1	0
Accel	0	0	0	2	0
Displ	180	1	0	3	64
Veloc	180	1	0	4	− 192
Accel	180	1	0	5	192
Displ	90	0.5	1	6	− 64

FIGURE 8-28

Boundary conditions and coefficients for a single-dwell polynomial application

$$s = h \left[64 \left(\frac{\theta}{\beta} \right)^3 - 192 \left(\frac{\theta}{\beta} \right)^4 + 192 \left(\frac{\theta}{\beta} \right)^5 - 64 \left(\frac{\theta}{\beta} \right)^6 \right] \qquad (8.22)$$

Figure 8-29 shows the $s\ v\ a\ j$ diagrams for this solution with its maximum values noted. Compare these acceleration and $s\ v\ a\ j$ curves to the double harmonic and cycloidal solutions to the same problem in Section 8.4 (Figures 8-23, p. 376, and 8-24, p. 377). Note that this sixth-degree polynomial function is as smooth as the double harmonic functions (Figure 8-24) and does not unnecessarily return the acceleration to zero at the top of the rise as does the cycloidal (Figure 8-23). The polynomial has a peak acceleration of 547 in/sec^2, which is less than that of either the cycloidal or double harmonic solution. This 3-4-5-6 polynomial is a superior solution to either of those presented for the same problem in Section 8.4 and is an example of how polynomial functions can be easily tailored to particular design specifications. The reader may open the file E08-08.cam in program DYNACAM to investigate this example in more detail.

SUMMARY This section has presented polynomial functions as the most versatile approach of those shown to virtually any cam design problems. It is only since the development and general availability of computers that these functions have become practical to use, as the computation to solve the simultaneous equations is often beyond hand calculation abilities. With the availability of a design aid to solve the equations such as program DYNACAM, polynomials have become a practical and preferable way to solve many cam design problems. **Spline functions**, of which polynomials are a subset, offer even more flexibility in meeting boundary constraints and other cam performance criteria.[5] [7] Space does not permit a detailed exposition of spline functions as applied to cam systems here. See the references for more information.

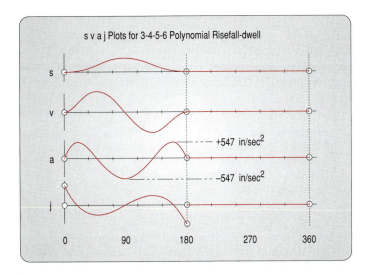

FIGURE 8-29

Sixth-degree 3-4-5-6 polynomial function for two-segment risefall, single-dwell cam

8.6   CRITICAL PATH MOTION (CPM)

Probably the most common application of **critical path motion** (CPM) specifications in production machinery design is the need for **constant velocity motion**. There are two general types of automated production machinery in common use, **intermittent motion** assembly machines and **continuous motion** assembly machines.

Intermittent motion assembly machines carry the manufactured goods from work station to work station, stopping the workpiece or subassembly at each station while another operation is performed upon it. The throughput speed of this type of automated production machine is typically limited by the dynamic forces which are due to accelerations and decelerations of the mass of the moving parts of the machine and its workpieces. The workpiece motion may be either in a straight line as on a conveyor or in a circle as on a rotary table as shown in Figure 8-21 (p.372).

Continuous motion assembly machines never allow the workpiece to stop and thus are capable of higher throughput speeds. All operations are performed on a moving target. Any tools which operate on the product have to "chase" the moving assembly line to do their job. Since the assembly line (often a conveyor belt or chain, or a rotary table) is moving at some constant velocity, there is a need for mechanisms to provide constant velocity motion, matched exactly to the conveyor, in order to carry the tools alongside for a long enough time to do their job. These cam driven "chaser" mechanisms must then return the tool quickly to its start position in time to meet the next part or subassembly on the conveyor (quick-return). There is a motivation in manufacturing to convert from intermittent motion machines to continuous motion in order to increase production rates. Thus there is considerable demand for this type of constant velocity mechanism. The cam-follower system is well suited to this problem, and the polynomial cam function is particularly adaptable to the task.

Polynomials Used for Critical Path Motion

EXAMPLE 8-9

Designing a Polynomial for Constant Velocity Critical Path Motion.

Problem: Consider the following statement of a critical path motion (CPM) problem:

Accelerate   the follower from zero to 10 in/sec
Maintain   a constant velocity of 10 in/sec for 0.5 sec
Decelerate   the follower to zero velocity
Return   the follower to start position
Cycle time   exactly 1 sec

Solution:

1   This unstructured problem statement is typical of real design problems as was discussed in Chapter 1. No information is given as to the means to be used to accelerate or decelerate the follower or even as to the portions of the available time to be used for those tasks. A little reflection will cause the engineer to recognize that the specification on total cycle time in effect defines the camshaft velocity to be its reciprocal or **one revolution per second**. Converted to appropriate units, this is an angular velocity of 2π rad/sec.

2   The constant velocity portion uses half of the total period of 1 sec in this example. The designer must next decide how much of the remaining 0.5 sec to devote to each other phase of the required motion.

3   The problem statement seems to imply that four segments are needed. Note that the designer has to somewhat arbitrarily select the lengths of the individual segments (except the constant velocity one). Some iteration may be required to optimize the result. Program DYNA-CAM makes the iteration process quick and easy, however.

4   Assuming four segments, the timing diagram in Figure 8-30 shows an acceleration phase, a constant velocity phase, a deceleration phase, and a return phase, labeled as segments 1 through 4.

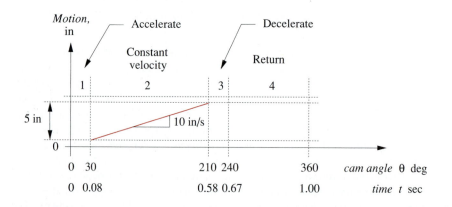

FIGURE 8-30

Constant velocity cam timing diagram

5 The segment angles (β's) are assumed, for a first approximation, to be 30° for segment 1, 180° for segment 2, 30° for segment 3, and 120° for segment 4 as shown in Figure 8-31. These angles may need to be adjusted in later iterations, except for segment 2 which is rigidly constrained in the specifications.

6 Figure 8-31 shows a tentative *s v a j* diagram. The solid circles indicate a set of boundary conditions which will constrain the continuous function to these specifications. These are for segment 1:

$$\text{when } \theta = 0°; \qquad s = 0, \qquad v = 0, \qquad none$$
$$\text{when } \theta = 30°; \qquad none, \qquad v = 10, \qquad a = 0 \qquad (a)$$

7 Note that the displacement at θ = 30° is left unspecified. The resulting polynomial function will provide us with the values of displacement at that point, which can then be used as a

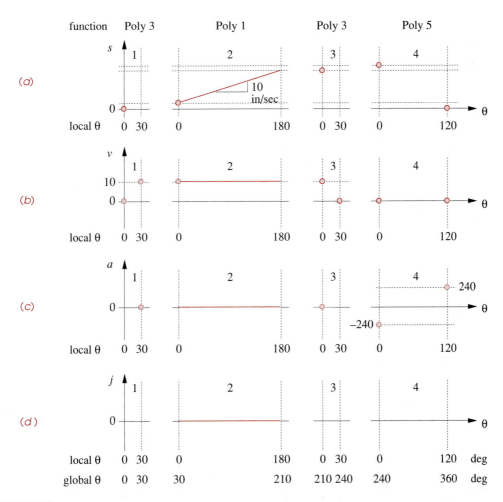

FIGURE 8-31

A possible set of boundary conditions for the four-segment constant velocity solution

boundary condition for the next segment, in order to make the overall functions continuous as required. The acceleration at $\theta = 30°$ must be zero in order to match that of the constant velocity segment 2. The acceleration at $\theta = 0$ is left unspecified. The resulting value will be used later to match the end of the last segment's acceleration.

8   Putting these four BCs for segment 1 into program DYNACAM yields a cubic function whose $s\,v\,a\,j$ plots are shown in Figure 8-32. Its equation is:

$$s = 0.83376\left(\frac{\theta}{\beta}\right)^{2} - 0.27792\left(\frac{\theta}{\beta}\right)^{3} \qquad (8.23a)$$

The maximum displacement occurs at $\theta = 30°$. This will be used as one BC for segment 2. The entire set for segment 2 is:

$$\text{when} \quad \theta = 30°; \qquad s = 0.556, \qquad v = 10$$
$$\text{when} \quad \theta = 210°; \qquad none, \qquad none \qquad (b)$$

9   Note that in the derivations and in the DYNACAM program each segment's local angles run from zero to the β for that segment. Thus, segment 2's local angles are 0° to 180°, which correspond to 30° to 210° globally in this example. We have left the displacement, velocity, and acceleration at the end of segment 2 unspecified. They will be determined by the computation.

10   Since this is a constant velocity segment, its integral, the displacement function, must be a polynomial of degree one, i.e., a straight line. If we specify more than two BCs we will get a function of higher degree than one which will pass through the specified endpoints but may also oscillate between them and deviate from the desired constant velocity. Thus we can *only* provide two BCs, a slope and an intercept, as defined in equation 8.2 (p. 355). But, we must

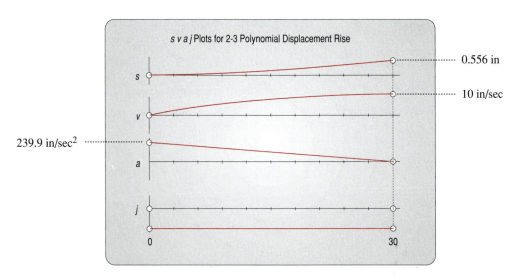

FIGURE 8-32

Segment one for the four-segment solution to the constant velocity problem provide at least one displacement boundary condition in order to compute the coefficient C_0 from equation 8.16 (p. 378). Specifying the two BCs at only one end of the interval is perfectly acceptable. The equation for segment 2 is:

$$s = 5\left(\frac{\theta}{\beta}\right) + 0.556 \qquad (8.23b)$$

11 Figure 8-33 shows the displacement and velocity plots of segment 2. The acceleration and jerk are both zero. The resulting displacement at $\theta = 210°$ is 5.556.

12 The displacement at the end of segment 2 is now known from its equation. The four boundary conditions for segment 3 are then:

$$\text{when} \quad \theta = 210°; \qquad s = 5.556, \qquad v = 10, \qquad a = 0$$
$$\text{when} \quad \theta = 240°; \qquad none, \qquad v = 0, \qquad none \qquad (c)$$

13 This generates a cubic displacement function as shown in Figure 8-34. Its equation is:

$$s = -0.27792\left(\frac{\theta}{\beta}\right)^3 + 0.83376\left(\frac{\theta}{\beta}\right) + 5.556 \qquad (8.23c)$$

14 The boundary conditions for the last segment 4 are now defined, as they must match those of the end of segment 3 and the beginning of segment 1. The displacement at the end of segment 3 is found from the computation in DYNACAM to be $s = 6.112$ at $\theta = 240°$ and the acceleration at that point is −239.9. We left the acceleration at the beginning of segment 1 unspecified. From the second derivative of the equation for displacement in that segment we find that the acceleration is 239.9 at $\theta = 0°$. The BCs for segment 4 are then:

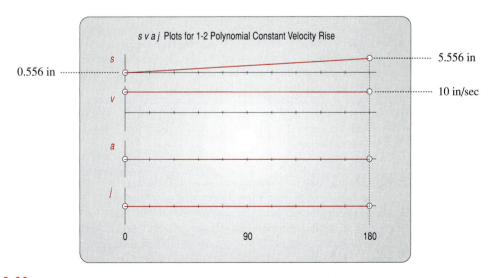

FIGURE 8-33

Segment two for the four-segment solution to the constant velocity problem

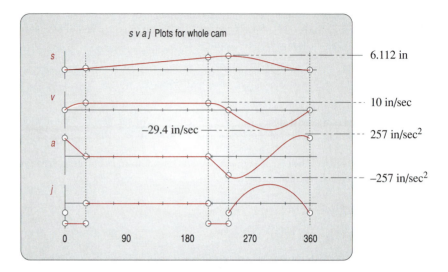

FIGURE 8-34

Four-segment solution to the constant velocity problem showing maximum values

$$
\begin{array}{llll}
\text{when} \quad \theta = 240°; & s = 6.112, & v = 0, & a = -239.9 \\
\text{when} \quad \theta = 360°; & s = 0, & v = 0, & a = 239.9
\end{array}
\qquad (d)
$$

15   The equation for segment 4 is then:

$$
s = -9.9894\left(\frac{\theta}{\beta}\right)^5 + 24.9735\left(\frac{\theta}{\beta}\right)^4 - 7.7548\left(\frac{\theta}{\beta}\right)^3 - 13.3413\left(\frac{\theta}{\beta}\right)^2 + 6.112 \qquad (8.23d)
$$

16   Figure 8-34 shows the *s v a j* plots for the complete cam. It obeys the fundamental law of cam design because the piecewise functions are continuous through the acceleration. The maximum value of acceleration is 257 in/sec². The maximum negative velocity is −29.4 in/sec. We now have four piecewise and continuous functions, equations 8.23, which will meet the performance specifications for this problem.

The reader may open the file E08-09.cam in program DYNACAM to investigate this example in more detail.

While this design is acceptable, it can be improved. One useful strategy in designing polynomial cams is to minimize the number of segments, provided that this does not result in functions of such high degree that they misbehave between boundary conditions. Another strategy is to always start with the segment for which you have the most information. In this example, the constant velocity portion is the most constrained and must be a separate segment, just as a dwell must be a separate segment. The rest of the cam motion exists only to return the follower to the constant velocity segment for the next cycle. If we start by designing the constant velocity segment, it may be possible to complete the cam with only one additional segment. We will now redesign this cam, to the same specifications but with only two segments as shown in Figure 8-35.

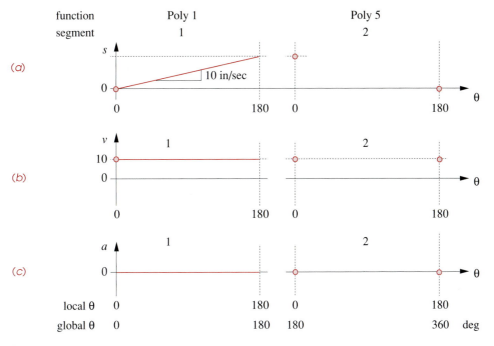

FIGURE 8-35

Boundary conditions for the two-segment constant velocity solution

✏️ **EXAMPLE 8-10**

Designing an Optimum Polynomial for Constant Velocity Critical Path Motion.

Problem: Redefine the problem statement of Example 8-9 to have only two segments.

Maintain   a constant velocity of 10 in/sec for 0.5 sec
Decelerate   and **accelerate** follower to constant velocity
Cycle time   exactly 1 sec

Solution:

1   The BCs for the first, constant velocity, segment will be similar to our previous solution except for the global values of its angles and the fact that we will start at zero displacement. They are:

$$\text{when} \quad \theta = 0°; \qquad s = 0, \qquad v = 10$$
$$\text{when} \quad \theta = 180°; \qquad none, \qquad none \qquad (a)$$

2   The displacement and velocity plots for this segment are identical to those in Figure 8-33 (p. 389) except that the displacement starts at zero. The equation for segment 1 is:

$$s = 5\left(\frac{\theta}{\beta}\right) \qquad (8.24a)$$

3   The program calculates the displacement at the end of segment 1 to be 5.00 in. This defines
     that BC for segment 2. The set of BCs for segment 2 is then:

$$\text{when} \quad \theta = 180°; \qquad s = 5.00, \qquad v = 10, \qquad a = 0$$
$$\text{when} \quad \theta = 360°; \qquad s = 0, \qquad v = 10, \qquad a = 0$$

(b)

The equation for segment 2 is:

$$s = -60\left(\frac{\theta}{\beta}\right)^5 + 150\left(\frac{\theta}{\beta}\right)^4 - 100\left(\frac{\theta}{\beta}\right)^3 + 5\left(\frac{\theta}{\beta}\right)^1 + 5 \qquad (8.24b)$$

4   The *s v a j* diagrams for this design are shown in Figure 8-36. Note that they are much
     smoother than the four-segment design. The maximum acceleration in this example is now
     230 in/sec², and the maximum negative velocity is –27.5 in/sec. These are both less than in
     the previous design of Example 8-9.

5   The fact that our displacement in this design contains negative values as shown in the *s* dia-
     gram of Figure 8-36 is of no concern. This is due to our starting with the beginning of the
     constant velocity portion as zero displacement. The follower has to go to a negative position
     in order to have distance to accelerate up to speed again. We will simply shift the displace-
     ment coordinates by that negative amount to make the cam. To do this, simply calculate the
     displacement coordinates for the cam. Note the value of the largest negative displacement.
     Add this value to the displacement boundary conditions for all segments and recalculate the
     cam functions with DYNACAM. (Do not change the BCs for the higher derivatives.) The fin-
     ished cam's displacement profile will be shifted up such that its minimum value will now be
     zero.

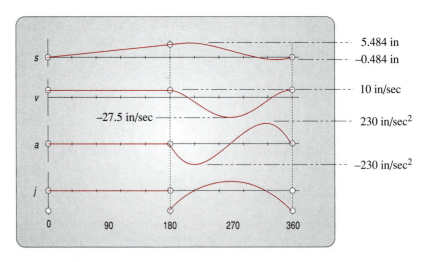

FIGURE 8-36

Two-segment solution to the constant velocity problem showing maximum values

So, not only do we now have a smoother cam but the dynamic forces and stored kinetic energy are both lower. Note that we did not have to make any assumptions about the portions of the available nonconstant velocity time to be devoted to speeding up or slowing down. This all happened automatically from our choice of only two segments and the specification of the minimum set of necessary boundary conditions. This is clearly a superior design to the previous attempt and is in fact an optimal polynomial solution to the given specifications. The reader is encouraged to read the file E08-10.cam into program DYNACAM to investigate this example in more detail.

Half-Period Harmonic Family Functions

The full-rise cycloidal and the full-rise modified sine functions are generally suited only to the double-dwell cases as they have zero acceleration at each end. However, pieces of these functions can be used to match other functions such as constant velocity segments in a similar fashion to that used to piece together the modified sine from two harmonics of different frequency. The full-rise functions mentioned above (except simple harmonic) contain one complete period in their velocity and acceleration. The simple harmonic (Figure 8-9, p. 360) does not have zero acceleration at its ends, but the modified sine's and cycloid's accelerations (Figure 8-12, p. 358) do start and end at zero. In order to match a nonzero velocity as in the example above, we could use half of any of these harmonic family functions and design them to match the desired constant velocity of the adjacent segment.

As an example of this approach, Figure 8-37 shows the *s v a j* functions for a **half-cycloid** rise function #1 which has zero velocity at the beginning of the interval and nonzero velocity at the end. Note that its displacement starts at zero and ends at some positive value, but its acceleration is zero at both extremes. This makes it possible to mate this function to a constant velocity segment and match both the desired velocity and its zero acceleration at the boundary. The total displacement required of the half-cycloid will "come out in the wash" when the required boundary conditions of velocity and duration are applied to the particular case. The equations for this half-cycloid #1 are:

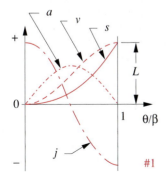
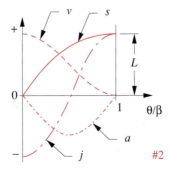

FIGURE 8-37

Half-cycloidal functions for use on a rise segment

$$s = L\left[\frac{\theta}{\beta} - \frac{1}{\pi}\sin\left(\pi\frac{\theta}{\beta}\right)\right] \tag{8.25a}$$

$$v = \frac{L}{\beta}\left[1 - \cos\left(\pi\frac{\theta}{\beta}\right)\right] \tag{8.25b}$$

$$a = \frac{\pi L}{\beta^2}\sin\left(\pi\frac{\theta}{\beta}\right) \tag{8.25c}$$

$$j = \frac{\pi^2 L}{\beta^3}\cos\left(\pi\frac{\theta}{\beta}\right) \tag{8.25d}$$

Figure 8-37 also shows the *s v a j* functions for the **half-cycloid** rise function #2 with nonzero velocity at the beginning of the interval and zero velocity at the end. The equations for this half-cycloid #2 are:

$$s = L\left[\frac{\theta}{\beta} + \frac{1}{\pi}\sin\left(\pi\frac{\theta}{\beta}\right)\right] \tag{8.26a}$$

$$v = \frac{L}{\beta}\left[1 + \cos\left(\pi\frac{\theta}{\beta}\right)\right] \tag{8.26b}$$

$$a = -\frac{\pi L}{\beta^2}\sin\left(\pi\frac{\theta}{\beta}\right) \tag{8.26c}$$

$$j = -\frac{\pi^2 L}{\beta^3}\cos\left(\pi\frac{\theta}{\beta}\right) \tag{8.26d}$$

For a fall instead of a rise, subtract the rise displacement expressions from the total rise *L* and negate all the higher derivatives.

To fit these functions to a particular constant velocity situation, solve either equation 8.25b or 8.26b (depending on which function is desired) for the value of *L* which results from the specification of the known constant velocity *v* to be matched at $\theta = \beta$ or $\theta = 0$. You will have to choose a value of β for the interval of this half-cycloid which is appropriate to the problem. In our example above, the value of $\beta = 30°$ used for the first segment of the four-piece polynomial could be tried as a first iteration. Once *L* and β are known, all the functions are defined.

The same approach can be taken with the **modified sine** and the **simple harmonic** functions. Either half of their full-rise functions can be sized to match with a constant velocity segment. The half-modified sine function mated with a constant velocity segment has the advantage of low peak velocity, useful with large inertia loads. When matched to a constant velocity, the half simple harmonic has the same disadvantage of infinite jerk as its full-rise counterpart does when matched to a dwell, so it is not recommended.

We will now solve the previous constant velocity example problem using half-cycloid, constant velocity, and full-fall modified sine functions.

✍️ EXAMPLE 8-11

Using Half-Cycloids to Match Constant Velocity Critical Path Motion.

Problem: Consider the same problem statement as Example 8-9.

Accelerate	the follower from zero to 10 in/sec
Maintain	a constant velocity of 10 in/sec for 0.5 sec
Decelerate	the follower to zero velocity
Return	the follower to start position
Cycle time	exactly 1 sec

Solution:

1   We must express the specified constant velocity in units of length per rad. The angular velocity is 2π rad/sec.

$$v = 10\frac{\text{in}}{\text{sec}}\left(\frac{1}{2\pi}\frac{\text{sec}}{\text{rad}}\right) = \frac{5}{\pi}\frac{\text{in}}{\text{rad}} \qquad (a)$$

2   The constant velocity portion uses half of the total period of 1 sec, or π rad, in this example. The designer must decide how much of the remaining 0.5 sec to devote to each other phase of the required motion. The segment angles (β's) are assumed, for a first approximation, to be 25° for segment 1, 180° for segment 2, 25° for segment 3, and 130° for segment 4. These angles may need to be adjusted in later iterations to balance and minimize the accelerations (except for segment 2 which is rigidly constrained in the specifications).

3   The segments will consist of:

Segment	β (deg)	β (rad)	Function	Motion
1	25	0.43633	Half-cycloid #1	Rise
2	180	3.14159	1° polynomial	Rise
3	25	0.43633	Half-cycloid #2	Rise
4	130	2.26893	Modified sine	Fall

4   To determine the total rise L of the half-cycloid needed to match the specified constant velocity, solve equation 8.25b for L at $\theta = \beta$ where it must match the constant velocity v.

$$L = \frac{\beta v}{1 - \cos\left(\pi\dfrac{\theta}{\beta}\right)} = \frac{0.43633\left(\dfrac{5}{\pi}\right)}{1 - \cos\left(\pi\dfrac{\beta}{\beta}\right)} = 0.34722 \qquad (b)$$

5   Substitute this value of L in equation 8.25a to get the displacement for the first segment:

$$s = 0.3472\left[\frac{\theta}{\beta} - \frac{1}{\pi}\sin\left(\pi\frac{\theta}{\beta}\right)\right] \qquad (c)$$

6   The constant velocity segment is found in the same way as was done in Example 8-9 (p. 386). The initial displacement for segment 2 in this case is the value of L, and the equation for segment 2 is:

$$s = 5\left(\frac{\theta}{\beta}\right) + 0.3472 \qquad\qquad (d)$$

The total lift within this segment is 5 in as before.

7   Segment 3 is a half-cycloid #2. Its coefficient L is identical to that of segment 1 because we used the same β for both. But, we must offset it by the sum of the displacement of segments 2 and 3 or $L + 5$. Program DYNACAM provides for the specification of this offset. The lift for this segment is 0.3472 and the offset is 5.3472. The equation for segment 3 (from equation 8.26a) is then:

$$s = 0.3472\left[\frac{\theta}{\beta} + \frac{1}{\pi}\sin\left(\pi\frac{\theta}{\beta}\right)\right] + 5.3472 \qquad\qquad (e)$$

8   Segment 4 is a full-period modified sine to return the follower from its maximum displacement of $h = 0.3472 + 5.0 + 0.3472 = 5.6944$. See equation 8.14 (p. 369).

9   The complete set of data needed to compute these functions in (or out of) DYNACAM is:

Segment	β (deg)	β (rad)	Function	Start (in)	Motion	Move (in)
1	25	0.43633	Half-cycloid #1	0	Rise	0.3472
2	180	3.14159	1° polynomial	0.3472	Rise	5.0000
3	25	0.43633	Half-cycloid #2	5.3472	Rise	0.3472
4	130	2.26893	Modified sine	5.6944	Fall	5.6944

10   The resulting $s\,v\,a\,j$ diagrams are shown in Figure 8-38. The peak acceleration is 241 in/sec^2 and peak velocity is -28 in/sec.

These results are nearly as low as the values from the two-segment polynomial solution in Example 8-10 (p. 391). The factor that makes this an inferior cam design to Example 8-10 is the unnecessary returns to zero in the acceleration waveform. This creates a more "ragged" jerk function which will increase vibration problems. The polynomial approach is superior to the other solutions presented in this case as it often is in cam design. The reader may open the file E08-11.cam in program DYNACAM to investigate this example in more detail.

8.7   SIZING THE CAM—PRESSURE ANGLE AND RADIUS OF CURVATURE

Once the $s\,v\,a\,j$ functions have been defined, the next step is to size the cam. There are two major factors which affect cam size, the **pressure angle** and the **radius of curvature**. Both of these involve either the **base circle radius** on the cam (R_b) when using flat-faced followers, or the **prime circle radius** on the cam (R_p) when using roller or curved followers.

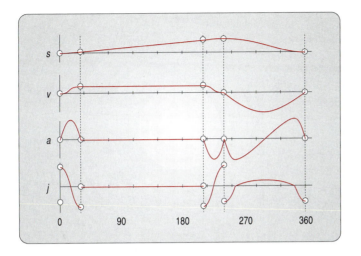

FIGURE 8-38

Half-cycloidal functions used as transitions to constant velocity (Example 8-11)

The base circle's and prime circle's centers are at the center of rotation of the cam. The base circle is defined as *the smallest circle which can be drawn tangent to the physical cam surface* as shown in Figure 8-39. All radial cams will have a base circle, regardless of the follower type used.

The prime circle is only applicable to cams with roller followers or radiused (mushroom) followers and is measured to the center of the follower. The **prime circle** is defined as *the smallest circle which can be drawn tangent to the locus of the centerline of the follower* as shown in Figure 8-39. *The locus of the centerline of the follower* is called the **pitch curve**. Cams with roller followers are in fact defined for manufacture with respect to the pitch curve rather than with respect to the cam's physical surface. Cams with flat-faced followers must be defined for manufacture with respect to their physical surface, as there is no pitch curve.

The process of creating the physical cam from the s diagram can be visualized conceptually by imagining the s diagram to be cut out of a flexible material such as rubber. The x axis of the s diagram represents the circumference of a circle, which could be either the **base circle**, or the **prime circle**, around which we will "wrap" our "rubber" s diagram. We are free to choose the initial length of our s diagram's x axis, though the height of the displacement curve is fixed by the cam displacement function we have chosen. In effect we will choose the base or prime circle radius as a design parameter and stretch the length of the s diagram's axis to fit the circumference of the chosen circle.

Pressure Angle—Roller Followers

The **pressure angle** is defined as shown in Figure 8-40. It is the complement of the transmission angle which was defined for linkages in previous chapters and has a similar meaning with respect to cam-follower operation. By convention, the pressure angle is used for cams, rather than the transmission angle. Force can only be transmitted from

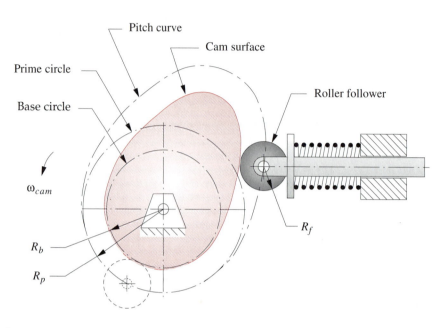

FIGURE 8-39

Base circle R_b, prime circle R_p, and pitch curve of a radial cam with roller follower

cam to follower or vice versa along the **axis of transmission** which is perpendicular to the **axis of slip**, or common tangent.

PRESSURE ANGLE   The **pressure angle** ϕ is *the angle between the direction of motion (velocity) of the follower and the direction of the axis of transmission.*[*] When $\phi = 0$, all the transmitted force goes into motion of the follower and none into slip velocity. When ϕ becomes 90° there will be no motion of the follower. As a rule of thumb, we would like the pressure angle to be between zero and about 30° for translating followers to avoid excessive side load on the sliding follower. If the follower is oscillating on a pivoted arm, a pressure angle up to about 35° is acceptable. Values of ϕ greater than this can increase the follower sliding or pivot friction to undesirable levels and may tend to jam a translating follower in its guides.

ECCENTRICITY   Figure 8-41 shows the geometry of a cam and translating roller follower in an arbitrary position. This shows the general case in that the axis of motion of the follower does not intersect the center of the cam. There is an **eccentricity** ϵ defined as *the perpendicular distance between the follower's axis of motion and the center of the cam.* Often this eccentricity ϵ will be zero, making it an **aligned follower**, which is the special case.

In the figure, the axis of transmission is extended to intersect effective link 1, which is the ground link. (See Section 8.0 and Figure 8-1, p. 347 for a discussion of effective links in cam systems.) This intersection is instant center $I_{2,4}$ (labeled B), which, by definition, has the same velocity in link 2 (the cam) and in link 4 (the follower). Because link 4 is in pure translation, all points on it have identical velocities $V_{follower}$, which are

[*] Dresner points out that this definition is only valid for single-degree-of-freedom systems. For multi-input systems, a more complicated definition and calculation of pressure angle (or transmission angle) is needed. For more information see Dresner, T. L., and K. W. Buffington. (1991). "Definition of Pressure and Transmission Angles Applicable to Multi-Input Mechanisms." *Journal of Mechanical Design,* **113**(4), p. 495.

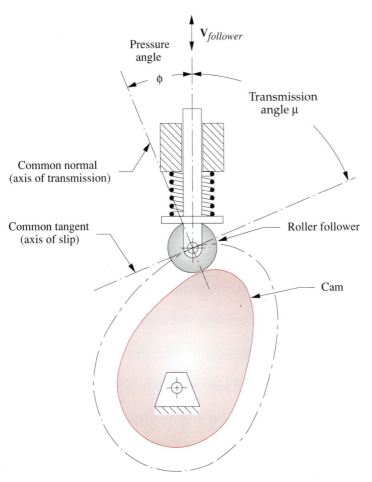

FIGURE 8-40

Cam pressure angle

equal to the velocity of $I_{2,4}$ in link 2. We can write an expression for the velocity of $I_{2,4}$ in terms of cam angular velocity and the radius b from cam center to $I_{2,4}$,

$$V_{I_{2,4}} = b\omega = \dot{s} \tag{8.27}$$

where s is the instantaneous displacement of the follower from the s diagram and s dot is its time derivative in units of length/sec. (Note that capital S V A J denote time-based variables rather than functions of cam angle.)

But $$\dot{s} = \frac{ds}{dt}$$

and $$\frac{ds}{dt}\frac{d\theta}{d\theta} = \frac{ds}{d\theta}\frac{d\theta}{dt} = \frac{ds}{d\theta}\omega = v\omega$$

so $$b\omega = v\omega$$

then $$b = v \tag{8.28}$$

This is an interesting relationship which says that the **distance b to the instant center I2,4 is numerically equal to the velocity of the follower** v in units of length per radian as derived in previous sections. We have reduced this expression to pure geometry, independent of the angular velocity ω of the cam.

Note that we can express the distance b in terms of the prime circle radius R_p and the eccentricity ε, by the construction shown in Figure 8-41. Swing the arc of radius R_p until it intersects the axis of motion of the follower at point D. This defines the length of line d from effective link 1 to this intersection. This is constant for any chosen prime circle radius R_p. Points A, C, and $I_{2,4}$ form a right triangle whose upper angle is the pressure angle ϕ and whose vertical leg is $(s + d)$, where s is the instantaneous displacement of the follower. From this triangle:

$$c = (s+d)\tan\phi$$

and

$$b = (s+d)\tan\phi + \varepsilon$$

(8.29a)

Then from equation 8.28,

$$v = (s+d)\tan\phi + \varepsilon$$

(8.29b)

and from triangle CDO_2,

$$d = \sqrt{R_P^2 - \varepsilon^2}$$

(8.29c)

Substituting equation 8.29c into equation 8.29b and solving for ϕ gives an expression for pressure angle in terms of displacement s, velocity v, eccentricity ε, and the prime circle radius R_p.

$$\phi = \arctan\frac{v - \varepsilon}{s + \sqrt{R_P^2 - \varepsilon^2}}$$

(8.29d)

The velocity v in this expression is in units of length/rad, and all other quantities are in compatible length units. We have typically defined s and v by this stage of the cam design process and wish to manipulate R_p and ε to get an acceptable maximum pressure angle ϕ. As R_p is increased, ϕ will be reduced. The only constraints against large values of R_p are the practical ones of package size and cost. Often there will be some upper limit on the size of the cam-follower package dictated by its surroundings. There will always be a cost constraint and bigger = heavier = more expensive.

Choosing a Prime Circle Radius

Both R_p and ε are within a transcendental expression in equation 8.29d, so they cannot be conveniently solved for directly. The simplest approach is to assume a trial value for R_p and an initial eccentricity of zero, and use program DYNACAM; your own program; or an equation solver such as *Matlab*, *TKSolver* or *Mathcad* to quickly calculate the values of ϕ for the entire cam, and then adjust R_p and repeat the calculation until an acceptable arrangement is found. Figure 8-42 shows the calculated pressure angles for a four-dwell cam. Note the similarity in shape to the velocity functions for the same cam in Figure 8-6 (p. 353), as that term is dominant in equation 8.29d.

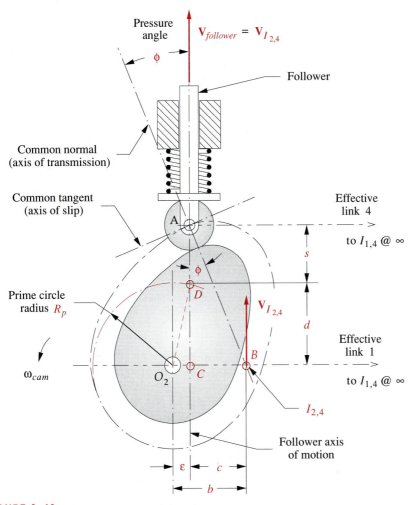

FIGURE 8-41

Geometry for the derivation of the equation for pressure angle

USING ECCENTRICITY   If a suitably small cam cannot be obtained with acceptable pressure angle, then eccentricity can be introduced to change the pressure angle. Using eccentricity to control the pressure angle has its limitations. For a positive ω, a positive value of eccentricity will *decrease the pressure angle on the rise* but will *increase it on the fall*. Negative eccentricity does the reverse.

This is of little value with a form-closed (groove or track) cam, as it is driving the follower in both directions. For a force-closed cam with spring return, you can sometimes afford to have a larger pressure angle on the fall than on the rise because the stored energy in the spring is attempting to speed up the camshaft on the fall, whereas the cam is storing that energy in the spring on the rise. The limit of this technique can be the degree of overspeed attained with a larger pressure angle on the fall. The resulting variations in cam angular velocity may be unacceptable.

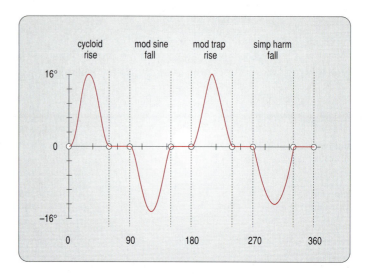

FIGURE 8-42

Pressure angle functions are similar in shape to velocity functions (See Figure 8-6, p. 353)

The most value gained from adding eccentricity to a follower comes in situations where the cam program is asymmetrical and significant differences exist (with no eccentricity) between maximum pressure angles on rise and fall. Introducing eccentricity can balance the pressure angles in this situation and create a smoother running cam.

If adjustments to R_p or ε do not yield acceptable pressure angles, the only recourse is to return to an earlier stage in the design process and redefine the problem. Less lift or more time to rise or fall will reduce the causes of the large pressure angle. Design is, after all, an iterative process.

Overturning Moment—Flat-Faced Follower

Figure 8-43 shows a translating, flat-faced follower running against a radial cam. The pressure angle can be seen to be zero for all positions of cam and follower. This seems to be giving us something for nothing, which can't be true. As the contact point moves left and right, the point of application of the force between cam and follower moves with it. There is an overturning moment on the follower associated with this off-center force which tends to jam the follower in its guides, just as did too large a pressure angle in the roller follower case. In this case, we would like to keep the cam as small as possible in order to minimize the moment arm of the force. Eccentricity will affect the average value of the moment, but the peak-to-peak variation of the moment about that average is unaffected by eccentricity. Considerations of too-large pressure angle do not limit the size of this cam, but other factors do. The minimum radius of curvature (see below) of the cam surface must be kept large enough to avoid undercutting. This is true regardless of the type of follower used.

Radius of Curvature—Roller Follower

The **radius of curvature** is a *mathematical property of a function*. Its value and use is not limited to cams but has great significance in their design. The concept is simple. No matter how complicated a curve's shape may be, nor how high the degree of the function which describes it, it will have an instantaneous radius of curvature at every point on the curve. These radii of curvature will have instantaneous centers (which may be at infinity), and the radius of curvature of any function is itself a function which can be computed and plotted. For example, the radius of curvature of a straight line is infinity everywhere; that of a circle is a constant value. A parabola has a constantly changing radius of curvature which approaches infinity along the parabola's asymptotes. A cubic curve will have radii of curvature that are sometimes positive (convex) and sometimes negative (concave). The higher the degree of a function, in general, the more potential variety in its radius of curvature.

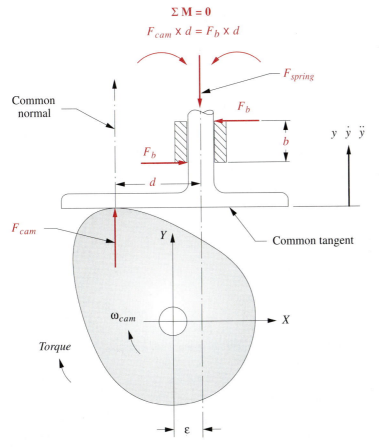

8

FIGURE 8-43

Overturning moment on a flat-faced follower

Cam contours are usually functions of high degree. When they are wrapped around their base or prime circles, they may have portions which are concave, convex, or flat. Infinitesimally short flats of infinite radius will occur at all inflection points on the cam surface where it changes from concave to convex or vice versa.

The radius of curvature of the finished cam is of concern regardless of the follower type, but the concerns are different for different followers. Figure 8-44 shows an obvious problem with a roller follower whose own (constant) radius of curvature R_f is too large to follow the locally smaller concave (negative) radius $-\rho$ on the cam.

A more subtle problem occurs when the roller follower radius R_f is larger than the smallest positive (convex) local radius $+\rho$ on the cam. This problem is called **undercutting** and is depicted in Figure 8-45. Recall that for a roller follower cam, the cam contour is actually defined as the locus of the center of the roller follower, or the **pitch curve**. The machinist is given these x,y coordinate data (on computer tape or disk) and also told the radius of the follower R_f. The machinist will then cut the cam with a cutter of the same effective radius as the follower, following the pitch curve coordinates with the center of the cutter.

Figure 8-45a shows the situation in which the follower (cutter) radius R_f is at one point exactly equal to the minimum convex radius of curvature of the cam $(+\rho_{min})$. The cutter creates a perfect sharp point, or **cusp**, on the cam surface. This cam will not run very well at speed! Figure 8-45b shows the situation in which the follower (cutter) radius is greater than the minimum convex radius of curvature of the cam. The cutter now undercuts or removes material needed for cam contours in different locations and also creates a sharp point or cusp on the cam surface. This cam no longer has the same displacement function you so carefully designed.

The rule of thumb is to keep the absolute value of the minimum radius of curvature ρ_{min} of the cam pitch curve preferably at least 2 to 3 times as large as the radius of the roller follower R_f.

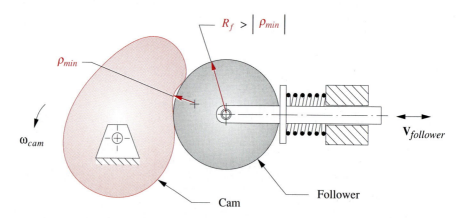

FIGURE 8-44

The result of using a roller follower larger than the one for which the cam was designed

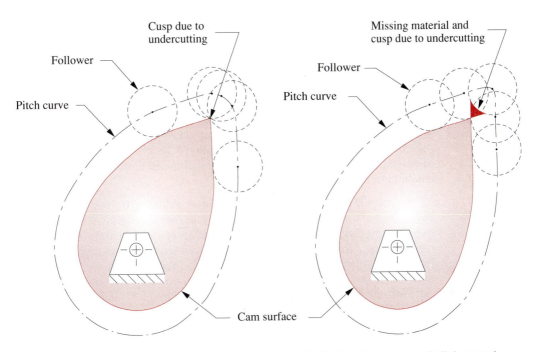

Cusp due to
undercutting

Follower

Pitch curve

Missing material and
cusp due to undercutting

Follower

Pitch curve

Cam surface

(*a*) Radius of curvature of pitch curve
equals the radius of the roller follower

(*b*) Radius of curvature of pitch curve is
less than the radius of the roller follower

8

FIGURE 8-45

Small positive radius of curvature can cause undercutting

$$|\rho_{min}| \gg R_f \tag{8.30}$$

A derivation for radius of curvature can be found in any calculus text. For our case of a roller follower, we can write the equation for the radius of curvature of the pitch curve of the cam as:

$$\rho_{pitch} = \frac{\left[(R_P + s)^2 + v^2\right]^{3/2}}{(R_P + s)^2 + 2v^2 - a(R_P + s)} \tag{8.31}$$

In this expression, s, v, and a are the displacement, velocity, and acceleration of the cam program as defined in a previous section. Their units are length, length/rad, and length/rad^2, respectively. R_p is the prime circle radius. **Do not confuse** this *prime circle radius* R_p with the *radius of curvature*, ρ_{pitch}. R_p is a **constant value** which you choose as a design parameter and ρ_{pitch} is the constantly changing radius of curvature which results from your design choices.

Also do not confuse R_p, the *prime circle radius* with R_f, the *radius of the roller follower*. See Figure 8-39 (p. 398) for definitions. You can choose the value of R_f to suit the problem, so you might think that it is simple to satisfy equation 8.30 by just selecting a roller follower with a small value of R_f. Unfortunately it is more complicated than that, as a small roller follower may not be strong enough to withstand the dynamic forces from the cam. The radius of the pin on which the roller follower pivots is substantially smaller than R_f because of the space needed for roller or ball bearings within the follower. Dynamic forces will be addressed in later chapters where we will revisit this problem.

We can solve equation 8.31 for ρ_{pitch} since we know s, v, and a for all values of θ and can choose a trial R_p. If the pressure angle has already been calculated, the R_p found for its acceptable values should be used to calculate ρ_{pitch} as well. If a suitable follower radius cannot be found which satisfies equation 8.30 for the minimum values of ρ_{pitch} calculated from equation 8.31, then further iteration will be needed, possibly including a redefinition of the cam specifications.

Program DYNACAM calculates ρ_{pitch} for all values of θ for a user supplied prime circle radius R_p. Figure 8-46 shows ρ_{pitch} for the four-dwell cam of Figure 8-6 (p. 353). Note that this cam has both positive and negative radii of curvature. The large values of radius of curvature are truncated at arbitrary levels on the plot as they are heading to infinity at the inflection points between convex and concave portions. Note that the radii of curvature go out to positive infinity and return from negative infinity or vice versa at these inflection points (perhaps after a round trip through the universe?).

Once an acceptable prime circle radius and roller follower radius are determined based on pressure angle and radius of curvature considerations, the cam can be drawn in finished form and subsequently manufactured. Figure 8-47 shows the profile of the four dwell cam from Figure 8-6. The cam surface contour is swept out by the envelope of follower positions just as the cutter will create the cam in metal. The sidebar shows the parameters for the design, which is an acceptable one. The ρ_{min} is 1.7 times R_f and the pressure angles are less than 30°. The contours on the cam surface appear smooth, with no sharp corners. Figure 8-48 shows the same cam with only one change. The radius of follower R_f has been made the same as the minimum radius of curvature, ρ_{min}. The sharp corners or cusps in several places indicate that undercutting has occurred. This has now become an **unacceptable cam**, *simply because of a roller follower that is too large.*

The coordinates for the cam contour, measured to the locus of the center of the roller follower, or the **pitch curve** as shown in Figure 8-47, are defined by the following expressions, referenced to the center of rotation of the cam. See Figure 8-42 (p. 402) for nomenclature. The subtraction of the cam input angle θ from 2π is necessary because the relative motion of the follower versus the cam is opposite to that of the cam versus the follower. In other words, to define the contour of the centerline of the follower's path around a stationary cam, we must move the follower (and also the cutter to make the cam) in the opposite direction of cam rotation.

$$x = \cos\lambda\sqrt{(d+s)^2 + \varepsilon^2}$$

$$y = \sin\lambda\sqrt{(d+s)^2 + \varepsilon^2} \tag{8.32}$$

where:

$$\lambda = (2\pi - \theta) - \arctan\left(\frac{\varepsilon}{d+s}\right)$$

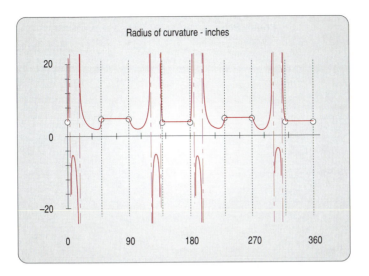

FIGURE 8-46

Radius of curvature of a four-dwell cam

Radius of Curvature—Flat-Faced Follower

The situation with a flat-faced follower is different to that of a roller follower. A negative radius of curvature on the cam cannot be accommodated with a flat-faced follower. The flat follower obviously cannot follow a concave cam. Undercutting will occur when the radius of curvature becomes negative if a cam with that condition is made.

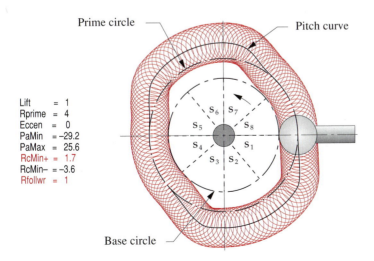

FIGURE 8-47

Radial plate cam profile is generated by the locus of the roller follower (or cutter)

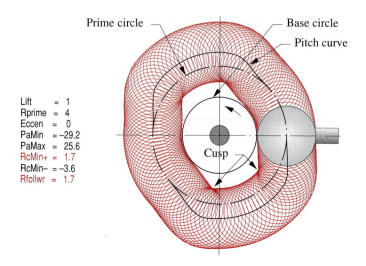

Lift   = 1
Rprime = 4
Eccen = 0
PaMin = −29.2
PaMax = 25.6
RcMin+ = 1.7
RcMin− = −3.6
Rfollwr = 1.7

FIGURE 8-48

Cusps formed by undercutting due to radius of follower $R_f \geq$ cam radius of curvature ρ

Figure 8-49 shows a cam and flat-faced follower in an arbitrary position. The origin of the global XY coordinate system is placed at the cam's center of rotation, and the X axis is defined parallel to the common tangent, which is the surface of the flat follower. The vector $\mathbf{r}$ is attached to the cam, rotates with it, and serves as the reference line to which the cam angle θ is measured from the X axis. The point of contact A is defined by the position vector $\mathbf{R}_A$. The instantaneous center of curvature is at C and the radius of curvature is ρ. R_b is the radius of the base circle and s is the displacement of the follower for angle θ. The eccentricity is ε.

We can define the location of contact point A from two vector loops (in complex notation).

$$\mathbf{R}_A = x + j(R_b + s)$$

and

$$\mathbf{R}_A = ce^{j(\theta+\alpha)} + j\rho$$

so:

$$ce^{j(\theta+\alpha)} + j\rho = x + j(R_b + s) \tag{8.33a}$$

Substitute the Euler equivalent (equation 4.4a, p. 155) in equation 8.33a and separate the real and imaginary parts.

real:

$$c\cos(\theta+\alpha) = x \tag{8.33b}$$

imaginary:

$$c\sin(\theta+\alpha) + \rho = R_b + s \tag{8.33c}$$

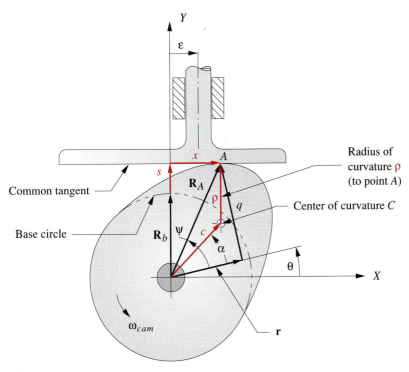

FIGURE 8-49

Geometry for derivation of radius of curvature and cam contour with flat-faced follower

The center of curvature C is **stationary** on the cam meaning that the magnitudes of c and ρ, and angle α do not change for small changes in cam angle θ. (These values are not constant but are at stationary values. Their first derivatives with respect to θ are zero, but their higher derivatives are not zero.)

Differentiating equation 8.33a with respect to θ then gives:

$$jce^{j(\theta+\alpha)} = \frac{dx}{d\theta} + j\frac{ds}{d\theta} \tag{8.34}$$

Substitute the Euler equivalent (equation 4.4a, p. 155) in equation 8.34 and separate the real and imaginary parts.

real:

$$-c\sin(\theta+\alpha) = \frac{dx}{d\theta} \tag{8.35}$$

imaginary:

$$c\cos(\theta+\alpha) = \frac{ds}{d\theta} = v \tag{8.36}$$

Inspection of equations 8.33b and 8.36 shows that:

$$x = v \tag{8.37}$$

This is an interesting relationship that says the x position of the contact point between cam and follower is numerically equal to the velocity of the follower in length/rad. This means that the v diagram gives a direct measure of the necessary minimum face width of the flat follower.

$$facewidth > v_{max} - v_{min} \tag{8.38}$$

If the velocity function is asymmetric, then a minimum-width follower will have to be asymmetric also, in order not to fall off the cam.

Differentiating equation 8.37 with respect to θ gives:

$$\frac{dx}{d\theta} = \frac{dv}{d\theta} = a \tag{8.39}$$

Equations 8.33c and 8.35 can be solved simultaneously and equation 8.39 substituted in the result to yield:

$$\rho = R_b + s + a \tag{8.40}$$

BASE CIRCLE   Note that equation 8.40 defines the radius of curvature in terms of the base circle radius and the displacement and acceleration functions from the $s\ v\ a\ j$ diagrams only. Because ρ cannot be allowed to become negative with a flat-faced follower, we can formulate a relationship from this equation which will predict the minimum base circle radius R_b needed to avoid undercutting. The only factor on the right side of equation 8.40 which can be negative is the acceleration, a. We have defined s to be always positive, as is R_b. Therefore, the worst case for undercutting will occur when a is at its **largest negative value**, a_{min}, whose value we know from the a diagram. The minimum base circle radius can then be defined as:

$$R_{b_{min}} > \rho_{min} - s_{@\,a_{min}} - a_{min} \tag{8.41}$$

Note that the value of s in this equation is taken at the cam angle θ corresponding to that of a_{min}. Because the value of a_{min} is negative and it is also negated in equation 8.41, it dominates the expression. To use this relationship, we must choose some minimum radius of curvature ρ_{min} for the cam surface as a design parameter. Since the hertzian contact stresses at the contact point are a function of local radius of curvature, that criterion can be used to select ρ_{min}. That topic is beyond the scope of this text and will not be further explored here. See reference 1 for further information on contact stresses.

CAM CONTOUR   For a flat-faced follower cam, the coordinates of the physical cam surface must be provided to the machinist as there is no pitch curve to work to. Figure 8-49 (p. 409) shows two orthogonal vectors, **r** and **q**, which define the cartesian coordinates of contact point A between cam and follower with respect to a rotating axis coordinate system embedded in the cam. Vector **r** is the rotating "x" axis of this embedded coordinate system. Angle ψ defines the position of vector $\mathbf{R}_A$ in this system. Two vector loop equations can be written and equated to define the coordinates of all points on the cam surface as a function of cam angle θ.

$$\mathbf{R}_A = x + j(R_b + s)$$

and

$$\mathbf{R}_A = re^{j\theta} + qe^{j\left(\theta + \frac{\pi}{2}\right)}$$

so:

$$re^{j\theta} + qe^{j\left(\theta + \frac{\pi}{2}\right)} = x + j(R_b + s) \qquad (8.42)$$

Divide both sides by $e^{j\theta}$:

$$r + jq = xe^{-j\theta} + j(R_b + s)e^{-j\theta} \qquad (8.43)$$

Separate into real and imaginary components and substitute v for x from equation 8.37:

real (x component):

$$r = (R_b + s)\sin\theta + v\cos\theta \qquad (8.44a)$$

imaginary (y component):

$$q = (R_b + s)\cos\theta - v\sin\theta \qquad (8.44b)$$

Equations 8.44 can be used to machine the cam for a flat-faced follower. These x,y components are in the rotating coordinate system that is embedded in the cam.

Note that none of the equations developed above for this case involve the **eccentricity**, ε. It is only a factor in cam size when a roller follower is used. It does not affect the geometry of a flat follower cam.

Figure 8-50 shows the result of trying to use a flat-faced follower on a cam with negative radius of curvature. If the follower could contact the cam at all the points necessary to control the follower to the function in the s diagram, the cam surface would be as developed by the envelope of straight lines. However, these loci of the follower face are cutting into cam contours which are needed for other cam angles. The line running through the forest of follower loci is the theoretical cam contour needed for this design. The undercutting can be clearly seen as the crescent-shaped missing pieces at four places between the cam contour and the follower loci.

SUMMARY The task of sizing a cam is an excellent example of the need for and value of iteration in design. Rapid recalculation of the relevant equations with a tool such as program DYNACAM makes it possible to quickly and painlessly arrive at an acceptable solution while balancing the often conflicting requirements of pressure angle and radius of curvature constraints. In any cam, either the pressure angle or radius of curvature considerations will dictate the minimum size of the cam. Both factors must be checked. The choice of follower type, either roller or flat-faced, makes a big difference in the cam geometry. Cam programs which generate negative radii of curvature are unsuited to the flat-faced type of follower unless very large base circles are used to force ρ to be positive everywhere.

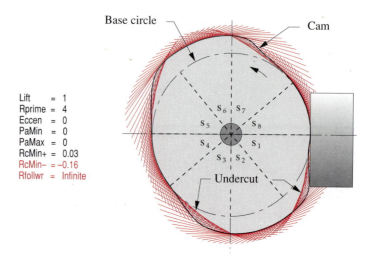

Lift   = 1
Rprime = 4
Eccen = 0
PaMin = 0
PaMax = 0
RcMin+ = 0.03
RcMin− = −0.16
Rfollwr = Infinite

FIGURE 8-50

Undercutting due to negative radius of curvature used with flat-faced follower

8.8   CAM MANUFACTURING CONSIDERATIONS

The preceding sections illustrate that there are a number of factors to consider when designing a cam. A great deal of care in design is necessary to obtain a good compromise of all factors, some of which conflict. Once the cam design is complete a whole new set of considerations must be dealt with that involve manufacturing the cam. After all, if your design cannot be successfully machined in metal in a way that truly represents the theoretical functions chosen, their benefits will not be realized. Unlike linkages, which are very easy to make, cams are a challenge to manufacture properly.

Cams are usually made from strong, hard materials such as **medium to high carbon steels** (case- or through-hardened) or **cast ductile iron** or **grey cast iron** (case-hardened). Cams for low loads and speeds or marine applications are sometimes made of brass or bronze. Even plastic cams are used in such applications as washing machine timers where the cam is merely tripping a switch at the right time. We will concentrate on the higher load-speed situations here for which steel or cast/ductile iron are the only practical choices. These materials range from fairly difficult to very difficult to machine depending on the alloy. At a minimum, a reasonably accurate milling machine is needed to make a cam. A computer controlled machining center is far preferable and is most often the choice for serious cam production.

Cams are typically milled with rotating cutters that in effect "tear" the metal away leaving a less than perfectly smooth surface at a microscopic level. For a better finish and better geometric accuracy, the cam can be ground after milling away most of the unneeded material. Heat treatment is usually necessary to get sufficient hardness to prevent rapid wear. Steel cams are typically hardened to about Rockwell Rc 50-55. Heat treatment introduces some geometric distortion. The grinding is usually done after heat treatment to correct the contour as well as to improve the finish.[*] The grinding step near-ly doubles the cost of an already expensive part, so it is often skipped in order to save money. A hardened but unground cam will have some heat distortion error despite accurate milling before hardening. There are several methods of cam manufacture in common use as shown in Table 8-3.

Geometric Generation

Geometric generation refers to the continuous "sweeping out" of a surface as in turning a cylinder on a lathe. This is perhaps the ideal way to make a cam because it creates a truly continuous surface with an accuracy limited only by the quality of the machine and tools used. Unfortunately there are very few types of cams that can be made by this method. The most obvious one is the eccentric cam (Figure 8-10, p. 359) which can be turned and ground on a lathe. A cycloid can also be geometrically generated. Very few other curves can. The presence of dwells makes it extremely difficult to apply this method. Thus, it is seldom used for cams. However, when it can be, as in the case of the eccentric cam of Figure 8-10, the resulting acceleration, though not perfect, is very close to the theoretical cosine wave as seen in Figure 8-11 (p. 359). This eccentric cam was made by turning and grinding on a high-quality lathe. This is the best that can be obtained in cam manufacture. Note that the displacement function is virtually perfect. The errors are only visible in the more sensitive acceleration function measurement.

Manual or NC Machining to Cam Coordinates (Plunge-Cutting)

Computer-aided manufacturing (CAM) has become the virtual standard for high accuracy machining in the United States. Numerical control (NC) machinery comes in many types. Lathes, milling machines, grinders, etc., are all available with on-board computers which control either the position of the workpiece, the tool, or both. The simplest type of NC machine moves the tool (or workpiece) to a specified x,y location and then drives the tool (say a drill) down through the workpiece to make a hole. This process is repeated as much as necessary to create the part. This simple process is referred to as NC to distinguish it from continuous numerical control (CNC).

This NC process is sometimes used for cam manufacture, and even for master cams as described below. It is, in fact, merely a computerized version of the old manual method of cam milling, which is often called **plunge-cutting** to refer to plunging the spinning milling cutter down through the workpiece. This is not the best way to machine a cam because it leaves "scallops" on the surface as shown in Figure 8-51, due to the fact

TABLE 8-3   Common Cam Manufacturing Methods

1	Geometric Generation
2	Manual or **N**umerical **C**ontrol (**NC**) machining to cam coordinates (plunge-cutting)
3	Continuous **N**umerical **C**ontrol (**CNC**) with **L**inear **I**nterpolation (**LI**)
4	Continuous **N**umerical **C**ontrol (**CNC**) with **C**ircular **I**nterpolation (**CI**)
5	**A**nalog **D**uplication (**AD**) of a hand-dressed master cam

[*] Some automotive camshafts are ground soft, then hardened and polished to remove the carburization film from the hardening process. The reason for this is to avoid the possibility of "burning" the surface during grinding which would locally anneal the material and lead to premature surface fatigue failure.

that the machinist can only *plunge* at a discrete number of positions around the cam. In effect, the displacement function which we developed has to be "discretized" or sampled at some finite number of places around the cam. Practicality limits this digitizing process to increments of about 1/2 to 1 degree. With an NC process the increment might be reduced to 1/4 or 1/10 degree. At some point "diminishing returns" will set in as the machine's ability to resolve positions spaced too closely will limit the accuracy. Standard milling machines can be expected to give accuracies in the 0.001 inch tolerance range. Tooling-quality machining centers, jig borers, and grinders can be as much as two to 20 times that accurate [0.000 5-in down to 0.000 050-in (50 millionths) tolerance].

The scallops that are left on the cam after plunge-cutting have to be removed by hand-dressing with files and grindstones. This obviously introduces more error. Even if the bottoms of the scallops at the sample increments were exactly correct, all points between are subject to the vagaries of hand work. The chance of exactly achieving the designed *s v a j* functions, especially the higher derivatives, with this manufacturing method is slight.

Continuous Numerical Control with Linear Interpolation

In a CNC machine, the tool is in constant contact with the workpiece, always cutting, while the computer controls the movement of the workpiece from position to position as stored in its memory. This is a **continuous cutting process** as opposed to the discrete one of NC. However, the cam displacement function must still be discretized or sampled at some angular increment. The common increments are 1/4, 1/2, and 1 degree. Since the machine only has information about the *x,y* locations of these 360, 720, or 1440 points around the cam, it has to figure out how to get from one point to another while cutting. The most common method used to "fill in" the missing data is **linear interpolation** (LI). The machine's computer calculates the straight line between each pair of data points and then drives the cutter (or workpiece) so as to stay as close to that line as it can. If it could do this perfectly (which it can't), we would get a piecewise continuous first-order approximation to the cam contour. This would introduce slope discontinuities which will create theoretically infinite pulses of acceleration. We would be back to

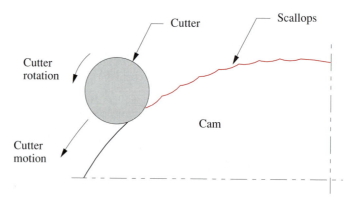

FIGURE 8-51

Plunge-cutting a cam leaves scallops on its surface

C
functic
celerat

Th
functic
which
The m
functio
smooth
functio
about 1
rather t

8.9

The car
stage o
without
The fol
the hop
sions.

Transl

There a
motion
motion
rocker c
when a
translati
to preve

our "bad cam by naive designer" of Example 8-1 having infinite acceleration regardless of the actual function selected.

An improvement can be made in this linear interpolation scheme by fitting a cubic spline curve to the cam coordinate data and then resampling this spline approximation at finer spacing down to the machine's angular resolution. This denser data set is then used to drive the cutter which still must traverse approximate straight-line paths between the close spaced data points. The curve fitting and resampling is typically done at the manufacturing stage.

Fortunately, the dynamics of the cutting process which are a function of speeds, feeds, tool sharpness, tool chatter, deflection of the spindle, etc., all conspire to prevent the formation of a series of distinct "flats" which would give the derivatives shown in Figure 8-8. Rather, the kind of acceleration curve that actually results from a cam which was milled (but not ground) on a very high-quality CNC machining center, using 1 degree linear interpolation, is as shown in Figure 8-52. The program is a simple harmonic eccentric with no dwells. The dynamic curves were measured with instrumentation on the roller follower while the cam was running at 600 rpm on a custom-designed cam dynamic test fixture (CDTF).[8] The actual displacement is quite true to the theoretical, but the acceleration has a significant amount of vibratory noise present which distorts the function from its theoretical cosine waveshape. The acceleration is shown in g's. Compare the peak-to-peak values in Figure 8-52 (8 g) with the same cam design made with geometric generation in Figure 8-11 (p. 359), (5 g). The error is of the order of 3 g's on a base of 5 g's. These errors in the acceleration are due to a combination of manufacturing factors as described above. It was found that the use of 1/4, 1/2 or 1 degree digitization increments on this size cam* made no statistically significant difference in the fidelity of the actual acceleration function to its theoretical waveform.[8]

The physical fidelity of the cam surface of the same 1 degree linear interpolated eccentric cam to a geometrically generated (turned and ground) reference cam can be seen in Figure 8-53 which is an enlarged section of a portion of the cams as measured to

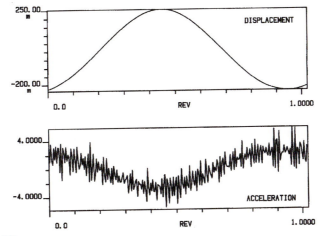

FIGURE 8-52

Displacement and acceleration of eccentric cam made with 1° linear interpolation CNC

* These cams were about 8 in (200 mm) in diameter. If the cam diameter is larger, then smaller angular increments of digitization will be needed as the distance along the pitch curve between data points for any angular increment increases linearly with prime circle diameter.

Conversely, the oscillating follower will keep the roller follower aligned in the same plane as the cam with no guiding beyond its own pivot. Also, the pivot friction in an oscillating follower typically has a small moment arm compared to the moment of the force from the cam on the follower arm. But, the friction force on a translating follower has a one-to-one geometric relationship with the cam force. This can have a larger parasitic effect on the system.

On the other hand, translating flat-faced followers are often deliberately arranged with their axis slightly out of the plane of the cam in order to create a rotation about their own axis due to the frictional moment resulting from the offset. The flat follower will then precess around its own axis and distribute the wear over its entire face surface. This is common practice in automotive valve cams that use flat-faced followers or "tappets."

Force or Form-Closed?

A form-closed (track or groove) cam is more expensive to make than a force-closed (open) cam simply because there are two surfaces to machine and grind. Also, heat treating will often distort the track of a form-closed cam, narrowing or widening it such that the roller follower will not fit properly. This virtually requires post heat-treat grinding for track cams in order to resize the slot. An open (force-closed) cam will also distort on heat-treating but may still be usable without grinding.

FOLLOWER JUMP   The principal advantage of a form-closed (track) cam is that it does not need a return spring, and thus can be run at higher speeds than a force-closed cam whose spring and follower mass will go into resonance at some speed, causing potentially destructive follower jump. This phenomenon will be investigated in Chapter 15 on cam dynamics. High-speed automobile and motorcycle racing engines often use form-closed (desmodromic) valve cam trains to allow higher engine rpm without incurring valve "float," or **follower jump**.

CROSSOVER SHOCK   Though the lack of a return spring can be an advantage, it comes, as usual, with a trade-off. In a form-closed (track) cam there will be **crossover shock** each time the acceleration changes sign. Crossover shock describes the impact force that occurs when the follower suddenly jumps from one side of the track to the other as the dynamic force (*ma*) reverses sign. There is no flexible spring in this system to absorb the force reversal as in the force-closed case. The high impact forces at crossover cause noise, high stresses, and local wear. Also, the roller follower has to reverse direction at each crossover which causes sliding and accelerates follower wear. Studies have shown that roller followers running against a well-lubricated open radial cam have slip rates of less than 1%.[9]

Radial or Axial Cam?

This choice is largely dictated by the overall geometry of the machine for which the cam is being designed. If the follower must move parallel to the camshaft axis, then an axial cam is dictated. If there is no such constraint, a radial cam is probably a better choice simply because it is a less complicated, thus cheaper, cam to manufacture.

Roller or Flat-Faced Follower?

The roller follower is a better choice from a cam design standpoint simply because it accepts negative radius of curvature on the cam. This allows more variety in the cam program. Also, for any production quantities, the roller follower has the advantage of being available from several manufacturers in any quantity from one to a million. For low quantities it is not usually economical to design and build your own custom follower. In addition, replacement roller followers can be obtained from suppliers on short notice when repairs are needed. Also, they are not particularly expensive even in small quantities.

Perhaps the largest users of flat-faced followers are automobile engine makers. Their quantities are high enough to allow any custom design they desire. It can be made or purchased economically in large quantity and can be less expensive than a roller follower in that case. Also with engine valve cams, a flat follower can save space over a roller. However, many manufacturers have switched to roller followers in automobile engines to reduce friction and improve fuel mileage. Diesel engines have long used roller followers (tappets) as have racers who "hop-up" engines for high performance.

Cams used in automated production line machinery use stock roller followers almost exclusively. The ability to quickly change a worn follower for a new one taken from the stockroom without losing much production time on the "line" is a strong argument in this environment. Roller followers come in several varieties (see Figure 8-5a, p. 351). They are based on roller or ball bearings. Plain bearing versions are also available for low-noise requirements. The outer surface, which rolls against the cam can be either cylindrical or spherical in shape. The "crown" on the spherical follower is slight, but it guarantees that the follower will ride near the center of a flat cam regardless of the accuracy of alignment of the axes of rotation of cam and follower. If a cylindrical follower is chosen and care is not taken to align the axes of cam and roller follower, the follower will ride on one edge and wear rapidly.

Commercial roller followers are typically made of high carbon alloy steel such as AISI 52100 and hardened to Rockwell Rc 60 - 62. The 52100 alloy is well suited to thin sections that must be heat-treated to a uniform hardness. Because the roller makes many revolutions for each cam rotation, its wear rate may be higher than that of the cam. Chrome plating the follower can markedly improve its life. Chrome is harder than steel at about Rc 70. Steel cams are typically hardened to a range of Rc 50 - 55.

To Dwell or Not to Dwell?

The need for a dwell is usually clear from the problem specifications. If the follower must be held stationary for any time, then a dwell is required. Some cam designers tend to insert dwells in situations where they are not specifically needed for follower stasis, in a mistaken belief that this is preferable to providing a rise-return motion when that is what is really needed. If the designer is attempting to use a double-dwell program in a single-dwell case, then perhaps his or her motivation to "let the vibrations settle out" by providing a "short dwell" at the end of the motion is justified. However, he or she probably should be using another cam program, perhaps a polynomial tailored to the specifications. Taking the acceleration to zero, whether for an instant or for a "short dwell," is generally unnecessary and undesirable. (See Examples 8-5, p. 375 , 8-6, p. 377, and 8-8,

p. 383) A dwell should be used only when the follower is required to be stationary for some measurable time. Moreover, if you do not need any dwell at all, consider using a linkage instead. They are a lot easier and cheaper to manufacture.

To Grind or Not to Grind?

Many production machinery cams are used as-milled, and not ground. Automotive valve cams are ground. The reasons are largely due to cost and quantity considerations as well as the high speeds of automotive cams. There is no question that a ground cam is superior to a milled cam. The question in each case is whether the advantage gained is worth the cost. In small quantities, as are typical of production machinery, grinding about doubles the cost of a cam. The advantages in terms of smoothness and quietness of operation, and of wear, are not in the same ratio as the cost difference. A well-machined cam can perform nearly as well as a well-ground cam and better than a poorly ground cam.[8]

Automotive cams are made in large quantity, run at very high speed, and are expected to last for a very long time with minimal maintenance. This is a very challenging specification. It is a great credit to the engineering of these cams that they very seldom fail in 100,000 miles or more of operation. These cams are made on specialized equipment which keeps the cost of their grinding to a minimum.

To Lubricate or Not to Lubricate?

Cams need lots of lubrication. Automotive cams are literally drowned in a flow of engine oil. Many production machine cams run immersed in an oil bath. These are reasonably happy cams. Others are not so fortunate. Cams which operate in close proximity to the product on an assembly machine in which oil would cause contamination of the product (food products, personal products) often are run dry. Camera mechanisms, which are full of linkages and cams, are often run dry. Lubricant would eventually find its way to the film.

Unless there is some good reason to eschew lubrication, a cam-follower should be provided with a generous supply of clean lubricant, preferably a hypoid-type oil containing additives for boundary lubrication conditions. The geometry of a cam-follower joint (half-joint) is among the worst possible from a lubrication standpoint. Unlike a journal bearing, which tends to trap a film of lubricant within the joint, the half joint is continually trying to squeeze the lubricant out of itself. This can result in a boundary, or mixed boundary/EHD* lubrication state in which some metal-to-metal contact will occur. Lubricant must be continually resupplied to the joint. Another purpose of the liquid lubricant is to remove the heat of friction from the joint. If run dry, significantly higher material temperatures will result, with accelerated wear and possible early failure.

8.10   REFERENCES

1   **McPhate, A. J., and L. R. Daniel**. (1962). "A Kinematic Analysis of Fourbar Equivalent Mechanisms for Plane Motion Direct Contact Mechanisms." *Proc. of Seventh Conference on Mechanisms*, Purdue University, pp. 61-65.

2   **Neklutin, C. N.** (1954). "Vibration Analysis of Cams." *Machine Design*, **26**, pp. 190-198.

* EHD = ElastoHydroDynamic.

3   **Wiederrich, J. L., and B. Roth**. (1978). "Design of Low Vibration Cam Profiles." *Cams and Cam Mechanisms*, Jones, J. R., ed. Institution of Mechanical Engineers: London, pp. 3-8.

4   **Chew, M., and C. H. Chuang**. (1995). "Minimizing Residual Vibrations in High Speed Cam-Follower Systems Over a Range of Speeds." *Journal of Mechanical Design*, **117**(1), p. 166.

5   **MacCarthy, B. L.** (1985). "Evaluation of Spline Functions for Use in Cam Design." *Proc Instn Mech Engrs*, **199**(C3), pp. 239-248.

6   **MacCarthy, B. L.** (1988). "Quintic Splines for Kinematic Design." *Computer-Aided Design*, **20**(7), pp. 406-415.

7   **Tsay, D. M., and C. O. Huey.** (1993). "Application of Rational B-Splines to the Synthesis of Cam-Follower Motion Programs." *Journal of Mechanical Design*, **115**(3), p. 621.

8   **Norton, R. L.** (1988). "Effect of Manufacturing Method on Dynamic Performance of Cams. " *Mechanism and Machine Theory*, **23**(3), pp. 191-208.

9   **Norton, R. L., et al.** (1988). "Analysis of the Effect of Manufacturing Methods and Heat Treatment on the Performance of Double Dwell Cams." *Mechanism and Machine Theory*, **23**(6), pp. 461-473.

8.11   PROBLEMS

Programs DYNACAM *and* MATRIX *may be used to solve these problems or to check your solution where appropriate.*

*8-1   Figure P8-1 shows the cam and follower from Problem 6-65. Using graphical methods, find and sketch the equivalent fourbar linkage for this position of the cam and follower.

8-2   Figure P8-1 shows the cam and follower from Problem 6-65. Using graphical methods, find the pressure angle at the position shown.

8-3   Figure P8-2 shows a cam and follower. Using graphical methods, find and sketch the equivalent fourbar linkage for this position of the cam and follower.

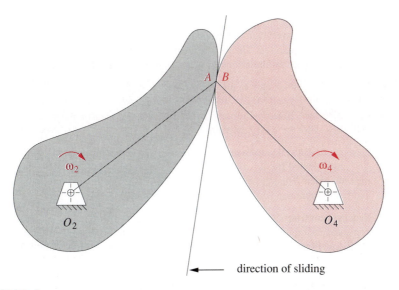

FIGURE P8-1

Problems 8-1 to 8-2. *Adapted from P. H. Hill and W. P. Rule. (1960). Mechanisms: Analysis and Design.*

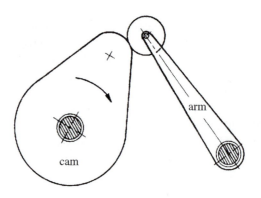

FIGURE P8-2

Problems 8-3 to 8-4

8-4   Figure P8-2 shows a cam and follower. Using graphical methods, find the pressure angle at the position shown.

8-5   Figure P8-3 shows a cam and follower. Using graphical methods, find and sketch the equivalent fourbar linkage for this position of the cam and follower.

8-6   Figure P8-3 shows a cam and follower. Using graphical methods, find the pressure angle at the position shown.

‡8-7 Design a double-dwell cam to move a follower from 0 to 2.5" in 60°, dwell for 120°, fall 2.5" in 30° and dwell for the remainder. The total cycle must take 4 sec. Choose suitable programs for rise and fall to minimize accelerations. Plot the *s v a j* diagrams.

‡8-8 Design a double-dwell cam to move a follower from 0 to 1.5" in 45°, dwell for 150°, fall 1.5" in 90° and dwell for the remainder. The total cycle must take 6 sec. Choose suitable programs for rise and fall to minimize velocities. Plot the *s v a j* diagrams.

‡8-9 Design a single-dwell cam to move a follower from 0 to 2" in 60°, fall 2" in 90° and dwell for the remainder. The total cycle must take 2 sec. Choose suitable programs for rise and fall to minimize accelerations. Plot the *s v a j* diagrams.

‡8-10 Design a three-dwell cam to move a follower from 0 to 2.5" in 40°, dwell for 100°, fall 1.5" in 90°, dwell for 20°, fall 1" in 30° and dwell for the remainder. The total cycle must take 10 sec. Choose suitable programs for rise and fall to minimize velocities. Plot the *s v a j* diagrams.

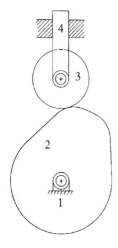

FIGURE P8-3

Problems 8-5 to 8-6

‡8-11 Design a four-dwell cam to move a follower from 0 to 2.5" in 40°, dwell for 100°, fall 1.5" in 90°, dwell for 20°, fall 0.5" in 30°, dwell for 40°, fall 0.5" in 30° and dwell for the remainder. The total cycle must take 15 sec. Choose suitable programs for rise and fall to minimize accelerations. Plot the *s v a j* diagrams.

‡8-12 Size the cam from Problem 8-7 for a 1" radius roller follower considering pressure angle and radius of curvature. Use eccentricity only if necessary to balance those functions. Plot both those functions. Draw the cam profile. Repeat for a flat-faced follower. Which would you use?

‡ These problems are suited to solution using program.DYNACAM, which is on the attached CD-ROM.

‡8-13 Size the cam from Problem 8-8 for a 1.5" radius roller follower considering pressure angle and radius of curvature. Use eccentricity only if necessary to balance those

functions. Plot both those functions. Draw the cam profile. Repeat for a flat-faced follower. Which would you use?

‡8-14   Size the cam from Problem 8-9 for a 0.5" radius roller follower considering pressure angle and radius of curvature. Use eccentricity only if necessary to balance those functions. Plot both those functions. Draw the cam profile. Repeat for a flat-faced follower. Which would you use?

‡8-15   Size the cam from Problem 8-10 for a 2" radius roller follower considering pressure angle and radius of curvature. Use eccentricity only if necessary to balance those functions. Plot both those functions. Draw the cam profile. Repeat for a flat-faced follower. Which would you use?

‡8-16   Size the cam from Problem 8-11 for a 0.5" radius roller follower considering pressure angle and radius of curvature. Use eccentricity only if necessary to balance those functions. Plot both those functions. Draw the cam profile. Repeat for a flat-faced follower. Which would you use?

‡8-17   A high friction, high inertia load is to be driven. We wish to keep peak velocity low. Combine segments of half period cycloidal displacements with a constant velocity segment on both rise and fall to reduce the maximum velocity below that obtainable with a full period modified sine acceleration alone (i.e., with no constant velocity portion). Rise is 1" in 90°, dwell for 60°, fall in 50°, dwell for remainder. Compare the two designs and comment. Use an ω of one for comparison.

‡8-18   A constant velocity of 0.4 in/sec must be matched for 1.5 sec. Then the follower must return to your choice of start point and dwell for 2 sec. The total cycle time is 6 sec. Design a cam for a follower radius of 0.75" and a maximum pressure angle of 30° absolute value.

‡8-19   A constant velocity of 0.25 in/sec must be matched for 3 sec. Then the follower must return to your choice of start point and dwell for 3 sec. The total cycle time is 12 sec. Design a cam for a follower radius of 1.25" and a maximum pressure angle of 35° absolute value.

‡8-20   A constant velocity of 2 in/sec must be matched for 1 second. Then the follower must return to your choice of start point. The total cycle time is 2.75 sec. Design a cam for a follower radius of 0.5" and a maximum pressure angle of 25° absolute value.

†8-21   Write a computer program or use an equation solver such as *Mathcad* or *TKSolver* to calculate and plot the *s v a j* diagrams for a modified trapezoidal acceleration cam function for any specified values of lift and duration. Test it using a lift of 20 mm over 60° at 1 rad/sec.

†8-22   Write a computer program or use an equation solver such as *Mathcad* or *TKSolver* to calculate and plot the *s v a j* diagrams for a modified sine acceleration cam function for any specified values of lift and duration. Test it using a lift of 20 mm over 60° at 1 rad/sec.

†8-23   Write a computer program or use an equation solver such as *Mathcad* or *TKSolver* to calculate and plot the *s v a j* diagrams for a cycloidal displacement cam function for any specified values of lift and duration. Test it using a lift of 20 mm over 60° at 1 rad/sec.

†8-24   Write a computer program or use an equation solver such as *Mathcad* or *TKSolver* to calculate and plot the *s v a j* diagrams for a 3-4-5 polynomial displacement cam

† These problems are suited to solution using *Mathcad, Matlab,* or *TKSolver* equation solver programs.

‡ These problems are suited to solution using program DYNACAM, which is on the attached CD-ROM..

function for any specified values of lift and duration. Test it using a lift of 20 mm over 60° at 1 rad/sec.

†8-25   Write a computer program or use an equation solver such as *Mathcad* or *TKSolver* to calculate and plot the *s v a j* diagrams for a 4-5-6-7 polynomial displacement cam function for any specified values of lift and duration. Test it using a lift of 20 mm over 60° at 1 rad/sec.

†8-26   Write a computer program or use an equation solver such as *Mathcad* or *TKSolver* to calculate and plot the *s v a j* diagrams for a simple harmonic displacement cam function for any specified values of lift and duration. Test it using a lift of 20 mm over 60° at 1 rad/sec.

†8-27   Write a computer program or use an equation solver such as *Mathcad* or *TKSolver* to calculate and plot the pressure angle and radius of curvature for a modified trapezoidal acceleration cam function for any specified values of lift, duration, eccentricity, and prime circle radius. Test it using a lift of 20 mm over 60° at 1 rad/sec, and determine the prime circle radius needed to obtain a maximum pressure angle of 20°. What is the minimum diameter of roller follower needed to avoid undercutting with these data?

†8-28   Write a computer program or use an equation solver such as *Mathcad* or *TKSolver* to calculate and plot the pressure angle and radius of curvature for a modified sine acceleration cam function for any specified values of lift, duration, eccentricity, and prime circle radius. Test it using a lift of 20 mm over 60° at 1 rad/sec, and determine the prime circle radius needed to obtain a maximum pressure angle of 20°. What is the minimum diameter of roller follower needed to avoid undercutting with these data?

†8-29   Write a computer program or use an equation solver such as *Mathcad* or *TKSolver* to calculate and plot the pressure angle and radius of curvature for a cycloidal displacement cam function for any specified values of lift, duration, eccentricity, and prime circle radius. Test it using a lift of 20 mm over 60° at 1 rad/sec, and determine the prime circle radius needed to obtain a maximum pressure angle of 20°. What is the minimum diameter of roller follower needed to avoid undercutting with these data?

†8-30   Write a computer program or use an equation solver such as *Mathcad* or *TKSolver* to calculate and plot the pressure angle and radius of curvature for a 3-4-5 polynomial displacement cam function for any specified values of lift, duration, eccentricity, and prime circle radius. Test it using a lift of 20 mm over 60° at 1 rad/sec, and determine the prime circle radius needed to obtain a maximum pressure angle of 20°. What is the minimum diameter of roller follower needed to avoid undercutting with these data?

†8-31   Write a computer program or use an equation solver such as *Mathcad* or *TKSolver* to calculate and plot the pressure angle and radius of curvature for a 4-5-6-7 polynomial displacement cam function for any specified values of lift, duration, eccentricity, and prime circle radius. Test it using a lift of 20 mm over 60° at 1 rad/sec, and determine the prime circle radius needed to obtain a maximum pressure angle of 20°. What is the minimum diameter of roller follower needed to avoid undercutting with these data?

† These problems are suited to solution using *Mathcad, Matlab,* or, *TKSolver* equation solver programs.

†8-32   Write a computer program or use an equation solver such as *Mathcad* or *TKSolver* to calculate and plot the pressure angle and radius of curvature for a simple harmonic displacement cam function for any specified values of lift, duration, eccentricity, and

prime circle radius. Test it using a lift of 20 mm over 60° at 1 rad/sec, and determine the prime circle radius needed to obtain a maximum pressure angle of 20°. What is the minimum diameter of roller follower needed to avoid undercutting with these data?

8-33   Derive equation 8.21 (p. 382) for the 4-5-6-7 polynomial function.

8-34   Derive an expression for the pressure angle of a barrel cam with zero eccentricity.

‡8-35   Design a radial plate cam to move a translating roller follower through 30 mm in 30°, dwell for 100°, fall 10 mm in 10°, dwell for 20°, fall 20 mm in 20°, and dwell for the remainder. Camshaft ω = 200 rpm. Minimize the follower's peak velocity and determine the minimum prime circle radius that will give a maximum 25° pressure angle. Determine the minimum radii of curvature on the pitch curve.

‡8-36   Repeat Problem 8-35, but minimize the follower's peak acceleration instead.

‡8-37   Repeat Problem 8-35, but minimize the follower's peak jerk instead.

‡8-38   Design a radial plate cam to lift a translating roller follower through 10 mm in 65°, return to 0 in 65° and dwell for the remainder. Camshaft ω = 3500 rpm. Minimize the cam size while not exceeding a 25° pressure angle. What size roller follower is needed?

‡8-39   Design a cam-driven quick-return mechanism for a 3:1 time ratio. The translating roller follower should move forward and back 50 mm and dwell in the back position for 80°. It should take one-third the time to return as to move forward. Camshaft ω = 100 rpm. Minimize the package size while maintaining a 25° maximum pressure angle. Draw a sketch of your design and provide *s v a j*, ϕ, and ρ diagrams.

‡8-40   Design a cam-follower system to drive a linear translating piston at constant velocity for 200° through a stroke of 100 mm at 60 rpm. Minimize the package size while maintaining a 25° maximum pressure angle. Draw a sketch of your design and provide *s v a j*, ϕ, and ρ diagrams.

8.12   PROJECTS

These larger-scale project statements deliberately lack detail and structure and are loosely defined. Thus, they are similar to the kind of "identification of need" or problem statement commonly encountered in engineering practice. It is left to the student to structure the problem through background research and to create a clear goal statement and set of task specifications before attempting to design a solution. This design process is spelled out in Chapter 1 and should be followed in all of these examples. All results should be documented in a professional engineering report. (See the bibliography of Chapter 1 for information on report writing.)

‡P8-1   A timing diagram for a halogen headlight filament insertion device is shown in Figure P8-4. Four points are specified. Point A is the start of rise. At B the grippers close to grab the filament from its holder. The filament enters its socket at C and is fully inserted at D. The high dwell from D to E holds the filament stationary while it is soldered in place. The follower returns to its start position from E to F. From F to A the follower is stationary while the next bulb is indexed into position. It is desirable to have low to zero velocity at point B where the grippers close on the fragile

‡ These problems are suited to solution using program DYNACAM, which is on the attached CD-ROM..

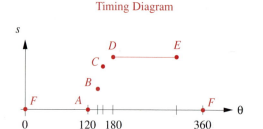

Timing Diagram

Cam angle, °	Point	s
120	A	0
140	B	2
150	C	3
180	D	3.5
300	E	3.5
360	F	0

Displacement Table

FIGURE P8-4

Data for cam design Project P8-1

filament. The velocity at C should not be so high as to "bend the filament in the breeze." Design and size a complete cam-follower system to do this job.

‡P8-2   A cam-driven pump to simulate human aortic pressure is needed to serve as a consistent, repeatable pseudo-human input to the operating room computer monitoring equipment, in order to test it daily. Figure P8-5 shows a typical aortic pressure curve and a pump pressure-volume characteristic. Design a cam to drive the piston and give as close an approximation to the aortic pressure curve shown as can be obtained without violating the fundamental law of cam design. Simulate the dichrotic notch as best you can.

‡P8-3   An athletic footwear manufacturer wants a device to test rubber heels for their ability to withstand millions of cycles of force similar to that which a walking human's foot applies to the ground. Figure P8-6 shows a typical walker's force-time function and a

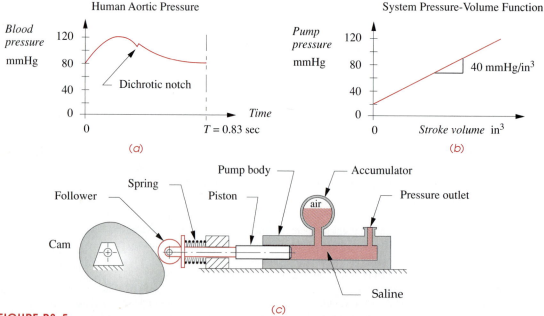

FIGURE P8-5

Data for cam design Project P8-2

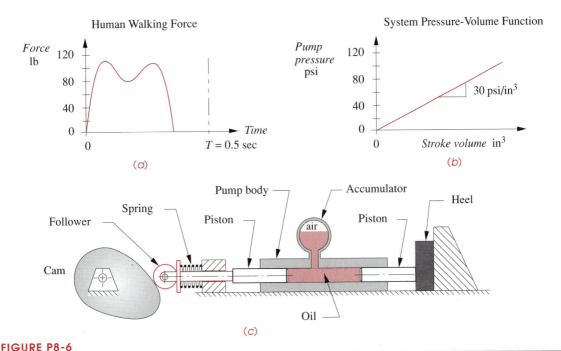

Human Walking Force

System Pressure-Volume Function

(a)

(b)

FIGURE P8-6

Data for cam design Project P8-3

8

pressure-volume curve for a piston-accumulator. Design a cam-follower system to drive the piston in a way that will create a force-time function on the heel similar to the one shown. Choose suitable piston diameters at each end.

‡P8-4 A fluorescent light bulb production machine moves 5500 lamps per hour through a 550°C oven on a chain conveyor which is in constant motion. The lamps are on 2-in centerlines. The bulbs must be sprayed internally with a tin oxide coating as they leave the oven, still hot. This requires a cam-driven device to track the bulbs at constant velocity for the 0.5 sec required to spray them. The spray guns will fit on a 6 x 10 in table. The spray creates hydrochloric acid, so all exposed parts must be resistant to that environment. The spray head transport device will be driven from the conveyor chain by a shaft having a 28-tooth sprocket in mesh with the chain. Design a complete spray gun transport assembly to these specifications.

‡P8-5 A 30-ft-tall drop tower is being used to study the shape of water droplets as they fall through air. A camera is to be carried by a cam-operated linkage which will track the droplet's motion from the 8-ft to the 10-ft point in its fall (measured from release point at the top of the tower). The drops are released every 1/2 sec. Every drop is to be filmed. Design a cam and linkage which will track these droplets, matching their velocities and accelerations in the 1-ft filming window.

‡P8-6 A device is needed to accelerate a 3000-lb vehicle into a barrier with constant velocity, to test its 5 mph bumpers. The vehicle will start at rest, move forward, and have constant velocity for the last part of its motion before striking the barrier with the specified velocity. Design a cam-follower system to do this. The vehicle will leave contact with your follower just prior to the crash.

‡ These problems are suited to solution using program DYNACAM, which is on the attached CD-ROM..

Chapter 9

GEAR TRAINS

Cycle and epicycle,
orb in orb
JOHN MILTON, PARADISE LOST

9.0   INTRODUCTION

The earliest known reference to gear trains is in a treatise by Hero of Alexandria (c. 100 B.C.). Gear trains are widely used in all kinds of mechanisms and machines, from can openers to aircraft carriers. Whenever a change in the speed or torque of a rotating device is needed, a gear train or one of its cousins, the belt or chain drive mechanism, will usually be used. This chapter will explore the theory of gear tooth action and the design of these ubiquitous devices for motion control. The calculations involved are trivial compared to those for cams or linkages. The shape of gear teeth has become quite standardized for good kinematic reasons which we will explore.

Gears of various sizes and styles are readily available from many manufacturers. Assembled gearboxes for particular ratios are also stock items. The kinematic design of gear trains is principally involved with the selection of appropriate ratios and gear diameters. A complete gear train design will necessarily involve considerations of strength of materials and the complicated stress states to which gear teeth are subjected. This text will not deal with the stress analysis aspects of gear design. There are many texts which do. Some are listed in the bibliography at the end of this chapter. This chapter will discuss the kinematics of gear tooth theory, gear types, and the kinematic design of gearsets and gear trains of simple, compound, reverted, and epicyclic types. Chain and belt drives will also be discussed. Examples of the use of these devices will be presented as well.

9.1   ROLLING CYLINDERS

The simplest means of transferring rotary motion from one shaft to another is a pair of rolling cylinders. They may be an external set of rolling cylinders as shown in Figure 9-1a or an internal set as in Figure 9-1b. Provided that sufficient friction is available at the rolling interface, this mechanism will work quite well. There will be no slip between the cylinders until the maximum available frictional force at the joint is exceeded by the demands of torque transfer.

A variation on this mechanism is what causes your car or bicycle to move along the road. Your tire is one rolling cylinder and the road the other (very large radius) one. Friction is all that prevents slip between the two, and it works well unless the friction coefficient is reduced by the presence of ice or other slippery substances. In fact, some early automobiles had rolling cylinder drives inside the transmission, as do some present-day snowblowers and garden tractors which use a rubber-coated wheel rolling against a steel disk to transmit power from the engine to the wheels.

A variant on the rolling cylinder drive is the flat or vee belt as shown in Figure 9-2. This mechanism also transfers power through friction and is capable of quite large power levels, provided enough belt cross section is provided. Friction belts are used in a wide variety of applications from small sewing machines to the alternator drive on your car, to multihorsepower generators and pumps. Whenever absolute phasing is not required and power levels are moderate, a friction belt drive may be the best choice. They are relatively quiet running, require no lubrication, and are inexpensive compared to gears and chain drives.

Both rolling cylinders and belt (or chain) drives have effective linkage equivalents as shown in Figure 9-3. These effective linkages are valid only for one instantaneous position but nevertheless show that these devices are just another variation of the four-bar linkage in disguise.

(a) External set

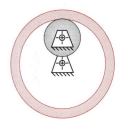

(b) Internal set

FIGURE 9-1

Rolling cylinders

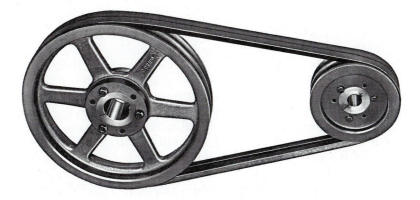

FIGURE 9-2

A two-groove vee belt drive   *Courtesy of T. B. Wood's Sons Co., Chambersburg, PA*

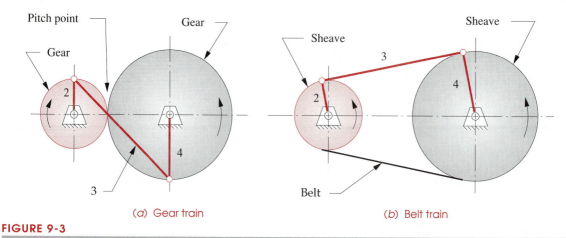

(a) Gear train   *(b)* Belt train

FIGURE 9-3

Gear and belt trains each have an equivalent fourbar linkage for any instantaneous position

The principal drawbacks to the rolling cylinder drive (or smooth belt) mechanism are its relatively low torque capability and the possibility of slip. Some drives require absolute phasing of the input and output shafts for timing purposes. A common example is the valve train drive in an automobile engine. The valve cams must be kept in phase with the piston motion or the engine will not run properly. A smooth belt or rolling cylinder drive from crankshaft to camshaft would not guarantee correct phasing. In this case some means of preventing slip is needed.

This usually means adding some meshing teeth to the rolling cylinders. They then become gears as shown in Figure 9-4 and are together called a *gearset*. When two gears are placed in mesh to form a gearset such as this one, it is conventional to refer to the smaller of the two gears as the *pinion* and to the other as the *gear*.

9.2   THE FUNDAMENTAL LAW OF GEARING

Conceptually, teeth of any shape will prevent gross slip. Old water-powered mills and windmills used wooden gears whose teeth were merely round wooden pegs stuck into the rims of the cylinders. Even ignoring the crudity of construction of these early examples of gearsets, there was no possibility of smooth velocity transmission because the geometry of the tooth "pegs" violated the **fundamental law of gearing** which, if followed, provides that *the angular velocity ratio between the gears of a gearset remains constant throughout the mesh.* A more complete and formal definition of this law is given on p. 436. The angular velocity ratio (m_V) referred to in this law is the same one that we derived for the fourbar linkage in Section 6.4 and equation 6.10 (p. 257). It is equal to the ratio of the radius of the input gear to that of the output gear.

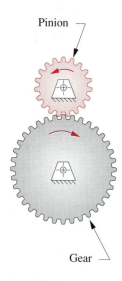

Pinion

Gear

FIGURE 9-4

An external gearset

$$m_V = \frac{\omega_{out}}{\omega_{in}} = \pm\frac{r_{in}}{r_{out}} = \pm\frac{d_{in}}{d_{out}} \qquad (9.1a)$$

$$m_T = \frac{\omega_{in}}{\omega_{out}} = \pm\frac{r_{out}}{r_{in}} = \pm\frac{d_{out}}{d_{in}} \qquad (9.1b)$$

The **torque ratio** (m_T) was shown in equation 6.12 (p. 259) to be the reciprocal of the velocity ratio (m_V); thus a gearset is essentially a device to exchange torque for velocity or vice versa. Since there are no applied forces as in a linkage, but only applied torques on the gears, the **mechanical advantage** m_A of a gearset is equal to its torque ratio m_T. The most common application is to reduce velocity and increase torque to drive heavy loads as in your automobile transmission. Other applications require an increase in velocity, for which a reduction in torque must be accepted. In either case, it is usually desirable to maintain a constant ratio between the gears as they rotate. Any variation in ratio will show up as oscillation in the output velocity and torque even if the input is constant with time.

The radii in equation 9.1 are those of the rolling cylinders to which we are adding the teeth. The positive or negative sign accounts for internal or external cylinder sets as defined in Figure 9-1 (p. 433). An external set reverses the direction of rotation between the cylinders and requires the negative sign. An internal gearset or a belt or chain drive will have the same direction of rotation on input and output shafts and require the positive sign in equation 9.1. The surfaces of the rolling cylinders will become the **pitch circles,** and their diameters the **pitch diameters** of the gears. The contact point between the cylinders lies on the line of centers as shown in Figure 9-3a, and this point is called the **pitch point**.

In order for the fundamental law of gearing to be true, the gear tooth contours on mating teeth must be conjugates of one another. There is an infinite number of possible conjugate pairs that could be used, but only a few curves have seen practical application as gear teeth. The **cycloid** still is used as a tooth form in watches and clocks, but most other gears use the **involute** curve for their shape.

The Involute Tooth Form

The involute is a curve which can be generated by unwrapping a taut string from a cylinder (called the evolute) as shown in Figure 9-5. Note the following about this involute curve:

The string is always tangent to the cylinder.

The center of curvature of the involute is always at the point of tangency of the string with the cylinder.

A tangent to the involute is then always normal to the string, the length of which is the instantaneous radius of curvature of the involute curve.

Figure 9-6 shows two involutes on separate cylinders in contact or "in mesh." These represent gear teeth. The cylinders from which the strings are unwrapped are called the **base circles** of the respective gears. Note that the base circles are necessarily smaller than the pitch circles, which are at the radii of the original rolling cylinders, r_p and r_g. The gear tooth must project both below and above the rolling cylinder surface (pitch circle) and the *involute only exists outside of the base circle*. The amount of tooth that sticks out above the pitch circle is the **addendum**, shown as a_p and a_g for pinion and gear, respectively. These are equal for standard, full-depth gear teeth.

The geometry at this tooth-tooth interface is similar to that of a cam-follower joint as was defined in Figure 8-40 (p. 399). There is a **common tangent** to both

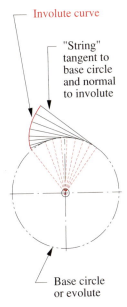

Involute curve

"String" tangent to base circle and normal to involute

Base circle or evolute

FIGURE 9-5

Development of the involute of a circle

TABLE 9-3
Standard Metric
Modules

Metric Module (mm)	Equivalent P_d (in^{-1})
0.3	84.67
0.4	63.50
0.5	50.80
0.8	31.75
1	25.40
1.25	20.32
1.5	16.93
2	12.70
3	8.47
4	6.35
5	5.08
6	4.23
8	3.18
10	2.54
12	2.12
16	1.59
20	1.27
25	1.02

Unequal-Addendum Tooth Forms

In order to avoid interference on small pinions, the tooth form can be changed from the standard, full-depth shapes of Figure 9-10 (p. 442) that have equal addenda on both pinion and gear to an involute shape with a longer addendum on the pinion and a shorter one on the gear called **profile-shifted gears**. The AGMA defines addendum modification coefficients, x_1 and x_2, which always sum to zero, being equal in magnitude and opposite in sign. The positive coefficient x_1 is applied to increase the pinion addendum, and the negative x_2 decreases the gear addendum by the same amount. The total tooth depth remains the same. This shifts the pinion dedendum circle outside its base circle and eliminates that noninvolute portion of pinion tooth below the base circle. The standard coefficients are ±0.25 and ±0.50, which add or subtract 25% or 50% of the standard addendum. The limit of this approach occurs when the pinion tooth becomes pointed.

There are secondary benefits to this technique. The pinion tooth becomes thicker at its base and thus stronger. The gear tooth is correspondingly weakened, but since a full-depth gear tooth is stronger than a full-depth pinion tooth, this shift brings them closer to equal strength. A disadvantage of unequal-addendum tooth forms is an increase in sliding velocity at the tooth tip. The percent sliding between the teeth is greater than with equal addendum teeth which increases tooth-surface stresses. Friction losses in the gear mesh are also increased by higher sliding velocities. Figure 9-13 shows the contours of profile-shifted involute teeth. Compare these to standard tooth shapes in Figure 9-10.

9.5   CONTACT RATIO

The contact ratio m_p defines the average number of teeth in contact at any one time as:

$$m_p = \frac{Z}{p_b} \qquad (9.6a)$$

where Z is the length of action from equation 9.2 (p. 437) and p_b is the base pitch from equation 9.4b (p. 441). Substituting equations 9.4b and 9.4d into 9.6a defines m_p in terms of p_d:

$$m_p = \frac{p_d Z}{\pi \cos \phi} \qquad (9.6b)$$

If the contact ratio is 1, then one tooth is leaving contact just as the next is beginning contact. This is undesirable because slight errors in the tooth spacing will cause oscillations in the velocity, vibration, and noise. In addition, the load will be applied at the tip of the tooth, creating the largest possible bending moment. At larger contact ratios than 1, there is the possibility of load sharing among the teeth. For contact ratios between 1 and 2, which are common for spur gears, there will still be times during the mesh when one pair of teeth will be taking the entire load. However, these will occur toward the center of the mesh region where the load is applied at a lower position on the tooth, rather than at its tip. This point is called the **highest point of single-tooth contact** (HPSTC). The minimum acceptable contact ratio for smooth operation is 1.2. A minimum contact ratio of 1.4 is preferred and larger is better. Most spur gearsets will have contact ratios between 1.4 and 2. Equation 9.6b shows that for smaller teeth (larger p_d) and larger pressure angle, the contact ratio will be larger.

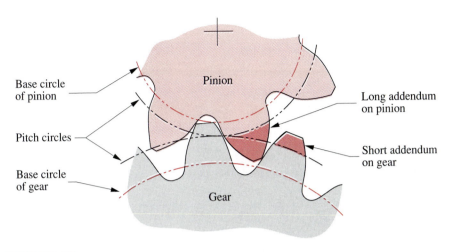

TABLE 9-4
Minimum Number of
Pinion Teeth
To Avoid Interference
Between a Full-Depth
Pinion and a Full-Depth
Rack

Pressure Angle (deg)	Minimum Number of Teeth
14.5	32
20	18
25	12

FIGURE 9-13

Profile-shifted teeth with long and short addenda to avoid interference and undercutting

EXAMPLE 9-1

Determining Gear Tooth and Gear Mesh Parameters.

Problem: Find the gear ratio, circular pitch, base pitch, pitch diameters, pitch radii, center distance, addendum, dedendum, whole depth, clearance, outside diameters, and contact ratio of a gearset with the given parameters. If the center distance is increased 2% what is the new pressure angle and increase in backlash?

Given: A 6 p_d, 20° pressure angle, 19-tooth pinion is meshed with a 37-tooth gear.

Assume: The tooth forms are standard AGMA full-depth involute profiles.

Solution:

1 The gear ratio is found from the tooth numbers on pinion and gear using equation 9.5b (p. 442).

$$m_G = \frac{N_g}{N_p} = \frac{37}{19} = 1.947 \qquad (a)$$

2 The circular pitch can be found either from equation 9.4a (p. 440) or 9.4c (p. 441).

$$p_c = \frac{\pi}{p_d} = \frac{\pi}{6} = 0.524 \text{ in} \qquad (b)$$

3 The base pitch measured on the base circle is (from equation 9.4b):

$$p_b = p_c \cos\phi = 0.524\cos(20°) = 0.492 \text{ in} \qquad (c)$$

TABLE 9-5
Minimum Number of
Pinion Teeth
To Avoid Interference
Between a 20° Full-Depth
Pinion and Full-Depth
Gears of Various Sizes

Minimum Pinion Teeth	Maximum Gear Teeth
17	1309
16	101
15	45
14	26
13	16

4   The pitch diameters and pitch radii of pinion and gear are found from equation 9.4c.

$$d_p = \frac{N_p}{P_d} = \frac{19}{6} = 3.167 \text{ in,} \qquad r_p = \frac{d_p}{2} = 1.583 \text{ in} \qquad (d)$$

$$d_g = \frac{N_g}{P_d} = \frac{37}{6} = 6.167 \text{ in,} \qquad r_g = \frac{d_g}{2} = 3.083 \text{ in} \qquad (e)$$

5   The nominal center distance C is the sum of the pitch radii:

$$C = r_p + r_g = 4.667 \text{ in} \qquad (f)$$

6   The addendum and dedendum are found from the equations in Table 9-1 (p. 441):

$$a = \frac{1.0}{P_d} = 0.167 \text{ in,} \qquad b = \frac{1.25}{P_d} = 0.208 \text{ in} \qquad (g)$$

7   The whole depth h_t is the sum of the addendum and dedendum.

$$h_t = a + b = 0.167 + 0.208 = 0.375 \text{ in} \qquad (h)$$

8   The clearance is the difference between dedendum and addendum.

$$c = b - a = 0.208 - 0.167 = 0.042 \text{ in} \qquad (i)$$

9   The outside diameter of each gear is the pitch diameter plus two addenda:

$$D_{o_p} = d_p + 2a = 3.500 \text{ in,} \qquad D_{o_g} = d_g + 2a = 6.500 \text{ in} \qquad (j)$$

10 The contact ratio is found from equations 9.2 (p. 437) and 9.6a (p. 444).

$$Z = \sqrt{(r_p + a_p)^2 - (r_p \cos\phi)^2} + \sqrt{(r_g + a_g)^2 - (r_g \cos\phi)^2} - C\sin\phi$$

$$= \sqrt{(1.583 + 0.167)^2 - (1.583\cos 20°)^2}$$

$$+ \sqrt{(3.083 + 0.167)^2 - (3.083\cos 20°)^2} - 4.667\sin 20° = 0.798 \text{ in}$$

$$m_p = \frac{Z}{p_b} = \frac{0.798}{0.492} = 1.62 \qquad (k)$$

11 If the center distance is increased from the nominal value due to assembly errors or other factors, the effective pitch radii will change by the same percentage. The gears' base radii will remain the same. The new pressure angle can be found from the changed geometry. For a 2% increase in center distance (1.02x):

$$\phi_{new} = \cos^{-1}\left(\frac{r_{base\ circle\ p}}{1.02 r_p}\right) = \cos^{-1}\left(\frac{r_p \cos\phi}{1.02 r_p}\right) = \cos^{-1}\left(\frac{\cos 20°}{1.02}\right) = 22.89° \qquad (l)$$

12 The change in backlash as measured at the pinion is found from equation 9.3 (p. 439).

$$\theta_B = 43\,200(\Delta C)\frac{\tan\phi}{\pi d} = 43\,200(0.02)(4.667)\frac{\tan(22.89°)}{\pi(3.167)} = 171 \text{ minutes of arc} \qquad (m)$$

9.6 GEAR TYPES

Gears are made in many configurations for particular applications. This section describes some of the more common types.

Spur, Helical, and Herringbone Gears

SPUR GEARS are ones in which the *teeth are parallel to the axis of the gear*. This is the simplest and least expensive form of gear to make. Spur gears can only be meshed if their axes are parallel. Figure 9-14 shows a spur gear.

HELICAL GEARS are ones in which the teeth are at a helix angle ψ with respect to the axis of the gear as shown in Figure 9-15a. Figure 9-16 shows a pair of opposite-hand[*] **helical gears** in mesh. Their axes are parallel. Two **crossed helical gears** of the same hand can be meshed with their axes at an angle as shown in Figure 9-17. The helix angles can be designed to accommodate any skew angle between the nonintersecting shafts.

Helical gears are more expensive than spur gears but offer some advantages. They run quieter than spur gears because of the smoother and more gradual contact between their angled surfaces as the teeth come into mesh. Spur gear teeth mesh along their entire face width at once. The sudden impact of tooth on tooth causes vibrations which are heard as a "whine" which is characteristic of spur gears but is absent with helicals. Also, for the same gear diameter and diametral pitch, a helical gear is stronger due to the slightly thicker tooth form in a plane perpendicular to the axis of rotation.

HERRINGBONE GEARS are formed by joining two helical gears of identical pitch and diameter but of opposite hand on the same shaft. These two sets of teeth are often cut on the same gear blank. The advantage compared to a helical gear is the internal cancellation of its axial thrust loads since each "hand" half of the herringbone gear has an oppositely directed thrust load. Thus no thrust bearings are needed other than to locate the shaft axially. This type of gear is much more expensive than a helical gear and tends to be used in large, high-power applications such as ship drives, where the frictional losses from axial loads would be prohibitive. A herringbone gear is shown in Figure 9-15b. Its face view is the same as the helical gear's.

EFFICIENCY The general definition of efficiency is *output power/input power* expressed as a percentage. A spur gearset can be 98 to 99% efficient. The helical gearset is less efficient than the spur gearset due to sliding friction along the helix angle. They

FIGURE 9-14

A spur gear
Courtesy of Martin Sprocket and Gear Co., Arlington, TX

FIGURE 9-16

Parallel axis helical gears
Courtesy of Martin Sprocket and Gear Co., Arlington, TX

[*] Helical gears are either right- or left-handed. Note that the gear of Figure 9-15a is left-handed because, if either face of the gear were placed on a horizontal surface, its teeth would slope up to the left.

Helix angle
ψ

(a) Helical gear

(b) Herringbone gear

FIGURE 9-15

A helical gear and a herringbone gear

TABLE 9-11   Reverted Gearsets and Errors in Ratio for Example 9-4

N_2	N_3	Ratio1	N_4	N_5	Ratio2	m_V	Error
22	39	1.773	22	39	1.773	3.142 562	-9.619 8 E-04
44	78	1.773	44	78	1.773	3.142 562	-9.619 8 E-04

5 The best reverted solution has an error in ratio of –9.619 8 E-04 (–0.030 62%) giving a ratio of 3.142 562 with gearsets of 22:39 and 22:39 teeth.

6 Note that imposing the additional constraint of reversion has reduced the number of possible solutions effectively to one (the two solutions in Table 9-11 differ by a factor of 2 in tooth numbers but have the same error) and the error is much greater than that of even the worst of the 11 nonreverted solutions in Table 9-10.

9.9   EPICYCLIC OR PLANETARY GEAR TRAINS

The conventional gear trains described in the previous sections are all one-degree-of-freedom (*DOF*) devices. Another class of gear train has wide application, the **epicyclic or planetary train**. This is a two-*DOF* device. Two inputs are needed to obtain a predictable output. In some cases, such as the automotive differential, one input is provided (the driveshaft) and two frictionally coupled outputs are obtained (the two driving wheels). In other applications such as automatic transmissions, aircraft engine to propeller reductions, and in-hub bicycle transmissions, two inputs are provided (one usually being a zero velocity, i.e., a fixed gear), and one controlled output results.

Figure 9-32a shows a conventional, one-*DOF* gearset in which link 1 is immobilized as the ground link. Figure 9-32b shows the same gearset with link 1 now free to rotate as an **arm** which connects the two gears. Now only the joint O_2 is grounded and the system *DOF* = 2. This has become an **epicyclic** train with a **sun gear** and a **planet gear** orbiting around the sun, held in orbit by the **arm**. Two inputs are required. Typically, the arm and the sun gear will each be driven in some direction at some velocity. In many cases, one of these inputs will be zero velocity, i.e., a brake applied to either the arm or the sun gear. Note that a zero velocity input to the arm merely makes a conventional train out of the epicyclic train as shown in Figure 9-32a. Thus the conventional gear train is simply a special case of the more complex epicyclic train, in which its arm is held stationary.

In this simple example of an epicyclic train, the only gear left to take an output from, after putting inputs to sun and arm, is the planet. It is a bit difficult to get a usable output from this orbiting gear as its pivot is moving. A more useful configuration is shown in Figure 9-33 to which a ring gear has been added. This **ring gear** meshes with the planet and pivots at O_2, so it can be easily tapped as the output member. Most planetary trains will be arranged with ring gears to bring the planetary motion back to a grounded pivot. Note how the sun gear, ring gear, and arm are all brought out as concentric hollow shafts so that each can be accessed to tap its angular velocity and torque either as an input or an output.

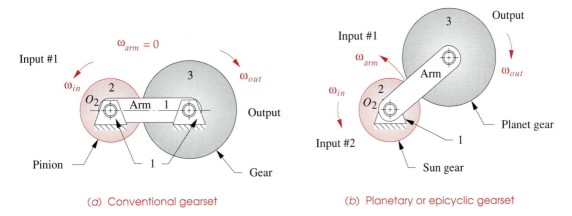

(a) Conventional gearset

(b) Planetary or epicyclic gearset

FIGURE 9-32

Conventional gearsets are special cases of planetary or epicyclic gearsets

Epicyclic trains come in many varieties. Levai[3] catalogued 12 possible types of basic epicyclic trains as shown in Figure 9-34. These basic trains can be connected together to create a larger number of trains having more degrees of freedom. This is done in automotive automatic transmissions as described in a later section.

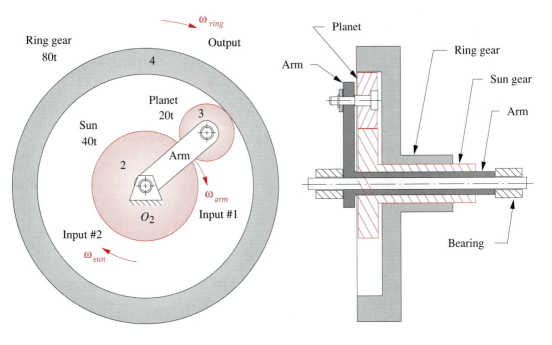

FIGURE 9-33

Planetary gearset with ring gear used as output

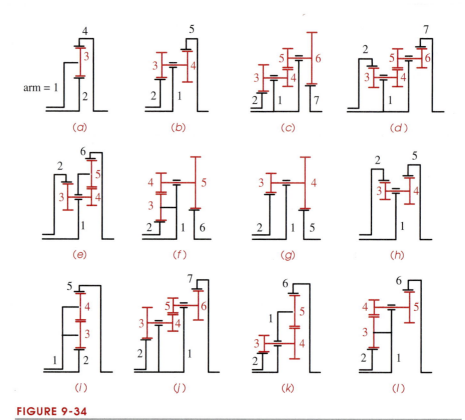

FIGURE 9-34

Levai's 12 possible epicyclic trains (3)

While it is relatively easy to visualize the power flow through a conventional gear train and observe the directions of motion for its member gears, it is very difficult to determine the behavior of a planetary train by observation. We must do the necessary calculations to determine its behavior and may be surprised at the often counterintuitive results. Since the gears are rotating with respect to the arm and the arm itself has motion, we have a velocity difference problem here which requires equation 6.5 (p. 243) be applied to this problem. Rewriting the velocity difference equation in terms of angular velocities specific to this system, we get:

$$\omega_{gear} = \omega_{arm} + \omega_{gear\,/\,arm} \tag{9.12}$$

Equations 9.12 and 9.5a (p. 442) are all that are needed to solve for the velocities in an epicyclic train, provided that the tooth numbers and two input conditions are known.

The Tabular Method

One approach to the analysis of velocities in an epicyclic train is to create a table which represents equation 9.12 for each gear in the train.

✎ EXAMPLE 9-5

Epicyclic Gear Train Analysis by the Tabular Method.

Problem: Consider the train in Figure 9-33 (p. 463) which has the following tooth numbers and initial conditions:

Sun gear	$N_2 = 40$-tooth external gear
Planet gear	$N_3 = 20$-tooth external gear
Ring gear	$N_4 = 80$-tooth internal gear
Input to arm	200 rpm clockwise
Input to sun	100 rpm clockwise

We wish to find the absolute output angular velocity of the ring gear.

Solution:

1 The solution table is set up with a column for each term in equation 9.12 and a row for each gear in the train. It will be most convenient if we can arrange the table so that meshing gears occupy adjacent rows. The table for this method, prior to data entry, is shown in Figure 9-35.

2 Note that the gear ratios are shown straddling the rows of gears to which they apply. The gear ratio column is placed next to the column containing the velocity differences $\omega_{gear/arm}$ because the gear ratios only apply to the velocity difference. The gear ratios **cannot be directly applied to the absolute velocities** in the ω_{gear} column.

3 The solution strategy is simple but is fraught with opportunities for careless errors. Note that we are solving a vector equation with scalar algebra and the signs of the terms denote the sense of the ω vectors which are all directed along the Z axis. Great care must be taken to get the signs of the input velocities and of the gear ratios correct in the table, or the answer will be wrong. Some gear ratios may be negative if they involve external gearsets, and some will be positive if they involve an internal gear. We have both types in this example.

4 The first step is to enter the known data as shown in Figure 9-36 which in this case are the arm velocity (in all rows) and the absolute velocity of gear 2 in column 1. The gear ratios can also be calculated and placed in their respective locations. Note that these ratios should be calculated for each gearset in a consistent manner, following the power flow through the train. That is, starting at gear 2 as the driver, it drives gear 3 directly. This makes its ratio $-N_2/N_3$, or input over output, not the reciprocal. *This ratio is negative because the gearset is external.* Gear 3 in turn drives gear 4 so its ratio is $+N_3/N_4$. *This is a positive ratio because of the internal gear.*

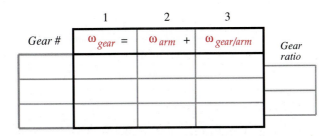

FIGURE 9-35

Table for the solution of planetary gear trains

Gear #	1 $\omega_{gear} =$	2 $\omega_{arm} +$	3 $\omega_{gear/arm}$	Gear Ratio
2	−100	−200		
				−40/20
3		−200		
				+20/80
4		−200		

FIGURE 9-36

Given data for planetary gear train from Example 9-5 placed in solution table

5   Once any one row has two entries, the value for its remaining column can be calculated from equation 9.12. Once any one value in the velocity difference column (column 3) is found, the gear ratios can be applied to calculate all other values in that column. Finally, the remaining rows can be calculated from equation 9.12 to yield the absolute velocities of all gears in column 1. These computations are shown in Figure 9-37 which completes the solution.

6   The overall train value for this example can be calculated from the table and is, from arm to ring gear +1.25:1 and from sun gear to ring gear +2.5:1.

In this example, the arm velocity was given. If it is to be found as the output, then it must be entered in the table as an unknown, x, and the equations solved for that unknown.

FERGUSON'S PARADOX   Epicyclic trains have several advantages over conventional trains among which are higher train ratios in smaller packages, reversion by default, and simultaneous, concentric, bidirectional outputs available from a single unidirectional input. These features make planetary trains popular as automatic transmissions in automobiles and trucks, etc.

The so-called **Ferguson's paradox** of Figure 9-38 illustrates all these features of the planetary train. It is a **compound epicyclic train** with one 20-tooth planet gear (gear 5) carried on the arm and meshing simultaneously with three sun gears. These sun gears have 100 teeth (gear 2), 99 teeth (gear 3), and 101 teeth (gear 4), respectively. The center distances between all sun gears and the planet are the same despite the slightly different pitch diameters of each sun gear. This is possible because of the properties of the involute tooth form as described in Section 9.2 (p. 434). Each sun gear will run smoothly with the planet gear. Each gearset will merely have a slightly different pressure angle.

Gear #	1 $\omega_{gear} =$	2 $\omega_{arm} +$	3 $\omega_{gear/arm}$	Gear Ratio
2	−100	−200	+100	
				−40/20
3	−400	−200	−200	
				+20/80
4	−250	−200	−50	

FIGURE 9-37

Solution for planetary gear train from Example 9-5

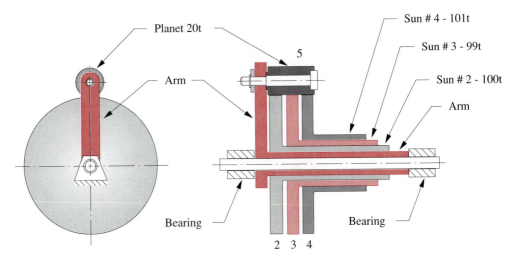

FIGURE 9-38

Ferguson's paradox compound planetary gear train

EXAMPLE 9-6

Analyzing Ferguson's Paradox by the Tabular Method.

Problem: Consider the Ferguson's paradox train in Figure 9-38 which has the following tooth numbers and initial conditions:

Sun gear #2	N_2 = 100-tooth external gear
Sun gear #3	N_3 = 99-tooth external gear
Sun gear #4	N_4 = 101-tooth external gear
Planet gear	N_5 = 20-tooth external gear
Input to sun #2	0 rpm
Input to arm	100 rpm counterclockwise

Sun gear 2 is fixed to the frame, thus providing one input (zero velocity) to the system. The arm is driven at 100 rpm counterclockwise as the second input. Find the angular velocities of the two outputs which are available from this compound train, one from gear 3 and one from gear 4, both of which are free to rotate on the main shaft.

Solution:

1   The tabular solution for this train is set up in Figure 9-39 which shows the given data. Note that the row for gear 5 is repeated for clarity in applying the gear ratio between gears 5 and 4.

2   The known input values of velocity are the arm angular velocity and the zero absolute velocity of gear 2.

3   The gear ratios in this case are all negative because of the external gear sets, and their values reflect the direction of power flow from gear 2 to 5, then 5 to 3 and 5 to 4 in the second branch.

	1	2	3	
Gear #	ω_{gear} =	ω_{arm} +	$\omega_{gear/arm}$	Gear Ratio
2	0	+100		
				−100/20
5		+100		
				−20/99
3		+100		
5		+100		
				−20/101
4		+100		

FIGURE 9-39

Given data for Ferguson's paradox planetary gear train from Example 9-6

4　Figure 9-40 shows the calculated values added to the table. Note that for a **counterclockwise** 100 rpm input to the arm, we get a **counterclockwise** 1 rpm output from gear 4 and a **clockwise** 1 rpm output from gear 3, simultaneously.

This result accounts for the use of the word **paradox** to describe this train. Not only do we get a much larger ratio (100:1) than we could from a conventional train with gears of 100 and 20 teeth, but we have our choice of output directions!

Automotive automatic transmissions use compound planetary trains, which are always in mesh, and which give different ratio forward speeds, plus reverse, by simply engaging and disengaging brakes on different members of the train. The brake provides zero velocity input to one train member. The other input is from the engine. The output is thus modified by the application of these internal brakes in the transmission according to the selection of the operator (**P**ark, **R**everse, **N**eutral, **D**rive, etc.).

	1	2	3	
Gear #	ω_{gear} =	ω_{arm} +	$\omega_{gear/arm}$	Gear Ratio
2	0	+100	−100	
				−100/20
5	+600	+100	+500	
				−20/99
3	−1.01	+100	−101.01	
5	+600	+100	+500	
				−20/101
4	+0.99	+100	−99.01	

FIGURE 9-40

Solution to Ferguson's paradox planetary gear train from example 9-6

The Formula Method

It is not necessary to tabulate the solution to an epicyclic train. The velocity difference formula can be solved directly for the train ratio. We can rearrange equation 9.12 (p. 464) to solve for the velocity difference term. Then, let ω_F represent the angular velocity of the first gear in the train (chosen at either end), and ω_L represent the angular velocity of the last gear in the train (at the other end).

For the first gear in the system:

$$\omega_{F/arm} = \omega_F - \omega_{arm} \qquad (9.13a)$$

For the last gear in the system:

$$\omega_{L/arm} = \omega_L - \omega_{arm} \qquad (9.13b)$$

Dividing the last by the first:

$$\frac{\omega_{L/arm}}{\omega_{F/arm}} = \frac{\omega_L - \omega_{arm}}{\omega_F - \omega_{arm}} = R \qquad (9.13c)$$

This gives an expression for the fundamental train value R which defines a velocity ratio for the train with the arm held stationary. The left-most side of equation 9.13c involves only the velocity difference terms which are relative to the arm. This fraction is equal to the ratio of the products of tooth numbers of the gears from first to last in the train as defined in equation 9.8b (p. 453) which can be substituted for the left-most side of equation 9.13c.

$$R = \pm\frac{\text{product of number of teeth on driver gears}}{\text{product of number of teeth on driven gears}} = \frac{\omega_L - \omega_{arm}}{\omega_F - \omega_{arm}} \qquad (9.14)$$

This equation can be solved for any one of the variables on the right side provided that the other two are defined as the two inputs to this two-*DOF* train. Either the velocities of the arm plus one gear must be known or the velocities of two gears, the first and last, as so designated, must be known. Another limitation of this method is that both the first and last gears chosen must be pivoted to ground (not orbiting), and there must be a path of meshes connecting them, which may include orbiting planet gears. Let us use this method to again solve the Ferguson's paradox of the previous example.

✏️EXAMPLE 9-7

Analyzing Ferguson's Paradox by the Formula Method

Problem: Consider the same Ferguson's paradox train as in Example 9-6 which has the following tooth numbers and initial conditions (See Figure 9-37):

Sun gear #2	$N_2 = 100$-tooth external gear
Sun gear #3	$N_3 = 99$-tooth external gear
Sun gear #4	$N_4 = 101$-tooth external gear
Planet gear	$N_5 = 20$-tooth external gear
Input to sun #2	0 rpm
Input to arm	100 rpm counterclockwise

Sun gear 2 is fixed to the frame, providing one input (zero velocity) to the system. The arm is driven at 100 rpm CCW as the second input. Find the angular velocities of the two outputs which are available from this compound train, one from gear 3 and one from gear 4, both of which are free to rotate on the main shaft.

Solution:

1   We will have to apply equation 9.14 twice, once for each output gear. Taking gear 3 as the last gear in the train with gear 2 as the first, we have:

$$N_2 = 100 \qquad N_3 = 99 \qquad N_5 = 20s$$

$$\omega_{arm} = +100 \qquad \omega_F = 0 \qquad \omega_L = ? \tag{a}$$

2   Substituting in equation 9.14 we get:

$$\left(-\frac{N_2}{N_5}\right)\left(-\frac{N_5}{N_3}\right) = \frac{\omega_L - \omega_{arm}}{\omega_F - \omega_{arm}}$$

$$\left(-\frac{100}{20}\right)\left(-\frac{20}{99}\right) = \frac{\omega_3 - 100}{0 - 100} \tag{b}$$

$$\omega_3 = -1.01$$

3   Now taking gear 4 as the last gear in the train with gear 2 as the first, we have:

$$N_2 = 100 \qquad N_4 = 101 \qquad N_5 = 20$$

$$\omega_{arm} = +100 \qquad \omega_F = 0 \qquad \omega_L = ? \tag{c}$$

4   Substituting in equation 9.14 we get:

$$\left(-\frac{N_2}{N_5}\right)\left(-\frac{N_5}{N_4}\right) = \frac{\omega_L - \omega_{arm}}{\omega_F - \omega_{arm}}$$

$$\left(-\frac{100}{20}\right)\left(-\frac{20}{101}\right) = \frac{\omega_4 - 100}{0 - 100} \tag{d}$$

$$\omega_4 = +0.99$$

These are the same results as were obtained with the tabular method.

9.10   EFFICIENCY OF GEAR TRAINS

The general definition of efficiency is *output power/input power.* It is expressed as a fraction (decimal %) or as a percentage. The efficiency of a conventional gear train (simple or compound) is very high. The power loss per gearset is only about 1 to 2% depending on such factors as tooth finish and lubrication. A gearset's basic efficiency is termed E_0. An external gearset will have an E_0 of about 0.98 or better and an external-internal gearset about 0.99 or better. When multiple gearsets are used in a conventional simple or compound train, the overall efficiency of the train will be the product of the efficiencies of all its stages. For example, a two-stage train with both gearset efficiencies of $E_0 = 0.98$ will have an overall efficiency of $\eta = 0.98^2 = 0.96$.

Epicyclic trains, if properly designed, can have even higher overall efficiencies than conventional trains. But, if the epicyclic train is poorly designed, its efficiency can be so low that it will generate excessive heat and may even be unable to operate at all. This strange result can come about if the orbiting elements (planets) in the train have high losses that absorb a large amount of "circulating power" within the train. It is possible for this circulating power to be much larger than the throughput power for which the train was designed, resulting in excessive heating or stalling. The computation of the overall efficiency of an epicyclic train is much more complicated than the simple multiplication indicated above that works for conventional trains. Molian[4] presents a concise derivation.

To calculate the overall efficiency η of an epicyclic train we need to define a basic ratio ρ which is related to the fundamental train value R defined in equation 9.13c:

$$\text{if } |R| \geq 1, \text{ then } \rho = R \text{ else } \rho = 1/R \qquad (9.15)$$

This constrains ρ to represent a speed increase rather than a decrease regardless of which way the gear train is intended to operate.

For the purpose of calculating torque and power in an epicyclic gear train, we can consider it to be a "black box" with three concentric shafts as shown in Figure 9-41. These shafts are labeled 1, 2, and arm and connect to either "end" of the gear train and to its arm, respectively. Two of these shafts can serve as inputs and the third as output in any combination. The details of the gear train's internal configuration are not needed if we know its basic ratio ρ and the basic efficiency E_0 of its gearsets. All the analysis is done relative to the arm of the train since the internal power flow and losses are only affected by rotation of shafts 1 and 2 with respect to the arm, not by rotation of the entire unit. We also model it as having a single planet gear for the purpose of determining E_0 on the assumption that the power and the losses are equally divided among all gears actually in the train. Counterclockwise torques and angular velocities are considered positive. Power is the product of torque and angular velocity, so a positive power is an input (torque and velocity in same direction) and negative power is an output.

If the gear train is running at constant speed or is changing speed too slowly to significantly affect its internal kinetic energy, then we can assume static equilibrium and the torques will sum to zero.

$$T_1 + T_2 + T_{arm} = 0 \qquad (9.16)$$

The sum of power in and out must also be zero, but the direction of power flow affects the computation. If the power flows from shaft 1 to shaft 2, then:

$$E_0 T_1 (\omega_1 - \omega_{arm}) + T_2 (\omega_2 - \omega_{arm}) = 0 \qquad (9.17a)$$

If the power flows from shaft 2 to shaft 1, then:

$$T_1 (\omega_1 - \omega_{arm}) + E_0 T_2 (\omega_2 - \omega_{arm}) = 0 \qquad (9.17b)$$

If the power flows from shaft 1 to 2, equations 9.16 and 9.17a are solved simultaneously to obtain the system torques. If the power flows in the other direction, then equations 9.16 and 9.17b are used instead. Substitution of equation 9.13c in combination with equation 9.15 introduces the basic ratio ρ and after simultaneous solution yields:

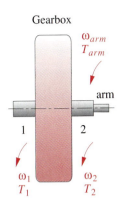

Gearbox

ω_{arm}
T_{arm}

arm

1   2

ω_1 ω_2
T_1 T_2

FIGURE 9-41

Generic epicyclic gear train

9

power flow from 1 to 2

$$T_1 = \frac{T_{arm}}{\rho E_0 - 1} \tag{9.18a}$$

$$T_2 = -\frac{\rho E_0 T_{arm}}{\rho E_0 - 1} \tag{9.18b}$$

power flow from 2 to 1

$$T_1 = \frac{E_0 T_{arm}}{\rho - E_0} \tag{9.19a}$$

$$T_2 = -\frac{\rho T_{arm}}{\rho - E_0} \tag{9.19b}$$

Once the torques are found, the input and output power can be calculated using the known input and output velocities (from a kinematic analysis as described in the previous section) and the efficiency then determined from *output power/input power*.

There are eight possible cases depending on which shaft is fixed, which shaft is input, and whether the basic ratio ρ is positive or negative. These cases are shown in Table 9-12[4] which includes expressions for the train efficiency as well as for the torques. Note that the torque on one shaft is always known from the load required to be driven or the power available from the driver, and this is needed to calculate the other two torques.

TABLE 9-12   Torques and Efficiencies in an Epicyclic Train (4)

Case	ρ	Fixed Shaft	Input Shaft	Train Ratio	T_1	T_2	T_{arm}	Efficiency (η)
1	$> +1$	2	1	$1-\rho$	$-\dfrac{T_{arm}}{1-\rho E_0}$	$\dfrac{\rho E_0 T_{arm}}{1-\rho E_0}$	T_{arm}	$\dfrac{\rho E_0 - 1}{\rho - 1}$
2	$> +1$	2	arm	$\dfrac{1}{1-\rho}$	T_1	$-\rho\dfrac{T_1}{E_0}$	$\left(\dfrac{\rho - E_0}{E_0}\right)T_1$	$\dfrac{E_0(\rho-1)}{\rho - E_0}$
3	$> +1$	1	2	$\dfrac{\rho-1}{\rho}$	$\dfrac{T_{arm}}{\rho E_0 - 1}$	$-\dfrac{\rho E_0 T_{arm}}{\rho E_0 - 1}$	T_{arm}	$\dfrac{\rho E_0 - 1}{E_0(\rho-1)}$
4	$> +1$	1	arm	$\dfrac{\rho}{\rho-1}$	$-\dfrac{E_0}{\rho}T_2$	T_2	$-\left(\dfrac{\rho - E_0}{\rho}\right)T_2$	$\dfrac{\rho-1}{\rho - E_0}$
5	≤ -1	2	1	$1-\rho$	$-\dfrac{T_{arm}}{1-\rho E_0}$	$\dfrac{\rho E_0 T_{arm}}{1-\rho E_0}$	T_{arm}	$\dfrac{\rho E_0 - 1}{\rho - 1}$
6	≤ -1	2	arm	$\dfrac{1}{1-\rho}$	T_1	$-\rho\dfrac{T_1}{E_0}$	$\left(\dfrac{\rho - E_0}{E_0}\right)T_1$	$\dfrac{E_0(\rho-1)}{\rho - E_0}$
7	≤ -1	1	2	$\dfrac{\rho-1}{\rho}$	$\dfrac{E_0 T_{arm}}{\rho - E_0}$	$-\dfrac{\rho T_{arm}}{\rho - E_0}$	T_{arm}	$\dfrac{\rho - E_0}{\rho - 1}$
8	≤ -1	1	arm	$\dfrac{\rho}{\rho-1}$	$-\dfrac{T_2}{\rho E_0}$	T_2	$-\left(\dfrac{\rho E_0 - 1}{\rho E_0}\right)T_2$	$\dfrac{E_0(\rho-1)}{\rho E_0 - 1}$

✍ EXAMPLE 9-8

Determining the Efficiency of an Epicyclic Gear Train.*

Problem: Find the overall efficiency of the epicyclic train shown in Figure 9-42.[5] The basic efficiency E_0 is 0.9928 and the gear tooth numbers are: $N_A = 82t$, $N_B = 84t$, $N_C = 86t$, $N_D = 82t$, $N_E = 82t$, and $N_F = 84t$. Gear A (shaft 2) is fixed to the frame, providing a zero velocity input. The arm is driven as the second input.

Solution:

1 Find the basic ratio ρ for the gear train using equations 9.14 (p. 467) and 9.15 (p. 471). Note that gears B and C have the same velocity as do gears D and E, so their ratios are 1 and thus are omitted.

$$\rho = \frac{N_F N_D N_B}{N_E N_C N_A} = \frac{84(82)(84)}{82(86)(82)} = \frac{1764}{1763} \cong 1.000567 \qquad (a)$$

2 The combination of $\rho > 1$, shaft 2 fixed and input to the arm corresponds to Case 2 in Table 9-12 giving an efficiency of:

$$\eta = \frac{E_0(\rho - 1)}{\rho - E_0} = \frac{0.9928(1.000567 - 1)}{1.000567 - 0.9928} = 0.073 = 7.3\% \qquad (b)$$

3 This is a very low efficiency which makes this gearbox essentially useless. About 93% of the input power is being circulated within the gear train and wasted as heat.

The above example points out a problem with epicyclic gear trains that have basic ratios near unity. They have low efficiency and are useless for transmission of power. Large speed ratios with high efficiency can only be obtained with trains having large basic ratios.[5]

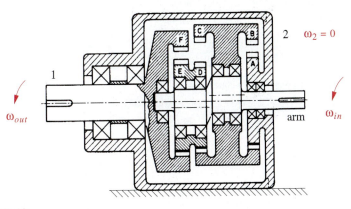

FIGURE 9-42

Epicyclic train for Example 9-8 (5)

* This example is adapted from reference [5].

9.11   TRANSMISSIONS

COMPOUND REVERTED GEAR TRAINS are commonly used in manual (nonautomatic) automotive transmissions to provide user-selectable ratios between the engine and the drive wheels for torque multiplication (mechanical advantage). These gearboxes usually have from three to six forward speeds and one reverse. Most modern transmissions of this type use helical gears for quiet operation. These gears are **not** moved into and out of engagement when shifting from one speed to another except for reverse. Rather, the desired ratio gears are selectively locked to the output shaft by synchromesh mechanisms as in Figure 9-43 which shows a four-speed, manually shifted, synchromesh automotive transmission.

The input shaft is at top left. The input gear is always in mesh with the left-most gear on the countershaft at the bottom. This countershaft has several gears integral with it, each of which meshes with a different output gear that is freewheeling on the output shaft. The output shaft is concentric with the input shaft, making this a reverted train, but the input and output shafts only connect through the gears on the countershaft except in "top gear" (fourth speed), for which the input and output shafts are directly coupled together with a synchromesh clutch for a 1:1 ratio.

The synchromesh clutches are beside each gear on the output shaft and are partially hidden by the shifting collars which move them left and right in response to the driver's hand on the shift lever. These clutches act to lock one gear to the output shaft at a time to provide a power path from input to output of a particular ratio. The arrows on the figure show the power path for third-speed forward, which is engaged. Reverse gear, on the lower right, engages an idler gear which is physically shifted into and out of mesh at standstill.

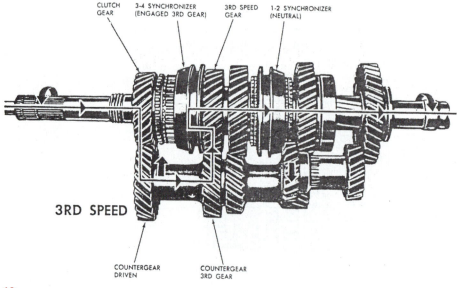

FIGURE 9-43

Four-speed manual synchromesh automobile transmission   *From Crouse, W. H. (1980). Automotive Mechanics, 8th ed.,* *McGraw-Hill, New York, NY, p. 480 Reprinted with permission.*

PLANETARY OR EPICYCLIC TRAINS are commonly used in automatic shifting automotive transmissions as shown in Figure 9-44. At the left is a turbinelike fluid coupling between engine and transmission, called a **torque converter**. This device allows sufficient slip in the coupling fluid to let the engine idle with the transmission engaged and the vehicle's wheels stopped. The engine-driven *impeller blades*, running in oil, transmit torque by pumping oil past a set of stationary *stator blades** and against the *turbine blades* attached to the transmission input shaft. This is one input to the multi-*DOF* transmission that consists of several stages of epicyclic trains. Automatic transmissions can have any number of ratios. Automotive examples typically have from two to five forward speeds. Truck and bus automatic transmissions may have more.

Three epicyclic gearsets can be seen near the center of the four-speed transmission in Figure 9-44. They are controlled by hydraulically operated multidisk clutches and brakes within the transmission that impart zero velocity (second) inputs to various elements of the train to create one of four forward velocity ratios plus reverse in this particular example. The clutches force zero relative velocity between the two elements engaged, and the brakes force zero absolute velocity on the element. Since all gears are in constant mesh, the transmission can be shifted under load by switching the internal brakes and clutches on and off. They are controlled by a combination of inputs that include driver selection (PRND), road speed, throttle position, engine load and speed, and other factors which are automatically monitored and computer controlled. Some modern transmission controllers use artificial intelligence techniques to learn and adapt to the operator's style of driving by automatically resetting the shift points for gentle or aggressive performance based on driving habits.

* The stator blades, which do not move, serve to redirect the flow of oil exiting the impeller blades to a more favorable angle relative to the turbine blades. This redirection of flow is responsible for the torque multiplication that gives the device its name, torque converter. Without the stator blades, it is just a *fluid coupling* which will transmit, but not multiply, the torque. In a torque converter, the maximum torque increase of about 2x occurs at stall when the transmission's turbine is stopped and the engine-driven impeller is turning, creating maximum slip between the two. This torque boost aids in accelerating the vehicle from rest when its inertia must be overcome. The transmitted torque decreases to zero at zero slip between impeller and turbine.

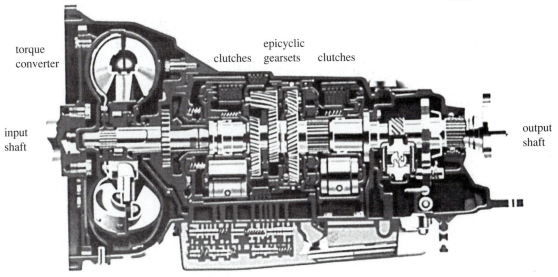

torque converter

input shaft

clutches

epicyclic gearsets

clutches

output shaft

control valves

FIGURE 9-44

Four-speed automatic automobile transmission *Courtesy of Mercedes Benz of North America Inc.*

Range	Clutch/Brake Activation				
	C_1	C_2	B_1	B_2	B_3
First	X			X	
Second	X				X
Third	X					X
Fourth	X	X			
Reverse		X	X		

(a) Schematic of 4-speed automatic transmission

(b) Clutch / brake activation table

FIGURE 9-45

Schematic of automatic transmission from Figure 9-44　*Adapted from reference (6)*

Figure 9-45a shows a schematic of the same transmission as in Figure 9-44. Its three epicyclic stages, two clutches (C_1, C_2), and three band brakes (B_1, B_2, B_3) are depicted. Figure 9-45b shows an activation table of the brake-clutch combinations for each speed ratio of this transmission.[6]

An historically interesting example of an epicyclic train used in a manually shifted gearbox is the Ford Model T transmission shown and described in Figure 9-46. Over 9 million were produced from 1909 to 1927, before the invention of the synchromesh mechanism shown in Figure 9-43 (p. 474). Conventional (compound-reverted) transmissions as used in most other automobiles of that era (and into the 1930's) were unaffec-

The input from the engine is to arm 2. Gear 6 is rigidly attached to the output shaft which drives the wheels.

There are two forward speeds. Low (1 : 2.75) is selected by engaging band brake B_2 to lock gear 7 to the frame. Clutch C is disengaged.

High (1 : 1) is selected by engaging clutch C which locks the input shaft directly to the output shaft.

Reverse (1 : –4) is obtained by engaging brake band B_1 to lock gear 8 to the frame. Clutch C is disengaged.

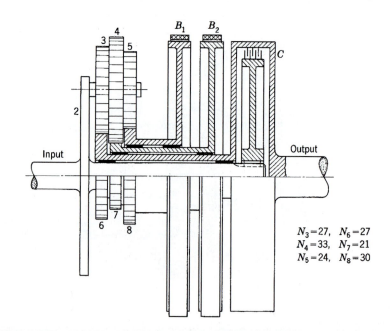

$N_3=27$,　$N_6=27$
$N_4=33$,　$N_7=21$
$N_5=24$,　$N_8=30$

FIGURE 9-46

Ford Model T epicyclic transmission *From R. M. Phelan. (1970). Fundamentals of Mechanical Design, 3rd ed., McGraw-Hill. NY.*

tionately known as "crashboxes," the name being descriptive of the noise made when shifting unsynchronized gears into and out of mesh while in motion. Henry Ford had a better idea. His Model T gears were in constant mesh. The two forward speeds and one reverse were achieved by engaging/disengaging a clutch and band brakes in various combinations via foot pedals. These provided second inputs to the epicyclic train which, like the Ferguson's paradox, gave bidirectional outputs, all without any "crashing" of gear teeth. This Model T transmission is the precursor to all modern automatic transmissions which replace the T's foot pedals with automated hydraulic operation of the clutches and brakes.

9.12 DIFFERENTIALS

A differential is a device that allows a difference in velocity (and displacement) between two elements. This requires a 2-*DOF* mechanism such as an epicyclic gear train. Perhaps the most common application of differentials is in the final drive mechanisms of wheeled land vehicles. When a four-wheeled vehicle turns, the wheels on the outside of the turn must travel farther than the inside wheels due to their different turning radii as shown in Figure 9-47. Without a differential mechanism between the inner and outer driving wheels, the tires must slip on the road surface for the vehicle to turn. If the tires have good traction, a nondifferentiated drive train will attempt to go in a straight line at all times and will fight the driver in turns. In a four-wheel drive (4WD) vehicle, an additional differential is needed between the front and rear wheels to allow the wheel velocities at each end of the vehicle to vary in proportion to the traction developed at either end of the vehicle under slippery conditions. Figure 9-48 shows a 4WD automotive chassis with its three differentials. In this example, the center differential is packaged with the transmission and front differential but effectively is in the driveshaft between the front and rear wheels as shown in Figure 9-47. Differentials are made with various gear types. For rear axle applications, a bevel gear epicyclic is commonly used as shown in Figure 9-49a and in Figure P9-3 (p. 482) in the problem section. For center and front differentials, helical or spur gear arrangements are often used as in Figure 9-49b and c.

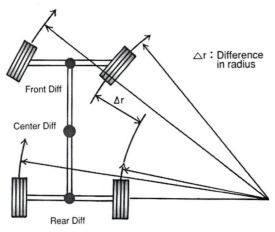

FIGURE 9-47

Turning behavior of a four-wheel vehicle *Courtesy of Tochigi Fuji Sangyo, Japan.*

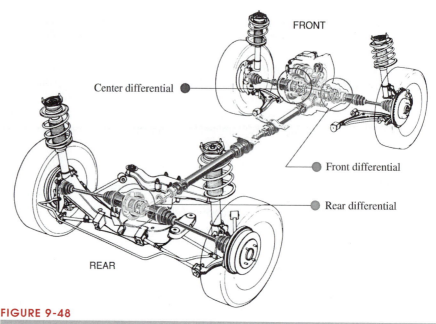

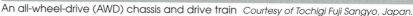

FRONT

Center differential

Front differential

Rear differential

REAR

(a)

FIGURE 9-48

An all-wheel-drive (AWD) chassis and drive train   *Courtesy of Tochigi Fuji Sangyo, Japan.*

(b)

(c)

FIGURE 9-49

Differentials
*Courtesy of Tochigi Fuji
Sangyo, Japan*

An epicyclic train used as a differential has one input and two outputs. Taking the rear differential in an automobile as an example, its input is from the driveshaft and its outputs are to the right and left wheels. The two outputs are coupled through the road via the traction (friction) forces between tires and pavement. The relative velocity between each wheel can vary from zero when both tires have equal traction and the car is not turning, to twice the epicyclic train's input speed when one wheel is on ice and the other has traction. Front or rear differentials split the torque equally between their two wheel outputs. Since power is the product of torque and angular velocity, and power out cannot exceed power in, the power is split between the wheels according to their velocities. When traveling straight ahead (both wheels having traction), half the power goes to each wheel. As the car turns, the faster wheel gets more power and the slower one less. When one wheel loses traction (as on ice), it gets *all* the power (50% torque x 200% speed), and the wheel with traction gets zero power (50% torque x 0% speed). This is why 4WD is needed in slippery conditions. The center differential splits the torque between front and rear in some proportion. If one end of the car loses traction, the other may still be able to control it provided it still has traction.

LIMITED SLIP DIFFERENTIALS   Because of their behavior when one wheel loses traction, various designs have been created to limit the slip between the two outputs under those conditions. These are called limited slip differentials and typically provide some type of friction device between the two output gears to transmit some torque but still allow slip for turning. Some use a fluid coupling between the gears, and others use spring-loaded friction disks or cones as can be seen in Figure 9-49a. Some use an electrically controlled clutch within the epicyclic train to lock it up on demand for off-road applications as shown in Figure 9-49c. The Torsen® differential uses wormsets whose resistance to backdriving (controlled by the choice of worm lead angle) provides some torque coupling between the outputs.

9.13 REFERENCES

1 **DilPare, A. L.** (1970). "A Computer Algorithm to Design Compound Gear Trains for Arbitrary Ratio." *J. of Eng. for Industry*, **93B**(1), pp. 196-200.

2 **Selfridge, R. G., and D. L. Riddle**. (1978). "Design Algorithms for Compound Gear Train Ratios." ASME Paper: 78-DET-62.

3 **Levai, Z.** (1968). "Structure and Analysis of Planetary Gear Trains." *Journal of Mechanisms*, **3**, pp. 131-148.

4 **Molian, S.** (1982). *Mechanism Design: An Introductory Text.* Cambridge University Press: Cambridge, p. 148.

5 **Auksmann, B., and D. A. Morelli**. (1963). "Simple Planetary-Gear System." ASME Paper: 63-WA-204.

6 **Pennestri, E., et al.** (1993). "A Catalog of Automotive Transmissions with Kinematic and Power Flow Analyses." *Proc. of 3rd Applied Mechanisms and Robotics Conference*, Cincinnati, pp. 57-1.

9.14 PROBLEMS

*†9-1 A 22-tooth gear has AGMA standard full-depth involute teeth with diametral pitch of 4. Calculate the pitch diameter, circular pitch, addendum, dedendum, tooth thickness, and clearance.

†9-2 A 40-tooth gear has AGMA standard full-depth involute teeth with diametral pitch of 10. Calculate the pitch diameter, circular pitch, addendum, dedendum, tooth thickness, and clearance.

†9-3 A 30-tooth gear has AGMA standard full-depth involute teeth with diametral pitch of 12. Calculate the pitch diameter, circular pitch, addendum, dedendum, tooth thickness, and clearance.

9-4 Using any available string, some tape, a pencil, and a drinking glass or tin can, generate and draw an involute curve on a piece of paper. With your protractor, show that all normals to the curve are tangent to the base circle.

*9-5 A spur gearset has pitch diameters of 4.5 and 12 in. What is the largest tooth size, in terms of diametral pitch, that can be used without having any interference and undercutting?

    a. For a 20° pressure angle.

    b. For a 25° pressure angle. (Note that diametral pitch need not be an integer.)

*†9-6 Design a simple, spur gear train for a ratio of –9:1 and diametral pitch of 8. Specify pitch diameters and numbers of teeth. Calculate the contact ratio.

*†9-7 Design a simple, spur gear train for a ratio of +8:1 and diametral pitch of 6. Specify pitch diameters and numbers of teeth. Calculate the contact ratio.

†9-8 Design a simple, spur gear train for a ratio of –7:1 and diametral pitch of 8. Specify pitch diameters and numbers of teeth. Calculate the contact ratio.

†9-9 Design a simple, spur gear train for a ratio of +6.5:1 and diametral pitch of 5. Specify pitch diameters and numbers of teeth. Calculate the contact ratio.

*†9-10 Design a compound, spur gear train for a ratio of –70:1 and diametral pitch of 10. Specify pitch diameters and numbers of teeth. Sketch the train to scale.

9

* Answers in Appendix F.

† These problems are suited to solution using *Mathcad, Matlab*, or *TKSolver* equation solver programs.

†9-11 Design a compound, spur gear train for a ratio of 50:1 and diametral pitch of 8. Specify pitch diameters and numbers of teeth. Sketch the train to scale.

*†9-12 Design a compound, spur gear train for a ratio of 150:1 and diametral pitch of 6. Specify pitch diameters and numbers of teeth. Sketch the train to scale.

†9-13 Design a compound, spur gear train for a ratio of −250:1 and diametral pitch of 9. Specify pitch diameters and numbers of teeth. Sketch the train to scale.

*†9-14 Design a compound, reverted, spur gear train for a ratio of 30:1 and diametral pitch of 10. Specify pitch diameters and numbers of teeth. Sketch the train to scale.

†9-15 Design a compound, reverted, spur gear train for a ratio of 40:1 and diametral pitch of 8. Specify pitch diameters and numbers of teeth. Sketch the train to scale.

*†9-16 Design a compound, reverted, spur gear train for a ratio of 75:1 and diametral pitch of 12. Specify pitch diameters and numbers of teeth. Sketch the train to scale.

†9-17 Design a compound, reverted, spur gear train for a ratio of 7:1 and diametral pitch of 4. Specify pitch diameters and numbers of teeth. Sketch the train to scale.

†9-18 Design a compound, reverted, spur gear train for a ratio of 12:1 and diametral pitch of 6. Specify pitch diameters and numbers of teeth. Sketch the train to scale.

*†9-19 Design a compound, reverted, spur gear transmission which will give two shiftable ratios of +3:1 forward and −4.5:1 reverse with diametral pitch of 6. Specify pitch diameters and numbers of teeth. Sketch the train to scale.

†9-20 Design a compound, reverted, spur gear transmission which will give two shiftable ratios of +5:1 forward and −3.5:1 reverse with diametral pitch of 6. Specify pitch diameters and numbers of teeth. Sketch the train to scale.

*†9-21 Design a compound, reverted, spur gear transmission which will give three shiftable ratios of +6:1, +3.5:1 forward and −4:1 reverse with diametral pitch of 8. Specify pitch diameters and numbers of teeth. Sketch the train to scale.

†9-22 Design a compound, reverted, spur gear transmission which will give three shiftable ratios of +4.5:1, +2.5:1 forward and −3.5:1 reverse with diametral pitch of 5. Specify pitch diameters and numbers of teeth. Sketch the train to scale.

†9-23 Design the rolling cones for a −3:1 ratio and a 60° included angle between the shafts. Sketch the train to scale.

†9-24 Design the rolling cones for a −4.5:1 ratio and a 40° included angle between the shafts. Sketch the train to scale.

TABLE P9-1   Data for Problem 9-25

Row	N_2	N_3	N_4	N_5	N_6	ω_2	ω_6	ω_{arm}
a	30	25	45	50	200	?	20	− 50
b	30	25	45	50	200	30	?	− 90
c	30	25	45	50	200	50	0	?
d	30	25	45	30	160	?	40	− 50
e	30	25	45	30	160	50	?	− 75
f	30	25	45	30	160	50	0	?

* Answers in Appendix F.

† These problems are suited to solution using *Mathcad, Matlab,* or *TKSolver* equation solver programs.

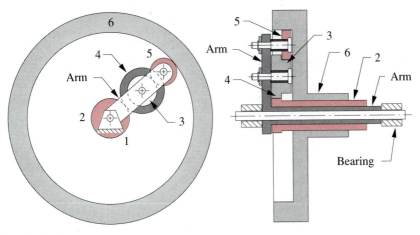

FIGURE P9-1

Planetary gearset for Problem 9-25

*†9-25 Figure P9-1 shows a compound planetary gear train (not to scale). Table P9-1 gives data for gear numbers of teeth and input velocities. For the row(s) assigned, find the variable represented by a question mark.

*†9-26 Figure P9-2 shows a compound planetary gear train (not to scale). Table P9-2 gives data for gear numbers of teeth and input velocities. For the row(s) assigned, find the variable represented by a question mark.

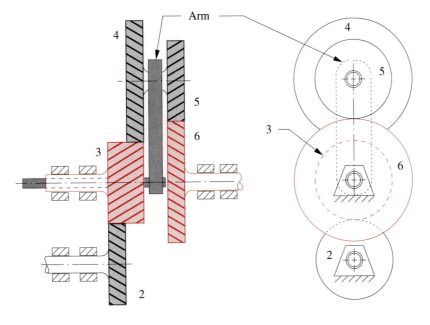

FIGURE P9-2

Compound planetary gear train for Problem 9-26

TABLE P9-2   Data for Problem 9-26

Row	N_2	N_3	N_4	N_5	N_6	ω_2	ω_6	ω_{arm}
a	50	25	45	30	40	?	20	− 50
b	30	35	55	40	50	30	?	− 90
c	40	20	45	30	35	50	0	?
d	25	45	35	30	50	?	40	− 50
e	35	25	55	35	45	30	?	− 75
f	30	30	45	40	35	40	0	?

*†9-27   Figure P9-3 shows a planetary gear train used in an automotive rear-end differential (not to scale). The car has wheels with a 15-inch rolling radius and is moving forward in a straight line at 50 mph. The engine is turning 2000 rpm. The transmission is in direct drive (1:1) with the driveshaft.

    a.   What is the rear wheels' rpm and the gear ratio between ring and pinion?

    b.   As the car hits a patch of ice, the right wheel speeds up to 800 rpm. What is the speed of the left wheel? Hint: The average of both wheels' rpm is a constant.

    c.   Calculate the fundamental train value of the epicyclic stage.

†9-28   Design a speed-reducing planetary gearbox to be used to lift a 5-ton load 50 ft with a motor that develops 20 lb-ft of torque at its operating speed of 1750 rpm. The available winch drum has no more than a 16-in diameter when full of its steel cable. The speed reducer should be no larger in diameter than the winch drum. Gears of no more than about 75 teeth are desired, and diametral pitch needs to be no smaller than

* Answers in Appendix F.

† These problems are
suited to solution using
Mathcad, Matlab, or
TKSolver equation solver
programs.

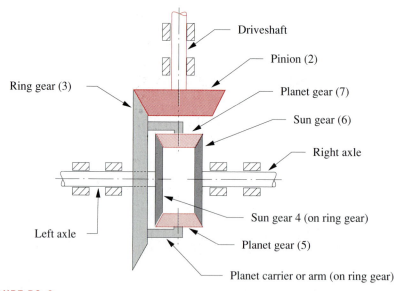

FIGURE P9-3

Automotive differential planetary gear train for Problem 9-27

6 to stand the stresses. Make multiview sketches of your design and show all calculations. How long will it take to raise the load with your design?

*†9-29   Determine all possible two-stage compound gear combinations that will give an approximation to the Naperian base 2.71828. Limit tooth numbers to between 18 and 80. Determine the arrangement that gives the smallest error.

†9-30   Determine all possible two-stage compound gear combinations that will give an approximation to 2π. Limit tooth numbers to between 15 and 90. Determine the arrangement that gives the smallest error.

†9-31   Determine all possible two-stage compound gear combinations that will give an approximation to $\pi/2$. Limit tooth numbers to between 20 and 100. Determine the arrangement that gives the smallest error.

†9-32   Determine all possible two-stage compound gear combinations that will give an approximation to $3\pi/2$. Limit tooth numbers to between 20 and 100. Determine the arrangement that gives the smallest error.

†9-33   Figure P9-4a shows a reverted clock train. Design it using 25° nominal pressure angle gears of 24 p_d having between 12 and 150 teeth. Determine the tooth numbers and nominal center distance. If the center distance has a manufacturing tolerance of ± 0.006 in, what will the pressure angle and backlash at the minute hand be at each extreme of the tolerance?

†9-34   Figure P9-4b shows a three-speed shiftable transmission. Shaft *F*, with the cluster of gears *E, G,* and *H,* is capable of sliding left and right to engage and disengage the

---

* Answers in Appendix F.

---

† These problems are suited to solution using *Mathcad, Matlab,* or *TKSolver* equation solver programs.

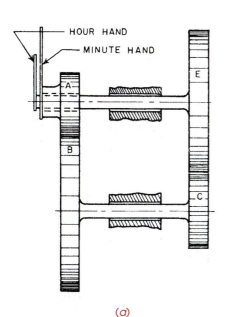

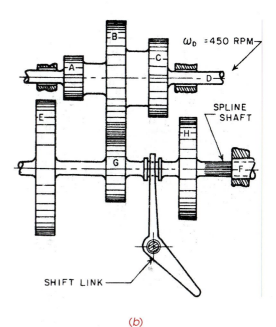

(a)   (b)

FIGURE P9-4

Problems 9-33 to 9-34 *From P. H. Hill and W. P. Rule. (1960). Mechanisms: Analysis and Design, with permission*

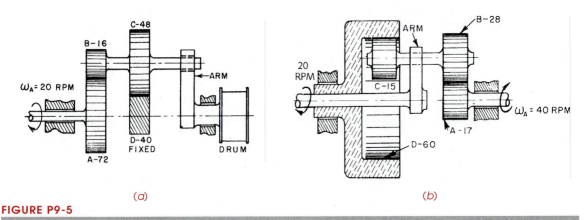

(a)   *(b)*

FIGURE P9-5

Problems 9-35 to 9-36 *From P. H. Hill and W. P. Rule. (1960). Mechanisms: Analysis and Design, with permission*

three gearsets in turn. Design the three reverted stages to give output speeds at shaft *F* of 150, 350, and 550 rpm for an input speed of 450 rpm to shaft *D*.

[†]9-35 Figure P9-5a shows a compound epicyclic train used to drive a winch drum. Gear *A* is driven at 20 rpm *CW* and gear *D* is fixed to ground. The tooth numbers are indicated in the figure. Determine the speed and direction of the drum. What is the efficiency of this train if the basic gearsets have $E_0 = 0.98$?

[†]9-36 Figure P9-5b shows a compound epicyclic train. The arm is driven *CCW* at 20 rpm. Gear *A* is driven *CW* at 40 rpm. The tooth numbers are indicated in the figure. Find the speed of the ring gear *D*.

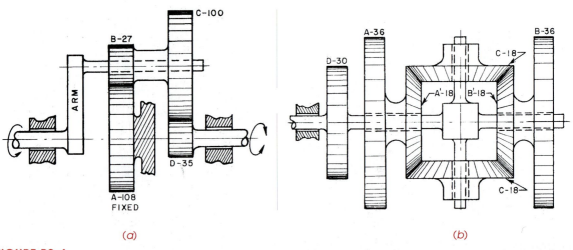

(a)   *(b)*

FIGURE P9-6

Problems 9-37 to 9-38 *From P. H. Hill and W. P. Rule. (1960). Mechanisms: Analysis and Design, with permission*

†9-37 Figure P9-6a shows an epicyclic train. The tooth numbers are indicated in the figure.
The arm is driven *CW* at 60 rpm and gear *A* on shaft 1 is fixed to ground. Find the
speed of gear *D* on shaft 2. What is the efficiency of this train if the basic gearsets
have $E_0 = 0.98$?

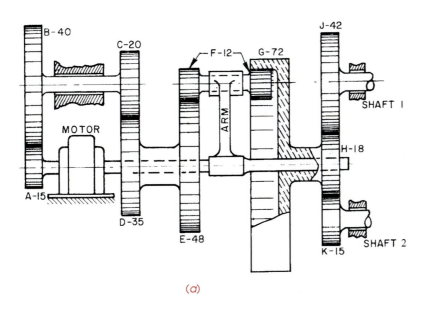

(a)

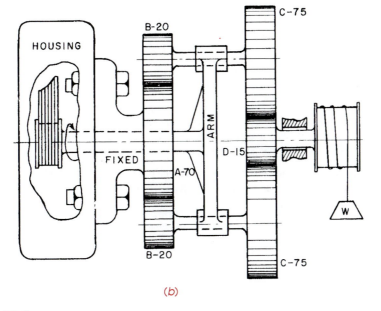

(b)

FIGURE P9-7

Problems 9-39 to 9-40 *From P. H. Hill and W. P. Rule. (1960). Mechanisms: Analysis and Design*

† These problems are
suited to solution using
Mathcad, Matlab, or
TKSolver equation solver
programs.

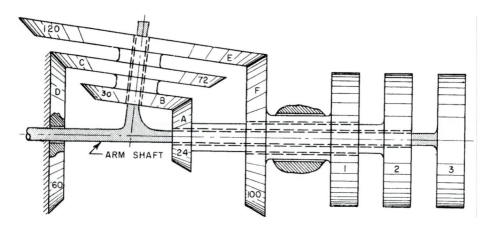

FIGURE P9-8

Problem 9-41   *From P. H. Hill and W. P. Rule. (1960). Mechanisms: Analysis and Design, with permission*

†9-38   Figure P9-6b shows a differential. Gear A is driven *CCW* at 10 rpm and gear B is driven *CW* at 24 rpm. The tooth numbers are indicated in the figure. Find the speed of gear D.

†9-39   Figure P9-7a shows a gear train containing both compound-reverted and epicyclic stages. The tooth numbers are indicated in the figure. The motor is driven *CCW* at 1750 rpm. Find the speeds of shafts 1 and 2.

†9-40   Figure P9-7b shows an epicyclic train used to drive a winch drum. The arm is driven at 250 rpm *CCW* and gear A, on shaft 2, is fixed to ground. The tooth numbers are indicated in the figure. Determine the speed and direction of the drum on shaft 1. What is the efficiency of this train if the basic gearsets have $E_0 = 0.98$?

†9-41   Figure P9-8 shows a compound epicyclic train. Gear 2 is driven at 800 rpm *CCW* and gear D is fixed to ground. The tooth numbers are indicated in the figure. Determine the speed and direction of gears 1 and 3.

†9-42   Figure P9-9a shows a compound epicyclic train. Shaft 1 is driven at 300 rpm *CCW* and gear A is fixed to ground. The tooth numbers are indicated in the figure. Determine the speed and direction of shaft 2.

†9-43   Figure P9-9b shows a compound epicyclic train. Shaft 1 is driven at 40 rpm. The tooth numbers are indicated in the figure. Determine the speed and direction of gears G and M.

†9-44   Calculate the ratios in the Model T transmission shown in Figure 9-46 (p. 476).

---

† These problems are suited to solution using *Mathcad, Matlab,* or *TKSolver* equation solver programs.

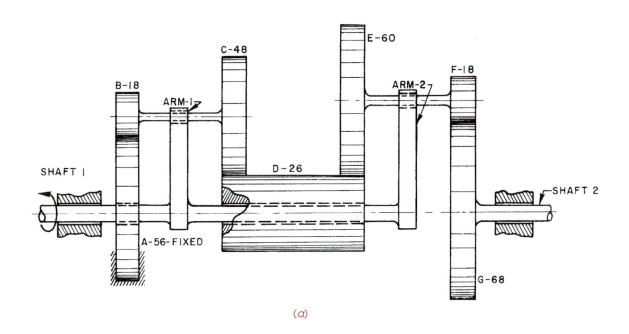

(a)

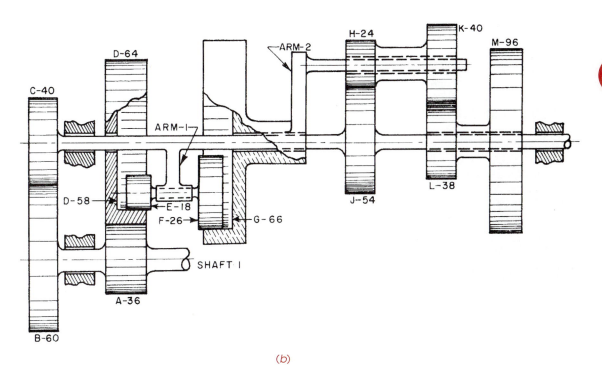

(b)

FIGURE P9-9

Problems 9-42 to 9-43 *From P. H. Hill and W. P. Rule. (1960). Mechanisms: Analysis and Design, with permission*

End
of
Part I

PART II

*The entire world of machinery ...
is inspired by the play of organs
of reproduction. The designer
animates artificial objects by
simulating the movements of
animals engaged in propagating
the species. Our machines are
Romeos of steel and Juliets of
cast iron.*

J. COHEN. (1966). *Human
Robots in Myth and Science,*
Allen & Unwin, London, p. 67.

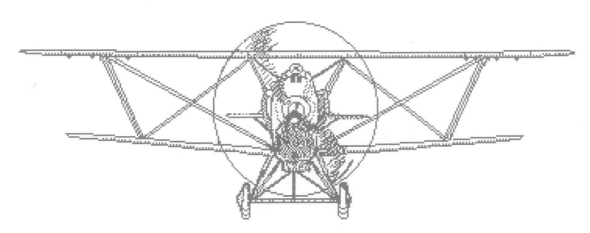

DYNAMICS OF MACHINERY

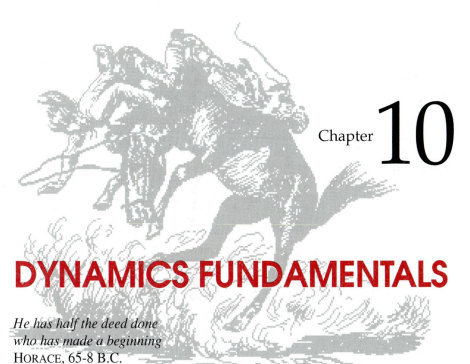

Chapter 10

DYNAMICS FUNDAMENTALS

He has half the deed done
who has made a beginning
HORACE, 65-8 B.C.

10.0 INTRODUCTION

Part I of this text has dealt with the **kinematics** of mechanisms while temporarily ignoring the forces present in those mechanisms. This second part will address the problem of determining the forces present in moving mechanisms and machinery. This topic is called **kinetics** or **dynamic force analysis**. We will start with a brief review of some fundamentals needed for dynamic analysis. It is assumed that the reader has had an introductory course in dynamics. If that topic is rusty, one can review it by referring to reference [1] or to any other text on the subject.

10.1 NEWTON'S LAWS OF MOTION

Dynamic force analysis involves the application of **Newton's** three **laws of motion** which are:

1 *A body at rest tends to remain at rest and a body in motion at constant velocity will tend to maintain that velocity unless acted upon by an external force.*

2 *The time rate of change of momentum of a body is equal to the magnitude of the applied force and acts in the direction of the force.*

3 *For every action force there is an equal and opposite reaction force.*

The second law is expressed in terms of rate of change of *momentum*, $\mathbf{M} = m\mathbf{v}$, where m is mass and $\mathbf{v}$ is velocity. Mass m is assumed to be constant in this analysis. The time rate of change of $m\mathbf{v}$ is $m\mathbf{a}$, where $\mathbf{a}$ is the acceleration of the mass center.

$$\mathbf{F} = m\mathbf{a} \tag{10.1}$$

$\mathbf{F}$ is the resultant of all forces on the system acting at the mass center.

We can differentiate between two subclasses of dynamics problems depending upon which quantities are known and which are to be found. The "**forward dynamics problem**" is the one in which we know everything about the external loads (forces and/or torques) being exerted on the system, and we wish to determine the accelerations, velocities, and displacements which result from the application of those forces and torques. This subclass is typical of the problems you probably encountered in an introductory dynamics course, such as determining the acceleration of a block sliding down a plane, acted upon by gravity. Given $\mathbf{F}$ and m, solve for $\mathbf{a}$.

The second subclass of dynamics problem, called the "**inverse dynamics problem**" is one in which we know the (desired) accelerations, velocities, and displacements to be imposed upon our system and wish to solve for the magnitudes and directions of the forces and torques which are necessary to provide the desired motions and which result from them. This inverse dynamics case is sometimes also called **kinetostatics**. Given $\mathbf{a}$ and m, solve for $\mathbf{F}$.

Whichever subclass of problem is addressed, it is important to realize that they are both dynamics problems. Each merely solves $\mathbf{F} = m\mathbf{a}$ for a different variable. To do so we must first review some fundamental geometric and mass properties which are needed for the calculations.

10.2   DYNAMIC MODELS

It is often convenient in dynamic analysis to create a simplified model of a complicated part. These models are sometimes considered to be a collection of point masses connected by massless rods. For a model of a rigid body to be dynamically equivalent to the original body, three things must be true:

1 *The mass of the model must equal that of the original body.*

2 *The center of gravity must be in the same location as that of the original body.*

3 *The mass moment of inertia must equal that of the original body.*

10.3   MASS

Mass is not weight! Mass is an invariant property of a rigid body. The weight of the same body varies depending on the gravitational system in which it sits. See Section 1.10 (p. 16) for a discussion of the use of proper mass units in various measuring systems. We will assume the mass of our parts to be constant in our calculations. For most earthbound machinery, this is a reasonable assumption. The rate at which an automobile or bulldozer loses mass due to fuel consumption, for example, is slow enough to be ignored when calculating dynamic forces over short time spans. However, this would not be a safe assumption for a vehicle such as the space shuttle, whose mass changes rapidly and drastically during liftoff.

When designing machinery, we must first do a complete kinematic analysis of our design, as described in Part I of this text, in order to obtain information about the accelerations of the moving parts. We next want to use Newton's second law to calculate the dynamic forces. But to do so we need to know the masses of all the moving parts which have these known accelerations. These parts do not exist yet! As with any design problem, we lack sufficient information at this stage of the design to accurately determine the best sizes and shapes of the parts. We must estimate the masses of the links and other parts of the design in order to make a first pass at the calculation. We will then have to iterate to better and better solutions as we generate more information. See Section 1.5 (p. 7) on the design process to review the use of iteration in design.

A first estimate of your parts' masses can be obtained by assuming some reasonable shapes and sizes for all the parts and choosing appropriate materials. Then calculate the volume of each part and multiply its volume by the material's **mass density** (not weight density) to obtain a first approximation of its mass. These mass values can then be used in Newton's equation. The densities of some common engineering materials can be found in Appendix B.

How will we know whether our chosen sizes and shapes of links are even acceptable, let alone optimal? Unfortunately, we will not know until we have carried the computations all the way through a complete stress and deflection analysis of the parts. It is often the case, especially with long, thin elements such as shafts or slender links, that the deflections of the parts under their dynamic loads will limit the design even at low stress levels. In some cases the stresses will be excessive.

We will probably discover that the parts fail under the dynamic forces. Then we will have to go back to our original assumptions about the shapes, sizes, and materials of these parts, redesign them, and repeat the force, stress, and deflection analyses. Design is, unavoidably, an **iterative process**.

The topic of stress and deflection analysis is beyond the scope of this text and will not be further discussed here. It is mentioned only to put our discussion of dynamic force analysis into context. We are analyzing these dynamic forces primarily to provide the information needed to do the stress and deflection analyses on our parts! It is also worth noting that, unlike a static force situation in which a failed design might be fixed by adding more mass to the part to strengthen it, to do so in a dynamic force situation can have a deleterious effect. More mass with the same acceleration will generate even higher forces and thus higher stresses! The machine designer often needs to remove mass (in the right places) from parts in order to reduce the stresses and deflections due to $\mathbf{F} = m\mathbf{a}$. Thus the designer needs to have a good understanding of both material properties and stress and deflection analysis to properly shape and size parts for minimum mass while maximizing the strength and stiffness needed to withstand the dynamic forces.

10.4 MASS MOMENT AND CENTER OF GRAVITY

When the mass of an object is distributed over some dimensions, it will possess a moment with respect to any axis of choice. Figure 10-1 shows a mass of general shape in an *xyz* axis system. A differential element of mass is also shown. The **mass moment (first moment of mass)** of the differential element is equal to the **product of its mass and its distance** from the axis of interest. With respect to the *x*, *y*, and *z* axes these are:

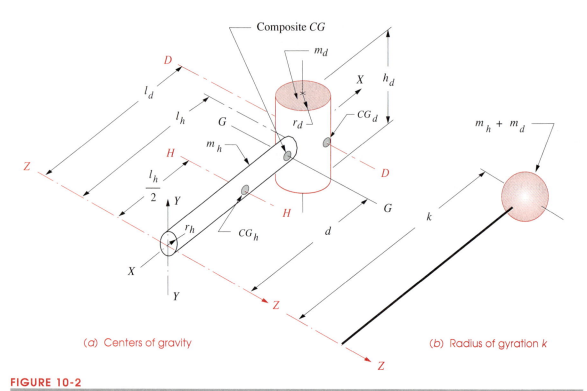

(a) Centers of gravity   (b) Radius of gyration k

FIGURE 10-2

Dynamic models, composite center of gravity, and radius of gyration of a mallet

While it is fairly intuitive to appreciate the physical significance of the first moment of mass, it is more difficult to do the same for the second moment, or moment of inertia.

Consider equation 10.4. It says that torque is proportional to angular acceleration, and the constant of proportionality is this moment of inertia, I. Picture a common hammer or mallet as depicted in Figure 10-2. The head, made of steel, has large mass compared to the light wooden handle. When gripped properly, at the end of the handle, the radius to the mass of the head is large. Its contribution to the total I of the mallet is proportional to the square of the radius from the axis of rotation (your wrist at axis ZZ) to the head. Thus it takes considerably more torque to swing (and thus angularly accelerate) the mallet when it is held properly than if held near the head. As a child you probably chose to hold a hammer close to its head because you lacked the strength to provide the larger torque it needed when held properly. You also found it ineffective in driving nails when held close to the head because you were unable to store very much **kinetic energy** in it. In a translating system kinetic energy is:

$$KE = \frac{1}{2}mv^2 \qquad (10.7a)$$

and in a rotating system kinetic energy is:

$$KE = \frac{1}{2}I\omega^2 \qquad (10.7b)$$

Thus the kinetic energy stored in the mallet is also proportional to its moment of inertia I and to ω^2. So, you can see that holding the mallet close to its head reduces the I and lowers the energy available for driving the nail.

Moment of inertia then is one indicator of the ability of the body to store rotational kinetic energy and is also an indicator of the amount of torque that will be needed to rotationally accelerate the body. Unless you are designing a device intended for the storage and transfer of large amounts of energy (punch press, drop hammer, rock crusher etc.) you will probably be trying to minimize the moments of inertia of your rotating parts. Just as mass is a measure of resistance to linear acceleration, moment of inertia is a measure of resistance to angular acceleration. A large I will require a large driving torque and thus a larger and more powerful motor to obtain the same acceleration. Later we will see how to make moment of inertia work for us in rotating machinery by using flywheels with large I. The units of moment of inertia can be determined by doing a unit balance on either equation 10.4 or equation 10.7 and are shown in Table 1-4 (p. 19). In the **ips** system they are lb-in-sec^2 or blob-in^2. In the **SI** system, they are N-m-sec^2 or kg-m^2.

10.6   PARALLEL AXIS THEOREM (TRANSFER THEOREM)

The moment of inertia of a body with respect to any axis *(ZZ)* can be expressed as the sum of its moment of inertia about an axis *(GG)* parallel to ZZ through its *CG*, and the square of the perpendicular distance between those parallel axes.

$$I_{ZZ} = I_{GG} + md^2 \qquad (10.8)$$

where ZZ and GG are parallel axes, GG goes through the *CG* of the body or assembly, m is the mass of the body or assembly, and d is the perpendicular distance between the parallel axes. This property is most useful when computing the moment of inertia of a complex shape which has been broken into a collection of simple shapes as shown in Figure 10-2a which represents a simplistic model of a mallet. The mallet is broken into two cylindrical parts, the handle and the head, which have masses m_h and m_d, and radii r_h and r_d, respectively. The expressions for the mass moments of inertia of a cylinder with respect to axes through its *CG* can be found in Appendix C and are for the handle about its *CG* axis *HH*:

$$I_{HH} = \frac{m_h\left(3r_h^2 + l_h^2\right)}{12} \qquad (10.9a)$$

and for the head about its *CG* axis *DD*:

$$I_{DD} = \frac{m_d\left(3r_d^2 + h_d^2\right)}{12} \qquad (10.9b)$$

Using the parallel axis theorem to transfer the moment of inertia to the axis ZZ at the end of the handle:

$$I_{ZZ} = \left[I_{HH} + m_h\left(\frac{l_h}{2}\right)^2\right] + \left[I_{DD} + m_d l_d^2\right] \qquad (10.9c)$$

10.7   RADIUS OF GYRATION

The **radius of gyration** of a body is defined as the radius at which the entire mass of the body could be concentrated such that the resulting model will have the same moment of inertia as the original body. The mass of this model must be the same as that of the original body. Let I_{ZZ} represent the mass moment of inertia about ZZ from equation 10.9c and m the mass of the original body. From the parallel axis theorem, a concentrated mass m at a radius k will have a moment of inertia:

$$I_{ZZ} = mk^2 \tag{10.10a}$$

Since we want I_{ZZ} to be equal to the original moment of inertia, the required **radius of gyration** at which we will concentrate the mass m is then:

$$k = \sqrt{\frac{I_{ZZ}}{m}} \tag{10.10b}$$

Note that this property of radius of gyration allows the construction of an even simpler dynamic model of the system in which all the system mass is concentrated in a "point mass" at the end of a massless rod of length k. Figure 10-2b shows such a model of the mallet in Figure 10-2a.

By comparing equation 10.10a with equation 10.8, it can be seen that the radius of gyration k will always be larger than the radius to the composite CG of the original body.

$$I_{CG} + md^2 = I_{ZZ} = mk^2 \qquad \therefore k > d \tag{10.10c}$$

Appendix C contains formulas for the moments of inertia and radii of gyration of some common shapes.

10.8   CENTER OF PERCUSSION

The **center of percussion** is a point on a body which, when struck with a force, will have associated with it another point called the **center of rotation** at which there will be a zero reaction force. You have probably experienced the result of "missing the center of percussion" when you hit a baseball or softball with the wrong spot on the bat. The "right place on the bat" to hit the ball is the center of percussion associated with the point that your hands grip the bat (the center of rotation). Hitting the ball at other than the center of percussion results in a stinging force being delivered to your hands. Hit the right spot and you feel no force (nor pain). The center of percussion is sometimes called the "sweet spot" on a bat, tennis racquet, or golf club. In the case of our mallet example, a center of percussion at the head corresponds to a center of rotation near the end of the handle, and the handle is usually contoured to encourage gripping it there.

The explanation of this phenomenon is quite simple. To make the example two dimensional and eliminate the effects of friction, consider a hockey stick of mass m lying on the ice as shown in Figure 10-3a. Strike it a sharp blow at point P with a force $\mathbf{F}$ perpendicular to the stick axis. The stick will begin to travel across the ice in complex planar motion, both rotating and translating. Its complex motion at any instant can be considered as the superposition of two components: pure translation of its center of gravity G in the direction of $\mathbf{F}$ and pure rotation about that point G. Set up an embedded coordi-

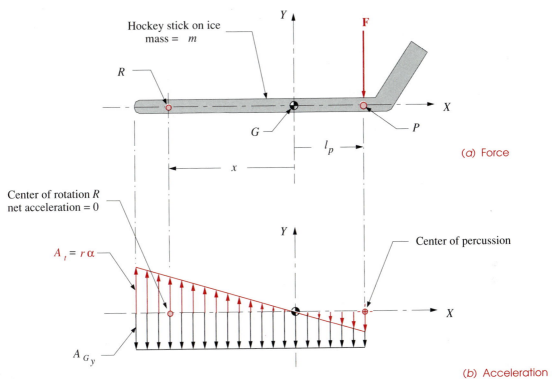

FIGURE 10-3

Center of percussion and center of rotation

nate system centered at G with the X axis along the stick in its initial position as shown. The translating component of acceleration of the CG resulting from the force **F** is (from Newton's law)

$$A_{G_y} = \frac{F}{m} \qquad (10.11a)$$

and the angular acceleration is:

$$\alpha = \frac{T}{I_G} \qquad (10.11b)$$

where I_G is its mass moment of inertia about the Z axis (out of the page) through the CG. But torque is also:

$$T = F l_P \qquad (10.11c)$$

where l_p is the distance along the X axis from point G to point P so:

$$\alpha = \frac{F l_P}{I_G} \qquad (10.11d)$$

10

The total linear acceleration at any point along the stick will be the sum of the linear acceleration A_{Gy} of the CG and the tangential component $(r\alpha)$ of the angular acceleration as shown in Figure 10-3b.

$$A_{y_{total}} = A_{G_y} + r\alpha$$

$$= \frac{F}{m} + x\left(\frac{F l_P}{I_G}\right) \tag{10.12}$$

where x is the distance to any point along the stick. Equation 10.12 can be set equal to zero and solved for the value of x for which the $r\alpha$ component exactly cancels the A_{Gy} component. This will be the **center of rotation** at which there is no translating acceleration, and thus no linear dynamic force. The solution for x when $A_{ytotal} = 0$ is:

$$x = -\frac{I_G}{m l_P} \tag{10.13a}$$

and substituting equation 10.10b:

$$x = -\frac{k^2}{l_P} \tag{10.13b}$$

where the radius of gyration k is calculated with respect to the axis ZZ through the CG.

Note that this relationship between the center of percussion and the center of rotation involves only geometry and mass properties. The magnitude of the applied force is irrelevant, but its location l_p completely determines x. Thus there is **not** just one center of percussion on a body. Rather there will be pairs of points. For every point (center of percussion) at which a force is applied there will be a corresponding center of rotation at which the reaction force felt will be zero. This center of rotation need not fall within the physical length of the body however. Consider the value of x predicted by equation 10.13b if you strike the body at its CG.

10.9   LUMPED PARAMETER DYNAMIC MODELS

Figure 10-4a shows a simple plate or disk cam driving a spring-loaded, roller follower. This is a force-closed system which depends on the spring force to keep the cam and follower in contact at all times. Figure 10-4b shows a lumped parameter model of this system in which all the **mass** which moves with the follower train is lumped together as m, all the springiness in the system is lumped within the **spring constant** k, and all the **damping** or resistance to movement is lumped together as a damper with coefficient c. The sources of mass which contribute to m are fairly obvious. The mass of the follower stem, the roller, its pivot pin, and any other hardware attached to the moving assembly all add together to create m. Figure 10-4c shows the free-body diagram of the system acted upon by the cam force F_c, the spring force F_s, and the damping force F_d. There will of course also be the effects of mass times acceleration on the system.

Spring Constant

We have been assuming all links and parts to be rigid bodies in order to do the kinematic analyses, but to do a more accurate force analysis we need to recognize that these bodies

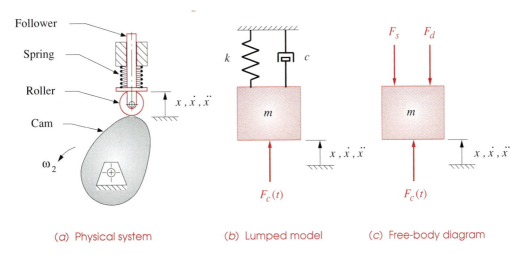

(a) Physical system   (b) Lumped model   (c) Free-body diagram

FIGURE 10-4

One-*DOF* lumped parameter model of a cam-follower system

are not truly rigid. The springiness in the system is assumed to be linear, thus describable by a spring constant k. A spring constant is defined as the force per unit deflection.

$$k = \frac{F_s}{x} \qquad (10.14)$$

The total spring constant k of the system is a combination of the spring constants of the actual coil spring, plus the spring constants of all other parts which are deflected by the forces. The roller, its pin, and the follower stem are all springs themselves as they are made of elastic materials. The spring constant for any part can be obtained from the equation for its deflection under the applied loading. Any deflection equation relates force to displacement and can be algebraically rearranged to express a spring constant. An individual part may have more than one k if it is loaded in several modes as, for example, a camshaft with a spring constant in bending and also one in torsion. We will discuss the procedures for combining these various spring constants in the system together into a combined, effective spring constant k in the next section. For now let us just assume that we can so combine them for our analysis and create an overall k for our lumped parameter model.

Damping

The friction, more generally called **damping**, is the most difficult parameter of the three to model. It needs to be a combination of all the damping effects in the system. These may be of many forms. **Coulomb friction** results from two dry or lubricated surfaces rubbing together. The contact surfaces between cam and follower and between the follower and its sliding joint can experience coulomb friction. It is generally considered to be independent of velocity magnitude but has a different, larger value when velocity is zero (static friction force F_{st} or *stiction*) than when there is relative motion between the parts (dynamic friction F_d). Figure 10-5a shows a plot of coulomb friction force versus relative velocity v at the contact surfaces. Note that friction always opposes motion, so

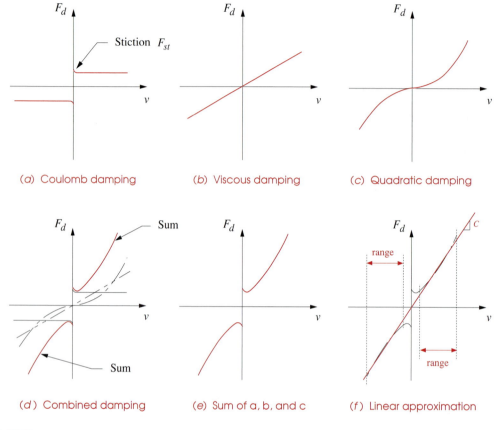

(a) Coulomb damping   (b) Viscous damping   (c) Quadratic damping

(d) Combined damping   (e) Sum of a, b, and c   (f) Linear approximation

FIGURE 10-5

Modeling damping

the friction force abruptly changes sign at $v = 0$. The stiction F_{st} shows up as a larger spike at zero v than the dynamic friction value F_d. Thus, this is a **nonlinear** friction function. It is multivalued at zero. In fact, at zero velocity, the friction force can be any value between $-F_{st}$ and $+F_{st}$. It will be whatever force is needed to balance the system forces and create equilibrium. When the applied force exceeds F_{st}, motion begins and the friction force suddenly drops to F_d. This nonlinear damping creates difficulties in our simple model since we want to describe our system with linear differential equations having known solutions.

Other sources of damping may be present besides coulomb friction. **Viscous damping** results from the shearing of a fluid (lubricant) in the gap between the moving parts and is considered to be a linear function of relative velocity as shown in Figure 10-5b. **Quadratic damping** results from the movement of an object through a fluid medium as with an automobile pushing through the air or a boat through the water. This factor is a fairly negligible contributor to a cam-follower's overall damping unless the speeds are very high or the fluid medium very dense. Quadratic damping is a function of the square of the relative velocity as shown in Figure 10-5c. The relationship of the dynamic damping force F_d as a function of relative velocity for all these cases can be expressed as:

$$F_d = cv|v|^{r-1} \qquad (10.15a)$$

where c is the constant damping coefficient, v is the relative velocity, and r is a constant which defines the type of damping.

For coulomb damping, $r = 0$ and:

$$F_d = \pm c \qquad (10.15b)$$

For viscous damping, $r = 1$ and:

$$F_d = cv \qquad (10.15c)$$

For quadratic damping, $r = 2$ and:

$$F_d = \pm cv^2 \qquad (10.15d)$$

If we combine these three forms of damping, their sum will look like Figure 10-5d and e. This is obviously a nonlinear function. But we can approximate it over a reasonably small range of velocity as a linear function with a slope c which is then a *pseudo-viscous damping coefficient*. This is shown in Figure 10-5f. While not an exact method to account for the true damping, this approach has been found to be acceptably accurate for a first approximation during the design process. The damping in these kinds of mechanical systems can vary quite widely from one design to the next due to different geometries, pressure or transmission angles, types of bearings, lubricants or their absence, etc. It is very difficult to accurately predict the level of damping (i.e., the value of c) in advance of the construction and testing of a prototype, which is the best way to determine the damping coefficient. If similar devices have been built and tested, their history can provide a good prediction. For the purpose of our dynamic modeling, we will assume *pseudo-viscous damping* and some value for c.

10.10 EQUIVALENT SYSTEMS

More complex systems than that shown in Figure 10-4 (p. 501) will have multiple masses, springs, and sources of damping connected together as shown in Figure 10-9. These models can be analyzed by writing dynamic equations for each subsystem and then solving the set of differential equations simultaneously. This allows a multi-degree-of-freedom analysis, with one-*DOF* for each subsystem included in the analysis. Koster[2] found in his extensive study of vibrations in cam mechanisms that a five-*DOF* model which included the effects of both torsional and bending deflection of the camshaft, backlash (see Section 10.2, p. 492) in the driving gears, squeeze effects of the lubricant, nonlinear coulomb damping, and motor speed variation gave a very good prediction of the actual, measured follower response. But he also found that a single-*DOF* model as shown in Figure 10-4 gave a reasonable simulation of the same system. We can then take the simpler approach and lump all the subsystems of Figure 10-9 together into a single-*DOF* **equivalent system** as shown in Figure 10-4. The combining of the various springs, dampers, and masses must be done carefully to properly approximate their dynamic interactions with each other.

10

There are only two types of variables active in any dynamic system. These are given the general names of *through variable* and *across variable*. These names are descriptive of their actions within the system. A **through variable** *passes through the system.* An **across variable** *exists across the system.* The power in the system is the product of the through and across variables. Table 10-1 lists the through and across variables for various types of dynamic systems.

We commonly speak of the voltage across a circuit and the current flowing through it. We also can speak of the velocity across a mechanical "circuit" or system and the force which flows through it. Just as we can connect electrical elements such as resistors, capacitors, and inductors together in series or parallel or a combination of both to make an electrical circuit, we can connect their mechanical analogs, dampers, springs, and masses together in series, parallel, or a combination thereof to make a mechanical system. Table 10-2 shows the analogs between three types of physical systems. The fundamental relationships between through and across variables in electrical, mechanical, and fluid systems are shown in Table 10-3.

Recognizing a series or parallel connection between elements in an electrical circuit is fairly straightforward, as their interconnections are easily seen. Determining how mechanical elements in a system are interconnected is more difficult as their interconnections are sometimes hard to see. The test for series or parallel connection is best done by examining the forces and velocities (or the integral of velocity, displacement) that exist in the particular elements. If two elements have the same force passing through them, they are in series. If two elements have the same velocity or displacement, they are in parallel.

Combining Dampers

DAMPERS IN SERIES   Figure 10-6a shows three dampers in series. The force passing through each damper is the same, and their individual displacements and velocities are different.

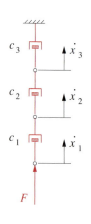

(a) Series

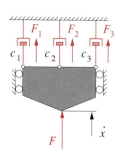

(b) Parallel

FIGURE 10-6

Dampers in series and in parallel

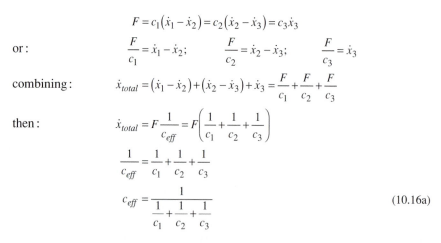

$$F = c_1(\dot{x}_1 - \dot{x}_2) = c_2(\dot{x}_2 - \dot{x}_3) = c_3\dot{x}_3$$

or :
$$\frac{F}{c_1} = \dot{x}_1 - \dot{x}_2; \qquad \frac{F}{c_2} = \dot{x}_2 - \dot{x}_3; \qquad \frac{F}{c_3} = \dot{x}_3$$

combining :
$$\dot{x}_{total} = (\dot{x}_1 - \dot{x}_2) + (\dot{x}_2 - \dot{x}_3) + \dot{x}_3 = \frac{F}{c_1} + \frac{F}{c_2} + \frac{F}{c_3}$$

then :
$$\dot{x}_{total} = F\frac{1}{c_{eff}} = F\left(\frac{1}{c_1} + \frac{1}{c_2} + \frac{1}{c_3}\right)$$

$$\frac{1}{c_{eff}} = \frac{1}{c_1} + \frac{1}{c_2} + \frac{1}{c_3}$$

$$c_{eff} = \frac{1}{\dfrac{1}{c_1} + \dfrac{1}{c_2} + \dfrac{1}{c_3}} \qquad (10.16a)$$

The reciprocal of the effective damping of the dampers in series is the sum of the reciprocals of their individual damping coefficients.

TABLE 10-1   Through and Across Variables in Dynamic Systems

System Type	Through Variable	Across Variable	Power Units
Electrical	Current (i)	Voltage (e)	ei = watts
Mechanical	Force (F)	Velocity (v)	Fv = (in-lb)/sec
Fluid	Flow (Q)	Pressure (P)	PQ = (in-lb)/sec

TABLE 10-2   Physical Analogs in Dynamic Systems

System Type	Energy Dissipator	Energy Storage	Energy Storage
Electrical	Resistor (R)	Capacitor (C)	Inductor (L)
Mechanical	Damper (c)	Mass (m)	Spring (k)
Fluid	Fluid resistor (R_f)	Accumulator (C_f)	Fluid inductor (L_f)

TABLE 10-3   Relationships Between Variables in Dynamic Systems

System Type	Resistance	Capacitance	Inductance
Electrical	$i = \dfrac{1}{R} e$	$i = C \dfrac{de}{dt}$	$i = \dfrac{1}{L} \int e\, dt$
Mechanical	$F = c\,v$	$F = m \dfrac{dv}{dt}$	$F = k \int v\, dt$
Fluid	$Q = \dfrac{1}{R_f} P$	$Q = C_f \dfrac{dP}{dt}$	$Q = \dfrac{1}{L_f} \int P\, dt$

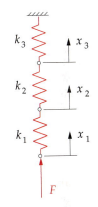

(a) Series

DAMPERS IN PARALLEL   Figure 10-6b shows three dampers in parallel. The force passing through each damper is different, and their displacements and velocities are all the same.

$$F = F_1 + F_2 + F_3$$
$$F = c_1 \dot{x} + c_2 \dot{x} + c_3 \dot{x}$$
$$F = (c_1 + c_2 + c_3) \dot{x}$$
$$F = c_{eff} \dot{x}$$
$$c_{eff} = c_1 + c_2 + c_3 \tag{10.16b}$$

The effective damping of the three is the sum of their individual damping coefficients.

Combining Springs

Springs are the mechanical analog of electrical inductors. Figure 10-7a shows three springs in series. The force passing through each spring is the same, and their individual

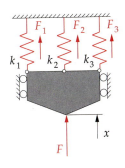

(b) Parallel

FIGURE 10-7

Springs in series and in parallel

to point A and attach it to the existing spring k, as shown in Figure 10-8b, or we can transfer an equivalent spring k_{eff} to point B and attach it to the existing mass m as shown in Figure 10-8c. In either case, for the lumped model to be equivalent to the original system, it must have the same energy in it.

Let's first find the effective mass that must be placed at point A to eliminate the lever. Equating the kinetic energies in the masses at points A and B:

$$\frac{1}{2}m_B v_B^2 = \frac{1}{2}m_{eff} v_A^2 \qquad (10.20a)$$

The velocities at each end of the lever can be related by the lever ratio:

$$v_A = \left(\frac{a}{b}\right)v_B$$

substituting:

$$m_B v_B^2 = m_{eff}\left(\frac{a}{b}\right)^2 v_B^2$$

$$m_{eff} = \left(\frac{b}{a}\right)^2 m_B \qquad (10.20b)$$

The effective mass varies from the original mass by the square of the lever ratio. Note that if the lever were instead a pair of gears of radii a and b, the result would be the same.

Now find the effective spring that would have to be placed at B to eliminate the lever. Equating the potential energies in the springs at points A and B:

$$\frac{1}{2}k_A x_A^2 = \frac{1}{2}k_{eff} x_B^2 \qquad (10.21a)$$

The deflection at B is related to the deflection at A by the lever ratio:

$$x_B = \left(\frac{b}{a}\right)x_A$$

substituting:

$$k_A x_A^2 = k_{eff}\left(\frac{b}{a}\right)^2 x_A^2$$

$$k_{eff} = \left(\frac{a}{b}\right)^2 k_A \qquad (10.21b)$$

The effective k varies from the original k by the square of the lever ratio. If the lever were instead a pair of gears of radii a and b, the result would be the same. So, gear or lever ratios can have a large effect on the lumped parameters' values in the simplified model.

Damping coefficients are also affected by the lever ratio. Figure 10-8d shows a damper and a mass at opposite ends of a lever. If the damper at A is to be replaced by a damper at B, then the two dampers must produce the same moment about the pivot, thus:

$$F_{d_A} a = F_{d_B} b \qquad (10.21c)$$

Substitute the product of the damping coefficient and velocity for force:

$$\left(c_A \dot{x}_A\right)a = \left(c_{B_{eff}} \dot{x}_B\right)b \qquad (10.21d)$$

The velocities at points A and B in Figure 10.8d can be related from kinematics:

$$\omega = \frac{\dot{x}_A}{a} = \frac{\dot{x}_B}{b}$$

$$\dot{x}_A = \dot{x}_B \frac{a}{b} \qquad (10.21e)$$

Substituting in equation 10.21d we get an expression for the effective damping coefficient at B resulting from a damper at A.

$$\left(c_A \dot{x}_B \frac{a}{b}\right)a = \left(c_{B_{eff}} \dot{x}_B\right)b$$

$$c_{B_{eff}} = c_A \left(\frac{a}{b}\right)^2 \qquad (10.21f)$$

Again, the square of the lever ratio determines the effective damping. The equivalent system is shown in Figure 10-8e.

EXAMPLE 10-1

Creating a Single-*DOF* Equivalent System Model of a Multielement Dynamic System.

Given: An automotive valve cam with translating flat follower, long pushrod, rocker arm, valve, and valve spring is shown in Figure 10-9a.

Problem: Create a suitable, approximate, single-*DOF*, lumped parameter model of the system. Define its effective mass, spring constant, and damping in terms of the individual elements' parameters.

Solution:

1   Break the system into individual elements as shown in Figure 10-9b. Each significant moving part is assigned a lumped mass element which has a connection to ground through a damper. There is also elasticity and damping within the individual elements, shown as connecting springs and dampers. The rocker arm is modeled as two lumped masses at its ends, connected with a rigid, massless rod for the crank and conrod of the slider-crank linkage. (See also Section 13.4, p. 614.) The breakdown shown represents a six-*DOF* model as there are six independent displacement coordinates, x_1 through x_6.

2   Define the individual spring constants of each element which represents the elasticity of a lumped mass from the elastic deflection formula for the particular part. For example, the pushrod is loaded in compression, so its relevant deflection formula and its k are,

$$x = \frac{Fl}{AE} \qquad \text{and} \qquad k_{pr} = \frac{F}{x} = \frac{AE}{l} \qquad (a)$$

where A is the cross-sectional area of the pushrod, l is its length, and E is Young's modulus for the material. The k of the tappet element will have the same expression. The expression for

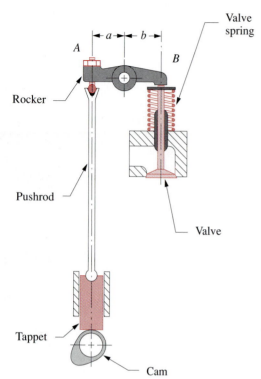

(a) Physical model

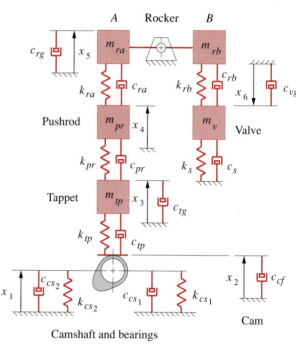

(b) Six-DOF model

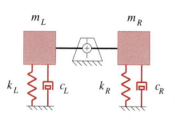

(c) One-DOF model with lever arm

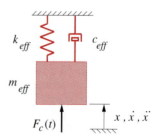

(d) One-DOF lumped model

FIGURE 10-9

Lumped parameter models of an overhead valve engine cam-follower system

the k of a helical coil compression spring, as used for the valve spring, can be found in any spring design manual or machine design text and is:

$$k_{sp} = \frac{d^4 G}{8D^3 N} \qquad (b)$$

where d is the wire diameter, D is the mean coil diameter, N is the number of coils, and G is the modulus of rupture of the material.

The rocker arm also acts as a spring, as it is a beam in bending. It can be modeled as a double cantilever beam with its deflection on each side of the pivot considered separately. These spring effects are shown in the model as if they were compression springs, but that is just schematic. They really represent the bending deflection of the rocker arms. From the deflection formula for a cantilever beam with concentrated load:

$$x = \frac{Fl^3}{3EI} \qquad \text{and} \qquad k_{ra} = \frac{3EI}{l^3} \qquad (c)$$

where I is the cross-sectional second moment of area of the beam, l is its length, and E is Young's modulus for the material. The spring constants of any other elements in a system can be obtained in similar fashion from their deflection formulas.

3 The dampers shown connected to ground represent the friction or viscous damping at the interfaces between the elements and the ground plane. The dampers between the masses represent the internal damping in the parts, which typically is quite small. These values will either have to be estimated from experience or measured in prototype assemblies.

4 The rocker arm provides a lever ratio which must be taken into account. The strategy will be to combine all elements on each side of the lever separately into two lumped parameter models as shown in Figure 10-9c, and then transfer one of those across the lever pivot to create one, single-*DOF* model as shown in Figure 10-9d.

5 The next step is to determine the types of connections, either series or parallel, between the elements. The masses are all in parallel as they each communicate their inertial force directly to ground and have independent displacements. On the left and right sides, respectively, the effective masses are:

$$m_L = m_{tp} + m_{pr} + m_{ra} \qquad\qquad m_R = m_{rb} + m_v \qquad (d)$$

Note that m_v includes about one-third of the spring's mass. The two springs shown representing the bending deflection of the camshaft split the force between them, so they are in parallel and thus add directly.

$$k_{cs} = k_{cs_1} + k_{cs_2} \qquad (e)$$

Note that, for completeness, the torsional deflection of the camshaft should also be included but is omitted in this example to reduce complexity. The combined camshaft spring constant and all the other springs shown on the left side are in series as they each have independent deflections and the same force passes through them all. The same is true of the springs on the right side. The effective spring constants for each side are then:

$$k_L = \frac{1}{\dfrac{1}{k_{cs}} + \dfrac{1}{k_{tp}} + \dfrac{1}{k_{pr}} + \dfrac{1}{k_{ra}}} \qquad\qquad k_R = \frac{1}{\dfrac{1}{k_{rb}} + \dfrac{1}{k_s}} \qquad (f)$$

The dampers are in a combination of series and parallel. The pair of dampers c_{cs1} and c_{cs2} shown supporting the camshaft represent the friction in the two camshaft bearings and are in parallel.

$$c_{cs} = c_{cs_1} + c_{cs_2} \qquad (g)$$

The ones representing internal damping are in series with one another and with the combined shaft damping.

$$c_{in_L} = \cfrac{1}{\cfrac{1}{c_{tp}} + \cfrac{1}{c_{pr}} + \cfrac{1}{c_{ra}} + \cfrac{1}{c_{cs}}} \qquad\qquad c_{in_R} = \cfrac{1}{\cfrac{1}{c_{rb}} + \cfrac{1}{c_s}} \qquad (h)$$

where c_{in_L} is all internal damping on the left side and c_{in_R} is all internal damping on the right side of the rocker arm pivot. The combined internal damping c_{in_L} goes to ground through c_{rg} and the combined internal damping c_{in_R} goes to ground through the valve spring c_s. These two series combinations are then in parallel with all the other dampers that go to ground. The combined dampings for each side of the system are then:

$$c_L = c_{tg} + c_{rg} + c_{in_L} \qquad\qquad c_R = c_{vg} + c_{in_R} \qquad (k)$$

6   The system can now be reduced to a single-*DOF* model with masses and springs lumped on either end of the rocker arm as shown in Figure 10-9c. We will bring the elements at point *B* across to point *A*. Note that we have reversed the sign convention across the pivot so that positive motion on one side results also in positive motion on the other. The damper, mass, and spring constant are affected by the square of the lever ratio as shown in equations 10.20 (p. 508) and 10.21 (p. 509).

$$m_{eff} = m_L + \left(\frac{b}{a}\right)^2 m_R$$

$$k_{eff} = k_L + \left(\frac{b}{a}\right)^2 k_R \qquad (m)$$

$$c_{eff} = c_L + \left(\frac{b}{a}\right)^2 c_R$$

These are shown in Figure 10-9d on the final, one-*DOF* lumped model of the system.

Note that this one-*DOF* model provides only a relatively crude approximation of this complex system's behavior. Even though it may be an oversimplification, it is nevertheless still useful as a first approximation and serves in this context as an example of the general method involved in modelling dynamic systems. A more complex model with multiple degrees of freedom will provide a better approximation of the dynamic system's behavior.

10.11 SOLUTION METHODS

Dynamic force analysis can be done by any of several methods. Two will be discussed here, **superposition** and **linear simultaneous equation solution**. Both methods require that the system be linear.

These dynamic force problems typically have a large number of unknowns and thus have multiple equations to solve. The method of superposition attacks the problem by solving for parts of the solution and then adding (superposing) the partial results together to get the complete result. For example, if there are two loads applied to the system, we solve independently for the effects of each load, and then add the results. In effect we solve an *N*-variable system by doing sequential calculations on parts of the problem. It can be thought of as a "serial processing" approach.

Another method writes all the relevant equations for the entire system as a set of linear simultaneous equations. These equations can then be solved simultaneously to obtain the results. This can be thought of as analogous to a "parallel processing" approach. A convenient approach to the solution of sets of simultaneous equations is to put them in a standard matrix form and use a numerical matrix solver to obtain the answers. Matrix solvers are built into most engineering and scientific pocket calculators. Some spreadsheet packages and equation solvers will also do a matrix solution. A brief introduction to matrix solution of simultaneous equations was presented in Section 5.5. Appendix A describes the use of the computer program MATRIX, included on CD-ROM with this text. This program allows the rapid calculation of the solution to systems of up to 40 simultaneous equations. Please refer to the sections in Chapter 5 to review these calculation procedures and Appendix *A* for program MATRIX. Reference [3] provides an introduction to matrix algebra.

We will use both superposition and simultaneous equation solution to solve various dynamic force analysis problems in the remaining chapters. Both have their place, and one can serve as a check on the results from the other. So it is useful to be familiar with more than one approach. Historically, superposition was the only practical method for systems involving large numbers of equations until computers became available to solve large sets of simultaneous equations. Now the simultaneous equation solution method is more popular.

10.12 THE PRINCIPLE OF D'ALEMBERT

Newton's second law (equations 10.1, p. 492 and 10.4, p. 495) are all that are needed to solve any dynamic force system by the newtonian method. Jean le Rond d'Alembert (1717-1783), a French mathematician, rearranged Newton's equations to create a "quasi-static" situation from a dynamic one. D'Alembert's versions of equations 10.1 and 10.4 are:

$$\sum \mathbf{F} - m\mathbf{a} = 0$$

$$\sum \mathbf{T} - I\alpha = 0 \tag{10.22}$$

All d'Alembert did was to move the terms from the right side to the left, changing their algebraic signs in the process as required. These are obviously still the same equations as 10.1 and 10.4, algebraically rearranged. The motivation for this algebraic manipulation was to make the system look like a statics problem in which, for equilibrium, all forces and torques must sum to zero. Thus, this is sometimes called a quasi-static problem when expressed in this form. The premise is that by placing an "inertia force" equal to $-ma$ and an "inertia torque" equal to $-I\alpha$ on our free-body diagrams, the system will then be in a state of "dynamic equilibrium" and can be solved by the familiar methods of statics. These inertia forces and torques are equal in magnitude, opposite in sense, and along the same line of action as ma and $I\alpha$. This was a useful and popular approach which made the solution of dynamic force analysis problems somewhat easier when graphical vector solutions were the methods of choice.

With the availability, literally in your pocket or on your desk, of calculators and computers which can solve the simultaneous equations for these problems, there is now little motivation to labor through the complicated tedium of a graphical force analysis. It is for this reason that graphical force analysis methods are not presented in this text. However, d'Alembert's concept of "inertia forces and torques" still has, at a minimum, historical value and, in many instances, can prove useful in understanding what is going on in a dynamic system. Moreover, the concept of inertia force has entered the popular lexicon and is often used in a lay context when discussing motion. Thus we present a simple example of its use here and will use it again in our discussion of dynamic force analysis later in this text where it helps us to understand some topics such as balancing and superposition.

The popular term **centrifugal force**, used by laypersons everywhere to explain why a mass on a rope keeps the rope taut when swung in a circle, is in fact a d'Alembert inertial force. Figure 10-10a shows such a mass, being rotated at the end of a flexible but inextensible cord at a constant angular velocity ω and constant radius r. Figure 10-10b shows "pure" free-body diagrams of both members in this system, the ground link (1) and the rotating link (2). The only real force acting on link 2 is the force of link 1 on 2, F_{12}. Since angular acceleration is zero in this example, the acceleration acting on the link is only the $r\omega^2$ component, which is a **centripetal acceleration**, i.e., directed *toward the center*. The force at the pin from Newton's equation 10.1 is then:

$$F_{12} = mr\omega^2 \qquad\qquad (10.23a)$$

Note that this force is directed toward the center, so it is a *centripetal* not a *centrifugal* (away from center) force. The force F_{21} which link 2 exerts on link 1 can be found from Newton's third law and is obviously equal and opposite to $\mathbf{F}_{12}$.

$$F_{21} = -F_{12} \qquad\qquad (10.23b)$$

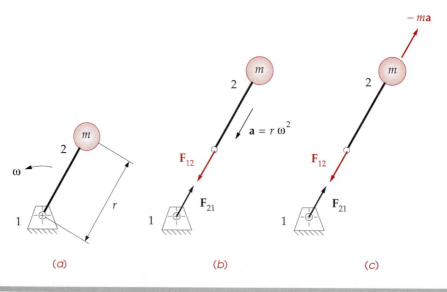

(a)   (b)   (c)

FIGURE 10-10

Centripetal and centrifugal forces

Thus it is the reaction force on link 1 which is centrifugal, not the force on link 2. Of course, it is this reaction force that your hand (link 1) feels, and this gives rise to the popular conception of something pulling centrifugally on the rotating weight. Now let us look at this through d'Alembert's eyes. Figure 10-10c shows another set of free-body diagrams done according to the principle of d'Alembert. Here we show a negative $m\mathbf{a}$ inertia force applied to the mass on link 2. The force at the pin from d'Alembert's equation is:

$$F_{12} - mr\omega^2 = 0$$

$$F_{12} = mr\omega^2 \qquad\qquad (10.23c)$$

Not surprisingly, the result is the same as equation 10.23a, as it must be. The only difference is that the free-body diagram shows an inertia force applied to the rotating mass on link 2. This is the centrifugal force of popular repute which takes the blame for keeping the cord taut.

Clearly, any problem can be solved for the right answer no matter how we may algebraically rearrange the correct equations. So, if it helps our understanding to think in terms of these inertia forces being applied to a dynamic system, we will do so. When dealing with the topic of balancing, this approach does, in fact, help to visualize the effects of the balance masses on the system.

10.13 ENERGY METHODS—VIRTUAL WORK

The newtonian methods of dynamic force analysis described above have the advantage of providing complete information about all interior forces at pin joints as well as about the external forces and torques on the system. One consequence of this fact is the relative complexity of their application which requires the simultaneous solution of large systems of equations. Other methods are available for the solution of these problems which are easier to implement but give less information. Energy methods of solution are of this type. Only the external, work producing, forces and torques are found by these methods. The internal joint forces are not computed. One chief value of the energy approach is its use as a quick check on the correctness of the newtonian solution for input torque. Usually we are forced to use the more complete newtonian solution in order to obtain force information at pin joints so that pins and links can be analyzed for failure due to stress.

The *law of conservation of energy* states that energy can neither be created nor destroyed, only converted from one form to another. Most machines are designed specifically to convert energy from one form to another in some controlled fashion. Depending on the efficiency of the machine, some portion of the input energy will be converted to heat which cannot be completely recaptured. But large quantities of energy will typically be stored temporarily within the machine in both potential and kinetic form. It is not uncommon for the magnitude of this internally stored energy, on an instantaneous basis, to far exceed the magnitude of any useful external work being done by the machine.

Work is defined as *the dot product of force and displacement.* It can be positive, negative, or zero and is a scalar quantity.

$$W = \mathbf{F} \cdot \mathbf{R} \tag{10.24a}$$

Since the forces at the pin joints between the links have no relative displacement associated with them, they do no work on the system, and thus will not appear in the work equation. The work done by the system plus losses is equal to the energy delivered to the system.

$$E = W + Losses \tag{10.24b}$$

Pin-jointed linkages with low-friction bearings at the pivots can have high efficiencies, above 95%. Thus it is not unreasonable, for a first approximation in designing such a mechanism, to assume the losses to be zero. **Power** is the time rate of change of energy:

$$P = \frac{dE}{dt} \tag{10.24c}$$

Since we are assuming the machine member bodies to be rigid, only a change of position of the *CG*s of the members will alter the stored potential energy in the system. The gravitational forces of the members in moderate- to high-speed machinery often tend to be dwarfed by the dynamic forces from the accelerating masses. For these reasons we will ignore the weights and the gravitational potential energy and consider only the kinetic energy in the system for this analysis. The time rate of change of the kinetic energy stored within the system for linear and angular motion, respectively, is then:

$$\frac{d\left(\frac{1}{2}m\mathbf{v}^2\right)}{dt} = m\mathbf{a} \cdot \mathbf{v} \tag{10.25a}$$

and:

$$\frac{d\left(\frac{1}{2}I\omega^2\right)}{dt} = I\alpha \cdot \omega \tag{10.25b}$$

These are, of course, expressions for power in the system, equivalent to:

$$P = \mathbf{F} \cdot \mathbf{v} \tag{10.25c}$$

and:

$$P = \mathbf{T} \cdot \omega \tag{10.25d}$$

The rate of change of energy in the system at any instant must balance between that which is externally supplied and that which is stored within the system (neglecting losses). Equation 10.25a and b represent change in the energy stored in the system, and equation 10.25c and d represent change in energy passing into or out of the system. In the absence of losses, these two must be equal in order to conserve energy. We can express this relationship as a summation of all the delta energies (or power) due to each moving element (or link) in the system.

$$\sum_{k=2}^{n} \mathbf{F}_k \cdot \mathbf{v}_k + \sum_{k=2}^{n} \mathbf{T}_k \cdot \omega_k = \sum_{k=2}^{n} m_k \mathbf{a}_k \cdot \mathbf{v}_k + \sum_{k=2}^{n} I_k \alpha_k \cdot \omega_k \qquad (10.26a)$$

The subscript k represents each of the n links or moving elements in the system, starting with link 2 because link 1 is the stationary ground link. Note that all the angular and linear velocities and accelerations in this equation must have been calculated, for all positions of the mechanism of interest, from a prior kinematic analysis. Likewise, the masses and mass moments of inertia of all moving links must be known.

If we use the principle of d'Alembert to rearrange this equation, we can more easily "name" the terms for discussion purposes.

$$\sum_{k=2}^{n} \mathbf{F}_k \cdot \mathbf{v}_k + \sum_{k=2}^{n} \mathbf{T}_k \cdot \omega_k - \sum_{k=2}^{n} m_k \mathbf{a}_k \cdot \mathbf{v}_k - \sum_{k=2}^{n} I_k \alpha_k \cdot \omega_k = 0 \qquad (10.26b)$$

The first two terms in equation 10.26b represent, respectively, the change in energy due to all **external forces** and all **external torques** applied to the system. These would include any forces or torques from other mechanisms which impinge upon any of these links and also includes the driving torque. The second two terms represent, respectively, the change in energy due to all **inertia forces** and all **inertia torques** present in the system. These last two terms define the change in stored kinetic energy in the system at each time step. The only unknown in this equation when properly set up is the **driving torque** (or driving force) to be supplied by the mechanism's motor or actuator. This driving torque (or force) is then the only variable which can be solved for with this approach. The internal joint forces are not present in the equation as they do no net work on the system.

Equation 10.26b is sometimes called the **virtual work equation**, which is something of a misnomer, as it is in fact a **power equation**. When this analysis approach is applied to a statics problem, there is no motion. The term **virtual work** comes from the concept of each force causing an infinitesimal, or virtual, displacement of the static system element to which it is applied over an infinitesimal delta time. The dot product of the force and the virtual displacement is the virtual work. In the limit, this becomes the instantaneous power in the system. We will present an example of the use of this method of virtual work in the next chapter along with examples of the newtonian solution applied to linkages in motion.

10.14 REFERENCES

1   **Beer, F. P., and E. R. Johnson.** (1984). *Vector Mechanics for Engineers, Statics and Dynamics,* McGraw-Hill Inc., New York.

2   **Koster, M. P.** (1974). *Vibrations of Cam Mechanisms.* Phillips Technical Library Series, Macmillan: London.

3   **Jennings, A.** (1977). *Matrix Computation for Engineers and Scientists,* John Wiley and Sons, New York.

10.15 PROBLEMS

*†10-1 The mallet shown in Figure 10-2 (p. 496) has the following specifications: The steel head has a 1-in diameter and is 3-in tall; the wood handle is 1.25-in diameter and 10-in long and necks down to 5/8-in wide where it enters the head. Find the location of its composite CG, and its moment of inertia and radius of gyration about axis ZZ. Assume the wood has a density equal to 0.9 times that of water.

*†10-2 Repeat Problem 10.1 using a wooden mallet head of 2-in diameter. Assume the wood has a density equal to 0.9 times that of water.

†10-3 Calculate the location of the composite CG, the mass moment of inertia and the radius of gyration with respect to the specified axis, for whichever of the following commonly available items that are assigned. (Note these are not short problems.)

    a. A good-quality writing pen, about the pivot point at which you grip it to write. (How does placing the cap on the upper end of the pen affect these parameters when you write?)

    b. Two table knives, one metal and one plastic, about the pivot axis when held for cutting. Compare the calculated results and comment on what they tell you about the dynamic usability of the two knives (ignore sharpness considerations).

    c. A ball-peen hammer (available for inspection in any university machine shop), about the center of rotation (after you calculate its location for the proper center of percussion).

    d. A baseball bat (see the coach) about the center of rotation (after you calculate its location for the proper center of percussion).

    e. A cylindrical coffee mug, about the handle hole.

*†10-4 Set up these equations in matrix form. Use program MATRIX, *Mathcad*, or a calculator which has matrix math capability to solve them.

$$
\begin{aligned}
a. \quad -5x &- 2y + 12z - w = -9 \\
x &+ 3y - 2z + 4w = 10 \\
-x &- y + z = -7 \\
3x &- 3y + 7z + 9w = -6
\end{aligned}
$$

$$
\begin{aligned}
b. \quad 3x &- 5y + 17z - 5w = -5 \\
-2x &+ 9y - 14z + 6w = 22 \\
-x &- y - 2w = 13 \\
4x &- 7y + 8z + 4w = -9
\end{aligned}
$$

†10-5 Figure P10-1 shows a bracket made of steel.

    a. Find the location of its centroid referred to point B.

    b. Find its mass moment of inertia I_{xx} about the X axis through point B.

    c. Find its mass moment of inertia I_{yy} about the Y axis through point B.

*†10-6 Two springs are connected in series. One has a k of 34 and the other a k of 3.4. Calculate their effective spring constant. Which spring dominates? Repeat with the two springs in parallel. Which spring dominates? (Use any unit system.)

†10-7 Repeat Problem 10-6 with $k_1 = 125$ and $k_2 = 25$. (Use any unit system.)

†10-8 Repeat Problem 10-6 with $k_1 = 125$ and $k_2 = 115$. (Use any unit system.)

* Answers in Appendix F.

† These problems are suited to solution using *Mathcad, Matlab,* or *TKSolver* equation solver programs.

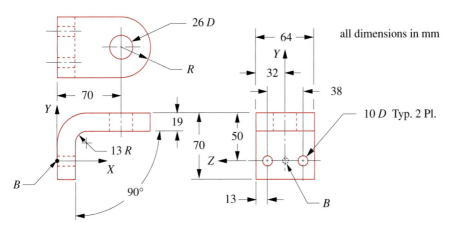

FIGURE P10-1

Problem 10-5

*†10-9 Two dampers are connected in series. One has a damping factor $c_1 = 12.5$ and the other, $c_2 = 1.2$. Calculate their effective damping constant. Which damper dominates? Repeat with the two dampers in parallel. Which damper dominates? (Use any unit system.)

†10-10 Repeat Problem 10-9 with $c_1 = 12.5$ and $c_2 = 2.5$. (Use any unit system.)

†10-11 Repeat Problem 10-9 with $c_1 = 12.5$ and $c_2 = 10$. (Use any unit system.)

*†10-12 A mass of $m = 2.5$ and a spring with $k = 42$ are attached to one end of a lever at a radius of 4. Calculate the effective mass and effective spring constant at a radius of 12 on the same lever. (Use any unit system.)

†10-13 A mass of $m = 1.5$ and a spring with $k = 24$ are attached to one end of a lever at a radius of 3. Calculate the effective mass and effective spring constant at a radius of 10 on the same lever. (Use any unit system.)

†10-14 A mass of m = 4.5 and a spring with $k = 15$ are attached to one end of a lever at a radius of 12. Calculate the effective mass and effective spring constant at a radius of 3 on the same lever. (Use any unit system.)

†10-15 Refer to Figure 10-9 and Example 10-1. The data for the valve train are:

Tappet is a solid cylinder 0.75 diameter by 1.25 long
Pushrod is a hollow tube. 0.375 outside diameter by 0.25 inside diameter by 12 long.
Rocker arm has an average cross section of 1 wide by 1.5 high
Length $a = 2$, $b = 3$.
Camshaft is 1 diameter by 3 between bearing supports, cam in center.
Valve spring $k = 200$
All parts are steel

Calculate the effective spring constant and effective mass of a single-DOF equivalent system placed on the cam side of the rocker arm. (Use ips unit system.)

†10-16 Figure P10-2 shows a cam-follower system. The dimensions of the solid, rectangular 2 x 2.5 in cross-section aluminum arm are given. The cutout for the 2-in dia by 1.5-in wide steel roller follower is 3-in long. Find the arm's mass, center of gravity location

---

* Answers in Appendix F.

† These problems are suited to solution using *Mathcad, Matlab,* or *TKSolver* equation solver programs.

10

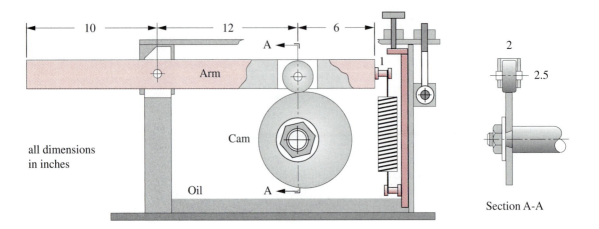

FIGURE P10-2

Problems 10-16, 10-17, 10-21 and 10-26

and mass moment of inertia about both its *CG* and the arm pivot. Create a linear, one-*DOF* lumped mass model of the dynamic system referenced to the cam-follower. Ignore damping.

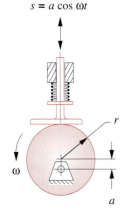

$s = a \cos \omega t$

FIGURE P10-3

Problems 10-18 to 10-19

---

* Answers in Appendix F.

---

† These problems are suited to solution using *Mathcad, Matlab,* or *TKSolver* equation solver programs.

†10-17 The cam in Figure P10-2 is a pure eccentric with eccentricity = 0.5 in and turns at 500 rpm. The spring has a rate of 123 lb/in and a preload of 173 lb. Use the method of virtual work to find the torque required to rotate the cam through one revolution. Use the data from the solution to Problem 10-16.

†10-18 The cam in Figure P10-3 is a pure eccentric with eccentricity = 20 mm and turns at 200 rpm. The mass of the follower is 1 kg. The spring has a rate of 10 N/m and a preload of 0.2 N. Use the method of virtual work to find the torque required to rotate the cam through one revolution.

†10-19 Repeat Problem 10-18 using a cam with a 20-mm, symmetric, double harmonic rise in 180° and double harmonic fall in 180°. See Chapter 8 for cam formulas.

*†10-20 A 3000 lb automobile has a final drive ratio of 1:3 and transmission gear ratios of 1:4, 1:3, 1:2, and 1:1 in first through fourth speeds, respectively. What is the effective mass of the vehicle as felt at the engine flywheel in each gear?

*†10-21 Determine the effective spring constant and effective preload of the spring in Figure P10-2 as reflected back to the cam-follower. See Problem 10-17 for additional data.

†10-22 What is the effective inertia of a load applied at the drum of Figure P9-5 (p. 484) as reflected back to gear *A*?

†10-23 What is the effective inertia of a load applied at the drum of Figure P9-7 (p. 485) as reflected back to the arm?

†10-24 Refer to Figure 10-8a (p. 507). Given $a = 100$ mm, $b = 150$ mm, $k_A = 2000$ N/m, and $m_B = 2$ kg, find the equivalent mass at point *A* and the equivalent spring at point *B*.

†10-25 Repeat Problem 10-24 with $a = 50$ mm, $b = 150$ mm, $k_A = 1000$ N/m, and $m_B = 3$ kg

†10-26 For the cam-follower arm in Figure P10-2, determine the location of its fixed pivot that will have zero reaction force when the cam applies its force to the follower.

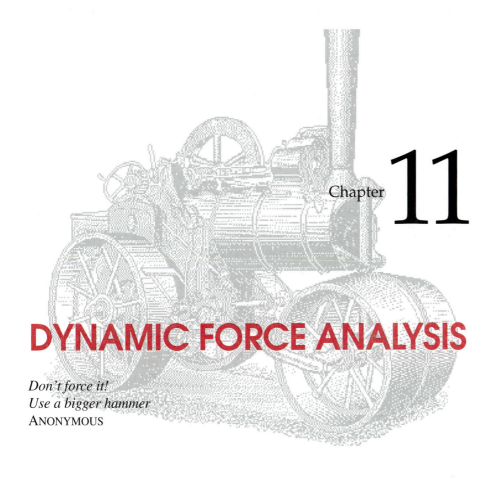

Chapter 11

DYNAMIC FORCE ANALYSIS

Don't force it!
Use a bigger hammer
ANONYMOUS

11.0 INTRODUCTION

When kinematic synthesis and analysis have been used to define a geometry and set of motions for a particular design task, it is logical and convenient to then use a **kinetostatic**, or **inverse dynamics,** solution to determine the forces and torques in the system. We will take that approach in this chapter and concentrate on solving for the forces and torques that result from, and are required to drive, our kinematic system in such a way as to provide the designed accelerations. Numerical examples are presented throughout this chapter. These examples are also provided as disk files for input to either program MATRIX or FOURBAR. These programs are described in Appendix A. The reader is encouraged to open the referenced files in these programs and investigate the examples in more detail. The file names are noted in the discussion of each example.

11.1 NEWTONIAN SOLUTION METHOD

Dynamic force analysis can be done by any of several methods. The one which gives the most information about forces internal to the mechanism requires only the use of Newton's law as defined in equations 10.1 (p. 492) and 10.4 (p. 495). These can be written as a summation of all forces and torques in the system.

$$\sum \mathbf{F} = m\mathbf{a} \qquad\qquad \sum \mathbf{T} = I_G \alpha \qquad\qquad (11.1a)$$

It is also convenient to separately sum force components in X and Y directions, with the coordinate system chosen for convenience. The torques in our two dimensional system are all in the Z direction. This lets us break the two vector equations into three scalar equations:

$$\sum F_x = ma_x \qquad \sum F_y = ma_y \qquad \sum T = I_G \alpha \qquad (11.1b)$$

These three equations must be written for each moving body in the system which will lead to a set of linear simultaneous equations for any system. The set of simultaneous equations can most conveniently be solved by a matrix method as was shown in Chapter 5. These equations do not account for the gravitational force (weight) on a link. If the kinematic accelerations are large compared to gravity, which is often the case, then the weight forces can be ignored in the dynamic analysis. If the machine members are very massive or moving slowly with small kinematic accelerations, or both, the weight of the members may need to be included in the analysis. The weight can be treated as an external force acting on the CG of the member at a constant angle.

11.2   SINGLE LINK IN PURE ROTATION

As a simple example of this solution procedure, consider the single link in pure rotation shown in Figure 11-1a. In any of these kinetostatic dynamic force analysis problems, the kinematics of the problem must first be fully defined. That is, the angular accelerations of all rotating members and the linear accelerations of the CGs of all moving members must be found for all positions of interest. The mass of each member and the mass moment of inertia I_G with respect to each member's CG must also be known. In addition there may be external forces or torques applied to any member of the system. These are all shown in the figure.

While this analysis can be approached in many ways, it is useful for the sake of consistency to adopt a particular arrangement of coordinate systems and stick with it. We present such an approach here which, if carefully followed, will tend to minimize the chances of error. The reader may wish to develop his or her own approach once the principles are understood. The underlying mathematics is invariant, and one can choose coordinate systems for convenience. The vectors which are acting on the dynamic system in any loading situation are the same at a particular time regardless of how we may decide to resolve them into components for the sake of computation. The solution result will be the same.

We will first set up a nonrotating, local coordinate system on each moving member, located at its CG. (In this simple example we have only one moving member.) All externally applied forces, whether due to other connected members or to other systems must then have their points of application located in this local coordinate system. Figure 11-1b shows a free-body diagram of the moving link 2. The pin joint at O_2 on link 2 has a force $\mathbf{F}_{12}$ due to the mating link 1, the x and y components of which are F_{12x} and F_{12y}. These subscripts are read "force of link 1 on 2" in the x or y direction. This subscript notation scheme will be used consistently to indicate which of the "action-reaction" pair of forces at each joint is being solved for.

Note: x,y is a local, nonrotating coordinate system (LNCS), attached to the link

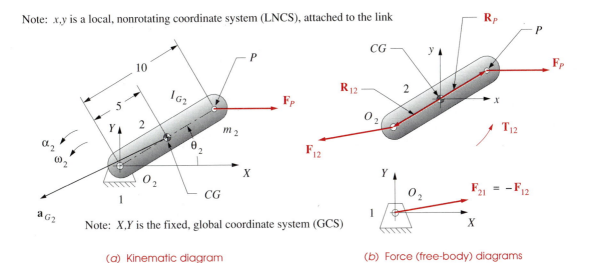

Note: X,Y is the fixed, global coordinate system (GCS)

(a) Kinematic diagram   (b) Force (free-body) diagrams

FIGURE 11-1

Dynamic force analysis of a single link in pure rotation

There is also an externally applied force $\mathbf{F}_P$ shown at point P, with components F_{Px} and F_{Py}. The points of application of these forces are defined by position vectors $\mathbf{R}_{12}$ and $\mathbf{R}_P$, respectively. These position vectors are defined with respect to the local coordinate system at the CG of the member. We will need to resolve them into x and y components. There will have to be a source torque available on the link to drive it at the kinematically defined accelerations. This is one of the unknowns to be solved for. The source torque is the torque delivered *from the ground to the driver link 2* and so is labeled $\mathbf{T}_{12}$. The other two unknowns in this example are the force components at the pin joint F_{12x} and F_{12y}.

We have three unknowns and three equations, so the system can be solved. Equations 11.1 can now be written for the moving link 2. Any applied forces or torques whose directions are known must retain the proper signs on their components. We will assume all unknown forces and torques to be positive. Their true signs will "come out in the wash."

$$\sum \mathbf{F} = \mathbf{F}_P + \mathbf{F}_{12} = m_2 \mathbf{a}_G$$
$$\sum \mathbf{T} = \mathbf{T}_{12} + (\mathbf{R}_{12} \times \mathbf{F}_{12}) + (\mathbf{R}_P \times \mathbf{F}_P) = I_G \alpha \qquad (11.2)$$

The force equation can be broken into its two components. The torque equation contains two cross product terms which represent torques due to the forces applied at a distance from the CG. When these cross products are expanded, the system of equations becomes:

$$F_{P_x} + F_{12_x} = m_2 a_{G_x}$$
$$F_{P_y} + F_{12_y} = m_2 a_{G_y} \tag{11.3}$$
$$T_{12} + \left(R_{12_x} F_{12_y} - R_{12_y} F_{12_x}\right) + \left(R_{P_x} F_{P_y} - R_{P_y} F_{P_x}\right) = I_G \alpha$$

This can be put in matrix form with the coefficients of the unknown variables forming the **A** matrix, the unknown variables the **B** vector, and the constant terms the **C** vector and then solved for **B**.

$$\qquad [\mathbf{A}] \qquad \times \quad [\mathbf{B}] \;=\; \qquad\qquad [\mathbf{C}]$$

$$\begin{bmatrix} 1 & 0 & 0 \\ 0 & 1 & 0 \\ -R_{12_y} & R_{12_x} & 1 \end{bmatrix} \times \begin{bmatrix} F_{12_x} \\ F_{12_y} \\ T_{12} \end{bmatrix} = \begin{bmatrix} m_2 a_{G_x} - F_{P_x} \\ m_2 a_{G_y} - F_{P_y} \\ I_G \alpha - \left(R_{P_x} F_{P_y} - R_{P_y} F_{P_x}\right) \end{bmatrix} \tag{11.4}$$

Note that the **A** matrix contains all the geometric information and the **C** matrix contains all the dynamic information about the system. The **B** matrix contains all the unknown forces and torques. We will now present a numerical example to reinforce your understanding of this method.

**EXAMPLE 11-1**

Dynamic Force Analysis of a Single Link in Pure Rotation. (See Figure 11-1)

Given: The 10-in-long link shown weighs 4 lb. Its *CG* is on the line of centers at the 5-in point. Its mass moment of inertia about its *CG* is 0.08 lb-in-sec^2. Its kinematic data are:

θ_2 deg	ω_2 rad/sec	α_2 rad/sec^2	a_{G_2} in/sec^2
30	20	15	2001 @ 208°

An external force of 40 lb at 0° is applied at point *P*.

Find: The **force** $\mathbf{F}_{12}$ at pin joint O_2 and the driving **torque** $\mathbf{T}_{12}$ needed to maintain motion with the given acceleration for this instantaneous position of the link.

Solution:

1   Convert the given weight to proper mass units, in this case blobs:

$$mass = \frac{weight}{g} = \frac{4\text{ lb}}{386\text{ in/sec}^2} = 0.0104 \text{ blobs} \tag{a}$$

2   Set up a local coordinate system at the *CG* of the link and draw all applicable vectors acting on the system as shown in the figure. Draw a free-body diagram as shown.

3   Calculate the *x* and *y* components of the position vectors $\mathbf{R}_{12}$ and $\mathbf{R}_P$ in this coordinate system:

$$\mathbf{R}_{12} = 5 \ in @ \angle 210°; \qquad R_{12_x} = -4.33, \qquad R_{12_y} = -2.50$$
$$\mathbf{R}_P = 5 \ in @ \angle 30°; \qquad R_{P_x} = +4.33, \qquad R_{P_y} = +2.50 \qquad (b)$$

4 Calculate the x and y components of the acceleration of the CG in this coordinate system:

$$\mathbf{a}_G = 2001 @ \angle 208°; \qquad a_{G_x} = -1766.78, \qquad a_{G_y} = -939.41 \qquad (c)$$

5 Calculate the x and y components of the external force at P in this coordinate system:

$$\mathbf{F}_P = 40 @ \angle 0°; \qquad F_{P_x} = 40, \qquad F_{P_y} = 0 \qquad (d)$$

6 Substitute these given and calculated values into the matrix equation 11.4

$$\begin{bmatrix} 1 & 0 & 0 \\ 0 & 1 & 0 \\ 2.50 & -4.33 & 1 \end{bmatrix} \times \begin{bmatrix} F_{12_x} \\ F_{12_y} \\ T_{12} \end{bmatrix} = \begin{bmatrix} (0.01)(-1766.78) - 40 \\ (0.01)(-939.41) - 0 \\ (0.08)(15) - \{(4.33)(0) - (2.5)(40)\} \end{bmatrix}$$
$$(e)$$

$$\begin{bmatrix} 1 & 0 & 0 \\ 0 & 1 & 0 \\ 2.50 & -4.33 & 1 \end{bmatrix} \times \begin{bmatrix} F_{12_x} \\ F_{12_y} \\ T_{12} \end{bmatrix} = \begin{bmatrix} -57.67 \\ -9.39 \\ 101.2 \end{bmatrix}$$

7 Solve this system either by inverting matrix **A** and premultiplying that inverse times matrix **C** using a pocket calculator such as the HP-15c or by inputting the values for matrices **A** and **C** to program MATRIX provided with this text.

Program MATRIX gives the following solution:

$$F_{12_x} = -57.67 \text{ lb}, \qquad F_{12_y} = -9.39 \text{ lb}, \qquad T_{12} = 204.72 \text{ lb-in} \qquad (f)$$

Converting the force to polar coordinates:

$$\mathbf{F}_{12} = 58.43 @ \angle 189.25° \qquad (g)$$

Read the disk file E011-01.mat into program MATRIX to exercise this example.

11.3 FORCE ANALYSIS OF A THREEBAR CRANK-SLIDE LINKAGE

When there is more than one link in the assembly, the solution simply requires that the three equations 11.1b be written for each link and then solved simultaneously. Figure 11-2a shows a threebar crank-slide linkage. This linkage has been simplified from the fourbar slider-crank (see Figure 11-4) by replacing the kinematically redundant slider block (link 4) with a half joint as shown. This linkage transformation reduces the number of links to three with no change in degree of freedom (see Section 2.9, p. 40). Only links 2 and 3 are moving. Link 1 is ground. Thus we should expect to have six equations in six unknowns (three per moving link).

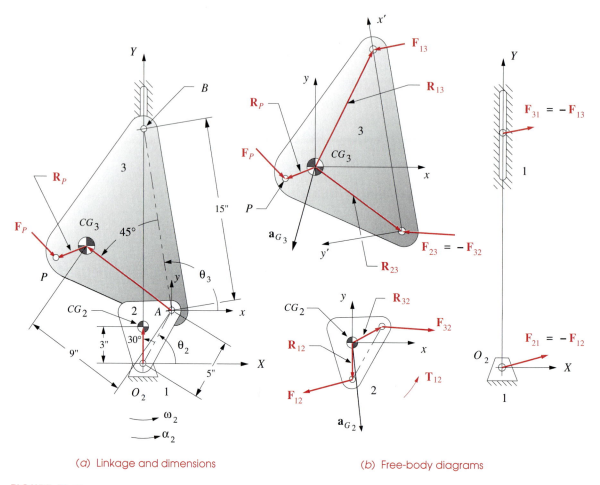

(a) Linkage and dimensions   (b) Free-body diagrams

FIGURE 11-2

Dynamic force analysis of a slider-crank linkage

Figure 11-2b shows the linkage "exploded" into its three separate links, drawn as free bodies. A kinematic analysis must have been done in advance of this dynamic force analysis in order to determine, for each moving link, its angular acceleration and the linear acceleration of its CG. For the kinematic analysis, only the link lengths from pin to pin were required. For a dynamic analysis the mass (m) of each link, the location of its CG, and its mass moment of inertia (I_G) about that CG are also needed.

The CG of each link is initially defined by a position vector rooted at one pin joint whose angle is measured with respect to the line of centers of the link in the local, rotating coordinate system (LRCS) x', y'. This is the most convenient way to establish the CG location since the link line of centers is the kinematic definition of the link. However, we will need to define the link's dynamic parameters and force locations with respect to a local, nonrotating coordinate system (LNCS) x, y located at its CG and which is always parallel to the global coordinate system (GCS) XY. The position vector locations of all attachment points of other links and points of application of external forces must be defined with respect to the link's LNCS. Note that these kinematic and applied force data must be available for all positions of the linkage for which a force analysis is desired. In the following discussion and examples, only one linkage position will be addressed. The process is identical for each succeeding position and only the calculations must be repeated. Obviously, a computer will be a valuable aid in accomplishing the task.

Link 2 in Figure 11-2b shows forces acting on it at each pin joint, designated $\mathbf{F}_{12}$ and $\mathbf{F}_{32}$. By convention their subscripts denote the force that the adjoining link is exerting *on* the link being analyzed; that is, $\mathbf{F}_{12}$ is the force of 1 *on* 2 and $\mathbf{F}_{32}$ is the force of 3 *on* 2. Obviously there is also an equal and opposite force at each of these pins which would be designated as $\mathbf{F}_{21}$ and $\mathbf{F}_{23}$, respectively. The choice of which of the members of these pairs of forces to be solved for is arbitrary. As long as proper bookkeeping is done, their identities will be maintained.

When we move to link 3, we maintain the same convention of showing forces acting *on* the link in its free-body diagram. Thus at instant center I_{23} we show $\mathbf{F}_{23}$ acting on link 3. However, because we showed force $\mathbf{F}_{32}$ acting at the same point on link 2, this introduces an additional unknown to the problem for which we need an additional equation. The equation is available from Newton's third law:

$$\mathbf{F}_{23} = -\mathbf{F}_{32} \tag{11.5}$$

Thus we are free to substitute the negative reaction force for any action force at any joint. This has been done on link 3 in the figure in order to reduce the unknown forces at that joint to one, namely $\mathbf{F}_{32}$. The same procedure is followed at each joint with one of the action-reaction forces arbitrarily chosen to be solved for and its negative reaction applied to the mating link.

The naming convention used for the position vectors ($\mathbf{R}_{ap}$) which locate the pin joints with respect to the *CG* in the link's nonrotating local coordinate system is as follows. The first subscript (*a*) denotes the adjoining link to which the position vector points. The second subscript (*p*) denotes the parent link to which the position vector belongs. Thus in the case of link 2 in Figure 11-2b, vector $\mathbf{R}_{12}$ locates the attachment point of link 1 to link 2, and $\mathbf{R}_{32}$ the attachment point of link 3 to link 2. Note that in some cases these subscripts will match those of the pin forces shown acting at those points, but where the negative reaction force has been substituted as described above, the subscript order of the force and its position vector will not agree. This can lead to confusion and must be carefully watched for typographical errors when setting up the problem.

Any external forces acting on the links are located in similar fashion with a position vector to a point on the line of application of the force. This point is given the same letter subscript as that of the external force. Link 3 in the figure shows such an external force $\mathbf{F}_P$ acting on it at point P. The position vector $\mathbf{R}_P$ locates that point with respect to the *CG*. It is important to note that the *CG* of each link is consistently taken as the point of reference for all forces acting on that link. Left to its own devices, an unconstrained body in complex motion will spin about its own *CG*; thus we analyze its linear acceleration at that point and apply the angular acceleration about the *CG* as a center.

Equations 11.1 are now written for each moving link. For link 2, with the cross products expanded:

$$F_{12_x} + F_{32_x} = m_2 a_{G_{2_x}}$$

$$F_{12_y} + F_{32_y} = m_2 a_{G_{2_y}}$$ (11.6a)

$$T_{12} + \left(R_{12_x} F_{12_y} - R_{12_y} F_{12_x} \right) + \left(R_{32_x} F_{32_y} - R_{32_y} F_{32_x} \right) = I_{G_2} \alpha_2$$

For link 3, with the cross products expanded, note the substitution of the reaction force $- \mathbf{F}_{32}$ for $\mathbf{F}_{23}$:

$$F_{13_x} - F_{32_x} + F_{P_x} = m_3 a_{G_{3_x}}$$

$$F_{13_y} - F_{32_y} + F_{P_y} = m_3 a_{G_{3_y}}$$ (11.6b)

$$\left(R_{13_x} F_{13_y} - R_{13_y} F_{13_x} \right) - \left(R_{23_x} F_{32_y} - R_{23_y} F_{32_x} \right) + \left(R_{P_x} F_{P_y} - R_{P_y} F_{P_x} \right) = I_{G_3} \alpha_3$$

Note also that $\mathbf{T}_{12}$, the source torque, only appears in the equation for link 2 as that is the driver crank to which the motor is attached. Link 3 has no externally applied torque but does have an external force $\mathbf{F}_P$ which might be due to whatever link 3 is pushing on to do its external work.

There are seven unknowns present in these six equations, F_{12x}, F_{12y}, F_{32x}, F_{32y}, F_{13x}, F_{13y}, and T_{12}. But, F_{13y} is due only to friction at the joint between link 3 and link 1. We can write a relation for the friction force f at that interface such as $f = \pm \mu N$, where $\pm \mu$ is a known coefficient of coulomb friction. The friction force always opposes motion. The kinematic analysis will provide the velocity of the link at the sliding joint. The direction of f will always be the opposite of this velocity. Note that μ is a nonlinear function which has a discontinuity at zero velocity; thus at the linkage positions where velocity is zero, the inclusion of μ in these linear equations is not valid. (See Figure 10-5a, p. 502.) In this example, the normal force N is equal to F_{13x} and the friction force f is equal to F_{13y}. For linkage positions with nonzero velocity, we can eliminate F_{13y} by substituting into equation 11.6b,

$$F_{13_y} = \mu F_{13_x}$$ (11.6c)

where the sign of F_{13y} is taken as the opposite of the sign of the velocity at that point. We are then left with six unknowns in equations 11.6 and can solve them simultaneously. We also rearrange equations 11.6a and 11.6b to put all known terms on the right side.

$$F_{12_x} + F_{32_x} = m_2 a_{G_{2_x}}$$

$$F_{12_y} + F_{32_y} = m_2 a_{G_{2_y}}$$

$$T_{12} + R_{12_x} F_{12_y} - R_{12_y} F_{12_x} + R_{32_x} F_{32_y} - R_{32_y} F_{32_x} = I_{G_2} \alpha_2$$ (11.6d)

$$F_{13_x} - F_{32_x} = m_3 a_{G_{3_x}} - F_{P_x}$$

$$\pm \mu F_{13_x} - F_{32_y} = m_3 a_{G_{3_y}} - F_{P_y}$$

$$\left(\pm \mu R_{13_x} - R_{13_y} \right) F_{13_x} - R_{23_x} F_{32_y} + R_{23_y} F_{32_x} = I_{G_3} \alpha_3 - R_{P_x} F_{P_y} + R_{P_y} F_{P_x}$$

Putting these six equations in matrix form we get:

$$
\begin{bmatrix}
1 & 0 & 1 & 0 & 0 & 0 \\
0 & 1 & 0 & 1 & 0 & 0 \\
-R_{12_y} & R_{12_x} & -R_{32_y} & R_{32_x} & 0 & 1 \\
0 & 0 & -1 & 0 & 1 & 0 \\
0 & 0 & 0 & -1 & \mu & 0 \\
0 & 0 & R_{23_y} & -R_{23_x} & \left(\mu R_{13_x} - R_{13_y}\right) & 0
\end{bmatrix}
\times
\begin{bmatrix}
F_{12_x} \\
F_{12_y} \\
F_{32_x} \\
F_{32_y} \\
F_{13_x} \\
T_{12}
\end{bmatrix}
=
$$

$$(11.7)$$

$$
\begin{bmatrix}
m_2 a_{G2_x} \\
m_2 a_{G2_y} \\
I_{G_2}\alpha_2 \\
m_3 a_{G3_x} - F_{P_x} \\
m_3 a_{G3_y} - F_{P_y} \\
I_{G_3}\alpha_3 - R_{P_x}F_{P_y} + R_{P_y}F_{P_x}
\end{bmatrix}
$$

This system can be solved by using program MATRIX or any other matrix solving calculator. As an example of this solution consider the following linkage data.

 **EXAMPLE 11-2**

Dynamic Force Analysis of a Threebar Crank-Slide Linkage with Half Joint. (See Figure 11-2, p. 526.)

Given: The 5-in long crank (link 2) shown weighs 2 lb. Its *CG* is at 3 in and 30° from the line of centers. Its mass moment of inertia about its *CG* is 0.05 lb-in-sec^2. Its acceleration is defined in its LNCS, *x,y*. Its kinematic data are:

θ_2 deg	ω_2 rad/sec	α_2 rad/sec^2	a_{G_2} in/sec^2
60	30	−10	2700.17 @ −89.4°

The coupler (link 3) is 15 in long and weighs 4 lb. Its *CG* is at 9 in and 45° from the line of centers. Its mass moment of inertia about its *CG* is 0.10 lb-in-sec^2. Its acceleration is defined in its LNCS, *x,y*. Its kinematic data are:

θ_3 deg	ω_3 rad/sec	α_3 rad/sec^2	a_{G_3} in/sec^2
99.59	−8.78	−136.16	3453.35 @ 254.4°

The sliding joint on link 3 has a velocity of 96.95 in/sec in the +*Y* direction.

There is an external force of 50 lb at −45°, applied at point *P* which is located at 2.7 in and 101° from the *CG* of link 3, measured in the link's embedded, rotating coordinate system or LRCS *x', y'* (origin at *A* and *x* axis from *A* to *B*). The coefficient of friction μ is 0.2.

11

Find: The **forces** F_{12}, F_{32}, F_{13} at the joints and the driving **torque** T_{12} needed to maintain motion with the given acceleration for this instantaneous position of the link.

Solution:

1 Convert the given weights to proper mass units, in this case blobs:

$$mass_{link2} = \frac{weight}{g} = \frac{2 \text{ lb}}{386 \text{ in/sec}^2} = 0.0052 \text{ blobs} \qquad (a)$$

$$mass_{link3} = \frac{weight}{g} = \frac{4 \text{ lb}}{386 \text{ in/sec}^2} = 0.0104 \text{ blobs} \qquad (b)$$

2 Set up a local, nonrotating xy coordinate system (LNCS) at the CG of each link, and draw all applicable position and force vectors acting within or on that system as shown in Figure 11-2. Draw a free-body diagram of each moving link as shown.

3 Calculate the x and y components of the position vectors R_{12}, R_{32}, R_{23}, R_{13}, and R_P in the LNCS coordinate system:

$$
\begin{array}{llll}
R_{12} = 3.00 \ @ \angle \ 270.0°; & R_{12_x} = \ 0, & R_{12_y} = -3.0 \\
R_{32} = 2.83 \ @ \angle \ 28.0°; & R_{32_x} = \ 2.500, & R_{32_y} = \ 1.333 \\
R_{23} = 9.00 \ @ \angle \ 324.5°; & R_{23_x} = \ 7.329, & R_{23_y} = -5.224 & (c) \\
R_{13} = 10.72 \ @ \angle \ 63.14°; & R_{13_x} = \ 4.843, & R_{13_y} = \ 9.563 \\
R_P = 2.70 \ @ \angle \ 201.0°; & R_{P_x} = -2.521, & R_{P_y} = -0.968
\end{array}
$$

These position vector angles are measured with respect to the LNCS which is always parallel to the global coordinate system (GCS), making the angles the same in both systems.

4 Calculate the x and y components of the acceleration of the CGs of all moving links in the global coordinate system:

$$a_{G_2} = 2700.17 \ @ \angle \ -89.4°; \qquad a_{G_{2x}} = 28.28, \quad a_{G_{2y}} = -2700$$

$$(d)$$

$$a_{G_3} = 3453.35 \ @ \angle \ 254.4°; \qquad a_{G_{3x}} = -930.82, \quad a_{G_{3y}} = -3325.54$$

5 Calculate the x and y components of the external force at P in the global coordinate system:

$$F_P = 50 @ \angle -45°; \qquad F_{P_x} = 35.36, \qquad F_{P_y} = -35.36 \qquad (e)$$

6 Substitute these given and calculated values into the matrix equation 11.7.

$$
\begin{bmatrix}
1 & 0 & 1 & 0 & 0 & 0 \\
0 & 1 & 0 & 1 & 0 & 0 \\
3 & 0 & -1.333 & 2.5 & 0 & 1 \\
0 & 0 & -1 & 0 & 1 & 0 \\
0 & 0 & 0 & -1 & 0.2 & 0 \\
0 & 0 & -5.224 & -7.329 & [(0.2)4.843-(9.563)] & 0
\end{bmatrix}
\times
\begin{bmatrix}
F_{12_x} \\
F_{12_y} \\
F_{32_x} \\
F_{32_y} \\
F_{13_x} \\
T_{12}
\end{bmatrix}
=
\tag{f}
$$

$$
\begin{bmatrix}
(0.005)(28.28) \\
(0.005)(-2700) \\
(0.05)(-10) \\
(0.01)(-930.82)-35.36 \\
(0.01)(-3325.54)-(-35.36) \\
(0.1)(-136.16)-(-2.521)(-35.36)+(-0.968)(35.36)
\end{bmatrix}
=
\begin{bmatrix}
0.141 \\
-13.500 \\
-.500 \\
-44.668 \\
2.105 \\
-136.987
\end{bmatrix}
$$

7 Solve this system either by inverting matrix **A** and premultiplying that inverse times matrix **C** using a pocket calculator such as the HP-15c, or by inputting the values for matrices **A** and **C** to program MATRIX provided with this text which gives the following solution:

$$
\begin{bmatrix}
F_{12_x} \\
F_{12_y} \\
F_{32_x} \\
F_{32_y} \\
F_{13_x} \\
T_{12}
\end{bmatrix}
=
\begin{bmatrix}
-39.232 \\
-10.336 \\
39.373 \\
-3.164 \\
-5.295 \\
177.590
\end{bmatrix}
\tag{g}
$$

Converting the forces to polar coordinates:

$$
\begin{aligned}
\mathbf{F}_{12} &= 40.57 \text{ lb } @ \angle 194.76° \\
\mathbf{F}_{32} &= 39.50 \text{ lb } @ \angle -4.60° \\
\mathbf{F}_{13} &= 5.40 \text{ lb } @ \angle 191.31°
\end{aligned}
\tag{h}
$$

Read the disk file E011-02.mat into program MATRIX to exercise this example.

11.4   FORCE ANALYSIS OF A FOURBAR LINKAGE

Figure 11-3a shows a fourbar linkage. All dimensions of link lengths, link positions, locations of the links' CGs, linear accelerations of those CGs, and link angular accelerations and velocities have been previously determined from a kinematic analysis. We now wish to find the forces acting at all the pin joints of the linkage for one or more positions. The procedure is exactly the same as that used in the above two examples. This linkage has three moving links. Equation 11.1 provides three equations for any link or rigid body in motion. We should expect to have nine equations in nine unknowns for this problem.

Figure 11-3b shows the free-body diagrams for all links, with all forces shown. Note that an external force $\mathbf{F}_P$ is shown acting on link 3 at point P. Also an external torque $\mathbf{T}_4$ is shown acting on link 4. These external loads are due to some other mechanism (de-

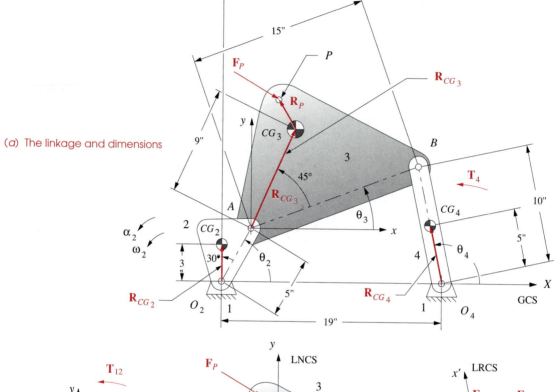

(a) The linkage and dimensions

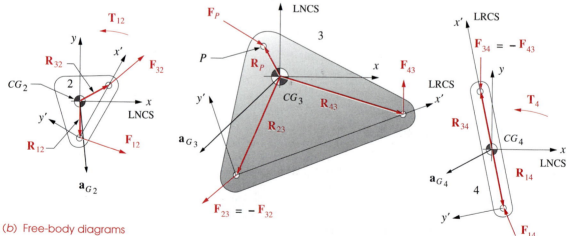

(b) Free-body diagrams

FIGURE 11-3

Dynamic force analysis of a fourbar linkage. (See also Figure P11-2, p. 561)

vice, person, thing, etc.) pushing or twisting against the motion of the linkage. Any link can have any number of external loads and torques acting on it. Only one external torque and one external force are shown here to serve as examples of how they are handled in the computation. (Note that a more complicated force system, if present, could also be reduced to the combination of a single force and torque on each link.)

To solve for the pin forces it is necessary that these applied external forces and torques be defined for all positions of interest. We will solve for one member of the pair of action-reaction forces at each joint, and also for the driving torque $\mathbf{T}_{12}$ needed to be supplied at link 2 in order to maintain the kinematic state as defined. The force subscript convention is the same as that defined in the previous example. For example, $\mathbf{F}_{12}$ is the force of 1 *on* 2 and $\mathbf{F}_{32}$ is the force of 3 *on* 2. The equal and opposite forces at each of these pins are designated $\mathbf{F}_{21}$ and $\mathbf{F}_{23}$, respectively. All the unknown forces in the figure are shown at arbitrary angles and lengths as their true values are still to be determined.

The linkage kinematic parameters are defined with respect to a global XY system (GCS) whose origin is at the driver pivot O_2 and whose X axis goes through link 4's fixed pivot O_4. The mass (m) of each link, the location of its CG, and its mass moment of inertia (I_G) about that CG are also needed. The CG of each link is initially defined within each link with respect to a local moving and rotating axis system (LRCS) embedded in the link because the CG is an unchanging physical property of the link. The origin of this x',y' axis system is at one pin joint and the x' axis is the line of centers of the link. The CG position within the link is defined by a position vector in this LRCS. The instantaneous location of the CG can easily be determined for each dynamic link position by adding the angle of the internal CG position vector to the current GCS angle of the link.

We need to define each link's dynamic parameters and force locations with respect to a local, moving, but nonrotating axis system (LNCS) x,y located at its CG as shown on each free-body diagram in Figure 11-3b. The position vector locations of all attachment points of other links and points of application of external forces must be defined with respect to this LNCS axis system. These kinematic and applied force data differ for each position of the linkage. In the following discussion and examples, only one linkage position will be addressed. The process is identical for each succeeding position.

Equations 11.1 (p. 522) are now written for each moving link. For link 2, the result is identical to that done for the slider-crank example in equation 11.6a (p. 528).

$$F_{12_x} + F_{32_x} = m_2 a_{G_{2x}}$$

$$F_{12_y} + F_{32_y} = m_2 a_{G_{2y}} \tag{11.8a}$$

$$T_{12} + \left(R_{12_x} F_{12_y} - R_{12_y} F_{12_x} \right) + \left(R_{32_x} F_{32_y} - R_{32_y} F_{32_x} \right) = I_{G_2} \alpha_2$$

For link 3, with substitution of the reaction force $-\mathbf{F}_{32}$ for $\mathbf{F}_{23}$, the result is similar to equation 11.6b with some subscript changes to reflect the presence of link 4.

$$F_{43_x} - F_{32_x} + F_{P_x} = m_3 a_{G_{3x}}$$

$$F_{43_y} - F_{32_y} + F_{P_y} = m_3 a_{G_{3y}} \tag{11.8b}$$

$$\left(R_{43_x} F_{43_y} - R_{43_y} F_{43_x} \right) - \left(R_{23_x} F_{32_y} - R_{23_y} F_{32_x} \right) + \left(R_{P_x} F_{P_y} - R_{P_y} F_{P_x} \right) = I_{G_3} \alpha_3$$

For link 4, substituting the reaction force $-\mathbf{F}_{43}$ for $\mathbf{F}_{34}$, a similar set of equations 11.1 can be written:

$$F_{14_x} - F_{43_x} = m_4 a_{G4_x}$$

$$F_{14_y} - F_{43_y} = m_4 a_{G4_y} \tag{11.8c}$$

$$\left(R_{14_x} F_{14_y} - R_{14_y} F_{14_x}\right) - \left(R_{34_x} F_{43_y} - R_{34_y} F_{43_x}\right) + T_4 = I_{G4}\alpha_4$$

Note again that $\mathbf{T}_{12}$, the source torque, only appears in the equation for link 2 as that is the driver crank to which the motor is attached. Link 3, in this example, has no externally applied torque (though it could have) but does have an external force $\mathbf{F}_P$. Link 4, in this example, has no external force acting on it (though it could have) but does have an external torque $\mathbf{T}_4$. (The driving link 2 could also have an externally applied force on it though it usually does not.) There are nine unknowns present in these nine equations, $F_{12x}, F_{12y}, F_{32x}, F_{32y}, F_{43x}, F_{43y}, F_{14x}, F_{14y},$ and T_{12}, so we can solve them simultaneously. We rearrange terms in equations 11.8 to put all known constant terms on the right side and then put them in matrix form.

$$\begin{bmatrix} 1 & 0 & 1 & 0 & 0 & 0 & 0 & 0 & 0 \\ 0 & 1 & 0 & 1 & 0 & 0 & 0 & 0 & 0 \\ -R_{12_y} & R_{12_x} & -R_{32_y} & R_{32_x} & 0 & 0 & 0 & 0 & 1 \\ 0 & 0 & -1 & 0 & 1 & 0 & 0 & 0 & 0 \\ 0 & 0 & 0 & -1 & 0 & 1 & 0 & 0 & 0 \\ 0 & 0 & R_{23_y} & -R_{23_x} & -R_{43_y} & R_{43_x} & 0 & 0 & 0 \\ 0 & 0 & 0 & 0 & -1 & 0 & 1 & 0 & 0 \\ 0 & 0 & 0 & 0 & 0 & -1 & 0 & 1 & 0 \\ 0 & 0 & 0 & 0 & R_{34_y} & -R_{34_x} & -R_{14_y} & R_{14_x} & 0 \end{bmatrix} \times \begin{bmatrix} F_{12_x} \\ F_{12_y} \\ F_{32_x} \\ F_{32_y} \\ F_{43_x} \\ F_{43_y} \\ F_{14_x} \\ F_{14_y} \\ T_{12} \end{bmatrix} =$$

$$\begin{bmatrix} m_2 a_{G2_x} \\ m_2 a_{G2_y} \\ I_{G2}\alpha_2 \\ m_3 a_{G3_x} - F_{P_x} \\ m_3 a_{G3_y} - F_{P_y} \\ I_{G3}\alpha_3 - R_{P_x} F_{P_y} + R_{P_y} F_{P_x} \\ m_4 a_{G4_x} \\ m_4 a_{G4_y} \\ I_{G4}\alpha_4 - T_4 \end{bmatrix}$$

(11.9)

This system can be solved by using program MATRIX or any matrix solving calculator. As an example of this solution consider the following linkage data.

✐EXAMPLE 11-3

Dynamic Force Analysis of a Fourbar Linkage. (See Figure 11-3, p. 532)

Given: The 5-in-long crank (link 2) shown weighs 1.5 lb. Its *CG* is at 3 in at +30° from the line of centers. Its mass moment of inertia about its *CG* is 0.4 lb-in-sec². Its kinematic data are:

θ_2 deg	ω_2 rad/sec	α_2 rad/sec²	a_{G_2} in/sec²
60	25	−40	1878.84 @ 273.66°

The coupler (link 3) is 15 in long and weighs 7.7 lb. Its *CG* is at 9 in at 45° off the line of centers. Its mass moment of inertia about its *CG* is 1.5 lb-in-sec². Its kinematic data are:

θ_3 deg	ω_3 rad/sec	α_3 rad/sec²	a_{G_3} in/sec²
20.92	−5.87	120.9	3646.1 @ 226.5°

There is an external force of 80 lb at 330° on link 3, applied at point *P* which is located 3 in at 100° from the *CG* of link 3. There is an external torque on link 4 of 120 lb-in. The ground link is 19 in long. The rocker (link 4) is 10 in long and weighs 5.8 lb. Its *CG* is at 5 in at 0° off the line of centers. Its mass moment of inertia about its *CG* is 0.8 lb-in-sec². Its kinematic data are:

θ_4 deg	ω_4 rad/sec	α_4 rad/sec²	a_{G_4} in/sec²
104.41	7.93	276.29	1416.8 @ 207.2°

Find: The forces $\mathbf{F}_{12}$, $\mathbf{F}_{32}$, $\mathbf{F}_{43}$, $\mathbf{F}_{14}$, at the joints and the driving torque $\mathbf{T}_{12}$ needed to maintain motion with the given acceleration for this instantaneous position of the link.

Solution:

1   Convert the given weight to proper mass units, in this case blobs:

$$mass_{link2} = \frac{weight}{g} = \frac{1.5 \text{ lb}}{386 \text{ in/sec}^2} = 0.004 \text{ blobs} \qquad (a)$$

$$mass_{link3} = \frac{weight}{g} = \frac{7.7 \text{ lb}}{386 \text{ in/sec}^2} = 0.020 \text{ blobs} \qquad (b)$$

$$mass_{link4} = \frac{weight}{g} = \frac{5.8 \text{ lb}}{386 \text{ in/sec}^2} = 0.015 \text{ blobs} \qquad (c)$$

2   Set up an LNCS *xy* coordinate system at the *CG* of each link, and draw all applicable vectors acting on that system as shown in the figure. Draw a free-body diagram of each moving link as shown.

3   Calculate the *x* and *y* components of the position vectors $\mathbf{R}_{12}$, $\mathbf{R}_{32}$, $\mathbf{R}_{23}$, $\mathbf{R}_{43}$, $\mathbf{R}_{34}$, $\mathbf{R}_{14}$, and $\mathbf{R}_P$ in the link's LNCS. $\mathbf{R}_{43}$, $\mathbf{R}_{34}$, and $\mathbf{R}_{14}$ will have to be calculated from the given link geometry data using the law of cosines and law of sines. Note that the current value of link 3's position angle (θ_3) in the GCS must be added to the angles of all position vectors before creating their *x,y* components in the LNCS if their angles were originally measured with respect to the link's embedded, local rotating coordinate system (LRCS).

11

$$R_{12} = 3.00 \ @ \ \angle \ 270.00°; \qquad R_{12_x} = 0.000, \qquad R_{12_y} = -3$$

$$R_{32} = 2.83 \ @ \ \angle \ 28.00°; \qquad R_{32_x} = 2.500, \qquad R_{32_y} = 1.333$$

$$R_{23} = 9.00 \ @ \ \angle \ 245.92°; \qquad R_{23_x} = -3.672, \qquad R_{23_y} = -8.217$$

$$R_{43} = 10.72 \ @ \ \angle \ -15.46°; \qquad R_{43_x} = 10.332, \qquad R_{43_y} = -2.858 \qquad (d)$$

$$R_{34} = 5.00 \ @ \ \angle \ 104.41°; \qquad R_{34_x} = -1.244, \qquad R_{34_y} = 4.843$$

$$R_{14} = 5.00 \ @ \ \angle \ 284.41°; \qquad R_{14_x} = 1.244, \qquad R_{14_y} = -4.843$$

$$R_P = 3.00 \ @ \ \angle \ 120.92°; \qquad R_{P_x} = -1.542, \qquad R_{P_y} = 2.574$$

4 Calculate the x and y components of the acceleration of the CGs of all moving links in the global coordinate system (GCS):

$$\mathbf{a}_{G_2} = 1878.84 @ \ \angle 273.66°; \qquad a_{G_{2_x}} = 119.94, \qquad a_{G_{2_y}} = -1875.01$$

$$\mathbf{a}_{G_3} = 3646.10 @ \ \angle 226.51°; \qquad a_{G_{3_x}} = -2509.35, \qquad a_{G_{3_y}} = -2645.23 \qquad (e)$$

$$\mathbf{a}_{G_4} = 1416.80 @ \ \angle 207.24°; \qquad a_{G_{4_x}} = -1259.67, \qquad a_{G_{4_y}} = -648.50$$

5 Calculate the x and y components of the external force at P in the GCS:

$$\mathbf{F}_{P3} = 80 \ @ \ \angle \ 330°; \qquad F_{P3_x} = 69.28, \qquad F_{P3_y} = -40.00 \qquad (f)$$

6 Substitute these given and calculated values into the matrix equation 11.9 (p. 534).

$$
\begin{bmatrix}
1 & 0 & 1 & 0 & 0 & 0 & 0 & 0 & 0 \\
0 & 1 & 0 & 1 & 0 & 0 & 0 & 0 & 0 \\
3 & 0 & -1.330 & 2.5 & 0 & 0 & 0 & 0 & 1 \\
0 & 0 & -1 & 0 & 1 & 0 & 0 & 0 & 0 \\
0 & 0 & 0 & -1 & 0 & 1 & 0 & 0 & 0 \\
0 & 0 & -8.217 & 3.673 & 2.861 & 10.339 & 0 & 0 & 0 \\
0 & 0 & 0 & 0 & -1 & 0 & 1 & 0 & 0 \\
0 & 0 & 0 & 0 & 0 & -1 & 0 & 1 & 0 \\
0 & 0 & 0 & 0 & 4.843 & 1.244 & 4.843 & 1.244 & 0
\end{bmatrix}
\times
\begin{bmatrix}
F_{12_x} \\
F_{12_y} \\
F_{32_x} \\
F_{32_y} \\
F_{43_x} \\
F_{43_y} \\
F_{14_x} \\
F_{14_y} \\
T_{12}
\end{bmatrix}
=
$$

$$(g)$$

$$
\begin{bmatrix}
(0.004)(119.94) \\
(0.004)(-1875.01) \\
(0.4)(-40) \\
(0.02)(-2509.35) - (69.28) \\
(0.02)(-2645.23) - (-40) \\
(1.5)(120.9) - \left[(-1.542)(-40) - (2.574)(69.28) \right] \\
(0.015)(-1259.67) \\
(0.015)(-648.50) \\
(0.8)(276.29) - (120)
\end{bmatrix}
=
\begin{bmatrix}
0.480 \\
-7.500 \\
-16.000 \\
-119.465 \\
-12.908 \\
298.003 \\
-18.896 \\
-9.727 \\
101.031
\end{bmatrix}
$$

7 Solve this system either by inverting matrix **A** and premultiplying that inverse times matrix **C** using a pocket calculator such as the HP-28, or by inputting the values for matrices **A** and **C** to program MATRIX provided with this text which gives the following solution:

$$
\begin{bmatrix}
F_{12_x} \\
F_{12_y} \\
F_{32_x} \\
F_{32_y} \\
F_{43_x} \\
F_{43_y} \\
F_{14_x} \\
F_{14_y} \\
T_{12}
\end{bmatrix}
=
\begin{bmatrix}
-117.65 \\
-107.84 \\
118.13 \\
100.34 \\
-1.34 \\
87.43 \\
-20.23 \\
77.71 \\
243.23
\end{bmatrix}
\tag{h}
$$

Converting the forces to polar coordinates:

$$
\begin{aligned}
\mathbf{F}_{12} &= \ 159.60 \text{ lb } @ \angle \ 222.52° \\
\mathbf{F}_{32} &= \ 154.99 \text{ lb } @ \angle \ \ \ 40.35° \\
\mathbf{F}_{43} &= \ \ \ 87.44 \text{ lb } @ \angle \ \ \ 90.88° \\
\mathbf{F}_{14} &= \ \ \ 80.30 \text{ lb } @ \angle \ 104.59°
\end{aligned}
\tag{i}
$$

8 The pin-force magnitudes in (*i*) are needed to size the pivot pins and links against failure and to select pivot bearings that will last for the required life of the assembly. The driving torque T_{12} defined in (*h*) is needed to select a motor or other device capable of supplying the power to drive the system. See Section 2.16 (p. 60) for a brief discussion of motor selection. Issues of stress calculation and failure prevention are beyond the scope of this text, but note that those calculations cannot be done until a good estimate of the dynamic forces and torques on the system has been made by methods such as those shown in this example.

This solves the linkage for one position. A new set of values can be put into the **A** and **C** matrices for each position of interest at which a force analysis is needed. Read the disk file E11-03.mat into program MATRIX to exercise this example. The disk file E11-03.4br can also be read into program FOURBAR which will run the linkage through a series of positions starting with the stated parameters as initial conditions. The linkage will slow to a stop and then run in reverse due to the negative acceleration. The matrix of equation (*g*) can be seen within FOURBAR using *Dynamics/Solve/Show Matrix*.

It is worth noting some general observations about this method at this point. The solution is done using cartesian coordinates of all forces and position vectors. Before being placed in the matrices, these vector components must be defined in the global coordinate system (GCS) or in nonrotating, local coordinate systems, parallel to the global coordinate system, with their origins at the links' *CGs* (LNCS). Some of the linkage parameters are normally expressed in such coordinate systems, but some are not, and so must be converted. The kinematic data should all be computed in the global system or in parallel, **nonrotating**, local systems placed at the *CGs* of individual links. Any external forces on the links must also be defined in the global system.

11

However, the position vectors that define intralink locations, such as the pin joints versus the CG, or which locate points of application of external forces versus the CG are defined in local, **rotating** coordinate systems embedded in the links (LRCS). Thus these position vectors must be redefined in a **nonrotating**, parallel system before being used in the matrix. An example of this is vector $\mathbf{R}_p$, which was initially defined as 3 in at 100° in link 3's embedded, **rotating** coordinate system. Note in the example above that its cartesian coordinates for use in the equations were calculated after adding the current value of θ_3 to its angle. This redefined $\mathbf{R}_p$ as 3 in at 120.92° in the **nonrotating** local system. The same was done for position vectors $\mathbf{R}_{12}$, $\mathbf{R}_{32}$, $\mathbf{R}_{23}$, $\mathbf{R}_{43}$, $\mathbf{R}_{34}$, and $\mathbf{R}_{14}$. In each case the **intralink angle** of these vectors (which is independent of linkage position) was added to the current link angle to obtain its position in the xy system at the link's CG. The proper definition of these position vector components is critical to the solution, and it is very easy to make errors in defining them.

To further confuse things, even though the position vector $\mathbf{R}_p$ is initially measured in the link's embedded, rotating coordinate system, the force $\mathbf{F}_p$, which it locates, is not. The force $\mathbf{F}_p$ is not part of the link, as is $\mathbf{R}_p$, but rather is part of the external world, so it is defined in the global system.

11.5   FORCE ANALYSIS OF A FOURBAR SLIDER-CRANK LINKAGE

The approach taken for the pin-jointed fourbar is equally valid for a fourbar slider-crank linkage. The principal difference will be that the slider block will have no angular acceleration. Figure 11-4 shows a fourbar slider-crank with an external force on the slider block, link 4. This is representative of the mechanism used extensively in piston pumps and internal combustion engines. We wish to determine the forces at the joints and the driving torque needed on the crank to provide the specified accelerations. A kinematic analysis must have previously been done in order to determine all position, velocity, and acceleration information for the positions being analyzed. Equations 11.1 are written for each link. For link 2:

$$F_{12_x} + F_{32_x} = m_2 a_{G2_x}$$

$$F_{12_y} + F_{32_y} = m_2 a_{G2_y} \tag{11.10a}$$

$$T_{12} + \left(R_{12_x} F_{12_y} - R_{12_y} F_{12_x}\right) + \left(R_{32_x} F_{32_y} - R_{32_y} F_{32_x}\right) = I_{G_2}\alpha_2$$

This is identical to equation 11.8a for the "pure" fourbar linkage. For link 3:

$$F_{43_x} - F_{32_x} = m_3 a_{G3_x}$$

$$F_{43_y} - F_{32_y} = m_3 a_{G3_y} \tag{11.10b}$$

$$\left(R_{43_x} F_{43_y} - R_{43_y} F_{43_x}\right) - \left(R_{23_x} F_{32_y} - R_{23_y} F_{32_x}\right) = I_{G_3}\alpha_3$$

This is similar to equation 11.8b, lacking only the terms involving $\mathbf{F}_p$ since there is no external force shown acting on link 3 of our example slider-crank. For link 4:

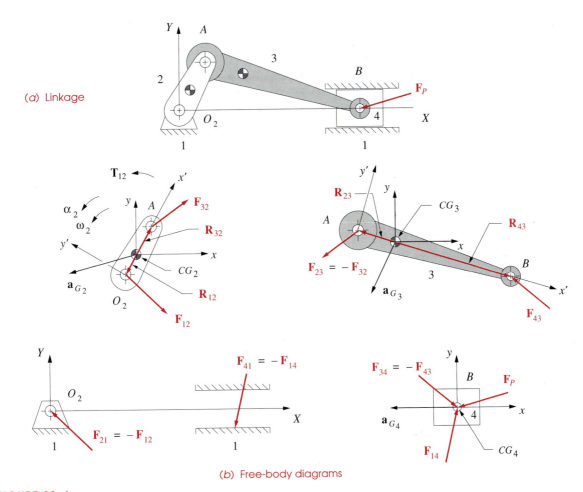

(a) Linkage

(b) Free-body diagrams

FIGURE 11-4

Dynamic force analysis of the fourbar slider-crank linkage

$$F_{14_x} - F_{43_x} + F_{P_x} = m_4 a_{G4_x}$$
$$F_{14_y} - F_{43_y} + F_{P_y} = m_4 a_{G4_y} \qquad (11.10c)$$

$$\left(R_{14_x} F_{14_y} - R_{14_y} F_{14_x}\right) - \left(R_{34_x} F_{43_y} - R_{34_y} F_{43_x}\right) + \left(R_{P_x} F_{P_y} - R_{P_y} F_{P_x}\right) = I_{G4} \alpha_4$$

These contain the external force $\mathbf{F}_P$ shown acting on link 4.

For the inversion of the slider-crank shown, the slider block, or piston, is in pure translation against the stationary ground plane; thus it can have no angular acceleration or angular velocity. Also, the position vectors in the torque equation (equation 11.10c) are all zero as the force $\mathbf{F}_P$ acts at the CG. Thus the torque equation for link 4 (third expression in equation 11.10c) is zero for this inversion of the slider-crank linkage. Its linear acceleration also has no y component.

$$\alpha_4 = 0, \qquad\qquad a_{G_{4_y}} = 0 \qquad\qquad (11.10d)$$

The only x directed force that can exist at the interface between links 4 and 1 is friction. Assuming coulomb friction, the x component can be expressed in terms of the y component of force at this interface. We can write a relation for the friction force f at that interface such as $f = \pm\mu N$, where $\pm\mu$ is a known coefficient of friction. The plus and minus signs on the coefficient of friction are to recognize the fact that the friction force always opposes motion. The kinematic analysis will provide the velocity of the link at the sliding joint. The sign of μ will always be the opposite of the sign of this velocity.

$$F_{14_x} = \pm\mu F_{14_y} \qquad\qquad (11.10e)$$

Substituting equations 11.10d and 11.10e into the reduced equation 11.10c yields:

$$\pm\mu F_{14_y} - F_{43_x} + F_{P_x} = m_4 a_{G_{4_x}}$$
$$F_{14_y} - F_{43_y} + F_{P_y} = 0 \qquad\qquad (11.10f)$$

This last substitution has reduced the unknowns to eight, F_{12x}, F_{12y}, F_{32x}, F_{32y}, F_{43x}, F_{43y}, F_{14y}, and T_{12}; thus we need only eight equations. We can now use the eight equations in 11.10a, b, and f to assemble the matrices for solution.

$$\begin{bmatrix} 1 & 0 & 1 & 0 & 0 & 0 & 0 & 0 \\ 0 & 1 & 0 & 1 & 0 & 0 & 0 & 0 \\ -R_{12_y} & R_{12_x} & -R_{32_y} & R_{32_x} & 0 & 0 & 0 & 1 \\ 0 & 0 & -1 & 0 & 1 & 0 & 0 & 0 \\ 0 & 0 & 0 & -1 & 0 & 1 & 0 & 0 \\ 0 & 0 & R_{23_y} & -R_{23_x} & -R_{43_y} & R_{43_x} & 0 & 0 \\ 0 & 0 & 0 & 0 & -1 & 0 & \pm\mu & 0 \\ 0 & 0 & 0 & 0 & 0 & -1 & 1 & 0 \end{bmatrix} \times \begin{bmatrix} F_{12_x} \\ F_{12_y} \\ F_{32_x} \\ F_{32_y} \\ F_{43_x} \\ F_{43_y} \\ F_{14_y} \\ T_{12} \end{bmatrix} =$$

$$(11.10g)$$

$$\begin{bmatrix} m_2 a_{G_{2_x}} \\ m_2 a_{G_{2_y}} \\ I_{G_2} \alpha_2 \\ m_3 a_{G_{3_x}} \\ m_3 a_{G_{3_y}} \\ I_{G_3} \alpha_3 \\ m_4 a_{G_{4_x}} - F_{P_x} \\ -F_{P_y} \end{bmatrix}$$

Solution of this matrix equation 11.10g plus equation 11.10e will yield complete dynamic force information for the fourbar slider-crank linkage.

11.6 FORCE ANALYSIS OF THE INVERTED SLIDER-CRANK

Another inversion of the fourbar slider-crank was also analyzed kinematically in Part I. It is shown in Figure 11-5. Link 4 does have an angular acceleration in this inversion. In fact, it must have the same angle, angular velocity, and angular acceleration as link 3 because they are rotationally coupled by the sliding joint. We wish to determine the forces at all pin joints and at the sliding joint as well as the driving torque needed to create the desired accelerations. Each link's joints are located by position vectors referenced to nonrotating local xy coordinate systems at each link's CG as before. The sliding joint is located by the position vector $\mathbf{R}_{43}$ to the center of the slider, point B. The instantaneous position of point B was determined from the kinematic analysis as length b referenced to instant center I_{23} (point A). See Sections 4.7 (p. 161), 6.7 (p. 276), and 7.3 (p. 315) to review the position, velocity, and acceleration analysis of this mechanism. Recall that this mechanism has a nonzero Coriolis component of acceleration. The force between link 3 and link 4 within the sliding joint is distributed along the unspecified length of the slider block. For this analysis the distributed force can be modeled as a force concentrated at point B within the sliding joint. We will neglect friction in this example.

The equations for links 2 and 3 are identical to those for the noninverted slider-crank (Equations 11.10a and b). The equations for link 4 are the same as equations 11.10c except for the absence of the terms involving $\mathbf{F}_p$ since no external force is shown acting on link 4 in this example. The slider joint can only transmit force from link 3 to link 4 or vice versa along a line perpendicular to the axis of slip. This line is called the axis of transmission. In order to guarantee that the force $\mathbf{F}_{34}$ or $\mathbf{F}_{43}$ is always perpendicular to the axis of slip, we can write the following relation:

$$\hat{\mathbf{u}} \cdot \mathbf{F}_{43} = 0 \qquad\qquad (11.11a)$$

which expands to:

$$u_x F_{43_x} + u_y F_{43_y} = 0 \qquad\qquad (11.11b)$$

The dot product of two vectors will be zero when the vectors are mutually perpendicular. The unit vector u *hat* is in the direction of link 3 which is defined from the kinematic analysis as θ_3.

$$u_x = \cos\theta_3, \qquad\qquad u_y = \sin\theta_3 \qquad\qquad (11.11c)$$

Equation 11.11 provides a tenth equation, but we have only nine unknowns, F_{12x}, F_{12y}, F_{32x}, F_{32y}, F_{43x}, F_{43y}, F_{14x}, F_{14y}, and T_{12}, so one of our equations is redundant. Since we must include equation 11.11, we will combine the torque equations for links 3 and 4 rewritten here in vector form and without the external force $\mathbf{F}_p$.

$$\left(\mathbf{R}_{43} \times \mathbf{F}_{43}\right) - \left(\mathbf{R}_{23} \times \mathbf{F}_{32}\right) = I_{G_3}\alpha_3 = I_{G_3}\alpha_4$$

$$(11.12a)$$

$$\left(\mathbf{R}_{14} \times \mathbf{F}_{14}\right) - \left(\mathbf{R}_{34} \times \mathbf{F}_{43}\right) = I_{G_4}\alpha_4$$

Note that the angular acceleration of link 3 is the same as that of link 4 in this linkage. Adding these equations gives:

$$\left(\mathbf{R}_{43} \times \mathbf{F}_{43}\right) - \left(\mathbf{R}_{23} \times \mathbf{F}_{32}\right) + \left(\mathbf{R}_{14} \times \mathbf{F}_{14}\right) - \left(\mathbf{R}_{34} \times \mathbf{F}_{43}\right) = \left(I_{G_3} + I_{G_4}\right)\alpha_4 \qquad (11.12b)$$

11

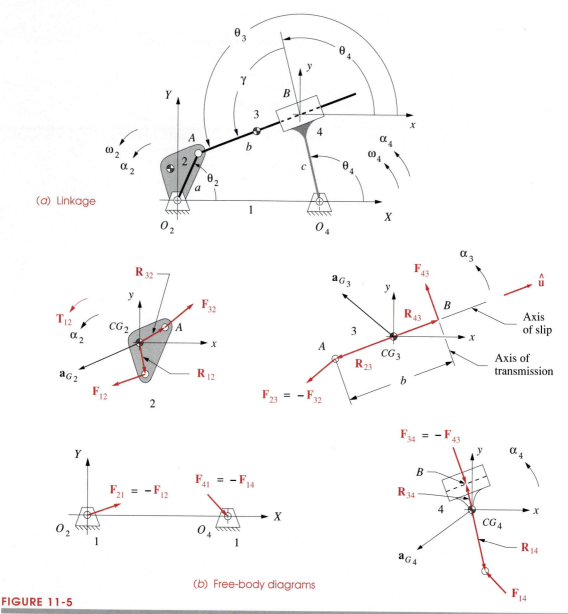

(a) Linkage

(b) Free-body diagrams

FIGURE 11-5

Dynamic forces in the inverted slider-crank fourbar linkage

Expanding and collecting terms:

$$\left(R_{43_x} - R_{34_x}\right)F_{43_y} + \left(R_{34_y} - R_{43_y}\right)F_{43_x} - R_{23_x}F_{32_y}$$

$$+ R_{23_y}F_{32_x} + R_{14_x}F_{14_y} - R_{14_y}F_{14_x} = \left(I_{G_3} + I_{G_4}\right)\alpha_4 \qquad (11.12c)$$

Equations 11.10a, 11.11b, 11.12c, and the four force equations from equations 11.10b and 11.10c (excluding the external force $\mathbf{F}_P$) give us nine equations in the nine unknowns which we can put in matrix form for solution.

$$
\begin{bmatrix}
1 & 0 & 1 & 0 & 0 & 0 & 0 & 0 & 0 \\
0 & 1 & 0 & 1 & 0 & 0 & 0 & 0 & 0 \\
-R_{12_y} & R_{12_x} & -R_{32_y} & R_{32_x} & 0 & 0 & 0 & 0 & 1 \\
0 & 0 & -1 & 0 & 1 & 0 & 0 & 0 & 0 \\
0 & 0 & 0 & -1 & 0 & 1 & 0 & 0 & 0 \\
0 & 0 & R_{23_y} & -R_{23_x} & \left(R_{34_y}-R_{43_y}\right) & \left(R_{43_x}-R_{34_x}\right) & -R_{14_y} & R_{14_x} & 0 \\
0 & 0 & 0 & 0 & -1 & 0 & 1 & 0 & 0 \\
0 & 0 & 0 & 0 & 0 & -1 & 0 & 1 & 0 \\
0 & 0 & 0 & 0 & u_x & u_y & 0 & 0 & 0
\end{bmatrix} \times
$$

$$
\begin{bmatrix}
F_{12_x} \\
F_{12_y} \\
F_{32_x} \\
F_{32_y} \\
F_{43_x} \\
F_{43_y} \\
F_{14_x} \\
F_{14_y} \\
T_{12}
\end{bmatrix}
=
\begin{bmatrix}
m_2 a_{G2_x} \\
m_2 a_{G2_y} \\
I_{G_2}\alpha_2 \\
m_3 a_{G3_x} \\
m_3 a_{G3_y} \\
\left(I_{G_3}+I_{G_4}\right)\alpha_4 \\
m_4 a_{G4_x} \\
m_4 a_{G4_y} \\
0
\end{bmatrix}
\tag{11.13}
$$

11.7 FORCE ANALYSIS—LINKAGES WITH MORE THAN FOUR BARS

This matrix method of force analysis can easily be extended to more complex assemblages of links. The equations for each link are of the same form. We can create a more general notation for equations 11.1 to apply them to any assembly of n pin-connected links. Let j represent any link in the assembly. Let $i = j - 1$ be the previous link in the chain and $k = j + 1$ be the next link in the chain; then, using the vector form of equations 11.1:

$$\mathbf{F}_{ij} + \mathbf{F}_{jk} + \sum \mathbf{F}_{ext_j} = m_j\, \mathbf{a}_{G_j} \tag{11.14a}$$

$$\left(\mathbf{R}_{ij} \times \mathbf{F}_{ij}\right)+\left(\mathbf{R}_{jk} \times \mathbf{F}_{jk}\right)+\sum \mathbf{T}_j +\left(\mathbf{R}_{ext_j} \times \sum \mathbf{F}_{ext_j}\right)= I_{G_j}\alpha_j \tag{11.14b}$$

where:

$$j = 2, 3, \ldots, n; \quad i = j-1; \quad k = j+1,\; j \neq n; \quad \text{if } j = n,\, k = 1$$

and

$$\mathbf{F}_{ji} = -\mathbf{F}_{ij}; \quad \mathbf{F}_{kj} = -\mathbf{F}_{jk} \tag{11.14c}$$

The sum of forces vector equation 11.14a can be broken into its two x and y component equations and then applied, along with the sum of torques equation 11.14b, to each of the links in the chain to create the set of simultaneous equations for solution. Any link

11

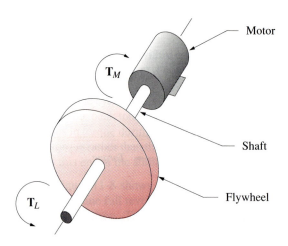

FIGURE 11-10

Flywheel on a driveshaft

$$E = \frac{1}{2}I\omega^2 \qquad (11.17)$$

where I is the moment of inertia of all rotating mass on the shaft. This includes the I of the motor rotor and of the linkage crank plus that of the flywheel. We want to determine how much I we need to add in the form of a flywheel to reduce the speed variation of the shaft to an acceptable level. We begin by writing Newton's law for the free-body diagram in Figure 11-10.

$$\sum T = I\alpha$$

$$T_L - T_M = I\alpha$$

but we want :

$$T_M = T_{avg}$$

so :

$$T_L - T_{avg} = I\alpha \qquad (11.18a)$$

substituting :

$$\alpha = \frac{d\omega}{dt} = \frac{d\omega}{dt}\left(\frac{d\theta}{d\theta}\right) = \omega\frac{d\omega}{d\theta}$$

gives :

$$T_L - T_{avg} = I\omega\frac{d\omega}{d\theta}$$

$$\left(T_L - T_{avg}\right)d\theta = I\omega\,d\omega \qquad (11.18b)$$

and integrating :

$$\int_{\theta@\omega_{min}}^{\theta@\omega_{max}} \left(T_L - T_{avg}\right)d\theta = \int_{\omega_{min}}^{\omega_{max}} I\omega\,d\omega \qquad (11.18c)$$

$$\int_{\theta@\omega_{min}}^{\theta@\omega_{max}} \left(T_L - T_{avg}\right)d\theta = \frac{1}{2}I\left(\omega_{max}^2 - \omega_{min}^2\right)$$

The left side of this expression represents the change in energy E between the maximum and minimum shaft ω's and is equal to the *area under the torque-time diagram*[*] (Figures 11-8, p. 548, and 11-11) between those extreme values of ω. The right side of equation 11.18c is the change in energy stored in the flywheel. The only way we can extract energy from the flywheel is to slow it down as shown in equation 11.17. Adding energy will speed it up. Thus it is impossible to obtain exactly constant shaft velocity in the face of changing energy demands by the load. The best we can do is to minimize the speed variation ($\omega_{max} - \omega_{min}$) by providing a flywheel with sufficiently large I.

✎ EXAMPLE 11-5

Determining the Energy Variation in a Torque-Time Function.

Given: An input torque-time function which varies over its cycle. Figure 11-11 shows the input torque curve from Figure 11-8. The torque is varying during the 360° cycle about its average value.

Find: The total energy variation over one cycle.

Solution:

1   Calculate the average value of the torque-time function over one cycle, which in this case is 70.2 lb-in. (Note that in some cases the average value may be zero.)

2   Note that the *integration on the left side of equation 11.18c is done with respect to the average line of the torque function, not with respect to the θ axis.* (From the definition of the

[*] There is often confusion between torque and energy because they appear to have the same units of *lb-in* (*in-lb*) or *N-m* (*m-N*). This leads some students to think that they are the same quantity, but they are not. Torque ≠ energy. The **integral** of torque with respect to angle, measured in radians, **is** equal to energy. This integral has the units of *in-lb-rad*. The radian term is usually omitted since it is in fact unity. Power in a rotating system is equal to torque x angular velocity (measured in *rad/sec*), and the power units are then *(in-lb-rad)/sec*. When power is integrated versus time to get energy, the resulting units are *in-lb-rad*, the same as the integral of torque versus angle. The radians are again usually dropped, contributing to the confusion.

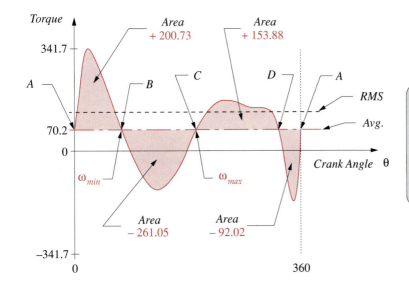

Areas of torque pulses in order over one cycle		
Order	Neg Area	Pos Area
1	– 261.05	200.73
2	– 92.02	153.88
Energy units are lb–in–rad		

FIGURE 11-11

Integrating the pulses above and below the average value in the input torque function

average, the sum of positive area above an average line is equal to the sum of negative area below that line.) The integration limits in equation 11.18 are from the shaft angle θ at which the shaft ω is a minimum to the shaft angle θ at which ω is a maximum.

3 The minimum ω will occur after the maximum positive energy has been delivered from the motor to the load, i.e., at a point (θ) where the summation of positive energy (area) in the torque pulses is at its largest positive value.

4 The maximum ω will occur after the maximum negative energy has been returned to the load, i.e., at a point (θ) where the summation of energy (area) in the torque pulses is at its largest negative value.

5 To find these locations in θ corresponding to the maximum and minimum ω's and thus find the amount of energy needed to be stored in the flywheel, we need to numerically integrate each pulse of this function from crossover to crossover with the average line. The crossover points in Figure 11-11 have been labeled A, B, C, and D. (Program FOURBAR does this integration for you numerically, using a trapezoidal rule.)

6 The FOURBAR program prints the table of areas shown in Figure 11-11. The positive and negative pulses are separately integrated as described above. Reference to the plot of the torque function will indicate whether a positive or negative pulse is the first encountered in a particular case. The first pulse in this example is a positive one.

7 The remaining task is to accumulate these pulse areas beginning at an arbitrary crossover (in this case point A) and proceeding pulse by pulse across the cycle. Table 11-1 shows this process and the result.

8 Note in Table 11-1 that the minimum shaft speed occurs after the largest accumulated positive energy pulse (+200.73 in-lb) has been delivered from the driveshaft to the system. This delivery of energy slows the motor down. The maximum shaft speed occurs after the largest accumulated negative energy pulse (–60.32 in-lb) has been received back from the system by the driveshaft. This return of stored energy will speed up the motor. The total energy variation is the algebraic difference between these two extreme values, which in this example is –261.05 in-lb. This negative energy coming out of the system needs to be absorbed by the flywheel and then returned to the system *during each cycle* to smooth the variations in shaft speed.

TABLE 11-1   Integrating the Torque Function

From	Δ Area = ΔE	Accum. Sum = E	
A to B	+200.73	+200.73	ω_{min}@B
B to C	–261.05	–60.32	ω_{max}@C
C to D	+153.88	+93.56	
D to A	–92.02	+1.54	
	Total Δ Energy	$= E @ \omega_{max} - E @ \omega_{min}$	
		$= (-60.32) - (+200.73) = -261.05 \text{ in-lb}$	

SIZING THE FLYWHEEL   We now must determine how large a flywheel is needed to absorb this energy with an acceptable change in speed. The change in shaft speed during a cycle is called its *fluctuation (Fl)* and is equal to:

$$Fl = \omega_{max} - \omega_{min} \qquad (11.19a)$$

We can normalize this to a dimensionless ratio by dividing it by the average shaft speed. This ratio is called the *coefficient of fluctuation (k)*.

$$k = \frac{\left(\omega_{max} - \omega_{min}\right)}{\omega_{avg}} \qquad (11.19b)$$

This *coefficient of fluctuation* is a design parameter to be chosen by the designer. It typically is set to a value between 0.01 and 0.05, which correspond to a 1 to 5% fluctuation in shaft speed. The smaller this chosen value, the larger the flywheel will have to be. This presents a design trade-off. A larger flywheel will add more cost and weight to the system, which factors have to be weighed against the smoothness of operation desired.

We found the required change in energy E by integrating the torque curve

$$\int_{\theta @ \omega_{min}}^{\theta @ \omega_{max}} \left(T_L - T_{avg}\right) d\theta = E \qquad (11.20a)$$

and can now set it equal to the right side of equation 11.18c (p. 550):

$$E = \frac{1}{2}I\left(\omega_{max}^2 - \omega_{min}^2\right) \qquad (11.20b)$$

Factoring this expression:

$$E = \frac{1}{2}I\left(\omega_{max} + \omega_{min}\right)\left(\omega_{max} - \omega_{min}\right) \qquad (11.20c)$$

If the torque-time function were a pure harmonic, then its average value could be expressed exactly as:

$$\omega_{avg} = \frac{\left(\omega_{max} + \omega_{min}\right)}{2} \qquad (11.21)$$

Our torque functions will seldom be pure harmonics, but the error introduced by using this expression as an approximation of the average is acceptably small. We can now substitute equations 11.19b and 11.21 into equation 11.20c to get an expression for the mass moment of inertia I_s of the system flywheel needed.

$$E = \frac{1}{2}I\left(2\omega_{avg}\right)\left(k\,\omega_{avg}\right)$$

$$I_s = \frac{E}{k\,\omega_{avg}^2} \qquad (11.22)$$

Equation 11.22 can be used to design the physical flywheel by choosing a desired coefficient of fluctuation k, and using the value of E from the numerical integration of

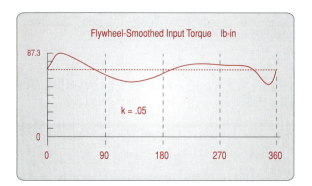

FIGURE 11-12

Input torque curve for the fourbar linkage in Figure 11-8 after smoothing with a flywheel

the torque curve (see Table 11-1, p. 552) and the average shaft ω to compute the needed system I_s. The physical flywheel's mass moment of inertia I_f is then set equal to the required system I_s. But if the moments of inertia of the other rotating elements on the same driveshaft (such as the motor) are known, the physical flywheel's required I_f can be reduced by those amounts.

The most efficient flywheel design in terms of maximizing I_f for minimum material used is one in which the mass is concentrated in its rim and its hub is supported on spokes, like a carriage wheel. This puts the majority of the mass at the largest radius possible and minimizes the weight for a given I_f. Even if a flat, solid circular disk flywheel design is chosen, either for simplicity of manufacture or to obtain a flat surface for other functions (such as an automobile clutch), the design should be done with an eye to reducing weight and thus cost. Since in general, $I = mr^2$, a thin disk of large diameter will need fewer pounds of material to obtain a given I than will a thicker disk of smaller diameter. Dense materials such as cast iron and steel are the obvious choices for a flywheel. Aluminum is seldom used. Though many metals (lead, gold, silver, platinum) are more dense than iron and steel, one can seldom get the accounting department's approval to use them in a flywheel.

Figure 11-12 shows the change in the input torque $\mathbf{T}_{12}$ for the linkage in Figure 11-8 after the addition of a flywheel sized to provide a coefficient of fluctuation of 0.05. The oscillation in torque about the unchanged average value is now 5%, much less than what it was without the flywheel. A much smaller horsepower motor can now be used because the flywheel is available to absorb the energy returned from the linkage during its cycle.

11.12 A LINKAGE FORCE TRANSMISSION INDEX

The transmission angle was introduced in Chapter 2 and used in subsequent chapters as an index of merit to predict the kinematic behavior of a linkage. A too-small transmission angle predicts problems with motion and force transmission in a fourbar linkage. Unfortunately, the transmission angle has limited application. It is only useful for fourbar linkages and then only when the input and output torques are applied to links that are pivoted to ground (i.e., the crank and rocker). When external forces are applied to the coupler link, the transmission angle tells nothing about the linkage's behavior.

Holte and Chase [1] define a joint-force index (JFI) which is useful as an indicator of any linkage's ability to smoothly transmit energy regardless of where the loads are applied on the linkage. It is applicable to higher-order linkages as well as to the fourbar linkage. The JFI at any instantaneous position is defined as the ratio of the maximum static force in any joint of the mechanism to the applied external load. If the external load is a force, then it is:

$$JFI = MAX \left| \frac{F_{ij}}{F_{ext}} \right| \qquad \text{for all pairs } i, j \qquad (11.23a)$$

If the external load is a torque, then it is:

$$JFI = MAX \left| \frac{F_{ij}}{T_{ext}} \right| \qquad \text{for all pairs } i, j \qquad (11.23b)$$

where, in both cases, F_{ij} is the force in the linkage joint connecting links i and j.

The F_{ij} are calculated from a static force analysis of the linkage. Dynamic forces can be much greater than static forces if speeds are high. However, if this static force transmission index indicates a problem in the absence of any dynamic forces, then the situation will obviously be worse at speed. The largest joint force at each position is used rather than a composite or average value on the assumption that high friction in any one joint is sufficient to hamper linkage performance regardless of the forces at other joints.

Equation 11.23a is dimensionless and so can be used to compare linkages of different design and geometry. Equation 11.23b has dimensions of reciprocal length, so caution must be exercised when comparing designs when the external load is a torque. Then the units used in any comparison must be the same, and the compared linkages should be similar in size.

Equations 11.23 apply to any one instantaneous position of the linkage. As with the transmission angle, this index must be evaluated for all positions of the linkage over its expected range of motion and the largest value of that set found. The peak force may move from pin to pin as the linkage rotates. If the external loads vary with linkage position, they can be accounted for in the calculation.

Holte and Chase suggest that the JFI be kept below a value of about 2 for linkages whose output is a force. Larger values may be tolerable especially if the joints are designed with good bearings that are able to handle the higher loads.

There are some linkage positions in which the JFI can become infinite or indeterminate as when the linkage reaches an immovable position, defined as the input link or input joint being inactive. This is equivalent to a stationary configuration as described in earlier chapters provided that the input joint is inactive in the particular stationary configuration. These positions need to be identified and avoided in any event, independent of the determination of any index of merit. In some cases the mechanism may be immovable but still capable of supporting a load. See reference [1] for more detailed information on these special cases.

11.13 PRACTICAL CONSIDERATIONS

This chapter has presented some approaches to the computation of dynamic forces in moving machinery. The newtonian approach gives the most information and is necessary in order to obtain the forces at all pin joints so that stress analyses of the members can be done. Its application is really quite straightforward, requiring only the creation of correct free-body diagrams for each member and the application of the two simple vector equations which express Newton's second law to each free-body. Once these equations are expanded for each member in the system and placed in standard matrix form, their solution (with a computer) is a trivial task.

The real work in designing these mechanisms comes in the determination of the shapes and sizes of the members. In addition to the kinematic data, the force computation requires only the masses, *CG* locations, and mass moments of inertia versus those *CGs* for its completion. These three geometric parameters completely characterize the member for dynamic modelling purposes. Even if the link shapes and materials are completely defined at the outset of the force analysis process (as with the redesign of an existing system), it is a tedious exercise to calculate the dynamic properties of complicated shapes. Current solids modelling *CAD* systems make this step easy by computing these parameters automatically for any part designed within them.

If, however, you are starting from scratch with your design, the *blank-paper syndrome* will inevitably rear its ugly head. A first approximation of link shapes and selection of materials must be made in order to create the dynamic parameters needed for a "first pass" force analysis. A stress analysis of those parts, based on the calculated dynamic forces, will invariably find problems that require changes to the part shapes, thus requiring recalculation of the dynamic properties and recomputation of the dynamic forces and stresses. This process will have to be repeated in circular fashion (*iteration*—see Chapter 1, p. 8) until an acceptable design is reached. The advantages of using a computer to do these repetitive calculations is obvious and cannot be overstressed. An equation solver program such as *TKSolver* or *Mathcad* will be a useful aid in this process by reducing the amount of computer programming necessary.

Students with no design experience are often not sure how to approach this process of designing parts for dynamic applications. The following suggestions are offered to get you started. As you gain experience, you will develop your own approach.

It is often useful to create complex shapes from a combination of simple shapes, at least for first approximation dynamic models. For example, a link could be considered to be made up of a hollow cylinder at each pivot end, connected by a rectangular prism along the line of centers. It is easy to calculate the dynamic parameters for each of these simple shapes and then combine them. The steps would be as follows (repeated for each link):

1 Calculate the volume, mass, *CG* location, and mass moments of inertia with respect to the local *CG* of each separate part of your built-up link. In our example link these parts would be the two hollow cylinders and the rectangular prism.

2 Find the location of the composite *CG* of the assembly of the parts into the link by the method shown in Section 11.4 (p. 531) and equation 11.3 (p. 524). See also Figure 11-2 (p. 526).

3 Use the *parallel axis theorem* (equation 10.8, p. 497) to transfer the mass moments of inertia of each part to the common, composite *CG* for the link. Then add the individual, transferred *I*'s of the parts to get the total *I* of the link about its composite *CG*. See Section 11.6 (p. 541).

Steps 1 to 3 will create the link geometry data for each link needed for the dynamic force analysis as derived in this chapter.

4 Do the dynamic force analysis.

5 Do a dynamic stress and deflection analysis of all parts.

6 Redesign the parts and repeat steps 1 to 5 until a satisfactory result is achieved.

Remember that lighter (lower mass) links will have smaller inertial forces on them and thus could have lower stresses despite their smaller cross sections. Also, smaller mass moments of inertia of the links can reduce the driving torque requirements, especially at higher speeds. But be cautious about the dynamic deflections of thin, light links becoming too large. We are assuming rigid bodies in these analyses. That assumption will not be valid if the links are too flexible. Always check the deflections as well as the stresses in your designs.

11.14 REFERENCES

1 **Holte, J. E., and T. R. Chase**. (1994). "A Force Transmission Index for Planar Linkage Mechanisms." *Proc. of 23rd Biennial Mechanisms Conference*, Minneapolis, MN, p. 377.

11.15 PROBLEMS

11-1 Draw free-body diagrams of the links in the geared fivebar linkage shown in Figure 4-11 (p. 165) and write the dynamic equations to solve for all forces plus the driving torque. Assemble the symbolic equations in matrix form for solution.

11-2 Draw free-body diagrams of the links in the sixbar linkage shown in Figure 4-12 (p. 167) and write the dynamic equations to solve for all forces plus the driving torque. Assemble the symbolic equations in matrix form for solution.

*†‡11-3 Table P11-1 shows kinematic and geometric data for several slider-crank linkages of the type and orientation shown in Figure P11-1. The point locations are defined as described in the text. For the row(s) in the table assigned, use the matrix method of Section 11.5 (p. 538) and program MATRIX, *Mathcad*, *Matlab*, *TKSolver*, or a matrix solving calculator to solve for forces and torques at the position shown. Also compute the shaking force and shaking torque. Consider the coefficient of friction μ between slider and ground to be zero. You may check your solution by opening the solution files (located in the Solutions folder on the CD-ROM) named P11-03x (where x is the row letter) into program SLIDER.

*†11-4 Repeat Problem 11-3 using the method of virtual work to solve for the input torque on link 2. Additional data for corresponding rows are given in Table P11-2.

*†11-5 Table P11-3 shows kinematic and geometric data for several pin-jointed fourbar linkages of the type and orientation shown in Figure P11-2. All have $\theta_1 = 0$. The

* Answers in Appendix F.

† These problems are suited to solution using *Mathcad, Matlab,* or *TKSolver* equation solver programs.

§ These problems are suited to solution using program SLIDER which is on the attached CD-ROM.

11

TABLE P11-1 Data for Problem 11-3 (See Figure P11-1 for Nomenclature)

Part 1   Lengths in inches, angles in degrees, mass in blobs, angular velocity in rad/sec

Row	link 2	link 3	offset	θ_2	ω_2	α_2	m_2	m_3	m_4
a.	4	12	0	45	10	20	0.002	0.020	0.060
b.	3	10	1	30	15	− 5	0.050	0.100	0.200
c.	5	15	-1	260	20	15	0.010	0.020	0.030
d.	6	20	1	− 75	−10	− 10	0.006	0.150	0.050
e.	2	8	0	135	25	25	0.001	0.004	0.014
f.	10	35	2	120	5	− 20	0.150	0.300	0.050
g.	7	25	-2	− 45	30	− 15	0.080	0.200	0.100

Part 2   Angular acceleration in rad/sec^2, moments of Inertia in blob-in^2, torque in lb-in

Row	I_2	I_3	Rg_2 mag	δ_2 ang	Rg_3 mag	δ_3 ang	F_{P3} mag	δF_{P3} ang	R_{P3} mag	δR_{P3} ang	T_3
a.	0.10	0.2	2	0	5	0	0	0	0	0	20
b.	0.20	0.4	1	20	4	− 30	10	45	4	30	− 35
c.	0.05	0.1	3	− 40	9	50	32	270	0	0	− 65
d.	0.12	0.3	3	120	12	60	15	180	2	60	− 12
e.	0.30	0.8	0.5	30	3	75	6	− 60	2	75	40
f.	0.24	0.6	6	45	15	135	25	270	0	0	− 75
g.	0.45	0.9	4	− 45	10	225	9	120	5	45	− 90

Part 3   Forces in lb, linear accelerations in inches/sec^2

Row	θ_3	α_3	ag_2 mag	ag_2 ang	ag_3 mag	ag_3 ang	ag_4 mag	ag_4 ang
a.	166.40	− 2.40	203.96	213.69	371.08	200.84	357.17	180
b.	177.13	34.33	225.06	231.27	589.43	200.05	711.97	180
c.	195.17	−134.76	1 200.84	37.85	2 088.04	43.43	929.12	0
d.	199.86	− 29.74	301.50	230.71	511.74	74.52	23.97	180
e.	169.82	113.12	312.75	−17.29	976.79	−58.13	849.76	0
f.	169.03	3.29	192.09	23.66	302.50	−29.93	301.92	0
g.	186.78	− 172.20	3 600.50	90.95	8 052.35	134.66	4 909.27	180

TABLE P11-2 Data for Problem 11-4

See also Table P11-1. Unit system is the same as in that table.

Row	ω_3	Vg_2 mag	Vg_2 ang	Vg_3 mag	Vg_3 ang	Vg_4 mag	Vg_4 ang	V_{P3} mag	V_{P3} ang
a.	− 2.43	20.0	135	35.24	152.09	35.14	180	35.24	152.09
b.	− 3.90	15.0	140	40.35	140.14	24.45	180	26.69	153.35
c.	1.20	60.0	310	89.61	−8.23	93.77	0	89.61	− 8.23
d.	0.83	30.0	315	69.10	191.15	63.57	180	70.63	191.01
e.	4.49	12.5	255	56.02	211.93	29.01	180	61.36	204.87
f.	0.73	30.0	255	60.89	210.72	38.46	180	60.89	210.72
g.	−5.98	120.0	0	211.46	61.31	166.14	0	208.60	53.19

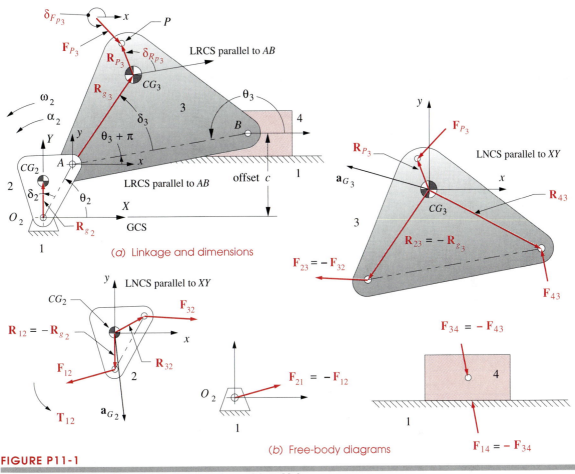

(a) Linkage and dimensions

(b) Free-body diagrams

FIGURE P11-1

Linkage geometry and free body diagrams for problem 11-3

point locations are defined as described in the text. For the row(s) in the table assigned, use the matrix method of Section 11.4 (p. 531) and program MATRIX or a matrix solving calculator to solve for forces and torques at the position shown. Also compute the shaking force and shaking torque. Work in any units system you prefer. You may check your solution by opening the solution files named P11-05x (where x is the row letter) into program FOURBAR.

*†11-6   Repeat Problem 11-5 using the method of virtual work to solve for the input torque on link 2. Additional data for corresponding rows are given in Table P11-4.

*‡11-7   For the row(s) assigned in Table P11-3 (a-f), input the associated disk file to program FOURBAR, calculate the linkage parameters for crank angles from zero to 360° by 5° increments, and design a steel disk flywheel to smooth the input torque using a coefficient of fluctuation of 0.05. Minimize the flywheel weight.

‡11-8   Figure P11-3 shows a fourbar linkage and its dimensions. The steel crank and rocker have uniform cross sections 1 in wide by 0.5 in thick. The aluminum coupler is 0.75

* Answers in Appendix F.

† These problems are suited to solution using *Mathcad, Matlab,* or *TKSolver* equation solver programs.

‡ These problems are suited to solution using program FOURBAR which is on the attached CD-ROM.

11

TABLE P11-3 Data for Problems 11-5 and 11-7

Part 1   Lengths in inches, angles in degrees, angular acceleration in rad/sec²

Row	link 2	link 3	link 4	link 1	θ_2	θ_3	θ_4	α_2	α_3	α_4
a.	4	12	8	15	45	24.97	99.30	20	75.29	244.43
b.	3	10	12	6	30	90.15	106.60	− 5	140.96	161.75
c.	5	15	14	2	260	128.70	151.03	15	78.78	53.37
d.	6	19	16	10	− 75	91.82	124.44	− 10	− 214.84	− 251.82
e.	2	8	7	9	135	34.02	122.71	25	71.54	− 14.19
f.	17	35	23	4	120	348.08	19.01	− 20	− 101.63	− 150.86
g.	7	25	10	19	100	4.42	61.90	− 15	− 17.38	−168.99

Part 2   Angular velocity in rad/sec, mass in blobs, moment of Inertia in blob-in², torque in lb-in

Row	ω_2	ω_3	ω_4	m_2	m_3	m_4	I_2	I_3	I_4	T_3	T_4
a.	20	− 5.62	3.56	0.002	0.02	0.10	0.10	0.20	0.50	− 15	25
b.	10	− 10.31	− 7.66	0.050	0.10	0.20	0.20	0.40	0.40	12	0
c.	20	16.60	14.13	0.010	0.02	0.05	0.05	0.10	0.13	− 10	20
d.	20	3.90	− 3.17	0.006	0.15	0.07	0.12	0.30	0.15	0	30
e.	20	1.06	5.61	0.001	0.04	0.09	0.30	0.80	0.30	25	40
f.	20	18.55	21.40	0.150	0.30	0.25	0.24	0.60	0.92	0	− 25
g.	20	4.10	16.53	0.080	0.20	0.12	0.45	0.90	0.54	0	0

Part 3   Lengths in inches, angles in degrees, linear accelerations in inches/sec²

Row	R_{g2} mag	R_{g2} ang	R_{g3} mag	R_{g3} ang	R_{g4} mag	R_{g4} ang	a_{g2} mag	a_{g2} ang	a_{g3} mag	a_{g3} ang
a.	2	0	5	0	4	30	801.00	222.14	1 691.49	208.24
b.	1	20	4	− 30	6	40	100.12	232.86	985.27	194.75
c.	3	− 40	9	50	7	0	1 200.84	37.85	3 120.71	22.45
d.	3	120	12	60	6	− 30	1 200.87	226.43	4 543.06	81.15
e.	0.5	30	3	75	2	− 40	200.39	341.42	749.97	295.98
f.	6	45	15	135	10	25	2 403.00	347.86	12 064.20	310.22
g.	4	− 45	10	225	4	45	1 601.12	237.15	2 562.10	−77.22

Part 4   Linear accelerations in inches/sec², forces in lb, lengths in inches, angles in degrees

Row	a_{g4} mag	a_{g4} ang	F_{P3} mag	δF_{P3} ang	R_{P3} mag	δR_{P3} ang	F_{P4} mag	δF_{P4} ang	R_{P4} mag	δR_{P4} ang
a.	979.02	222.27	0	0	0	0	40	− 30	8	0
b.	1 032.32	256.52	4	30	10	45	15	− 55	12	0
c.	1 446.58	316.06	0	0	0	0	75	45	14	0
d.	1 510.34	2.15	2	45	15	180	20	270	16	0
e.	69.07	286.97	9	0	6	− 60	16	60	7	0
f.	4 820.72	242.25	0	0	0	0	23	0	23	0
g.	1 284.55	−41.35	12	− 60	9	120	32	20	10	0

11

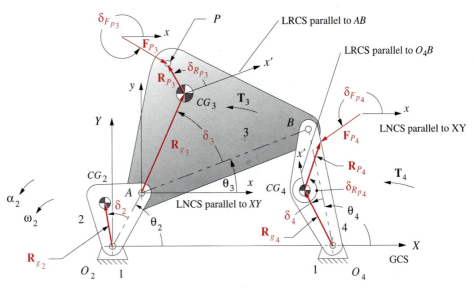

(a) The linkage and dimensions

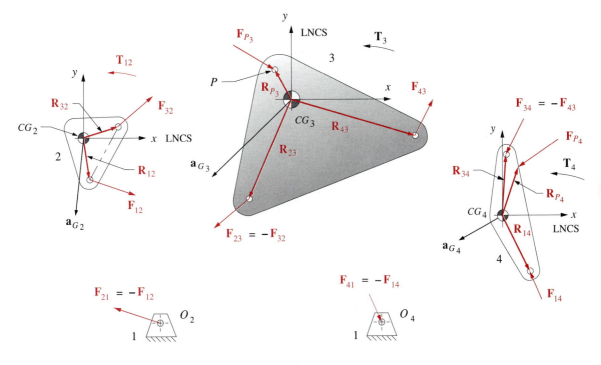

(b) Free-body diagrams

FIGURE P11-2

Linkage geometry and free-body diagrams for Problem 11-5

11

TABLE P11-4 Data for Problem 11-6

Row	Vg_2 mag	Vg_2 ang	Vg_3 mag	Vg_3 ang	Vg_4 mag	Vg_4 ang	V_{P_3} mag	V_{P_3} ang	V_{P_4} mag	V_{P_4} ang
a.	40.00	135.00	54.44	145.19	14.23	219.30	54.44	145.19	41.39	−160.80
b.	10.00	140.00	21.46	14.74	45.94	56.60	122.10	40.04	130.51	29.68
c.	60.00	−50.00	191.94	299.70	98.91	241.03	191.94	−60.30	296.73	−118.97
d.	60.00	135.00	94.36	353.80	19.03	4.44	152.51	−3.13	67.86	26.38
e.	10.00	255.00	42.89	223.13	11.22	172.71	37.01	−140.37	48.41	−155.86
f.	120.00	255.00	618.05	211.39	213.98	134.01	618.03	−148.61	692.08	116.52
g.	80.00	145.00	118.29	205.52	66.10	196.90	154.85	−152.36	217.15	164.33

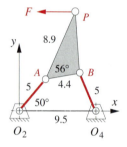

$F \leftarrow \circ P$

8.9

A 56° B

5   4.4   5

50°

O_2 O_4

9.5

Dimensions in inches

FIGURE P11-3

Problem 11-8

in thick. In the instantaneous position shown, the crank O_2A has $\omega = 40$ rad/sec and $\alpha = -20$ rad/sec². There is a horizontal force at P of $F = 50$ lb. Find all pin forces and the torque needed to drive the crank at this instant.

‡11-9   Figure P11-4a shows a fourbar linkage and its dimensions in meters. The steel crank and rocker have uniform cross sections of 50 mm wide by 25 mm thick. The aluminum coupler is 25 mm thick. In the instantaneous position shown, the crank O_2A has $\omega = 10$ rad/sec and $\alpha = 5$ rad/sec². There is a vertical force at P of $F = 100$ N. Find all pin forces and the torque needed to drive the crank at this instant.

‡11-10   Figure P11-4b shows a fourbar linkage and its dimensions in meters. The steel crank and rocker have uniform cross sections of 50 mm wide by 25 mm thick. The aluminum coupler is 25 mm thick. In the instantaneous position shown, the crank O_2A has $\omega = 15$ rad/sec and $\alpha = -10$ rad/sec². There is a horizontal force at P of $F = 200$ N. Find all pin forces and the torque needed to drive the crank at this instant.

‡11-11   Figure P11-5a shows a fourbar linkage and its dimensions in meters. The steel crank, coupler, and rocker have uniform cross sections of 50 mm wide by 25 mm thick. In the instantaneous position shown, the crank O_2A has $\omega = 15$ rad/sec and $\alpha = -10$ rad/

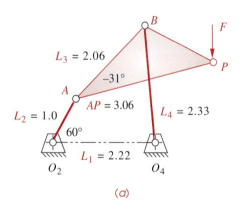

$L_3 = 2.06$

B

F

−31°

A

P

$L_2 = 1.0$

$AP = 3.06$

$L_4 = 2.33$

60°

$L_1 = 2.22$

O_2 O_4

(a)

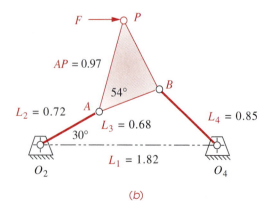

$F \longrightarrow P$

$AP = 0.97$

54° B

$L_2 = 0.72$ A

$L_3 = 0.68$ $L_4 = 0.85$

30°

$L_1 = 1.82$

O_2 O_4

(b)

FIGURE P11-4

Problem 11-9 to 11-10

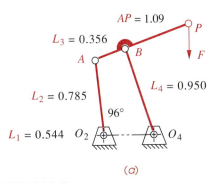

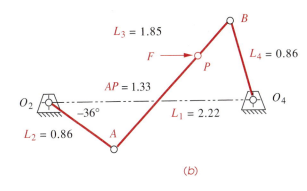

(a)   (b)

FIGURE P11-5

Problems 11-11 to 11-12

sec^2. There is a vertical force at P of $F = 500$ N. Find all pin forces and the torque needed to drive the crank at this instant.

*†‡11-12 Figure P11-5b shows a fourbar linkage and its dimensions in meters. The steel crank, coupler, and rocker have uniform cross sections of 50 mm diameter. In the instantaneous position shown, the crank O_2A has $\omega = -10$ rad/sec and $\alpha = 10$ rad/sec^2. There is a horizontal force at P of $F = 300$ N. Find all pin forces and the torque needed to drive the crank at this instant.

*†‡11-13 Figure P11-6 shows a water jet loom laybar drive mechanism driven by a pair of Grashof crank rocker fourbar linkages. The crank rotates at 500 rpm. The laybar is carried between the coupler-rocker joints of the two linkages at their respective instant centers $I_{3,4}$. The combined weight of the reed and laybar is 29 lb. A 540-lb beat-up force from the cloth is applied to the reed as shown. The steel links have a 2 X 1 in uniform cross section. Find the forces on the pins for one revolution of the crank. Find the torque-time function required to drive the system.

†11-14 Figure P11-7 shows a crimping tool. Find the force F_{hand} needed to generate a 2000-lb F_{crimp}. Find the pin forces. What is this linkage's joint force transmission index (JFI) in this position?

†§11-15 Figure P11-8 shows a walking beam conveyor mechanism that operates at slow speed (25 rpm). The boxes being pushed each weigh 50 lb. Determine the pin forces in the linkage and the torque required to drive the mechanism through one revolution. Neglect the masses of the links.

†§11-16 Figure P11-9 shows a surface grinder table drive that operates at 120 rpm. The crank radius is 22 mm, the coupler is 157 mm, and its offset is 40 mm. The mass of table and workpiece combined is 50 kg. Find the pin forces, slider side loads, and driving torque over one revolution.

†§11-17 Figure P11-10 shows a power hacksaw that operates at 50 rpm. The crank is 75 mm, the coupler is 170 mm, and its offset is 45 mm. Find the pin forces, slider side loads, and driving torque over one revolution for a cutting force of 250 N in the forward direction and 50 N during the return stroke.

* Answers in Appendix F.

† These problems are suited to solution using *Mathcad, Matlab,* or *TKSolver* equation solver programs.

‡ These problems are suited to solution using program FOURBAR which is on the attached CD-ROM.

§ These problems are suited to solution using the *Working Model* program, which is on the attached CD-ROM.

11

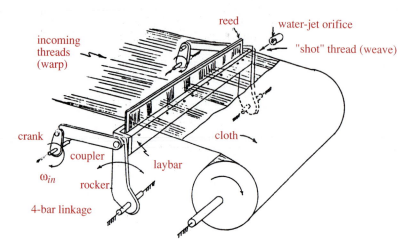

(*a*) Warp, weave, laybar, reed, and laybar drive for a water-jet loom

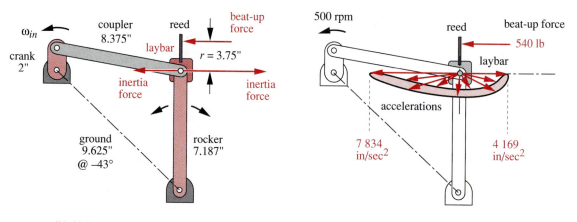

(*b*) Linkage, laybar, reed, and dimensions   (*c*) Acceleration on laybar and force on reed

FIGURE P11-6

Fourbar linkage for laybar drive, showing forces and accelerations on laybar

†§11-18   Figure P11-11 shows a paper roll off-loading station. The paper rolls have a 0.9-m OD, 0.22-m ID, are 3.23 m long, and have a density of 984 kg/m^3. The forks that support the roll are 1.2 m long. The motion is slow so inertial loading can be neglected. Find the force required of the air cylinder to rotate the roll through 90°.

†11-19   Derive an expression for the relationship between flywheel mass and the dimensionless parameter radius/thickness (r/t) for a solid disk flywheel of moment of inertia I. Plot this function for an arbitrary value of I and determine the optimum r/t ratio to minimize flywheel weight for that I.

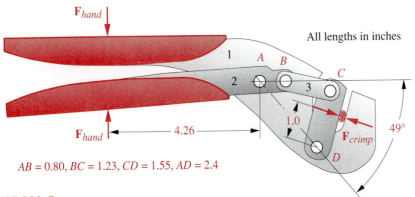

F_{hand}

All lengths in inches

1   A B

2   3   C

1.0

F_{hand} |———— 4.26 ————→|

F_{crimp} 49°

D

$AB = 0.80, BC = 1.23, CD = 1.55, AD = 2.4$

FIGURE P11-7

Problem 11-14

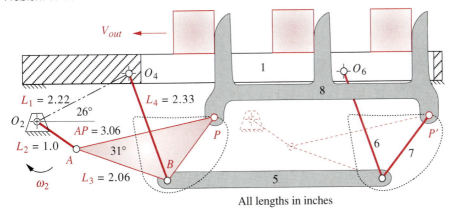

V_{out}

1   O_6

O_4

$L_1 = 2.22$ $L_4 = 2.33$ 8

26°

O_2 $AP = 3.06$

$L_2 = 1.0$ 31°   P   6

A   B   7

ω_2 $L_3 = 2.06$ 5   P'

All lengths in inches

FIGURE P11-8

Problem 11-15

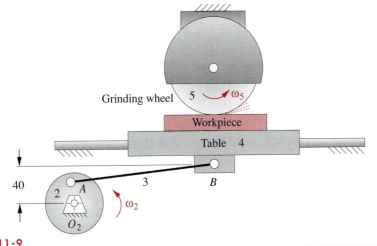

Grinding wheel   5 ω_5

Workpiece

Table   4

40   3   B

2   A

O_2 ω_2

FIGURE P11-9

Problem 11-16

† These problems are suited to solution using *Mathcad*, *Matlab*, or *TKSolver* equation solver programs.

§ These problems are suited to solution using the *Working Model* program, which is on the attached CD-ROM.

11

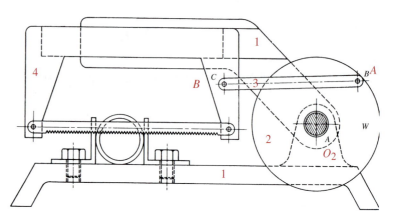

FIGURE P11-10

Problem 11-17 *From P. H. Hill and W. P. Rule. (1960). Mechanisms: Analysis and Design, with permission.*

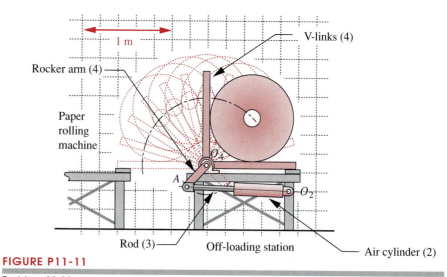

FIGURE P11-11

Problem 11-18

11.16 PROJECTS

The following problem statement applies to all the projects listed.

These larger-scale project statements deliberately lack detail and structure and are loosely defined. Thus, they are similar to the kind of "identification of need" or problem statement commonly encountered in engineering practice. It is left to the student to structure the problem through **background research** *and to create a* **clear goal statement** *and set of* **performance specifications** *before attempting to design a solution. This design process is spelled out in Chapter 1 and should be followed in all of these examples. All results should be documented in a professional engineering report. See the Bibliography in Chapter 1 for references on report writing.*

Some of these project problems are based on the kinematic design projects in Chapter 3. Those kinematic devices can now be designed more realistically with consideration of the dynamic forces that they generate. The strategy in most of the following project problems is to keep the dynamic pin forces and thus the shaking forces to a minimum and also keep the input torque-time curve as smooth as possible to minimize power requirements. **All these problems can be solved with a pin-jointed fourbar linkage.** *This fact will allow you to use program* FOURBAR *to do the kinematic and dynamic computations on a large number and variety of designs in a short time. There are infinities of viable solutions to these problems.* **Iterate to find the best one!** *All links must be designed in detail as to their geometry (mass, moment of inertia, etc.). An equation solver such as* Mathcad *or* TKSolver *will be useful here. Determine all pin forces, shaking force, shaking torque, and input horsepower required for your designs.*

P11-1 Project P3-1 stated that:

The tennis coach needs a better tennis ball server for practice. This device must fire a sequence of standard tennis balls from one side of a standard tennis court over the net such that they land and bounce within each of the three court areas defined by the court's white lines. The order and frequency of a ball's landing in any one of the three court areas must be random. The device should operate automatically and unattended except for the refill of balls. It should be capable of firing 50 balls between reloads. The timing of ball releases should vary. For simplicity, a motor-driven pin-jointed linkage design is preferred.

This project asks you to design such a device to be mounted upon a tripod stand of 5-foot height. Design it, and the stand, for stability against tipover due to the shaking forces and shaking torques which should also be minimized in the design of your linkage. Minimize the input torque required.

P11-2 Project P3-9 stated that:

The "Save the Skeet" foundation has requested a more humane skeet launcher be designed. While they have not yet succeeded in passing legislation to prevent the wholesale slaughter of these little devils, they are concerned about the inhumane aspects of the large accelerations imparted to the skeet as it is launched into the sky for the sportsperson to shoot it down. The need is for a skeet launcher that will smoothly accelerate the clay pigeon onto its desired trajectory.

Design a skeet launcher to be mounted upon a child's "little red wagon." Control your design parameters so as to minimize the shaking forces and torques so that the wagon will remain as nearly stationary as possible during the launch of the clay pigeon.

P11-3   Project P3-10 stated that:

The coin-operated "kid bouncer" machines found outside supermarkets typically provide a very unimaginative rocking motion to the occupant. There is a need for a superior "bouncer" which will give more interesting motions while remaining safe for small children.

Design this device for mounting in the bed of a pickup truck. Keep the shaking forces to a minimum and the input torque-time curve as smooth as possible.

P11-4   Project P3-15 stated that:

NASA wants a zero-G machine for astronaut training. It must carry one person and provide a negative 1-g acceleration for as long as possible.

Design this device and its mounting hardware to the ground plane minimizing the dynamic forces and driving torque.

P11-5   Project P3-16 stated that:

The Amusement Machine Co. Inc. wants a portable "WHIP" ride which will give two or four passengers a thrilling but safe ride and which can be trailed behind a pickup truck from one location to another.

Design this device and its mounting hardware to the truck bed minimizing the dynamic forces and driving torque.

P11-6   Project P3-17 stated that:

The Air Force has requested a pilot training simulator which will give potential pilots exposure to G forces similar to those they will experience in dogfight maneuvers.

Design this device and its mounting hardware to the ground plane minimizing the dynamic forces and driving torque.

P11-7   Project P3-18 stated that:

Cheers needs a better "mechanical bull" simulator for their "yuppie" bar in Boston. It must give a thrilling "bucking bronco" ride but be safe.

Design this device and its mounting hardware to the ground plane minimizing the dynamic forces and driving torque.

P11-8   Gargantuan Motors Inc. is designing a new light military transport vehicle. Their current windshield wiper linkage mechanism develops such high shaking forces when run at its highest speed that the engines are falling out! Design a superior windshield wiper mechanism to sweep the 20-lb armored wiper blade through a 90° arc while minimizing both input torque and shaking forces. The wind load on the blade, perpendicular to the windshield, is 50 lb. The coefficient of friction of the wiper blade on glass is 0.9.

sure and correct in
convenient to use t
rotating imbalance

12.1 STATIC

Despite its name, s
of concern are du
static balance is
d'Alembert inertia

This, of course, is s

Another nam
masses which are
essentially a two-d
this criterion, and
ley on a shaft, a b
peller, an individu
nominator among
to the radial direct
bile tire and wheel
in the axial directi
statically balanced
under that topic.

Figure 12-1a
to statically balan
m_2 concentrated a
These point masse
ported on massles
required amount a
some location R_b i

Assume that t
erations of the ma
tia forces will be c
system is rotating,
we "stop the actio
both arbitrary and
its origin at the ce
system. Writing v

Note that the
it does not matter
cannot be balanced
the ω^2 terms cance

P11-9 The Army's latest helicopter gunship is to be fitted with the Gatling gun, which fires 50-mm-diameter, 2-cm-long spent uranium slugs at a rate of 10 rounds per second. The reaction (recoil) force may upset the chopper's stability. A mechanism is needed which can be mounted to the frame of the helicopter and which will provide a synchronous shaking force, 180° out of phase with the recoil force pulses, to counteract the recoil of the gun. Design such a linkage and minimize its torque and power drawn from the aircraft's engine. Total weight of your device should also be minimized.

P11-10 Steel pilings are universally used as foundations for large buildings. These are often driven into the ground by hammer blows from a "pile driver." In certain soils (sandy, muddy) the piles can be "shaken" into the ground by attaching a "vibratory driver" which imparts a vertical, dynamic shaking force at or near the natural frequency of the pile-earth system. The pile can literally be made to "fall into the ground" under optimal conditions. Design a fourbar linkage-based pile shaker mechanism which, when its ground link is firmly attached to the top of a piling (supported from a crane hook), will impart a dynamic shaking force that is predominantly directed along the piling's long, vertical axis. Operating speed should be in the vicinity of the natural frequency of the pile-earth system.

P11-11 Paint can shaker mechanisms are common in paint and hardware stores. While they do a good job of mixing the paint, they are also noisy and transmit their vibrations to the shelves and counters. A better design of the paint can shaker is possible using a balanced fourbar linkage. Design such a portable device to sit on the floor (not bolted down) and minimize the shaking forces and vibrations while still effectively mixing the paint.

P11-12 Convertible automobiles are once again popular. While offering the pleasure of open-air motoring, they offer little protection to the occupants in a rollover accident. Permanent roll bars are ugly and detract from the open feeling of a true convertible. An automatically deployable roll bar mechanism is needed that will be out of sight until needed. In the event that sensors in the vehicle detect an imminent rollover, the mechanism should deploy within 250 ms. Design a collapsible/deployable roll bar mechanism to retrofit to the convertible of your choice.

It is always good practice to first statically balance all individual components that go into an assembly, if possible. This will reduce the amount of dynamic imbalance that must be corrected in the final assembly and also reduce the bending moment on the shaft. A common example of this situation is the aircraft turbine which consists of a number of circular turbine wheels arranged along a shaft. Since these spin at high speed, the inertia forces due to any imbalance can be very large. The individual wheels are statically balanced before being assembled to the shaft. The final assembly is then dynamically balanced.

Some devices do not lend themselves to this approach. An electric motor rotor is essentially a spool of copper wire wrapped in a complex pattern around the shaft. The mass of the wire is not uniformly distributed either rotationally or longitudinally, so it will not be balanced. It is not possible to modify the windings' local mass distribution after the fact without compromising electrical integrity. Thus the entire rotor imbalance must be countered in the two correction planes after assembly.

Consider the system of three lumped masses arranged around and along the shaft in Figure 12-3. Assume that, for some reason, they cannot be individually statically balanced within their own planes. We then create two correction planes labeled A and B. In this design example, the unbalanced masses m_1, m_2, m_3 and their radii R_1, R_2, R_3 are known along with their angular locations θ_1, θ_2, and θ_3. We want to dynamically balance the system. A three-dimensional coordinate system is applied with the axis of rotation in the Z direction. Note that the system has again been stopped in an arbitrary freeze-frame position. Angular acceleration is assumed to be zero. The summation of forces is:

$$-m_1\mathbf{R}_1\omega^2 - m_2\mathbf{R}_2\omega^2 - m_3\mathbf{R}_3\omega^2 - m_A\mathbf{R}_A\omega^2 - m_B\mathbf{R}_B\omega^2 = 0 \qquad (12.4a)$$

Dividing out the ω^2 and rearranging we get:

$$m_A\mathbf{R}_A + m_B\mathbf{R}_B = -m_1\mathbf{R_1} - m_2\mathbf{R_2} - m_3\mathbf{R}_3 \qquad (12.4b)$$

Breaking into x and y components:

$$m_A R_{A_x} + m_B R_{B_x} = -m_1 R_{1_x} - m_2 R_{2_x} - m_3 R_{3_x}$$
$$m_A R_{A_y} + m_B R_{B_y} = -m_1 R_{1_y} - m_2 R_{2_y} - m_3 R_{3_y} \qquad (12.4c)$$

Equations 12.4c have four unknowns in the form of the $m\mathbf{R}$ products at plane A and the $m\mathbf{R}$ products at plane B. To solve we need the sum of the moments equation which we can take about a point in one of the correction planes such as point O. The moment arm z distances of each force measured from plane A are labeled l_1, l_2, l_3, l_B in the figure; thus

$$\left(m_B\mathbf{R}_B\omega^2\right)l_B = -\left(m_1\mathbf{R}_1\omega^2\right)l_1 - \left(m_2\mathbf{R}_2\omega^2\right)l_2 - \left(m_3\mathbf{R}_3\omega^2\right)l_3 \qquad (12.4d)$$

Dividing out the ω^2, breaking into x and y components, and rearranging:

The moment in the XZ plane (i.e., about the Y axis) is:

$$m_B R_{B_x} = \frac{-\left(m_1 R_{1_x}\right)l_1 - \left(m_2 R_{2_x}\right)l_2 - \left(m_3 R_{3_x}\right)l_3}{l_B} \qquad (12.4e)$$

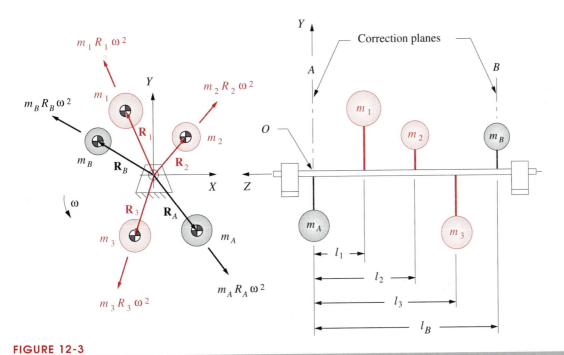

FIGURE 12-3

Two-plane dynamic balancing

The moment in the YZ plane (i.e., about the X axis) is:

$$m_B R_{B_y} = \frac{-\left(m_1 R_{1_y}\right)l_1 - \left(m_2 R_{2_y}\right)l_2 - \left(m_3 R_{3_y}\right)l_3}{l_B}$$
(12.4f)

These can be solved for the $m\mathbf{R}$ products in x and y directions for correction plane B which can then be substituted into equation 12.4c to find the values needed in plane A. Equations 12.2d and 12.2e can then be applied to each correction plane to find the angles at which the balance masses must be placed and the $m\mathbf{R}$ products needed in each plane. The physical counterweights can then be designed consistent with the constraints outlined in the section on static balance. Note that the radii R_A and R_B do not have to be the same value.

 **EXAMPLE 12-2**

Dynamic Balancing.

Given: The system shown in Figure 12-3 has the following data:

$m_1 = 1.2$ kg	$R_1 = 1.135$ m @ $\angle 113.4°$
$m_2 = 1.8$ kg	$R_2 = 0.822$ m @ $\angle 48.8°$
$m_3 = 2.4$ kg	$R_3 = 1.04$ m @ $\angle 251.4°$

The z distances in meters from plane A are:

$$l_1 = 0.854, \qquad l_2 = 1.701, \qquad l_3 = 2.396, \qquad l_B = 3.097$$

Find: The mass-radius products and their angular locations needed to dynamically balance the system using the correction planes A and B.

Solution:

1   Resolve the position vectors into xy components in the arbitrary coordinate system associated with the freeze-frame position of the linkage chosen for analysis.

$$R_1 = 1.135 \, @ \angle 113.4°; \qquad R_{1_x} = -0.451, \qquad R_{1_y} = +1.042$$
$$R_2 = 0.822 \, @ \angle 48.8°; \qquad R_{2_x} = +0.541, \qquad R_{2_y} = +0.618 \qquad (a)$$
$$R_3 = 1.04 \, @ \angle 251.4°; \qquad R_{3_x} = -0.332, \qquad R_{3_y} = -0.986$$

2   Solve equations 12.4e for summation of moments about point O.

$$m_B R_{B_x} = \frac{-\left(m_1 R_{1_x}\right)l_1 - \left(m_2 R_{2_x}\right)l_2 - \left(m_3 R_{3_x}\right)l_3}{l_B}$$

$$= \frac{-1.2(-0.451)(0.854) - 1.8(0.541)(1.701) - 2.4(-0.332)(2.396)}{3.097} = 0.230 \qquad (b)$$

$$m_B R_{B_y} = \frac{-\left(m_1 R_{1_y}\right)l_1 - \left(m_2 R_{2_y}\right)l_2 - \left(m_3 R_{3_y}\right)l_3}{l_B}$$

$$= \frac{-1.2(1.042)(0.854) - 1.8(0.618)(1.701) - 2.4(-0.986)(2.396)}{3.097} = 0.874 \qquad (c)$$

3   Solve equations 12.2d and 12.2e for the mass radius product in plane B.

$$\theta_B = \arctan \frac{0.874}{0.230} = 75.27°$$

$$m_B R_B = \sqrt{(0.230)^2 + (0.874)^2} = 0.904 \ \text{kg-m} \qquad (d)$$

4   Solve equations 12.4c for forces in x and y directions.

$$m_A R_{A_x} = -m_1 R_{1_x} - m_2 R_{2_x} - m_3 R_{3_x} - m_B R_{B_x}$$
$$m_A R_{A_y} = -m_1 R_{1_y} - m_2 R_{2_y} - m_3 R_{3_y} - m_B R_{B_y}$$

$$m_A R_{A_x} = -1.2(-0.451) - 1.8(0.541) - 2.4(-0.332) - 0.230 = 0.134 \qquad (e)$$
$$m_A R_{A_y} = -1.2(1.042) - 1.8(0.618) - 2.4(-0.986) - 0.874 = -0.870$$

5 Solve equations 12.2d and 12.2e for the mass radius product in plane A.

$$\theta_A = \arctan \frac{-0.870}{0.134} = -81.25°$$

$$m_A R_A = \sqrt{(0.134)^2 + (-0.870)^2} = 0.880 \ \text{kg-m}$$

(f)

6 These mass-radius products can be obtained with a variety of shapes appended to the assembly in planes A and B. Many shapes are possible. As long as they provide the required mass-radius products at the required angles in each correction plane, the system will be dynamically balanced.

So, when the design is still on the drawing board, these simple analysis techniques can be used to determine the necessary sizes and locations of balance masses for any assembly in pure rotation for which the mass distribution is defined. This two-plane balance method can be used to dynamically balance any system in pure rotation, and all such systems should be balanced unless the purpose of the device is to create shaking forces or moments.

12.3 BALANCING LINKAGES

Many methods have been devised to balance linkages. Some achieve a complete balance of one dynamic factor, such as shaking force, at the expense of other factors such as shaking moment or driving torque. Others seek an optimum arrangement that collectively minimizes (but does not zero) shaking forces, moments, and torques for a best compromise. Lowen and Berkof, [1] and Lowen, Tepper, and Berkof [2] give comprehensive reviews of the literature on this subject up to 1983. Additional work has been done on the problem since that time, some of which is noted in the references at the end of this chapter.

Complete balance of any mechanism can be obtained by creating a second "mirror image" mechanism connected to it so as to cancel all dynamic forces and moments. Certain configurations of multicylinder internal combustion engines do this. The pistons and cranks of some cylinders cancel the inertial effects of others. We will explore these engine mechanisms in Chapter 14. However, this approach is expensive and is only justified if the added mechanism serves some second purpose such as increasing power, as in the case of additional cylinders in an engine. Adding a "dummy" mechanism whose only purpose is to cancel dynamic effects is seldom economically justifiable.

Most practical linkage balancing schemes seek to minimize or eliminate one or more of the dynamic effects (forces, moments, torques) by redistributing the mass of the existing links. This typically involves adding counterweights and/or changing the shapes of links to relocate their CGs. More elaborate schemes add geared counterweights to some links in addition to redistributing their mass. As with any design endeavor, there are trade-offs. For example, elimination of shaking forces usually increases the shaking moment and driving torque. We can only present a few approaches to this problem in the space available. The reader is directed to the literature for information on other methods.

12

Complete Force Balance of Linkages

The rotating links (cranks, rockers) of a linkage can be individually balanced by the rotating balance methods described in the previous section. The effects of the couplers, which are in complex motion, are more difficult to compensate for. Note that the process of statically balancing a rotating link, in effect, forces its mass center (*CG*) to be at its fixed pivot and thus stationary. In other words the condition of **static balance** can also be **defined as** one of *making the mass center stationary*. A coupler has no fixed pivot, and thus its mass center is, in general, always in motion.

Any mechanism, no matter how complex, will have, for every instantaneous position, a single, overall, *global mass center* located at some particular point. We can calculate its location knowing only the link masses and the locations of the *CG*s of the individual links at that instant. The global mass center normally will change position as the linkage moves. If we can somehow force this global mass center to be stationary, we will have a state of static balance for the overall linkage.

The Berkof-Lowen method of linearly independent vectors[3] provides a means to calculate the magnitude and location of counterweights to be placed on the rotating links which will make the global mass center stationary for all positions of the linkage. Placement of the proper balance masses on the links will cause the dynamic forces on the fixed pivots to always be equal and opposite, i.e., a couple, thus creating static balance ($\Sigma F = 0$ but $\Sigma M \neq 0$) in the moving linkage.

This method works for any *n*-link planar linkage having a combination of revolute (pin) and prismatic (slider) joints, provided that there exists a path to the ground from every link which only contains revolute joints.[4] In other words, if all possible paths from any one link to the ground contain sliding joints, then the method fails. Any linkage of *n* links that meets the above criterion can be balanced by the addition of *n*/2 balance weights, each on a different link.[4] We will apply the method from reference [3] to a fourbar linkage.

Figure 12-4 shows a fourbar linkage with its overall global mass center located by the position vector $\mathbf{R}_t$. The individual *CG*s of the links are located *in the global system* by position vectors $\mathbf{R}_2, \mathbf{R}_3$, and $\mathbf{R}_4$ (magnitudes R_2, R_3, R_4), rooted at its origin, the crank pivot O_2. The link lengths are defined by position vectors labeled $\mathbf{L}_1, \mathbf{L}_2, \mathbf{L}_3, \mathbf{L}_4$ (magnitudes l_1, l_2, l_3, l_4), and the local position vectors which locate the *CG*s *within each link* are $\mathbf{B}_2, \mathbf{B}_3, \mathbf{B}_4$ (magnitudes b_2, b_3, b_4). The angles of the vectors $\mathbf{B}_2, \mathbf{B}_3, \mathbf{B}_4$ are ϕ_2, ϕ_3, ϕ_4 measured internal to the links with respect to the links' lines of centers $\mathbf{L}_2, \mathbf{L}_3, \mathbf{L}_4$. The instantaneous link angles which locate $\mathbf{L}_2, \mathbf{L}_3, \mathbf{L}_4$ in the global system are $\theta_2, \theta_3, \theta_4$. The total mass of the system is simply the sum of the individual link masses:

$$m_t = m_2 + m_3 + m_4 \tag{12.5a}$$

The total mass moment about the origin must be equal to the sum of the mass moments due to the individual links:

$$\sum M_{O_2} = m_t \mathbf{R}_t = m_2 \mathbf{R}_2 + m_3 \mathbf{R}_3 + m_4 \mathbf{R}_4 \tag{12.5b}$$

The position of the global mass center is then:

$$\mathbf{R}_t = \frac{m_2 \mathbf{R}_2 + m_3 \mathbf{R}_3 + m_4 \mathbf{R}_4}{m_t} \tag{12.5c}$$

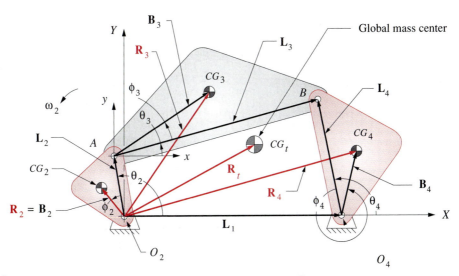

FIGURE 12-4

Static (force) balancing a fourbar linkage

and from the linkage geometry:

$$\mathbf{R}_2 = b_2\, e^{j(\theta_2 + \phi_2)} = b_2\, e^{j\theta_2} e^{j\phi_2}$$

$$\mathbf{R}_3 = l_2\, e^{j\theta_2} + b_3\, e^{j(\theta_3 + \phi_3)} = l_2\, e^{j\theta_2} + b_3\, e^{j\theta_3} e^{j\phi_3} \qquad (12.5d)$$

$$\mathbf{R}_4 = l_1\, e^{j\theta_1} + b_4\, e^{j(\theta_4 + \phi_4)} = l_1\, e^{j\theta_1} + b_4\, e^{j\theta_4} e^{j\phi_4}$$

We can solve for the location of the global mass center for any link position for which we know the link angles θ_2, θ_3, θ_4. We want to make this position vector $\mathbf{R}_t$ be a constant. The first step is to substitute equations 12.5d into 12.5b,

$$m_t\mathbf{R}_t = m_2\!\left(b_2\, e^{j\theta_2} e^{j\phi_2}\right) + m_3\!\left(l_2\, e^{j\theta_2} + b_3\, e^{j\theta_3} e^{j\phi_3}\right) + m_4\!\left(l_1\, e^{j\theta_1} + b_4\, e^{j\theta_4} e^{j\phi_4}\right) \qquad (12.5e)$$

and rearrange to group the constant terms as coefficients of the time-dependent terms:

$$m_t\mathbf{R}_t = \left(m_4 l_1\, e^{j\theta_1}\right) + \left(m_2 b_2 e^{j\phi_2} + m_3 l_2\right) e^{j\theta_2} + \left(m_3 b_3\, e^{j\phi_3}\right) e^{j\theta_3} + \left(m_4 b_4\, e^{j\phi_4}\right) e^{j\theta_4} \qquad (12.5f)$$

Note that the terms in parentheses are all constant with time. The only time-dependent terms are the ones containing θ_2, θ_3, and θ_4.

We can also write the vector loop equation for the linkage,

$$l_2\, e^{j\theta_2} + l_3\, e^{j\theta_3} - l_4\, e^{j\theta_4} - l_1\, e^{j\theta_1} = 0 \qquad (12.6a)$$

and solve it for one of the unit vectors that define a link direction, say link 3:

$$e^{j\theta_3} = \frac{\left(l_1\, e^{j\theta_1} - l_2\, e^{j\theta_2} + l_4\, e^{j\theta_4}\right)}{l_3} \qquad (12.6b)$$

Substitute this into equation 12.5f to eliminate the θ_3 term and rearrange:

$$m_t \mathbf{R}_t = \left(m_2 b_2 e^{j\phi_2} + m_3 l_2 \right) e^{j\theta_2} + \frac{1}{l_3} \left(m_3 b_3\, e^{j\phi_3} \right) \left(l_1\, e^{j\theta_1} - l_2\, e^{j\theta_2} + l_4\, e^{j\theta_4} \right)$$

$$+ \left(m_4 b_4\, e^{j\phi_4} \right) e^{j\theta_4} + \left(m_4 l_1\, e^{j\theta_1} \right) \qquad (12.7a)$$

and collect terms:

$$m_t \mathbf{R}_t = \left(m_2 b_2 e^{j\phi_2} + m_3 l_2 - m_3 b_3 \frac{l_2}{l_3} e^{j\phi_3} \right) e^{j\theta_2} + \left(m_4 b_4\, e^{j\phi_4} + m_3 b_3 \frac{l_4}{l_3} e^{j\phi_3} \right) e^{j\theta_4}$$

$$+ m_4 l_1\, e^{j\theta_1} + m_3 b_3 \frac{l_1}{l_3} e^{j\phi_3}\, e^{j\theta_1} \qquad (12.7b)$$

This expression gives us the tool to force $\mathbf{R}_t$ to be a constant and make the linkage mass center stationary. For that to be so, the terms in parentheses which multiply the only two time-dependent variables, θ_2 and θ_4, must be forced to be zero. (The fixed link angle θ_1 is a constant.) Thus the requirement for linkage force balance is:

$$\left(m_2 b_2 e^{j\phi_2} + m_3 l_2 - m_3 b_3 \frac{l_2}{l_3} e^{j\phi_3} \right) = 0$$

$$(12.8a)$$

$$\left(m_4 b_4\, e^{j\phi_4} + m_3 b_3 \frac{l_4}{l_3} e^{j\phi_3} \right) = 0$$

Rearrange to isolate one link's terms (say link 3) on one side of each of these equations:

$$m_2 b_2 e^{j\phi_2} = m_3 \left(b_3 \frac{l_2}{l_3} e^{j\phi_3} - l_2 \right)$$

$$(12.8b)$$

$$m_4 b_4\, e^{j\phi_4} = -m_3 b_3 \frac{l_4}{l_3} e^{j\phi_3}$$

We now have two equations involving three links. The parameters for any one link can be assumed and the other two solved for. A linkage is typically first designed to satisfy the required motion and packaging constraints before this force balancing procedure is attempted. In that event, the link geometry and masses are already defined, at least in a preliminary way. A useful strategy is to leave the link 3 mass and *CG* location as originally designed and calculate the necessary masses and *CG* locations of links 2 and 4 to satisfy these conditions for balanced forces. Links 2 and 4 are in pure rotation, so it is straightforward to add counterweights to them in order to move their *CG*s to the necessary locations. With this approach, the right sides of equations 12.8b are reducible to numbers for a designed linkage. We want to solve for the mass radius products $m_2 b_2$ and $m_4 b_4$ and also for the angular locations of the *CG*s within the links. Note that the angles ϕ_2 and ϕ_4 in equation 12.8 are measured with respect to the lines of centers of their respective links.

Equations 12.8b are vector equations. Substitute the Euler identity (equation 4.4a, p. 155) to separate into real and imaginary components, and solve for the x and y components of the mass-radius products.

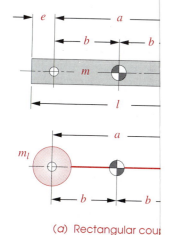

(a) Rectangular cou[...]

FIGURE 12-10

Making the coupler link a physical per[...]

12-10a as a uniform rectangular bar of [...]
as a "dogbone." These are only two c[...]
to be at the pivot pins, connected by a [...]
es will be in pure rotation either as pa[...]
accomplished by adding mass as indi[...]

The three requirements for dynam[...]
and are: equal mass, same *CG* locatio[...]
second of these are easily satisfied by [...]
ment can be stated in terms of radius [...]
equation 10.10 (p. 498).

Taking each lump separately as if [...]
each of length *b*, the moment of inerti[...]

and $I = $[...]

then $k = $[...]

For the link configuration in Figur[...]
sions have the following dimensionless[...]

$$\frac{e}{h} = \frac{1}{2}\sqrt{3\Big(}$$

$$\left(m_2 b_2\right)_x = m_3 \left(b_3 \frac{l_2}{l_3}\cos\phi_3 - l_2 \right)$$

$$\left(m_2 b_2\right)_y = m_3 \left(b_3 \frac{l_2}{l_3}\sin\phi_3 \right)$$

(12.8c)

$$\left(m_4 b_4\right)_x = -m_3 b_3 \frac{l_4}{l_3}\cos\phi_3$$

$$\left(m_4 b_4\right)_y = -m_3 b_3 \frac{l_4}{l_3}\sin\phi_3$$

(12.8d)

These components of the *mR* product needed to force balance the linkage represent the entire amount needed. If links 2 and 4 are already designed with some individual unbalance (the *CG* not at pivot), then the existing *mR* product of the unbalanced link must be subtracted from that found in equations 12.8c and 12.8d in order to determine the size and location of additional counterweights to be added to those links. As we did with the balance of rotating links, any combination of mass and radius that gives the desired product is acceptable. Use equations 12.2d and 12.2e to convert the cartesian *mR* products in equations 12.8c and 12.8d to polar coordinates in order to find the magnitude and angle of the counterweight's *mR* vector. Note that the angle of the *mR* vector for each link will be referenced to that link's line of centers. Design the shape of the physical counterweights to be put on the links as discussed in Section 12.1 (p. 571).

12.4 EFFECT OF BALANCING ON SHAKING AND PIN FORCES

Figure 12-5 shows a fourbar linkage[*] to which balance masses have been added in accord with equations 12.8. Note the counterweights placed on links 2 and 4 at the calculated locations for complete force balance. Figure 12-6a shows a polar plot of the shaking forces of this linkage without the balance masses. The maximum is 462 lb at 15°. Figure 12-6b shows the shaking forces after the balance masses are added. The shaking forces are reduced to essentially zero. The small residual forces seen in Figure 12-6b are due to computational round-off errors—the method gives theoretically exact results.

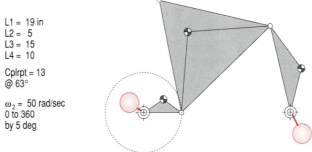

L1 = 19 in
L2 = 5
L3 = 15
L4 = 10

Cplrpt = 13
@ 63°

ω_2 = 50 rad/sec
0 to 360
by 5 deg

FIGURE 12-5

A balanced fourbar linkage showing balance masses applied to links 2 and 4

[*] Open the disk file F12-05.4br in program FOURBAR to see more detail on this linkage and its balancing.

12

Note, l
ing is depen
extra mass
square of th
computes th
us as possil
shown that a
is a good co
duce the tor
balance and

FIGURE 12-9

An inline fourbar
linkage (6)
with optimally
located circular
counterweights. (5)

12.6   BA

The shaking
and the shak

where T_{21} is
O_4 (i.e., link
age, the mag
means of ma
ment require
cial configura

Many te
age mass con
with minimiz
tuation in kin
tribution of m
ing a flywhee
input torque
nestri [12] sho
ment, and inp
counterweigh
of parameters
shaking mome
nificant comp
all here. The

Berkof's r
and useful ever
their respective
tive constraint
must have a sh
line by adding i

For compl
must be reconf
alent to a lumpe

where e defines the length of the material that must be added at each end to satisfy equation 12.11b.

For the link configuration in Figure 12-10b, the length e of the added material of width h needed to make it a physical pendulum can be found from

$$A\left(\frac{e}{h}\right)^3 + B\left(\frac{e}{h}\right)^2 + C\left(\frac{e}{h}\right) + D = 0$$

(12.13)

where : $A = 8$

$$B = 12\left(\frac{a}{c}\right) + 24$$

$$C = 24\left(\frac{a}{c}\right) + 26$$

$$D = -2\left(\frac{a}{c}\right)^3 + 13\left(\frac{a}{c}\right) + 12\pi - 10$$

The second step is to force balance the linkage with its modified coupler using the method of Section 12.3 (p.579) and define the required counterweights on links 2 and 4. With the shaking forces eliminated, the shaking moment is a free vector, as is the input torque.

Then as the third step, the shaking moment can be counteracted by adding geared inertia counterweights to links 2 and 4 as shown in Figure 12-11. These must turn in the opposite direction to the links, so they require a gear ratio of –1. Such an inertia counterweight can balance any planar moment that is proportional to an angular acceleration and does not introduce any net inertia forces to upset the force balance of the linkage.

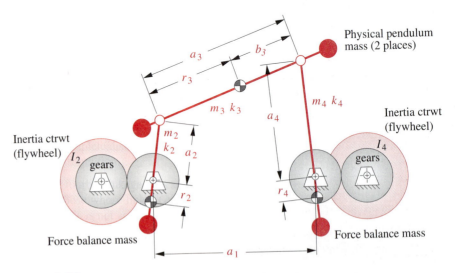

FIGURE 12-11

Completely force and moment balanced inline fourbar linkage with physical pendulum coupler and inertia counterweights on rotating links

Trade-offs include increased input torque and larger pin forces resulting from the torque required to accelerate the additional rotational inertia. There can also be large loads on the gear teeth and impact when torque reversals take up the gearsets' backlash, causing noise.

The shaking moment of an inline fourbar linkage is derived in reference [6] as

$$M_s = \sum_{i=2}^{4} A_i \alpha_i$$
(12.14)

where: $A_2 = -m_2\left(k_2^2 + r_2^2 + a_2 r_2\right)$

$A_3 = -m_3\left(k_3^2 + r_3^2 - a_3 r_3\right)$

$A_4 = -m_4\left(k_4^2 + r_4^2 + a_4 r_4\right)$

α_i is the angular acceleration of link i. The other variables are defined in Figure 12-11.

Adding the effects of the two inertia counterweights gives

$$M_s = \sum_{i=2}^{4} A_i \alpha_i + I_2 \alpha_2 + I_4 \alpha_4$$
(12.15)

The shaking moment can be forced to zero if

$I_2 = -A_2$

$I_4 = -A_4$
(12.16)

$A_3 = 0, \quad \text{or} \quad k_3^2 = r_3(a_3 - r_3)$

This leads to a set of five design equations that must be satisfied for complete force and moment balancing of an inline fourbar linkage.[*]

$$m_2 r_2 = -m_3 b_3 \left(\frac{a_2}{a_3}\right)$$
(12.17a)

$$m_4 r_4 = -m_3 r_3 \left(\frac{a_4}{a_3}\right)$$
(12.17b)

$$k_3^2 = r_3 b_3$$
(12.17c)

$$I_2 = m_2\left(k_2^2 + r_2^2 + a_2 r_2\right)$$
(12.17d)

$$I_4 = m_4\left(k_4^2 + r_4^2 + a_4 r_4\right)$$
(12.17e)

Equations 12.17a and 12.17b are the force-balance criteria of equation 12.8 written for the inline linkage case. Equation 12.17c defines the coupler as a physical pendulum. Equations 12.17d and 12.17e define the mass moments of inertia required for the two inertia counterweights. Note that if the linkage is run at constant angular velocity, α_2 will be zero in equation 12.14 and the inertia counterweight on link 2 can be omitted.

[*] These components of the mR product needed to force balance the linkage represent the entire amount needed. If links 2 and 4 are already designed with some individual unbalance (i.e., the CG not at pivot), then the existing mR product of the unbalanced link must be subtracted from that found in equations 12.17a and 12.17b in order to determine the size and location of additional counterweights to be added to those links.

12

12.7   MEASURING AND CORRECTING IMBALANCE

While we can do a great deal to ensure balance when designing a machine, variations and tolerances in manufacturing will preclude even a well-balanced design from being in perfect balance when built. Thus there is need for a means to measure and correct the imbalance in rotating systems. Perhaps the best example assembly to discuss is that of the automobile tire and wheel, with which most readers will be familiar. Certainly the design of this device promotes balance, as it is essentially cylindrical and symmetrical. If manufactured to be perfectly uniform in geometry and homogeneous in material, it should be in perfect balance as is. But typically it is not. The wheel (or rim) is more likely to be close to balanced, as manufactured, than is the tire. The wheel is made of a homogeneous metal and has fairly uniform geometry and cross section. The tire, how-ever, is a composite of synthetic rubber elastomer and fabric cord or metal wire. The whole is compressed in a mold and steam-cured at high temperature. The resulting ma-terial varies in density and distribution, and its geometry is often distorted in the process of removal from the mold and cooling.

STATIC BALANCING   After the tire is assembled to the wheel, the assembly must be balanced to reduce vibration at high speeds. The simplest approach is to statically bal-ance it, though it is not really an ideal candidate for this approach as it is thick axially compared to its diameter. To do so it is typically suspended in a horizontal plane on a cone through its center hole. A bubble level is attached to the wheel, and weights are placed at positions around the rim of the wheel until it sits level. These weights are then attached to the rim at those points. This is a single-plane balance and thus can only can-cel the unbalanced forces. It has no effect on any unbalanced moments due to uneven distribution of mass along the axis of rotation. It also is not very accurate.

DYNAMIC BALANCING   The better approach is to dynamically balance it. This requires a dynamic balancing machine be used. Figure 12-12 shows a schematic of such a device used for balancing wheels and tires or any other rotating assembly. The assem-bly to be balanced is mounted temporarily on an axle, called a mandrel, which is sup-ported in bearings within the balancer. These two bearings are each mounted on a sus-pension which contains a transducer that measures dynamic force. A common type of force transducer contains a piezoelectric crystal which delivers a voltage proportional to the force applied. This voltage is amplified electronically and delivered to circuitry or software which can compute its peak magnitude and the phase angle of that peak with respect to some time reference signal. The reference signal is supplied by a shaft encod-er on the mandrel which provides a short duration electrical pulse once per revolution in exactly the same angular location. This encoder pulse triggers the computer to begin processing the force signal. The encoder may also provide some large number of addi-tional pulses equispaced around the shaft circumference (often 1024). These are used to trigger the recording of each data sample from the transducers in exactly the same loca-tion around the shaft and to provide a measure of shaft velocity via an electronic counter.

The assembly to be balanced is then "spun up" to some angular velocity, usually with a friction drive contacting its circumference. The drive torque is then removed and the drive motor stopped, allowing the assembly to "freewheel." (This is to avoid mea-suring any forces due to imbalances in the drive system.) The measuring sequence is begun, and the dynamic forces at each bearing are measured simultaneously and their waveforms stored. Many cycles can be measured and averaged to improve the quality

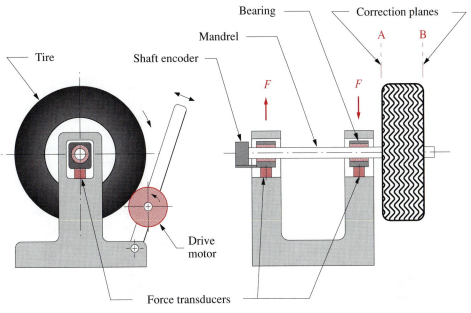

FIGURE 12-12

A dynamic wheel balancer

of the measurement. Because forces are being measured at two locations displaced along the axis, both summation of moment and summation of force data are computed.

The force signals are sent to a built-in computer for processing and computation of the needed balance masses and locations. The data needed from the measurements are the magnitudes of the peak forces and the angular locations of those peaks with respect to the shaft encoder's reference angle (which corresponds to a known point on the wheel). The axial locations of the wheel rim's inside and outside edges (the correction planes) with respect to the balance machine's transducer locations are provided to the machine's computer by operator measurement. From these data the net unbalanced force and net unbalanced moment can be calculated since the distance between the measured bearing forces is known. The mass-radius products needed in the correction planes on each side of the wheel can then be calculated from equations 12.3 (p. 574) in terms of the mR product of the balance weights. The correction radius is that of the wheel rim. The balance masses and angular locations are calculated for each correction plane to put the system in dynamic balance. Weights having the needed mass are clipped onto the inside and outside wheel rims (which are the correction planes in this case), at the proper angular locations. The result is a fairly accurately dynamically balanced tire and wheel.

12.8 REFERENCES

1   **Lowen, G. G., and R. S. Berkof**. (1968). "Survey of Investigations into the Balancing of Linkages." *J. Mechanisms*, **3**(4), pp. 221-231.

2   **Lowen, G. G., et al.** (1983). "Balancing of Linkages—An Update." *Journal of Mechanism and Machine Theory*, **18**(3), pp. 213-220.

3   **Berkof, R. S., and G. G. Lowen**. (1969). "A New Method for Completely Force Balancing Simple Linkages." *Trans. ASME J. of Eng. for Industry, (*February) pp. 21-26.

4   **Tepper, F. R., and G. G. Lowen**. (1972). "General Theorems Concerning Full Force Balancing of Planar Linkages by Internal Mass Redistribution." *Trans ASME J. of Eng. for Industry*, **94 series B**(3), pp. 789-796.

5   **Weiss, K., and R. G. Fenton**. (1972). "Minimum Inertia Weight." *Mech. Chem. Engng. Trans. I.E. Aust.*, **MC8**(1), pp. 93-96.

6   **Berkof, R. S., and G. G. Lowen**. (1971). "Theory of Shaking Moment Optimization of Force-Balanced Four-Bar Linkages." *Trans. ASME J. of Eng. for Industry, (*February*)*, pp. 53-60.

7   **Berkof, R. S.** (1972). "Complete Force and Moment Balancing of Inline Four-Bar Linkages." *J. Mechanism and Machine Theory*, **8** (August)**,** pp. 397-410.

8   **Hockey, B. A.** (1971). "An Improved Technique for Reducing the Fluctuation of Kinetic Energy in Plane Mechanisms." *J. Mechanisms*, **6**, pp. 405-418.

9   **Hockey, B. A.** (1972). "The Minimization of the Fluctuation of Input Torque in Plane Mechanisms." *Mechanism and Machine Theory*, **7**, pp. 335-346.

10   **Berkof, R. S.** (1979). "The Input Torque in Linkages." *Mechanism and Machine Theory*, **14,** pp. 61-73.

11   **Lee, T. W., and C. Cheng.** (1984). "Optimum Balancing of Combined Shaking Force, Shaking Moment, and Torque Fluctuations in High Speed Linkages." *Trans. ASME J. Mechanisms, Transmission, Automation and Design.*, 106, pp. 242-251.

12   **Qi, N. M., and E. Pennestri**. (1991). "Optimum Balancing of Fourbar Linkages." *Mechanism and Machine Theory*, **26**(3), pp. 337-348.

13   **Porter, B., et al.** (1994). "Genetic Design of Dynamically Optimal Fourbar Linkages." *Proc. of 23rd Biennial Mechanisms Conference*, Minneapolis, MN, p. 413.

14   **Bagci, C.** (1975). "Shaking Force and Shaking Moment Balancing of the Plane Slider-Crank Mechanism." *Proc. of The 4th OSU Applied Mechanism Conference*, Stillwater, OK, pp. 25-1.

12.9 PROBLEMS

*†12-1   A system of two coplanar arms on a common shaft, as shown in Figure 12-1 (p. 573), is to be designed. For the row(s) assigned in Table P12-1, find the shaking force of the linkage when run unbalanced at 10 rad/sec and design a counterweight to statically balance the system. Work in any consistent units system you prefer.

†12-2   The minute hand on Big Ben weighs 40 lb and is 10 ft long. Its *CG* is 4 ft from the pivot. Calculate the *mR* product and angular location needed to statically balance this link and design a physical counterweight, positioned close to the center. Select material and design the detailed shape of the counterweight which is of 2 in uniform thickness in the *Z* direction.

†12-3   A "V for victory" advertising sign is being designed to be oscillated about the apex of the V, on a billboard, as the rocker of a fourbar linkage. The angle between the legs of the V is 20°. Each leg is 8 ft long and 1.5 ft wide. Material is 0.25 in-thick aluminum. Design the V link for static balance.

†12-4   A three-bladed ceiling fan has 1.5 ft by 0.25 ft equispaced rectangular blades that nominally weigh 2 lb each. Manufacturing tolerances will cause the blade weight to vary up to plus or minus 5%. The mounting accuracy of the blades will vary the location of the *CG* versus the spin axis by plus or minus 10% of the blades' diame-

---

* Answers in Appendix F.

---

† These problems are suited to solution using *Mathcad, Matlab,* or *TKSolver* equation solver programs.

TABLE P12-1 Data for Problem 12-1

Row	m_1	m_2	R_1	R_2
a.	0.20	0.40	1.25 @ 30°	2.25 @ 120°
b.	2.00	4.36	3.00 @ 45°	9.00 @ 320°
c.	3.50	2.64	2.65 @ 100°	5.20 @ –60°
d.	5.20	8.60	7.25 @ 150°	6.25 @ 220°
e.	0.96	3.25	5.50 @ –30°	3.55 @ 120°

ters. Calculate the weight of the largest steel counterweight needed at a 2-in radius to statically balance the worst-case blade assembly if the minimum radius of the blades is 6 in.

*†12-5 A system of three noncoplanar weights is arranged on a shaft generally as shown in Figure 12-3 (p. 577). For the dimensions from the row(s) assigned in Table P12-2, find the shaking forces and shaking moment when run unbalanced at 100 rpm and specify the mR product and angle of the counterweights in correction planes A and B needed to dynamically balance the system. The correction planes are 20 units apart. Work in any consistent units system you prefer.

*†12-6 A wheel and tire assembly has been run at 100 rpm on a dynamic balancing machine as shown in Figure 12-12 (p. 591). The force measured at the left bearing had a peak of 5 lb at a phase angle of 45° with respect to the zero reference angle on the tire. The force measured at the right bearing had a peak of 2 lb at a phase angle of –120° with respect to the reference zero on the tire. The center distance between the two bearings on the machine is 10 in. The left edge of the wheel rim is 4 in from the centerline of the closest bearing. The wheel is 7 in wide at the rim. Calculate the size and location with respect to the tire's zero reference angle, of balance weights needed on each side of the rim to dynamically balance the tire assembly. The wheel rim diameter is 15 in.

*†12-7 Repeat Problem 12-6 for measured forces of 6 lb at a phase angle of –60° with respect to the reference zero on the tire, measured at the left bearing, and 4 lb at a phase angle of 150° with respect to the reference zero on the tire, measured at the right bearing. The wheel diameter is 16 in.

TABLE P12-2 Data for Problem 12-5

Row	m_1	m_2	m_3	l_1	l_2	l_3	R_1	R_2	R_3
a.	0.20	0.40	1.24	2	8	17	1.25 @ 30°	2.25 @ 120°	5.50 @ – 30°
b.	2.00	4.36	3.56	5	7	16	3.00 @ 45°	9.00 @ 320°	6.25 @ 220°
c.	3.50	2.64	8.75	4	9	11	2.65 @ 100°	5.20 @ – 60°	1.25 @ 30°
d.	5.20	8.60	4.77	7	12	16	7.25 @ 150°	6.25 @ 220°	9.00 @ 320°
e.	0.96	3.25	0.92	1	3	18	5.50 @ 30°	3.55 @ 120°	2.65 @ 100°

* Answers in Appendix F.

† These problems are suited to solution using *Mathcad, Matlab,* or *TKSolver* equation solver programs.

12

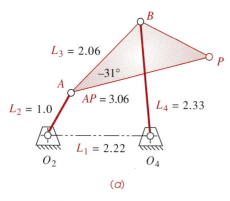

FIGURE P12-1

Problem 12-9

---

† These problems are suited to solution using *Mathcad, Matlab,* or *TKsolver* equation solver programs.

---

‡ These problems are suited to solution using program FOURBAR which is on the attached CD-ROM.

*†‡12-8   Table P11-3 (p. 560) shows the geometry and kinematic data for several fourbar linkages.

a.   For the row(s) from Table P11-3 assigned in this problem, calculate the size and angular locations of the counterbalance mass-radius products needed on links 2 and 4 to completely force balance the linkage by the method of Berkof and Lowen. Check your manual calculation with program FOURBAR.

b.   Calculate the input torque for the linkage both with and without the added balance weights and compare the results. Use program FOURBAR.

*†12-9   Link 2 in Figure P12-1 rotates at 500 rpm. The links are steel with cross sections of 1 x 2 in. Half of the 29-lb weight of the laybar and reed are supported by the linkage at point B. Design counterweights to force balance the linkage and determine its change in peak torque versus the unbalanced condition. See Problem 11-13 (p. 563) for more information on the overall mechanism.

†‡12-10   Figure P12-2a shows a fourbar linkage and its dimensions in meters. The steel crank and rocker have uniform cross sections of 50 mm wide by 25 mm thick. The aluminum coupler is 25 mm thick. The crank O_2A rotates at a constant speed of ω = 40 rad/sec. Design counterweights to force balance the linkage and determine its change in peak torque versus the unbalanced condition.

†‡12-11   Figure P12-2b shows a fourbar linkage and its dimensions in meters. The steel crank and rocker have uniform cross sections of 50 mm wide by 25 mm thick. The aluminum coupler is 25 mm thick. The crank O_2A rotates at a constant speed of ω = 50 rad/sec. Design counterweights to force balance the linkage and determine its change in peak torque versus the unbalanced condition.

†12-12   Write a computer program or use an equation solver such as *Mathcad, Matlab,* or *TKSolver* to solve for the mass-radius products that will force balance any fourbar linkage for which the geometry and mass properties are known.

†12-13   Figure P12-3 shows a system with two weights on a rotating shaft. W_1 = 15 lb @ 0° at a 6-in radius and W_2 = 20 lb @ 270° at a 5-in radius. Determine the magnitudes and angles of the balance weights needed to dynamically balance the system. The balance weight in plane 3 is placed at a radius of 5 in and in plane 4 of 8 in.

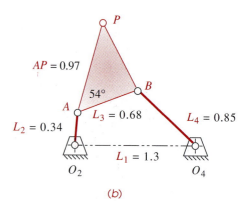

(a)   (b)

FIGURE P12-2

Problems 12-10 to 12-11

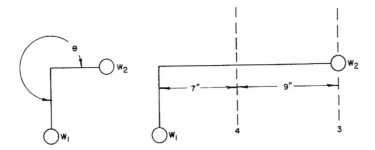

FIGURE P12-3

Problem 12-13

[†]12-14   Figure P12-4 shows a system with two weights on a rotating shaft. $W_1 = 15$ lb @ 30° at a 4-in radius and $W_2 = 20$ lb @ 270° at a 6-in radius. Determine the radii and angles of the balance weights needed to dynamically balance the system. The balance weight in plane 3 weighs 15 lb and in plane 4 weighs 30 lb.

[†]12-15   Figure P12-5 shows a system with two weights on a rotating shaft. $W_1 = 10$ lb @ 90° at a 3-in radius and $W_2 = 15$ lb @ 240° at a 3-in radius. Determine the magnitudes and angles of the balance weights needed to dynamically balance the system. The balance weights in planes 3 and 4 are placed at a 3-in radius.

[†]12-16   Figure P12-6 shows a system with three weights on a rotating shaft. $W_1 = 9$ lb @ 90° at a 4-in radius, $W_2 = 9$ lb @ 225° at a 6-in radius, and $W_3 = 6$ lb @ 315° at a 10-in radius. Determine the magnitudes and angles of the balance weights needed to dynamically balance the system. The balance weights in planes 4 and 5 are placed at a 3-in radius.

[†]12-17   Figure P12-7 shows a system with three weights on a rotating shaft. $W_2 = 10$ lb @ 90° at a 3-in radius, $W_3 = 10$ lb @ 180° at a 4-in radius, and $W_4 = 8$ lb @ 315° at a 4-in radius. Determine the magnitudes and angles of the balance weights needed to dynamically balance the system. The balance weight in plane 1 is placed at a radius of 4 in and in plane 5 of 3 in.

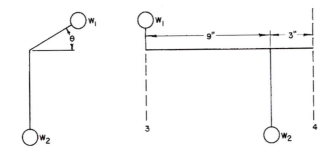

FIGURE P12-4

Problem 12-14

[†] These problems are suited to solution using *Mathcad, Matlab,* or *TKSolver* equation solver programs.

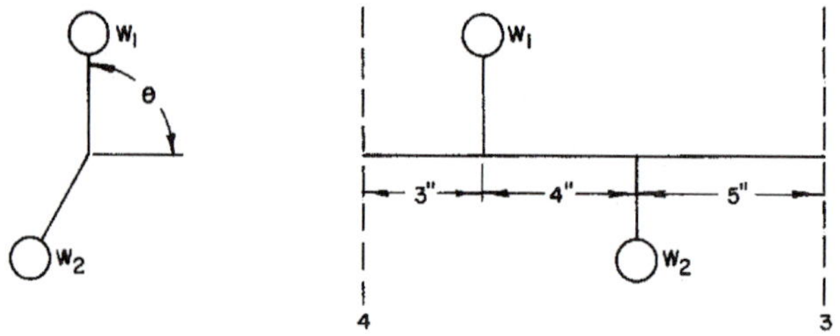

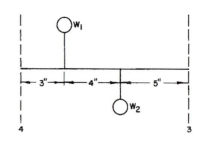

FIGURE P12-5

Problem 12-15

†12-18 The 400-mm-dia steel roller in Figure P12-8 has been tested on a dynamic balancing machine at 100 rpm and shows an unbalanced force of $F_1 = 0.291$ N @ $\theta_1 = 45°$ in the x-y plane at 1 and $F_4 = 0.514$ N @ $\theta_4 = 210°$ in the x-y plane at 4. Determine the angular locations and required diameters of 25-mm-deep holes drilled radially inward from the surface in planes 2 and 3 to dynamically balance the system.

†12-19 The 500-mm-dia steel roller in Figure P12-8 has been tested on a dynamic balancing machine at 100 rpm and shows an unbalanced force of $F_1 = 0.23$ N @ $\theta_1 = 30°$ in the x-y plane at 1 and $F_4 = 0.62$ N @ $\theta_4 = 135°$ in the x-y plane at 4. Determine the angular locations and required diameters of 25-mm-deep holes drilled radially inward from the surface in planes 2 and 3 to dynamically balance the system.

†‡12-20 The linkage in Figure P12-9a has rectangular steel links of 20 x 10 mm cross section similar to that shown in Figure 12-10a (p. 587). Design the necessary balance weights and other features necessary to completely eliminate the shaking force and shaking moment. State all assumptions.

†‡12-21 Repeat Problem 12-20 using links configured as in Figure 12-10b with the same cross section but having "dogbone" end diameters of 50 mm.

†‡12-22 The linkage in Figure P12-9b has rectangular steel links of 20 x 10 mm cross section similar to that shown in Figure 12-10a (p. 587). Design the necessary balance weights and other features necessary to completely eliminate the shaking force and shaking moment. State all assumptions.

†‡12-23 Repeat Problem 12-22 using steel links configured as in Figure 12-10b with a 20 x 10 mm cross section and having "dogbone" end diameters of 50 mm.

† These problems are suited to solution using *Mathcad, Matlab,* or *TKsolver* equation solver programs.

‡ These problems are suited to solution using program FOURBAR which is on the attached CD-ROM.

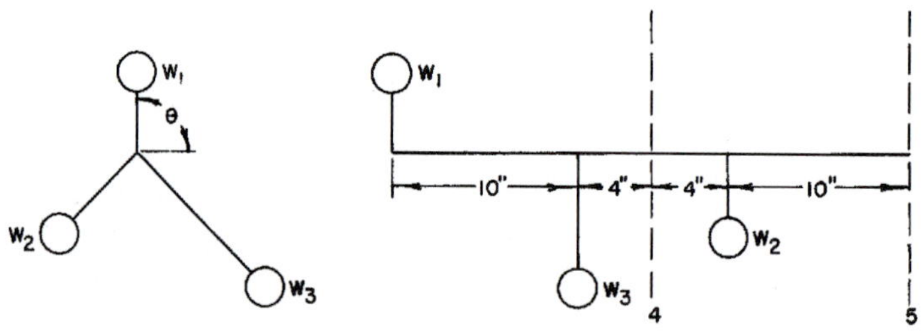

FIGURE P12-6

Problem 12-16

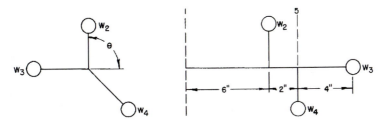

FIGURE P12-7

Problem 12-17

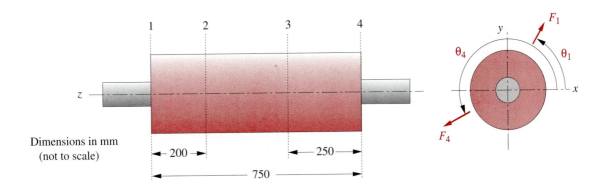

FIGURE P12-8

Problems 12-18 and 12-19

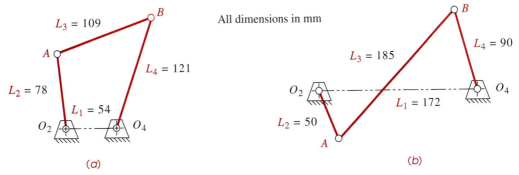

(a)   All dimensions in mm   (b)

FIGURE P12-9

Problems 12-20 to 12-23

Chapter 13

ENGINE DYNAMICS

*I have always thought that the substitution of the
internal combustion machine for the horse marked
a very gloomy milestone in the progress of mankind.*
WINSTON S. CHURCHILL

13.0   INTRODUCTION

The previous chapters have introduced analysis techniques for the determination of dynamic forces, moments, and torques in machinery. Shaking forces, moments, and their balancing have also been discussed. We will now attempt to integrate all these dynamic considerations into the design of a common device, the slider-crank linkage as used in the internal combustion engine. This deceptively simple mechanism will be found to be actually quite complex in terms of the dynamic considerations necessary to its design for high-speed operation. Thus it will serve as an excellent example of the application of the dynamics concepts just presented. We will not address the thermodynamic aspects of the internal combustion engine beyond defining the combustion forces which are necessary to drive the device. Many other texts, such as those listed in the bibliography at the end of this chapter, deal with the very complex thermodynamic and fluid dynamic aspects of this ubiquitous device. We will concentrate only on its kinematics and mechanical dynamics aspects. It is not our intention to make an "engine designer" of the student so much as to apply dynamic principles to a realistic design problem of general interest and also to convey the complexity and fascination involved in the design of an apparently simple dynamic device.

Some students may have had the opportunity to disassemble and service an internal combustion engine, but many will have never done so. Thus we will begin with very fundamental descriptions of engine design and operation. The program ENGINE, supplied with this text, is designed to reinforce and amplify the concepts presented. It will perform all the tedious computations necessary to provide the student with dynamic in-

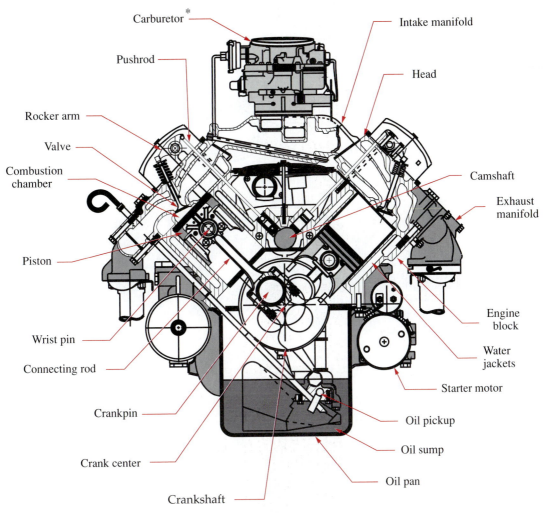

FIGURE 13-1

Cutaway cross section of a vee-eight engine
Adapted from a drawing by Lane Thomas, Western Carolina University, Dept. of Industrial Education, with permission.

formation for design choices and trade-offs. The student is encouraged to use this program concurrent with a reading of the text. Many examples and illustrations within the text will be generated with this program and reference will frequently be made to it. A user manual for program ENGINE is provided in Appendix A which can be read or referred to at any time in order to gain familiarity with the program's operation. Examples used in Chapters 13 and 14 which deal with engine dynamics are built into program ENGINE for student observation and exercise. They can be found on a drop-down menu in that program. Other example engine files for program ENGINE are on the CD-ROM.

* Carburetors are being replaced by fuel injection systems on automotive and other engines that are required to meet increasingly stringent exhaust emission control regulations in the United States. Fuel injection gives better control over the fuel-air mixture than a carburetor.

13

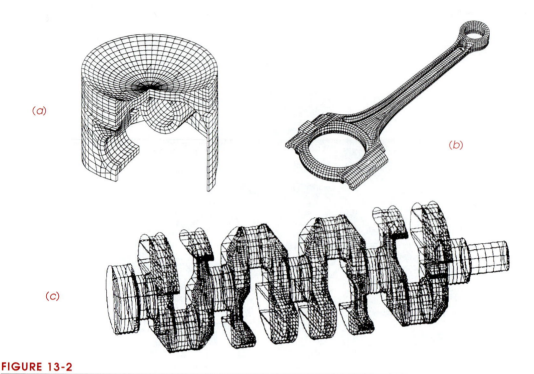

FIGURE 13-2

Finite-element models of an engine piston (a), connecting rod (b), and crankshaft (c) *Courtesy of General Motors Co.*

13.1   ENGINE DESIGN

Figure 13-1 (p. 599) shows a detailed cross section of an internal combustion engine. The basic mechanism consists of a crank, a connecting rod (coupler), and piston (slider). Since this figure depicts a **multicylinder vee-eight** engine configuration, there are four cranks arranged on a crankshaft, and eight sets of connecting rods and pistons, four in the left bank of cylinders and four in the right bank. Only two piston-connecting rod assemblies are visible in this view, both on a common crankpin. The others are behind those shown. Figure 13-2 shows finite-element models of a piston, connecting rod and crankshaft for a four-cylinder inline engine. The most usual arrangement is an inline engine with cylinders all in a common plane. Three-, four-, five-, and six-cylinder **inline engines** are in production the world over. **Vee engines** in four-, six-, eight-, ten-, and twelve-cylinder versions are also in production, with vee six and vee eight being the most popular vee configurations. The geometric arrangements of the crankshaft and cylinders have a significant effect on the dynamic condition of the engine. We will explore these effects of multicylinder arrangements in the next chapter. At this stage we wish to deal only with the design of a **single-cylinder** engine. After optimizing the geometry and dynamic condition of one cylinder, we will be ready to assemble combinations of cylinders into multicylinder configurations.

A schematic of the basic **one-cylinder slider-crank** mechanism and the terminology for its principal parts are shown in Figure 13-3. Note that it is "back-driven" com-

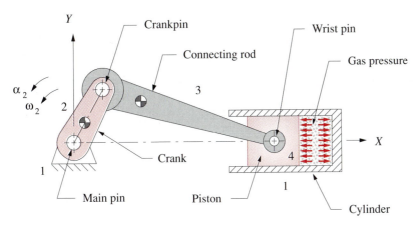

Fourbar slider-crank mechanism for single-cylinder internal combustion engine

pared to the linkages we have been analyzing in previous chapters. That is, the explosion of the combustible mixture in the cylinder drives the piston to the left in Figure 13-3 or down in Figure 13-4, turning the crank. The crank torque that results is ultimately delivered to the drive wheels of the vehicle through a transmission (see Section 9.11, p. 474) to propel the car, motorcycle, or other device. The same slider-crank mechanism can also be used "forward-driven," by motor-driving the crank and taking the output energy from the piston end. It is then called a **piston pump** and is used to compress air, pump gasoline and well water, etc.

In the internal combustion engine of Figure 13-3, it should be fairly obvious that at most we can only expect energy to be delivered from the exploding gases to the crank during the power stroke of the cycle. The piston must return from bottom dead center (BDC) to top dead center (TDC) on its own momentum before it can receive another push from the next explosion. In fact, some rotational kinetic energy must be stored in the crankshaft merely to carry it through the TDC and BDC points as the moment arm for the gas force at those points is zero. This is why an internal combustion engine must be "spun-up" with a hand crank, pull rope, or starter motor to get it running.

There are two common combustion cycles in use in internal combustion engines, the **Clerk two-stroke cycle** and the **Otto four-stroke cycle**, named after their nineteenth century inventors. The four-stroke cycle is most common in automobile, truck, and stationary gasoline engines. The two-stroke cycle is used in motorcycles, outboard motors, chain saws, and other applications where its better power-to-weight ratio outweighs its drawbacks of higher pollution levels and poor fuel economy compared to the four-stroke.

FOUR-STROKE CYCLE   The **Otto four-stroke cycle** is shown in Figure 13-4. It takes four full strokes of the piston to complete one Otto cycle. A piston stroke is defined as its travel from TDC to BDC or the reverse. Thus there are two strokes per 360° crank revolution and it takes 720° of crankshaft rotation to complete one four-stroke cycle. This engine requires at least two valves per cylinder, one for intake and one for exhaust. For discussion, we can start the cycle at any point as it repeats every two crank

13

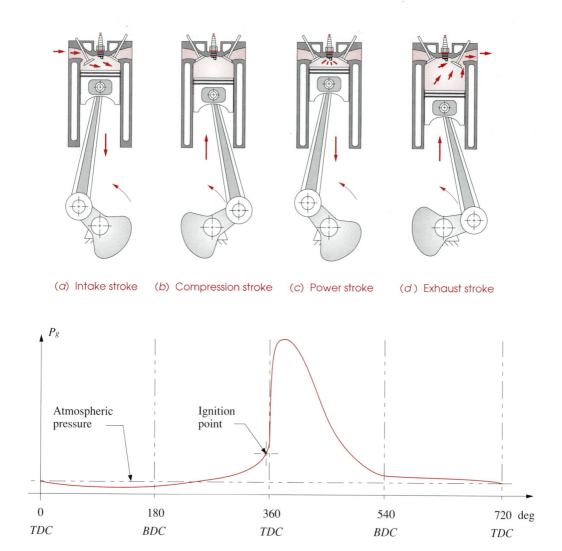

(a) Intake stroke   (b) Compression stroke   (c) Power stroke   (d) Exhaust stroke

(e) The gas pressure curve

FIGURE 13-4

The Otto four-stroke combustion cycle

revolutions. Figure 13-4a shows the **intake stroke** which starts with the piston at TDC. A mixture of fuel and air is drawn into the cylinder from the induction system (the fuel injectors, or the carburetor and intake manifold in Figure 13-1, p. 599) as the piston descends to BDC, increasing the volume of the cylinder and creating a slight negative pressure.

During the **compression stroke** in Figure 13-4b, all valves are closed and the gas is compressed as the piston travels from BDC to TDC. Slightly before TDC, a spark is ignited to explode the compressed gas. The pressure from this explosion builds very quickly and pushes the piston down from TDC to BDC during the **power stroke** shown in Figure 13-4c. The exhaust valve is opened and the piston's **exhaust stroke** from BDC to TDC (Figure 13-4d) pushes the spent gases out of the cylinder into the exhaust manifold (see also Figure 13-1) and thence to the catalytic converter for cleaning before being dumped out the tailpipe. The cycle is then ready to repeat with another intake stroke. The valves are opened and closed at the right times in the cycle by a camshaft which is driven in synchrony with the crankshaft by gears, chain, or toothed belt drive. (See Figure 9-24, p. 452.) Figure 13-4e shows the gas pressure curve for one cycle. With a one-cylinder Otto cycle engine, power is delivered to the crankshaft, at most, 25% of the time as there is only 1 power stroke per 2 revolutions.

TWO-STROKE CYCLE   The **Clerk two-stroke cycle** is shown in Figure 13-5. This engine does not need any valves, though to increase its efficiency it is sometimes provided with a passive (pressure differential operated) one at the intake port. It does not have a camshaft or valve train or cam drive gears to add weight and bulk to the engine. As its name implies, it requires only two-strokes, or 360°, to complete its cycle. There is a passageway, called a transfer port, between the combustion chamber above the piston and the crankcase below. There is also an exhaust port in the side of the cylinder. The piston acts to sequentially block or expose these ports as it moves up and down. The crankcase is sealed and mounts the carburetor on it, serving also as the intake manifold.

Starting at TDC (Figure 13-5a), the two-stroke cycle proceeds as follows: The spark plug ignites the fuel-air charge, compressed on the previous revolution. The expansion of the burning gases drives the piston down, delivering torque to the crankshaft. Partway down, the piston uncovers the exhaust port, allowing the burnt (and also some unburned) gases to begin to escape to the exhaust system.

As the piston descends (Figure 13-5ba), it compresses the charge of fuel-air mixture in the sealed crankcase. The piston blocks the intake port preventing blowback through the carburetor. As the piston clears the transfer port in the cylinder wall, its downward motion pushes the new fuel-air charge up through the transfer port to the combustion chamber. The momentum of the exhaust gases leaving the chamber on the other side helps pull in the new charge as well.

The piston passes BDC (Figure 13-5c) and starts up, pushing out the remaining exhaust gases. The exhaust port is closed by the piston as it ascends, allowing compression of the new charge. As the piston approaches TDC, it exposes the intake port (Figure 13-5d), sucking a new charge of air and fuel into the expanded crankcase from the carburetor. Slightly before TDC, the spark is ignited and the cycle repeats as the piston passes TDC.

Clearly, this Clerk cycle is not as efficient as the Otto cycle in which each event is more cleanly separated from the others. Here there is much mixing of the various phases of the cycle. Unburned hydrocarbons are exhausted in larger quantities. This accounts for the poor fuel economy and dirty emissions of the Clerk engine.[*] It is nevertheless popular in applications where low weight is paramount.

Lubrication is also more difficult in the two-stroke engine than in the four-stroke as the crankcase is not available as an oil sump. Thus the lubricating oil must be mixed with the fuel. This further increases the emissions problem compared to the Otto cycle engine which burns raw gasoline and pumps its lubricating oil separately throughout the engine.

---

[*]   Research and development is under way to clean up the emissions of the two-stroke engine by using fuel injection and compressed air scavenging of the cylinders. These efforts may yet bring this potentially more powerful engine design into compliance with air quality specifications.

13

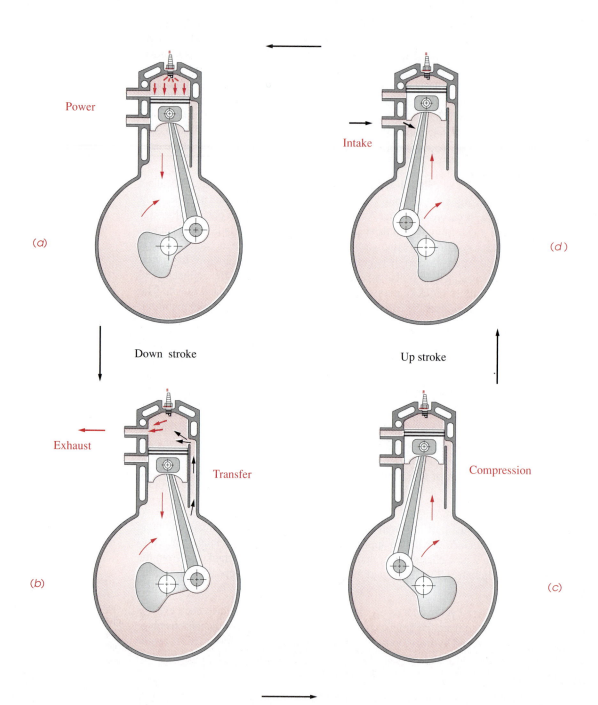

Power

Intake

(a)

(d)

Down stroke

Up stroke

Exhaust

Transfer

Compression

(b)

(c)

FIGURE 13-5

The Clerk two-stroke combustion cycle

DIESEL CYCLE   The **diesel cycle** can be either two-stroke or four-stroke. It is a **compression-ignition** cycle. No spark is needed to ignite the air-fuel mixture. The air is compressed in the cylinder by a factor of about 14 to 15 (versus 8 to 10 in the spark engine), and a low volatility fuel is injected into the cylinder just before TDC. The heat of compression causes the explosion. Diesel engines are larger and heavier than spark ignition engines for the same power output because the higher pressures and forces at which they operate require stronger, heavier parts. Two-stroke cycle Diesel engines are quite common. Diesel fuel is a better lubricant than gasoline.

GAS FORCE   In all the engines discussed here, the usable **output torque** is created from the explosive gas pressure generated within the cylinder either once or twice per two revolutions of the crank, depending on the cycle used. The magnitude and shape of this explosion pressure curve will vary with the engine design, stroke cycle, fuel used, speed of operation, and other factors related to the thermodynamics of the system. For our purpose of analyzing the mechanical dynamics of the system, we need to keep the gas pressure function consistent while we vary other design parameters in order to compare the results of our mechanical design changes. For this purpose, program ENGINE has been provided with a built-in **gas pressure curve** whose peak value is about 600 psi and whose shape is similar to the curve from a real engine. Figure 13-6 shows the **gas force curve** that results from the built-in gas pressure function in program ENGINE applied to a piston of particular area, for both two- and four-stroke engines. Changes in piston area will obviously affect the gas force magnitude for this consistent pressure function, but no changes in engine design parameters input to this program will change its built-in gas pressure curve. To see this gas force curve, run program ENGINE and select any one of the example engines from the pull-down menu. Then calculate and plot Gas Force.

13.2   SLIDER-CRANK KINEMATICS

In Chapters 4, 6, 7, and 11 we developed general equations for the exact solution of the positions, velocities, accelerations, and forces in the pin-jointed fourbar linkage, and also for two inversions of the **slider-crank linkage**, using vector equations. We could again apply that method to the analysis of the "standard" slider-crank linkage, used in the majority of internal combustion engines, as shown in Figure 13-7. Note that its slider motion has been aligned with the X axis. This is a "nonoffset" slider-crank, because the slider axis extended passes through the crank pivot. It also has its slider block translating against the stationary ground plane; thus there will be no Coriolis component of acceleration (see Section 7.3, p. 308).

The simple geometry of this particular inversion of the slider-crank mechanism allows a very straightforward approach to the exact analysis of its slider's position, velocity, and acceleration, using only plane trigonometry and scalar equations. Because of this method's simplicity and to present an alternative solution approach we will analyze this device again.

Let the crank radius be r and the conrod length be l. The angle of the crank is θ, and the angle that the conrod makes with the X axis is ϕ. For any constant crank angular velocity ω, the crank angle $\theta = \omega t$. The instantaneous piston position is x. Two right triangles rqs and lqu are constructed. Then from geometry:

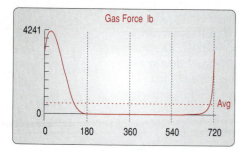

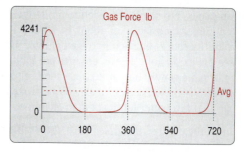

(a) Otto four-stroke cycle   (b) Clerk two-stroke cycle

FIGURE 13-6

Gas force functions in the two-stroke and four-stroke cycle engines

$$q = r\sin\theta = l\sin\phi$$
$$\theta = \omega t$$
$$\sin\phi = \frac{r}{l}\sin\omega t$$
(13.1a)

$$s = r\cos\omega t$$
$$u = l\cos\phi$$
$$x = s + u = r\cos\omega t + l\cos\phi$$
(13.1b)

$$\cos\phi = \sqrt{1 - \sin^2\phi} = \sqrt{1 - \left(\frac{r}{l}\sin\omega t\right)^2}$$
(13.1c)

$$x = r\cos\omega t + l\sqrt{1 - \left(\frac{r}{l}\sin\omega t\right)^2}$$
(13.1d)

Equation 13.1d is an exact expression for the piston position x as a function of r, l, and ωt. This can be differentiated versus time to obtain exact expressions for the velocity and acceleration of the piston. For a steady-state analysis we will assume ω to be constant.

$$\dot{x} = -r\omega\left[\sin\omega t + \frac{r}{2l}\frac{\sin 2\omega t}{\sqrt{1 - \left(\frac{r}{l}\sin\omega t\right)^2}}\right]$$
(13.1e)

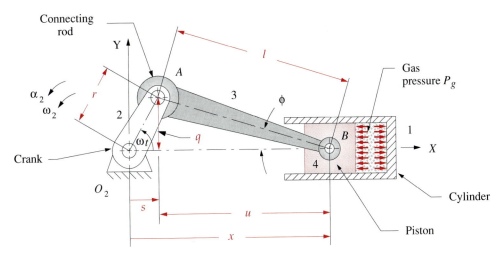

Note:
Link 3 can be considered as a 2-force member for this gas-force analysis because the inertia forces are being temporarily ignored. They will be superposed later.

(a) The linkage geometry

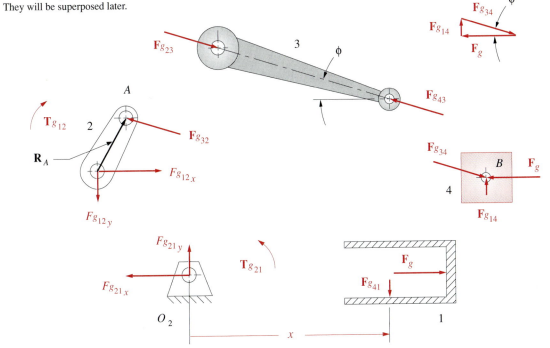

(b) Free-body diagrams

FIGURE 13-7

Fourbar slider-crank linkage position and gas force analysis. (See Figure 13-12, p. 618 for the inertia force analysis.)

13

$$\ddot{x} = -r\omega^2 \left\{ \cos\omega t - \frac{r\left[l^2\left(1 - 2\cos^2\omega t\right) - r^2\sin^4\omega t\right]}{\left[l^2 - \left(r\sin\omega t\right)^2\right]^{\frac{3}{2}}} \right\} \qquad (13.1f)$$

Equations 13.1 can easily be solved with a computer for all values of ωt needed. But, it is rather difficult to look at equation 13.1f and visualize the effects of changes in the design parameters r and l on the acceleration. It would be useful if we could derive a simpler expression, even if approximate, that would allow us to more easily predict the results of design decisions involving these variables. To do so, we will use the binomial theorem to expand the radical in equation 13.1d for piston position to put the equations for position, velocity, and acceleration in simpler, approximate forms which will shed some light on the dynamic behavior of the mechanism.

The general form of the binomial theorem is:

$$(a+b)^n = a^n + na^{n-1}b + \frac{n(n-1)}{2!}a^{n-2}b^2 + \frac{n(n-1)(n-2)}{3!}a^{n-3}b^3 + \cdots \qquad (13.2a)$$

The radical in equation 13.1d is:

$$\sqrt{1 - \left(\frac{r}{l}\sin\omega t\right)^2} = \left[1 - \left(\frac{r}{l}\sin\omega t\right)^2\right]^{\frac{1}{2}} \qquad (13.2b)$$

where, for the binomial expansion:

$$a = 1 \qquad\qquad b = -\left(\frac{r}{l}\sin\omega t\right)^2 \qquad\qquad n = \frac{1}{2} \qquad (13.2c)$$

It expands to:

$$1 - \frac{1}{2}\left(\frac{r}{l}\sin\omega t\right)^2 + \frac{1}{8}\left(\frac{r}{l}\sin\omega t\right)^4 - \frac{1}{16}\left(\frac{r}{l}\sin\omega t\right)^6 + \cdots \qquad (13.2d)$$

or: $$1 - \left(\frac{r^2}{2l^2}\right)\sin^2\omega t + \left(\frac{r^4}{8l^4}\right)\sin^4\omega t - \left(\frac{r^6}{16l^6}\right)\sin^6\omega t + \cdots \qquad (13.2e)$$

Each nonconstant term contains the **crank-conrod ratio** r/l to some power. Applying some engineering common sense to the depiction of the slider-crank in Figure 13-7a (p. 607), we can see that if r/l were greater than 1 the crank could not make a complete revolution. In fact if r/l even gets close to 1, the piston will hit the fixed pivot O_2 before the crank completes its revolution. If r/l is as large as 1/2, the transmission angle ($\pi/2 - \phi$) will be too small (see Sections 3.3, p. 81 and 4.10, p.169) and the linkage will not run well. A practical upper limit on the value of r/l is about 1/3. Most slider-crank linkages will have this **crank-conrod ratio** somewhere between 1/3 and 1/5 for smooth operation. If we substitute this practical upper limit of $r/l = 1/3$ into equation 13.2e, we get:

$$1-\left(\frac{1}{18}\right)\sin^2\omega t+\left(\frac{1}{648}\right)\sin^4\omega t-\left(\frac{1}{11,664}\right)\sin^6\omega t+\cdots$$

(13.2f)

$$1-0.05556\sin^2\omega t+0.00154\sin^4\omega t-0.00009\sin^6\omega t+\cdots$$

Clearly we can drop all terms after the second with very small error. Substituting this approximate expression for the radical in equation 13.1d gives an approximate expression for piston displacement with only a fraction of one percent error.

$$x\doteq r\cos\omega t+l\left[1-\left(\frac{r^2}{2l^2}\right)\sin^2\omega t\right]$$

(13.3a)

Substitute the trigonometric identity:

$$\sin^2\omega t=\frac{1-\cos 2\omega t}{2}$$

(13.3b)

and simplify:

$$x\doteq l-\frac{r^2}{4l}+r\left(\cos\omega t+\frac{r}{4l}\cos 2\omega t\right)$$

(13.3c)

Differentiate for velocity of the piston (with constant ω):

$$\dot{x}\doteq-r\omega\left(\sin\omega t+\frac{r}{2l}\sin 2\omega t\right)$$

(13.3d)

Differentiate again for acceleration (with constant ω):

$$\ddot{x}\doteq-r\omega^2\left(\cos\omega t+\frac{r}{l}\cos 2\omega t\right)$$

(13.3e)

The process of binomial expansion has, in this particular case, led us to Fourier series approximations of the exact expressions for the piston displacement, velocity, and acceleration. Fourier[*] showed that any periodic function can be approximated by a series of sine and cosine terms of integer multiples of the independent variable. Recall that we dropped the fourth, sixth, and subsequent power terms from the binomial expansion, which would have provided cos $4\omega t$, cos $6\omega t$, etc., terms in this expression. These multiple angle functions are referred to as the **harmonics** of the fundamental cos ωt term. The cos ωt term repeats once per crank revolution and is called the fundamental frequency or the **primary component**. The second harmonic, cos $2\omega t$, repeats twice per crank revolution and is called the **secondary component**. The higher harmonics were dropped when we truncated the series. The constant term in the displacement function is the **DC component** or **average value**. The complete function is the sum of its harmonics. The Fourier series form of the expressions for displacement and its derivatives lets us see the relative contributions of the various harmonic components of the functions. This approach will prove to be quite valuable when we attempt to dynamically balance an engine design.

Program ENGINE calculates the position, velocity, and acceleration of the piston according to equations 13.3c, d, and e. Figure 13-8a, b, and c shows these functions for

[*] Baron Jean Baptiste Joseph Fourier (1768-1830) published the description of the mathematical series which bears his name in *The Analytic Theory of Heat* in 1822. The Fourier series is widely used in harmonic analysis of all types of physical systems. Its general form is:

$$y=\frac{a_0}{2}+\left(a_1\cos x+b_1\sin x\right)$$
$$+\left(a_2\cos 2x+b_2\sin 2x\right)$$
$$+\cdots+\left(a_n\cos nx+b_n\sin nx\right)$$

13

this example engine in the program as plotted for constant crank ω over two full revolutions. The acceleration curve shows the effects of the second harmonic term most clearly because that term's coefficient is larger than its correspondent in either of the other two functions. The fundamental ($-\cos \omega t$) term gives a pure harmonic function with a period of 360°. This fundamental term dominates the function as it has the largest coefficient in equation 13.3e. The flat top and slight dip in the positive peak acceleration of Figure 13-8c is caused by the $\cos 2\omega t$ second harmonic adding or subtracting from the fundamental. Note the very high value of peak acceleration of the piston even at the midrange engine speed of 3400 rpm. It is 747 g's! At 6000 rpm this increases to nearly 1300 g's. This is a moderately sized engine, of 3-in (76 mm) bore and 3.54-in (89 mm) stroke, with 25-in³ (400-cc) displacement per cylinder.

SUPERPOSITION   We will now analyze the dynamic behavior of the single-cylinder engine based on the approximate kinematic model developed above. Since we have several sources of dynamic excitation to deal with, we will use the method of superposition to separately analyze them and then combine their effects. We will first consider the **forces and torques** which are due to the presence of the **explosive gas forces** in the cylinder, which drive the engine. Then we will analyze the **inertia forces and torques** that result from the high-speed motion of the elements. The total force and torque state of the machine at any instant will be the sum of these components. Finally we will look at the **shaking forces and torques** on the ground plane and the **pin forces** within the linkage that result from the combination of applied and dynamic forces on the system.

13.3   GAS FORCE AND GAS TORQUE

The **gas force** is due to the gas pressure from the exploding fuel-air mixture impinging on the top of the piston surface as shown in Figure 13-3 (p. 601). Let F_g = gas force, P_g = gas pressure, A_p = area of piston, and B = bore of cylinder, which is also equal to the piston diameter. Then:

$$\mathbf{F}_g = -P_g A_p \,\hat{\mathbf{i}}; \qquad\qquad A_p = \frac{\pi}{4}B^2$$

$$\mathbf{F}_g = -\frac{\pi}{4}P_g B^2 \,\hat{\mathbf{i}}$$

(13.4)

The negative sign is due to the choice of engine orientation in the coordinate system of Figure 13-3. The **gas pressure** P_g in this expression is a function of crank angle ωt and is defined by the thermodynamics of the engine. A typical **gas pressure curve** for a four-stroke engine is shown in Figure 13-4 (p. 602). The **gas force curve** shape is identical to that of the gas pressure curve as they differ only by a constant multiplier, the piston area A_p. Figure 13-6 (p. 606) shows the approximation of the gas force curve used in program ENGINE for both four- and two-stroke engines.

The **gas torque** in Figure 13-9 is due to the gas force acting at a moment arm about the crank center O_2 in Figure 13-7 (p. 607). This moment arm varies from zero to a maximum as the crank rotates. The distributed gas force over the piston surface has been resolved to a single force acting through the mass center of link 4 in the free-body diagrams of Figure 13-7b. The concurrent force system at point B is resolved in the vector diagram showing that:

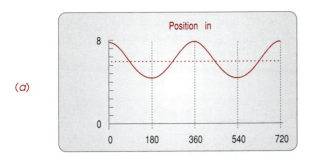

(a)

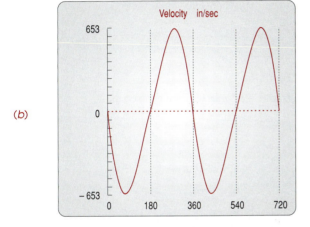

(b)

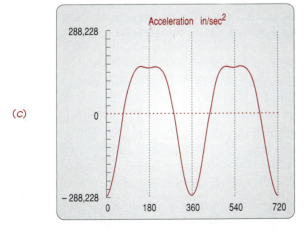

(c)

1 Cylinder
4 Stroke Cycle
Bore = 3.00 in
Stroke = 3.54
B/S   = 0.85
L/R   = 3.50
RPM = 3400

FIGURE 13-8

Position, velocity, and acceleration functions for a single-cylinder engine

$$\mathbf{F}_{g14} = F_g \tan\phi \,\hat{\mathbf{j}} \qquad\qquad (13.5a)$$

$$\mathbf{F}_{g34} = F_g \,\hat{\mathbf{i}} - F_g \tan\phi \,\hat{\mathbf{j}} \qquad\qquad (13.5b)$$

13

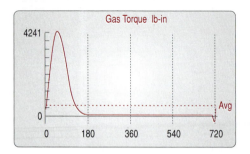

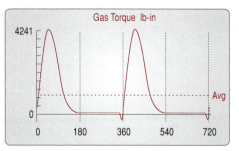

(a) Otto four-stroke cycle   (b) Clerk two-stroke cycle

FIGURE 13-9

Gas torque functions in the two-stroke and four-stroke cycle engines

From the free-body diagrams in Figure 13-7 we can see that:

$$\mathbf{F}_{g41} = -\mathbf{F}_{g14}$$

$$\mathbf{F}_{g43} = -\mathbf{F}_{g34}$$

$$\mathbf{F}_{g23} = -\mathbf{F}_{g43}$$

$$\mathbf{F}_{g32} = -\mathbf{F}_{g23}$$

so:

$$\mathbf{F}_{g32} = -\mathbf{F}_{g34} = -F_g\,\hat{\mathbf{i}} + F_g\tan\phi\,\hat{\mathbf{j}} \qquad (13.5c)$$

The **driving torque** $\mathbf{T}_{g21}$ at link 2 due to the gas force can be found from the cross product of the position vector to point A and the force at point A.

$$\mathbf{T}_{g21} = \mathbf{R}_A \times \mathbf{F}_{g32} \qquad (13.6a)$$

This expression can be expanded and will involve the crank length r and the angles θ and ϕ as well as the gas force $\mathbf{F}_g$. Note from the free-body diagram for link 1 that we can also express the torque in terms of the forces $\mathbf{F}_{g14}$ or $\mathbf{F}_{g41}$ which act always perpendicular to the motion of the slider (neglecting friction), and the distance x, which is their instantaneous moment arm about O_2. The reaction torque $\mathbf{T}_{g12}$ due to the gas force trying to rock the ground plane is:

$$\mathbf{T}_{g12} = F_{g41} \cdot x\ \hat{\mathbf{k}} \qquad (13.6b)$$

If you have ever abruptly opened the throttle of a running automobile engine while working on it, you probably noticed the engine move to the side as it rocked in its mounts from the reaction torque. The driving torque $\mathbf{T}_{g21}$ is the negative of this reaction torque.

$$\mathbf{T}_{g21} = -\mathbf{T}_{g12}$$

$$\mathbf{T}_{g21} = -F_{g41} \cdot x\ \hat{\mathbf{k}} \qquad (13.6c)$$

and:

$$F_{g14} = -F_{g41}$$

so:

$$\mathbf{T}_{g21} = F_{g14} \cdot x\ \hat{\mathbf{k}} \qquad (13.6d)$$

Equation 13.6d gives us an expression for **gas torque** which involves the displacement of the piston x for which we have already derived equation 13.3a. Substituting equation 13.3a for x and the magnitude of equation 13.5a for F_{g14} we get:

$$\mathbf{T}_{g21} = \left(F_g \tan\phi \right) \left[l - \frac{r^2}{4l} + r\left(\cos\omega t + \frac{r}{4l}\cos 2\omega t \right) \right] \hat{\mathbf{k}} \qquad (13.6e)$$

Equation 13.6e contains the conrod angle ϕ as well as the independent variable, crank angle ωt. We would like to have an expression which involves only ωt. We can substitute an expression for $\tan\phi$ generated from the geometry of Figure 13-7a.

$$\tan\phi = \frac{q}{u} = \frac{r\sin\omega t}{l\cos\phi} \qquad (13.7a)$$

Substitute equation 13.1c for $\cos\phi$:

$$\tan\phi = \frac{r\sin\omega t}{l\sqrt{1 - \left(\dfrac{r}{l}\sin\omega t \right)^2}} \qquad (13.7b)$$

The radical in the denominator can be expanded using the binomial theorem as was done in equations 13.2 (p. 608), and the first two terms retained for a good approximation to the exact expression,

$$\frac{1}{\sqrt{1 - \left(\dfrac{r}{l}\sin\omega t \right)^2}} \cong 1 + \frac{r^2}{2l^2}\sin^2\omega t \qquad (13.7c)$$

giving:

$$\tan\phi \cong \frac{r}{l}\sin\omega t \left(1 + \frac{r^2}{2l^2}\sin^2\omega t \right) \qquad (13.7d)$$

Substitute this into equation 13.6e for the gas torque:

$$\mathbf{T}_{g21} \cong F_g \left[\frac{r}{l}\sin\omega t \left(1 + \frac{r^2}{2l^2}\sin^2\omega t \right) \right] \left[l - \frac{r^2}{4l} + r\left(\cos\omega t + \frac{r}{4l}\cos 2\omega t \right) \right] \hat{\mathbf{k}} \qquad (13.8a)$$

Expand this expression and neglect any terms containing the conrod crank ratio r/l raised to any power greater than one since these will have very small coefficients as was seen in equation 13.2. This results in a simpler, but even more approximate expression for the gas torque:

$$\mathbf{T}_{g21} \cong F_g\, r\sin\omega t \left(1 + \frac{r}{l}\cos\omega t \right) \qquad (13.8b)$$

Note that the **exact value** of this **gas torque** can always be calculated from equations 13.1d, 13.5a and 13.6d in combination, or from the expansion of equation 13.6a, if you require a more accurate answer. For design purposes the approximate equation 13.8b will usually be adequate. Program ENGINE calculates the gas torque using equation 13.8b and its built-in gas pressure curve to generate the gas force function. Plots of the gas torque for two- and four-stroke cycles are shown in Figure 13-9. Note the simi-

13

larity in shape to that of the gas force curve in Figure 13-6 (p. 606). Note also that the two-stroke has theoretically twice the power available as the four-stroke, all other factors equal, because there are twice as many torque pulses per unit time. The poorer efficiency of the two-stroke reduces this theoretical advantage, however.

13.4   EQUIVALENT MASSES

To do a complete dynamic force analysis on any mechanism we need to know the geometric properties (mass, center of gravity, mass moment of inertia) of the moving links as was discussed in previous chapters (see Sections 10.3 to 10.8 and Chapter 11). This is easy to do if the link already is designed in detail and its dimensions are known. When designing the mechanism from scratch, we typically do not yet know that level of detail about the links' geometries. But we must nevertheless make some estimate of their geometric parameters in order to begin the iteration process which will eventually converge on a detailed design.

In the case of this slider-crank mechanism, the **crank** is in **pure rotation** and the **piston** is in **pure translation**. By assuming some reasonable geometries and materials we can make approximations of their dynamic parameters. Their kinematic motions are easily determined. We have already derived expressions for the piston motion in equations 13.3 (p. 609). Further, if we balance the rotating crank, as described and recommended in the previous chapter, then the *CG* of the crank will be motionless at its center O_2 and will not contribute to the dynamic forces. We will do this in a later section.

The conrod is in complex motion. To do an exact dynamic analysis as was derived in Section 11.5, we need to determine the linear acceleration of its *CG* for all positions. At the outset of the design, the conrod's *CG* location is not accurately defined. To "bootstrap" the design, we need a simplified model of this connecting rod which we can later refine, as more dynamic information is generated about our engine design. The requirements for a dynamically equivalent model were stated in Section 10.2 and are repeated here as Table 13-1 for your convenience.

If we could model our still-to-be-designed conrod as two, lumped, point masses, concentrated one at the crankpin (point *A* in Figure 13-7, p. 607), and one at the wrist pin (point *B* in Figure 13-7), we would at least know what the motions of these lumps are. The lump at *A* would be in pure rotation as part of the crank, and the lump at point *B* would be in pure translation as part of the piston. These lumped, point masses have no dimension and are assumed to be connected with a magical, massless but rigid rod.[*]

DYNAMICALLY EQUIVALENT MODEL   Figure 13-10a shows a typical conrod. Figure 13-10b shows a generic two-mass model of the conrod. One mass m_t is located at distance l_t from the *CG* of the original rod, and the second mass m_p at distance l_p from the *CG*. The mass of the original part is m_3, and its moment of inertia about its *CG* is I_{G3}. Expressing the three requirements for dynamic equivalence from Table 13-1 mathematically in terms of these variables, we get:

$$m_p + m_t = m_3 \tag{13.9a}$$

$$m_p l_p = m_t l_t \tag{13.9b}$$

$$m_p l_p^2 + m_t l_t^2 = I_{G3} \tag{13.9c}$$

[*] These lumped mass models have to be made with *very* special materials. *Unobtainium 206* has the property of infinite mass density, thus occupies no space and can be used for "point masses." *Unobtainium 208* has infinite stiffness and zero mass and thus can be used for rigid but "massless rods."

There are four unknowns in these three equations, m_p, l_p, m_t, l_t, which means we must choose a value for any one variable to solve the system. Let us choose the distance l_t equal to the distance to the wrist pin, l_b, as shown in Figure 13-10c. This will put one mass at a desired location. Solving equations 13.9a and 13.9b simultaneously with that substitution gives expressions for the two lumped masses:

$$m_p = m_3 \frac{l_b}{l_p + l_b}$$

$$m_b = m_3 \frac{l_p}{l_p + l_b}$$

(13.9d)

Substituting equation 13.9d into 13.9c gives a relation between l_p and l_b:

$$m_3 \frac{l_b}{l_p + l_b} l_p^2 + m_3 \frac{l_p}{l_p + l_b} l_b^2 = I_{G_3} = m_3 l_p l_b$$

(13.9e)

$$l_p = \frac{I_{G_3}}{m_3 l_b}$$

Please refer to Section 10.8 and equation 10.13 which define the *center of percussion* and its geometric relationship to a corresponding *center of rotation*. Equation 13.9e is the same as equation 11.13 (except for sign which is due to an arbitrary choice of the link's orientation in the coordinate system). The distance l_p is the location of the center of percussion corresponding to a center of rotation at l_b. Thus our second mass m_p must be placed at the link's **center of percussion** P (using point B as its center of rotation) to obtain exact dynamic equivalence. The masses must be as defined in equation 13.9d.

The geometry of the typical conrod, as shown in Figures 13-2 (p. 600) and 13-10a, is large at the crankpin end (A) and small at the wrist pin end (B). This puts the CG close to the "big end." The center of percussion P will be even closer to the big end than is the CG. For this reason, we can place the second lumped mass, which belongs at P, at point A with relatively small error in our dynamic model's accuracy. This approximate model is adequate for our initial design calculations. Once a viable design geometry is established, we will have to do a complete and exact force analysis with the methods of Chapter 11 before considering the design complete.

Making this substitution of distance l_a for l_p and renaming the lumped masses at those distances m_{3a} and m_{3b}, to reflect both their identity with link 3 and with points A and B, we rewrite equations 13.9d.

13

TABLE 13-1   Requirements for Dynamic Equivalence

1   The mass of the model must equal that of the original body.

2   The center of gravity must be in the same location as that of the original body.

3   The mass moment of inertia must equal that of the original body.

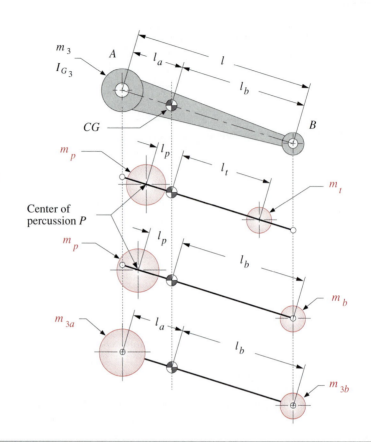

(a) Original connecting rod

(b) Generic two mass model

Center of percussion P

(c) Exact dynamic model

(d) Approximate model

FIGURE 13-10

Lumped mass dynamic models of a connecting rod

let

$$l_p = l_a$$

then :

$$m_{3a} = m_3 \frac{l_b}{l_a + l_b} \qquad (13.10a)$$

and :

$$m_{3b} = m_3 \frac{l_a}{l_a + l_b} \qquad (13.10b)$$

These define the amounts of the total conrod mass to be placed at each end, to approximately dynamically model that link. Figure 13-10d shows this dynamic model. In the absence of any data on the shape of the conrod at the outset of a design, preliminary dynamic force information can be obtained by using the rule of thumb of placing two-thirds of the conrod's mass at the crankpin and one-third at the wrist pin.

STATICALLY EQUIVALENT MODEL   We can create a similar lumped mass model of the crank. Even though we intend to balance the crank before we are done, for generality we will initially model it *unbalanced* as shown in Figure 13-11. Its *CG* is located at some distance r_{G2} from the pivot, O_2, on the line to the crankpin, A. We would like to

---

TABLE 13-2   Requirements for Static Equivalence

1   The mass of the model must equal that of the original body.

2   The center of gravity must be in the same location as that of the original body.

---

model it as a lumped mass at A on a massless rod pivoted at O_2. If our principal concern is with a steady-state analysis, then the crank velocity ω will be held constant. An absence of angular acceleration on the crank allows a statically equivalent model to be used because the equation $\mathbf{T} = I\alpha$ will be zero regardless of the value of I. **A statically equivalent model** needs only to have equivalent mass and equivalent first moments as shown in Table 13-2. The moments of inertia need not match. We model it as two lumped masses, one at point A and one at the fixed pivot O_2. Writing the two requirements for static equivalence from Table 13-2:

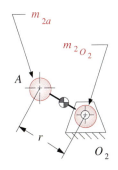

$$m_2 = m_{2a} + m_{2O_2}$$

$$m_{2a}r = m_2 r_{G_2} \qquad (13.11)$$

$$m_{2a} = m_2 \frac{r_{G_2}}{r}$$

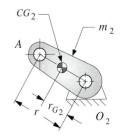

The lumped mass m_{2a} can be placed at point A to represent the unbalanced crank. The second lumped mass, at the fixed pivot O_2, is not necessary to any calculations as that point is stationary.

These simplifications lead to the lumped parameter model of the slider-crank linkage shown in Figure 13-12. The crankpin, point A, has two masses concentrated at it, the equivalent mass of the crank m_{2a} and the portion of conrod m_{3a}. Their sum is m_A. At the wrist pin, point B, two masses are also concentrated, the piston mass m_4 and the remaining portion of the conrod mass m_{3b}. Their sum is m_B. This model has masses which are either in pure rotation (m_A) or in pure translation (m_B), so it is very easy to dynamically analyze.

FIGURE 13-11

Statically equivalent lumped mass model of a crank

$$m_A = m_{2a} + m_{3a}$$

$$m_B = m_{3b} + m_4 \qquad (13.12)$$

VALUE OF MODELS   *The value of constructing simple, lumped mass models of complex systems increases with the complexity of the system being designed.* It makes little sense to spend large amounts of time doing sophisticated, detailed analyses of designs which are so ill-defined at the outset that their conceptual viability is as yet unproven. Better to get a reasonably approximate and rapid answer that tells you the concept needs to be rethought, than to spend a greater amount of time reaching the same conclusion to more decimal places.

13.5   INERTIA AND SHAKING FORCES

The simplified, lumped mass model of Figure 13-12 can be used to develop expressions for the forces and torques due to the accelerations of the masses in the system. The method of d'Alembert is of value in visualizing the effects of these moving masses on the sys-

13

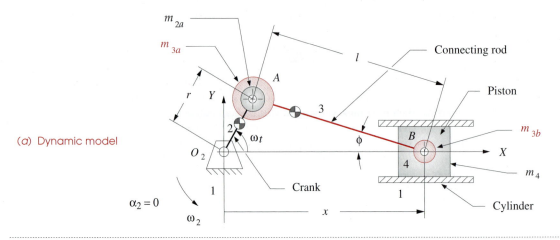

(a) Dynamic model

$\alpha_2 = 0$

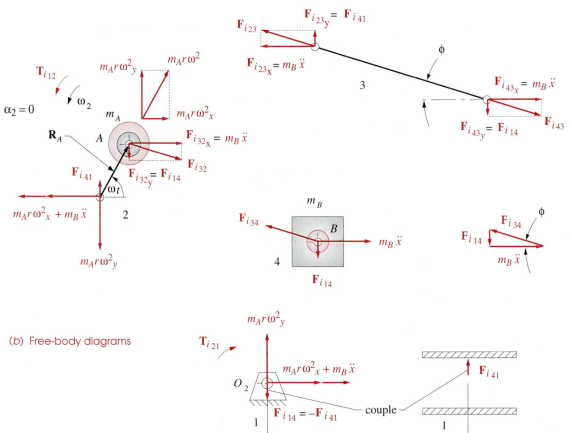

(b) Free-body diagrams

FIGURE 13-12

Lumped mass dynamic model of the slider-crank—arrows show vector direction and sense, labels show magnitude

13

tem and on the ground plane. Accordingly, the free-body diagrams of Figure 13-12b show the d'Alembert inertia forces acting on the masses at points A and B. Friction is again ignored. The acceleration for point B is given in equation 13.3e. The acceleration of point A in pure rotation is obtained by differentiating the position vector $\mathbf{R}_A$ twice, assuming a constant crankshaft ω, which gives:

$$\mathbf{R}_A = r\cos\omega t\,\hat{\mathbf{i}} + r\sin\omega t\,\hat{\mathbf{j}}$$

$$\mathbf{a}_A = -r\omega^2\cos\omega t\,\hat{\mathbf{i}} - r\omega^2\sin\omega t\,\hat{\mathbf{j}}$$

(13.13)

The total inertia force $\mathbf{F}_i$ is the sum of the centrifugal (inertia) force at point A and the inertia force at point B.

$$\mathbf{F}_i = -m_A\,\mathbf{a}_A - m_B\,\mathbf{a}_B \qquad (13.14a)$$

Breaking it into x and y components:

$$F_{i_x} = -m_A\left(-r\omega^2\cos\omega t\right) - m_B\,\ddot{x} \qquad (13.14b)$$

$$F_{i_y} = -m_A\left(-r\omega^2\sin\omega t\right) \qquad (13.14c)$$

Note that only the x component is affected by the acceleration of the piston. Substituting equation 13.3e into equation 13.14b:

$$F_{i_x} \cong -m_A\left(-r\omega^2\cos\omega t\right) - m_B\left[-r\omega^2\left(\cos\omega t + \frac{r}{l}\cos 2\omega t\right)\right]$$

$$F_{i_y} = -m_A\left(-r\omega^2\sin\omega t\right)$$

(13.14d)

Notice that the x directed inertia forces have primary components at crank frequency and secondary (second harmonic) forces at twice crank frequency. There are also small-magnitude, higher, even harmonics which we truncated in the binomial expansion of the piston displacement function. The force due to the rotating mass at point A has only a primary component.

The **shaking force** was defined in Section 11.8 to be *the sum of all forces acting on the ground plane*. From the free-body diagram for link 1 in Figure 13-12:

$$\sum F_{s_x} \cong -m_A\left(r\omega^2\cos\omega t\right) - m_B\left[r\omega^2\left(\cos\omega t + \frac{r}{l}\cos 2\omega t\right)\right]$$

$$\sum F_{s_y} = -m_A\left(r\omega^2\sin\omega t\right) + F_{i_{41}} - F_{i_{41}}$$

(13.14e)

Note that the side force of the piston on the cylinder wall $F_{i_{41}}$ is cancelled by an equal and opposite force $F_{i_{14}}$ passed through the connecting rod and crankshaft to the main pin at O_2. These two forces create a couple that provides the shaking torque. The shaking force $\mathbf{F}_s$ is equal to the negative of the inertia force.

$$\mathbf{F}_s = -\mathbf{F}_i \qquad (13.14f)$$

13

Note that the gas force from equation 13.4 does not contribute to the shaking force. Only inertia forces and external forces are felt as shaking forces. The gas force is an internal force which is cancelled within the mechanism. It acts equally and oppositely on both the piston top and the cylinder head as shown in Figure 13-7 (p. 607).

Program ENGINE calculates the shaking force at constant ω, for any combination of linkage parameters input to it. Figure 13-13 shows the shaking force plot for the same unbalanced built-in example engine as shown in the acceleration plot (Figure 13-8c, p. 611). The linkage orientation is the same as in Figure 13-12 (p. 618) with the x axis horizontal. The x component is larger than the y component due to the high acceleration of the piston. The forces are seen to be quite large despite this being a relatively small (0.4 liter per cylinder) engine running at moderate speed (3400 rpm). We will soon investigate techniques to reduce or eliminate this shaking force from the engine. It is an undesirable feature which creates noise and vibration.

13.6   INERTIA AND SHAKING TORQUES

The **inertia torque** results from the action of the inertia forces at a moment arm. The inertia force at point A in Figure 13-12 has two components, radial and tangential. The radial component has no moment arm. The tangential component has a moment arm of crank radius r. If the crank ω is constant, the mass at A will not contribute to inertia torque. The inertia force at B has a nonzero component perpendicular to the cylinder wall except when the piston is at TDC or BDC. As we did for the gas torque, we can express the inertia torque in terms of the couple $-\mathbf{F}_{i41}$, $\mathbf{F}_{i41}$ whose forces act always perpendicular to the motion of the slider (neglecting friction), and the distance x, which is their instantaneous moment arm (see Figure 13-12 on p. 618). The inertia torque is:

$$\mathbf{T}_{i_{21}} = \left(F_{i_{41}} \cdot x \right)\ \hat{\mathbf{k}} = \left(-F_{i_{14}} \cdot x \right)\ \hat{\mathbf{k}} \tag{13.15a}$$

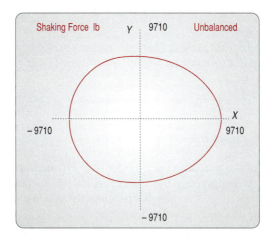

FIGURE 13-13

Shaking force in an unbalanced slider-crank linkage

Substituting for F_{i14} (see Figure 13-12b, p. 618) and for x, (see equation 13.3a. p. 609) we get:

$$\mathbf{T}_{i_{21}} \cong -(-m_B \ddot{x} \tan \phi) \left[l - \frac{r^2}{4l} + r \left(\cos \omega t + \frac{r}{4l} \cos 2\omega t \right) \right] \hat{\mathbf{k}} \qquad (13.15b)$$

We previously developed expressions for x *double dot* (equation 13.3e, p. 609) and $\tan \phi$ (equation 13.7d, p. 613) which can now be substituted.

$$
\begin{aligned}
\mathbf{T}_{i_{21}} \cong m_B &\left[-r\omega^2 \left(\cos \omega t + \frac{r}{l} \cos 2\omega t \right) \right] \\
&\cdot \left[\frac{r}{l} \sin \omega t \left(1 + \frac{r^2}{2l^2} \sin^2 \omega t \right) \right] \\
&\cdot \left[l - \frac{r^2}{4l} + r \left(\cos \omega t + \frac{r}{4l} \cos 2\omega t \right) \right] \hat{\mathbf{k}}
\end{aligned}
\qquad (13.15c)
$$

Expanding this and then dropping all terms with coefficients containing r/l to powers higher than one gives the following approximate equation for inertia torque with constant shaft ω :

$$\mathbf{T}_{i_{21}} \cong -m_B r^2 \omega^2 \sin \omega t \left(\frac{r}{2l} + \cos \omega t + \frac{3r}{2l} \cos 2\omega t \right) \hat{\mathbf{k}} \qquad (13.15d)$$

This contains products of sine and cosine terms. Putting it entirely in terms of harmonics will be instructive, so substitute the identities:

$$2 \sin \omega t \cos 2\omega t = \sin 3\omega t - \sin \omega t$$
$$2 \sin \omega t \cos \omega t = \sin 2\omega t$$

to get: $\qquad \mathbf{T}_{i_{21}} \cong \frac{1}{2} m_B r^2 \omega^2 \left(\frac{r}{2l} \sin \omega t - \sin 2\omega t - \frac{3r}{2l} \sin 3\omega t \right) \hat{\mathbf{k}} \qquad (13.15e)$

This shows that the **inertia torque** has a third harmonic term as well as a first and second. The second harmonic is the dominant term as it has the largest coefficient because r/l is always less than 2/3.

The **shaking torque** is equal to the inertia torque.

$$\mathbf{T}_s = \mathbf{T}_{i_{21}} \qquad (13.15f)$$

Program ENGINE calculates the inertia torque from equation 13.15e. Figure 13-14 shows a plot of the inertia torque for this built-in example engine. Note the dominance of the second harmonic. The ideal magnitude for the inertia torque is zero, as it is parasitic. Its average value is always zero, so *it contributes nothing to the net driving torque*. It merely creates large positive and negative oscillations in the total torque which increase vibration and roughness. We will soon investigate means to reduce or eliminate this inertia and shaking torque in our engine designs. It is possible to cancel their effects by proper arrangement of the cylinders in a multicylinder engine, as will be explored in the next chapter.

13

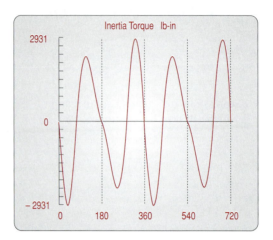

FIGURE 13-14

Inertia torque in the slider-crank linkage

13.7   TOTAL ENGINE TORQUE

The total engine torque is the sum of the gas torque and the inertia torque.

$$\mathbf{T}_{total} = \mathbf{T}_g + \mathbf{T}_i \tag{13.16}$$

The gas torque is less sensitive to engine speed than is the inertia torque, which is a function of ω^2. So the relative contributions of these two components to the total torque will vary with engine speed. Figure 13-15a shows the total torque for this example engine plotted by program ENGINE for an idle speed of 800 rpm. Compare this to the gas torque plot of the same engine in Figure 13-9a (p. 612). The inertia torque component is negligible at this slow speed compared to the gas torque component. Figure 13-15c shows the same engine run at 6000 rpm. Compare this to the plot of inertia torque in Figure 13-14. The inertia torque component is dominating at this high speed. At the midrange speed of 3400 rpm (Figure 13-15b), a mix of both components is seen.

13.8   FLYWHEELS

We saw in Section 11.11 (p. 548) that large oscillations in the torque-time function can be significantly reduced by the addition of a flywheel to the system. The single-cylinder engine is a prime candidate for the use of a flywheel. The intermittent nature of its power strokes makes one mandatory as it will store the kinetic energy needed to carry the piston through the Otto cycle's exhaust, intake, and compression strokes during which work must be done on the system. Even the two-stroke engine needs a flywheel to drive the piston up on the compression stroke.

The procedure for designing an engine flywheel is identical to that described in Section 11.11 for the fourbar linkage. The total torque function for one revolution of the crank is integrated, pulse by pulse, with respect to its average value. These integrals represent energy fluctuations in the system. The maximum change in energy under the torque curve during one cycle is the amount needed to be stored in the flywheel. Equation 11.20c expresses this relationship. Program ENGINE does the numerical integration of the total torque function and presents a table similar to the one shown in Figure 11-11 (p. 551). These data and the designer's choice of a coefficient of fluctuation k (see equation 11.19) are all that are needed to solve equations 11.20 and 11.21 for the required moment of inertia of the flywheel.

The calculation must be done at some average crank ω. Since the typical engine operates over a wide range of speeds, some thought needs to be given to the most appropriate speed to use in the flywheel calculation. The flywheel's stored kinetic energy is proportional to ω^2 (see equation 11.17). Thus at high speeds a flywheel can have a small moment of inertia and still be effective. The slowest operating speed will require the largest flywheel and should be the one used in the computation of required flywheel size.

Program ENGINE plots the flywheel-smoothed total torque for a user-supplied coefficient of fluctuation k. Figure 13-16 shows the smoothed torque functions for $k = 0.05$ corresponding to the unsmoothed ones in Figure 13-15. Note that the smoothed curves shown for each engine speed are what would result with the flywheel size necessary to obtain that coefficient of fluctuation at that speed. In other words, the flywheel applied to the 800-rpm engine is much larger than the one on the 6000-rpm engine, in these plots. Compare corresponding rows (speeds) between Figures 13-15 and 13-16 to see the effect of the addition of a flywheel. But do not directly compare parts a, b, and c within Figure 13-16 as to the amount of smoothing since the flywheel sizes used are different at each operating speed.

An engine flywheel is usually designed as a flat disk, bolted to one end of the crankshaft. One flywheel face is typically used for the clutch to run against. The clutch is a friction device which allows disconnection of the engine from the drive train (the wheels of a vehicle) when no output is desired. The engine can then remain running at idle speed with the vehicle or output device stopped. When the clutch is engaged, all engine torque is transmitted through it, by friction, to the output shaft.

13.9   PIN FORCES IN THE SINGLE-CYLINDER ENGINE

In addition to calculating the overall effects on the ground plane of the dynamic forces present in the engine, we also need to know the magnitudes of the forces at the pin joints. These forces will dictate the design of the pins and the bearings at the joints. Though we were able to lump the mass due to both conrod and piston, or conrod and crank, at points A and B for an overall analysis of the linkage's effects on the ground plane, we cannot do so for the pin force calculations. This is because the pins feel the effect of the conrod pulling on one "side" and the piston (or crank) pulling on the other "side" of the pin as shown in Figure 13-17. Thus we must separate the effects of the masses of the links joined by the pins.

We will calculate the effect of each component due to the various masses, and the gas force and then superpose them to obtain the complete pin force at each joint. We need a bookkeeping system to keep track of all these components. We have already used some subscripts for these forces, so we will retain them and add others. The resultant bearing loads have the following components:

1   The gas force components, with the subscript g, as in F_g.

2   The inertia force due to the piston mass, with subscript ip, as in F_{ip}.

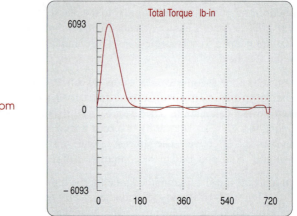

(a) 800 rpm

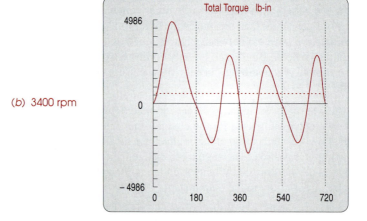

(b) 3400 rpm

1 Cylinder
4 Stroke Cycle
Bore = 3.00 in
Stroke = 3.54
B/S   = 0.85
L/R   = 3.50

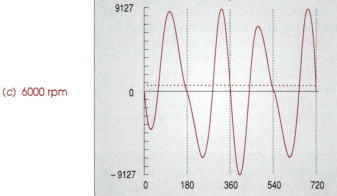

(c) 6000 rpm

FIGURE 13-15

The total torque function's shape and magnitude vary with crankshaft speed

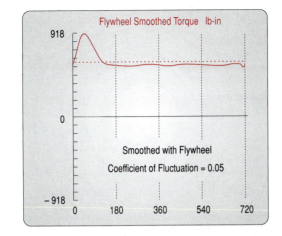

(a) 800 rpm

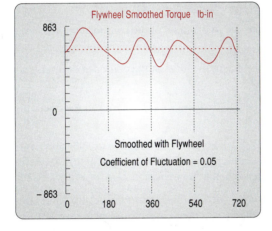

(b) 3400 rpm

1 Cylinder
4 Stroke Cycle
Bore = 3.00 in
Stroke = 3.54
B/S   = 0.85
L/R   = 3.50

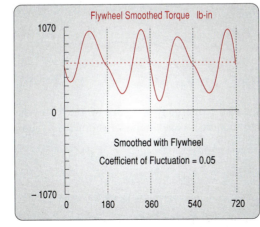

(c) 6000 rpm

FIGURE 13-16

The total torque function's shape and magnitude vary with crankshaft speed

13

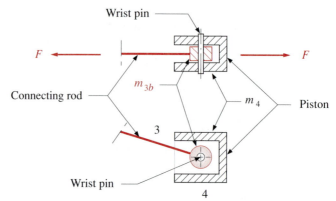

FIGURE 13-17

Forces on a pivot pin

3   The inertia force due to the mass of the conrod at the wrist pin, with subscript *iw*, as in F_{iw}.

4   The inertia force due to the mass of the conrod at the crank pin, with subscript *ic*, as in F_{ic}.

5   The inertia force due to the mass of the crank at the crank pin, with subscript *ir*, as in F_{ir}.

The link number designations will be added to each subscript in the same manner as before, indicating the link from which the force is coming as the first number and the link being analyzed as the second. (See Section 11.2 for further discussion of this notation.)

Figure 13-18 shows the free-body diagrams for the forces due only to the acceleration of the mass of the piston, m_4. Those components are:

$$\mathbf{F}_{ip41} = -m_4 a_B \tan\phi \,\hat{\mathbf{j}} \tag{13.17a}$$

$$\mathbf{F}_{ip34} = m_4 a_B \,\hat{\mathbf{i}} - m_4 a_B \tan\phi \,\hat{\mathbf{j}} \tag{13.17b}$$

$$\mathbf{F}_{ip32} = -\mathbf{F}_{ip34} \tag{13.17c}$$

$$\mathbf{F}_{ip21} = \mathbf{F}_{ip32} \tag{13.17d}$$

Figure 13-19 shows the free-body diagrams for the forces due only to the acceleration of the mass of the conrod located at the wrist pin, m_{3b}. Those components are:

$$\mathbf{F}_{iw41} = -m_{3b} a_B \tan\phi \,\hat{\mathbf{j}} \tag{13.18a}$$

$$\mathbf{F}_{iw34} = \mathbf{F}_{iw41} \tag{13.18b}$$

$$\mathbf{F}_{iw32} = -m_{3b} a_B \,\hat{\mathbf{i}} + m_{3b} a_B \tan\phi \,\hat{\mathbf{j}} \tag{13.18c}$$

$$\mathbf{F}_{iw21} = \mathbf{F}_{iw32} \tag{13.18d}$$

Figure 13-20a shows the free-body diagrams for the forces due only to the acceleration of the mass of the conrod located at the crankpin, m_{3a}. That component is:

$$\mathbf{F}_{ic32} = -m_{3a}\mathbf{a}_A \tag{13.19a}$$

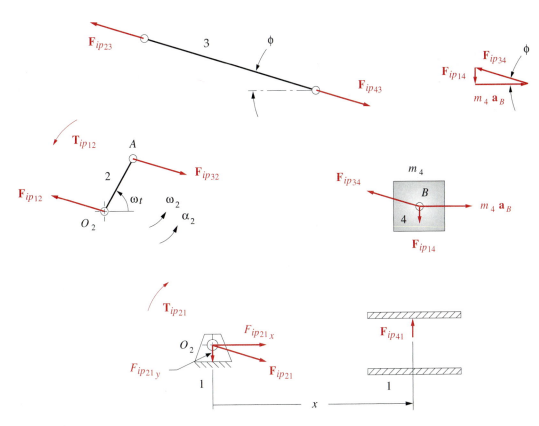

FIGURE 13-18

Free-body diagrams for forces due to the piston mass

Substitute equation 13.13 (p. 619) for $\mathbf{a}_A$:

$$\mathbf{F}_{ic32} = m_{3a}r\omega^2\left(\cos\omega t\ \hat{\mathbf{i}} + \sin\omega t\ \hat{\mathbf{j}}\right) \tag{13.19b}$$

Figure 13-20b shows the free-body diagrams for the forces due only to the acceleration of the lumped mass of the crank at the crankpin, m_{2a}. These affect only the main pin at O_2. That component is:

$$\mathbf{F}_{ir21} = m_{2a}r\omega^2\left(\cos\omega t\ \hat{\mathbf{i}} + \sin\omega t\ \hat{\mathbf{j}}\right) \tag{13.19c}$$

The gas force components were shown in Figure 13-7 (p. 607) and defined in equations 13.5 (p. 611).

We can now sum the components of the forces at each pin joint. For the sidewall force $\mathbf{F}_{41}$ of the piston against the cylinder wall:

$$\begin{aligned}
\mathbf{F}_{41} &= \mathbf{F}_{g41} + \mathbf{F}_{ip41} + \mathbf{F}_{iw41} \\
&= -F_g\tan\phi\ \hat{\mathbf{j}} - m_4 a_B\tan\phi\ \hat{\mathbf{j}} - m_{3b}a_B\tan\phi\ \hat{\mathbf{j}} \\
&= -\left[(m_4 + m_{3b})a_B + F_g\right]\tan\phi\ \hat{\mathbf{j}}
\end{aligned} \tag{13.20}$$

13

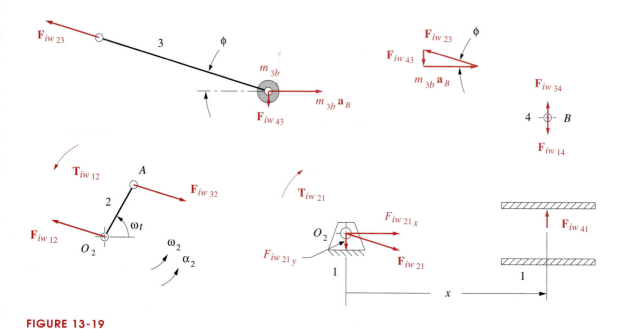

FIGURE 13-19

Free-body diagrams for forces due to the conrod mass concentrated at the wrist pin

The total force $\mathbf{F}_{34}$ on the wrist pin is:

$$\mathbf{F}_{34} = \mathbf{F}_{g34} + \mathbf{F}_{ip34} + \mathbf{F}_{iw34}$$

$$= \left(F_g \,\hat{\mathbf{i}} - F_g \tan\phi \,\hat{\mathbf{j}} \right) + \left(m_4 a_B \,\hat{\mathbf{i}} - m_4 a_B \tan\phi \,\hat{\mathbf{j}} \right) + \left(-m_{3b} a_B \tan\phi \,\hat{\mathbf{j}} \right) \qquad (13.21)$$

$$= \left(F_g + m_4 a_B \right) \hat{\mathbf{i}} - \left[F_g + \left(m_4 + m_{3b} \right) a_B \right] \tan\phi \,\hat{\mathbf{j}}$$

The total force $\mathbf{F}_{32}$ on the crankpin is:

$$\mathbf{F}_{32} = \mathbf{F}_{g32} + \mathbf{F}_{ip32} + \mathbf{F}_{iw32} + \mathbf{F}_{ic32}$$

$$= \left(-F_g \,\hat{\mathbf{i}} + F_g \tan\phi \,\hat{\mathbf{j}} \right) + \left(-m_4 a_B \,\hat{\mathbf{i}} + m_4 a_B \tan\phi \,\hat{\mathbf{j}} \right)$$

$$+ \left(-m_{3b} a_B \,\hat{\mathbf{i}} + m_{3b} a_B \tan\phi \,\hat{\mathbf{j}} \right) + \left[m_{3a} r\omega^2 \left(\cos\omega t \,\hat{\mathbf{i}} + \sin\omega t \,\hat{\mathbf{j}} \right) \right] \qquad (13.22)$$

$$= \left[m_{3a} r\omega^2 \cos\omega t - \left(m_{3b} + m_4 \right) a_B - F_g \right] \hat{\mathbf{i}}$$

$$+ \left\{ m_{3a} r\omega^2 \sin\omega t + \left[\left(m_{3b} + m_4 \right) a_B + F_g \right] \tan\phi \right\} \hat{\mathbf{j}}$$

The total force $\mathbf{F}_{21}$ on the main pin is:

$$\mathbf{F}_{21} = \mathbf{F}_{32} + \mathbf{F}_{ir21} \qquad (13.23)$$

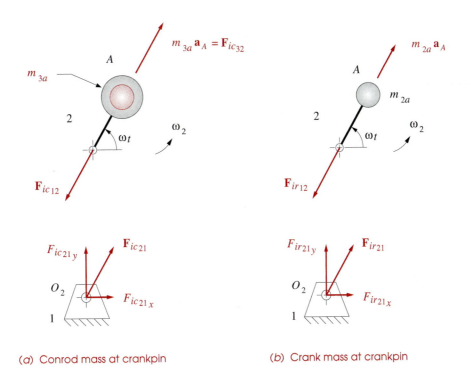

(a) Conrod mass at crankpin   (b) Crank mass at crankpin

FIGURE 13-20

Free-body diagrams for forces due to masses at the crankpin

Note that, unlike the inertia force in equation 13.14 (p. 619), which was unaffected by the gas force, these pin forces **are** a function of the gas force as well as of the $-ma$ forces. Engines with larger piston diameters will experience greater pin forces as a result of the explosion pressure acting on their larger piston area.

Program ENGINE calculates the pin forces on all joints using equations 13.20 to 13.23. Figure 13-21 shows the wrist-pin force on the same unbalanced engine example as shown in previous figures, for three engine speeds. The "bow tie" loop is the inertia force and the "teardrop" loop is the gas force portion of the force curve. An interesting trade-off occurs between the gas force components and the inertia force components of the pin forces. At a low speed of 800 rpm (Figure 13-21a), the gas force dominates as the inertia forces are negligible at small ω. The peak wrist-pin force is then about 4200 lb. At high speed (6000 rpm), the inertia components dominate and the peak force is about 4500 lb (Figure 13-21c). But at a midrange speed (3400 rpm), the inertia force cancels some of the gas force and the peak force is only about 3200 lb (Figure 13-21b). These plots show that the pin forces can be quite large even in a moderately sized (0.4 liter/cylinder) engine. The pins, links, and bearings all have to be designed to withstand hundreds of millions of cycles of these reversing forces without failure.

Figure 13-22 shows further evidence of the interaction of the gas forces and inertia forces on the crankpin and the wrist pin. Figures 13-22a and 13-22c show the variation in the inertia force component on the crankpin and wrist pin, respectively, over one full

13

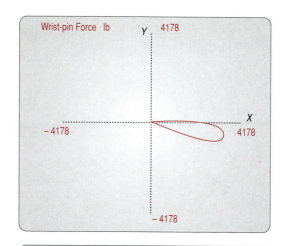

(*a*) 800 rpm

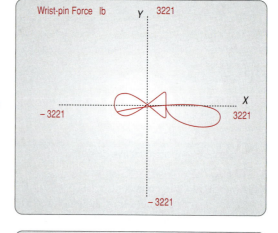

(*b*) 3400 rpm

1 Cylinder
4 Stroke Cycle
Bore = 3.00 in
Stroke = 3.54
B/S   = 0.85
L/R   = 3.50

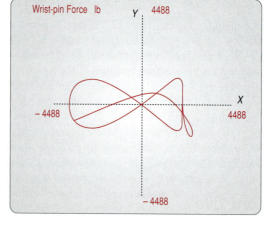

(*c*) 6000 rpm

FIGURE 13-21

Forces on the wrist pin of the single-cylinder engine at various speeds

revolution of the crank as the engine speed is increased from idle to redline. Figure 13-22b and d show the variation in the total force on the same respective pins with both the inertia and gas force components included. These two plots show only the first 90° of crank revolution where the gas force in a four-stroke cylinder occurs. Note that the gas force and inertia force components counteract one another resulting in one particular speed where the total pin force is a minimum during the power stroke. This is the same phenomenon as seen in Figure 13-21.

13.10 BALANCING THE SINGLE-CYLINDER ENGINE

The derivations and figures in the preceding sections have shown that significant forces are developed both on the pivot pins and on the ground plane due to the gas forces and the inertia and shaking forces. Balancing will not have any effect on the gas forces, which are internal, but it can have a dramatic effect on the inertia and shaking forces. The main pin force can be reduced, but the crankpin and wrist pin forces will be unaffected by any crankshaft balancing done. Figure 13-13 (p. 620) shows the unbalanced shaking force as felt on the ground plane of our 0.4-liter-per-cylinder example engine from program ENGINE. It is about 9700 lb even at the moderate speed of 3400 rpm. At 6000 rpm it increases to over 30 000 lb. The methods of Chapter 12 can be applied to this mechanism to balance the members in pure rotation and reduce these large shaking forces.

Figure 13-23a shows the dynamic model of our slider-crank with the conrod mass lumped at both crankpin A and wrist pin B. We can consider this single-cylinder engine to be a single-plane device, thus suitable for static balancing (see Section 13.1). It is straightforward to statically balance the crank. We need a balance mass at some radius, 180° from the lumped mass at point A whose mr product is equal to the product of the mass at A and its radius r. Applying equation 13.2 to this simple problem we get:

$$m_{bal}r_{bal} = -m_A r_A \tag{13.24}$$

Any combination of mass and radius which gives this product, placed at 180° from point A will balance the crank. For simplicity of example, we will use a balance radius equal to r. Then a mass equal to m_A placed at A' will exactly balance the rotating masses. The CG of the crank will then be at the fixed pivot O_2 as shown in Figure 13-23a. In a real crankshaft, actually placing the counterweight at this large a radius would not work. The balance mass has to be kept close to the centerline to clear the piston at BDC. Figure 13-2c shows the shape of typical crankshaft counterweights.

Figure 13-24a shows the shaking force from the same engine as in Figure 13-13 after the crank has been exactly balanced in this manner. The Y component of the shaking force has been reduced to zero and the x component to about 3300 lb at 3400 rpm. This is a factor of three reduction over the unbalanced engine. Note that the only source of Y directed inertia force is the rotating mass at point A of Figure 13-23 (see equations 13.14, p. 619). What remains after balancing the rotating mass is the force due to the acceleration of the piston and conrod masses at point B of Figure 13-23 which are in linear translation along the X axis, as shown by the inertia force $-m_B\mathbf{a}_B$ at point B in that figure.

To completely eliminate this reciprocating unbalanced shaking force would require the introduction of another reciprocating mass, which oscillates 180° out of phase with the piston. Adding a second piston and cylinder, properly arranged, can accomplish this.

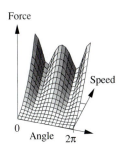

(a) Crankpin inertia force

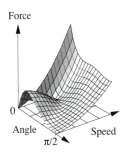

(b) Crankpin total force

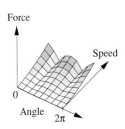

(c) Wrist-pin inertia force

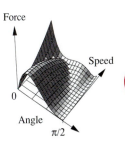

(d) Wrist-pin total force

FIGURE 13-22

Pin force variation

13

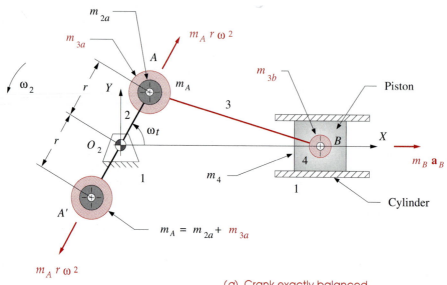

(a) Crank exactly balanced

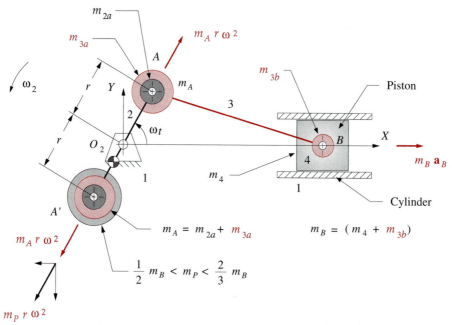

(b) Crank overbalanced

FIGURE 13-23

Balancing and overbalancing the single-cylinder engine

One of the principal advantages of multicylinder engines is their ability to reduce or elim-inate the shaking forces. We will investigate this in the next chapter.

In the single-cylinder engine, there is no way to completely eliminate the recipro-cating unbalance with a single, rotating counterweight, but we can reduce the shaking force still further. Figure 13-23b shows an additional amount of mass m_P added to the counterweight at point A'. (Note that the crank's CG has now moved away from the fixed pivot.) This extra balance mass creates an additional inertia force $(-m_P r \omega^2)$ as is shown, broken into x and y components, in the figure. The y component is not opposed by any other inertia forces present, but the x component will always be opposite to the reciprocating inertia force at point B. Thus this extra mass, m_P, which *overbalances the crank*, will reduce the x directed shaking force at the expense of adding back some y directed shaking force. This is a useful trade-off as the direction of the shaking force is usually of less concern than is its magnitude. Shaking forces create vibrations in the sup-porting structure which are transmitted through it and modified by it. As an example, it is unlikely that you could define the direction of a motorcycle engine's shaking forces by feeling their resultant vibrations in the handlebars. But you **will** detect an increase in the magnitude of the shaking forces from the larger amplitude of vibration they cause in the cycle frame.

The correct amount of additional "overbalance" mass needed to minimize the peak shaking force, regardless of its direction, will vary with the particular engine design. It will usually be between one-half and two-thirds of the reciprocating mass at point B (pis-ton plus conrod at wrist pin), if placed at the crank radius r. Of course, once this mass-radius product is determined, it can be achieved with any combination of mass and radi-us. Figure 13-24b shows the minimum shaking force achieved for this engine with the addition of 65.5% of the mass at B acting at radius r. The shaking force has now been reduced to 1700 lb at 3400 rpm, which is **18% of its original unbalanced value**. The

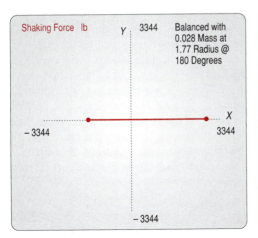

(a) Crank exactly balanced

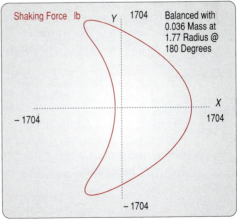

(b) Crank overbalanced

1 Cylinder
4 Stroke Cycle
Bore   = 3.00 in
Stroke = 3.54
B/S   = 0.85
L/R   = 3.50
RPM = 3400

13

FIGURE 13-24

Effects of balancing and overbalancing on the shaking force in the slider-crank linkage

benefits of balancing, and of overbalancing in the case of the single-cylinder engine, should now be obvious.

13.11 DESIGN TRADE-OFFS AND RATIOS

In the design of any system or device, no matter how simple, there will always be conflicting demands, requirements, or desires which must be traded off to achieve the best design compromise. This single-cylinder engine is no exception. There are two dimensionless design ratios which can be used to characterize an engine's dynamic behavior in a general way. The first is the crank/conrod ratio r/l, introduced in Section 13.2 (p. 605), or its inverse, the **conrod/crank ratio** l/r. The second is the **bore/stroke ratio** B/S.

Conrod/Crank Ratio

The crank/conrod ratio r/l appears in all the equations for acceleration, forces, and torques. In general, the smaller the r/l ratio, the smoother will be the acceleration function and thus all other factors which it influences. Program ENGINE uses the inverse of this ratio as an input parameter. The **conrod/crank ratio** l/r must be greater than about two to obtain acceptable transmission angles in the slider-crank linkage. The ideal value for l/r from a kinematic standpoint would be infinity as that would result in the piston acceleration function being a pure harmonic. The second and all subsequent harmonic terms in equations 13.3 (p. 619) would be zero in this event, and the peak value of acceleration would be a minimum. However, an engine this tall would not package very well, and package considerations often dictate the maximum value of the l/r ratio. Most engines will have an l/r ratio between three and five which values give acceptable smoothness in a reasonably short engine.

Bore/Stroke Ratio

The bore B of the cylinder is essentially equal to the diameter of the piston. (There is a small clearance.) The stroke S is defined as the distance travelled by the piston from TDC to BDC and is twice the crank radius, $S = 2r$. The bore appears in the equation for gas force (equation 13.4, p. 610) and thus also affects gas torque. The crank radius appears in every equation. An engine with a B/S ratio of 1 is referred to as a "square" engine. If B/S is larger than 1, it is "oversquare"; if less, "undersquare." The designer's choice of this ratio can have a significant effect on the dynamic behavior of the engine. Assuming that the displacement, or stroke volume V, of the engine has been chosen and is to remain constant, this displacement can be achieved with an infinity of combinations of bore and stroke ranging from a "pancake" piston with tiny stroke to a "pencil" piston with very long stroke.

$$V = \frac{\pi B^2}{4} S \qquad\qquad (13.25)$$

There is a classic design trade-off here between B and S for a constant stroke volume V. A large bore and small stroke will result in high gas forces which will affect pin forces adversely. A large stroke and small bore will result in high inertia forces which will affect pin forces (as well as other forces and torques) adversely. So there should be an optimum value for the B/S ratio in each case, which will minimize these adverse effects. Most production engines have B/S ratios in the range of about 0.75 to 1.5.

It is left as an exercise for the student to investigate the effects of variation in the B/S and l/r ratios on forces and torques in the system. Program ENGINE will demonstrate the effects of changes made independently to each of these ratios, while all other design parameters are held constant. The student is encouraged to experiment with the program to gain insight into the role of these ratios in the dynamic performance of the engine.

Materials

There will always be a strength/weight trade-off. The forces in this device can be quite high, due both to the explosion and to the inertia of the moving elements. We would like to keep the masses of the parts as low as possible as the accelerations are typically very high, as seen in Figure 13-8c (p. 611). But the parts must be strong enough to withstand the forces, so materials with good strength-to-weight ratios are needed. Pistons are usually made of an aluminum alloy, either cast or forged. Conrods are most often cast iron or forged steel, except in very small engines (lawn mower, chain saw, motorcycle) where they may be aluminum alloy. High-performance engines may have titanium connecting rods. Crankshafts are usually cast iron or forged steel, and wrist pins are of hardened steel tubing or rod. Plain bearings of a special soft, nonferrous metal alloy called babbitt are usually used. In the four-stroke engine these are pressure lubricated with oil pumped through drilled passageways in the block, crankshaft, and connecting rods. In the two-stroke engine, the fuel carries the lubricant to these parts. Engine blocks are cast iron or cast aluminum alloy. The chrome-plated steel piston rings seal and wear well against cast iron cylinders. Most aluminum blocks are fitted with cast iron liners around the cylinder bores. Some are unlined and made of a high-silicon aluminum alloy which is specially cooled after casting to precipitate the hard silicon in the cylinder walls for wear resistance.

13.12 BIBLIOGRAPHY

1   **Heywood, J. B.** (1988). *Internal Combustion Engine Fundamentals.* McGraw-Hill: New York.

2   **Taylor, C. F.** (1966). *The Internal Combustion Engine in Theory and Practice.* MIT Press: Cambridge, MA.

13.13 PROBLEMS

*†13-1   A slider-crank linkage has $r = 3$ and $l = 12$. It has an angular velocity of 200 rad/sec at time $t = 0$. Its initial crank angle is zero. Calculate the piston acceleration at $t = 1$ sec. Use two methods, the exact solution, and the approximate Fourier series solution and compare the results.

†13-2   Repeat Problem 13-1 for $r = 4$ and $l = 15$ and $t = 0.9$ sec.

*†13-3   A slider-crank linkage has $r = 3$ and $l = 12$, and a piston bore $B = 2$. The peak gas pressure in the cylinder occurs at a crank angle of 10° and is 1000 pressure units. Calculate the gas force and gas torque at this position.

13

* Answers in Appendix F.

† These problems are suited to solution using *Mathcad, Matlab,* or *TKSolver* equation solver programs.

†13-4   A slider-crank linkage has $r = 4$ and $l = 15$, and a piston bore $B = 3$. The peak gas pressure in the cylinder occurs at a crank angle of 5° and is 600 pressure units. Calculate the gas force and gas torque at this position.

*†13-5   Repeat Problem 13-3 using the exact method of calculation of gas torque and compare its result to that obtained by the approximate expression in equation 13.8b (p. 613). What is the percent error?

†13-6   Repeat Problem 13-4 using the exact method of calculation of gas torque and compare its result to that obtained by the approximate expression in equation 13.8b (p. 613). What is the percent error?

*†13-7   A connecting rod of length $l = 12$ has a mass $m_3 = 0.020$. Its mass moment of inertia is 0.620. Its CG is located at 0.4 l from the crankpin, point A.

   a.   Calculate an exact dynamic model using two lumped masses, one at the wrist pin, point B, and one at whatever other point is required. Define the lumped masses and their locations.
   b.   Calculate an approximate dynamic model using two lumped masses, one at the wrist pin, point B, and one at the crankpin, point A. Define the lumped masses and their locations.
   c.   Calculate the error in the mass moment of inertia of the approximate model as a percentage of the original mass moment of inertia.

†13-8   Repeat Problem 13-7 for these data: $l = 15$, mass $m_3 = 0.025$, mass moment of inertia is 1.020. Its CG is located at 0.25 l from the crankpin, point A.

*†13-9   A crank of length $r = 3.5$ has a mass $m_2 = 0.060$. Its mass moment of inertia about its pivot is 0.300. Its CG is located at 0.30 r from the main pin, point O_2. Calculate a statically equivalent two-lumped mass dynamic model with the lumps placed at the main pin and crankpin. What is the percent error in the model's moment of inertia about the crank pivot.

†13-10   Repeat Problem 13-9 for a crank length $r = 4$, a mass $m_2 = 0.050$, a mass moment of inertia about its pivot of 0.400. Its CG is located at 0.40 r from the main pin, point O_2.

*†13-11   Combine the data from problems 13-7 and 13-9. Run the linkage at a constant 2000 rpm. Calculate the inertia force and inertia torque at $\omega t = 45°$. Piston mass = 0.012.

†13-12   Combine the data from problems 13-7 and 13-10. Run the linkage at a constant 3000 rpm. Calculate the inertia force and inertia torque at $\omega t = 30°$. Piston mass = 0.019.

†13-13   Combine the data from problems 13-8 and 13-9. Run the linkage at a constant 2500 rpm. Calculate the inertia force and inertia torque at $\omega t = 24°$. Piston mass = 0.023.

*†13-14   Combine the data from problems 13-8 and 13-10. Run the linkage at a constant 4000 rpm. Calculate the inertia force and inertia torque at $\omega t = 18°$. Piston mass = 0.015.

†13-15   Combine the data from problems 13-7 and 13-9. Run the linkage at a constant 2000 rpm. Calculate the pin forces at $\omega t = 45°$. Piston mass = 0.022. $F_g = 300$.

†13-16   Combine the data from problems 13-7 and 13-10. Run the linkage at a constant 3000 rpm. Calculate the pin forces at $\omega t = 30°$. Piston mass = 0.019. $F_g = 600$.

13

* Answers in Appendix F.

† These problems are suited to solution using Mathcad, Matlab, or TKSolver equation solver programs.

†13-17 Combine the data from problems 13-8 and 13-9. Run the linkage at a constant 2500 rpm. Calculate the pin forces at $\omega t = 24°$. Piston mass = 0.032. $F_g = 900$.

†13-18 Combine the data from problems 13-8 and 13-10. Run the linkage at a constant 4000 rpm. Calculate the pin forces at $\omega t = 18°$. Piston mass = 0.014. $F_g = 1200$.

*†‡13-19 Using the data from Problem 13-11:

    a. Exactly balance the crank and recalculate the inertia force.
    b. Overbalance the crank with approximately two-thirds of the mass at the wrist pin placed at radius $-r$ on the crank and recalculate the inertia force.
    c. Compare these results to those for the unbalanced crank.

†‡13-20 Repeat Problem 13-19 using the data from Problem 13-12.

†‡13-21 Repeat Problem 13-19 using the data from Problem 13-13.

†‡13-22 Repeat Problem 13-19 using the data from Problem 13-14.

†13-23 Combine the necessary equations to develop expressions that show how each of these dynamic parameters varies as a function of the crank/conrod ratio alone:

    a. Piston acceleration
    b. Inertia force
    c. Inertia torque
    d. Pin forces

Plot the functions. Check your conclusions with program ENGINE.

Hint: *Consider all other parameters to be temporarily constant. Set the crank angle to some value such that the gas force is nonzero.*

†13-24 Combine the necessary equations to develop expressions that show how each of these dynamic parameters varies as a function of the bore/stroke ratio alone:

    a. Gas force
    b. Gas torque
    c. Inertia force
    d. Inertia torque
    e. Pin forces

Plot the functions. Check your conclusions with program ENGINE.

Hint: *Consider all other parameters to be temporarily constant. Set the crank angle to some value such that the gas force is nonzero.*

†13-25 Develop an expression to determine the optimum bore/stroke ratio to minimize the wrist pin force. Plot the function.

†‡13-26 Use program ENGINE, your own computer program, or an equation solver to calculate the maximum value and the polar-plot shape of the force on the main pin of a 1-in³ displacement, single-cylinder engine with bore = 1.12838 in for the following situations.

    a. Piston, conrod and crank masses = 0
    b. Piston mass = 1 blob, conrod and crank masses = 0
    c. Conrod mass = 1 blob, piston and crank masses = 0
    d. Crank mass = 1 blob, conrod and piston masses = 0

* Answers in Appendix F.

† These problems are suited to solution using Mathcad, Matlab, or TKSolver equation solver programs.

‡ These problems are suited to solution using program ENGINE which is on the attached CD-ROM.

13

Place the *CG* of the crank at 0.5 *r* and the conrod at 0.33 *l*. Compare and explain the differences in the main pin force under these different conditions with reference to the governing equations.

†13-27   Repeat Problem 13-26 for the crankpin.

†13-28   Repeat Problem 13-26 for the wrist pin.

†‡13-29   Use program ENGINE, your own computer program, or an equation solver to calculate the maximum value and the polar plot shape of the force on the main pin of a 1 in^3 single-cylinder engine with bore = 1.12838 in for the following situations.

   a.   Engine unbalanced.
   b.   Crank exactly balanced against mass at crankpin.
   c.   Crank optimally overbalanced against masses at crankpin and wrist pin.

Piston, conrod, and crank masses = 1. Place the *CG* of the crank at 0.5 *r* and the conrod at 0.33 *l*. Compare and explain the differences in the main pin force under these conditions with reference to the governing equations.

†‡13-30   Repeat Problem 13-29 for the crankpin force.

†‡13-31   Repeat Problem 13-29 for the wrist pin force.

†‡13-32   Repeat Problem 13-29 for the shaking force.

13.14 PROJECTS

These are loosely structured design problems intended for solution using program EN-GINE. All involve the design of a single-cylinder engine and differ only in the specific data for the engine. The general problem statement is:

Design a single-cylinder engine for a specified displacement and stroke cycle. Optimize the conrod/crank ratio and bore/stroke ratio to minimize shaking forces, shaking torque, and pin forces, also considering package size. Design your link shapes and calculate realistic dynamic parameters (mass, CG location, moment of inertia) for those links using the methods shown in Chapter 11 and Section 12.12. Dynamically model the links as described in this chapter. Balance or overbalance the linkage as needed to achieve these results. Overall smoothness of total torque is desired. Design and size a minimum weight flywheel by the method of Chapter 12 to smooth the total torque. Write an engineering report on your design.

P13-1   Two-stroke cycle with a displacement of 0.125 liters.

P13-2   Four-stroke cycle with a displacement of 0.125 liters.

P13-3   Two-stroke cycle with a displacement of 0.25 liters.

P13-4   Four-stroke cycle with a displacement of 0.25 liters.

P13-5   Two-stroke cycle with a displacement of 0.50 liters.

P13-6   Four-stroke cycle with a displacement of 0.50 liters.

P13-7   Two-stroke cycle with a displacement of 0.75 liters.

P13-8   Four-stroke cycle with a displacement of 0.75 liters.

13

† These problems are suited to solution using *Mathcad, Matlab,* or *TKSolver* equation solver programs.

‡ These problems are suited to solution using program ENGINE which is on the attached CD-ROM.

Chapter 14

MULTICYLINDER ENGINES

Look long on an engine,
it is sweet to the eyes
MACKNIGHT BLACK

14.0 INTRODUCTION

The previous chapter discussed the design of the slider-crank mechanism as used in the single-cylinder internal combustion engine and piston pumps. We will now extend the design to multicylinder configurations. Some of the problems with shaking forces and torques can be alleviated by proper combination of multiple slider-crank linkages on a common crankshaft. Program ENGINE, included with this text, will calculate the equations derived in this chapter and allow the student to exercise many variations of an engine design in a short time. Some examples are provided as disk files to be read into the program. These are noted in the text. The student is encouraged to investigate these examples with program ENGINE in order to develop an understanding of and insight to the subtleties of this topic. A user manual for program ENGINE is provided in Appendix A which can be read or referred to out of sequence, with no loss in continuity, in order to gain familiarity with the program's operation.

As with the single-cylinder case, we will not address the thermodynamic aspects of the internal combustion engine beyond the definition of the combustion forces necessary to drive the device presented in the previous chapter. We will concentrate on the engine's kinematic and mechanical dynamics aspects. It is not our intention to make an "engine designer" of the student so much as to apply dynamic principles to a realistic design problem of general interest and also to convey the complexity and fascination involved in the design of a more complicated dynamic device than the single-cylinder engine.

14

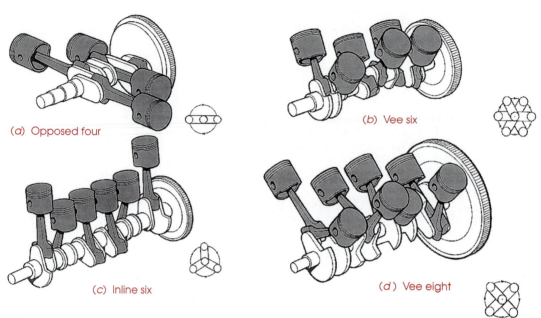

(a) Opposed four

(b) Vee six

(c) Inline six

(d) Vee eight

FIGURE 14-1

Various multicylinder engine configurations
Illustrations copyright Eaglemoss Publications/Car Care Magazine. Reprinted with permission.

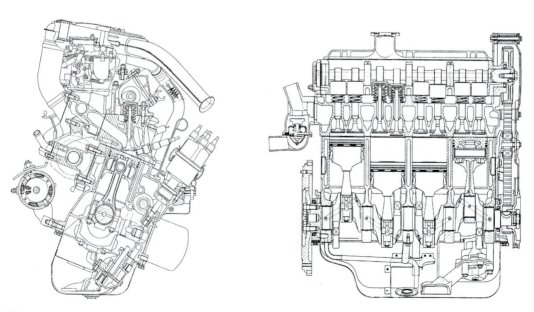

FIGURE 14-2

Cutaway views of a four-stroke, four-cylinder inline engine. *Courtesy of FIAT Corporation, Italy*

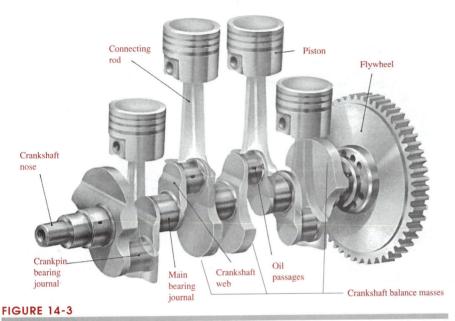

FIGURE 14-3

Crankshaft from an inline four-cylinder engine with pistons, connecting rods, and flywheel
Illustration copyright Eaglemoss Publications/Car Care Magazine. Reprinted with permission.

14.1 MULTICYLINDER ENGINE DESIGNS

Multicylinder engines are designed in a wide variety of configurations from the simple inline arrangement to vee, opposed, and radial arrangements some of which are shown in Figure 14-1. These arrangements may use any of the stroke cycles discussed in Chapter 13, Clerk, Otto, or Diesel.

INLINE ENGINES The most common and simplest arrangement is an inline engine with its cylinders all in a common plane as shown in Figure 14-2. Two-,[*] three-,[*] four, five, and six-cylinder **inline engines** are in common use. Each cylinder will have its individual slider-crank mechanism consisting of a crank, conrod, and piston. The cranks are formed together in a common **crankshaft** as shown in Figure 14-3. Each cylinder's crank on the crankshaft is referred to as a **crank throw**. These crank throws will be arranged with some **phase angle** relationship one to the other, in order to stagger the motions of the pistons in time. It should be apparent from the discussion of shaking forces and balancing in the previous chapter that we would like to have pistons moving in opposite directions to one another at the same time in order to cancel the reciprocating inertial forces. The optimum phase angle relationships between the crank throws will differ depending on the number of cylinders and the stroke cycle of the engine. There will usually be one (or a small number of) viable crank throw arrangements for a given engine configuration to accomplish this goal. The engine in Figure 14-2 is a four-stroke cycle, four-cylinder, inline engine with its crank throws at 0, 180, 180, and 0° phase angles which we will soon see are optimum for this engine. Figure 14-3 shows the crankshaft, connecting rods and pistons for the same design of engine as in Figure 14-2.

14

[*] Mainly in motorcycles and boats.

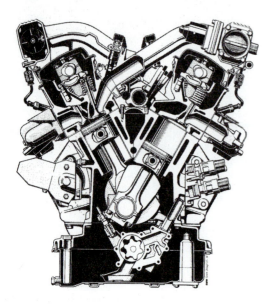

Engine power and torque curves

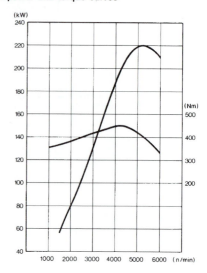

FIGURE 14-4

Cross section of a BMW 5-liter V-12 engine and its power and torque curves. *Courtesy of BMW of North America Inc.*

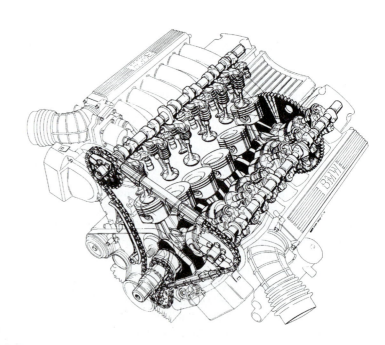

FIGURE 14-5

Cutaway view of a 5-liter BMW V-12 engine. *Courtesy of BMW of North America Inc.*

14

VEE ENGINES in two-,[*] four-,[*] six-, eight-, ten-,[†] and twelve-cylinder[‡] versions are produced, with vee six and vee eight being the most common configurations. Figure 14-4 shows a cross section and Figure 14-5 a cutaway of a 60° vee -twelve engine. **Vee engines** can be thought of as *two inline engines grafted together onto a common crankshaft*. The two "inline" portions, or **banks**, are arranged with some **vee angle** between them. Figure 14-1d shows a vee-eight engine. Its crank throws are at 0, 90, 270, and 180° respectively. A vee eight's vee angle is 90°. The geometric arrangements of the crankshaft (phase angles) and cylinders (vee angle) have a significant effect on the dynamic condition of the engine. We will soon explore these relationships in detail.

OPPOSED ENGINES are essentially vee engines with a vee angle of 180°. The pistons in each bank are on opposite sides of the crankshaft as shown in Figure 14-6. This arrangement promotes cancellation of inertial forces and is popular in aircraft engines.[§] It has also been used in some automotive applications.[‖]

RADIAL ENGINES have their cylinders arranged radially around the crankshaft in nearly a common plane. These were common on World War II vintage aircraft as they allowed large displacements, and thus high power, in a compact form whose shape was well suited to that of an airplane. Typically air cooled, the cylinder arrangement allowed good exposure of all cylinders to the airstream. Large versions had multiple rows of radial cylinders, rotationally staggered to allow cooling air to reach the back rows. The gas turbine jet engine has rendered these radial aircraft engines obsolete.

ROTARY ENGINES were an interesting variant on the aircraft radial engine. Similar in appearance and cylinder arrangement to the radial engine, the anomaly was that the crankshaft was the stationary ground plane. The propeller was attached to the crankcase (block) which rotated around the crankshaft! It is a kinematic inversion of the radial engine. At least it didn't need a flywheel.

Many other configurations of engines have been tried over the century of development of this ubiquitous device. The bibliography at the end of this chapter contains several references which describe other engine designs, the usual, unusual, and exotic. We will begin our detailed exploration of multicylinder engine design with the simplest configuration, the inline engine, and then progress to the vee and opposed versions.

[*] Mainly in motorcycles, and boats.

[†] Honda, Chrysler.

[‡] BMW, Jaguar, Mercedes.

[§] Continental six-cylinder aircraft engine.

[‖] Original VW "Beetle" four cylinder, Subaru four, Honda motorcycle six, Ferrari twelve, Porsche six, the ill-fated Corvair six, and the short-lived Tucker (Continental) six, among others.

14

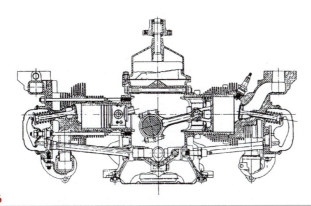

FIGURE 14-6

Chevrolet Corvair horizontally opposed six-cylinder engine
Courtesy of Chevrolet Division, General Motors Corp.

14.2   THE CRANK PHASE DIAGRAM

Fundamental to the design of any multicylinder engine (or piston pump) is the arrangement of crank throws on the crankshaft. We will use the four-cylinder inline engine as an example. Many choices are possible for the crank phase angles in the four-cylinder engine. We will start, for example, with the one that seems most obvious from a commonsense standpoint. There are 360° in any crankshaft. We have four cylinders, so an arrangement of 0, 90, 180, and 270° seems appropriate. The **delta phase angle** between throws is then 90°. In general, for maximum cancellation of inertia forces, which have a period of one revolution, the optimum delta phase angle will be:

$$\Delta\phi_{\text{inertia}} = \frac{360°}{n} \qquad (14.1)$$

where n is the number of cylinders.

We must establish some convention for the measurement of these phase angles which will be:

1   The first (front) cylinder will be number 1 and its phase angle will always be zero. It is the reference cylinder for all others.

2   The phase angles of all other cylinders will be measured with respect to the crank throw for cylinder 1.

3   Phase angles are measured internal to the crankshaft, that is, with respect to a rotating coordinate system embedded in the first crank throw.

4   Cylinders will be numbered consecutively from front to back of the engine.

The phase angles are defined in a **crank phase diagram** as shown in Figure 14-7 for a four-cylinder, inline engine. Figure 14-7a shows the crankshaft with the throws numbered clockwise around the axis. The shaft is rotating counterclockwise. The pistons are oscillating horizontally in this diagram, along the x axis. Cylinder 1 is shown with its piston at top dead center (TDC). Taking that position as the starting point for the abscissas (thus time zero) in Figure 14-7b, we plot the velocity of each piston for two revolutions of the crank (to accommodate one complete four-stroke cycle). Piston 2 arrives at TDC 90° after piston 1 has left. Thus we say that cylinder 2 lags cylinder 1 by 90 degrees. By convention a *lagging event is defined as having a negative phase angle*, shown by the clockwise numbering of the crank throws. The velocity plots clearly show that each cylinder arrives at TDC (zero velocity) 90° later than the one before it. Negative velocity on the plots in Figure 14-7b indicates piston motion to the left (down stroke) in Figure 14-7a; positive velocity indicates motion to the right (up stroke).

For the discussion in this chapter we will assume counterclockwise rotation of all crankshafts, and all phase angles will thus be negative. We will, however, omit the negative signs on the listings of phase angles with the understanding that they follow this convention.

Figure 14-7 shows the timing of events in the cycle and is a necessary and useful aid in defining our crankshaft design. However, it is not necessary to go to the trouble of drawing the correct sinusoidal shapes of the velocity plots to obtain the needed information. All that is needed is a schematic indication of the relative positions within the cy-

14

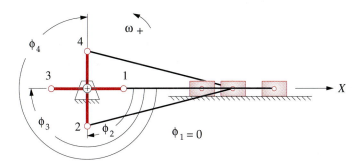

(a) Crankshaft phase angles

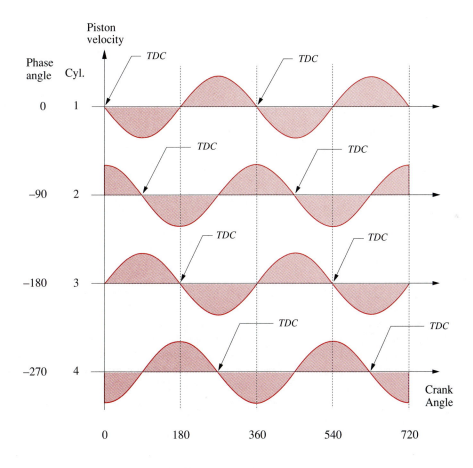

(b) The crank phase diagram

FIGURE 14-7

Crank phase angles and the phase diagram

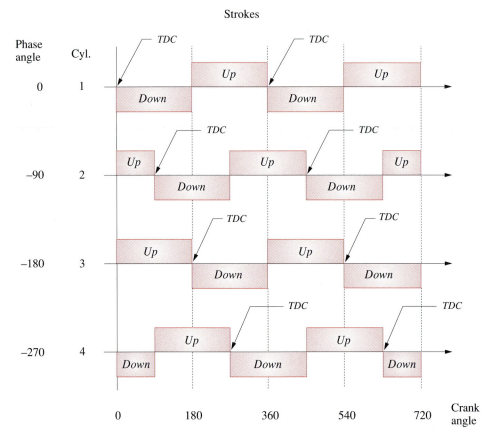

FIGURE 14-8

The schematic crank phase diagram

cle of the ups and downs of the various cylinders. This same information is conveyed by the simplified crank phase diagram shown in Figure 14-8. Here the piston motions are represented by rectangular blocks with a negative block arbitrarily used to denote a piston downstroke and a positive one a piston upstroke. It is strictly schematic. The positive and negative values of the blocks imply nothing more than that stated. Such a schematic crank phase diagram can (and should) be drawn for any proposed arrangement of crankshaft phase angles. To draw it, simply shift each cylinder's blocks to the right by its phase angle with respect to the first cylinder.

14.3   SHAKING FORCES IN INLINE ENGINES

We want to determine the overall shaking force which results from our chosen crankshaft phase angle arrangement. The individual cylinders will each contribute to the total shaking force. We can superpose their effects, taking their phase shifts into account. Equation 13.14e (p. 619) defined the shaking force for one cylinder with the crankshaft rotating at constant ω.

for $\alpha = 0$:

$$\mathbf{F_s} \cong \left[m_A\, r\omega^2 \cos\omega t + m_B r\omega^2 \left(\cos\omega t + \frac{r}{l}\cos 2\omega t \right) \right] \hat{\mathbf{i}} + \left[m_A r\omega^2 \sin\omega t \right] \hat{\mathbf{j}} \qquad (14.2a)$$

This expression is for an unbalanced crank. In multicylinder engines each crank throw on the crankshaft is at least counterweighted to eliminate the shaking force effects of the combined mass m_A of crank and conrod assumed concentrated at the crankpin. (See Section 13.10 and equation 13.24, p. 631.) Sometimes the crank throws in a multicylinder engine are also overbalanced, although to a lesser extent than for a one-cylinder engine.[*] The need for overbalancing is less if the crankshaft phase angles are arranged to cancel the effects of the reciprocating masses at the wrist pins. This is possible except in some two-cylinder, four-stroke, inline engines.

If we provide balance masses with an mr product equal to $m_A r_A$ on each crank throw as shown in Figure 14-3 (p. 641), the terms in equation 14.2a which include m_A will be eliminated, reducing it to:

$$\mathbf{F_s} \cong m_B r\omega^2 \left(\cos\omega t + \frac{r}{l}\cos 2\omega t \right) \hat{\mathbf{i}} \qquad (14.2b)$$

Recall that these are approximate expressions which exclude all harmonics above the second.

We will assume that all cylinders in the engine are of equal displacement and that all pistons and all conrods are interchangeable. This is desirable both for dynamic balance and for lower production costs. If we let the crank angle ωt represent the instantaneous position of the reference crank throw for cylinder 1, the corresponding positions of the other cranks can be defined by their phase angles as shown in Figure 14-7 (p. 645). The total shaking force for a multicylinder inline engine is then:

$$\mathbf{F_s} \cong m_B r\omega^2 \sum_{i=1}^{n} \left[\cos(\omega t - \phi_i) + \frac{r}{l}\cos 2(\omega t - \phi_i) \right] \hat{\mathbf{i}} \qquad (14.2c)$$

where n = number of cylinders and $\phi_1 = 0$. Substitute the identity:

$$\cos(a - b) = \cos a \cos b + \sin a \sin b$$

and factor to:

$$\mathbf{F_s} \cong m_B r\omega^2 \left[\begin{array}{c} \cos\omega t \sum\limits_{i=1}^{n} \cos\phi_i + \sin\omega t \sum\limits_{i=1}^{n} \sin\phi_i \\[2ex] + \frac{r}{l}\left(\cos 2\omega t \sum\limits_{i=1}^{n} \cos 2\phi_i + \sin 2\omega t \sum\limits_{i=1}^{n} \sin 2\phi_i \right) \end{array} \right] \hat{\mathbf{i}} \qquad (14.2d)$$

The ideal value for the shaking force is zero. This expression can only be zero for all values of ωt if:

14

[*] A 90° vee-eight engine typically has approximately $m_B/2$ of overbalance mass added per crank throw.

$$\sum_{i=1}^{n} \cos\phi_i = 0 \qquad\qquad \sum_{i=1}^{n} \sin\phi_i = 0 \qquad\qquad (14.3a)$$

$$\sum_{i=1}^{n} \cos 2\phi_i = 0 \qquad\qquad \sum_{i=1}^{n} \sin 2\phi_i = 0 \qquad\qquad (14.3b)$$

Thus, there are some combinations of phase angles ϕ_i which will cause cancellation of the shaking force through the second harmonic. If we wish to cancel higher harmonics, we could replace those harmonics' terms truncated from the Fourier series representation and find that the fourth and sixth harmonics will be cancelled if:

$$\sum_{i=1}^{n} \cos 4\phi_i = 0 \qquad\qquad \sum_{i=1}^{n} \sin 4\phi_i = 0 \qquad\qquad (14.3c)$$

$$\sum_{i=1}^{n} \cos 6\phi_i = 0 \qquad\qquad \sum_{i=1}^{n} \sin 6\phi_i = 0 \qquad\qquad (14.3d)$$

Equations 14.3 provide us with a convenient predictor of the shaking force behavior of any proposed inline engine design. Program ENGINE calculates equations 14.3 and displays a table of their values. Note that **both the sine and cosine summations of any multiple of the phase angles must be zero for that harmonic of the shaking force to be zero**. The calculation for our example of a four-cylinder engine with phase angles of $\phi_1 = 0$, $\phi_2 = 90$, $\phi_3 = 180$, $\phi_4 = 270°$ in Table 14-1 shows that the shaking forces are zero for the first, second, and sixth harmonics and nonzero for the fourth. So, our common-sense choice in this instance has proven a good one as far as shaking forces are concerned. The coefficients of the fourth and sixth harmonic terms are minuscule, so their contributions, if any, can be ignored. The primary component is of most concern, because of its potential magnitude. The secondary (second harmonic) term is less critical than the primary as it is multiplied by r/l which is generally less than 1/3. An unbalanced secondary harmonic of shaking force is undesirable but can be lived with, especially if the engine is of small displacement (less than about 1/2 liter per cylinder).

TABLE 14-1   Force Balance State of a 4-Cylinder Inline Engine with a 0, 90, 180, 270° Crankshaft

Primary forces:	$\displaystyle\sum_{i=1}^{n} \sin\phi_i = 0$	$\displaystyle\sum_{i=1}^{n} \cos\phi_i = 0$
Secondary forces:	$\displaystyle\sum_{i=1}^{n} \sin 2\phi_i = 0$	$\displaystyle\sum_{i=1}^{n} \cos 2\phi_i = 0$
Fourth harmonic forces:	$\displaystyle\sum_{i=1}^{n} \sin 4\phi_i = 0$	$\displaystyle\sum_{i=1}^{n} \cos 4\phi_i = 4$
Sixth harmonic forces:	$\displaystyle\sum_{i=1}^{n} \sin 6\phi_i = 0$	$\displaystyle\sum_{i=1}^{n} \cos 6\phi_i = 0$

14

To see more detail on the results of this 0, 90, 180, 270 inline four-cylinder engine configuration, run program ENGINE, select the configuration from the *Example* pull-down menu, and then *Plot* the shaking force. See Appendix A for more detailed instructions on the use of program ENGINE.

14.4   INERTIA TORQUE IN INLINE ENGINES

The inertia torque for a single-cylinder engine was defined in Section 13.6 and equation 13.15e (p. 612). We are concerned with reducing this inertia torque, preferably to zero, because it adds to the gas torque to form the total torque. (See Section 13.7, p. 622.) Inertia torque adds nothing to the net driving torque as its average value is always zero, but it does create large oscillations in the total torque which detracts from its smoothness. Inertia torque oscillations can be masked to a degree with the addition of sufficient flywheel to the system, or their external, net effect can be cancelled by the proper choice of phase angles. However, the torque oscillations, even if hidden from the outside observer, or made to sum to zero, are still present within the crankshaft and can lead to torsional fatigue failure if the part is not properly designed. (See also Figure 14-22, p. 671.) The approximate one-cylinder inertia torque equation for three harmonics is:

$$\mathbf{T}_{i21} \cong \frac{1}{2} m_B r^2 \omega^2 \left(\frac{r}{2l} \sin \omega t - \sin 2\omega t - \frac{3r}{2l} \sin 3\omega t \right) \hat{\mathbf{k}} \qquad (14.4a)$$

Summing for all cylinders and including their phase angles:

$$\mathbf{T}_{i21} \cong \frac{1}{2} m_B r^2 \omega^2 \sum_{i=1}^{n} \left[\frac{r}{2l} \sin(\omega t - \phi_i) - \sin 2(\omega t - \phi_i) - \frac{3r}{2l} \sin 3(\omega t - \phi_i) \right] \hat{\mathbf{k}} \qquad (14.4b)$$

Substitute the identity:

$$\sin(a - b) = \sin a \cos b - \cos a \sin b$$

and factor to:

$$\mathbf{T}_{i21} \cong \frac{1}{2} m_B r^2 \omega^2 \begin{bmatrix} \dfrac{r}{2l} \left(\sin \omega t \displaystyle\sum_{i=1}^{n} \cos \phi_i - \cos \omega t \displaystyle\sum_{i=1}^{n} \sin \phi_i \right) \\[2ex] - \left(\sin 2\omega t \displaystyle\sum_{i=1}^{n} \cos 2\phi_i - \cos 2\omega t \displaystyle\sum_{i=1}^{n} \sin 2\phi_i \right) \\[2ex] - \dfrac{3r}{2l} \left(\sin 3\omega t \displaystyle\sum_{i=1}^{n} \cos 3\phi_i - \cos 3\omega t \displaystyle\sum_{i=1}^{n} \sin 3\phi_i \right) \end{bmatrix} \hat{\mathbf{k}} \qquad (14.4c)$$

This can only be zero for all values of ωt if:

14

$$\sum_{i=1}^{n} \sin \phi_i = 0 \qquad\qquad \sum_{i=1}^{n} \cos \phi_i = 0 \qquad (14.5a)$$

$$\sum_{i=1}^{n} \sin 2\phi_i = 0 \qquad\qquad \sum_{i=1}^{n} \cos 2\phi_i = 0 \qquad (14.5b)$$

$$\sum_{i=1}^{n} \sin 3\phi_i = 0 \qquad\qquad \sum_{i=1}^{n} \cos 3\phi_i = 0 \qquad (14.5c)$$

Equations 14.5 provide us with a convenient predictor of the inertia torque behavior of any proposed inline engine design. Calculation for our example of a four-cylinder engine with phase angles of $\phi_1 = 0$, $\phi_2 = 90$, $\phi_3 = 180$, $\phi_4 = 270°$ shows that the inertia torque components are zero for the first, second, and third harmonics. So, our current example is a good one for inertia torques as well.

14.5   SHAKING MOMENT IN INLINE ENGINES

We were able to consider the single-cylinder engine to be a single plane, or two dimensional, device and thus could statically balance it. The multicylinder engine is three dimensional. Its multiple cylinders are distributed along the axis of the crankshaft. Even though we may have cancellation of the shaking forces, there may still be unbalanced moments in the plane of the engine block. We need to apply the criteria for dynamic balance. (See Section 12.2 and equation 12.3, p. 574.) Figure 14-9 shows a schematic of an inline four-cylinder engine with crank phase angles of $\phi_1 = 0$, $\phi_2 = 90$, $\phi_3 = 180$, $\phi_4 = 270°$. The spacing between the cylinders is normally uniform. We can sum moments in the plane of the cylinders about any convenient point such as one end of the crankshaft,

$$\sum M_L = \sum_{i=1}^{n} z_i \mathbf{F}_{\mathbf{s}_i} \, \hat{\mathbf{j}} \qquad (14.6a)$$

where $\mathbf{F}_{si}$ is the shaking force and z_i is the moment arm of the ith cylinder. Substituting equation 14.2d for $\mathbf{F}_{si}$:

$$\sum M_L \cong m_B r \omega^2 \left[\begin{array}{l} \cos \omega t \sum_{i=1}^{n} z_i \cos \phi_i + \sin \omega t \sum_{i=1}^{n} z_i \sin \phi_i \\[2mm] + \dfrac{r}{l}\left(\cos 2\omega t \sum_{i=1}^{n} z_i \cos 2\phi_i + \sin 2\omega t \sum_{i=1}^{n} z_i \sin 2\phi_i \right) \end{array} \right] \hat{\mathbf{j}} \qquad (14.6b)$$

This expression can only be zero for all values of ωt if:

$$\sum_{i=1}^{n} z_i \cos \phi_i = 0 \qquad\qquad \sum_{i=1}^{n} z_i \sin \phi_i = 0 \qquad (14.7a)$$

$$\sum_{i=1}^{n} z_i \cos 2\phi_i = 0 \qquad\qquad \sum_{i=1}^{n} z_i \sin 2\phi_i = 0 \qquad (14.7b)$$

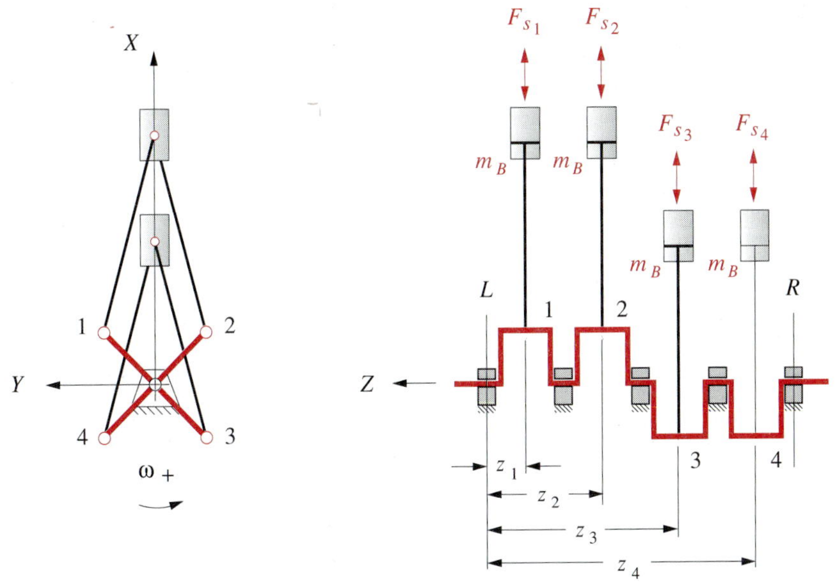

FIGURE 14-9

Moment arms of the shaking moment

These will guarantee no shaking moments through the second harmonic. We can extend this to higher harmonics as we did for the shaking force.

$$\sum_{i=1}^{n} z_i \cos 4\phi_i = 0 \qquad \sum_{i=1}^{n} z_i \sin 4\phi_i = 0 \qquad (14.7c)$$

$$\sum_{i=1}^{n} z_i \cos 6\phi_i = 0 \qquad \sum_{i=1}^{n} z_i \sin 6\phi_i = 0 \qquad (14.7d)$$

Note that both the sine and cosine summations of any multiple of the phase angles must be zero for that harmonic of the shaking moment to be zero. The calculation for our example of a four-cylinder engine with phase angles of $\phi_1 = 0$, $\phi_2 = 90$, $\phi_3 = 180$, $\phi_4 = 270°$, and an assumed cylinder spacing of one length unit ($z_1 = 1$) in Table 14-2, shows that the shaking moments are not zero for any of these harmonics. So, our choice of phase angles which is a good one for shaking forces and torques fails the test for zero shaking moments. The coefficients of the fourth and sixth harmonic terms in the moment equations are minuscule, so they can be ignored. The secondary (second harmonic) term is less critical than the primary as it is multiplied by r/l which is generally less than 1/3. An unbalanced secondary harmonic of shaking moment is undesirable but can be tolerated, especially if the engine is of small displacement (less than about 1/2 liter per cylinder). The primary component is of most concern, because of its magnitude. If we wish to use this crankshaft configuration, we will need to apply a balancing technique to the engine as described in a later section to at least eliminate the primary moment. A large shaking moment is undesirable as it will cause the engine to **pitch** forward and back

14

TABLE 14-2   Moment Balance State of a 4-Cylinder, Inline Engine with a 0, 90, 180, 270° Crankshaft, and $z_1 = 1$, $z_2 = 2$, $z_3 = 3$, $z_4 = 4$

Primary moments:	$\sum_{i=1}^{n} z_i \sin \phi_i = -2$	$\sum_{i=1}^{n} z_i \cos \phi_i = -2$
Secondary moments:	$\sum_{i=1}^{n} z_i \sin 2\phi_i = 0$	$\sum_{i=1}^{n} z_i \cos 2\phi_i = -2$
Fourth harmonic moments:	$\sum_{i=1}^{n} z_i \sin 4\phi_i = 0$	$\sum_{i=1}^{n} z_i \cos 4\phi_i = 6$
Sixth harmonic moments:	$\sum_{i=1}^{n} z_i \sin 6\phi_i = 0$	$\sum_{i=1}^{n} z_i \cos 6\phi_i = -2$

(like a bucking bronco) as the moment oscillates from positive to negative in the plane of the cylinders. *Do not confuse this shaking moment with the shaking torque* which acts to **roll** the engine back and forth about the axis of the crankshaft.

Figure 14-10 shows the primary and secondary components of the shaking moment for this example engine for two revolutions of the crank. Each is a pure harmonic of zero average value. The total moment is the sum of these two components. Use example engine #6 in program ENGINE to view these functions. See Appendix A for help with the program.

14.6   EVEN FIRING

The inertial forces, torques, and moments are only one set of criteria which need to be considered in the design of multicylinder engines. Gas force and gas torque considerations are equally important. In general, it is desirable to create a firing pattern among the cylinders that is evenly spaced in time. If the cylinders fire unevenly, vibrations will be created which may be unacceptable. Smoothness of the power pulses is desired. The power pulses depend on the stroke cycle. If the engine is a two-stroke, there will be one power pulse per revolution in each of its n cylinders. The optimum delta phase angle between the cylinders' crank throws for evenly spaced power pulses will then be:

$$\Delta\phi_{two\ stroke} = \frac{360°}{n} \qquad (14.8a)$$

For a four-stroke engine there will be one power pulse in each cylinder every two revolutions. The optimum delta phase angle of the crank throws for evenly spaced power pulses will then be:

$$\Delta\phi_{four\ stroke} = \frac{720°}{n} \qquad (14.8b)$$

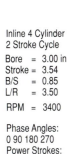

Inline 4 Cylinder
2 Stroke Cycle

Bore = 3.00 in
Stroke = 3.54
B/S   = 0.85
L/R   = 3.50
RPM = 3400

Phase Angles:
0 90 180 270
Power Strokes:
0 90 180 270

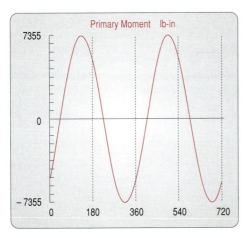

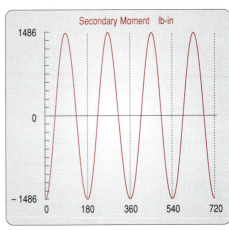

FIGURE 14-10

Primary and secondary moments in the 0, 90, 180, 270° crankshaft four-cylinder engine

Compare equations 14.8a and 14.8b to equation 14.1 (p. 644) which defined the optimum delta phase angle for cancellation of inertia forces. A two-stroke engine can have both even firing and inertia balance, but a four-stroke engine has a conflict between these two criteria. Thus some design trade-offs will be necessary to obtain the best compromise between these factors in the four-stroke case.

Two-Stroke Cycle Engine

To determine the firing pattern of an engine design we must return to the crank phase diagram. Figure 14-11 reproduces Figure 14-8 and adds new information to it. It shows the power pulses for a **two-stroke cycle, four-cylinder engine** with the $\phi_i = 0, 90, 180, 270°$ phase angle crank configuration. Note that each cylinder's negative block in Figure 14-11 is shifted to the right by its phase angle with respect to the reference cylinder 1. In this schematic representation, only the negative blocks on the diagram are available for power pulses as they represent the downstroke of the piston. By convention, cylinder one fires first, so its negative block at 0° is labeled **Power**. The other cylinders may be fired in any order, but their power pulses should be as evenly spaced as possible across the interval.

The available power pulse spacings are dictated by the crank phase angles. There may be more than one firing order which will give even firing, especially with large numbers of cylinders. In this simple example the firing order 1, 2, 3, 4 will work as it will provide successive power pulses every 90° across the interval. The **power stroke angles** ψ_i are the angles in the cycle at which the cylinders fire. They are defined by the crankshaft phase angles and the choice of firing order in combination and in this example are $\psi_i = 0, 90, 180,$ and $270°$. In general, ψ_i are not equal to ϕ_i. Their correspondence with the phase angles in this example results from choosing the consecutive firing order 1, 2, 3, 4.

14

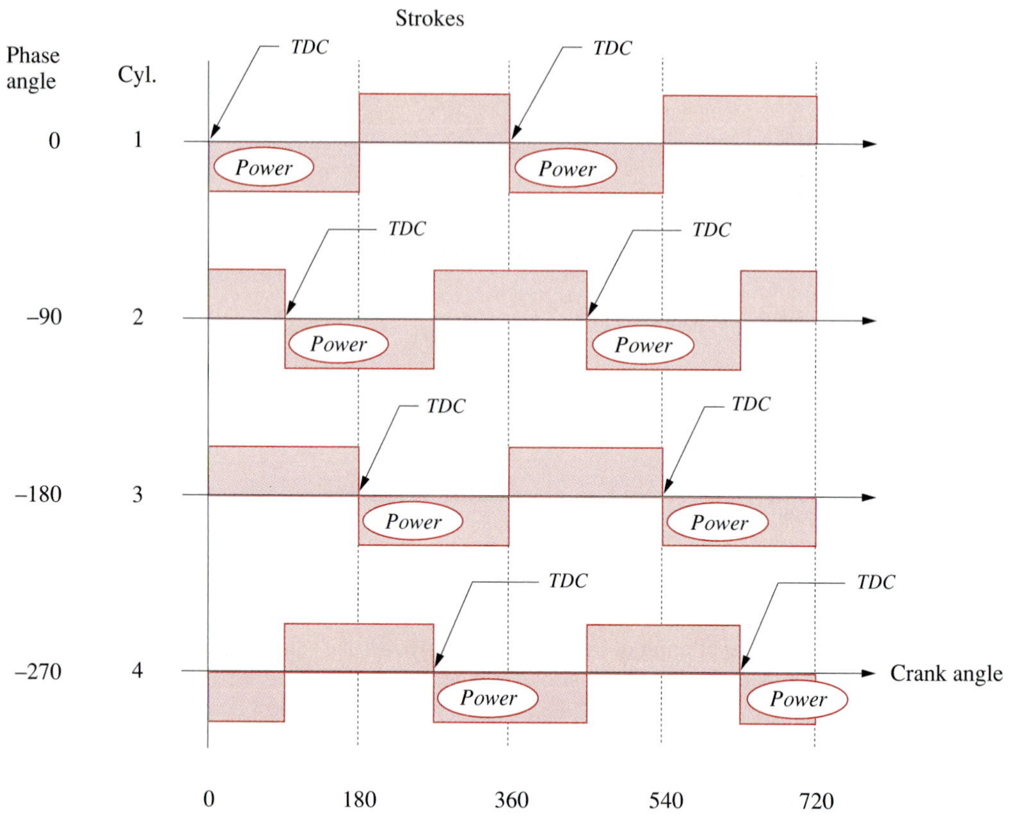

FIGURE 14-11

Two-stroke inline four-cylinder engine crank phase diagram with 0, 90, 180, 270° crankshaft phase angles

For a **two-stroke engine**, the power stroke angles ψ_i must be *between* 0 *and* 360°. We always want them to be evenly spaced in that interval with a delta power stroke angle defined by equation 14.8c. For our four-cylinder, two-stroke engine, ideal power stroke angles are then $\psi_i = 0, 90, 180, 270°$, which we have achieved in this example.

We define the **delta power stroke angle** differently for each stroke cycle. For the two-stroke engine:

$$\Delta\psi_{two\ stroke} = \frac{360°}{n} \qquad (14.8c)$$

For the four-stroke engine:

$$\Delta\psi_{four\ stroke} = \frac{720°}{n} \qquad (14.8d)$$

The gas torque for a one-cylinder engine was defined in equation 13.8b (p. 613). The combined gas torque for all cylinders must sum the contributions of n cylinders, each phase-shifted by its power stroke angle ψ_i:

$$\mathbf{T}_{g21} \cong F_g r \sum_{i=1}^{n} \left\{ \sin(\omega t - \psi_i) \left[1 + \frac{r}{l} \cos(\omega t - \psi_i) \right] \right\} \hat{\mathbf{k}} \qquad (14.9)$$

Figure 14-12 shows the gas torque, inertia torque, and shaking force for this two-stroke four-cylinder engine plotted from program ENGINE. The shaking moment components are shown in Figure 14-10 (p. 653). Except for the unbalanced shaking moments, this design is otherwise acceptable. The inertia force and inertia torque are both zero which is ideal. The gas torque consists of uniformly shaped and spaced pulses across the interval, four per revolution. Note that program ENGINE plots two full revolutions to accommodate the four-stroke case; thus eight power pulses are seen. Open the file F14-12.eng in the program to exercise this example.

Four-Stroke Cycle Engine

Figure 14-13 shows a crank phase diagram for the *same crankshaft design* as in Figure 14-11 except that it is designed as a *four-stroke cycle engine*. There is now only one power stroke every 720° for each cylinder. The second negative block for each cylinder must be used for the intake stroke. Cylinder 1 is again fired first. An evenly spaced pattern of power pulses among the other cylinders is again desired but is now not possible with this crankshaft. Whether the firing order is 1, 3, 4, 2 or 1, 2, 4, 3 or 1, 4, 2, 3, or any other chosen, there will be both gaps and overlaps in the power pulses. The first firing order listed, 1, 3, 4, 2, has been chosen for this example. This results in the set of power stroke angles $\psi_i = 0, 180, 270, 450°$. These **power stroke angles** define the points in the **720° cycle** where each cylinder fires. Thus for a four-stroke engine, the power stroke angles ψ_i must be between 0 and 720°. We would like them to be evenly spaced in that interval with a delta angle defined by equation 14.8d. For our four-cylinder, four-stroke engine, the ideal power stroke angles would then be $\psi_i = 0, 180, 360, 540°$. We clearly have not achieved them in this example. Figure 14-14 shows the resulting gas torque. Open the file F14-14.eng in program ENGINE to exercise this example.

The uneven firing in Figure 14-14 is obvious. This uneven gas torque will be perceived by the operator of any vehicle containing this engine as rough running and vibration, especially at idle speed. At higher engine speeds the flywheel will tend to mask this roughness, but flywheels are ineffective at low speeds. It is this fact that causes most engine designers to *favor even firing over elimination of inertia effects* in their selection of crankshaft phase angles. The inertia force, torque, and moment are all functions of engine speed squared. But, as engine speed increases the magnitude of these factors, the same speed is also increasing the flywheel's ability to mask their effects. Not so with gas-torque roughness due to uneven firing. It is bad at all speeds and the flywheel won't hide it at low speed.

We therefore must reject this crankshaft design for our four-stroke, four-cylinder engine. Equation 14.8b (p. 652) indicates that we need a delta phase angle $\Delta\phi_i = 180°$ in our crankshaft to obtain even firing. We need four crank throws, and all crank phase angles must be less than 360°. So, we must repeat some angles if we use a delta phase angle of 180°. One possibility is $\phi_i = 0, 180, 0, 180°$ for the four crank throws.[*] The crank phase diagram for this design is shown in Figure 14-15. The power strokes can

14

---

[*] Note that 0, 180, 360, 540, modulo 360 is the same as 0, 180, 0, 180.

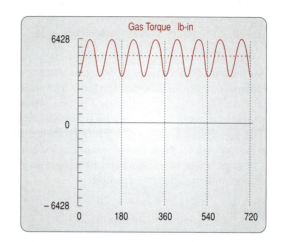

(a)

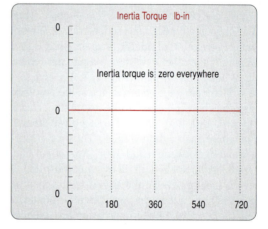

(b)

Inline 4 Cylinder
2 Stroke Cycle

Bore = 3.00 in
Stroke = 3.54
B/S = 0.85
L/R = 3.50

RPM = 3400

Phase Angles:
0 90 180 270
Power Strokes:
0 90 180 270

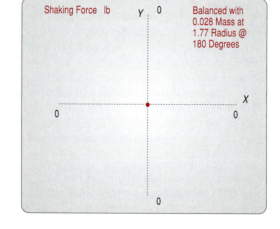

(c)

FIGURE 14-12

Torque and shaking force in the two-stroke four-cylinder inline engine

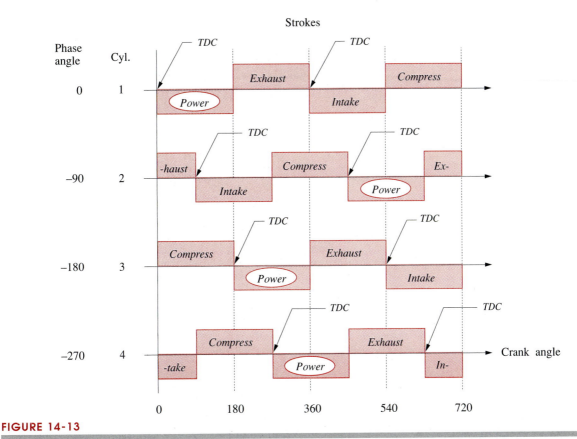

FIGURE 14-13

Four-stroke inline four-cylinder engine crank phase diagram with a 0, 90, 180, 270° crankshaft phase angles

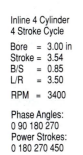

Inline 4 Cylinder
4 Stroke Cycle

Bore = 3.00 in
Stroke = 3.54
B/S = 0.85
L/R = 3.50

RPM = 3400

Phase Angles:
0 90 180 270
Power Strokes:
0 180 270 450

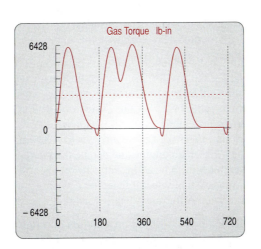

FIGURE 14-14

Uneven firing four-stroke, four-cylinder inline engine with a 0, 90, 180, 270° crankshaft

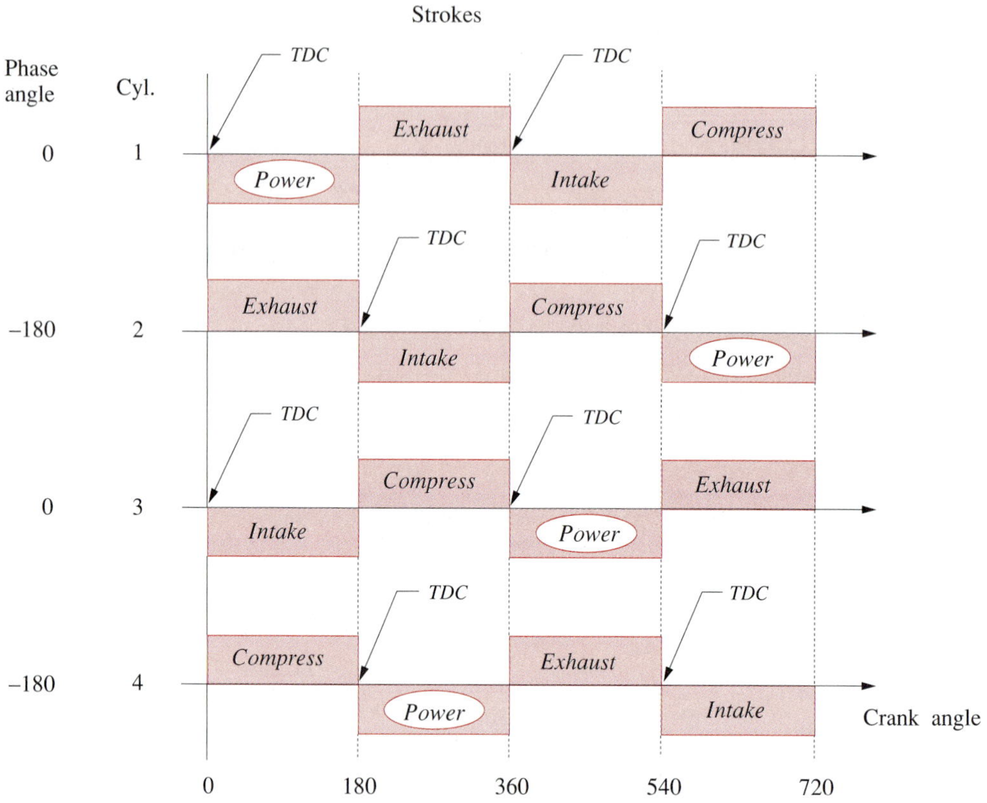

FIGURE 14-15

Even firing four-stroke, four-cylinder engine crank phase diagram with 0, 180, 0, 180° crankshaft phase angles

now be evenly spaced over 720°. A firing order of 1, 4, 3, 2 has been chosen which gives the desired sequence of power stroke angles, $\psi_i = 0, 180, 360, 540°$. (Note that a firing order of 1, 2, 3, 4 would also work with this engine.)[*]

The inertial balance condition of this design must now be checked with equations 14.3, 14.5 and 14.7 (pp. 648-650). These show that the primary inertia force is zero, but the primary moment, secondary force, secondary moment, and inertia torque are all non-zero as shown in Table 14-3. So, this even-firing design has compromised the very good state of inertia balance of the previous design in order to achieve even firing. The inertia torque variations can be masked by a flywheel. The secondary forces and moments are relatively small in a small engine and can be tolerated. The nonzero primary moment is a problem which needs to be addressed. To see the results of this engine configuration, run program ENGINE and select it from the *Example* pull-down menu. Then plot the results. See Appendix A for more detailed instructions on the use of program ENGINE.

We shall soon discuss ways to counter an unbalanced moment with the addition of balance shafts, but there is a more direct approach available in this example. Figure 14-16 shows that the shaking moment is due to the action of the individual cylinders' inertial forces acting at moment arms about some center. If we consider that center to be

[*] Note the pattern of acceptable firing orders (FO). Write two revolutions worth of any acceptable FO, as in 1, 4, 3, 2, 1, 4, 3, 2. Any set of four successive numbers in this sequence, *either forward or backward*, is an acceptable FO. If we require the first to be cylinder 1, then the only other possibility here is the backward set 1, 2, 3, 4.

TABLE 14-3   Force and Moment Balance State of a 4-Cylinder, Inline Engine with a 0, 180, 0, 180° Crankshaft, and $z_1 = 0$, $z_2 = 1$, $z_3 = 2$, $z_4 = 3$

Primary forces:	$\displaystyle\sum_{i=1}^{n} \sin\phi_i = 0$	$\displaystyle\sum_{i=1}^{n} \cos\phi_i = 0$
Secondary forces:	$\displaystyle\sum_{i=1}^{n} \sin 2\phi_i = 0$	$\displaystyle\sum_{i=1}^{n} \cos 2\phi_i = 4$
Fourth harmonic forces:	$\displaystyle\sum_{i=1}^{n} \sin 4\phi_i = 0$	$\displaystyle\sum_{i=1}^{n} \cos 4\phi_i = 4$
Sixth harmonic forces:	$\displaystyle\sum_{i=1}^{n} \sin 6\phi_i = 0$	$\displaystyle\sum_{i=1}^{n} \cos 6\phi_i = 4$
Primary moments:	$\displaystyle\sum_{i=1}^{n} z_i \sin\phi_i = 0$	$\displaystyle\sum_{i=1}^{n} z_i \cos\phi_i = -2$
Secondary moments:	$\displaystyle\sum_{i=1}^{n} z_i \sin 2\phi_i = 0$	$\displaystyle\sum_{i=1}^{n} z_i \cos 2\phi_i = 6$
Fourth harmonic moments:	$\displaystyle\sum_{i=1}^{n} z_i \sin 4\phi_i = 0$	$\displaystyle\sum_{i=1}^{n} z_i \cos 4\phi_i = 6$
Sixth harmonic moments:	$\displaystyle\sum_{i=1}^{n} z_i \sin 6\phi_i = 0$	$\displaystyle\sum_{i=1}^{n} z_i \cos 6\phi_i = 6$

point C in the middle of the engine, it should be apparent that any primary force-balanced crankshaft design which is mirror symmetric about a transverse plane through point C would also have balanced primary moments as long as all cylinder spacings were uniform and all inertial forces equal. Figure 14-16a shows the 0, 180, 0, 180° crankshaft which is not mirror symmetric. The couple $\mathbf{F}_{s1}\Delta z$ due to cylinder pairs 1, 2 has the same magnitude and sense as the couple $\mathbf{F}_{s3}\Delta z$ due to cylinders 3 and 4, so they add. Figure 14-16b shows the 0, 180, 180, 0° crankshaft which is *mirror symmetric*. The couple $\mathbf{F}_{s1}\Delta z$ due to cylinder pairs 1, 2 has the same magnitude but opposite sense to the couple $\mathbf{F}_{s3}\Delta z$ due to cylinders 3, 4, so they cancel. We can then achieve both even firing and balanced primary moments by changing the sequence of crank throw phase angles to $\phi_i = 0, 180, 180, 0°$ which is *mirror symmetric*.

The crank phase diagram for this design is shown in Figure 14-17. The power strokes can still be evenly spaced over 720°. A firing order of 1, 3, 4, 2 has been chosen which gives the same desired sequence of power stroke angles, $\psi_i = 0, 180, 360, 540°$. (Note that a firing order of 1, 2, 4, 3 would also work with this engine.)[*] Equations 14.3, 14.5, and 14.7 and Table 14-4 show that the primary inertia force and primary moment are both now zero, but the secondary force, secondary moment, and inertia torque are still nonzero.

[*] In carbureted, inline engines a nonconsecutive firing order (i.e., not 1, 2, 3, 4) is usually preferred so that adjacent cylinders near the ends of the engine do not fire sequentially. This allows the intake manifold more time to recharge locally between intake strokes. With fuel injection, this is less critical.

14

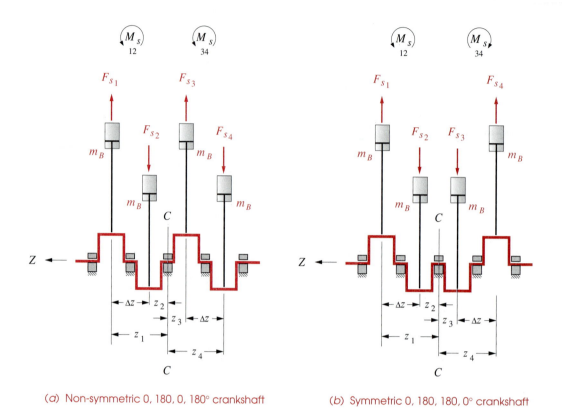

(a) Non-symmetric 0, 180, 0, 180° crankshaft   (b) Symmetric 0, 180, 180, 0° crankshaft

FIGURE 14-16

Mirror symmetric crankshafts cancel primary moments

This $\phi_i = 0, 180, 180, 0°$ crankshaft is considered the best design trade-off and is the one universally used in these four-cylinder inline, four-stroke production engines. Figures 14-2 (p. 640) and 14-3 (p. 641) show such a four-cylinder design. Inertia balance is sacrificed to gain even firing for the reasons cited above. Figure 14-18 shows the gas torque, inertia torque, and total torque for this design. Figure 14-19 shows the secondary shaking moment, secondary shaking force component, and a polar plot of the total shaking force for this design. Note that Figures 14-19b and c are just different views of the same parameter. The polar plot of the shaking force in Figure 14-19c is a view of the shaking force looking at the end of the crankshaft axis with the piston motion horizontal. The cartesian plot in Figure 14-19b shows the same force on a time axis. Since the primary component is zero, this total force is due only to the secondary component. We will soon discuss ways to eliminate these secondary forces and moments.

To see the results of this engine configuration, run program ENGINE and select it from the *Examples* pull-down menu. Then *Plot* or *Print* the results. See Appendix A for more detailed instructions on the use of the program.

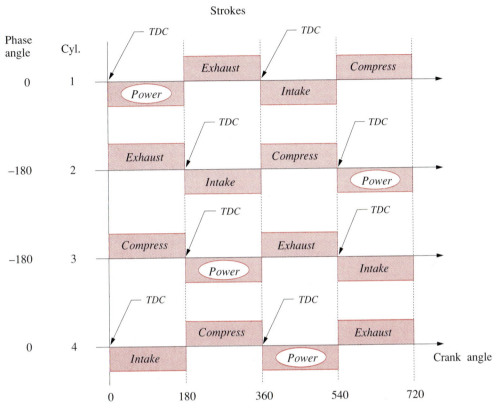

FIGURE 14-17

Even firing four-stroke, four-cylinder engine crank phase diagram with a mirror symmetric 0, 180, 180, 0° crankshaft

14.7   VEE ENGINE CONFIGURATIONS

The same design principles which apply to inline engines also apply to vee and opposed configurations. Even firing takes precedence over inertia balance and mirror symmetry of the crankshaft balances primary moments. In general, a vee engine will have similar inertia balance to that of the inline engines from which it is constructed. A vee six is essentially two three-cylinder inline engines on a common crankshaft, a vee eight is two four-cylinder inlines, etc. The larger number of cylinders allows more power pulses to be spaced out over the cycle for a smoother (and larger average) gas torque. The existence of a **vee angle** between the two inline engines introduces an additional phase shift of the inertial and gas events which is analogous to, but independent of, the phase angle effects. This vee angle is the designer's choice, but there are good and poor choices. The same criteria of even firing and inertia balance apply to its selection.

The **vee angle** 2γ is defined as shown in Figure 14-20. Each bank is offset by its **bank angle** γ referenced to the central X axis of the engine. The crank angle ωt is measured from the X axis. Cylinder one in the right bank is the reference cylinder. Events in each bank will now be phase shifted by its bank angle as well as by the crankshaft phase

14

(a)

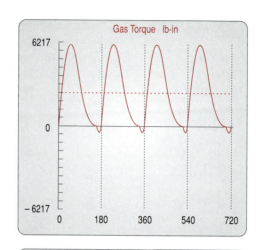

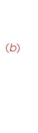
(b)

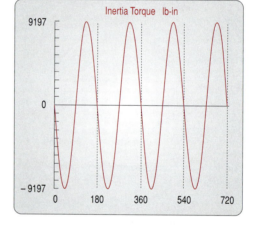

Inline 4 Cylinder
4 Stroke Cycle

Bore = 3.00 in
Stroke = 3.54
B/S   = 0.85
L/R   = 3.50

RPM = 3400

Phase Angles:
0 180 180 0
Power Strokes:
0 180 360 540

(c)

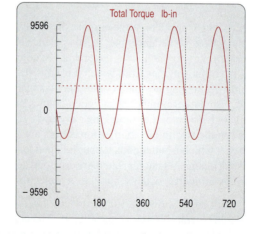

14

FIGURE 14-18

Torque in the four-stroke, four-cylinder inline engine with a 0, 180, 180, 0° crankshaft

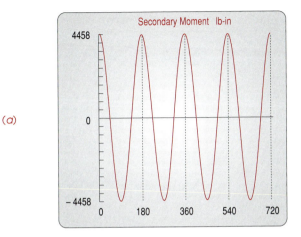

(a)

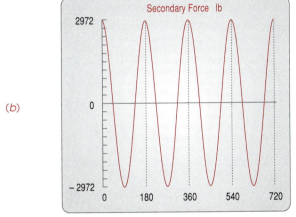

(b)

Inline 4 Cylinder
4 Stroke Cycle

Bore = 3.00 in
Stroke = 3.54
B/S   = 0.85
L/R   = 3.50

RPM = 3400

Phase Angles:
0 180 180 0
Power Strokes:
0 180 360 540

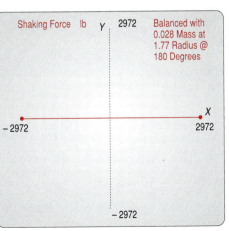

(c)

14

FIGURE 14-19

Shaking forces and moments in the four-stroke, four-cylinder 0, 180, 180, 0° engine

TABLE 14-4   Force and Moment Balance State of a 4-Cylinder, Inline Engine
             with a 0, 180, 180, 0° Crankshaft, and $z_1 = 0$, $z_2 = 1$, $z_3 = 2$, $z_4 = 3$

Primary forces:	$\displaystyle\sum_{i=1}^{n} \sin\phi_i = 0$	$\displaystyle\sum_{i=1}^{n} \cos\phi_i = 0$
Secondary forces:	$\displaystyle\sum_{i=1}^{n} \sin 2\phi_i = 0$	$\displaystyle\sum_{i=1}^{n} \cos 2\phi_i = 4$
Fourth harmonic forces:	$\displaystyle\sum_{i=1}^{n} \sin 4\phi_i = 0$	$\displaystyle\sum_{i=1}^{n} \cos 4\phi_i = 4$
Sixth harmonic forces:	$\displaystyle\sum_{i=1}^{n} \sin 6\phi_i = 0$	$\displaystyle\sum_{i=1}^{n} \cos 6\phi_i = 4$
Primary moments:	$\displaystyle\sum_{i=1}^{n} z_i \sin\phi_i = 0$	$\displaystyle\sum_{i=1}^{n} z_i \cos\phi_i = 0$
Secondary moments:	$\displaystyle\sum_{i=1}^{n} z_i \sin 2\phi_i = 0$	$\displaystyle\sum_{i=1}^{n} z_i \cos 2\phi_i = 6$
Fourth harmonic moments:	$\displaystyle\sum_{i=1}^{n} z_i \sin 4\phi_i = 0$	$\displaystyle\sum_{i=1}^{n} z_i \cos 4\phi_i = 6$
Sixth harmonic moments:	$\displaystyle\sum_{i=1}^{n} z_i \sin 6\phi_i = 0$	$\displaystyle\sum_{i=1}^{n} z_i \cos 6\phi_i = 6$

angles. These two phase shifts will superpose. Taking any one cylinder in either bank as an example, let its instantaneous crank angle be represented by:

$$\theta = (\omega t - \phi_i) \tag{14.10a}$$

Consider first a two-cylinder vee engine with one cylinder in each bank and with both sharing a common crank throw. The shaking force for a single cylinder in the direction of piston motion **u** *hat* with θ measured from the piston axis is:

$$\mathbf{F_s} \cong m_B r \omega^2 \left(\cos\theta + \frac{r}{l}\cos 2\theta \right) \hat{\mathbf{u}} \tag{14.10b}$$

The total shaking force is the vector sum of the contributions from each bank.

$$\mathbf{F_s} = \mathbf{F_{s_L}} + \mathbf{F_{s_R}} \tag{14.10c}$$

We now want to refer the crank angle to the central X axis. The shaking forces for the right (R) and left (L) banks, in the planes of the respective cylinder banks can then be expressed as:

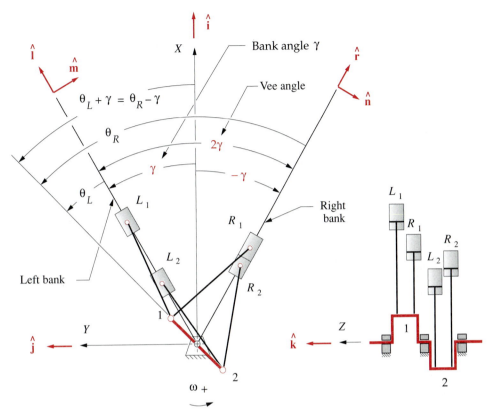

FIGURE 14-20

Vee-engine geometry

$$\mathbf{F}_{s_R} \cong m_B r \omega^2 \left[\cos(\theta+\gamma) + \frac{r}{l}\cos 2(\theta+\gamma) \right] \hat{\mathbf{r}}$$

(14.10d)

$$\mathbf{F}_{s_L} \cong m_B r \omega^2 \left[\cos(\theta-\gamma) + \frac{r}{l}\cos 2(\theta-\gamma) \right] \hat{\mathbf{l}}$$

Note that the bank angle γ is added to or subtracted from the crank angle for each cylinder bank to reference it to the central X axis. The forces are still directed along the planes of the cylinder banks. Substitute the identities:

$$\cos(\theta+\gamma) = \cos\theta\cos\gamma - \sin\theta\sin\gamma$$

(14.10e)

$$\cos(\theta-\gamma) = \cos\theta\cos\gamma + \sin\theta\sin\gamma$$

to get:

14

$$\mathbf{F}_{s_R} \cong m_B r \omega^2 \left[\begin{array}{l} \cos\theta\cos\gamma - \sin\theta\sin\gamma \\ + \dfrac{r}{l}\left(\cos 2\theta\cos 2\gamma - \sin 2\theta\sin 2\gamma\right) \end{array} \right] \hat{\mathbf{r}}$$

$$\mathbf{F}_{s_L} \cong m_B r \omega^2 \left[\begin{array}{l} \cos\theta\cos\gamma + \sin\theta\sin\gamma \\ + \dfrac{r}{l}\left(\cos 2\theta\cos 2\gamma + \sin 2\theta\sin 2\gamma\right) \end{array} \right] \hat{\mathbf{i}}$$

(14.10f)

Now, to account for the possibility of multiple cylinders, phase shifted within each bank, substitute equation 14.10a for θ and replace the sums of angle terms with products from the identities:

$$\cos(\omega t - \phi_i) = \left(\cos\omega t\cos\phi_i + \sin\omega t\sin\phi_i\right)$$

$$\sin(\omega t - \phi_i) = \left(\sin\omega t\cos\phi_i - \cos\omega t\sin\phi_i\right)$$

(14.10g)

After much manipulation, the expressions for the contributions from right and left banks reduce to:

for the right bank:

$$\mathbf{F}_{s_R} \cong m_B r \omega^2 \left[\begin{array}{l} \left(\cos\omega t\cos\gamma - \sin\omega t\sin\gamma\right)\displaystyle\sum_{i=1}^{n/2}\cos\phi_i \\[2ex] + \left(\cos\omega t\sin\gamma + \sin\omega t\cos\gamma\right)\displaystyle\sum_{i=1}^{n/2}\sin\phi_i \\[2ex] + \dfrac{r}{l}\left(\cos 2\omega t\cos 2\gamma - \sin 2\omega t\sin 2\gamma\right)\displaystyle\sum_{i=1}^{n/2}\cos 2\phi_i \\[2ex] + \dfrac{r}{l}\left(\cos 2\omega t\sin 2\gamma + \sin 2\omega t\cos 2\gamma\right)\displaystyle\sum_{i=1}^{n/2}\sin 2\phi_i \end{array} \right] \hat{\mathbf{r}}$$

(14.10h)

and for the left bank:

$$\mathbf{F}_{s_L} \cong m_B r \omega^2 \left[\begin{array}{l} \left(\cos\omega t\cos\gamma + \sin\omega t\sin\gamma\right)\displaystyle\sum_{i=n/2+1}^{n}\cos\phi_i \\[2ex] - \left(\cos\omega t\sin\gamma - \sin\omega t\cos\gamma\right)\displaystyle\sum_{i=n/2+1}^{n}\sin\phi_i \\[2ex] + \dfrac{r}{l}\left(\cos 2\omega t\cos 2\gamma + \sin 2\omega t\sin 2\gamma\right)\displaystyle\sum_{i=n/2+1}^{n}\cos 2\phi_i \\[2ex] - \dfrac{r}{l}\left(\cos 2\omega t\sin 2\gamma - \sin 2\omega t\cos 2\gamma\right)\displaystyle\sum_{i=n/2+1}^{n}\sin 2\phi_i \end{array} \right] \hat{\mathbf{i}}$$

(14.10i)

14

The summations in equations 14.10h and i give a set of sufficient criteria for **zero shaking force** through the second harmonic for each bank, similar to those for the inline engine in equation 14.3 (p. 648). We can resolve the shaking forces for each bank into components along and normal to the central X axis of the vee engine:

$$F_{s_x} = \left(F_{s_L} + F_{s_R}\right)\cos\gamma \; \hat{\mathbf{i}}$$

$$F_{s_y} = \left(F_{s_L} - F_{s_R}\right)\sin\gamma \; \hat{\mathbf{j}} \qquad (14.10j)$$

$$\mathbf{F_s} = F_{s_x}\,\hat{\mathbf{i}} + F_{s_y}\,\hat{\mathbf{j}}$$

Equation 14.10j provides additional opportunities for cancellation of shaking forces beyond the choice of phase angles; e.g., even with nonzero values of F_{sL} and F_{sR}, if γ is 90°, then the x component of the shaking force will be zero. Also, if $F_{sL} = F_{sR}$, then the y component of the shaking force will be zero for any γ. This situation obtains for the case of a horizontally opposed engine (see Section 14.8, p. 674). With some vee or opposed engines it is possible to get cancellation of shaking force components even when the summations in equation 14.10 are not all zero.

The **shaking moment** equations are easily formed from the shaking force equations by multiplying each term in the summations by the moment arm as was done in equations 14.6 (p. 650). The moments exist within each bank and their vectors will be orthogonal to the respective cylinder planes. For the right bank we define a moment unit vector **n** *hat* perpendicular to the **r** *hat*-Z plane in Figure 14-20 (p. 665). For the left bank we define a moment unit vector **m** *hat* perpendicular to the **l** *hat*-Z plane in Figure 14-20.

$$\mathbf{M_{s_R}} \doteq m_B r\omega^2 \begin{bmatrix} \left(\cos\omega t\cos\gamma - \sin\omega t\sin\gamma\right)\displaystyle\sum_{i=1}^{n/2} z_i\cos\phi_i \\[2mm] + \left(\cos\omega t\sin\gamma + \sin\omega t\cos\gamma\right)\displaystyle\sum_{i=1}^{n/2} z_i\sin\phi_i \\[2mm] + \dfrac{r}{l}\left(\cos2\omega t\cos2\gamma - \sin2\omega t\sin2\gamma\right)\displaystyle\sum_{i=1}^{n/2} z_i\cos2\phi_i \\[2mm] + \dfrac{r}{l}\left(\cos2\omega t\sin2\gamma + \sin2\omega t\cos2\gamma\right)\displaystyle\sum_{i=1}^{n/2} z_i\sin2\phi_i \end{bmatrix} \hat{\mathbf{n}} \qquad (14.11a)$$

$$\mathbf{M_{s_L}} \doteq m_B r\omega^2 \begin{bmatrix} \left(\cos\omega t\cos\gamma + \sin\omega t\sin\gamma\right)\displaystyle\sum_{i=n/2+1}^{n} z_i\cos\phi_i \\[2mm] - \left(\cos\omega t\sin\gamma - \sin\omega t\cos\gamma\right)\displaystyle\sum_{i=n/2+1}^{n} z_i\sin\phi_i \\[2mm] + \dfrac{r}{l}\left(\cos2\omega t\cos2\gamma + \sin2\omega t\sin2\gamma\right)\displaystyle\sum_{i=n/2+1}^{n} z_i\cos2\phi_i \\[2mm] - \dfrac{r}{l}\left(\cos2\omega t\sin2\gamma - \sin2\omega t\cos2\gamma\right)\displaystyle\sum_{i=n/2+1}^{n} z_i\sin2\phi_i \end{bmatrix} \hat{\mathbf{m}} \qquad (14.11b)$$

14

The summations provide a set of sufficient criteria for **zero shaking moment** through the second harmonic for each bank, similar to those found for the inline engine in equation 14.7. Resolving the shaking moments for each bank into components along and normal to the central X axis of the vee engine gives:

$$M_{s_x} = \left(M_{s_L} - M_{s_R}\right)\sin\gamma$$

$$M_{s_y} = \left(-M_{s_L} - M_{s_R}\right)\cos\gamma \qquad (14.11c)$$

$$\mathbf{M_s} = M_{s_x}\,\hat{\mathbf{i}} + M_{s_y}\,\hat{\mathbf{j}}$$

Equation 14.11c allows possible cancellation of shaking moment components for some vee or opposed configurations even when the summations in equation 14.11 are not all zero; e.g., if γ is 90°, then the y component of the shaking moment is zero.

The **inertia torques** from the right and left banks of a vee engine are:

$$\mathbf{T}_{i21_R} \cong \frac{1}{2}m_B r^2 \omega^2 \left[\begin{array}{l} \dfrac{r}{2l}\left(\sin(\omega t + \gamma)\displaystyle\sum_{i=1}^{n/2}\cos\phi_i - \cos(\omega t + \gamma)\displaystyle\sum_{i=1}^{n/2}\sin\phi_i\right) \\[2em] -\left(\sin 2(\omega t + \gamma)\displaystyle\sum_{i=1}^{n/2}\cos 2\phi_i - \cos 2(\omega t + \gamma)\displaystyle\sum_{i=1}^{n/2}\sin 2\phi_i\right) \\[2em] -\dfrac{3r}{2l}\left(\sin 3(\omega t + \gamma)\displaystyle\sum_{i=1}^{n/2}\cos 3\phi_i - \cos 3(\omega t + \gamma)\displaystyle\sum_{i=1}^{n/2}\sin 3\phi_i\right) \end{array}\right]\hat{\mathbf{k}} \qquad (14.12a)$$

$$\mathbf{T}_{i21_L} \cong \frac{1}{2}m_B r^2 \omega^2 \left[\begin{array}{l} \dfrac{r}{2l}\left(\sin(\omega t - \gamma)\displaystyle\sum_{i=n/2+1}^{n}\cos\phi_i - \cos(\omega t - \gamma)\displaystyle\sum_{i=n/2+1}^{n}\sin\phi_i\right) \\[2em] -\left(\sin 2(\omega t - \gamma)\displaystyle\sum_{i=n/2+1}^{n}\cos 2\phi_i - \cos 2(\omega t - \gamma)\displaystyle\sum_{i=n/2+1}^{n}\sin 2\phi_i\right) \\[2em] -\dfrac{3r}{2l}\left(\sin 3(\omega t - \gamma)\displaystyle\sum_{i=n/2+1}^{n}\cos 3\phi_i - \cos 3(\omega t - \gamma)\displaystyle\sum_{i=n/2+1}^{n}\sin 3\phi_i\right) \end{array}\right]\hat{\mathbf{k}} \qquad (14.12b)$$

Add the contributions from each bank for the total. For **zero inertia torque** through the third harmonic in a vee engine it is sufficient (but not necessary) that:

$$\sum_{i=1}^{n/2} \sin\phi_i = 0 \qquad \sum_{i=1}^{n/2} \cos\phi_i = 0 \qquad \sum_{i=n/2+1}^{n} \sin\phi_i = 0 \qquad \sum_{i=n/2+1}^{n} \cos\phi_i = 0$$

$$\sum_{i=1}^{n/2} \sin 2\phi_i = 0 \qquad \sum_{i=1}^{n/2} \cos 2\phi_i = 0 \qquad \sum_{i=n/2+1}^{n} \sin 2\phi_i = 0 \qquad \sum_{i=n/2+1}^{n} \cos 2\phi_i = 0 \qquad (14.12c)$$

$$\sum_{i=1}^{n/2} \sin 3\phi_i = 0 \qquad \sum_{i=1}^{n/2} \cos 3\phi_i = 0 \qquad \sum_{i=n/2+1}^{n} \sin 3\phi_i = 0 \qquad \sum_{i=n/2+1}^{n} \cos 3\phi_i = 0$$

Note that when equations 14.12a and b are added, particular combinations of ϕ_i and γ may cancel the inertia torque even when some terms of equation 14.12c are nonzero.

The **gas torque** is:

$$\mathbf{T}_{g21} \cong F_g r \sum_{i=1}^{n} \left(\sin\left[\omega t - (\psi_i + \gamma_k)\right] \left\{ 1 + \frac{r}{l} \cos\left[\omega t - (\psi_i + \gamma_k)\right] \right\} \right) \hat{\mathbf{k}} \qquad (14.13)$$

where the left bank has a *bank angle,* $\gamma_k = +\gamma$ and the right bank a *bank angle,* $\gamma_k = -\gamma$.

It is possible to design a vee engine which has as many crank throws as cylinders, but, for several reasons, this is not always done. The principal advantage of a vee engine over an inline of the same number of cylinders is its more compact size and greater stiffness. It can be about half the length of a comparable inline engine (at the expense of greater width), provided that the crankshaft is designed to accommodate two conrods per crank throw as shown in Figure 14-21. Cylinders in opposite banks then share a crank throw. One bank of cylinders is shifted along the crankshaft axis by the thickness of a conrod. The shorter, wider cylinder block and the shorter crankshaft are much stiffer in both torsion and bending than are those for an inline engine with the same number of cylinders. Figure 14-22 shows computer simulations of several bending and one torsional mode of vibration for a four throw crankshaft. The deflections are greatly exaggerated. The necessarily contorted shape of a crankshaft makes it difficult to control these deflections by design. If excessive in magnitude, they can lead to structural failure.

As an example we will now design the crankshaft for a four-stroke cycle, vee-eight engine. We could put two $\phi_i = 0, 180, 180, 0°$ four-cylinder engines together on one such crankshaft and have the same balance conditions as the four-stroke, four-cylinder engine designed in the previous section (primaries balanced, secondaries unbalanced). However, the motivation for choosing that crankshaft for the four-cylinder engine was the need to space the four available power pulses evenly across the cycle. Equation 14.8b (p. 652) then dictated a 180° delta phase angle for that engine. Now we have eight cylinders available and equation 14.8b defines a delta phase angle of 90° for optimum power pulse spacing. This means we could use the $\phi_i = 0, 90, 180, 270°$ crankshaft designed for the two-stroke four-cylinder engine shown in Figure 14-11 (p. 654), and take advantage of its better inertia balance condition as well as achieve even firing in the four-stroke eight-cylinder vee engine.

14

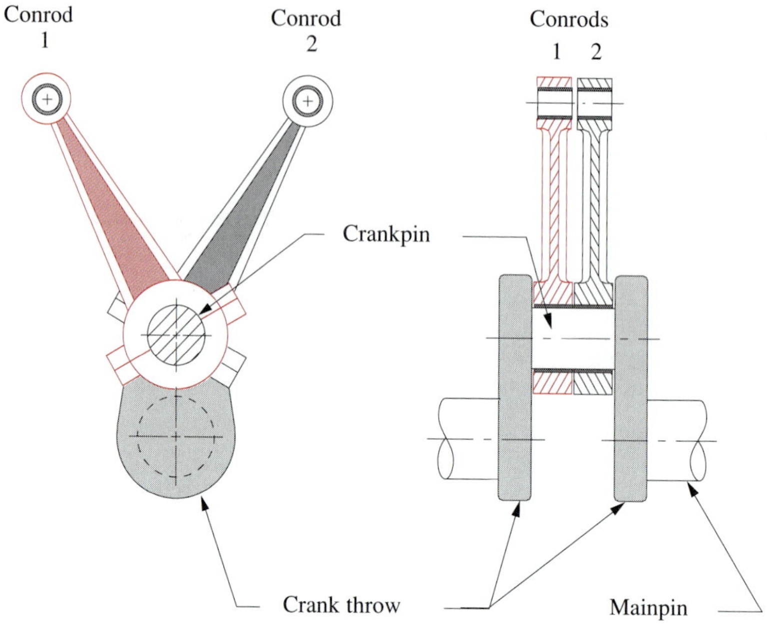

FIGURE 14-21

Two connecting rods on a common crank throw

The $\phi_i = 0, 90, 180, 270°$ four-cylinder crankshaft has all inertia factors equal to zero except for the primary and secondary moments. We learned that arranging the crank throws with mirror symmetry about the midplane would balance the primary moment. Some thought and/or sketches will reveal that it is not possible to obtain this mirror symmetry with any four throw, 90° delta phase angle crankshaft arrangement. However, just as rearranging the order of the crank throws from $\phi_i = 0, 180, 0, 180°$ to $\phi_i = 0, 180, 180, 0°$ had an effect on the shaking moments, rearranging this crankshaft's throw-order will as well. A crankshaft of $\phi_i = 0, 90, 270, 180°$ has all inertia factors equal to zero except for the primary moment. The secondary moment is now gone.[*] This is an advantage worth taking. We will use this crankshaft for the vee eight and deal with the primary moment later.

Figure 14-23a shows the crank phase diagram for the right bank of a vee-eight engine with a $\phi_i = 0, 90, 270, 180°$ crankshaft. Figure 14-23b shows the crank phase diagram for the second (left) bank which is identical to that of the right bank (as it must be since they share crank throws), but it is *shifted to the right by the vee angle 2γ*. Note that in Figure 14-20 (p. 665), the two pistons are driven by conrods on a common crank throw with positive ω, and the piston in the right bank will reach TDC before the one in the left bank. Thus as we show it, the left bank's piston motions lag those of the right bank. Lagging events occur later in time, so we must shift the second (left) bank rightward by the vee angle on the crank phase diagram.

We would like to shift the second bank of cylinders such that its power pulses are evenly spaced among those of the first bank. A little thought (and reference to equation

[*] The explanation for this is quite simple. Equation 14.7b (p. 650) shows that second moments are a function of twice the phase angles and the cylinder moment arms. If you double the values of the original 0, 90, 180, 270° phase angle sequence and modulo them with 360, you get 0, 180, 0, 180° which is not mirror symmetric. Doubling the new phase angle sequence of 0, 90, 270, 180°, modulo 360 gives 0, 180, 180, 0° which is mirror symmetric. It is this symmetry of the doubled phase angles that causes cancellation of the second harmonic of the shaking moment.

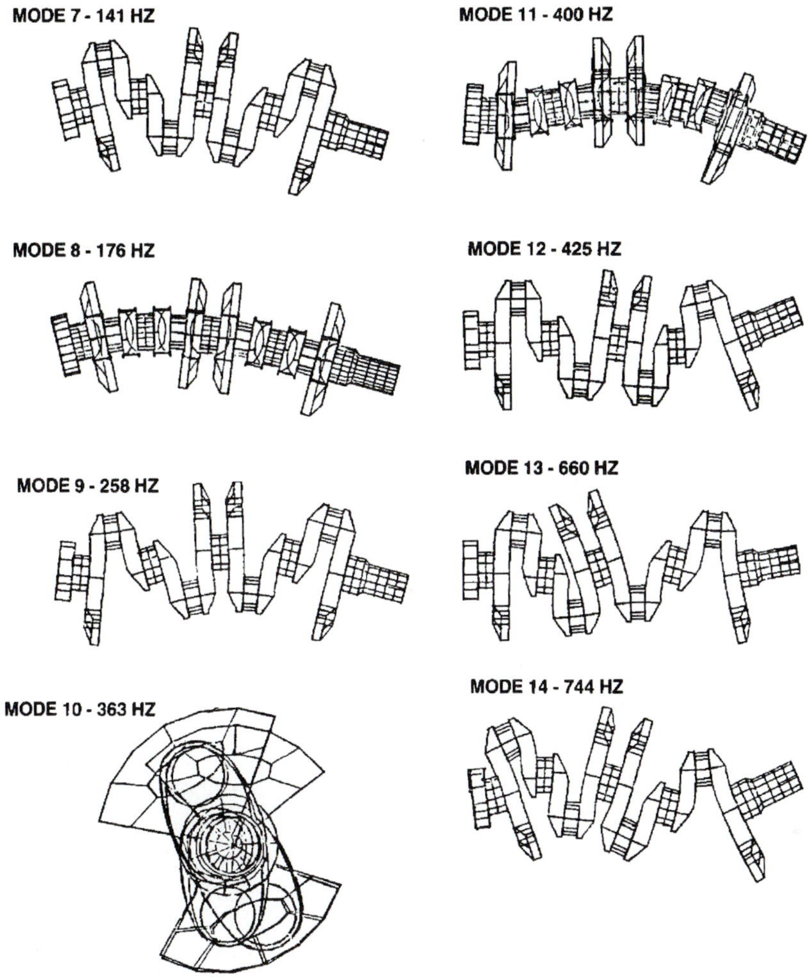

MODE 7 - 141 HZ

MODE 11 - 400 HZ

MODE 8 - 176 HZ

MODE 12 - 425 HZ

MODE 9 - 258 HZ

MODE 13 - 660 HZ

MODE 10 - 363 HZ

MODE 14 - 744 HZ

FIGURE 14-22

Bending and torsional modes of vibration in a four-throw crankshaft
Courtesy of Chevrolet Division, General Motors Corp.

14.8b, p. 652) should reveal that, in this example, each four-cylinder bank has potential-ly $720/4 = 180°$ between power pulses. Our chosen crank throws are spaced at 90° in-crements. A 90° bank angle will be optimum in this case as the phase angles and bank angles will add to create an effective spacing of 180°. Every vee engine design will have one or more optimum vee angles that will give approximately even firing with any par-ticular set of crank phase angles.

Several firing orders are possible with this many cylinders. Vee engines are often arranged to fire cylinders in opposite banks successively to balance the fluid flow de-mands in the intake manifold. Our cylinders are numbered from front to back, first down the right bank and then down the left bank. The firing order shown in Figure 14-23b is

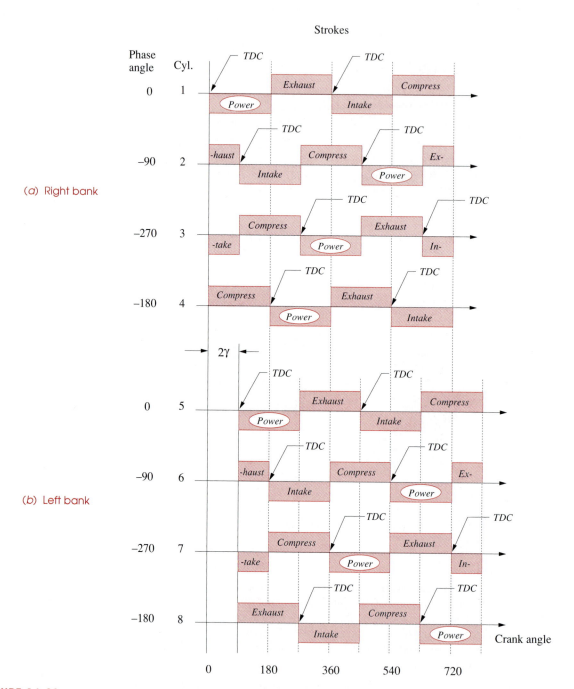

Strokes

FIGURE 14-23

Four-stroke vee-eight crank phase diagram with 0, 90, 270, 180° crankshaft phase angles

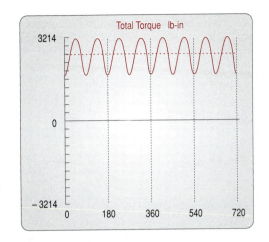

90 deg Vee 8 Cylinder
4 Stroke Cycle

Bore = 2.50 in
Stroke = 2.55
B/S   = 0.98
L/R   = 3.50

RPM = 3400

Phase Angles:
0 90 270 180 0 90 270 180
Power Strokes:
0 90 180 270 360 450 540 630

FIGURE 14-24

Total torque in the 90° V-8 engine with 0, 90, 270, 180° crankshaft phase angles

1, 5, 4, 3, 7, 2, 6, 8 and results in power stroke angles ψ_i = 0, 90, 180, 270, 360, 450, 540, 630°. This will clearly give even firing with a power pulse every 90°.

Figure 14-24 shows the total torque for this engine design. The total torque in this case is equal to the gas torque as the inertia torque is zero. Table 14-5 and Figure 14-25 show the only significant unbalanced inertial component in this engine to be the primary moment, which is quite large. The fourth harmonic terms have negligible coefficients in the Fourier series, and we have truncated them from the equations. We will address the balancing of this primary moment in a later section of this chapter.

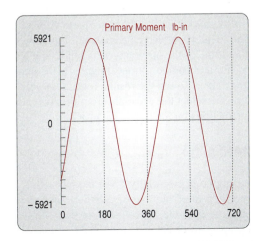

90 deg Vee 8 Cylinder
4 Stroke Cycle

Bore = 2.50 in
Stroke = 2.55
B/S   = 0.98
L/R   = 3.50

RPM = 3400

Phase Angles:
0 90 270 180 0 90 270 180
Power Strokes:
0 90 180 270 360 450 540 630

FIGURE 14-25

Unbalanced primary moment in the 90° V-8 engine with a 0, 90, 270, 180° crankshaft

TABLE 14-5   Force and Moment Balance State of an 8-Cylinder, Vee Engine
              with a 0, 90, 270, 180° Crankshaft, and $z_1 = 1$, $z_2 = 2$, $z_3 = 3$, $z_4 = 4$

Primary forces:	$\sum_{i=1}^{n} \sin\phi_i = 0$	$\sum_{i=1}^{n} \cos\phi_i = 0$
Secondary forces:	$\sum_{i=1}^{n} \sin 2\phi_i = 0$	$\sum_{i=1}^{n} \cos 2\phi_i = 0$
Fourth harmonic forces:	$\sum_{i=1}^{n} \sin 4\phi_i = 0$	$\sum_{i=1}^{n} \cos 4\phi_i = 8$
Sixth harmonic forces:	$\sum_{i=1}^{n} \sin 6\phi_i = 0$	$\sum_{i=1}^{n} \cos 6\phi_i = 0$
Primary moments:	$\sum_{i=1}^{n} z_i \sin\phi_i = -4$	$\sum_{i=1}^{n} z_i \cos\phi_i = -2$
Secondary moments:	$\sum_{i=1}^{n} z_i \sin 2\phi_i = 0$	$\sum_{i=1}^{n} z_i \cos 2\phi_i = 0$
Fourth harmonic moments:	$\sum_{i=1}^{n} z_i \sin 4\phi_i = 0$	$\sum_{i=1}^{n} z_i \cos 4\phi_i = 28$
Sixth harmonic moments:	$\sum_{i=1}^{n} z_i \sin 6\phi_i = 0$	$\sum_{i=1}^{n} z_i \cos 6\phi_i = 0$

Any vee cylinder configuration may have one or more desirable vee angles which will give both even firing and acceptable inertia balance. However, vee engines of fewer than twelve cylinders will not be completely balanced by means of their crankshaft configuration. The desirable vee angles will typically be an integer multiple (including one) or submultiple of the optimum delta phase angle as defined in equations 14.8 (pp. 652 and 654) for that engine. Ninety degrees is the optimum vee angle (2γ) for an eight-cylinder vee engine. To see the results for this vee-eight engine configuration, run program ENGINE and select the vee eight from the *Examples* pull-down menu. See Appendix A for instructions on the use of the program.

14.8   OPPOSED ENGINE CONFIGURATIONS

An opposed engine is essentially a vee engine with a 180° vee angle. The advantage, particularly with a small number of cylinders such as two or four, is the relatively good balance condition possible. A four-stroke opposed twin[*] with 0, 180° crank has even firing plus primary force balance, though the primary moment and all higher harmonics of force and moment are nonzero. A four-stroke, opposed four-cylinder engine (flat four) with a 0, 180, 180, 0°, four-throw crank has primary force balance but must fire its cyl-

[*] As in the BMW motorcycle.

inders in pairs, so its firing pattern looks like a twin. A four-stroke flat four with a two-throw, 0, 180° crank will have even firing and the same balance condition as the inline four with a 0, 180, 180, 0° crank. Program ENGINE will calculate the parameters for opposed as well as vee and inline configurations.

14.9 BALANCING MULTICYLINDER ENGINES

With a sufficient number (n) of cylinders, properly arranged, an engine can be inherently balanced. In a two-stroke engine with its crank throws arranged for even firing, all harmonics of shaking force will be balanced except those whose harmonic number is a multiple of n. In a four-stroke engine with its crank throws arranged for even firing, all harmonics of shaking force will be balanced except those whose harmonic number is a multiple of $n/2$. Primary shaking moment components will be balanced if the crankshaft is mirror symmetric about the central transverse plane. A four-stroke inline configuration then requires at least six cylinders to be inherently balanced through the fourth harmonic. We have seen that an inline four with a 0, 180, 180, 0° crankshaft has nonzero secondary forces and moments as well as nonzero inertia torque. The inline six with a mirror symmetric crank of $\phi_i = 0, 240, 120, 120, 240, 0°$ will have zero shaking forces and moments through the fourth harmonic, though the inertia torque's third harmonic will still be present. To see the results of this six-cylinder inline engine configuration, run program ENGINE and select the inline six from the *Examples* pull-down menu.

A vee twelve is then the smallest vee engine with this inherent state of near perfect balance, as it is two inline sixes on a common crankshaft. We have seen that vee engines take on the balance characteristics of the inline banks from which they are made. Equations 14.10 and 14.11 (pp. 664 to 668) introduced no new criteria for balance in the vee engine over those already defined in equations 14.3 (p. 648) and 14.5 (p. 650) for shaking force and moment balance in the inline engine. Open the file BMWV12.ENG in program ENGINE to see the results for a vee-twelve engine. The common vee-eight engine with crankshaft phase angles of $\phi_i = 0, 90, 270, 180°$ has an unbalanced primary moment as does the inline four from which it is made. It is an example in program ENGINE.

A vee-six engine with 0, 240, 120° crankshaft has unbalanced primary and secondary moments as does the three-cylinder inline from which it is made. This vee six needs a 120° vee angle for proper balance. To reduce engine width, vee sixes are most often made with a 60° vee angle which gives even firing with a 6-throw crank. Some use 90° vee angles to allow their assembly on the same production-line tooling as 90° vee eights, but 90° vee sixes will run rough due to uneven firing unless the crankshaft is redesigned to shift (or splay) the two conrods on each pin by 30°. This results in a more complicated 6-throw crankshaft but gives an even firing, but dynamically unbalanced engine.

Unbalanced inertia torques can be smoothed with a flywheel as was shown in Section 13.10 (p. 631) for the single-cylinder engine. Note that even an engine having zero inertial torque may require a flywheel to smooth its variations in gas torque. The total torque function should be used to determine the energy variations to be absorbed by a flywheel as it contains both gas torque and inertia torque (if any). The method of Section 11.11 (p. 548) also applies to calculation of the flywheel size needed in an engine, based on its variation in the total torque function. Program ENGINE will compute the areas under the total torque pulses needed for the calculation. See the referenced sections for the proper flywheel design procedure.

14

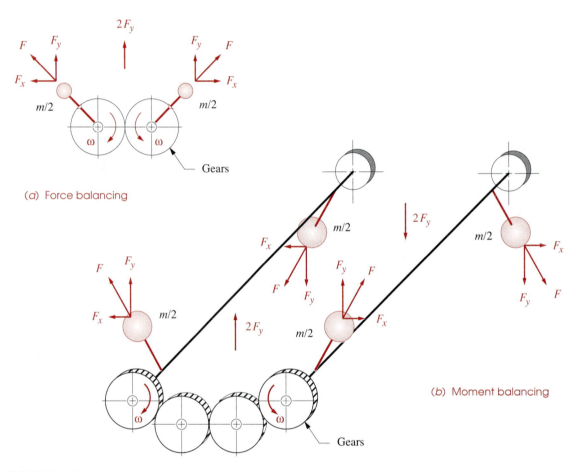

FIGURE 14-26

Counterrotating eccentric masses can balance forces and moments

Unbalanced shaking forces and shaking moments can be cancelled by the addition of one or more rotating balance shafts within the engine. To cancel the primary components requires two balance shafts rotating at crank speed, one of which can be the crankshaft itself. To cancel the secondary components usually requires at least two balance shafts rotating at twice crank speed, gear or chain driven from the crankshaft. Figure 14-26a shows a pair of counterrotating shafts which are fitted with eccentric masses arranged 180° out of phase.[*] As shown, the unbalanced centrifugal forces from the equal, unbalanced masses will add to give a shaking force component in the vertical direction of twice the unbalanced force from each mass, while their horizontal components will exactly cancel. Pairs of counterrotating eccentrics can be arranged to provide a harmonically varying force in any one plane. The harmonic frequency will be determined by the rotational speed of the shafts.

If we arrange two pairs of eccentrics, with one pair displaced some distance along the shaft from the other, and also rotated 180° around the shaft from the first, as shown

* This is called a Lanchester balancer after its English inventor who developed it for aircraft engines prior to World War I (c. 1913). It is still used in various kinds of machinery as well as in engines to cancel inertia forces.

14

in Figure 14-26b, we will get a harmonically varying couple in one plane. There will be cancellation of the forces in one direction and summation in an orthogonal direction.

Thus to cancel the shaking moment in any plane, we can arrange a pair of shafts, each containing two eccentric masses displaced along those shafts, 180° out of phase, and gear the shafts together to rotate in opposite directions at any multiple of crankshaft speed. To cancel the shaking force as well, it is only necessary to provide sufficient additional unbalanced mass in one of the pairs of eccentric masses to generate a shaking force opposite to that of the engine, over and above that needed to generate the forces of the couple.

In an inline engine, the unbalanced forces and moments are all confined to the single plane of the cylinders as they are due entirely to the reciprocating masses assumed concentrated at the wrist pin. (We are assuming that all crank throws are exactly balanced rotationally to cancel the effects of mass at the crankpin.) In a vee engine, however, the shaking forces and moments have both x and y components as shown in equations 14.10 and 14.11 (pp. 664 to 668) and in Figure 14-20 (p. 665). Each bank's piston's shaking effects are acting within the plane of that bank's cylinders, and the bank angle γ is used to resolve them into x and y components. Figure 14-27 shows the two-dimensional shaking force present in a two-cylinder 90° vee engine. The two conrods share a common crank throw at a phase angle of zero in this particular engine. The force due to each piston is confined to the reciprocating plane (bank) of that piston, but the vee angle between the cylinder banks creates the pattern shown when the primary and secondary components of each piston force are added together in their proper phase relationship.

The 90° vee eight engine with a 0, 90, 270, 180° crankshaft, which has only an unbalanced primary moment, presents a special case. The 90° angle between the banks results in equal horizontal and vertical components of the primary shaking moment which reduces it to a couple of constant magnitude rotating about the crank axis at crankshaft speed as shown in Figure 14-28, which is a view looking at the end of the crankshaft axis.

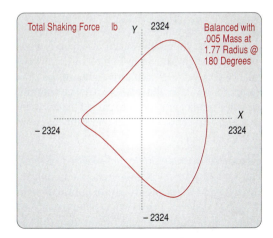

90 deg Vee 2 Cylinder
4 Stroke Cycle

Bore = 3.00 in
Stroke = 3.54
B/S   = 0.85
L/R   = 3.50

RPM = 3400

Phase Angles:
0   0
Power Strokes:
0   270

Total Shaking Force lb   Y   2324   Balanced with .005 Mass at 1.77 Radius @ 180 Degrees

− 2324   X   2324

− 2324

FIGURE 14-27

Shaking force in a 90° vee twin engine (looking end-on to the crankshaft axis)

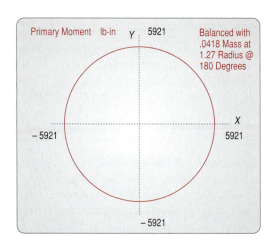

90 deg Vee 8 Cylinder
4 Stroke Cycle
Bore   = 2.50 in
Stroke = 2.55
B/S   = 0.98
L/R   = 3.50

RPM   = 3400

Phase Angles:
0 90 270 180 0 90 270 180
Power Strokes:
0 90 180 270 360 450 540 630

FIGURE 14-28

Primary moment in the 90° vee-eight engine (looking end-on to the crankshaft axis)

With this vee-eight engine, the primary moment can be balanced by merely adding two eccentric counterweights of proper size and opposite orientation to the crankshaft alone. No independent, second balance shaft is needed in the 90° vee-eight engine with this crankshaft. The 180°-out-of-phase counterweights are typically placed near the ends of the crankshaft to obtain the largest moment arm possible and thus reduce their size.

The shaking force of the 90° vee twin depicted in Figure 14-27 (p. 677) has a rotating primary component of constant magnitude that also can be cancelled with counterweights on the crankshaft alone. However, its second harmonic is planar (in the YZ plane). To cancel it requires a pair of twice-speed balance shafts as shown in Figure 14-26a (p. 676).

The 60° vee-six engine with a 0, 240, 120, 60, 300, 180° crankshaft has unbalanced primary and secondary moments, each of which is a constant magnitude rotating vector like that of the vee eight shown in Figure 14-28. The primary component can be completely balanced by adding counterweights to the crankshaft as in the 90° vee eight. Some manufacturers[*] also add a single balance shaft in the valley of the vee six, driven by gears at twice crankshaft speed to cancel the constant magnitude second harmonic of the shaking moment. This combination results in complete cancellation of the shaking moment in this engine.

Calculation of the magnitude and location of the eccentric balance masses needed to cancel any shaking forces or moments is a straightforward exercise in **static balancing** (for forces) and **two-plane dynamic balancing** (for moments) as discussed in Sections 12.1 (p. 571) and 12.2 (p. 574). The unbalanced forces and moments for the particular engine configuration are calculated from the appropriate equations in this chapter. Two correction planes must be selected along the length of the balance shafts/crankshaft being designed. The magnitude and angular locations of the balance masses can then be calculated by the methods described in the noted sections of Chapter 12.

[*] General Motors Corp., Buick V6.

Secondary Balance in the Four Cylinder Inline Engine

The four-cylinder inline engine with a 0, 180, 180, 0° crankshaft is one of the most wide-ly used engines in the automobile industry. As described in a previous section, this en-gine suffers from unbalanced secondary force, moment and torque. If the displacement of the engine is less than about 2.0 liters, then the magnitudes of the secondary forces may be small enough to be ignored, especially if the engine mounts provide good vibra-tion isolation of the engine from the passenger compartment. Above that displacement, objectionable noise, vibration, and harshness (NVH) may be heard and felt by the pas-sengers at certain engine speeds where the frequency of the engine's second harmonic coincides with one of the natural frequencies of the body structure. Then some balanc-ing is needed in the engine to avoid customer dissatisfaction.

The shaking force for an inline engine is given by equation 14.2d (p. 647). Taking only the second harmonic term and applying the relevant factors from Table 14-4 (p. 664) for this engine gives

$$\mathbf{F}_{s_2} \cong m_B r \omega^2 \frac{4r}{l} \cos 2\omega t \ \hat{\mathbf{i}} \tag{14.14}$$

The shaking torque for an inline engine is given by equation 14.4c (p. 649) in com-bination with equation 13.15f (p. 621). Taking only the second harmonic term and ap-plying the relevant factors from Table 14-4 for this engine gives

$$\mathbf{T}_{s_2} \cong 2 m_B r^2 \omega^2 \sin 2\omega t \ \hat{\mathbf{k}} \tag{14.15}$$

The principle of the Lanchester balancer, shown in Figure 14-26a, can be used to counteract the secondary forces by driving the two counterrotating balance shafts at twice crankshaft speed with chains and/or gears. Figure 14-29 shows such an arrange-ment as applied to a Mitsubishi 2.6L, four-cylinder engine.[*]

H. Nakamura [1] improved on Lanchester's 1913 design by arranging the balance shafts within the engine so as to cancel the second harmonic of the inertia torque as well as the secondary inertia force. But, his arrangement does not affect the unbalanced sec-ondary shaking moment. In fact, it is designed to impart zero net moment about a trans-verse axis to either balance shaft in order to minimize bending moments on the shafts, and so reduce bearing loads and friction losses. This feature is the subject and principal claim of Nakamura's patent on this design. [2]

Figure 14-30a shows a schematic of a conventional Lanchester balancer arranged with the two counterrotating balance shafts with their centers in a single horizontal plane transverse to the vertical plane of piston motion.

The balance force from the two balance shafts combined is

$$\mathbf{F}_{bal} \cong -8 m_{bal} r_{bal} \omega^2 \cos 2\omega t \ \hat{\mathbf{i}} \tag{14.16}$$

where m_{bal} and r_{bal} are the mass and radius of one balance weight.

Figure 14-30b shows Nakamura's arrangement of the balance shafts with one situ-ated above the other in separate horizontal planes. The vertical offset $x_1 - x_2$ between the shafts, in combination with the oppositely directed but equal magnitude horizontal

14

---
[*] Also Chrysler and Porsche under license from Mitsubishi.

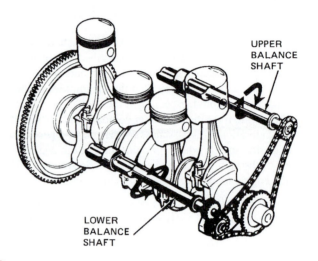

UPPER
BALANCE
SHAFT

LOWER
BALANCE
SHAFT

FIGURE 14-29

Balance shafts used to eliminate the secondary unbalance in the four-cylinder inline engine
Courtesy of the Chrysler Corporation

components of the counterweights' centrifugal forces, creates a time-varying couple *about the crankshaft axis* defined as:

$$\mathbf{T}_{bal} \cong -4m_{bal}r_{bal}\omega^2 \left(x_1 - x_2\right)\sin 2\omega t - \left(y_1 + y_2\right)\cos 2\omega t \;\; \hat{\mathbf{k}} \qquad (14.17)$$

where x and y refer to the coordinates of the shaft centers referenced to the crankshaft center, and the subscripts 1 and 2 refer, respectively, to the balance shaft turning in the same and opposite directions as the crankshaft.

The vertical components of the balance weights' centrifugal forces still add to provide force balance as in equation 14.16. The torque in equation 14.17 will have opposite sense to the shaking torque if the upper shaft turns in the same direction, and the lower shaft in the opposite direction, of the crankshaft.

For force balance, equations 14.14 and 14.16 must sum to zero,

$$m_B r \omega^2 \frac{4r}{l} \cos 2\omega t - 8m_{bal}r_{bal}\omega^2 \cos 2\omega t = 0$$

or
$$m_{bal}r_{bal} = \frac{r}{2l}m_B r \qquad (14.18)$$

which defines the mass radius product needed for the balance mechanism.

For torque balance, equations 14.15 and 14.17 must sum to zero.

$$2m_B r^2 \omega^2 \sin 2\omega t - 4m_{bal}r_{bal}\omega^2 \left(x_1 - x_2\right)\sin 2\omega t - \left(y_1 + y_2\right)\cos 2\omega t = 0 \qquad (14.19a)$$

Substitute equation 14.18 in 14.19a.

$$2m_B r^2 \sin 2\omega t - 4\frac{r}{2l}m_B r\left(x_1 - x_2\right)\sin 2\omega t - \left(y_1 + y_2\right)\cos 2\omega t = 0 \qquad (14.19b)$$

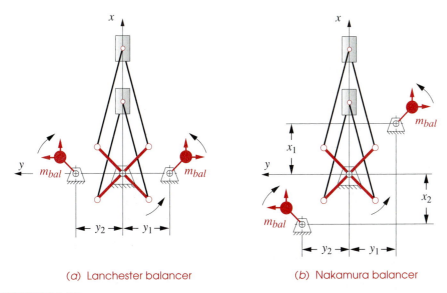

(a) Lanchester balancer   (b) Nakamura balancer

FIGURE 14-30

Two types of secondary balancer mechanisms for the four-cylinder inline engine

For this equation to be zero for all ωt,

$$y_2 = -y_1$$

$$x_1 - x_2 = l$$

(14.19c)

So, if the balance shafts are arranged symmetrically with respect to the piston plane at any convenient locations y_1 and $-y_1$, and the distance $x_1 - x_2$ is made equal to the length of the connecting rod l, then the second harmonic of the inertia torque will be completely cancelled. Since the second harmonic is the only nonzero component of inertia torque in this engine as can be seen in Figure 14-18b (p. 662), it will now be completely balanced for shaking force and shaking torque (but not shaking moment).

There is also significant oscillation of the gas torque in a four cylinder engine as shown in Figure 14-18a. The gas torque is 180° out of phase with the inertia torque as can be seen in Figure 14-18b and so provides some natural cancellation as shown in the total torque curve of Figure 14-18c. The magnitude of the gas torque varies with engine load and so cannot itself be cancelled with any particular balance shaft geometry for all conditions. However, one engine speed and load condition can be selected as representative of the majority of typical driving conditions, and the balance system geometry altered to give an optimum reduction of total engine torque under those conditions. Nakamura estimates that gas torque magnitude is about 30% of the inertia torque under typical driving conditions and so suggests a value of $x_1 - x_2 = 0.7l$ for the best overall reduction of total torque oscillation in this engine. Note that the average value of the driving torque is not affected by balancing because the average torque of any rotating balance system is always zero.

14

14.10 REFERENCES

1   **Nakamura, H.** (1976). "A Low Vibration Engine with Unique Counter-Balance Shafts." SAE Paper: 760111.

2   **Nakamura, H., et al.** (1977). "Engine Balancer." Assignee: Mitsubishi Corp.: U.S. Patent 4,028,963.

14.11 BIBLIOGRAPHY

Crouse, W. H. (1970). *Automotive Engine Design*, McGraw-Hill Inc., New York.

Jennings, G. (1979). "A Short History of Wonder Engines," *Cycle Magazine*, May 1979, p. 68ff.

Setright, L. J. K. (1975). *Some Unusual Engines*, Mechanical Engineering Publications Ltd., The Inst. of Mech. Engr., London.

Thomson, W. (1978). *Fundamentals of Automotive Balance*, Mechanical Engineering Publications Ltd., London.

14.12 PROBLEMS

14-1   Draw a crank phase diagram for a three-cylinder inline engine with a 0, 120, 240° crankshaft and determine all possible firing orders for:

    a.   Four-stroke cycle   b.   Two-stroke cycle

    Select the best arrangement to give even firing for each stroke cycle.

14-2   Repeat Problem 14-1 for an inline four-cylinder engine with a 0, 90, 270, 180° crankshaft.

14-3   Repeat Problem 14-1 for a 45° vee, four-cylinder engine with a 0, 90, 270, 180° crankshaft.

14-4   Repeat Problem 14-1 for a 45° vee, two-cylinder engine with a 0, 90° crankshaft.

14-5   Repeat Problem 14-1 for a 90° vee, two-cylinder engine with a 0, 180° crankshaft.

14-6   Repeat Problem 14-1 for a 180° opposed, two-cylinder engine with a 0, 180° crankshaft.

14-7   Repeat Problem 14-1 for a 180° opposed, four-cylinder engine with a 0, 180, 180, 0° crankshaft.

[†]14-8   Calculate the shaking force, torque, and moment balance conditions through the second harmonic for the engine design in Problem 14-1.

[†]14-9   Calculate the shaking force, torque, and moment balance conditions through the second harmonic for the engine design in Problem 14-2.

[†]14-10   Calculate the shaking force, torque, and moment balance conditions through the second harmonic for the engine design in Problem 14-3.

[†]14-11   Calculate the shaking force, torque, and moment balance conditions through the second harmonic for the engine design in Problem 14-4.

[†]14-12   Calculate the shaking force, torque, and moment balance conditions through the second harmonic for the engine design in Problem 14-5.

14

[†] These problems are suited to solution using *Mathcad, Matlab,* or *TKSolver* equation solver programs.

†14-13 Calculate the shaking force, torque, and moment balance conditions through the second harmonic for the engine design in Problem 14-6.

†14-14 Calculate the shaking force, torque, and moment balance conditions through the second harmonic for the engine design in Problem 14-7.

14-15 Derive expressions, in general terms, for the magnitude and angular location with respect to the first crank throw, of the mass-radius products needed on the crankshaft to balance the shaking moment in a 90° vee-eight engine with a 0, 90, 270, 180° crankshaft.

14-16 Repeat Problem 14-15 for a 90° vee six with a 0, 240, 120° crankshaft.

14-17 Repeat Problem 14-15 for a 90° vee four with a 0, 180° crankshaft.

†14-18 Design a pair of Nakamura balance shafts to cancel the shaking force and reduce torque oscillations in the engine shown in Figure 14-18 (p. 662).*

14.13 PROJECTS

These are loosely structured design problems intended for solution using program EN-GINE. All involve the design of one or more multicylinder engines and differ mainly in the specific data for the engine. The general problem statement is:

Design a multicylinder engine for a specified displacement and stroke cycle. Optimize the conrod/crank ratio and bore/stroke ratio to minimize shaking forces, shaking torque, and pin forces, also considering package size. Design your link shapes and calculate realistic dynamic parameters (mass, CG location, moment of inertia) for those links using the methods shown in Chapters 10 through 13. Dynamically model the links as described in those chapters. Balance or overbalance the linkage as needed to achieve the desired results. Choose crankshaft phase angles to optimize the inertial balance of the engine. Choose a firing order and determine the power stroke angles to optimize even firing. Trade off inertia balance if necessary to achieve even firing. Choose an optimum vee angle if appropriate to the particular problem. Overall smoothness of total torque is desired. Design and size a minimum weight flywheel by the method of Chapter 11 to smooth the total torque. Write an engineering report on your design and analysis.

P14-1 Two-stroke cycle inline twin with a displacement of 1 liter.

P14-2 Four-stroke cycle inline twin with a displacement of 1 liter.

P14-3 Two-stroke cycle vee twin with a displacement of 1 liter.

P14-4 Four-stroke cycle vee twin with a displacement of 1 liter.

P14-5 Two-stroke cycle opposed twin with a displacement of 1 liter.

P14-6 Four-stroke cycle opposed twin with a displacement of 1 liter.

P14-7 Two-stroke cycle vee four with a displacement of 2 liters.

P14-8 Four-stroke cycle vee four with a displacement of 2 liters.

P14-9 Two-stroke cycle opposed four with a displacement of 2 liters.

P14-10 Four-stroke cycle opposed four with a displacement of 2 liters.

P14-11 Two-stroke inline five cylinder with a displacement of 2.5 liters.

P14-12 Four-stroke inline five cylinder with a displacement of 2.5 liters.

† These problems are suited to solution using *Mathcad, Matlab,* or *TKSolver* equation solver programs.

* More information on this engine design can be found in program ENGINE where it is one of the examples.

14

P14-13　Two-stroke cycle vee six with a displacement of 3 liters.

P14-14　Four-stroke cycle vee six with a displacement of 3 liters.

P14-15　Two-stroke cycle opposed six with a displacement of 3 liters.

P14-16　Four-stroke cycle opposed six with a displacement of 3 liters.

P14-17　Two-stroke inline seven cylinder with a displacement of 3.5 liters.

P14-18　Four-stroke inline seven cylinder with a displacement of 3.5 liters.

P14-19　Two-stroke inline eight cylinder with a displacement of 4 liters.

P14-20　Four-stroke inline eight cylinder with a displacement of 4 liters.

P14-21　Two-stroke vee ten cylinder with a displacement of 5 liters.

P14-22　Four-stroke vee ten cylinder with a displacement of 5 liters.

P14-23　Four-stroke W-6 with a displacement of 5 liters. (A W engine has three banks of cylinders on a common crank.)

P14-24　Four-stroke W-9 with a displacement of 5 liters. (A W engine has three banks of cylinders on a common crank.)

P14-25　Four-stroke W-12 with a displacement of 5 liters. (A W engine has three banks of cylinders on a common crank.)

P14-26　Design a family of vee engines to be built on the same assembly line. They must all have the same vee angle and use the same pistons, connecting rods, and strokes. Their crankshafts can each be different. Four configurations are needed: vee four, vee six, vee eight, and vee ten. All will have the same single-cylinder displacement of 0.5 liters. Optimize the single-cylinder configuration from which the multicylinder engines will be constructed for bore/stroke ratio and conrod crank ratio. Then assemble this cylinder design into the above configurations. Find the best compromise of vee angle to provide a good mix of balance and even firing in all engines.

P14-27　Repeat Project P14-26 for a family of W engines: W-3, W-6, W-9, and W-12. The inter-bank angles must be the same for all models. See the built-in example W-12 engine in program ENGINE for more information on this unusual W configuration.

P14-28　In recent years some automobile manufacturers have made unusual vee configurations such as the VW-Audi VR15 which is a 15° vee six. Obtain detailed information on this engine design and then analyze it with program ENGINE. Write a report that explains why the manufacturer chose this unusual arrangement and justify your conclusions with sound engineering analysis.

P14-29　Design an inline six- and an inline five-cylinder engine of the same displacement, say 2.5 liters. Analyze their dynamics with program ENGINE. Write an engineering report to explain why such manufacturers as Audi, Volvo, and Acura have chosen a five-cylinder inline over a six cylinder of comparable torque and power output.

P14-30　Ferrari has produced vee-twelve engines in both 60° vee and horizontally opposed configurations. Design 3-liter versions of each and compare their dynamics. Write a report that explains why the manufacturer chose these arrangements and justify your conclusions with sound engineering analysis.

P14-31　Design and compare a 3-liter 90° vee six, 60° vee six, inline six, and 180° opposed six, all of which designs are in volume production. Explain their advantages and disadvantages and justify your conclusions with sound engineering analysis.

Chapter 15

CAM DYNAMICS

The universe is full of magical things
patiently waiting for our wits to grow sharper
EDEN PHILLPOTS

15.0 INTRODUCTION

Chapter 8 presented the kinematics of cams and followers and methods for their design. We will now extend the study of cam-follower systems to include considerations of the dynamic forces and torques developed. While the discussion in this chapter is limited to examples of cams and followers, the principles and approaches presented are applicable to most dynamic systems. The cam-follower system can be considered a useful and convenient example for the presentation of topics such as creating lumped parameter dynamic models and defining equivalent systems as described in Chapter 10. These techniques as well as the discussion of natural frequencies, effects of damping, and analogies between physical systems will be found useful in the analysis of all dynamic systems regardless of type.

In Chapter 10 we discussed the two approaches to dynamic analysis, commonly called the forward and the inverse dynamics problems. The forward problem assumes that all the forces acting on the system are known and seeks to solve for the resulting displacements, velocities, and accelerations. The inverse problem is, as its name says, the inverse of the other. The displacements, velocities, and accelerations are known, and we solve for the dynamic forces that result. In this chapter we will explore the application of both methods to cam-follower dynamics. Section 15.1 explores the forward solution. Section 15.3 will present the inverse solution. Both are instructive in this application of a force-closed (spring-loaded) cam-follower system and will each be discussed in this chapter.

15

TABLE 15-1 Notation Used in This Chapter

c = damping coefficient
c_c = critical damping coefficient
k = spring constant
F_c = force of cam on follower
F_s = force of spring on follower
F_d = force of damper on follower
m = mass of moving elements
t = time in seconds
T_c = torque on camshaft
θ = camshaft angle, in degrees or radians
ω = camshaft angular velocity, rad/sec
ω_d = damped circular natural frequency, rad/sec
ω_f = forcing frequency, rad/sec
ω_n = undamped circular natural frequency, rad/sec
x = follower displacement, length units
$\dot{x} = v$ = follower velocity, length/sec
$\ddot{x} = a$ = follower acceleration, length/sec^2
ζ = damping ratio

15.1 DYNAMIC FORCE ANALYSIS OF THE FORCE-CLOSED CAM-FOLLOWER

Figure 15-1a shows a simple plate or disk cam driving a spring-loaded, roller follower. This is a force-closed system which depends on the spring force to keep the cam and follower in contact at all times. Figure 15-1b shows a lumped parameter model of this system in which all the **mass** which moves with the follower train is lumped together as m, all the springiness in the system is lumped within the **spring constant** k, and all the **damping** or resistance to movement is lumped together as a damper with coefficient c. The sources of mass which contribute to m are fairly obvious. The mass of the follower stem, the roller, its pivot pin, and any other hardware attached to the moving assembly all add together to create m. Figure 15-1c shows the free-body diagram of the system acted upon by the cam force F_c, the spring force F_s, and the damping force F_d. There will of course also be the effects of mass times acceleration on the system.

Undamped Response

Figure 15-2 shows an even simpler lumped parameter model of the same system as in Figure 15-1 but which omits the damping altogether. This is referred to as a *conservative model* since it conserves energy with no losses. This is not a realistic or safe assumption in this case but will serve a purpose in the path to a better model which will include the damping. The free-body diagram for this mass-spring model is shown in Figure 15-2c. We can write Newton's equation for this one-*DOF* system:

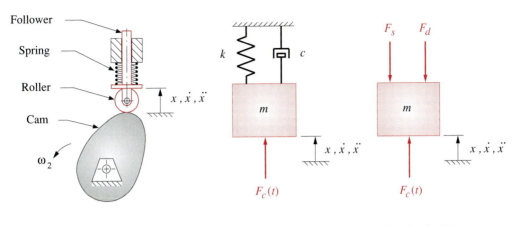

(a) Physical system   (b) Lumped model   (c) Free-body diagram

FIGURE 15-1

One-*DOF* lumped parameter model of a cam-follower system including damping

$$\sum F = ma = m\ddot{x}$$

$$F_c(t) - F_s = m\ddot{x}$$

From equation 10.14:

$$F_s = kx$$

then:

$$m\ddot{x} + kx = F_c(t) \tag{15.1a}$$

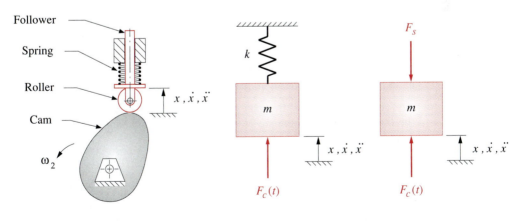

(a) Physical system   (b) Lumped model   (c) Free-body diagram

FIGURE 15-2

One-*DOF* lumped parameter model of a cam-follower system without damping

15

This is a second-order ordinary differential equation (ODE) with constant coefficients. The complete solution will consist of the sum of two parts, the transient (homogeneous) and the steady state (particular). The homogeneous ODE is,

$$m\ddot{x} + k\,x = 0$$

$$\ddot{x} = -\frac{k}{m}x$$

(15.1b)

which has the well-known solution,

$$x = A\cos\omega t + B\sin\omega t$$

(15.1c)

where A and B are the constants of integration to be determined by the initial conditions. To check the solution, differentiate it twice, assuming constant ω, and substitute in the homogeneous ODE.

$$-\omega^2\left(A\cos\omega t + B\sin\omega t\right) = -\frac{k}{m}\left(A\cos\omega t + B\sin\omega t\right)$$

This is a solution provided that:

$$\omega^2 = \frac{k}{m} \qquad\qquad \omega_n = \sqrt{\frac{k}{m}}$$

(15.1d)

The quantity ω_n (rad/sec) is called the *circular natural frequency* of the system and is the frequency at which the system wants to vibrate if left to its own devices. This represents the *undamped natural frequency* since we ignored damping. The *damped natural frequency* will be slightly lower than this value. Note that ω_n is a function only of the physical parameters of the system m and k; thus it is completely determined and unchanging with time once the system is built. By creating a one-*DOF* model of the system, we have limited ourselves to one natural frequency which is an "average" natural frequency usually close to the lowest, or fundamental, frequency.

Any real physical system will also have higher natural frequencies which in general will not be integer multiples of the fundamental. In order to find them we need to create a multi-degree-of-freedom model of the system. The fundamental tone at which a bell rings when struck is its natural frequency defined by this expression. The bell also has overtones which are its other, higher, natural frequencies. The fundamental frequency tends to dominate the transient response of the system.[1]

The circular natural frequency ω_n (rad/sec) can be converted to cycles per second (hertz) by noting that there are 2π radians per revolution and one revolution per cycle:

$$f_n = \frac{1}{2\pi}\omega_n \quad \text{hertz}$$

(15.1e)

The constants of integration, A and B in equation 15.1c, depend on the initial conditions. A general case can be stated as,

When $t = 0$,   let $x = x_0$ and $v = v_0$,   where x_0 and v_0 are constants

which gives a general solution to the homogeneous ODE 15.1b of:

$$x = x_0\cos\omega_n t + \frac{v_0}{\omega_n}\sin\omega_n t$$

(15.1f)

Equation 15.1f can be put into polar form by computing the magnitude and phase angle:

$$X_0 = \sqrt{x_0^2 + \left(\frac{v_0}{\omega_n}\right)^2} \qquad\qquad \phi = \arctan\left(\frac{v_0}{x_0\omega_n}\right)$$

then:

$$x = X_0 \cos(\omega_n t - \phi) \tag{15.1g}$$

Note that this is a pure harmonic function whose amplitude X_0 and phase angle ϕ are a function of the initial conditions and the natural frequency of the system. It will oscillate forever in response to a single, transitory input if there is truly no damping present.

Damped Response

If we now reintroduce the damping of the model in Figure 15-1b and draw the free-body diagram as shown in Figure 15-1c, the summation of forces becomes:

$$F_c(t) - F_d - F_s = m\ddot{x} \tag{15.2a}$$

Substituting equations 10.14 (p. 501) and 10.15c (p. 503):

$$m\ddot{x} + c\dot{x} + kx = F_c(t) \tag{15.2b}$$

HOMOGENEOUS SOLUTION   We again separate this differential equation into its homogeneous and particular components. The homogeneous part is:

$$\ddot{x} + \frac{c}{m}\dot{x} + \frac{k}{m}x = 0 \tag{15.2c}$$

The solution to this ODE is of the form:

$$x = Re^{st} \tag{15.2d}$$

where R and s are constants. Differentiating versus time:

$$\dot{x} = Rse^{st}$$
$$\ddot{x} = Rs^2e^{st}$$

and substituting in equation 15.2c:

$$Rs^2e^{st} + \frac{c}{m}Rse^{st} + \frac{k}{m}Re^{st} = 0$$

$$\left(s^2 + \frac{c}{m}s + \frac{k}{m}\right)Re^{st} = 0 \tag{15.2e}$$

For this solution to be valid either R or the expression in parentheses must be zero as e^{st} is never zero. If R were zero, then the assumed solution, in equation 15.2d, would also be zero and thus not be a solution. Therefore, the quadratic equation in parentheses must be zero.

15

$$\left(s^2 + \frac{c}{m}s + \frac{k}{m}\right) = 0 \qquad (15.2f)$$

This is called the characteristic equation of the ODE and its solution is:

$$s = \frac{-\dfrac{c}{m} \pm \sqrt{\left(\dfrac{c}{m}\right)^2 - 4\dfrac{k}{m}}}{2}$$

which has the two roots:

$$s_1 = -\frac{c}{2m} + \sqrt{\left(\frac{c}{2m}\right)^2 - \frac{k}{m}}$$

$$(15.2g)$$

$$s_2 = -\frac{c}{2m} - \sqrt{\left(\frac{c}{2m}\right)^2 - \frac{k}{m}}$$

These two roots of the characteristic equation provide two independent terms of the homogeneous solution:

$$x = R_1 e^{s_1 t} + R_2 e^{s_2 t} \qquad \text{for } s_1 \neq s_2 \qquad (15.2h)$$

If $s_1 = s_2$, then another form of solution is needed. The quantity s_1 will equal s_2 when:

$$\sqrt{\left(\frac{c}{2m}\right)^2 - \frac{k}{m}} = 0 \qquad \text{or:} \qquad \frac{c}{2m} = \sqrt{\frac{k}{m}}$$

and:

$$c = 2m\sqrt{\frac{k}{m}} = 2m\omega_n = c_c \qquad (15.2i)$$

This particular value of c is called the **critical damping** and is labeled c_c. The system will behave in a unique way when critically damped, and the solution must be of the form:

$$x = R_1 e^{s_1 t} + R_2 t e^{s_2 t} \qquad \text{for } s_1 = s_2 = -\frac{c}{2m} \qquad (15.2j)$$

It will be useful to define a dimensionless ratio called the **damping ratio** ζ which is the actual damping divided by the critical damping.

$$\zeta = \frac{c}{c_c}$$

$$(15.3a)$$

$$\zeta = \frac{c}{2m\omega_n}$$

and then:

15

$$\zeta\omega_n = \frac{c}{2m} \qquad (15.3b)$$

The damped natural frequency ω_d is slightly less than the undamped natural frequency ω_n and is:

$$\omega_d = \sqrt{\frac{k}{m} - \left(\frac{c}{2m}\right)^2} \qquad (15.3c)$$

We can substitute equations 15.1d and 15.3b into equations 15.2g to get an expression for the characteristic equation in terms of dimensionless ratios:

$$s_{1,2} = -\omega_n\zeta \pm \sqrt{\left(\omega_n\zeta\right)^2 - \omega_n{}^2}$$

$$\qquad (15.4a)$$

$$s_{1,2} = \omega_n\left(-\zeta \pm \sqrt{\zeta^2 - 1}\right)$$

This shows that the system response is determined by the damping ratio ζ which dictates the value of the discriminant. There are three possible cases:

CASE 1:	$\zeta > 1$	Roots real and unequal	
CASE 2:	$\zeta = 1$	Roots real and equal	(15.4b)
CASE 3:	$\zeta < 1$	Roots complex conjugate	

Let's consider the response of each of these cases separately.

CASE 1: $\zeta > 1$ *overdamped*

The solution is of the form in equation 15.2h and is:

$$x = R_1 e^{\left(-\zeta + \sqrt{\zeta^2 - 1}\right)\omega_n t} + R_2 e^{\left(-\zeta - \sqrt{\zeta^2 - 1}\right)\omega_n t} \qquad (15.5a)$$

Note that since $\zeta > 1$, both exponents will be negative making x the sum of two decaying exponentials as shown in Figure 15-3. This is the transient response of the system to a disturbance and dies out after a time. There is no oscillation in the output motion. An example of an overdamped system which you have probably encountered is the tone arm on a good-quality record turntable with a "cueing" feature. The tone arm can be lifted up, then released, and it will slowly "float" down to the record. This is achieved by putting a large amount of damping in the system, at the arm pivot. The arm's motion follows an exponential decay curve such as in Figure 15-3.

CASE 2: $\zeta = 1$ *critically damped*

The solution is of the form in equation 15.2j and is:

$$x = R_1 e^{-\omega_n t} + R_2 t e^{-\omega_n t} = \left(R_1 + R_2 t\right)e^{-\omega_n t} \qquad (15.5b)$$

This is the product of a linear function of time and a decaying exponential function and can take several forms depending on the values of the constants of integration, R_1

15

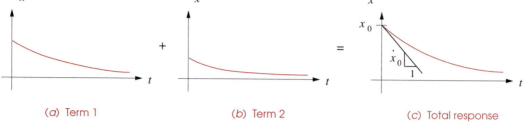

(a) Term 1   (b) Term 2   (c) Total response

FIGURE 15-3

Transient response of an overdamped system

and R_2, which in turn depend on initial conditions. A typical transient response might look like Figure 15-4. This is the transient response of the system to a disturbance, which response dies out after a time. There is fast response but no oscillation in the output motion. An example of a critically damped system is the suspension system of a new sports car in which the damping is usually made close to critical in order to provide crisp handling response without either oscillating or being slow to respond. A critically damped system will, when disturbed, return to its original position within one bounce. It may overshoot but will not oscillate and will not be sluggish.

CASE 3: $\zeta < 1$ *underdamped*

The solution is of the form in equation 15.2h and s_1, s_2 are complex conjugate. Equation 15.4a can be rewritten in a more convenient form as:

$$s_{1,2} = \omega_n \left(-\zeta \pm j\sqrt{1-\zeta^2} \right); \qquad j = \sqrt{-1} \qquad (15.5c)$$

Substituting in equation 15.2h:

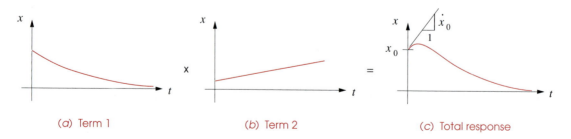

(a) Term 1   (b) Term 2   (c) Total response

FIGURE 15-4

Transient response of a critically damped system

15

$$x = R_1 e^{\left(-\zeta + j\sqrt{1-\zeta^2}\right)\omega_n t} + R_2 e^{\left(-\zeta - j\sqrt{1-\zeta^2}\right)\omega_n t}$$

and noting that:
$$y^{a+b} = y^a y^b$$

$$x = R_1\left[e^{-\zeta\omega_n t}e^{\left(j\sqrt{1-\zeta^2}\right)\omega_n t}\right] + R_2\left[e^{-\zeta\omega_n t}e^{\left(-j\sqrt{1-\zeta^2}\right)\omega_n t}\right]$$

factor:

$$x = e^{-\zeta\omega_n t}\left[R_1 e^{\left(j\sqrt{1-\zeta^2}\right)\omega_n t} + R_2 e^{\left(-j\sqrt{1-\zeta^2}\right)\omega_n t}\right] \tag{15.5d}$$

Substitute the Euler identity from equation 4.4a (p. 155):

$$x = e^{-\zeta\omega_n t}\left\{ \begin{array}{l} R_1\left[\cos\left(\sqrt{1-\zeta^2}\,\omega_n t\right) + j\sin\left(\sqrt{1-\zeta^2}\,\omega_n t\right)\right] \\ \\ + R_2\left[\cos\left(\sqrt{1-\zeta^2}\,\omega_n t\right) - j\sin\left(\sqrt{1-\zeta^2}\,\omega_n t\right)\right] \end{array} \right\}$$

and simplify: $\tag{15.5e}$

$$x = e^{-\zeta\omega_n t}\left\{(R_1 + R_2)\left[\cos\left(\sqrt{1-\zeta^2}\,\omega_n t\right) + (R_1 - R_2)j\sin\left(\sqrt{1-\zeta^2}\,\omega_n t\right)\right]\right\}$$

Note that R_1 and R_2 are just constants yet to be determined from the initial conditions, so their sum and difference can be denoted as some other constants:

$$x = e^{-\zeta\omega_n t}\left\{A\left[\cos\left(\sqrt{1-\zeta^2}\,\omega_n t\right) + B\sin\left(\sqrt{1-\zeta^2}\,\omega_n t\right)\right]\right\} \tag{15.5f}$$

We can put this in polar form by defining the magnitude and phase angle as:

$$X_0 = \sqrt{A^2 + B^2} \qquad\qquad \phi = \arctan\frac{B}{A}$$

then: $\tag{15.5g}$

$$x = X_0 e^{-\zeta\omega_n t}\cos\left[\left(\sqrt{1-\zeta^2}\,\omega_n t\right) - \phi\right]$$

This is the product of a harmonic function of time and a decaying exponential function where X_0 and f are the constants of integration determined by the initial conditions.

Figure 15-5 shows the transient response for this **underdamped case**. The response overshoots and oscillates before finally settling down to its final position. Note that if the damping ratio ζ is zero, equation 15.5g reduces to equation 15.1g which is a pure harmonic.

15

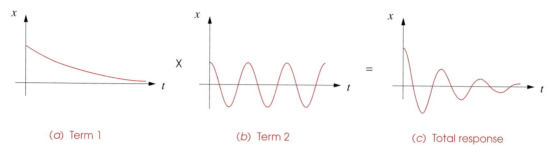

(a) Term 1   (b) Term 2   (c) Total response

FIGURE 15-5

Transient response of an underdamped system

An example of an **underdamped system** is a diving board which continues to oscillate after the diver has jumped off, finally settling back to zero position. *Many real systems in machinery are underdamped, including the typical cam-follower system.* This often leads to **vibration problems**. It is not usually a good solution simply to add damping to the system as this causes heating and is very energy inefficient. It is better to design the system to avoid the vibration problems.

PARTICULAR SOLUTION   Unlike the homogeneous solution which is always the same regardless of the input, the particular solution to equation 15.2b (p. 689) will depend on the forcing function $F_c(t)$ which is applied to the cam-follower from the cam. In general the output displacement x of the follower will be a function of similar shape to the input function but will lag the input function by some phase angle. It is quite reasonable to use a sinusoidal function as an example since any periodic function can be represented as a Fourier series of sine and cosine terms of different frequencies (see equation 13.2, p. 608, equation 13.3, p. 609, and the footnote on p. 609).

Assume the forcing function to be:

$$F_c(t) = F_0 \sin \omega_f t \tag{15.6a}$$

where F_0 is the amplitude of the force and ω_f is its circular frequency. Note that ω_f is unrelated to ω_n or ω_d and may be any value. The system equation then becomes:

$$m\ddot{x} + c\dot{x} + kx = F_0 \sin \omega_f t \tag{15.6b}$$

The solution must be of harmonic form to match this forcing function and we can try the same form of solution as used for the homogeneous solution.

$$x_f(t) = X_f \sin(\omega_f t - \psi) \tag{15.6c}$$

where :

X_f = amplitude

ψ = phase angle between applied force and displacement

ω_f = angular velocity of forcing function

The factors X_f and ψ are not constants of integration here. They are constants determined by the physical characteristics of the system and the forcing function's frequency and magnitude. They have nothing to do with the initial conditions. To find their values, differentiate the assumed solution twice, substitute in the ODE, and get:

$$X_f = \frac{F_0}{\sqrt{\left(k - m\omega_f^2\right)^2 + \left(c\omega_f\right)^2}}$$

$$\psi = \arctan\left[\frac{c\omega_f}{\left(k - m\omega_f^2\right)^2}\right]$$

(15.6d)

Substitute equations 15.1d, 15.2i, and 15.3a and put in dimensionless form:

$$\frac{X_f}{\left(\dfrac{F_0}{k}\right)} = \frac{1}{\sqrt{\left[1 - \left(\dfrac{\omega_f}{\omega_n}\right)^2\right]^2 + \left(2\zeta\dfrac{\omega_f}{\omega_n}\right)^2}}$$

$$\psi = \arctan\left[\frac{2\zeta\dfrac{\omega_f}{\omega_n}}{1 - \left(\dfrac{\omega_f}{\omega_n}\right)^2}\right]$$

(15.6e)

The ratio ω_f / ω_n is called the **frequency ratio**. Dividing X_f by the static deflection F_0 / k creates the **amplitude ratio** which defines the relative dynamic displacement compared to the static.

COMPLETE RESPONSE The complete solution to our system differential equation for a sinusoidal forcing function is the sum of the homogeneous and particular solutions:

$$x = X_0 e^{-\zeta\omega_n t}\cos\left[\left(\sqrt{1 - \zeta^2}\,\omega_n t\right) - \phi\right] + X_f\sin\left(\omega_f t - \psi\right)$$

(15.7)

The homogeneous term represents the **transient response** of the system which will die out in time but is reintroduced any time the system is again disturbed. The **particular term** represents the **forced response** or **steady-state response** to a sinusoidal forcing function which will continue as long as the forcing function is present.

Note that the solution to this equation, shown in equations 15.5 and 15.6, depends only on two ratios, the damping ratio ζ which relates the actual damping relative to the critical damping, and the *frequency ratio* ω_f / ω_n which relates the forcing frequency to the natural frequency of the system. Koster[1] found that a typical value for the damping ratio in cam-follower systems is $\zeta = 0.06$, so they are underdamped and can **resonate** if operated at frequency ratios close to 1.

The initial conditions for the specific problem are applied to equation 15.7 to determine the values of X_0 and ϕ. Note that these constants of integration are contained within the homogeneous part of the solution.

15.2   RESONANCE

The natural frequency (and its overtones) are of great interest to the designer as they define the frequencies at which the system will **resonate**. The single-D*OF* lumped parameter systems shown in Figures 15-1 and 15-2 (p. 687) are the simplest possible to describe a dynamic system, yet they contain all the basic dynamic elements. Masses and springs are energy storage elements. A mass stores kinetic energy, and a spring stores potential energy. The damper is a dissipative element. It uses energy and converts it to heat. Thus all the losses in the model of Figure 15-1 occur through the damper.

These are "pure" idealized elements which possess only their own characteristics. That is, the spring has no damping and the damper no springiness, etc. Any system which contains more than one energy storage device such as a mass and a spring will possess at least one natural frequency. If we excite the system at its natural frequency, we will set up the condition called resonance in which the energy stored in the system's elements will oscillate from one element to the other at that frequency. The result can be violent oscillations in the displacements of the movable elements in the system as the energy moves from potential to kinetic form and vice versa.

Figure 15-6 shows a plot of the amplitude and phase angle of the displacement response X of the system to a sinusoidal input forcing function at various frequencies ω_f. In our case, the forcing frequency is the angular velocity at which the cam is rotating. The plot normalizes the forcing frequency as the frequency ratio ω_f / ω_n. The response amplitude X is normalized by dividing the dynamic deflection x by the static deflection F_0 / k that the same force amplitude would create on the system. Thus at a frequency of zero, the output is one, equal to the static deflection of the spring at the amplitude of the input force . As the forcing frequency increases toward the natural frequency ω_n, the amplitude of the output motion, for zero damping, increases rapidly and becomes theoretically infinite when $\omega_f = \omega_n$. Beyond this point the amplitude decreases rapidly and asymptotically toward zero at high frequency ratios. It is obvious that we must avoid driving this system at or near its natural frequency. The addition of damping reduces the amplitude of vibration at the natural frequency, but very large damping ratios (> 0.6) are needed to keep the output amplitude less than or equal to the input amplitude. This is much more damping than is found in cam-follower systems and most machinery. One result of operation of an underdamped cam-follower system near ω_n can be **follower jump**. The system of follower mass and spring can oscillate violently at its natural frequency and leave contact with the cam. When it does reestablish contact, it may do so with severe impact loads that can quickly fail the materials.

The designer has a degree of control over resonance in that the system's mass m and stiffness k can be tailored to move its natural frequency away from any required operating frequencies. A common rule of thumb is to design the system to have a fundamental natural frequency ω_n at least ten times the highest forcing frequency expected in service, thus keeping all operation well below the resonance point. This is often difficult to achieve in mechanical systems. One tries to achieve the largest ratio ω_n / ω_f possible nevertheless, even if it is less than 10.

For cams with finite jerk having a dwell after the rise of the same order of duration as the rise, Koster[1] defines a dimensionless ratio,

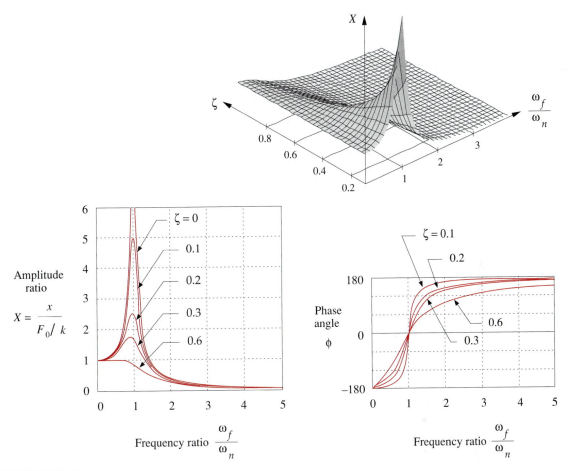

FIGURE 15-6

Amplitude ratio and phase angle of system response

$$\tau = \frac{T_n}{t_r} \qquad\qquad T_n = \frac{1}{\omega_n} \qquad\qquad (15.8)$$

where T_n is the period of the follower system's natural frequency and t_r is the time to complete the rise portion of the follower motion. To minimize the effects of the transient vibrations on the system, this ratio should be as small as possible and always less than 1. A small value of τ is equivalent to a large ratio of ω_n / ω_f. The error in acceleration of the follower due to vibrations during the dwell period will be proportional to the first power of τ, and the error in follower position will be proportional to the third power of τ, for values of $\tau < 0.5$. If the dwell's duration is about as long or longer than that of the rise, these transient vibrations will tend to die out by the end of the dwell. The next fall or rise will again provide an input to the system and cause a new transient response.

15

Thus, the response of a cam-follower system of this type will be dominated by the recurring transient responses rather than by the forced, or steady-state, response. It is important to adhere to the fundamental law of cam design and use cam programs with finite jerk in order to minimize these residual vibrations in the follower system. Koster[1] reports that the cycloidal and 3-4-5 polynomial programs both gave low residual vibrations in the double-dwell cam. The modified sine acceleration program will also give good results. All these have finite jerk. (See Section 8.8 and Table 8-4, p. 421.)

Some thought and observation of equation 15.1d (p. 688) will show that we would like our system members to be both light (low m) and stiff (high k) to get high values for ω_n and thus low values for τ. Unfortunately, the lightest materials are seldom also the stiffest. Aluminum is one-third the weight of steel but is also about a third as stiff. Titanium is about half the weight of steel but also about half as stiff. Some of the exotic composite materials such as carbon fiber/epoxy offer better stiffness-to-weight ratios but their cost is high and processing is difficult. Another job for *Unobtainium 208*!

Note in Figure 15-6 (p. 697) that the amplitude of vibration at large frequency ratios approaches zero. So, if the system can be brought up to speed through the resonance point without damage and then kept operating at a large frequency ratio, the vibration will be minimal. An example of systems designed to be run this way are large devices that must run at higher speed such as electrical power generators. Their large mass creates a lower natural frequency than their required operating speeds. They are "run up" as quickly as possible through the resonance region to avoid damage from their vibrations and "rundown" quickly through resonance when stopping them. They also have the advantage of long duty cycles of constant speed operation in the safe frequency region between infrequent starts and stops.

15.3   KINETOSTATIC FORCE ANALYSIS OF THE FORCE-CLOSED CAM-FOLLOWER

The analysis in the previous sections is of the "**forward dynamic analysis**" type as defined in Section 10.1, p. 491. That is, the applied force [$F_c(t)$ in equation 15.2b, p. 689] is presumed to be known, and the equation is solved for the resulting displacement x from which its derivatives can also be determined. Cam-follower force analysis problems, like linkage analysis problems, lend themselves well to solution by the **inverse dynamics**, or **kinetostatics**, approach. Here the displacement and its derivatives are defined from the kinematic design of the cam based on an assumed angular velocity ω of the cam, and we want to solve for the force $F_c(t)$. Equation 15.2b can be solved directly for the force in a spring-loaded cam-follower system provided that values for mass m, spring constant k, preload F_{pl}, and damping factor c are known in addition to the displacement, velocity, and acceleration functions.

The designer has a large degree of control over the system spring constant k_{eff} as it tends to be dominated by the k_s of the physical return spring. The elasticities of the follower parts also contribute to the overall system k_{eff} but are usually much stiffer than the physical spring. If the follower stiffness is in series with the return spring, as it often is, equation 10.17c (p. 506) shows that the softest spring in series will dominate the effective spring constant. Thus the return spring will virtually determine the overall k unless some parts of the follower train have similarly low stiffness.

The designer will choose or design the return spring and thus can specify both its k and the amount of preload to be introduced at assembly. Preload of a spring occurs when it is compressed (or extended if an extension spring) from its *free length* to its initial assembled length. This is a necessary and desirable situation as we want some residual force on the follower even when the cam is at its lowest displacement. This will help maintain good contact between the cam and follower at all times. This spring preload F_{pl} adds a constant term to equation 15.2b which becomes:

$$F_c(t) = m\ddot{x} + c\dot{x} + kx + F_{pl} \tag{15.9}$$

The value of m is determined from the effective mass of the system as lumped in the single-*DOF* model of Figure 15-1 (p. 687). The value of c for most cam-follower systems can be estimated for a first approximation to be about 0.05 to 0.15 of the critical damping c_c as defined in equation 15.2i (p. 690). Koster[1] found that a typical value for the damping ratio in cam-follower systems is $\zeta = 0.06$.

Calculating the damping c based on an assumed value of ζ requires specifying a value for the overall system k and for its effective mass. The choice of k will affect both the natural frequency of the system for a given mass and the available force to keep the joint closed. Some iteration will probably be needed to find a good compromise. A selection of data for commercially available helical coil springs is provided in Appendix D. Note in equation 15.9 that the terms involving acceleration and velocity can be either positive or negative. The terms involving the spring parameters k and F_{pl} are the only ones that are always positive. So, to keep the overall function always positive requires that the spring force terms be large enough to counteract any negative values in the other terms. Typically, the acceleration is larger numerically than the velocity, so the negative acceleration usually is the principal cause of a negative force F_c.

The principal concern in this analysis is to keep the cam force always positive in sign as its direction is defined in Figure 15-1. The cam force is shown as positive in that figure. In a force-closed system the cam can only push on the follower. It cannot pull. The follower spring is responsible for providing the force needed to keep the joint closed during the negative acceleration portions of the follower motion. The damping force also can contribute, but the spring must supply the bulk of the force to maintain contact between the cam and follower. If the force F_c goes negative at any time in the cycle, the follower and cam will part company, a condition called **follower jump**. When they meet again, it will be with large and potentially damaging impact forces. The follower jump, if any, will occur near the point of maximum negative acceleration. Thus we must select the spring constant and preload to guarantee a positive force at all points in the cycle. In automotive engine valve cam applications follower jump is also called *valve float*, because the valve (follower) "floats" above the cam, also periodically impacting the cam surface. This will occur if the cam rpm is increased to the point that the larger negative acceleration makes the follower force negative. The "redline" maximum engine rpm often indicated on its tachometer is to warn of impending valve float above that speed which will damage the cam and follower.

Program DYNACAM allows the iteration of equation 15.9 to be done quickly and painlessly for any cam whose kinematics have been defined in that program. The program's *Dynamics* button will solve equation 15.9 for all values of camshaft angle, using the displacement, velocity, and acceleration functions previously calculated for that cam design in the program. The program requires values for the effective system mass m, effective spring constant k, preload F_{pl}, and the assumed value of the damping ratio ζ. These values need to be determined for the model by the designer using the methods described in Sections 10.9 (p. 500) and 10.10 (p. 503). The calculated force at the cam-follower interface can then be plotted or its values printed in tabular form. The system's natural frequency is also reported when the tabular force data are printed.

✍ EXAMPLE 15-1

Kinetostatic Force Analysis of a Force-Closed (Spring-loaded) Cam-Follower System.

Given: A translating roller follower as shown in Figure 15-1 is driven by a force-closed radial plate cam which has the following program:

Segment 1:   Rise 1 inch in 50° with modified sine acceleration
Segment 2:   Dwell for 40°
Segment 3:   Fall 1 inch in 50° with cycloidal displacement
Segment 4:   Dwell for 40°
Segment 5:   Rise 1 inch in 50° with 3-4-5 polynomial displacement
Segment 6:   Dwell for 40°
Segment 7:   Fall 1 inch in 50° with 4-5-6-7 polynomial displacement
Segment 8:   Dwell for 40°
Camshaft angular velocity is 18.85 rad/sec.
Follower effective mass is 0.0738 in-lb-sec^2 (blobs).
Damping is 15% of critical ($\zeta = 0.15$).

Problem: Compute the necessary spring constant and spring preload to maintain contact between cam and follower and calculate the dynamic force function for the cam. Calculate the system natural frequency with the selected spring. Keep the pressure angle under 30°.

Solution:

1   Calculate the kinematic data (follower displacement, velocity, acceleration, and jerk) for the specified cam functions. The acceleration for this cam is shown in Figure 15-7 and has a maximum value of 3504 in/sec^2. See Chapter 8 to review this procedure.

2   Calculate the pressure angle and radius of curvature for trial values of prime circle radius, and size the cam to control these values. Figure 15-8 shows the pressure angle function and Figure 15-9 the radii of curvature for this cam with a prime circle radius of 4 in and zero eccentricity. The maximum pressure angle is 29.2° and the minimum radius of curvature is 1.7 in. Figure 8-47 (p. 407) shows the finished cam profile. See Chapter 8 to review these calculations.

3   With the kinematics of the cam defined, we can address its dynamics. To solve equation 15.9 for cam force, we must assume values for the spring constant k and the preload F_{pl}. The value of c can be calculated from equation 15.3a using the given mass m, the damping factor ζ, and assumed k. The kinematic parameters are known.

4   Program DYNACAM does this computation for you. The dynamic force that results from an assumed k of 150 lb/in and a preload of 75 lb is shown in Figure 15-10a. The damping coefficient $c = 0.998$. Note that the force dips below the zero axis in two places during negative acceleration. These are locations of follower jump. The follower has left the cam during the

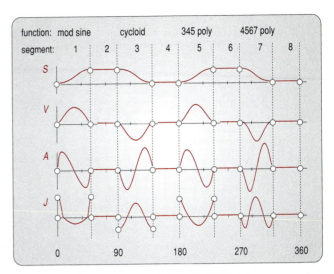

Segment Number	Function used	Start Angle	End Angle	Delta Angle
1	ModSine rise	0	50	50
2	Dwell	50	90	40
3	Cycloid fall	90	140	50
4	Dwell	140	180	40
5	345 poly rise	180	230	50
6	Dwell	230	270	40
7	4567 poly fall	270	320	50
8	Dwell	320	360	40

(a) Cam program specifications

(b) Plots of cam-follower's *S V A J* diagrams

FIGURE 15-7

S V A J diagrams for Examples 15-1 and 15-2 (p. 705)

fall because the spring does not have enough available force to keep the follower in contact with the rapidly falling cam. Open the file E15-01.cam in DYNACAM and provide the specified *k* and F_{pl} to see this example. Another iteration is needed to improve the design.

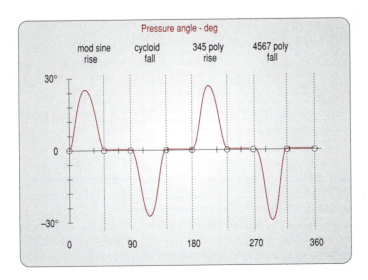

FIGURE 15-8

Pressure angle plot for Examples 15-1 and 15-2 (p. 705)

15

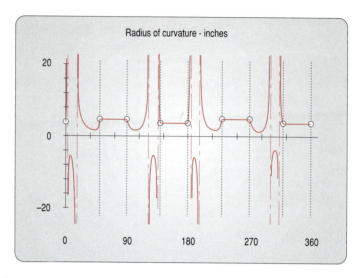

FIGURE 15-9

Radius of curvature of a four-dwell cam for Examples 15-1 and 15-2 (p. 705)

5   Figure 15-10b shows the dynamic force for the same cam with a spring constant of $k = 200$
     lb/in and a preload of 150 lb. The damping coefficient $c = 1.153$. This additional force has
     lifted the function up sufficiently to keep it positive everywhere. There is no follower jump
     in this case. The maximum force during the cycle is 400.4 lb. A margin of safety has been
     provided by keeping the minimum force comfortably above the zero line at 36.9 lb. Run ex-
     ample #5 in the program and provide the specified spring constant and preload values to see this
     example.

6   The fundamental natural frequency, both undamped and damped, can be calculated for the
     system from equations 15.1d (p. 688) and 15.3c (p. 691) and are:

$$\omega_n = 52.06 \text{ rad/sec}; \qquad\qquad \omega_d = 51.98 \text{ rad/sec}$$

15.4   KINETOSTATIC FORCE ANALYSIS OF THE FORM-CLOSED CAM-FOLLOWER

Section 8.1 described two types of joint closure used in cam-follower systems, **force closure** and **form closure**. Force closure uses an open joint and requires a spring or other
force source to maintain contact between the elements. Form closure provides a geometric constraint at the joint such as the cam groove shown in Figure 15-11a or the conjugate cams of Figure 15-11b. No spring is needed to keep the follower in contact with
these cams. The follower will run against one side or the other of the groove or conjugate pair as necessary to provide both positive and negative forces. Since there is no
spring in this system, its dynamic force equation 15.9 (p. 699) simplifies to:

$$F_c(t) = m\ddot{x} + c\dot{x} \qquad\qquad (15.10)$$

15

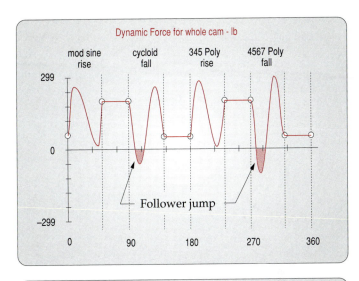

(a) Insufficient spring force allows follower jump

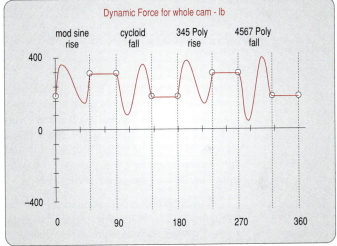

(b) Sufficient spring force keeps the dynamic force positive

FIGURE 15-10

Dynamic forces in a force-closed cam-follower system

Note that there is now only one energy storage element in the system (the mass), so, theoretically, resonance is not possible. There is no natural frequency for it to resonate at. This is the chief advantage of a form-closed system over a force-closed one. Follower jump will not occur, short of complete failure of the parts, no matter how fast the system is run. This arrangement is sometimes used in high-performance or racing engine valve trains to allow higher redline engine speeds without valve float. In engine valve trains, a form-closed cam-follower valve train is called a *desmodromic* system.

As with any design, there are trade-offs. While the form-closed system typically allows higher operating speeds than a comparable force-closed system, it is not free of all

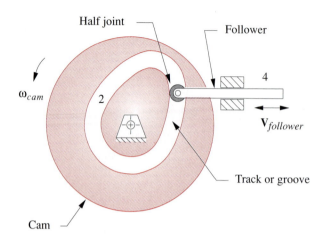

(a) Form-closed cam with translating follower

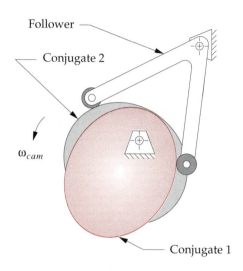

(b) Conjugate cams on common shaft

FIGURE 15-11

Form-closed cam-follower systems

vibration problems. Even though there is no physical return spring in the system, the follower train, the camshaft, and all other parts still have their own spring constants which combine to provide an overall effective spring constant for the system as shown in Section 15.2 (p. 696). The positive side is that this spring constant will typically be quite large (stiff) since properly designed follower parts are designed to be stiff. The effective natural frequency will then be high (see equation 15.1d, p. 688) and possibly well above the forcing frequency as desired.

Another problem with form-closed cams, especially the grooved or track type shown in Figure 15-11a, is the phenomenon of **crossover shock**. Every time the acceleration of the follower changes sign, the inertial force also does so. This causes the follower to abruptly shift from one side of the cam groove to the other. There cannot be zero clearance between the roller follower and the groove and still have it operate. Even if the clearance is very small, there will still be an opportunity for the follower to develop some velocity in its short trip across the groove, and it will impact the other side. Track cams of the type shown in Figure 15-11a typically fail at the points where the acceleration reverses sign, due to many cycles of crossover shock. Note also that the roller follower has to reverse direction every time it crosses over to the other side of the groove. This causes significant follower slip and high wear on the follower compared to an open, force-closed cam where the follower will have less than 1% slip.

Because there are two cam surfaces to machine and because the cam track, or groove, must be cut and ground to high precision to control the clearance, form-closed cams tend to be more expensive to manufacture than force-closed cams. Track cams usually must be ground after heat treatment to correct the distortion of the groove resulting from the high temperatures. Grinding significantly increases cost. Many force-closed cams are not ground after heat treatment and are used as-milled. Though the conjugate cam approach avoids the groove tolerance and heat treat distortion problems, there are still two matched cam surfaces to be made per cam. Thus, the desmodromic cam's dynamic advantages come at a significant cost premium.

We will now repeat the cam design of Example 15-1, modified for desmodromic operation. This is simple to do with program DYNACAM. We will merely specify the spring constant and preload values to be zero, which then assumes that the follower train is a rigid body. A more accurate result can be obtained by calculating and using the effective spring constant of the combination of parts in the follower train, once their geometries and materials are defined. The dynamic forces will now be negative as well as positive, but a form-closed cam can both push and pull.

 **EXAMPLE 15-2**

Dynamic Force Analysis of a Form-Closed (Desmodromic) Cam-Follower System.

Given: A translating roller follower as shown in Figure 15-11a is driven by a form-closed radial plate cam which has the following program:

Segment 1:   Rise 1 inch in 50° with modified sine acceleration
Segment 2:   Dwell for 40°
Segment 3:   Fall 1 inch in 50° with cycloidal displacement
Segment 4:   Dwell for 40°
Segment 5:   Rise 1 inch in 50° with 3-4-5 polynomial displacement
Segment 6:   Dwell for 40°
Segment 7:   Fall 1 inch in 50° with 4-5-6-7 polynomial displacement
Segment 8:   Dwell for 40°
Camshaft angular velocity is 18.85 rad/sec.
Follower effective mass is 0.0738 in-lb-sec^2 (blobs).
Damping is 15% of critical ($\zeta = 0.15$).

Problem: Compute the dynamic force function for the cam. Keep the pressure angle under 30°.

Solution:

1 Calculate the kinematic data (follower displacement, velocity, acceleration, and jerk) for the specified cam functions. The acceleration for this cam is shown in Figure 15-7 (p. 701) and has a maximum value of 3504 in/sec^2. See Chapter 8 to review this procedure.

2 Calculate radius of curvature and pressure angle for trial values of prime circle radius, and size the cam to control these values. Figure 15-8 shows the pressure angle function and Figure 15-9 the radii of curvature for this cam with a prime circle radius of 4 in and zero eccentricity. The maximum pressure angle is 29.2° and the minimum radius of curvature is 1.7 in. Figure 8-47 (p. 407) shows the finished cam profile. See Chapter 8 to review these calculations.

3 With the kinematics of the cam defined, we can address its dynamics. To solve equation 15.10 (p. 702) for the cam force, we assume zero values for the spring constant k and the preload F_{pl}. The value of c is assumed to be the same as in the previous example, 1.153. The kinematic parameters are known.

15

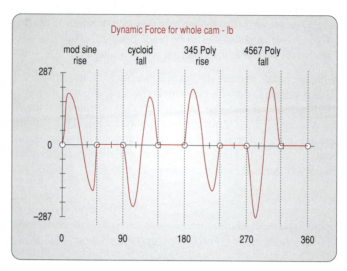

FIGURE 15-12

Dynamic force in a form-closed cam-follower system

4   Program DYNACAM does this computation for you. The dynamic force that results is shown in Figure 15-12. Note that the force is now more nearly symmetric about the axis and its peak absolute value is 289 lb. Crossover shock will occur each time the follower force changes sign. Open the file E15-02.cam in DYNACAM to see this example.

Compare the dynamic force plots for the force-closed system (Figure 15-10b, p. 703) and the form-closed system (Figure 15-12). The absolute peak force magnitude on either side of the track in the form-closed cam is less than that on the spring-loaded one. This shows the penalty that the spring imposes on the system in order to keep the joint closed. Thus, either side of the cam groove will experience lower stresses than will the open cam, except for the areas of crossover shock mentioned on p. 704.

15.5   CAMSHAFT TORQUE

The kinetostatic analysis assumes that the camshaft will operate at some constant speed ω. As we saw in the case of the fourbar linkage in Chapter 11 and with the slider-crank mechanism in Chapter 13, the input torque must vary over the cycle if the shaft velocity is to remain constant. The torque can be easily calculated from the power relationship, ignoring losses.

$$\text{Power in} = \text{Power out}$$
$$T_c \omega = F_c v \tag{15.11}$$
$$T_c = \frac{F_c v}{\omega}$$

15

Once the cam force has been calculated from either equation 15.9 or 15.10, the camshaft torque T_c is easily found since the follower velocity v and camshaft ω are both known. Figure 15-13a shows the camshaft input torque needed to drive the force-closed cam designed in Example 15-1 (p. 700). Figure 15-13b shows the camshaft input torque needed to drive the form-closed cam designed in Example 15-2 (p. 705). Note that the torque required to drive the force-closed (spring-loaded) system is significantly higher than that needed to drive the form-closed (track) cam. The spring force is also extracting a penalty here as energy must be stored in the spring during the rise portions which will tend to slow the camshaft. This stored energy is then returned to the camshaft during the fall portions, tending to speed it up. The spring loading causes larger oscillations in the torque.

(*a*) Force-closed (spring-loaded) cam-follower

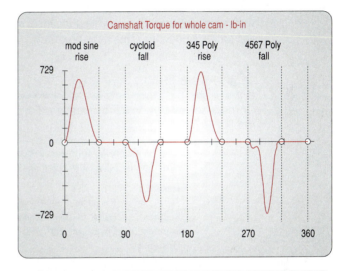

(*b*) Form-closed (desmodromic) cam-follower

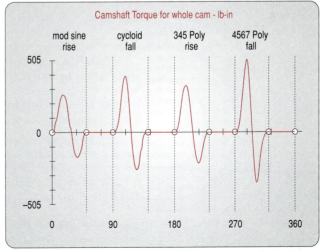

FIGURE 15-13

Input torque in force- and form-closed cam-follower systems

15

A flywheel can be sized and fitted to the camshaft to smooth these variations in torque just as was done for the fourbar linkage in Section 11.11 (p. 548). See that section for the design procedure. Program DYNACAM integrates the camshaft torque function pulse by pulse and prints those areas to the screen. These energy data can be used to calculate the required flywheel size for any selected coefficient of fluctuation.

One useful way to compare alternate cam designs is to look at the torque function as well as at the dynamic force. A smaller torque variation will require a smaller motor and/ or flywheel and will run more smoothly. Three different designs for a single-dwell cam were explored in Chapter 8. (See Examples 8-5, p. 375, 8-6, p. 377, and 8-8, p. 383.) All had the same lift and duration but used different cam functions. One was a double harmonic, one cycloidal, and one a sixth-degree polynomial. On the basis of their kinematic results, principally acceleration magnitude, we found that the polynomial design was superior. We will now revisit this cam as an example and compare its dynamic force and torque among the same three programs.

✎EXAMPLE 15-3

Comparison of Dynamic Torques and Forces Among Three Alternate Designs of the Same Cam.

Given: A translating roller follower as shown in Figure 15-1 (p. 687) is driven by a force-closed radial plate cam which has the following program:

Design 1
Segment 1:   Rise 1 inch in 90° double harmonic displacement
Segment 2:   Fall 1 inch in 90° double harmonic displacement
Segment 3:   Dwell for 180°
Design 2
Segment 1:   Rise 1 inch in 90° cycloidal displacement
Segment 2:   Fall 1 inch in 90° cycloidal displacement
Segment 3:   Dwell for 180°
Design 3
Segment 1:   Rise 1 inch in 90° and fall 1 inch in 90° with polynomial displacement. (A single polynomial can create both rise and fall.)
Segment 2:   Dwell for 180°

Camshaft angular velocity is 15 rad/sec. Follower effective mass is 0.0738 in-lb-sec^2 (blobs). Damping is 15% of critical ($\zeta = 0.15$).

Find: The dynamic force and torque functions for the cam. Compare their peak magnitudes for the same prime circle radius.

Solution: Note that these are the same kinematic cam designs as are shown in Figures 8-24 (p. 377), 8-23 (p. 376), and 8-29 (p. 385).

1   Calculate the kinematic data (follower displacement, velocity, acceleration, and jerk) for each of the specified cam designs. See Chapter 8 to review this procedure.

2   Calculate the radius of curvature and pressure angle for trial values of prime circle radius, and size the cam to control these values. A prime circle radius of 3 in gives acceptable pressure angles and radii of curvature. See Chapter 8 to review these calculations.

15

3   With the kinematics of the cam defined, we can address its dynamics. To solve equation 15.1a (p. 687) for the cam force, we will assume a value of 50 lb/in for the spring constant k and adjust the preload F_{pl} for each design to obtain a minimum dynamic force of about 10 lb. For design 1 this requires a spring preload of 28 lb; for design 2, 15 lb; and for design 3, 10 lb.

4   The value of damping c is calculated from equation 15.2i (p. 690). The kinematic parameters x, v, and a are known from the prior analysis.

5   Program DYNACAM will do these computations for you. The dynamic forces that result from each design are shown in Figure 15-14 and the torques in Figure 15-15. Note that the force is largest for design 1 at 82 lb peak and least for design 3 at 53 lb peak. The same ranking holds for the torques which range from 96 lb-in for design 1 to 52 lb-in for design 3. These represent reductions of 35% and 46% in the dynamic loading due to a change in the kinematic design. Not surprisingly, the sixth-degree polynomial design which had the lowest acceleration also has the lowest forces and torques and is the clear winner. Open the files E08-05.cam, E08-06.cam, and E08-08.cam in program DYNACAM to see these results.

15.6   MEASURING DYNAMIC FORCES AND ACCELERATIONS

As described in previous sections, cam-follower systems tend to be underdamped. This allows significant oscillations and vibrations to occur in the follower train. Section 8.8 (p. 412) discussed the effects of manufacturing errors on the actual accelerations of cam-followers and showed dynamic measurements of acceleration. Dynamic forces and accelerations can be measured fairly easily in operating machinery. Compact, piezoelectric force and acceleration transducers are available that have frequency response ranges in the high thousands of hertz. Strain gages provide strain measurements that are proportional to force and have bandwidths of a kilohertz or better.

Figure 15-16 shows acceleration and force curves as measured on the follower train of a single overhead camshaft (SOHC) valve train in a 1.8-liter four-cylinder inline engine. [2] The nonfiring engine was driven by an electric motor on a dynamometer. The camshaft is turning at 500, 2000, and 3000 rpm (1000, 4000, and 6000 crankshaft rpm), respectively, in the three plots of Figure 15-16a, b, and c. Acceleration was measured with a piezoelectric accelerometer attached to the head of one intake valve, and the force was calculated from the output of strain gages placed on the rocker arm for that intake valve. The theoretical follower acceleration curve (as designed) is superposed on the measured acceleration curve. All acceleration measurements are converted to units of mm/deg^2 (i.e., normalized to camshaft speed) to allow comparison with one another and with the theoretical acceleration curve.

At 500 camshaft rpm, the measured acceleration closely matches the theoretical acceleration curve with some minor oscillations due to spring vibration. At 2000 camshaft rpm, significant oscillation in the measured acceleration is seen during the first positive and in the negative acceleration phase. This is due to the valve spring vibrating at its natural frequency in response to excitation by the cam. This is called "spring surge" and is a significant factor in valve spring fatigue failure. At 3000 camshaft rpm, the spring surge is still present but is less prominent as a percentage of total acceleration. The frequency content of the cam's forcing function passed through the first natural fre-

15

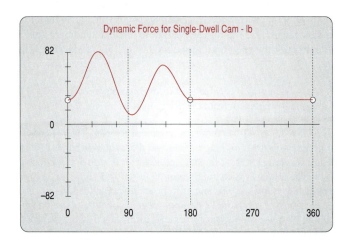

(a) Double harmonic rise - double harmonic fall

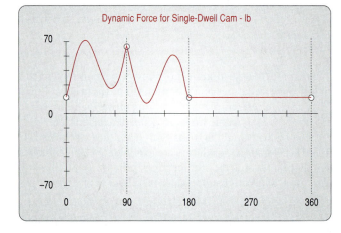

(b) Cycloidal rise - cycloidal fall

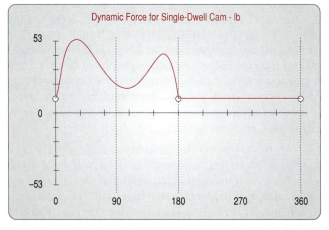

(c) Sixth-degree polynomial

FIGURE 15-14

Dynamic forces in three different designs of a single-dwell cam

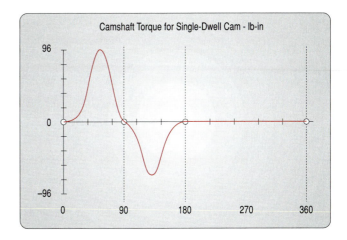

(a) Double harmonic rise - double harmonic fall

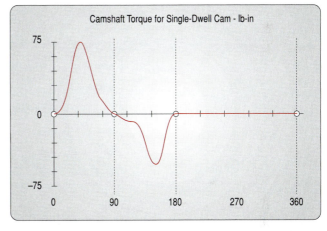

(b) Cycloidal rise - cycloidal fall

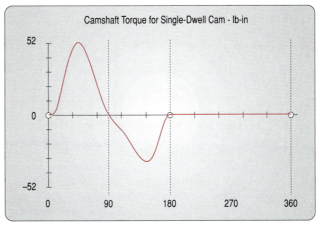

(c) Sixth-degree polynomial

FIGURE 15-15

15

Dynamic input torque in three different designs of a single-dwell cam

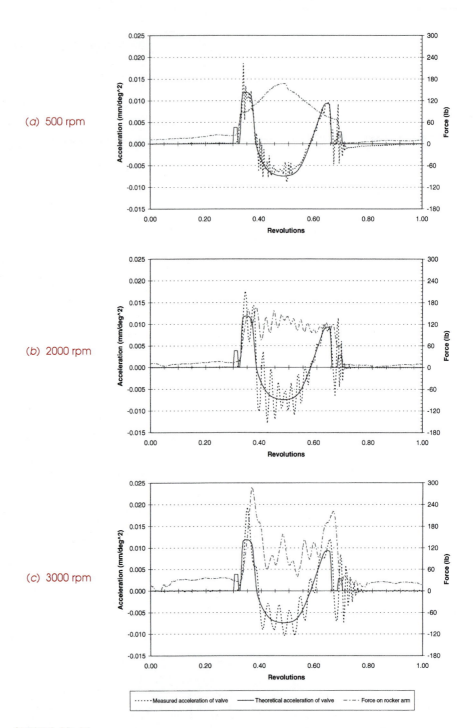

(a) 500 rpm

(b) 2000 rpm

(c) 3000 rpm

········ Measured acceleration of valve   ——— Theoretical acceleration of valve   — · — Force on rocker arm

FIGURE 15-16

Valve acceleration and rocker arm force in a single overhead cam valve train

15

quency of the valve spring at about 2000 camshaft rpm, causing the spring to resonate. The same effects can be seen in the rocker arm force. Everything in a machine tends to sympathetically vibrate at its own natural frequency when excited by any input forcing function. Sensitive transducers such as accelerometers will pick up these vibrations as they are transmitted through the structure.

15.7 PRACTICAL CONSIDERATIONS

Koster[1] proposes some general rules for the design of cam-follower systems for high-speed operation based on his extensive dynamic modeling and experimentation.

To minimize the positional error and residual acceleration:

1 Keep the total lift of the follower to a minimum.

2 If possible, arrange the follower spring to preload all pivots in a consistent direction to control backlash in the joints.

3 Keep the duration of rises and falls as long as possible to minimize τ.

4 Keep follower train mass low and follower train stiffness high to increase natural frequency.

5 Any lever ratios present will change the effective stiffness of the system by the square of the ratio. Try to keep lever ratios close to 1.

6 Make the camshaft as stiff as possible **both in torsion and in bending**. This is perhaps the most important factor in controlling follower vibration.

7 Reduce pressure angle by increasing the cam pitch circle diameter.

8 Use low backlash or antibacklash gears in the camshaft drive train.

15.8 REFERENCES

1 **Koster, M. P.** (1974). *Vibrations of Cam Mechanisms.* Phillips Technical Library Series, Macmillan Press Ltd.: London.

2 **Norton, R. L., et al.** (1998). "Analyzing Vibrations in an IC Engine Valve Train." SAE Paper: 980570.

15.9 BIBLIOGRAPHY

Barkan, P., and R. V. McGarrity. (1965). "A Spring Actuated, Cam-Follower System: Design Theory and Experimental Results." *ASME J. Engineering for Industry (August)*, pp. 279-286.

Chen, F. Y. (1975). "A Survey of the State of the Art of Cam System Dynamics." *Mechanism and Machine Theory*, **12**, pp. 210 - 224.

Chen, F. Y. (1982). *Mechanics and Design of Cam Mechanisms.* Pergamon Press: New York, p. 520.

Freudenstein, F. (1959). "On the Dynamics of High-Speed Cam Profiles." *Int. J. Mech. Sci.*, **1**, pp. 342 - 349.

Freudenstein, F., et al. (1969). "Dynamic Response of Mechanical Systems." IBM: New York Scientific Center, Report No. 320-2967.

Hrones, J. A. (1948). "An Analysis of the Dynamic Forces in a Cam System." *Trans ASME*, **70**, pp. 473-482.

15

Johnson, A. R. (1965). "Motion Control for a Series System of N Degrees of Freedom Using Numerically Derived and Evaluated Equations." *ASME J. Eng. Industry*, pp. 191-204.

Knight, B. A., and H. L. Johnson. (1966). "Motion Analysis of Flexible Cam-Follower Systems." ASME Paper: 66-Mech-3.

Matthew, G. K., and D. Tesar. (1975). "Cam System Design: The Dynamic Synthesis and Analysis of the One Degree of Freedom Model." *Mechanisms and Machine Theory*, **11**, pp. 247 - 257.

Matthew, G. K., and D. Tesar. (1975). "The Design of Modeled Cam Systems Part I: Dynamic Synthesis and Design Chart for the Two-Degree-of-Freedom Model." *Journal of Engineering for Industry* (November), pp. 1175-1180.

Midha, A., and D. A. Turic. (1980). "On the Periodic Response of Cam Mechanisms with Flexible Follower and Camshaft." *J. Dyn. Sys. Meas. Control*, **102**(December), pp. 225-264.

Rothbart, H. A. (1956). *Cams: Design, Dynamics, and Accuracy.* John Wiley & Sons: New York, p. 350.

15.10 PROBLEMS

Program DYNACAM *may be used to solve these problems where applicable. Where units are unspecified, work in any consistent units system you wish. Appendix D contains some pages from a catalog of commercially available helical coil springs to aid in designing realistic solutions to these problems.*

*‡15-1   Design a double-dwell cam to move a 2-in-dia roller follower of mass = 2.2 bl from 0 to 2.5 inches in 60° with modified sine acceleration, dwell for 120°, fall 2.5 inches in 30° with cycloidal motion, and dwell for the remainder. The total cycle must take 4 sec. Size a return spring and specify its preload to maintain contact between cam and follower. Calculate and plot the dynamic force and torque. Assume damping of 0.2 times critical. Repeat for a form-closed cam. Compare the dynamic force, torque, and natural frequency for the form-closed design and the force-closed design.

*‡15-2   Design a double-dwell cam to move a 2-in-dia roller follower of mass = 1.4 bl from 0 to 1.5 inches in 45° with 3-4-5 polynomial motion, dwell for 150°, fall 1.5 inches in 90° with 4-5-6-7 polynomial motion, and dwell for the remainder. The total cycle must take 6 sec. Size a return spring and specify its preload to maintain contact between cam and follower. Calculate and plot the dynamic force and torque. Assume damping of 0.1 times critical. Repeat for a form-closed cam. Compare the dynamic force, torque, and natural frequency for the form-closed design and the force-closed design.

*‡15-3   Design a single-dwell cam to move a 2-in-dia roller follower of mass = 3.2 bl from 0 to 2 inches in 60°, fall 2 inches in 90°, and dwell for the remainder. The total cycle must take 5 sec. Use a seventh-degree polynomial. Size a return spring and specify its preload to maintain contact between cam and follower. Calculate and plot the dynamic force and torque. Assume damping of 0.15 times critical. Repeat for a form-closed cam. Compare the dynamic force, torque, and natural frequency for the form-closed design and the force-closed design.

*‡15-4   Design a three-dwell cam to move a 2-in-dia roller follower of mass = 0.4 bl from 0 to 2.5 inches in 40°, dwell for 100°, fall 1.5 inches in 90°, dwell for 20°, fall 1 inch in 30°, and dwell for the remainder. The total cycle must take 10 sec. Choose suitable programs for rise and fall to minimize dynamic forces and torques. Size a return spring and specify its preload to maintain contact between cam and follower. Calculate and plot the dynamic force and torque. Assume damping of 0.12 times critical. Repeat for a form-closed cam. Compare the dynamic force, torque, and natural frequency for the form-closed design and the force-closed design.

---

* Answers in Appendix F.

---

† These problems are suited to solution using *Mathcad, Matlab,* or *TKSolver* equation solver programs.

---

‡ These problems are suited to solution using program DYNACAM which is on the attached CD-ROM.

15

*‡15-5   Design a four-dwell cam to move a 2-in-dia roller follower of mass = 1.25 bl from 0 to 2.5 inches in 40°, dwell for 100°, fall 1.5 inches in 90°, dwell for 20°, fall 0.5 inches in 30°, dwell for 40°, fall 0.5 inches in 30°, and dwell for the remainder. The total cycle must take 15 sec. Choose suitable programs for rise and fall to minimize dynamic forces and torques. Size a return spring and specify its preload to maintain contact between cam and follower. Calculate and plot the dynamic force and torque. Assume damping of 0.18 times critical. Repeat for a form-closed cam. Compare the dynamic force, torque, and natural frequency for the form-closed design and the force-closed design.

*†15-6   A mass-spring damper system as shown in Figure 15-1b (p. 687) has the values shown in Table P15-1. Find the undamped and damped natural frequencies and the value of critical damping for the system(s) assigned.

†15-7   Figure P15-1 shows a cam-follower system. The dimensions of the solid, rectangular 2 x 2.5 in cross-section aluminum arm are given. The cutout for the 2-in-diameter , 1.5-in-wide steel roller follower is 3 in long. Find the arm's mass, center of gravity location, and mass moment of inertia about both its *CG* and the arm pivot. Create a linear, one-*DOF* lumped mass model of the dynamic system referenced to the cam-follower and calculate the cam-follower force for one revolution. The cam is a pure eccentric with eccentricity = 0.5 in and turns at 500 rpm. The spring has a rate of 123 lb/in and a preload of 173 lb. Ignore damping.

†‡15-8   Repeat Problem 15-7 for a double-dwell cam to move the roller follower from 0 to 2.5 inches in 60° with modified sine acceleration, dwell for 120°, fall 2.5 inches in 30° with cycloidal motion, and dwell for the remainder. Cam speed is 100 rpm. Choose a suitable spring rate and preload to maintain follower contact. Select a spring from Appendix D. Assume a damping ratio of 0.10.

†‡15-9   Repeat Problem 15-7 for a double-dwell cam to move the roller follower from 0 to 1.5 inches in 45° with 3-4-5 polynomial motion, dwell for 150°, fall 1.5 inches in 90° with 4-5-6-7 polynomial motion, and dwell for the remainder. Cam speed is 250 rpm. Choose a suitable spring rate and preload to maintain follower contact. Select a spring from Appendix D. Assume a damping ratio of 0.15.

TABLE P15-1
Problem 15-6

	m	k	c
a.	1.2	14	1.1
b.	2.1	46	2.4
c.	30.0	2	0.9
d.	4.5	25	3.0
e.	2.8	75	7.0
f.	12.0	50	14.0

*   Answers in Appendix F..

† These problems are suited to solution using *Mathcad, Matlab,* or *TKSolver* equation solver programs.

‡ These problems are suited to solution using program DYNACAM which is on the attached CD-ROM.

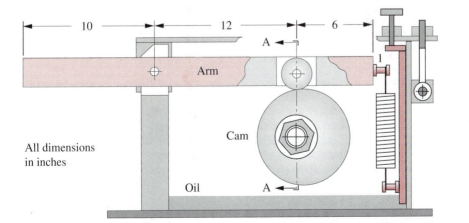

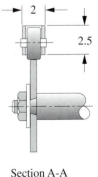

All dimensions in inches

Cam

Arm

Oil

Section A-A

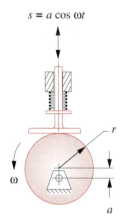

$s = a \cos \omega t$

FIGURE P15-2

Problems 15-12 to 15-14

†‡15-10 Repeat Problem 15-7 for a single-dwell cam to move the follower from 0 to 2 inches in 60°, fall 2 inches in 90°, and dwell for the remainder. Use a seventh-degree polynomial. Cam speed is 250 rpm. Choose a suitable spring rate and preload to maintain follower contact. Select a spring from Appendix D. Assume a damping ratio of 0.15.

†‡15-11 Repeat Problem 15-7 for a double-dwell cam to move the roller follower from 0 to 2 inches in 45° with cycloidal motion, dwell for 150°, fall 2 inches in 90° with modified sine motion, and dwell for the remainder. Cam speed is 200 rpm. Choose a suitable spring rate and preload to maintain follower contact. Select a spring from Appendix D. Assume a damping ratio of 0.15.

†‡15-12 The cam in Figure P15-2 is a pure eccentric with eccentricity = 20 mm and turns at 200 rpm. The mass of the follower is 1 kg. The spring has a rate of 10 N/m and a preload of 0.2 N. Find the follower force over one revolution. Assume a damping ratio of 0.10. If there is follower jump, respecify the spring rate and preload to eliminate it.

†‡15-13 Repeat Problem 15-12 using a cam with a 20-mm symmetric double harmonic rise and fall (180° rise -180° fall). See Chapter 8 for cam formulas.

‡15-14 Repeat Problem 15-12 using a cam with a 20-mm 3-4-5-6 polynomial rise and fall (180° rise -180° fall). See Chapter 8 for cam formulas.

‡15-15 Design a double-dwell cam to move a 50-mm-dia roller follower of mass = 2 kg from 0 to 45 mm in 60° with modified sine acceleration, dwell for 120°, fall 45 mm in 90° with 3-4-5 polynomial motion, and dwell for the remainder. The total cycle must take 1 sec. Size a return spring and specify its preload to maintain contact between cam and follower. Select a spring from Appendix D. Calculate and plot the dynamic force and torque. Assume damping of 0.25 times critical. Repeat for a form-closed cam. Compare the dynamic force, torque, and natural frequency for the form-closed design and the force-closed design.

‡15-16 Design a single-dwell cam using polynomials to move a 50-mm-dia roller follower of mass = 10 kg from 0 to 25 mm in 60°, fall 25 mm in 90°, and dwell for the remainder. The total cycle must take 2 sec. Size a return spring and specify its preload to maintain contact between cam and follower. Select a spring from Appendix D. Calculate and plot the dynamic force and torque. Assume damping of 0.15 times critical. Repeat for a form-closed cam. Compare the dynamic force, torque, and natural frequency for the form-closed design and the force-closed design.

‡15-17 Design a four-dwell cam to move a 50-mm-dia roller follower of mass = 3 kg from 0 to 40 mm in 40°, dwell for 100°, fall 20 mm in 90°, dwell for 20°, fall 10 mm in 30°, dwell for 40°, fall 10 mm in 30°, and dwell for the remainder. The total cycle must take 10 sec. Choose suitable programs for rise and fall to minimize dynamic forces and torques. Size a return spring and specify its preload to maintain contact between cam and follower. Calculate and plot the dynamic force and torque. Assume damping of 0.25 times critical. Repeat for a form-closed cam. Compare the dynamic force, torque, and natural frequency for the form-closed design and the force-closed design.

‡15-18 Design a cam to drive an automotive valve train whose effective mass is 0.2 kg. ζ = 0.3. Valve stroke is 12 mm. Roller follower is 10 mm diameter. The open-close event occupies 160° of camshaft revolution; dwell for remainder. Use one or two polynomials for the rise-fall event. Select a spring constant and preload to avoid jump to 3500 rpm. Fast opening and closing and maximum open time are desired.

---

† These problems are suited to solution using *Mathcad, Matlab,* or *TKSolver* equation solver programs.

---

‡ These problems are suited to solution using program DYNACAM which is on the attached CD-ROM.

15

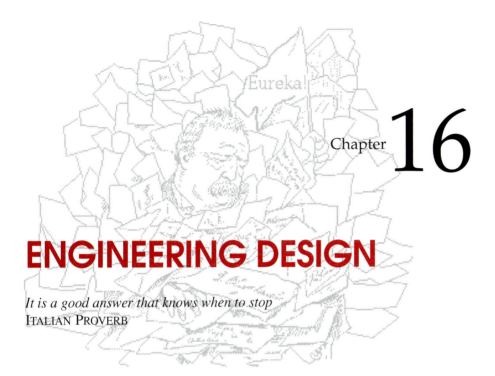

Chapter 16

ENGINEERING DESIGN

It is a good answer that knows when to stop
ITALIAN PROVERB

16.0 INTRODUCTION

Of all the myriad activities that the practicing engineer engages in, the one that is at once the most challenging and potentially the most satisfying is that of design. Doing calculations to analyze a clearly defined and structured problem, no matter how complex, may be difficult, but the exercise of creating something from scratch, to solve a problem that is often poorly defined, is *very* difficult. The sheer pleasure and joy at conceiving a viable solution to such a design problem is one of life's great satisfactions for anyone, engineer or not. The preceding chapters have attempted to present the particular subject matter in a way and context that will not only enhance the student's fundamental understanding of the topic but also encourage and promote his or her creative efforts toward the solution of design problems.

As a closing to a book that began with a discussion of a design process in Chapter 1, it is perhaps appropriate to present something of a "case study" of the design process to conclude the discussion. Some years ago, a very creative engineer of the author's acquaintance, George A. Wood Jr., heard a presentation by another creative engineer of the author's acquaintance, Keivan Towfigh, about one of his designs. Years later, Mr. Wood himself wrote a short paper about creative engineering design in which he reconstructed Mr. Towfigh's presumed creative process when designing the original invention. Both Mr. Wood and Mr. Towfigh have kindly consented to the reproduction of that paper here. It serves, in this author's opinion, as an excellent example and model for the student of engineering design to consider when pursuing his or her own design career.

16.1   A DESIGN CASE STUDY

Educating for Creativity in Engineering[1]

by George A. Wood Jr.

One facet of engineering, as it is practiced in industry, is the creative process. Let us define creativity as Rollo May does in his book, The Courage to Create.[2] It is "the process of bringing something new into being." Much of engineering has little to do with creativity in its fullest sense. Many engineers choose not to enter into creative enterprise, but prefer the realms of analysis, testing and product or process refinement. Many others find their satisfaction in management or business roles and are thus removed from engineering creativity as we shall discuss it here.

From the outset, I wish to note that the less creative endeavors are no less important or satisfying to many engineers than is the creative experience to those of us with the will to create. It would be a false goal for all engineering schools to assume that their purpose was to make all would-be engineers creative and that their success should be measured by the "creative quotient" of their graduates.

On the other hand, for the student who has a creative nature, a life of high adventure awaits if he can find himself in an academic environment which recognizes his needs, enhances his abilities and prepares him for a place in industry where his potential can be realized.

In this talk I will review the creative process as I have known it personally and witnessed it in others. Then I shall attempt to indicate those aspects of my training that seemed to prepare me best for a creative role and how this knowledge and these attitudes toward a career in engineering might be reinforced in today's schools and colleges.

During a career of almost thirty years as a machine designer, I have seen and been a part of a number of creative moments. These stand as the high points of my working life. When I have been the creator I have felt great elation and immense satisfaction. When I have been with others at their creative moments I have felt and been buoyed up by their delight. To me, the creative moment is the greatest reward that the profession of engineering gives.

Let me recount an experience of eight years ago when I heard a paper given by a creative man about an immensely creative moment. At the First Applied Mechanisms Conference in Tulsa, Oklahoma, was a paper entitled The Four-Bar Linkage as an Adjustment Mechanism.[3] It was nestled between two "how to do it" academic papers with graphs and equations of interest to engineers in the analysis of their mechanism problems. This paper contained only one very elementary equation and five simple illustrative figures; yet, I remember it now more clearly than any other paper I have ever heard at mechanism conferences. The author was Keivan Towfigh and he described the application of the geometric characteristics of the instant center of the coupler of a four bar mechanism.

His problem had been to provide a simple rotational adjustment for the oscillating mirror of an optical galvanometer. To accomplish this, he was required to rotate the entire galvanometer assembly about an axis through the center of the mirror and perpendicular to the pivot axis of the mirror. High rigidity of the system after adjustment was essential with very limited space available and low cost required, since up to sixteen of these galvanometer units were used in the complete instrument.

His solution was to mount the galvanometer elements on the coupler link of a one-piece, flexure hinged, plastic four bar mechanism so designed that the mirror center was at the instant center of the linkage at the midpoint of its adjustment. (See Fig 4.) It is about this particular geometric point (see Fig 1.) That pure rotation occurs and with proper selection of linkage dimensions this condition of rotation without translation could be made to hold sufficiently accurately for the adjustment angles required.

Unfortunately, this paper was not given the top prize by the judges of the conference. Yet, it was, indirectly, a description of an outstandingly creative moment in the life of a creative man.

Let us look at this paper together and build the steps through which the author probably progressed in the achievement of his goal. I have never seen Mr. Towfigh since, and I shall therefore describe a generalized creative process which may be incorrect in some details but which, I am sure, is surprisingly close to the actual story he would tell.

The galvanometer problem was presented to Mr. Towfigh by his management. It was, no doubt, phrased something like this: "In our new model, we must improve the stability of the adjustment of the equipment but keep the cost down. Space is critical and low weight is too. The overall design must be cleaned up, since customers like modern, slim-styled equipment and we'll lose sales to others if we don't keep ahead of them on all points. Our industrial designer has this sketch that all of us in sales like and within which you should be able to make the mechanism fit."

Then followed a list of specifications the mechanism must meet, a time when the new model should be in production and, of course, the request for some new feature that would result in a strong competitive edge in the marketplace.

I wish to point out that the galvanometer adjustment was probably only one hoped-for improvement among many others. The budget and time allowed were little more than enough needed for conventional redesign, since this cost must be covered by the expected sales of the resulting instrument. For every thousand dollars spent in engineering, an equivalent increase in sales or reduction in manufacturing cost must be realized at a greater level than the money will bring if invested somewhere else.

In approaching this project, Mr. Towfigh had to have a complete knowledge of the equipment he was designing. He had to have run the earlier models himself. He must have adjusted the mirrors of existing machines many times. He had to be able to visualize the function of each element in the equipment in its most basic form.

(research)

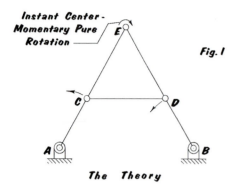

Instant Center-
Momentary Pure
Rotation

Fig. 1

The   Theory

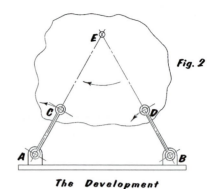

Fig. 2

The   Development

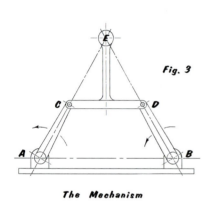

Fig. 3

The   Mechanism

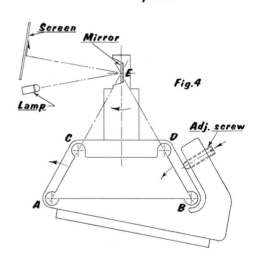

Fig. 4

Screen   Mirror   Lamp   Adj. screw

The Final Product of Keivan Towfigh

(ideation)

(frustration)

(incubation)

(Eureka!)

Secondly, he had to ask himself (as if he were the customer) what operational and maintenance requirements would frustrate him most. He had to determine which of these might be improved within the design time available. In this case he focused on the mirror adjustment. He considered the requirement of rotation without translation. He determined the maximum angles that would be necessary and the allowable translation that would not affect the practical accuracy of the equipment. He recognized the desirability of a one screw adjustment. He spent a few hours thinking of all the ways he had seen of rotating an assembly about an arbitrary point. He kept rejecting each solution as it came to him as he felt, in each case, that there was a better way. His ideas had too many parts, involved slides, pivots, too many screws, were too vibration sensitive or too large.

He thought about the problem that evening and at other times while he proceeded with the design of other aspects of the machine. He came back to the problem several times during the next few days. His design time was running out. He was a mechanism specialist and visualized a host of cranks and bars moving the mirrors. Then one day, probably after a period when he had turned his attention elsewhere, on rethinking of the adjustment device an image of the system based on one of the elementary characteristics of a four bar mechanism came to him.

I feel certain that this was a visual image, as clear as a drawing on paper. It was probably not complete but involved two inspirations. First was the characteristics of the instant center. (See Figs 1, 2, 3.) Second was the use of flexure hinge joints which led to a one-piece plastic molding. (See Fig 4.) I am sure that at this moment he had a feeling that this solution was right. He knew it with certainty. The whole of his engineering background told him. He was elated. He was filled with joy. His pleasure was not because of the knowledge that his superiors would be impressed or that his security in the company would be enhanced. It was the joy of personal victory, the awareness that he had conquered.

The creative process has been documented before by many others far more qualified to analyze the working of the human mind than I. Yet I would like to address, for the remaining minutes, how education can enhance this process and help more engineers, designers and draftsmen extend their creative potential.

The key elements I see in creativity that have greatest bearing on the quality that results from the creative effort are visualization and basic knowledge that gives strength to the feeling that the right solution has been achieved. There is no doubt in my mind that the fundamental mechanical principles that apply in the area in which the creative effort is being made must be vivid in the mind of the creator. The words that he was given in school must describe real elements that have physical, visual significance. F = ma must bring a picture to his mind vivid enough to touch.

If a person decides to be a designer, his training should instill in him a continuing curiosity to know how each machine he sees works. He must note its elements and mentally see them function together even when they are not moving. I feel that this kind of solid, basic knowledge couples with physical experience to build ever more critical levels at which one accepts a tentative solution as "right."

(analysis)

It should be noted that there have been times for all of us when the inspired "right" solution has proven wrong in the long run. That this happens does not detract from the process but indicates that creativity is based on learning and that failures build toward a firmer judgment base as the engineer matures. These failure periods are only negative, in the growth of a young engineer, when they result in the fear to accept a new challenge and beget excessive caution which then stifles the repetition of the creative process.

What would seem the most significant aspects of an engineering curriculum to help the potentially creative student develop into a truly creative engineer?

First is a solid, basic knowledge in physics, mathematics, chemistry and those subjects relating to his area of interest. These fundamentals should have physical meaning to the student and a vividness that permits him to explain his thoughts to the untrained layman. All too often technical words are used to cover cloudy concepts. They serve the ego of the user instead of the education of the listener.

Second is the growth of the student's ability to visualize. The creative designer must be able to develop a mental image of that which he is inventing. The editor of the book Seeing with the Mind's Eye,[4] by Samuels and Samuels, says in the preface:

". . . visualization is the way we think. Before words, images were. Visualization is the heart of the bio-computer. The human brain programs and self-programs through its images. Riding a bicycle, driving a car, learning to read, baking a cake, playing golf - all skills are acquired through the image making process. Visualization is the ultimate consciousness tool."

Obviously, the creator of new machines or products must excel in this area.

To me, a course in Descriptive Geometry is one part of an engineer's training that enhances one's ability to visualize theoretical concepts and graphically reproduce the result. This ability is essential when one sets out to design a piece of new equipment. First, he visualizes a series of complete machines with gaps where the problem or unknown areas are. During this time, a number of directions the development could take begin to form. The best of these images are recorded on paper and then are reviewed with those around him until, finally, a basic concept emerges.

The third element is the building of the student's knowledge of what can be or has been done by others with different specialized knowledge than he has. This is the area to which experience will add throughout his career as long as he maintains an enthusiastic curiosity. Creative engineering is a building process. No one can develop a new concept involving principles about which he has no knowledge. The creative engineer looks at problems in the light of what he has seen, learned and experienced and sees new ways for combining these to fill a new need.

Fourth is the development of the ability of the student to communicate his knowledge to others. This communication must involve not only skills with the techniques used by technical people but must also include the ability to share engineering concepts with untrained shop workers, business people and the general public. The engineer will seldom gain the opportunity to develop a concept if he cannot pass on to those around him his enthusiasm and confidence in the idea. Frequently, truly ingenious ideas are lost because the creator cannot transfer his vivid image to those who might finance or market it.

Fifth is the development of a student's knowledge of the physical result of engineering. The more he can see real machines doing real work, the more creative he can be as a designer. The engineering student should be required to run tools, make products, adjust machinery and visit factories. It is through this type of experience that judgement grows as to what makes a good machine, when approximation will suffice and where optimization should halt.

It is often said that there has been so much theoretical development in engineering during the past few decades that the colleges and universities do not have time for the basics I have outlined above. It is suggested that industry should fill in the practice areas that colleges have no time for, so that the student can be exposed to the latest technology. To some degree I understand and sympathize with this approach, but I feel that there is a negative side that needs to be recognized. If a potentially creative engineer leaves col-

lege without the means to achieve some creative success as he enters his first job, his enthusiasm for creative effort is frustrated and his interest sapped long before the most enlightened company can fill in the basics. Therefore, a result of the "basics later" approach often is to remove from the gifted engineering student the means to express himself visually and physically. Machine design tasks therefore become the domain of the graduates of technical and trade schools and the creative contribution by many a brilliant university student to products that could make all our lives richer is lost.

As I said at the start, not all engineering students have the desire, drive and enthusiasm that are essential to creative effort. Yet I feel deeply the need for the enhancement of the potential of those who do. That expanding technology makes course decisions difficult for both student and professor is certainly true. The forefront of academic thought has a compelling attraction for both the teacher and the learner. Yet I feel that the development of strong basic knowledge, the abilities to visualize, to communicate, to respect what has been done, to see and feel real machinery, need not exclude or be excluded by the excitement of the new. I believe that there is a curriculum balance that can be achieved which will enhance the latent creativity in all engineering and science students. It can give a firm basis for those who look towards a career of mechanical invention and still include the excitement of new technology.

I hope that this discussion may help in generating thought and providing some constructive suggestions that may lead more engineering students to find the immense satisfaction of the creative moment in the industrial environment. In writing this paper I have spent considerable time reflecting on my years in engineering and I would close with the following thought. For those of us who have known such times during our careers, the successful culminations of creative efforts stand among our most joyous hours.

16.2 CLOSURE

Mr. Wood's description of his creative experiences in engineering design and the educational factors which influenced them closely parallel this author's experience as well. The student is well advised to follow his prescription for a thorough grounding in the fundamentals of engineering and communication skills. A most satisfying career in the design of machinery can result.

16.3 REFERENCES

1. **Wood, G. A. Jr.** (1977). "Educating for Creativity in Engineering," Presented at the 85th Annual Conf. ASEE, Univ. of No. Dakota.

2. **May, R**. (1976). *The Courage to Create*, Bantam Books, New York.

3. **Towfigh, K.** (1969). "The Four-Bar Linkage as an Adjustment Mechanism," 1st Applied Mechanism Conf., Oklahoma State Univ., Tulsa Okla., pp. 27-1– 27-4.

4. **Samuels and Samuels.** (1975). *Seeing with the Mind's Eye: the History, Techniques and Uses of Visualization,* Random House, New York.

THE
END

COMPUTER PROGRAMS

I really hate this damned machine;
I wish that they would sell it.
It never does quite what I want
But only what I tell it.
FROM THE FORTUNE DATABASE, BERKELEY UNIX

A.0 INTRODUCTION

In addition to the student version of the commercial simulation program *Working Model*, there are seven custom computer programs provided on the CD-ROM with this text: programs FOURBAR, FIVEBAR, SIXBAR, SLIDER, MATRIX, DYNACAM, and ENGINE. These are student editions of the programs. Professional versions with extended capabilities are also available. Contact the author at rlnorton@wpi.edu or visit his faculty page at the web site http://www.wpi.edu/ for more information. Programs FOURBAR, FIVEBAR, SIXBAR, and SLIDER are based on the mathematics derived in Chapters 4 to 7 and 10 to 11 and use the equations presented therein to solve for position, velocity, and acceleration in linkages of the variety described in the particular program's name. Program DYNACAM is a cam design program based on the mathematics derived in Chapters 8 and 15. Program ENGINE is based on the mathematics derived in Chapters 13 and 14. Program MATRIX is a general linear simultaneous equation solver. All have similar choices for display of output data in the form of tables and plots. All the programs are designed to be user friendly and reasonably "crashproof." The author encourages users to email reports of any "bugs" or problems encountered in their use to him at rlnorton@wpi.edu.

Learning Tools

All the custom programs provided with this text are designed to be learning tools to aid in the understanding of the relevant subject matter and *are specifically not intended to be used for commercial purposes in the design of hardware* **and should not be so used**. It is quite possible to obtain inappropriate (but mathematically correct) results to any problem solved with these programs, due to incorrect or inappropriate input of data. In other words, the user is expected to understand the kinematic and dynamic theory underlying the program's structure and to also understand the mathematics on which the program's algorithms are based. This information on the underlying theory and mathematics is derived and described in the noted chapters of this text. Most equations used in the programs are presented or derived in this textbook.

Disclaimer

Commercial software for use in design or analysis needs to have built-in safeguards against the possibility of the user providing incorrect, inappropriate, or ridiculous values for input variables, in order to guard against erroneous results due to user ignorance or inexperience. **The student editions of the custom programs provided with this text are not commercial software and deliberately do not contain such safeguards against improper input data**, on the premise that to do so would "short circuit" the student's learning process. We learn most from our failures. These programs provide a consequence-free environment to explore failure of your designs "on paper" and in the process come to a more thorough and complete understanding of the subject matter. **The author and publisher are not responsible for any damages which may result from the use or misuse of these programs.**

Brute Force and Ignorance

The very rapid computation speed of these programs allows the student to explore a much larger number and variety of potential solutions to more realistic and comprehensive problems than could be accomplished using only hand calculator solutions of these complicated systems of equations. This is both an advantage and a danger. The advantage is that the student can use the programs like a "flight simulator" to "fly" potential design solutions through their paces with no consequences from a "crash" of the design not yet built. If the student diligently attempts to interpret the program's results and relate them to the relevant theory, a more thorough understanding of the kinematics and dynamics can result. On the other hand, there is a great temptation to use these programs with "brute force and ignorance" (BFI) to somewhat randomly try solutions without regard to what the theory and equations are telling you and hope that somehow a usable solution will "pop out." The student who succumbs to this temptation will not obtain much benefit from the exercise and will probably have a poor design result. This situation is probably best summed up in the following comment from a student who had suffered through the course in kinematics using these programs to design solutions to three project problems like those listed at the end of Chapter 3.

> . . . The computer, with its immense benefits for the engineer, can also be a hindrance if one does not first develop a thorough understanding of the theory upon which a particular program is based. An over-reliance on the computer can leave one "computer smart" and "engineering stupid." The BFI approach becomes increasingly tempting when it can be employed with such ease. . . . *BRIAN KIMBALL*

Smart student! Use these computer programs wisely. Avoid *Brute Force* and *Ignorance*! **Engineer** your solutions and understand the theory behind them.

A.1   GENERAL INFORMATION

Hardware Requirements

These programs work on any computer that runs Windows 3.1, Windows 95, or Windows NT. Different versions of the programs are provided for each operating system. A CD-ROM drive is needed, as is a hard disk drive. If your computer does not read a CD-ROM, your school's computer center can probably convert it to a usable format for you.

Operating System Requirements

These custom programs are written in Microsoft Visual Basic and are compiled in two versions: *Pgmname16.exe* for the Windows 3.1 (16 bit) operating system and *Pgmname32.exe* for the Windows 95 and NT (32 bit) operating systems. Be sure to use the version suitable to your computer's hardware and software.

Installing the Software

The CD-ROM contains the executable program files plus all necessary Dynamic Link Library (DLL) and other ancillary files needed to run the programs. Run the SETUP file from the individual program's folder on the CD-ROM to automatically decompress and install all of its files on your hard drive. A folder and icon will be created with that program's name. The program can then be run by double clicking on the icon or on the Pgmname16.exe or Pgmname32.exe file itself. In Windows 95 or NT the program name will appear in the list under the *Start* button's *Program* menu after installation and can be run from there.

How to Use This Manual

This manual is intended to be used while running the programs. To see a screen referred to, bring it up within the program to follow its discussion.

A.2   GENERAL PROGRAM OPERATION

All seven programs in the set have similar features and operate in a consistent way. For example, all printing and plotting functions are selected from identical screens common to all programs. Opening and saving files are done identically in all programs. These common operations will be discussed in this section independent of the particular program. Later sections will address the unique features and operations of each program.

Note that *student editions* of these programs are supplied with this book at no charge and carry a limited-term license restricted to use in course work for up to two semesters (9 months). If you continue to use the program after that time, you are expected to register it and pay the shareware fee defined on the *Registration* screen. If you wish to use the program for the benefit of a company or for any commercial purpose, then you must obtain the professional edition of the same program. The student editions are not to be used commercially. The professional editions typically offer more features than the student editions.

A

Running the Programs (All Programs)

At start-up, a splash screen appears which identifies the program version, revision number, and revision date. Click the button labeled *Start* or press the *Enter* key to run the program. A *Disclaimer* screen next appears which defines the registered owner and allows the printing of a registration form if the software is as yet unregistered. A registration form can be accessed and printed from this screen.

The next screen, the *Title* screen, allows the input of any user and/or project identification desired. This information must be provided to proceed and is used to identify all plots and printouts from this program session. The second box on the *Title* screen allows any desired file name to be supplied for storing data to disk. This name defaults to **Model1** and may be changed at this screen and/or when later writing the data to disk. The third box allows the typing of a starting design number for the first design. This design number defaults to 1 and is automatically incremented each time you change the basic design during this program session. It is used only to identify plots, data files, and printouts so they can be grouped, if necessary, at a later date. When the *Done* button on the *Title* screen is clicked, the *Home* screen appears.

The *Home Screen* (All Programs)

All program actions start and end at the *Home* screen which has several pull-down menus and buttons, some of which commands (*File*, *New*, *Open*, *Save*, *Save As*, *Units*, *About*, *Plot*, *Print*, *Quit*) are common to all programs. These will be described below.

General User Actions Possible Within a Program (All Programs)

The programs are constructed to allow operation from the keyboard or the mouse or with any combination of both input devices. Selections can be made either with the mouse or, if a button is highlighted (showing a dotted square within the button), the *Enter* key will activate the button as if it had been clicked with the mouse. Text boxes are provided where you need to type in data. These have a yellow background. In general, what you type in any text box is not accepted until you hit the *Enter* key or move off that box with the *Tab* key or the mouse. This allows you to retype or erase with no effect until you leave the text box. You can move between available input fields with the *Tab* key (and backup with *Shift-Tab*) on most screens. If you are in doubt as to the order in which to input the data on any screen, try using the *Tab* key as it will take you to each needed entry field in a sensible order. You can then type or mouse click to input the desired data in that field. Remember that a yellow background means typed input data is expected. Boxes with a cyan background provide information back to you but cannot be typed in.

Other information required from you is selected from drop-down menus or lists. These have a white background. Some lists allow you to type in a value different than any provided in the available list of selections. If you type an inappropriate response, it will simply ignore you or choose the closest value to your request. Typing the first few letters of a listed selection will sometimes cause it to be selected. Double clicking on a selectable item in a list will often provide a shortcut.

Units (All Programs)

The *Units* menu defines several units systems to choose from. *Note that all programs work entirely in pure numbers without regard to units.* It is your responsibility to ensure that the data as input are in some consistent units system. **No units conversion is done within the programs**. The *Units* menu selection that you make has only one effect, namely to change the labels displayed on various input and output parameters within the program. You mix units systems at your own peril.

Examples (Most Programs)

Most of the programs have an *Examples* pull-down menu on the *Home* screen which provides some number of example mechanisms that will demonstrate the program's capability. Selecting an example from this menu will cause calculation of the mechanism and open a screen to allow viewing the results. In some cases you may need to hit a button marked *Calculate*, *Run*, or *Animate* on the presented screen to see the results. Some programs also provide access to examples from various screens.

Creating New, Saving, and Opening Files (*File* - All Programs)

The standard *Windows* functions for creating new files, saving your work to disk, and opening a previously saved file are all accessible from the pull-down menu labeled *File* on each program's *Home* screen. Selecting *New* from this menu will zero any data you may have already created within the program, but before doing so will give warning and prompt you to save the data to disk.

The *Save* and *Save As* selections on the *File* menu prompt you to provide a file name and disk location to save your current model data to disk. The data are saved in a custom format and with a three-character suffix unique to the particular program. You should use the recommended suffix on these files as that will allow the program to see them on the disk when you want to open them later. If you forget to add the suffix when saving a file, you can still recover the file.

Selecting *Open* from the *File* menu prompts you to pick a file from those available in the disk directory that you choose. If you do not see any files with the program's suffix, use the pull-down menu within the *Open File* dialog box to choose *Show All Files* and you will then see them. They will read into the program properly with or without the suffix in their name as long as they were saved from the same program. If you are a former user of the DOS versions of these programs, files saved from the older version of a given program can also be opened in its *Windows* version, though they will not have as much information as is needed in the new programs.

Copying Screens to Clipboard or Printer (*Copy* - All Programs)

Any screen can be copied as a graphic to the clipboard by using the standard *Windows* keyboard combo of *Alt-PrintScrn*. It will then be available for pasting into any compatible *Windows* program such as *Word* or *Powerpoint* that is running concurrently in *Windows*. Most screens also provide a button to dump the screen image to an attached laser printer. However, the quality of that printed image may be less than could be obtained

A

by copying and pasting the image into a program such as *Word* or *Powerpoint* and then printing it from that program. It seems that *Visual Basic* does not print graphics as well as some other *Windows* applications.. NOTE: *In some cases the plotted functions may not print properly. If so, copy the screen to clipboard, paste into Word and print from Word.*

Printing to Screen, Printer, and Exporting Disk Files (*Print* Button)

Selecting the *Print* button from the *Home* screen will open the *Print Select* screen (see Figure A-1) containing lists of variables that may be printed. Buttons on the left of this screen can be clicked to direct the printed output to one of *Screen*, *Printer*, or *Disk*. This choice defaults to *Screen* and so must be clicked each time the screen is opened to obtain either of the other options. The output is different with each of these selections.

Selecting *Screen* will result in a scrollable screen window full of the requested data. Scrolling will allow you to view all data requested serially. This data screen can be copied to the clipboard or dumped to a printer as described above, but this clip or dump will typically show only a portion of the available data, i.e., one screenful.

Selecting *Printer* as the output device will cause the entire selection of data to print to an available printer. Only some of the sidebar information shown on the screen display will be included in this printout.

Selecting *Disk* as the output device will cause your selections to be sent to the file of your choice in an ASCII text format (tab delimited) that can be opened in a spreadsheet program such as *Lotus 123* or *Excel*. You can then do further calculations or plotting of data within the spreadsheet program.

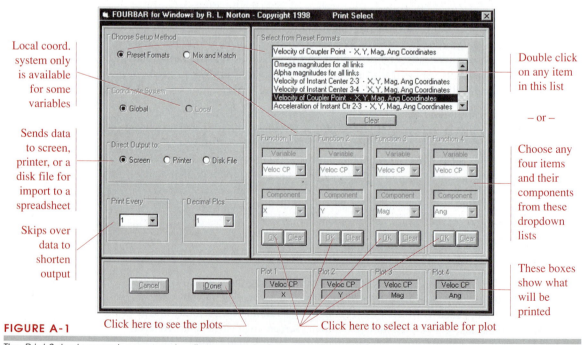

FIGURE A-1

The *Print Select* screen is common to all programs (not all programs allow component selection)

The *Print Select* screen has two modes for data selection, *Preset Formats* and *Mix and Match*. The former provides preselected sets of four variables for printing. Selecting *Mix and Match* allows you to pick any four of the available variables for printing. You must print four variables at a time in either mode. Depending on the program, you may be able to select other ancillary parameters such as the number of decimal places or the frequency of data to be printed.

Plotting Data (*Plot* Button)

The *Plot* button on the *Home* screen brings up the *PlotType* screen (see Figure A-2) which is the same in all programs. Variables in these programs can be plotted in one of several formats, three cartesian (see below) and one polar (see below). This screen allows a choice among these four "flavors" of plots shown as plot-style icons. The first icon (upper left) provides four functions plotted on cartesian axes in four separate windows. The second icon (upper right) plots two functions on cartesian axes in separate windows. The third icon (lower left) allows one to four functions to be plotted on common cartesian axes in a single window. This choice is of value to show a single function full screen or to overlay multiple functions. (Be advised, however, that multiple functions will scale to the largest function of the set, so if there are large differences in magnitude between the members of the set, it may be difficult to see and interpret the smaller ones.) The fourth icon (lower right) provides a polar plot of one selected function. You may select any of these four plot styles by clicking on its icon or on the *Select* button above it and then clicking *Done*. Double clicking on a plot icon will bring up the next screen immediately.

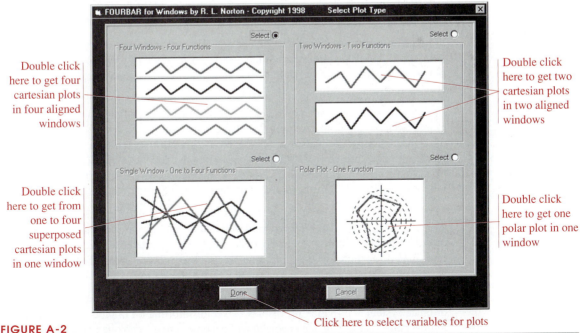

FIGURE A-2

The *PlotType* screen is common to all programs that allow plotting

CARTESIAN PLOTS depict a dependent variable versus an independent variable on cartesian (x, y) axes. In these programs, the independent variable shown on the x axis may be either time or angle, depending on the calculation choice made in the particular program. The variable for the y axis is selected from the plot menu. Angular velocities and torques are vectors but are directed along the z axis in a two-dimensional system. So their magnitudes can be plotted on cartesian axes and compared because their directions are constant, known, and the same.

POLAR PLOTS Plots of linear velocities, linear accelerations, and forces require a different treatment than the cartesian plots used for the angular vector parameters. Their directions are not the same and vary with time or input angle. One way to represent these linear vectors is to make two cartesian plots, one for magnitude and one for angle of the vector at each time or angle step. Alternatively, the x and y components of the vector at each time or angle step can be presented as a pair of cartesian plots. Either of these approaches requires two plots per vector and has the disadvantage of being difficult to interpret. A better method for vectors that act on a moving point (such as a force on a moving pin) can be to make a polar plot with respect to a local, nonrotating axis coordinate system (LNCS) attached at the moving point. This local, nonrotating x, y axis system translates with the point as it moves but remains always parallel to the global axis system X,Y. By plotting the vectors on this moving axis system we can see both their magnitude and direction at each time or angle step, since we are attaching the roots of all the vectors to the moving point at which they act.

In some of the programs, polar plots can be paused between the plotting of each vector. Without a pause, the plot can occur too quickly for the eye to detect the order in which they are drawn. When a mouse click is required between the drawing of each vector, their order is easily seen. With each pause, the current value of the independent variable (time or angle) as well as the magnitude and angle of the vector are displayed.

The programs also allow alternate presentations of polar plots, showing just the vectors, just the envelope of the path of the vector tips, or both. A plot that connects the tips of the vectors with a line (its envelope) is sometimes called a **hodograph**.

SELECTING PLOT VARIABLES Choosing any one of the four plot types from the *Plot Type* screen brings up a *Plot Select* screen which is essentially the same in all programs. (See Figure A-3.) As with the *Print Select* screen, two arrangements for selecting the functions to be plotted are provided, *Preset Formats* and *Mix and Match*. The former provides preselected collections of functions, and the latter allows you to select up to four functions from those available on the pull-down menus. In some cases you will also have to select the component of the function desired, i.e., x, y, *mag*, or *angle*.

PLOT ALIGNMENT Some of the *Plot Select* screens offer a choice of two further plot style variants labeled *Aligned* and *Annotated*. The aligned style places multiple plots in exact phase relationship, one above the other. The annotated style does not align the plots but allows more variety in their display such as fills and grids. The data displayed is the same in each case.

COORDINATE SYSTEMS For particular variables in some programs, a choice of coordinate system is provided for display of vector information in plots. The *Coordinate System* panel on the *Plot Select* screen will become active when one of these variables is selected. Then either the *Global* or *Local* button can be clicked. (It defaults to *Global*.)

A

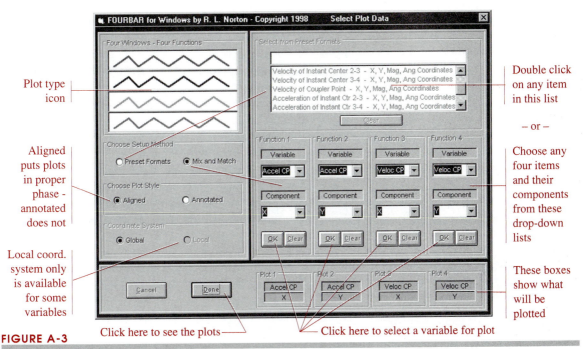

FIGURE A-3

The *PlotSelect* screens are common to all programs. This shows one of four styles of *PlotSelect* screens

GLOBAL COORDINATES   The *Global* choice in the *Coordinate System* panel refers all angles to the *XY* axes of Figure A-6 (p. 737). For polar plots the vectors shown with the *Global* choice actually are drawn in a local, nonrotating coordinate system (LNCS) that remains parallel to the global system such as x_1, y_1 at point A and x_2, y_2 at point B in Figure A-1. The LNCS x_2, y_2 at point B behaves in the same way as the LNCS x_1, y_1 at point A; that is, it travels with point B but remains parallel to the world coordinate system *X,Y* at all times.

LOCAL COORDINATES   The coordinate system x',y' also travels with point B as its origin, but is embedded in link 4 and rotates with that link, continuously changing its orientation with respect to the global coordinate system *X,Y* making it an LRCS. Each link has such an LRCS but not all are shown in the figure. The *Local* choice in the *Coordinate System* panel uses the LRCS for each link to allow the plotting and printing of the tangential and radial components of acceleration or force on a link. This is of value if, for example, a bending stress analysis of the link is wanted. The dynamic force components perpendicular to the link due to the product of the link mass and tangential acceleration will create a bending moment in the link. The radial component will create tension or compression.

PLOTTING   Once your selections are made and are shown in the cyan boxes at the lower right of the *Plot Select* screen, the *Done* button will become available. Clicking it will bring up the plots that you selected. Figure A-4 shows examples of the four plot types available. From this *Plot* screen you may copy to the clipboard for pasting into another application or dump the *Plot* screen to a printer. The *Select Another* button returns you to the previous *Plot Select* screen. *Done* returns you to the *Home* screen.

A

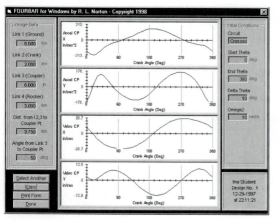

(a) Four aligned plots in separate windows

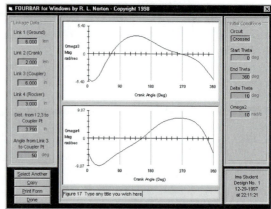

(b) Two aligned plots in separate windows

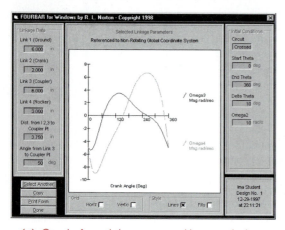

(c) One to four plots superposed in one window

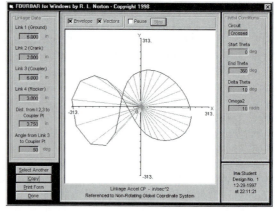

(d) Single polar plot

FIGURE A-4

The four styles of plots available in all programs. Sidebar information is different in each program

The *About* Menu (All Programs)

The *About* pull-down menu on the *Home* screen will display a splash screen containing information on the edition and revision of your copy of the program. The *Disclaimer* and *Registration* form can also be accessed from this menu.

Exiting a Program (All Programs)

Choosing either the *Quit* button or *Quit* on the *File* pull-down menu on the *Home* screen will exit the program. If the current data has not been saved since it was last changed, it will prompt you to save the model using an appropriate suffix. In all cases, it will ask you to confirm that you want to quit. If you choose yes, the program will terminate and any unsaved data will be gone at that point.

Support (All Programs)

Please notify the author of any bugs via email to rlnorton@wpi.edu.

A.3   PROGRAM FOURBAR

FOURBAR *for Windows* is a linkage design and analysis program intended for use by students, engineers and other professionals who are knowledgeable in or are learning the art and science of linkage design. It is assumed that the user knows how to determine whether a linkage design is good or bad and whether it is suitable for the application for which it is intended. The program will calculate the kinematic and dynamic data associated with any linkage design, but cannot substitute for the engineering judgment of the user. The linkage theory and mathematics on which this program is based are documented in Chapters 4 to 7, 10, and 11 of this textbook. Please consult them for explanations of the theory and mathematics involved.

The FOURBAR *Home Screen*

Initially, only the *Input* and *Quit* buttons are active on the *Home* screen. Typically, you will start a linkage design with the *Input* button, but for a quick look at a linkage as drawn by the program, one of the examples under the *Example* pull-down menu can be selected and it will draw a linkage. If you activate one of these examples, when you return from the *Animate* screen you will find all the other buttons on the *Home* screen to be active. We will address each of these buttons in due course below.

Input Data (FOURBAR *Input* Screen)

Figure A-5 defines the input parameters, link numbering, and the axis system used in program FOURBAR. The link lengths needed are ground link 1, input link 2, coupler link 3, and output link 4, defined by their pin-to-pin distances and labeled a, b, c, d in the figure. The X axis is constrained to lie along link 1, defined by the instant centers $I_{1,2}$ and $I_{1,4}$ which are also labeled, respectively, O_2 and O_4 in the figure. Instant center $I_{1,2}$, the driver crank pivot, is the origin of the global coordinate system.

It might seem overly restrictive to force the X axis to lie on link 1 in this "aligned system." Many linkages will have their ground link at some angle other than zero. However, reorienting the linkage after designing and analyzing it merely involves rotating the piece of paper on which it is drawn to the desired final angle of the ground link. More formally, it means a rotation of the coordinate system through the negative of the angle of the ground link. In effect, the actual final angle of the ground link must be subtracted from all angles of links and vectors calculated in the aligned axis system.

In addition to the link lengths, you must supply the location of one coupler point on link 3 to find that point's coupler curve positions, velocities, and accelerations. This point is located by a position vector rooted at $I_{2,3}$ (point A) and directed to the coupler point P of interest which can be anywhere on link 3. The program requires that you input the polar coordinates of this vector which are labeled p and δ_3 in Figure A-5. The program asks for the distance from $I_{2,3}$ to the coupler point, which is p, and the angle the coupler point

A

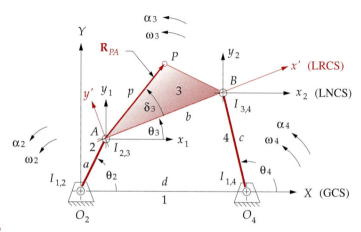

FIGURE A-5

Linkage data for program FOURBAR. (Open the file F_A-05.5br in FOURBAR to see this linkage)

makes with link 3 which is δ_3. Note that angle δ_3 is not referenced to either the global coordinate system (GCS) *X,Y* or to the local nonrotating coordinate system (LNCS) *x,y* at point *A*. Rather, it is referenced to the line *AB* which is the pin-to-pin edge of link 3 (LRCS). Angle δ_3 is a property of link 3 and is embedded in it. The angle which locates vector $\mathbf{R}_{CA}$ in the *x,y* coordinate system is the sum of angle δ_3 and angle θ_3. This addition is done in the program, after θ_3 is calculated for each position of the input crank. Also see Section 4.5 (p. 152). The coordinate system, dimensions, angles, and nomenclature in Figure A-5 are consistent with those of Figure 4-6 (p. 154) which were used in the derivation of the equations used in program FOURBAR.

Calculation (FOURBAR, FIVEBAR, SIXBAR, and SLIDER *Input* Screens)

Basic data for a linkage design is defined on the *Input* screen shown in Figure A-6, which is activated by selecting the *Input* button on the *Home* screen. When you open this screen for the first time, it will have default data for all parameters. The linkage geometry is defined in the *Linkage Data* panel on the left side of the screen. You may change these by typing over the data in the yellow text boxes.

The open or crossed circuit of the linkage is selected in a panel at the lower left of the screen in Figure A-6. Select the type of calculation desired, one of *Angle Steps, Time Steps,* or *One Position* from the *Calculation Mode* in the upper-right panel. The start, finish, and delta step information is different for each of these calculation methods, and the input text box labels in the *Initial Conditions* panel will change based on your choice. Type the desired initial, delta, and final conditions as desired.

ONE POSITION   will calculate position, velocity, and acceleration for any one specified input position θ_2, input angular velocity ω_2, and angular acceleration α_2.

ANGLE STEPS   assumes that the angular acceleration α_2 of input link 2 is zero, making ω_2 constant. The values of initial and final crank angle θ_2, angle step $\Delta\theta_2$, and the constant input crank velocity ω_2 are requested. The program will compute all linkage pa-

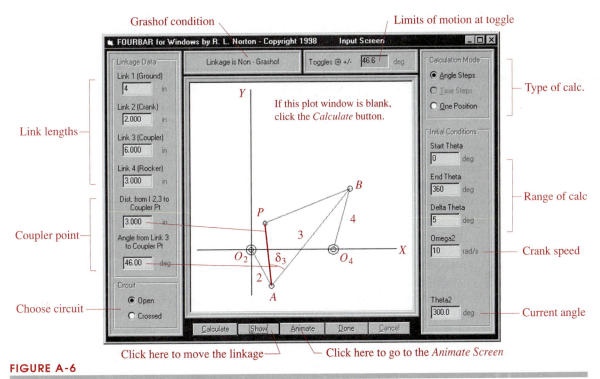

FIGURE A-6

Input Data screen for program FOURBAR. Corresponding screens in FIVEBAR, SIXBAR, and SLIDER are similar

rameters for each angle step. This is a steady-state analysis and is suitable for either Grashof or non-Grashof linkages provided that the total linkage excursion is limited in the latter case.

TIME STEPS requires input of a start time, finish time, and a time step, all in seconds. The value for α_2 (which must be either a constant or zero) and the initial position θ_2 and initial velocity ω_2 of link 2 at time zero must also be supplied. The program will then calculate all linkage parameters for each time step by applying the specified acceleration, which of course will change the angular velocity of the driver link with time. This is a transient analysis. The linkage will make as many revolutions of the driver crank as is necessary to run for the specified time. This choice is more appropriate for Grashof linkages, unless very short time durations are specified, as a non-Grashof linkage will quickly reach its toggle positions.

Note that a combination of successive **Time Step, Crank Angle,** and **Time Step** analyses can be used to simulate the start-up, steady-state, and deceleration phases, respectively, of a system for a complete analysis.

CALCULATE The *Calculate* button will compute all data for your linkage and show it in an arbitrary position in the linkage window on the *Input* screen. If at any time the white linkage window is blank, the *Calculate* button will bring back the image. The *Show* button will move the linkage through its range in "giant steps."

A

After you have calculated the linkage, the *Animate* and *Done* buttons on the *Input* screen will become available. The *Animate* button takes you to the *Animate* screen where you can run the linkage through any range of motion to observe its behavior. You can also change any of the linkage parameters on the *Animate* screen and then recalculate the results there with the *Recalc* button. The *Done* button on either the *Input* or *Animate* screen returns you to the *Home* screen. The *Plot* and *Print* buttons will now be available as well as the *Animate* button which will send you to the *Animate* screen.

CALCULATION ERRORS   If a position is encountered which cannot be reached by the links (in either the angle step or time step calculations), the mathematical result will be an attempt to take the square root of a negative number. The program will then show a dialog box with the message *Links do not connect for Theta2=xx* and present three choices: Abort, Retry, or Ignore. **Abort** will terminate the calculation at this step and return you to the *Input* screen. **Retry** will set the calculated parameters to zero at the current position and attempt to continue the computation at the next step, reporting successive problems as they occur. **Ignore** will continue the calculation for the entire excursion, setting the calculated parameters to zero at any subsequent positions with problems but will not present any further error messages. If a linkage is non-Grashof and you request calculation for angles that it cannot reach, then you will trip this error sequence. Choosing Ignore will force the calculation to completion, and you can then observe the possible motions of the linkage in the linkage window of the *Input* screen with the *Show* button.

GRASHOF CONDITION   Once the calculation is done, the linkage's Grashof condition is displayed in a panel at the top left of the screen. If the linkage is non-Grashof, the angles at which it reaches toggle positions are displayed in a second panel at top right. This information can be used to reset the initial conditions to avoid tripping the "links cannot connect" error.

Animation (FOURBAR)

The *Animate* button on the *Home* screen brings up the *Animate* screen as shown in Figure A-7. Its *Run* button activates the linkage and runs it through the range of motion defined in its most recent calculation. The Grashof condition is reported at the upper right corner of the screen. The number of cycles for the animation can be typed in the *Cycles* box at the lower right of the screen.

A time delay (defaulted to 0) can be set with a drop-down menu below the *Cycles* box. Any number of cycles not on its list can be typed in the box. This time-delay feature is provided to accommodate variations in speed among computers. If your computer is very fast, the animation may occur too rapidly to be seen. If so, selecting larger positive numbers for the time delay will slow the animation. For slow computers, negative numbers will speed the animation to a degree but can make it jerky. The *Step* button moves the linkage one increment of the independent variable at a time.

Text boxes in the *Linkage Data* panel allow the linkage geometry to be changed without returning to the *Input* screen. The initial conditions can be redefined in the panel on the left of this screen. The *Open-Crossed* selection can be switched, but the *Calculation Type* can only be changed on the *Input* screen. After any such change, the linkage must be recalculated with the *Recalc* button and then rerun.

A

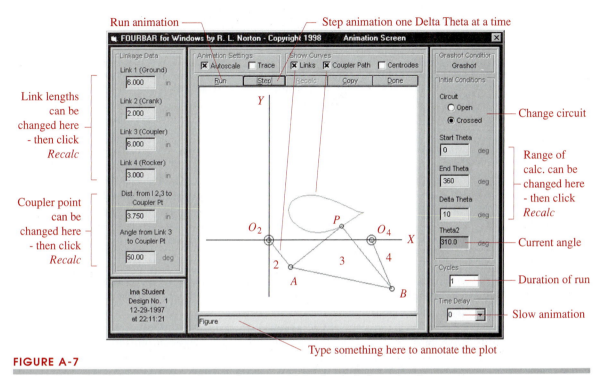

Run animation — ——— Step animation one Delta Theta at a time

Link lengths can be changed here – then click *Recalc*

Coupler point can be changed here – then click *Recalc*

Change circuit

Range of calc. can be changed here – then click *Recalc*

Current angle

Duration of run

Slow animation

Type something here to annotate the plot

FIGURE A-7

FOURBAR *Animation Screen.* Programs FIVEBAR, SIXBAR, and SLIDER have similar animation screens

Two panels at the top center of the *Animate* screen provide switches to change the animation display. In the *Show Curves* panel, displays of *Links*, *Coupler Path*, and *Centrodes* can be turned on and off in subsequent animations.

CENTRODES Only the FOURBAR program calculates and draws the fixed and moving centrodes (the loci of the instant centers as defined in Section 6.5 (p. 263) and shown in Figure 6-15a, p. 266). Different colors are used to distinguish the fixed from the moving centrode. The centrodes are drawn with their point of common tangency located at the first position calculated. Thus, you can orient them anywhere by your choice of start angle for the calculation.

AUTOSCALE can be turned on or off in the *Animation Settings* panel. The linkage animation plot is normally autoscaled to fit the screen based on the size of the linkage and its coupler curves (but not of the centrodes as they can go to infinity). You may want to turn off autoscaling when you wish to print two plots of different linkages at the same scale for later manual superposition. Turning off autoscaling will retain the most recent scale factor used. When on, it will rescale each plot to fit the screen.

TRACE Turning *Trace* on keeps all positions of the linkage visible on the screen so that the pattern of motion can be seen. Turning *Trace* off erases all prior positions, showing only the current position as it cycles the linkage through all positions giving a dynamic view of linkage behavior.

A

Dynamics (FOURBAR, FIVEBAR, SIXBAR, and SLIDER *Dynamics* Screen)

The *Dynamics* button on the *Home* screen brings up a screen that allows input of data on link masses, mass moments of inertia, *CG* locations, and any external forces or torques acting on the links. The location of the *CG* of each moving link is defined in the same way as the coupler point, by a position vector whose root is at the low-numbered instant center of each link. That is, for link 2 it is $I_{1,2}$; for link 3, $I_{2,3}$; and for link 4, $I_{1,4}$. These vectors for a fourbar linkage are shown in Figure A-2a (p. 731) labeled R_{CG_i} where *i* is the link number. Note that the angle of each *CG* vector is measured with respect to a *local rotating coordinate system* (LRCS) whose origin is at the aforementioned instant center and whose *x* axis lies along the line of centers of the link. For example, in Figure A-8a, link 2's *CG* vector is 3 in at 30°, link 3's is 9 in at 45°, and link 4's is 5 in at 0°. The program will automatically create the necessary position vectors $\mathbf{R}_{12}, \mathbf{R}_{32}, \mathbf{R}_{23}, \mathbf{R}_{43},$ $\mathbf{R}_{34},$ and $\mathbf{R}_{14}$ needed for the dynamic force equations as shown in the free-body diagrams in Figure A-8b. These position vectors are, of course, recalculated in the *nonrotating local coordinate systems* (LNCS) at the links' *CG*s for each new position of the linkage as the link angles change.

The masses and mass moments of inertia with respect to the CGs of the moving links are also required. Any external forces or torques which are applied to links 3 or 4 are typed in the appropriate boxes on the *Dynamics* screen as shown in Figure A-9. The direction angle of any external force must be measured with respect to the *global coordinate system* (GCS). The program will assume that this angle remains constant for all positions of the linkage analyzed. You must also supply the magnitude and direction of the position vector $\mathbf{R}_P$ which locates any point on the force vector $\mathbf{F}_P$. This $\mathbf{R}_P$ vector is measured in the *rotating, local coordinate system* (LRCS) embedded in the link, as were the *CG*s of the links. $\mathbf{R}_P$ is *not* measured in the global system. The program takes care of the resolution of these $\mathbf{R}_P$ vectors, for each position of the linkage, into coordinates in the nonrotating local coordinate system at the *CG*. Note that if you wish to account for the gravitational force on a heavy link, you may do so by applying that weight as an external force acting through the link's *CG* at 270° in the global system ($\mathbf{R}_P = 0$).

For other linkages such as the fivebar, sixbar, and fourbar slider-crank, the dynamic data input is similar. The only difference is the number of links for which mass property data and possible external forces and torques must be supplied.

After solving (by clicking the *Solve* button), clicking the *Show Matrix* button will display the dynamics matrix for the linkage. The results of the dynamics calculations are automatically stored for later plotting and printing. The menus on the *Plot* and *Print* screens will expand to include forces and torques for all links.

Balancing (FOURBAR Only)[*]

* Program ENGINE also allows balancing but its *Balance Screen* is completely different than in FOURBAR and so will be discussed separately in the section on the ENGINE program.

The *Balance* button on the FOURBAR *Home* screen brings up the *Balance* screen shown in Figure A-10, which immediately displays the mass-radius products needed on links 2 and 4 to force balance it and reduce the shaking force to zero. If you place the total amounts of the calculated mass-radius products on the rotating links, the shaking force will become zero and the shaking torque will increase. A partial balance condition can be specified by reducing the balance masses and accepting some nonzero shaking force in return for a smaller increase in torque.

A

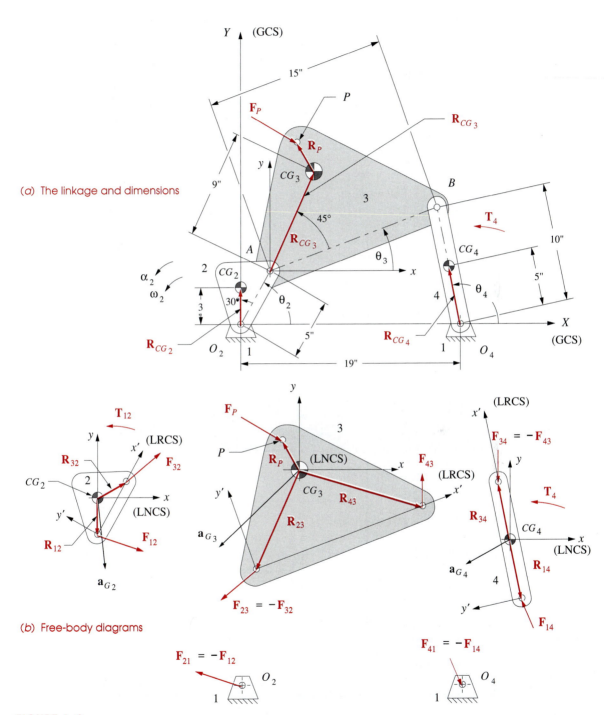

(a) The linkage and dimensions

(b) Free-body diagrams

FIGURE A-8

Definition of data needed for dynamics calculations in Program FOURBAR

A

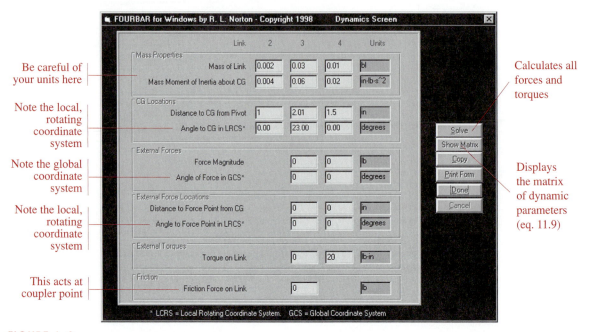

Be careful of
your units here

Note the local,
rotating
coordinate
system

Note the global
coordinate
system

Note the local,
rotating
coordinate
system

This acts at
coupler point

Calculates all
forces and
torques

Displays
the matrix
of dynamic
parameters
(eq. 11.9)

FIGURE A-9

Dynamic Data Input screen for program FOURBAR. Screens for programs FIVEBAR, SIXBAR, and SLIDER are similar

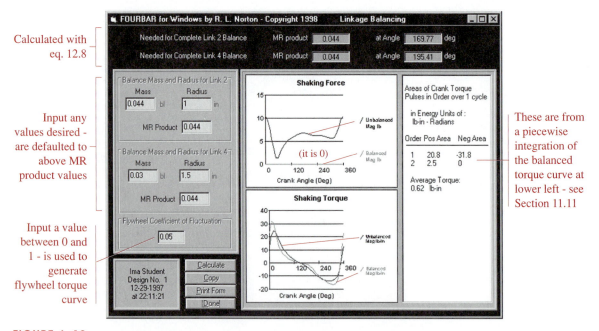

Calculated with
eq. 12.8

Input any
values desired -
are defaulted to
above MR
product values

Input a value
between 0 and
1 - is used to
generate
flywheel torque
curve

These are from
a piecewise
integration of
the balanced
torque curve at
lower left - see
Section 11.11

FIGURE A-10

Linkage Balancing screen in program FOURBAR only. Other programs do allow flywheel calculations

A

The FOURBAR *Balance* screen in Figure A-10 displays two mini-plots that superpose the shaking forces and shaking torques before and after balancing. Effects on the shaking force and torque from changes in the amount of balance mass-radius product placed on each link can be immediately seen in these plots. The energy in each pulse of the torque-time curve is also displayed in a sidebar on the right of this screen for use in a flywheel sizing calculation. See Section 11.11 (p. 548) for a discussion of the meaning and use of these data. The program calculates a smoothed torque function by multiplying the raw torque by the coefficient of fluctuation specified in the box at lower left.

Cognates (FOURBAR Only)

The *Cognates* pull-down menu allows switching among the three cognates which create the same coupler curve. Switching among them requires recalculation of all kinematic and dynamic parameters via the *Input*, *Dynamics*, and *Balance* buttons. The previously used mass property data is retained but can be changed easily by selecting the *Dynamics* button. The *Cayley Diagram* menu pick under *Cognates* displays that diagram of all three cognates. See Chapter 3. Whenever linkage data are changed on the *Input Screen* and recalculated, the program automatically calculates the dimensions of that linkage's two cognates. These can be switched to, calculated, and investigated at any time.

Synthesis (FOURBAR Only)

This pull-down menu allows selection of two- or three-position synthesis of a linkage, each with a choice of two methods. See Chapter 5 for a discussion of these methods and derivations of the equations used. When the linkage is synthesized, its link geometry is automatically put into the input sheet and recalculation is then required.

Other

See Section A.2, General Program Operation (p. 727) for information on *New, Open, Save, Save As, Plot, Print, Units*, and *Quit* functions.

A.4   PROGRAM FIVEBAR

FIVEBAR *for Windows* is a linkage design and analysis program intended for use by students, engineers, and other professionals who are knowledgeable in or are learning the art and science of linkage design. It is assumed that the user knows how to determine whether a linkage design is good or bad and whether it is suitable for the application for which it is intended. The program will calculate the kinematic and dynamic data associated with any geared fivebar linkage design, but cannot substitute for the engineering judgment of the user. The linkage theory and mathematics on which this program is based are documented in Chapters 4 to 7, and 11 of this textbook. Please consult them for explanations of the theory and mathematics involved.

The FIVEBAR *Home* Screen

Initially, only the *Input* and *Quit* buttons are active on the *Home* screen. Typically, you will start a linkage design with the *Input* button, but for a quick look at a linkage as drawn

A

by the program, one of the examples under the *Example* pull-down menu can be selected and it will draw a linkage. If you activate one of these examples, when you return from the *Animate* screen you will find all the other buttons on the *Home* screen to be active. We will address each of these buttons in due course below.

Input Data (The *Input* Screen)

Because this program deals with the geared fivebar mechanism (GFBM), it requires more input data than for the fourbar mechanism. Five link lengths must be supplied as shown in Figure A-11: driver crank (link 2), first coupler (link 3), second coupler (link 4), driven crank (link 5), and ground (link 1). Two other linkage parameters must be defined as input, namely, the gear ratio (λ) and the phase angle (ϕ). The phase angle is defined as the angle of link 5 when link 2 is at 0° as shown in Figure A-11. Note that the gear ratio as shown in the figure is a *negative ratio* because the external gears turn in opposite directions. The addition of an idler gear will create a positive gear ratio. It is also worth noting that the gear ratios defined in the *ZNH Atlas of Geared Fivebar Linkages* (Appendix E) are the reciprocal of the gear ratio in program FIVEBAR. So, when transferring data from this atlas to the program, the gear ratio must be reciprocated in order to get the same linkage as shown in the atlas. Otherwise the coupler curve will be a mirror image of the one in the atlas.

The coupler point *P* is defined in the same way as in the fourbar linkage. A coupler point can only be placed on link 3 in this program. If you want a coupler point on link 4, mirror your linkage and renumber the links to put the coupler point on link 3.

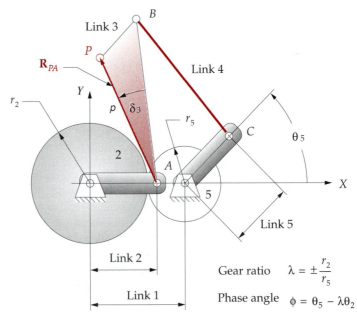

Gear ratio $\lambda = \pm\dfrac{r_2}{r_5}$

Phase angle $\phi = \theta_5 - \lambda\theta_2$

FIGURE A-11

Input data for program FIVEBAR. (Open the disk file F_A-11.5br in FIVEBAR to see this linkage)

A

Basic data for the fivebar linkage is defined on the *Input* screen which is similar to that for FOURBAR shown in Figure A-6 (p. 737). The *Input* screen is activated by selecting the *Input* button on the *Home* screen. When you open this screen for the first time, it will have default data for all link parameters. You may change any of these by typing over the data in the yellow text boxes.

Select the type of calculation desired in the upper right corner of the screen, one of *Angle Steps, Time Steps*, or *One Position*. The *Calculate* button will compute all data for your linkage and show it in an arbitrary position on the screen. If the white linkage display window is blank, the *Calculate* button will bring back the image. See the discussion of calculations for the FOURBAR program in Section A.3 (p. 735). They are similar in FIVEBAR.

After you have calculated the linkage, the *Animate* and *Done* buttons on the *Input* screen will become available. *Animate* takes you to the *Animate screen* where you can run the linkage through any range of motion to observe its behavior. You can also change any of the linkage parameters on the *Animate* screen and then recalculate the results with the *Recalc* button. The *Done* button on either the *Input* or *Animate* screen returns you to the *Home* screen. The *Plot* and *Print* buttons will now be available as well as the *Animate* button which returns you to the *Animate* screen.

Animation (FIVEBAR)

In program FIVEBAR, the *Animation* screen and its features are essentially similar to those of program FOURBAR. The only exception is the lack of a centrode selection in FIVEBAR. See the *Animation* discussion for FOURBAR in Section A-3 for more information.

Dynamics (FIVEBAR *Dynamics* Screen)

Input data for dynamics calculation in FIVEBAR are similar to that for program FOURBAR with the addition of one more link. However, linkage balancing is not available in FIVEBAR. See the discussion of dynamics calculations for program FOURBAR in Section A-3 for more information.

Other

See Section A.2, General Program Operation (p. 727) for information on *New, Open, Save, Save As, Plot, Print, Units,* and *Quit* functions.

A.5   PROGRAM SIXBAR

SIXBAR *for Windows* is a linkage design and analysis program intended for use by students, engineers, and other professionals who are knowledgeable in or are learning the art and science of linkage design. It is assumed that the user knows how to determine whether a linkage design is good or bad and whether it is suitable for the application for which it is intended. The program will calculate the kinematic and dynamic data associated with any linkage design but cannot substitute for the engineering judgment of the user. The linkage theory and mathematics on which this program is based are document-

A

ed in Chapters 4 to 10, and 11 of this textbook. Please consult them for explanations of the theory and mathematics involved.

The SIXBAR program is generally similar to the FOURBAR program. It will analyze the **Watt's II** and the **Stephenson's III** linkage isomers as defined in Figure 2-14 (p. 45). These are two of the five distinct sixbar isomers. The **Watt's II** mechanism is shown in Figure A-12 with the program's input parameters defined. The **Stephenson's III** mechanism is shown in Figure A-5 (p. 736) with its input parameters defined. Note that the program divides the sixbar into two stages of fourbar linkages. Stage 1 is the left half of the mechanism as shown in Figures A-12 and A-5. Stage 2 is the right half. The X axis of the global coordinate system is defined by instant centers $I_{1,2}$ and $I_{1,4}$ with its origin at $I_{1,2}$. The third fixed pivot $I_{1,6}$ can be anywhere in the plane. Its coordinates must be supplied as input.

The SIXBAR *Home* Screen

Initially, only the *Input* and *Quit* buttons are active on the *Home* screen. Typically, you will start a linkage design with the *Input* button, but for a quick look at a linkage as drawn by the program, one of the examples under the *Example* pull-down menu can be selected and it will draw a linkage. If you activate one of these examples, when you return from the *Animate* screen you will find all the other buttons on the *Home* screen to be active. We will address each of these buttons in due course below.

The *Home* screen's *Examples* pull-down menu includes both Watt's and Roberts' straight-line fourbar linkage stages driven by dyads (making them sixbars), a single-dwell sixbar linkage similar to that of Example 3-13 (p. 126) and Figure 3-31 (p. 127), and a double-dwell sixbar linkage that uses an alternate approach to that of Example 3-14 (p. 130) and Figure 3-32 (p. 129).

Input Data (SIXBAR *Input* Screen)

Much of the basic data for the linkage design is defined on the *Input* screen which is activated by selecting the *Input* button on the *Home* screen. When you open this screen for the first time, it will have default data for all link parameters. You may change any of these by typing over the data in the yellow text boxes. A choice of Watt's or Stephenson's linkage must be made on the *Input* screen. The link information differs for the Watt's and Stephenson's linkages.

WATT'S II LINKAGE　 For the Watt's linkage (Figure A-12) the stage 1 data is: crank, first coupler, first rocker, and ground link segment from instant centers $I_{1,2}$ to $I_{1,4}$. These correspond to links 2, 3a, 4a, and 1a, respectively, as labeled in the Figure A-12. The stage 2 data are: second crank, second coupler, second rocker, corresponding respectively to links 4b, 5a, and 6 in Figure A-12. The angle δ_4 that the second crank (4b) makes with the first rocker (4a) is also requested. Note that this angle obeys the right-hand rule as do all angles in these programs.

Two coupler points are allowed to be defined in this linkage, one on link 3 and one on link 5. The method of location is by polar coordinates of a position vector embedded in the link as was done for the fourbar and fivebar linkages. The first coupler point C is on link 3 and is defined in the same way as in FOURBAR. Program SIXBAR requires the

A

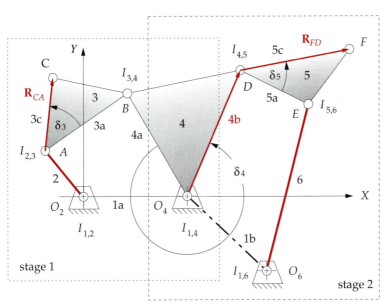

FIGURE A-12

Input data for program SIXBAR—a Watt's II linkage

length (3c) of its position vector $\mathbf{R}_{CA}$ and the angle δ_3 which that vector makes with line 3a in Figure A-12. The second coupler point F is on link 5 and is defined with a position vector $\mathbf{R}_{FD}$ rooted at instant center $I_{4,5}$. The program requires the length (5c) of this position vector $\mathbf{R}_{FD}$ and the angle δ_5 which that vector makes with line 5a. The X and Y components of the location of the third fixed pivot $I_{1,6}$ are also needed. These are with respect to the global X,Y axis system whose origin is at $I_{1,2}$.

STEPHENSON'S III LINKAGE   The stage 1 data for the Stephenson's linkage is similar to that of the Watt's linkage. The first stage's crank, first coupler, first rocker, and ground link segment from instant centers $I_{1,2}$ to $I_{1,4}$ correspond to links 2, 3a, 4, and 1a, respectively, in Figure A-13. The link lengths in stage 2 of the linkage are: second coupler and second rocker corresponding, respectively, to links 5a and 6 in Figure A-13.

Note that, unlike the Watt's linkage, there is no "second crank" in the Stephenson's linkage because the second stage is driven by the coupler (link 3) of the first stage. In this program link 5 is constrained to be connected to link 3 at link 3's defined coupler point C which then becomes instant center $I_{3,5}$. The data for this are requested in similar format to the Watt's linkage, namely, the length of the position vector $\mathbf{R}_{CA}$ (line 3c) and its angle δ_3. The second coupler point location on link 5 is defined, as before, by position vector $\mathbf{R}_{FC}$ with length 5c and angle δ_5.

The X and Y components of the location of the third fixed pivot $I_{1,6}$ are required. These are with respect to the global X,Y axis system.

Select the type of calculation desired in the upper right corner of the screen, one of *Angle Steps, Time Steps,* or *One Position.* The *Calculate* button will compute all data for your linkage and show it in an arbitrary position on the screen. If at any time the link-

A

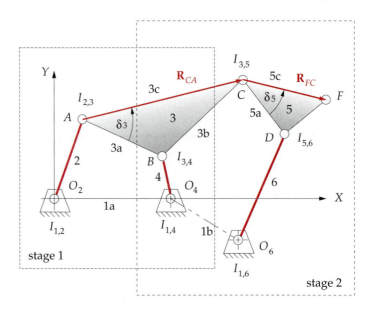

FIGURE A-13

Input data for program SIXBAR—a Stephenson's III linkage

age display window is blank, the *Calculate* button will bring back the image. See the description of calculations for program FOURBAR in Section A.3 (p. 735). They are similar in SIXBAR.

After you have calculated the linkage, the *Animate* and *Done* buttons on the *Input* screen will become available. *Animate* takes you to the *Animate* screen where you can run the linkage through any range of motion to observe its behavior. You can also change any of the linkage parameters on the *Animate* screen and then recalculate the results with the *Recalc* button. The *Done* button on either the *Input* or *Animate* screen returns you to the *Home* screen. The *Plot* and *Print* buttons will now be available as well as the *Animate* button which returns you to the *Animate* screen.

Animation (SIXBAR)

The *Animation* screen and its features in SIXBAR are essentially similar to those of Program FOURBAR. The only exception is the lack of a *Centrode* selection in SIXBAR. See the Animation discussion for FOURBAR in Section A.3 for more information.

Dynamics (SIXBAR *Dynamics* Screen)

Input data for dynamics calculation in SIXBAR are similar to that for program FOURBAR with the addition of two links. See the discussion of dynamics calculations for program FOURBAR in Section A.3 for more information.

Other

See Section A.2, General Program Operation (p. 727) for information on *New, Open, Save, Save As, Plot, Print, Units*, and *Quit* functions.

A.6   PROGRAM SLIDER

SLIDER *for Windows* is a linkage design and analysis program intended for use by students, engineers, and other professionals who are knowledgeable in or are learning the art and science of linkage design. It is assumed that the user knows how to determine whether a linkage design is good or bad and whether it is suitable for the application for which it is intended. The program will calculate the kinematic and dynamic data associated with any linkage design, but cannot substitute for the engineering judgment of the user. The linkage theory and mathematics on which this program is based are documented in Chapters 4 to 10, and 11 of this textbook. Please consult them for explanations of the theory and mathematics involved.

The Slider Home Screen

Initially, only the *Input* and *Quit* buttons are active on the *Home* screen. Typically, you will start a linkage design with the *Input* button, but for a quick look at a linkage as drawn by the program, one of the examples under the *Example* pull-down menu can be selected and it will draw a linkage. If you activate one of these examples, when you return from the *Animate* screen you will find all the other buttons on the *Home* screen to be active. We will address each of these buttons below.

Input Data (SLIDER *Input* Screen)

Basic data for the sixbar linkage is defined on the *Input* screen, which is activated by selecting the *Input* button on the *Home* screen. When you open this screen for the first time, it will have default data for all link parameters. You may change any of these by typing over the data in the yellow text boxes.

Figure A-14 defines the input parameters for the fourbar slider-crank linkage. The link lengths needed are input link 2 and coupler link 3, defined by their pin-to-pin distances and labeled a and b in the figure. The X axis lies along the line d, through instant center $I_{1,2}$ (point O_2) and parallel to the direction of motion of slider 4. Instant center $I_{1,2}$, the driver crank pivot, is the origin of the global coordinate system. The slider offset c is the perpendicular distance from the X axis to the sliding axis. Slider position d will be calculated for all positions of the linkage.

In addition to the link lengths, you must supply the location of one coupler point on link 3 to find that point's coupler curve positions, velocities, and accelerations. This point is located by a position vector rooted at $I_{2,3}$ (point A) and directed to the coupler point P of interest which can be anywhere on link 3. The program requires that you input the polar coordinates of this vector which are labeled p and δ_3 in Figure A-14. The program needs the distance from $I_{2,3}$ to the coupler point p and the angle δ_3 that the coupler point makes with link 3. Note that angle δ_3 is not referenced to either the global coordinate system

A

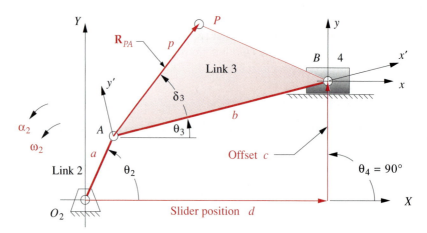

FIGURE A-14

Input data for Program SLIDER

X,Y or to the local nonrotating coordinate system x,y at point A. Rather, it is referenced to the line AB which is the pin-to-pin edge of link 3. Angle δ_3 is a property of link 3 and is embedded in it. The angle which locates vector $\mathbf{R}_{CA}$ in the x,y coordinate system is the sum of angle δ_3 and angle θ_3. This addition is done in the program, after θ_3 is calculated for each position of the input crank.

The definition of θ_3 is different in program SLIDER than in Figure 4-9 (p. 160) and the derivations in Section 4.6 (p. 159). This was done to maintain consistency of input data among all the linkage programs in this package. Actually, within program SLIDER, the angle shown in Figure A-14 that you input as θ_3 is converted using the law of cosines to the one shown in Figure 4-9 before the calculations are done with the equations from Section 4.6.

Calculation (SLIDER *Input* Screen)

See the description of calculations for program FOURBAR in Section A.3 (p. 735) for more information. They are similar in SLIDER. Select the type of calculation desired in the upper right corner of the screen, one of *Angle Steps, Time Steps,* or *One Position.* The *Calculate* button will compute all data for your linkage and show it in an arbitrary position on the screen. If at any time the white linkage window is blank, the *Calculate* button will bring back the image.

After you have calculated the linkage, the *Animate* and *Done* buttons on the *Input* screen will become available. *Animate* takes you to the *Animate* screen where you can run the linkage through any range of motion to observe its behavior. You can also change any of the linkage parameters on the *Animate* screen and then recalculate the results with the *Recalc* button. The *Done* button on either the *Input* or *Animate* screen returns you to the *Home* screen. The *Plot* and *Print* buttons will now be available as well as the *Animate* button which returns you to the *Animate* screen.

A

Animation (SLIDER *Animation* Screen)

The *Animation* screen and it features in SLIDER are essentially similar to those of program FOURBAR. An exception is the lack of a *Centrode* selection in SLIDER. See the discussion of the *Animation* screen for FOURBAR in Section A.3 for more information.

Dynamics (SLIDER *Dynamics* Screen)

Input data for dynamics calculation in SLIDER are similar to that for program FOURBAR. See the discussion of dynamics calculations for Program FOURBAR in Section A.3 for more information.

Other

See Section A.2, General Program Operation (p. 727) for information on *New, Open, Save, Save As, Plot, Print, Units,* and *Quit* functions.

A.7   PROGRAM DYNACAM

DYNACAM *for Windows* is a cam design and analysis program intended for use by students, engineers, and other professionals who are knowledgeable in the art and science of cam design. It is assumed that the user knows how to determine whether a cam design is good or bad and whether it is suitable for the application for which it is intended. The program will calculate the kinematic and dynamic data associated with any cam design but cannot substitute for the engineering judgment of the user. The cam theory and mathematics on which this program is based are shown in Chapter 8 and 15 of this text. Please consult them for complete explanations of the theory and mathematics involved.

The DYNACAM *Home* Screen

Initially, only the *SVAJ* and *Quit* buttons are active on the *Home* screen. Typically, you will start a cam design with the *SVAJ* button, but for a quick look at a cam as drawn by the program, one of the examples under the *Example* pull-down menu can be selected and it will draw a cam profile. If you activate one of these examples, when you return from the *Cam Profile* screen you will find all the other buttons on the *Home* screen to be active. We will address each of these buttons in due course below.

Input Data (DYNACAM *Input* Screen)

Much of the basic data for the cam design is defined on the *Input* screen shown in Figure A-15, which is activated by selecting the *SVAJ* button on the *Home* screen. When you open this screen for the first time, it will be nearly blank, with only one segment's row visible. (Note that the built-in examples can also be accessed from this form at its upper right corner.) If you selected an example cam from the pull-down menu on the *Home* screen, making the *Input* screen nonblank, please now select the *Clear All* button on the *Input* screen to zero all the data and blank the screen, in order to better follow the presentation below. We will proceed with the explanation as if you were typing your data into an initially blank *Input* screen.

A

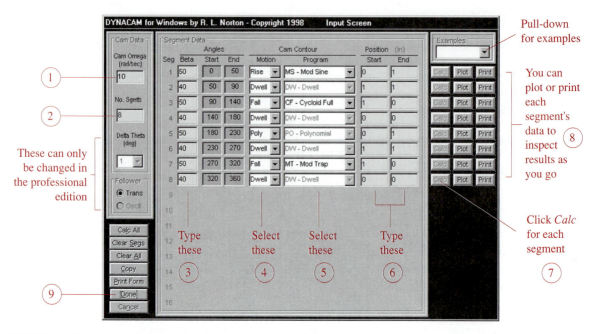

FIGURE A-15

S V A J input screen for program DYNACAM

If you use the *Tab* button, it will lead you through the steps needed to input all data. On a blank *Input* screen, *Tab* first to the *Cam Omega* box in the upper left corner and type in the speed of the camshaft. *Tab* again (or mouse click if you prefer) to the *Number of Segments* box and type in any number desired between 1 and 16. That number of rows will immediately become visible on the *Input* screen. Note that some of this screen's choices such as *Osclt* (oscillating arm follower) and *Delta Theta* are disabled in the student edition of the program.

Another *Tab* should put your cursor in the box for the *Beta* (segment duration angle) of segment 1. Type any desired angle (in degrees). Successive *Tabs* will take you to each *Beta* box to type in the desired angles. All *Betas* must, of course, sum to 360 degrees. If they do not, a warning will appear.

As you continue to *Tab* (or click your mouse in the appropriate box, if you prefer), you will arrive at the boxes for *Motion* selection. These boxes offer a pull-down selection of *Rise, Fall, Poly,* and *Dwell.* You may select from the pull-down menus with the mouse, or you can type the first letter of each word to select them. Rise, fall, and dwell have obvious meanings. The Poly choice indicates that you wish to create a customized polynomial function for that segment, and this will later cause a new screen to appear on which you will define the boundary conditions of your desired polynomial function.

The next set of choices that you will *Tab* or mouse click to are the *Program* pull-downs. These provide a menu of standard cam functions such as *Modified Trapezoid, Modified Sine,* and *Cycloid.* Also included are portions of functions such as the first and

A

second halves of cycloids and simple harmonics that can be used to assemble piecewise continuous functions for special situations. See Chapter 8 for additional information.

After you have selected the desired *Program* functions for each segment, you will *Tab* or *Click* to the *Position Start* and *End* boxes. *Start* in this context refers to the beginning displacement position for the follower in the particular segment, and *End* for its final position. You may begin at the "top" or "bottom" of the displacement "hill" as you wish, but be aware that the range of position values of the follower must be from a zero value to some positive value over the whole cam. In other words, you cannot include any anticipated base or prime circle radius in these position data. These position numbers represent the excursion of the so-called *S* diagram (displacement) of the cam and cannot include any prime circle information (which will be input later).

As each row's (segment's) input data are completed, the *Calc* button for that row will become enabled. Clicking on this button will cause that segment's *S, V, A,* and *J* data to be calculated and stored. After the *Calc* button has been clicked for any row, the *Plot* and *Print* buttons for that segment will become available. Clicking on these buttons will bring up a plot or a printed table of data for *S, V, A, J* data for that segment only. More detailed plots and printouts can be obtained later from the *Home* screen.

Polynomial Functions

If any of your segments specified a *Poly* motion, clicking the *Calc* button will bring up the *Boundary Condition* screen shown in Figure A-16. The cursor will be in the box for *Number of Conditions Requested.* Type the number of boundary conditions (BCs) de-

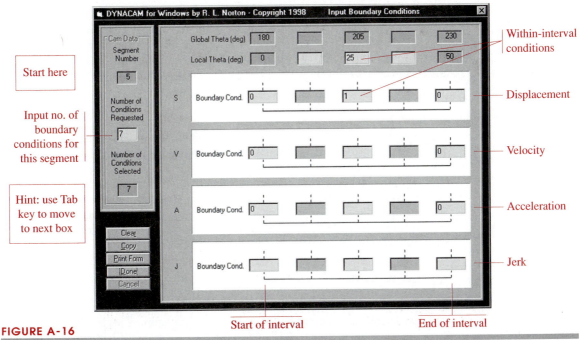

FIGURE A-16

Boundary Condition Input screen for polynomial functions in program DYNACAM—3-4-5-6 single-dwell function shown

A

sired, which must be between 2 and 20 inclusive. When you hit *Enter* or *Tab* or mouse click away from this box, the rest of the screen will activate, allowing you to type in the desired values of BCs. Note that the start and end values of position that you typed on the *Input* screen are already entered in their respective *S* boundary condition boxes at the beginning and end of the segment. Type your additional end of interval conditions on *V*, *A*, and *J* as desired. If you also need some BCs within the interval, click or tab to one of the boxes in the row labeled *Local Theta* at the top of the screen and type in the value of the angle at which you wish to provide a BC. That column will activate and you may type whatever additional BCs you need.

The box labeled *Number of Conditions Selected* monitors the BC count, and when it matches the *Number of Conditions Requested,* the *Done* button becomes available. Note that what you type in any (yellow) text box is not accepted until you hit *Enter* or move off that box with the *Tab* key or the mouse, allowing you to retype or erase with no effect until you leave the text box. (This is generally true throughout the program.)

Selecting the *Done* button from the BC screen calculates the coefficients of the polynomial by a Gauss-Jordan reduction method with partial pivoting. All computations are done in double precision for accuracy. If an inconsistent set of conditions is sent to the solver, an error message will appear. If the solution succeeds, it calculates *s v a j* for the segment. When finished, it brings up a summary screen that shows the BCs you selected and also the coefficients of the polynomial equation that resulted. You may at this point want to print this screen to the printer or copy and paste it into another document for your records. You will only be able to reconstruct it later by again defining the BCs and recalculating the polynomial.

Back to the *Input* Screen

Completing a polynomial function returns you to the *Input* screen. When all segments are calculated, select the *Done* button on the *Input* screen (perhaps after copying it to the clipboard or printing it to the printer with the appropriate buttons). This will bring up the *Continuity Check* screen.

Continuity Check Screen

This screen provides a visual check on the continuity of the cam design at the segment interfaces. The values of each function at the beginning and end of each segment are grouped together for easy viewing. The fundamental law of cam design requires that the *S, V,* and *A* functions be continuous (see Chapter 8). This will be true if the boundary values for those functions shown grouped as pairs are equal. If this is not true, then a warning dialog box will appear when the *Done* button is clicked. This will not prevent you from proceeding but will suggest you return to the *Input Data* screen to correct the problem. Then it will return you to the *Home* screen.

Sizing the Cam

Once the *S V A J* functions have been defined to your satisfaction, it remains to size the cam and determine its pressure angles and radii of curvature. This is done from the *Size Cam* screen shown in Figure A-17, which is accessible from the button of that name on

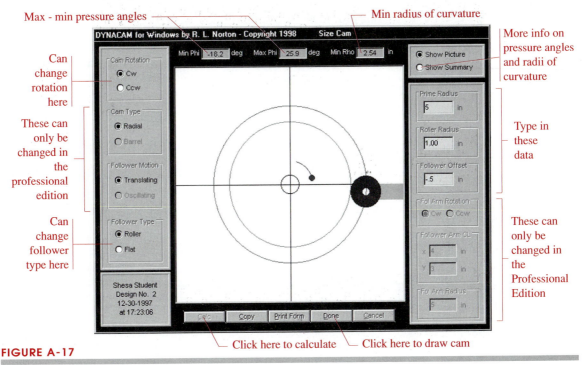

FIGURE A-17

Size Cam screen from DYNACAM

the *Home* screen. The *Size Cam* screen allows the cam rotation direction and follower type (flat or roller) to be set. The cam type (radial or barrel) and follower motion (translating or oscillating) are only available in the professional version. The prime circle radius and follower offset for a translating follower can be typed into their boxes. When a change is made to any of these parameters, the schematic image of cam and follower is updated.

Select the *Calc* button to compute the cam size parameters. The max and min pressure angles will appear in the boxes at the top of the screen. For more information, select the *Show Summary* button at the upper right. This will change the schematic image to a summary of pressure angle and radius of curvature information. When you hit the *Done* button, the cam profile will be drawn.

Drawing the Cam

The cam profile shown when leaving the *Size Cam* screen can be independently accessed at any time from the *Home* screen with the *Draw Cam* button. The *Draw Cam* screen shown in Figure A-18, allows the cam rotation direction, prime circle radius, roller radius, and offset (if available) to be changed. The cam size data will be recalculated and the new profile displayed.

In the cam profile drawing, a curved arrow indicates the direction of cam rotation. The initial position of a roller follower at cam angle $\theta = 0$ is shown as a filled circle with rectangular stem, and of a flat-faced follower as a filled rectangle. Any eccentricity

A

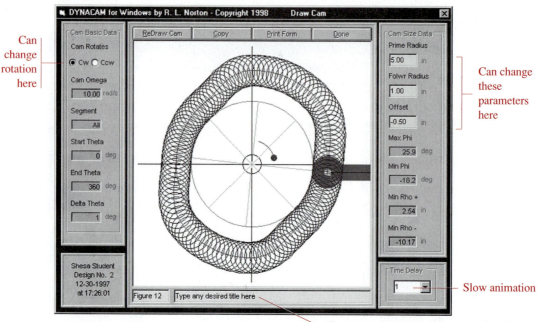

Can change rotation here

Can change these parameters here

Slow animation

Type something here to annotate the plot

FIGURE A-18

Cam Profile screen in program DYNACAM

shows as a shift up or down of the filled follower with respect to the *X* axis through the cam center. The smallest filled circle on the cam centerline represents the camshaft. The smallest unfilled circle is the base circle. The prime circle may be somewhat obscured by the repeated drawings of the follower diameter, which sweep out the cam surface. The pitch curve is drawn along the locus of the roller follower centers. The radial lines that form pieces of pie within the base circle represent the segments of the cam. If the cam turns counterclockwise, the radial lines are numbered clockwise around the circumference and vice versa.

Follower Dynamics (DYNACAM Only)

When the cam has been sized, the *Dynamics* button on the *Home* screen will become available. This button brings up the *Dynamics* screen shown in Figure A-19. Text boxes are provided for typing in values of the effective mass of the follower system, the effective spring constant, the spring preload, and a damping factor. By effective mass is meant the mass of the entire follower system as reflected back to the cam-follower roller centerline or cam contact point. Any link ratios between the cam-follower and any physical masses must be accounted for in calculating the effective mass. Likewise the effective spring in the system must be reflected back to the follower. The damping is defined by the damping ratio zeta, as defined for second-order vibrating systems. See Chapter 10 for further information.

The journal diameter and the coefficients of friction are used for calculating the friction torque on the shaft. The *Start New* or *Accumulate* switch allows you to either make

Undamped natural frequency — 2nd row: max, min force & max torque — Damping coeff. — Damped natural frequency

Mass reflected to follower

Spring reflected to follower

Usually an assumed value

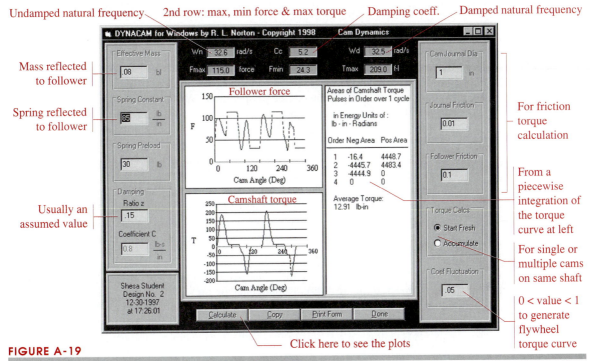

For friction torque calculation

From a piecewise integration of the torque curve at left

For single or multiple cams on same shaft

0 < value < 1 to generate flywheel torque curve

Click here to see the plots

FIGURE A-19

Cam Dynamics screen in DYNACAM

a fresh torque calculation or accumulate the torques for several cams running on a common shaft. The energy information in the window can be used to calculate a flywheel needed for any coefficient of fluctuation chosen as described in Section 11.11 (p.548). The program calculates a smoothed torque function by multiplying the raw camshaft torque by the coefficient of fluctuation specified in the box at lower right of the screen.

Other

See Section A.2, General Program Operation (p. 727) for information on *New, Open, Save, Save As, Plot, Print, Units,* and *Quit* functions.

A.8   PROGRAM ENGINE

ENGINE *for Windows* is an internal combustion (IC) engine design and analysis program intended for use by students, engineers, and other professionals who are knowledgeable in the art and science of engineering design. It is assumed that the user knows how to determine whether a design is good or bad and whether it is suitable for the application for which it is intended. The program will calculate the kinematic and dynamic data associated with any engine design, but cannot substitute for the engineering judgment of the user. The theory and mathematics on which this program is based are shown in Chapters 13 and 14 of this textbook Please consult them for an explanation of the theory and mathematics involved.

A

The ENGINE *Home* Screen

Initially, only the *Input* and *Quit* buttons are active on the *Home* screen. Typically, you will start a design with the *Input* button, but before doing so, one of the examples under the *Example* pull-down menu can be selected. If you activate one of these examples, it will calculate the result and take you to the *Engine Data* screen (see Figure A-22). Click on the *Run* button to see the engine in operation. When you return to the *Home* screen you will find all the other buttons to be active. We will address each of these buttons below.

Single-Cylinder Engine (*Input One-Cylinder Data* Screen)

This program can be thought of as being used in two stages, each of which roughly corresponds to the topics in Chapters 13 and 14, respectively. That is, the *Input* and *Balance* buttons on the *Home* screen deal with single-cylinder engines (Chapter 13), and the *Assemble* button with multicylinder engines (Chapter 14). The *Flywheel* button can be used with either a single- or multicylinder design.

THE *INPUT* BUTTON   on the *Home* screen brings up the *Input One-Cylinder Data* screen shown in Figure A-20. Default data are present in all entry boxes. Device type can be set to either an IC engine or compressor. The only difference between these two choices is the magnitude of the cylinder pressure used in the calculations. The default pressure for either device can also be changed in the box at lower right. The stroke cycle can be specified as 2 or 4 stroke. Two operating speeds must be specified, idle and redline. The midrange speed is the average of the two speeds supplied. One of three calculation speeds must be selected from the *Idle, Midrange,* and *Redline* options. The number of cylinders can be specified in the *No. Cylinders* box, and *Engine Type* can be selected from its pull-down menu. At this screen, only one cylinder of the engine is being calculated, so the choices of number of cylinders and engine type are not used until later when the engine is assembled. Choices made on this screen will be carried forward to the *Assemble* screen and also can be changed there if desired. The *CG* locations of crank and conrod, expressed as a percent of length are typed in boxes on the left. The masses of crank, conrod, and piston, must be supplied (in proper units) in the boxes at the right.

Text boxes in the *Engine Geometry Panel* on the left allow typing of the cylinder displacement, bore, stroke or crank radius, and L/R ratio. Cylinder displacement is the primary variable in this program. Any displacement volume can be achieved with an infinity of combinations of bore and stroke. To resolve this indeterminacy, bore is arbitrarily given precedence over stroke when the displacement is changed. That is, a change of displacement will force a change in stroke, leaving the bore unchanged. Changing the bore will force a change in stroke, keeping the specified displacement. Either the stroke or the crank radius may be changed, but either will force the other to change accordingly since stroke is always twice the crank radius, so only one need be input. The bore/stroke ratio is calculated and displayed at upper right.

The diameters of main pins and crankpins are used only to calculate an estimated friction torque in combination with an estimate of friction coefficient in the engine. The friction torque is calculated by multiplying the user-specified coefficient of friction by the forces calculated at the piston-cylinder interface and at the main pin and crankpin

A

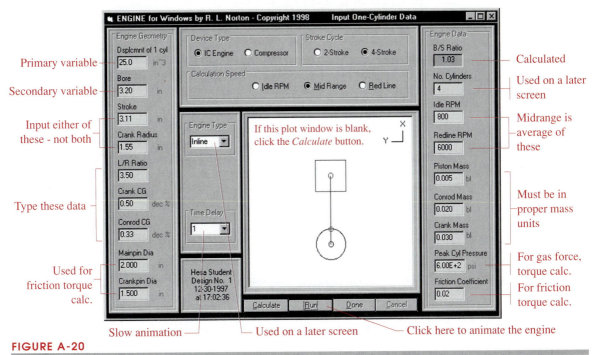

Primary variable

Secondary variable

Input either of these - not both

Type these data

Used for friction torque calc.

Slow animation

Used on a later screen

Click here to animate the engine

Calculated

Used on a later screen

Midrange is average of these

Must be in proper mass units

For gas force, torque calc.

For friction torque calc.

FIGURE A-20

Program ENGINE *Input Screen* for one-cylinder engine data

journals. These last two friction forces are multiplied by the journal radii supplied by the user to obtain friction torque. The piston friction creates friction torque through the geometry described in the gas torque equation 13.8 (p. 613). The other torque values (gas torque, inertia torque, and total torque) are **not** reduced in the program by the amount of the calculated friction torque, which is at best a crude estimate.

CALCULATION When all data are supplied, a click of the *Calculate* button will cause the piston position, velocity, and acceleration; the inertia forces and torques; and pin forces plus gas force and gas torque to be calculated for two revolutions of the crankshaft. At this screen, these data are computed only for one cylinder of the engine, regardless of how many cylinders were specified. The gas force and gas torque calculation is based on a built-in gas pressure curve similar to the one shown in Figure 13-4e (p. 602). This gas pressure function in the program is kept the same at all engine speeds as discussed in Section 13.1 (p. 600). Though this is not accurate in a thermodynamic sense, it is both necessary and appropriate for the purpose of comparing designs based solely on their kinematic and dynamic factors.

Calculations are done for all parameters at 3° increments over two crankshaft revolutions, giving 241 data points per variable. When calculations are complete, the plot window will show a schematic of the single-cylinder engine that can be animated with the *Run* button. The bore, stroke and conrod dimensions are to scale in the animation. The *Time Delay* value can be increased to slow the animation.

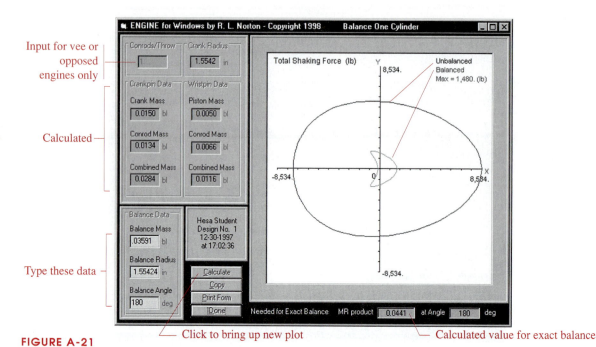

Input for vee or opposed engines only

Calculated

Type these data

Click to bring up new plot   Calculated value for exact balance

FIGURE A-21

Single-cylinder *Balance* screen from program ENGINE

The *Done* button returns you to the *Home* screen where you can use the *Plot* and *Print* buttons to display the results of the single-cylinder calculations. See Section A.2 (p. 727) for information on the *Print* and *Plot* screens.

Balancing the Crank (ENGINE *Balance* Screen)

The *Balance* button on the *Home* screen brings up the *Balance* screen shown in Figure A-21. The bottom of this screen displays the mass-radius product needed on the crank to cancel the primary component of unbalance due to the mass at the crankpin. The unbalanced shaking force is displayed as a hodograph in a plot window. Information on the amounts of mass estimated to be located at the crankpin and wrist pin is displayed in the sidebar. These data provide enough information to determine the counterweight parameters needed to either exactly balance or optimally overbalance the single-cylinder engine. Three text boxes at the bottom left of the screen allow input of the desired mass, radius, and angle of the counterweight proposed to be placed on the crankshaft.

Clicking the *Calculate* button recomputes the shaking force and torque with the added counterweight and superposes a hodograph of the new, balanced shaking force on the plot of the unbalanced force at the same scale so the improvement can be seen. See Chapter 13 for a discussion of the meaning and use of these data. Note that if the engine design has enough cylinders to allow a crankshaft arrangement that will cancel the inertial forces, then there may be no advantage to overbalancing the crank. But, for a single-cylinder engine and some two-cylinder engines (twins), overbalancing the crank can significantly reduce the shaking force.

A

Click on an example to activate it - Warning! - will overwrite previous data

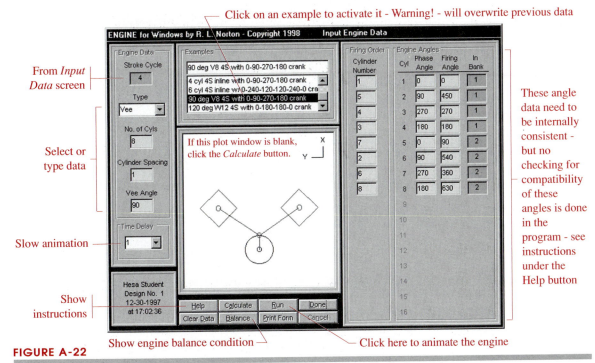

From *Input Data* screen

Select or type data

Slow animation

Show instructions

Show engine balance condition

Click here to animate the engine

These angle data need to be internally consistent - but no checking for compatibility of these angles is done in the program - see instructions under the Help button

FIGURE A-22

The *Input Engine Data* screen in program ENGINE is used to assemble multicylinder engines

Assembling a Multicylinder Engine (*Engine Data* Screen)

Once the single-cylinder configuration is satisfactorily designed and balanced, the engine can be assembled. Figure A-22 shows the *Engine Data* screen used for assembly of a multicylinder engine. Its *Help* button brings up the information shown in Figure A-23, which points out that the crank phase diagram (called a **Power Chart** in program ENGINE) as defined in Section 14.2 (p. 644) and shown in Figures 14-8, 14-11, 14-13, 14-15, 14-17, and 14-23 (pp. 646-672) should be defined and drawn manually before proceeding with the assembly of the engine. The assumptions and conventions used in the program to number the cylinders and banks are also stated on these screens. These are the same conventions as defined in Section 14.7 (p. 661). The upper limit stated in item 7 of Figure A-23 for the acceptable range of power stroke angles will be either 360° for a two-stroke engine or 720° (as shown) for a four-stroke engine.

The firing order, the crankshaft phase angles, and the angles at which the cylinders fire (firing angles)—all angles in cylinder order—are typed in the boxes on the right of the *Engine Data* screen based on an arrangement according to the rules in Figure A-23. Note that **the program does not do any internal check on the compatibility of these data**. It will accept any combination of phase angles, firing orders, and power stroke angles you provide. It is up to you to ensure that these data are compatible and realistic.

The program also needs the distance between cylinders which is defaulted to one. This information is used to define the z_i in equation 14.11 (p. 667). If you want to com-

+++++++++++++++++ Warning +++++++++++++++++
Before proceeding you need to work out, on paper, the crank's
Phase Angles, the engine's Firing Order, and the Power Stroke
Angles (the angles at which the cylinders fire).

This program uses the Phase Angles and Firing Angles
and Vee Angle to calculate the results. Firing Order is used as a pointer.
If you change only the Firing Order, it will make the results incorrect.
It is your responsibility to ensure that all these factors agree!

This program requires that these rules be followed when designing
and numbering your crankshaft phase and firing angles, and vee angle:

1 - The engine's front cylinder in the right bank is always number 1
2 - Number one cylinder is the reference cylinder for all others
3 - Number one cylinder always has a Crank Phase Angle of zero degrees
4 - Number one cylinder always fires first in Firing Order
5 - Thus the Power Stroke Angle for number one is also zero degrees
6 - Crank Phase Angles are always between 0 and 360 degrees
7 - Firing Angles are between 0 and 720 deg. and in ascending order
8 - Phase Angles shift to the right versus cylinder 1 on the Crank Phase Diagram
9 - The Vee Angle shifts the second bank to the Right on the Crank Phase Diagram
10 - Cylinders are numbered first down right bank and then down left bank

You must draw your crank phase diagram to decide these values before proceeding.

FIGURE A-23

Rules for assembling a multicylinder engine in program ENGINE

pute the correct magnitude of the shaking moment, the actual cylinder spacing of your design needs to be supplied in the box at the left of the screen in Figure A-22. If you only want to compare two or more engine designs on the basis of relative shaking moment, then the default value of one will suffice. The computation sums moments about the first cylinder, so its moment arm is always zero. The small offset between a vee engine's banks, due to having two conrods per crank throw, is ignored in the moment calculation.

Charts (Pull-Down Menu)

Figure A-24 shows the *Charts* pull-down menu on the *Home* screen. Several summaries of engine information are available there. The *Force Balance*, *Moment Balance*, and *Torque Balance* menu selections display the results of engine balance condition calculations, i.e., equations 14.3a and 14.3b (p. 648) for force balance; 14.7a and 14.7b (p. 650) for moment balance; and 14.5a, 14.5b, and 14.5c (p. 650) for torque balance. (These charts are also accessible from the *Engine Data* screen via its *Balance* button.)

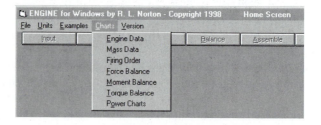

FIGURE A-24

Charts available from pull-down menu in program ENGINE

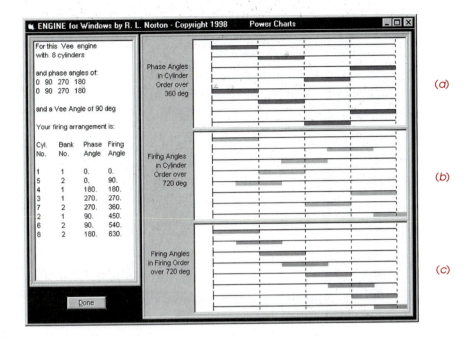

FIGURE A-25

Phase angle and power charts for a vee-eight engine from program ENGINE. Compare to Figure 14-23 (p. 672)

The **Power Chart** selection presents the three plots shown in Figure A-25. Figure A-25a is a plot of the crankshaft phase angles over 360° for the 90° vee eight engine designed in Section 14.7 (p. 661). Figure A-25b shows the power stroke angles in cylinder order over 720° for this vee eight. This is the same as the crank phase diagram shown in Figure 14-23 (p. 672) and indicates when the power pulses will occur in each cylinder during the cycle. Figure A-25c shows the power stroke angles in firing order over 720°. This third plot rearranges the order of the cylinders on the power chart to match that of the selected firing order. If these three factors, *crankshaft phase angles, firing order,* and *firing angles* have all been correctly chosen for compatibility, this third power chart plot will appear *as a staircase*. The power pulses should occur in equispaced steps across the cycle if even firing has been achieved. Thus these power chart plots can serve as a check on the correct choices of these factors. Note that the program only draws the boxes on the crank phase diagrams and power charts which represent the power strokes of each cylinder. The other piston strokes shown in Figures 14-8, 14-11, 14-13, 14-15, 14-17, and 14-23 (pp. 646-672) are not drawn.

Flywheel Calculations

Figure A-26 shows the *Flywheel Data* screen from program ENGINE. The only input to this screen is the flywheel coefficient of fluctuation which must be between 0 and 1. When the *Calc* button is clicked, it does the same pulse-by-pulse integration of the total torque curve as is done in program FOURBAR. See Section 11.11 and Figure 11-16 (p. 548). The areas under the pulses of the total torque curve are a measure of energy variation in the cycle. The flywheel size is based on these energy variations as defined in Sec-

A

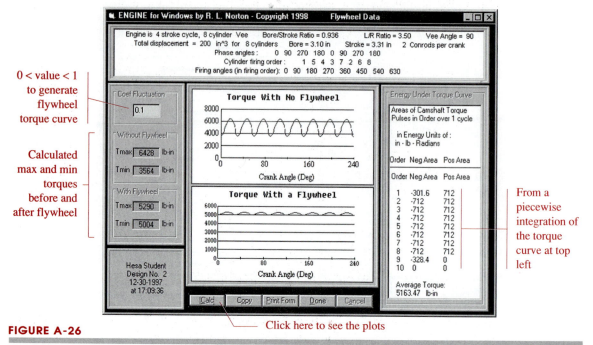

0 < value < 1 to generate flywheel torque curve

Calculated max and min torques before and after flywheel

From a piecewise integration of the torque curve at top left

FIGURE A-26

Flywheel Data screen from program ENGINE for a vee-eight engine

Click here to see the plots

tion 11.11. The total torque curve for this example vee-eight engine is shown in the upper plot of Figure A-26. The table of areas under its pulses is shown in the table at the right of the screen. Note that for a properly designed, even firing, multicylinder engine, the absolute values of the areas under each pulse will be equal. In that case the energy value needed for the flywheel calculation is simply the area under one pulse. These data can be used in equation 11.22 (p. 553) to compute the needed flywheel moment of inertia based on a chosen coefficient of fluctuation k. Program ENGINE then calculates a smoothed total torque function by multiplying the raw torque by the specified coefficient of fluctuation. The lower plot in Figure A-26 shows the flywheel smoothed torque for the example vee-eight engine with $k = 0.1$. The torque function with the flywheel added is nearly a constant value, which is desirable. The pulses due to the individual cylinder's explosions have been effectively masked by the flywheel. The maximum and minimum values of torque with and without the flywheel are shown in boxes at the left of the screen. Data on the engine design are displayed in the panel at the top of the screen.

A.9   PROGRAM MATRIX

Program MATRIX *for Windows* solves a matrix equation of the form $\mathbf{A} \times \mathbf{B} = \mathbf{C}$. $\mathbf{A}$ must be square and can have up to 16 rows and 16 columns. $\mathbf{C}$ must be a vector with as many elements as $\mathbf{A}$ has rows. When the data have all been typed in and checked, click the *Solve* button and it will compute the terms of vector $\mathbf{B}$. The results can be printed to screen, printer, or a disk file in the same manner as described in Section A.2 (p. 727). If the matrix is singular or the solution fails for other reasons, an error message will be returned. No plotting or animation is provided in this program.

A

Appendix **B**

MATERIAL PROPERTIES

For selected engineering materials. Many other alloys are available.

The following tables contain approximate values for strengths and other specifications of a variety of engineering materials compiled from various sources. In some cases the data are minimum recommended values and in other cases are data from a single test specimen. These data are suitable for use in the engineering exercises contained in this text but should not be considered as statistically valid representations of specifications for any particular alloy or material. The designer should consult the materials' manufacturers for more accurate and up-to-date strength information on materials used in engineering applications or conduct independent tests of the selected materials to determine their ultimate suitability to any application.

Table B-1   Physical Properties of Some Engineering Materials

Data from Various Sources.[*]   These Properties are Essentially Similar for All Alloys of the Particular Material

Material	Modulus of Elasticity E		Modulus of Rigidity G		Poisson's Ratio ν	Weight Density γ	Mass Density ρ	Specific Gravity
	Mpsi	GPa	Mpsi	GPa		lb/in^3	Mg/m^3	
Aluminum Alloys	10.4	71.7	3.9	26.8	0.34	0.10	2.8	2.8
Beryllium Copper	18.5	127.6	7.2	49.4	0.29	0.30	8.3	8.3
Brass, Bronze	16.0	110.3	6.0	41.5	0.33	0.31	8.6	8.6
Copper	17.5	120.7	6.5	44.7	0.35	0.32	8.9	8.9
Iron, Cast, Gray	15.0	103.4	5.9	40.4	0.28	0.26	7.2	7.2
Iron, Cast, Ductile	24.5	168.9	9.4	65.0	0.30	0.25	6.9	6.9
Iron, Cast, Malleable	25.0	172.4	9.6	66.3	0.30	0.26	7.3	7.3
Magnesium Alloys	6.5	44.8	2.4	16.8	0.33	0.07	1.8	1.8
Nickel Alloys	30.0	206.8	11.5	79.6	0.30	0.30	8.3	8.3
Steel, Carbon	30.0	206.8	11.7	80.8	0.28	0.28	7.8	7.8
Steel, Alloys	30.0	206.8	11.7	80.8	0.28	0.28	7.8	7.8
Steel, Stainless	27.5	189.6	10.7	74.1	0.28	0.28	7.8	7.8
Titanium Alloys	16.5	113.8	6.2	42.4	0.34	0.16	4.4	4.4
Zinc Alloys	12.0	82.7	4.5	31.1	0.33	0.24	6.6	6.6

[*] *Properties of Some Metals and Alloys,* International Nickel Co., Inc., NY; *Metals Handbook,* American Society for Metals, Materials Park, OH.

Table B-2   Mechanical Properties for Some Wrought-Aluminum Alloys

Data from Various Sources.[*]   Approximate Values. Consult Manufacturers for More Accurate Information

Wrought-Aluminum Alloy	Condition	Tensile Yield Strength (2% offset)		Ultimate Tensile Strength		Fatigue Strength at 5E8 cycles		Elongation over 2 in	Brinell Hardness
		kpsi	MPa	kpsi	MPa	kpsi	MPa	%	-HB
1100	Sheet annealed	5	34	13	90			35	23
	Cold rolled	22	152	24	165			5	44
2024	Sheet annealed	11	76	26	179			20	-
	Heat treated	42	290	64	441	20	138	19	-
3003	Sheet annealed	6	41	16	110			30	28
	Cold rolled	27	186	29	200			4	55
5052	Sheet annealed	13	90	28	193			25	47
	Cold rolled	37	255	42	290			7	77
6061	Sheet annealed	8	55	18	124			25	30
	Heat treated	40	276	45	310	14	97	12	95
7075	Bar annealed	15	103	33	228			16	60
	Heat treated	73	503	83	572	14	97	11	150

[*] *Properties of Some Metals and Alloys,* International Nickel Co., Inc., NY; *Metals Handbook,* American Society for Metals, Materials Park, OH.

Table B-3 Mechanical Properties for Some Carbon Steels

Data from Various Sources.[*] Approximate Values. Consult Manufacturers for More Accurate Information

SAE / AISI Number	Condition	Tensile Yield Strength (2% offset)		Ultimate Tensile Strength		Elongation over 2 in	Brinell Hardness
		kpsi	MPa	kpsi	MPa	%	-HB
1010	Hot rolled	26	179	47	324	28	95
	Cold rolled	44	303	53	365	20	105
1020	Hot rolled	30	207	55	379	25	111
	Cold rolled	57	393	68	469	15	131
1030	Hot rolled	38	259	68	469	20	137
	Normalized @ 1650°F	50	345	75	517	32	149
	Cold rolled	64	441	76	524	12	149
	Q&T @ 1000°F	75	517	97	669	28	255
	Q&T @ 800°F	84	579	106	731	23	302
	Q&T @ 400°F	94	648	123	848	17	495
1035	Hot rolled	40	276	72	496	18	143
	Cold rolled	67	462	80	552	12	163
1040	Hot rolled	42	290	76	524	18	149
	Normalized @ 1650°F	54	372	86	593	28	170
	Cold rolled	71	490	85	586	12	170
	Q&T @ 1200°F	63	434	92	634	29	192
	Q&T @ 800°F	80	552	110	758	21	241
	Q&T @ 400°F	86	593	113	779	19	262
1045	Hot rolled	45	310	82	565	16	163
	Cold rolled	77	531	91	627	12	179
1050	Hot rolled	50	345	90	621	15	179
	Normalized @ 1650°F	62	427	108	745	20	217
	Cold rolled	84	579	100	689	10	197
	Q&T @ 1200°F	78	538	104	717	28	235
	Q&T @ 800°F	115	793	158	1 089	13	444
	Q&T @ 400°F	117	807	163	1 124	9	514
1060	Hot rolled	54	372	98	676	12	200
	Normalized @ 1650°F	61	421	112	772	18	229
	Q&T @ 1200°F	76	524	116	800	23	229
	Q&T @ 1000°F	97	669	140	965	17	277
	Q&T @ 800°F	111	765	156	1 076	14	311
1095	Hot rolled	66	455	120	827	10	248
	Normalized @ 1650°F	72	496	147	1 014	9	13
	Q&T @ 1200°F	80	552	130	896	21	269
	Q&T @ 800°F	112	772	176	1 213	12	363
	Q&T @ 600°F	118	814	183	1 262	10	375

[*] *SAE Handbook,* Society of Automotive Engineers, Warrendale PA; *Metals Handbook,* American Society for Metals, Materials Park, OH.

A

Table B-4   Mechanical Properties of Some Cast-Iron Alloys

Data from Various Sources.* Approximate Values. Consult Manufacturers for More Accurate Information

Cast-Iron Alloy	Condition	Tensile Yield Strength (2% offset)		Ultimate Tensile Strength		Compressive Strength		Brinell Hardness
		kpsi	MPa	kpsi	MPa	kpsi	MPa	HB
Gray Cast Iron—Class 20	As cast	–	–	22	152	83	572	156
Gray Cast Iron—Class 30	As cast	–	–	32	221	109	752	210
Gray Cast Iron—Class 40	As cast	–	–	42	290	140	965	235
Gray Cast Iron—Class 50	As cast	–	–	52	359	164	1 131	262
Gray Cast Iron—Class 60	As cast	–	–	62	427	187	1 289	302
Ductile Iron 60-40-18	Annealed	47	324	65	448	52	359	160
Ductile Iron 65-45-12	Annealed	48	331	67	462	53	365	174
Ductile Iron 80-55-06	Annealed	53	365	82	565	56	386	228
Ductile Iron 120-90-02	Q & T	120	827	140	965	134	924	325

* *Properties of Some Metals and Alloys,* International Nickel Co., Inc., NY; *Metals Handbook,* American Society for Metals, Materials Park, OH.

Table B-5   Properties of Some Engineering Plastics

Data from Various Sources.* Approximate Values. Consult Manufacturers for More Accurate Information

Material	Approximate Modulus of Elasticity E †		Ultimate Tensile Strength		Ultimate Compressive Strength		Elongation over 2 in	Max Temp	Specific Gravity
	Mpsi	GPa	kpsi	MPa	kpsi	MPa	%	°F	
ABS	0.3	2.1	6.0	41.4	10.0	68.9	5-25	160-200	1.05
20-40% glass filled	0.6	4.1	10.0	68.9	12.0	82.7	3	200-230	1.30
Acetal	0.5	3.4	8.8	60.7	18.0	124.1	60	220	1.41
20-30% glass filled	1.0	6.9	10.0	68.9	18.0	124.1	7	185-220	1.56
Acrylic	0.4	2.8	10.0	68.9	15.0	103.4	5	140-190	1.18
Fluoroplastic (PTFE)	0.2	1.4	5.0	34.5	6.0	41.4	100	350-330	2.10
Nylon 6/6	0.2	1.4	10.0	68.9	10.0	68.9	60	180-300	1.14
Nylon 11	0.2	1.3	8.0	55.2	8.0	55.2	300	180-300	1.04
20-30% glass filled	0.4	2.5	12.8	88.3	12.8	88.3	4	250-340	1.26
Polycarbonate	0.4	2.4	9.0	62.1	12.0	82.7	100	250	1.20
10-40% glass filled	1.0	6.9	17.0	117.2	17.0	117.2	2	275	1.35
HMW Polyethylene	0.1	0.7	2.5	17.2	–	–	525	–	0.94
Polyphenylene Oxide	0.4	2.4	9.6	66.2	16.4	113.1	20	212	1.06
20-30% glass filled	1.1	7.8	15.5	106.9	17.5	120.7	5	260	1.23
Polypropylene	0.2	1.4	5.0	34.5	7.0	48.3	500	250-320	0.90
20-30% glass filled	0.7	4.8	7.5	51.7	6.2	42.7	2	300-320	1.10
Impact Polystryrene	0.3	2.1	4.0	27.6	6.0	41.4	2-80	140-175	1.07
20-30% glass filled	0.1	0.7	12.0	82.7	16.0	110.3	1	180-200	1.25
Polysulfone	0.4	2.5	10.2	70.3	13.9	95.8	50	300-345	1.24

* *Modern Plastics Encyclopedia,* McGraw-Hill, New York; *Machine Design Materials Reference Issue,* Penton Publishing, Cleveland, OH.
† Most plastics do not obey Hooke's Law. These apparent moduli of elasticity vary with time and temperature.

A

Appendix C

GEOMETRIC PROPERTIES

DIAGRAMS AND FORMULAS TO CALCULATE THE FOLLOWING PARAMETERS FOR SEVERAL COMMON GEOMETRIC SOLIDS

V = volume

m = mass

C_g = location of center of mass

I_x = second moment of mass about x axis = $\int \left(y^2 + z^2\right) dm$

I_y = second moment of mass about y axis = $\int \left(x^2 + z^2\right) dm$

I_z = second moment of mass about z axis = $\int \left(x^2 + y^2\right) dm$

k_x = radius of gyration about x axis

k_y = radius of gyration about y axis

k_z = radius of gyration about z axis

A

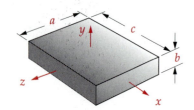

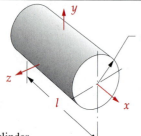

(a) Rectangular prism

$$V = abc \qquad m = V \cdot \text{mass density}$$

$$x_{Cg} @ \frac{c}{2} \qquad y_{Cg} @ \frac{b}{2} \qquad z_{Cg} @ \frac{a}{2}$$

$$I_x = \frac{m(a^2 + b^2)}{12} \qquad I_y = \frac{m(a^2 + c^2)}{12} \qquad I_z = \frac{m(b^2 + c^2)}{12}$$

$$k_x = \sqrt{\frac{I_x}{m}} \qquad k_y = \sqrt{\frac{I_y}{m}} \qquad k_z = \sqrt{\frac{I_z}{m}}$$

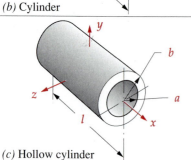

(b) Cylinder

$$V = \pi r^2 l \qquad m = V \cdot \text{mass density}$$

$$x_{Cg} @ \frac{l}{2} \qquad y_{Cg} \text{ on axis} \qquad z_{Cg} \text{ on axis}$$

$$I_x = \frac{mr^2}{2} \qquad I_y = I_z = \frac{m(3r^2 + l^2)}{12}$$

$$k_x = \sqrt{\frac{I_x}{m}} \qquad k_y = k_z = \sqrt{\frac{I_y}{m}}$$

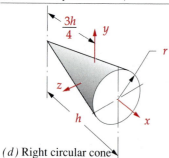

(c) Hollow cylinder

$$V = \pi(b^2 - a^2)l \qquad m = V \cdot \text{mass density}$$

$$x_{Cg} @ \frac{l}{2} \qquad y_{Cg} \text{ on axis} \qquad z_{Cg} \text{ on axis}$$

$$I_x = \frac{m(a^2 + b^2)}{2} \qquad I_y = I_z = \frac{m(3a^2 + 3b^2 + l^2)}{12}$$

$$k_x = \sqrt{\frac{I_x}{m}} \qquad k_y = k_z = \sqrt{\frac{I_y}{m}}$$

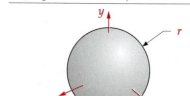

(d) Right circular cone

$$V = \pi \frac{r^2 h}{3} \qquad m = V \cdot \text{mass density}$$

$$x_{Cg} @ \frac{3h}{4} \qquad y_{Cg} \text{ on axis} \qquad z_{Cg} \text{ on axis}$$

$$I_x = \frac{3}{10} mr^2 \qquad I_y = I_z = \frac{m(12r^2 + 3h^2)}{80}$$

$$k_x = \sqrt{\frac{I_x}{m}} \qquad k_y = k_z = \sqrt{\frac{I_y}{m}}$$

(e) Sphere

$$V = \frac{4}{3} \pi r^3 \qquad m = V \cdot \text{mass density}$$

$$x_{Cg} \text{ at center} \qquad y_{Cg} \text{ at center} \qquad z_{Cg} \text{ at center}$$

$$I_x = I_y = I_z = \frac{2}{5} mr^2$$

$$k_x = k_y = k_z = \sqrt{\frac{I_y}{m}}$$

A

Appendix D

SPRING DATA

The following catalog pages of helical compression and extension spring data provided courtesy of *Hardware Products Co., Chelsea, Massachusetts.*

A

COMPRESSION SPRINGS

Will go in hole In.	7/16			1/2				5/8				3/4				7/8			
Wire Dia. In.	.031	.047	.062	.047	.062	.078	.094	.047	.062	.078	.094	.062	.078	.094	.125	.062	.078	.094	.125
7/16 Catalog No.	247	248	249																
Price Code	HB	HB	HC																
lbs./in.	12	55	180																
Max. Defl.	.32	.23	.16																
1/2 Catalog No.	250	251	252	283	284	285	286												
Price Code	HB	HB	HC	HB	HB	HE	HE												
lbs./in.	10	47	150	37	110	320	840												
Max. Defl.	.37	.27	.19	.29	.22	.15	.10												
5/8 Catalog No.	253	254	255	287	288	289	290	331	332	333	334								
Price Code	HB	HB	HC	HB	HB	HE	HE	HB	HB	HD	HE								
lbs./in.	7.9	36	175	29	85	240	610	20	54	140	320								
Max. Defl.	.47	.35	.25	.38	.29	.20	.14	.43	.35	.27	.20								
3/4 Catalog No.	256	257	258	291	292	293	294	335	336	337	338	375	376	377	378				
Price Code	HB	HB	HC	HB	HB	HE	HE	HB	HB	HD	HE	HD	HD	HF	HG				
lbs./in.	6.4	29	90	23	68	185	470	16	43	105	250	32	78	170	650				
Max. Defl.	.58	.44	.32	.48	.37	.26	.18	.53	.44	.34	.26	.59	.40	.32	.19				
7/8 Catalog No.	259	260	261	295	296	297	298	339	340	341	342	379	380	381	382	419	420	421	422
Price Code	HB	HB	HC	HB	HB	HE	HE	HB	HB	HD	HE	HD	HD	HF	HG	HF	HF	HJ	HK
lbs./in.	5.4	24	75	19	56	155	384	13	36	90	204	27	65	140	520	21	49	100	350
Max. Defl.	.68	.52	.39	.57	.44	.32	.23	.64	.53	.42	.32	.58	.48	.39	.24	.62	.53	.44	.30
1 Catalog No.	262	263	264	299	300	301	302	343	344	345	346	383	384	385	386	423	424	425	426
Price Code	HB	HB	HC	HB	HC	HE	HE	HB	HC	HD	HE	HD	HD	HF	HG	HF	HF	HJ	HK
lbs./in.	4.7	21	65	17	48	130	320	11	31	77	170	23	55	115	430	18	42	86	290
Max. Defl.	.79	.60	.45	.66	.51	.37	.27	.74	.62	.49	.38	.68	.57	.46	.29	.73	.63	.53	.36
1¼ Catalog No.	265	266	267	303	304	305	306	347	348	349	350	387	388	389	390	427	428	429	430
Price Code	HB	HC	HC	HC	HC	HF	HF	HC	HC	HD	HE	HE	HE	HF	HG	HG	HG	HJ	HK
lbs./in.	3.7	16	50	13	38	100	245	9.0	24	59	130	18	42	89	320	14	32	66	220
Max. Defl.	1.0	.77	.58	.84	.66	.49	.35	.94	.80	.64	.49	.88	.74	.60	.39	.94	.81	.69	.47
1½ Catalog No.	268	269	270	307	308	309	310	351	352	353	354	391	392	393	394	431	432	433	434
Price Code	HB	HC	HC	HC	HC	HF	HF	HD	HD	HE	HF	HE	HE	HF	HG	HG	HG	HJ	HK
lbs./in.	3.1	14	41	11	31	83	200	7.4	20	48	105	15	34	72	260	12	26	53	175
Max. Defl.	1.2	.94	.70	1.0	.81	.60	.43	1.1	.98	.78	.61	1.1	.91	.74	.48	1.1	1.0	.85	.59
1¾ Catalog No.	271	272	273	311	312	313	314	355	356	357	358	395	396	397	398	435	436	437	438
Price Code	HC	HD	HE	HC	HD	HF	HF	HD	HD	HC	HF	HE	HE	HF	HG	HG	HG	HJ	HK
lbs./in.	2.6	11	35	9.1	26	70	170	6.2	17	41	90	12.4	29	61	216	9.9	22	45	147
Max. Defl.	1.4	1.1	.84	1.2	.96	.71	.52	1.3	1.1	.93	.73	1.3	1.1	.89	.58	1.35	1.2	1.0	.71
2 Catalog No.	274	275	276	315	316	317	318	359	360	361	362	399	400	401	402	439	440	441	442
Price Code	HD	HD	HE	HC	HD	HF	HF	HD	HD	HE	HF	HE	HE	HF	HG	HG	HG	HJ	HL
lbs./in.	2.3	10	30	7.9	23	60	145	5.4	14	35	77	11	25	52	185	8.6	19	38	125
Max. Defl.	1.6	1.3	.96	1.4	1.1	.82	.60	1.5	1.3	1.1	.85	1.4	1.2	1.0	.68	1.5	1.4	1.2	.83
3 Catalog No.	277	278	279	319	320	321	322	363	364	365	366	403	404	405	406	443	444	445	446
Price Code	HD	HE	HE	HF	HF	HG	HG	HD	HD	HF	HG	HF	HF	HG	HK	HJ	HJ	HK	HL
lbs./in.	1.5	6.6	20	5.2	15	39	94	3.6	9.4	23	50	7	16	34	115	5.6	12	25	80
Max. Defl.	2.4	1.9	1.4	2.1	1.7	1.2	.93	2.4	2.0	1.6	1.3	2.2	1.9	1.6	1.0	2.4	2.1	1.8	1.3
4 Catalog No.	280	281	282	323	324	325	326	367	368	369	370	407	408	409	410	447	448	449	450
Price Code	HE	HE	HE	HE	HF	HG	HG	HE	HE	HF	HG	HF	HF	HG	HL	HJ	HJ	HL	HM
lbs./in.	1.1	4.9	15	3.9	11	29	69	2.6	6.9	17	37	5.2	12	25	86	4.2	9.2	18	59
Max. Defl.	3.3	2.6	2.0	2.8	2.3	1.7	1.2	3.2	2.7	2.2	1.8	3.0	2.6	2.1	1.4	3.2	2.8	2.5	1.8
6 Catalog No.				327	328	329	330	371	372	373	374	411	412	413	414	451	452	453	454
Price Code				HF	HF	HG	HG	HF	HF	HG	HG	HG	HG	HJ	HN	HK	HK	HL	HO
lbs./in.				2.5	7.0	17	45	1.8	4.6	11	24	3.4	7.9	16	56	2.7	6.0	12	38
Max. Defl.				4.4	3.5	2.5	2.	4.8	4.2	3.4	2.7	4.6	3.9	3.3	2.2	4.9	4.3	3.8	2.7
8 Catalog No.												415	416	417	418	455	456	457	458
Price Code.												HJ	HJ	HK	HO	HL	HL	HM	HP
lbs./in.												2.6	6	11	40	2.0	4.5	8.9	28
Max. Defl.												6.1	5.2	4.5	3.0	6.5	5.8	5.1	3.7
Maximum Load	3.7	12.7	29	11	25	45	88	8.3	19	38	66	15.8	31.2	54	125	13.4	26.3	45	105
Will work free over	.347	.315	.285	.375	.345	.313	.281	.505	.475	.443	.411	.585	.554	.522	.460	.700	.670	.638	.576
Pitch	.195	.141	.128	.173	.151	.141	.141	.259	.214	.188	.177	.284	.240	.217	.204	.371	.306	.268	.239
Solid Stress (000 omitted)	125	118	113	118	113	109	105	118	113	109	105	113	109	105	99	113	109	105	99

Note (in 3/4 and 7/8 header region): Pounds per inch figure is a constant for each spring, and represents the number of pounds required to compress the spring 1". To compress the spring ½" or ¼" requires ½ or ¼ of this value.

Maximum Deflection is the amount spring deflects to give the maximum load. This value subtracted from the free length gives the solid or compressed length.

NOTE: Stock springs can be ordered in stainless steel or plated. Prices quoted upon request.

A

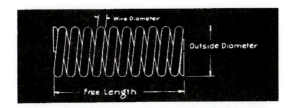

Spring diagram: Wire Diameter, Outside Diameter, Free Length

Will go in hole In.		1	1	1	1	1 1/4	1 1/4	1 1/4	1 1/2	1 1/2	1 1/2	2	2	2	3	3	3	4	4	4	6	6
Wire Dia. In.		.078	.094	.125	.187	.094	.125	.187	.125	.187	.250	.187	.250	.375	.250	.375	.500	.375	.500	.750	.750	1.000
1 Catalog No.		459	460	461	462																	
Price Code		HK	HL	HL	HR																	
lbs./in.		34	67	210	1500																	
Max. Defl.		.67	.58	.41	.19																	
1¼ Catalog No.		463	464	465	466	499	500	501														
Price Code		HL	HM	HM	HS	HN	HN	HM														
lbs./in.		26	52	160	1100	35	100	600														
Max. Defl.		.87	.76	.55	.26	.85	.67	.37														
1½ Catalog No.		467	468	469	470	502	503	504	526	527	528											
Price Code		HL	HM	HN	HS	HN	HO	HT	HR	HX	HAC											
lbs./in.		21	42	130	870	29	82	460	60	300	1200											
Max. Defl.		1.0	.93	.69	.34	1.1	.84	.48	.95	.60	.35											
1¾ Catalog No.		471	472	473	474	505	506	507	529	530	531											
Price Code		HL	HM	HN	HS	HN	HO	HT	HR	HX	HAC											
lbs./in.		18	35	108	712	24	68	379	50	244	960											
Max. Defl.		1.3	1.1	.83	.41	1.3	1.0	.59	1.1	.74	.44											
2 Catalog No.		475	476	477	478	508	509	510	532	533	534	553	554	555								
Price Code		HM	HN	HO	HT	HO	HP	HU	HS	HZ	HAE	HAA	HAG	HZZ								
lbs./in.		16	30	93	600	21	59	320	43	200	800	115	390	3000								
Max. Defl.		1.4	1.3	.97	.49	1.4	1.2	.70	1.3	.87	.53	1.1	.77	.34								
3 Catalog No.		479	480	481	482	511	512	513	535	536	537	556	557	558	577	578	579					
Price Code		HN	HO	HP	HZ	HP	HR	HAA	HT	HAD	HAL	HAE	HAN	HZZ	HAR	HZZ	HZZ					
lbs./in.		10	19	59	370	13	37	200	27	130	480	73	230	1650	105	560	2300					
Max. Defl.		2.2	2.0	1.5	.79	2.2	1.8	1.1	2.1	1.4	.89	1.8	1.3	.61	1.8	1.1	.64					
4 Catalog No.		483	484	485	486	514	515	516	538	539	540	559	560	561	580	581	582	598	599	610		
Price Code		HP	HR	HS	HAC	HS	HT	HAD	HW	HAG	HAQ	HAJ	HAR	HZZ	HAT	HZZ	HZZ	HZZ	HZZ	HZZ		
lbs./in.		7.4	14	43	270	9.9	27	144	20	93	340	53	170	1150	76	390	1500	210	720	4600		
Max. Defl.		3.0	2.7	2.1	1.1	3.0	2.5	1.5	2.8	1.9	1.2	2.5	1.8	.88	2.5	1.6	.96	2.1	1.4	.8		
6 Catalog No.		487	488	489	490	517	518	519	541	542	543	562	563	564	583	584	585	600	601	611	616	621
Price Code		HR	HT	HU	HAD	HU	HW	HAE	HX	HAJ	HAT	HAM	HAW	HZZ	HAZ	HZZ	HZZ	HZZ	HZZ	HZZ	HZZ	HZZ
lbs./in.		4.9	9.4	28	175	6.5	18	93	13	60	220	34	105	710	49	240	920	130	430	2840	850	3500
Max. Defl.		4.6	4.1	3.2	1.7	4.7	3.9	2.4	4.3	3.0	1.9	3.8	2.8	1.4	4.0	2.6	1.6	3.4	2.4	1.3	1.9	1.4
8 Catalog No.		491	492	493	494	520	521	522	544	545	546	565	566	567	586	587	588	602	603	612	617	622
Price Code		HS	HU	HW	HAL	HW	HX	HAM	HAA	HAW	HAW	HAR	HAZ	HZZ	HBD	HZZ	HZZ	HZZ	HZZ	HZZ	HZZ	HZZ
lbs./in.		3.6	7.0	21	125	4.8	13	68	9.6	44	160	25	79	510	36	175	660	95	310	2050	630	2500
Max. Defl.		6.2	5.6	4.3	2.3	6.3	5.2	3.2	5.9	4.1	2.7	5.2	3.9	1.9	5.4	3.6	2.2	4.5	3.4	1.8	2.7	2.0
12 Catalog No.		495	496	497	498	523	524	525	547	548	549	568	569	570	589	590	591	604	605	613	618	623
Price Code		HT	HW	HZ	HAP	HX	HAA	HAR	HAC	HAU	HBA	HAZ	HBE	HZZ	HBK	HZZ	HZZ	HZZ	HZZ	HZZ	HZZ	HZZ
lbs./in.		2.4	4.6	14	84	3.2	8.7	45	6.3	29	105	16	52	330	23	110	420	61	195	1325	400	1580
Max. Defl.		9.4	8.4	6.5	3.5	9.5	7.9	5.0	8.9	6.2	4.1	8.0	5.9	3.0	8.3	5.5	3.5	7.3	5.3	2.6	4.3	3.1
16 Catalog No.									550	551	552	571	572	573	592	593	594	606	607	614	619	624
Price Code									HAE	HAW	HBD	HAZ	HBG	HZZ	HBL	HZZ	HZZ	HZZ	HZZ	HZZ	HZZ	HZZ
lbs./in.									4.7	21	74	12	38	240	17	83	310	45	145	975	300	1170
Max. Defl.									11.9	8.5	5.6	10.7	8.0	4.1	11.3	7.5	4.0	10	7.3	3.8	6.1	4.3
24 Catalog No.												574	575	576	595	596	597	608	609	615	620	625
Price Code												HBA	HBL	HZZ	HBP	HZZ	HZZ	HZZ	HZZ	HZZ	HZZ	HZZ
lbs./in.												7.8	23.4	150	11.4	54	200	29	94	640	175	760
Max. Defl.												16.3	12.1	7.0	17.1	11.5	7.3	15.2	11.1	5.8	9.4	6.5
Maximum Load		23	39	90	295	30	69	224	57	180	428	131	307	1000	195	624	1470	449	1040	3700	2000	4800
Will work free over		.784	.752	.690	.565	1.00	.940	.815	1.19	1.06	.940	1.52	1.39	1.14	2.33	2.08	1.83	3.08	2.83	2.25	4.25	3.75
Pitch		.382	.328	.279	.268	.481	.384	.327	.516	.403	.388	.596	.518	.514	.917	.741	.736	1.08	.969	1.00	1.3	1.4
Solid Stress (000 omitted)		109	105	99	90	105	99	90	99	90	85	90	85	77	85	77	73	77	73	70	70	65

Pounds per inch figure is a constant for each spring, and represents the number of pounds required to compress the spring 1". To compress the spring ½" or ¼" requires ½ or ¼ of this value.

Maximum Deflection is the amount spring deflects to give the maximum load. This value subtracted from the free length gives the solid or compressed length.

NOTE: Stock springs can be ordered in stainless steel or plated. Prices quoted upon request.

Hardware Products Company, Inc.

A

EXTENSION SPRINGS

Regular hook　　　Regular loop

← Length →

ORDER BY:
SE LENGTH x O.D. x WIRE DIA.

SPECIFY HOOKS OR LOOPS

The figures given for "Maximum Extension" and "lbs. per inch" are for a spring 1" long. For other lengths multiply the "Maximum Extension" and divide the "lbs. per inch" by the length in inches. The "Maximum Load" and "Initial Tension" remain constant for any length.

Example: A spring ½" diam. .062" wire and 4" long will have a safe maximum extension of 3.2" and it will require 4 lbs. to deflect it 1 in. The spring will hold approximately 3.3 lbs. before it starts to extend, and will hold a maximum of 16.1 lbs. without permanent stretch. If 8.5 lbs. is hung on the spring it will deflect 1.3". 8.5 lbs. minus 3.3 lbs. divided by 4 lbs. per inch equals 1.3".

FOR QUICK DELIVERY
Call 617-884-9410

NOTE: Stock springs can be ordered in stainless steel or plated. Prices quoted upon request.

Outside dia.	Wire dia.	Catalog No.	Price code	Safe maximum load in pounds	Safe maximum extension - in.	Approx. initial tension in pounds	Pound per inch extension	Stress at max. load (000 omitted)	Weight per foot (lbs.)
1/8	.012	01	EHE	.6	1.9	.07	.27	100	.012
	.016	02	EHD	1.3	.9	.2	1.2	93	.015
	.023	03	EHD	4.2	.35	.9	9.0	90	.02
5/32	.012	04	EHE	.47	3.5	.01	.12	100	.015
	.016	05	EHD	1.1	1.7	.15	.55	93	.019
	.023	06	EHD	3.2	.7	.5	3.9	90	.027
3/16	.016	07	EHD	.87	2.5	.1	.3	93	.024
	.023	08	EHD	2.6	1.0	.4	2.2	90	.032
	.031	09	EHD	6.5	.45	1.5	10.7	88	.04
7/32	.016	10	EHD	.75	4.0	.01	.18	93	.028
	.023	11	EHD	2.3	1.6	.32	1.2	90	.039
	.031	12	EHD	5.5	.7	1.0	6.5	88	.048
1/4	.023	13	EHD	1.8	1.9	.26	.8	90	.044
	.031	14	EHD	4.7	1.0	.75	3.8	88	.055
	.047	15	EHE	16.0	.3	3.5	40.0	83	.082
5/16	.023	16	EHE	1.5	3.5	.16	.38	90	.058
	.031	17	EHE	3.6	1.6	.55	1.9	88	.072
	.047	18	EHE	12.5	.9	2.2	10.8	83	.108
3/8	.031	19	EHE	2.9	2.5	.37	1.0	88	.084
	.047	20	EHE	10.5	.9	1.7	9.5	83	.13
	.062	21	EHF	23.0	.39	5.3	45.0	79	.16
7/16	.031	22	EHF	2.5	3.5	.26	.63	88	.105
	.047	23	EHF	8.5	1.2	1.4	5.7	83	.163
	.062	24	EHF	20.0	.6	4.3	26.0	79	.2
1/2	.047	25	EHF	7.3	1.6	1.1	3.7	83	.18
	.062	26	EHF	17.0	.8	3.3	16.0	79	.23
	.078	27	EHG	34.0	.45	8.0	57.0	77	.28
	.094	28	EHJ	57.0	.25	16.0	160.0	74	.32
5/8	.047	29	EHG	6.0	3.0	.7	1.7	83	.24
	.062	30	EHG	13.3	1.4	2.1	7.6	79	.3
	.078	31	EHG	27.0	.9	5.2	23.0	77	.37
	.094	32	EHJ	45.0	.4	11.0	73.0	74	.44
3/4	.062	33	EHG	10.5	2.2	1.5	4.1	79	.36
	.078	34	EHJ	22.0	1.3	3.5	14.0	77	.46
	.094	35	EHJ	36.0	.7	8.0	38.0	74	.51
	.125	36	EHK	85.0	.3	22.0	180.0	69	.64
7/8	.062	37	EHK	9.2	3.3	1.1	2.4	79	.4
	.078	38	EHK	18.0	1.7	2.6	8.7	77	.59
	.094	39	EHL	31.0	1.0	6.0	25.0	74	.64
	.125	40	EHM	72.0	.5	17.0	107.0	69	.8
1	.078	41	EHL	16.0	2.5	2.0	5.5	77	.67
	.094	42	EHL	26.0	1.5	4.5	13.7	74	.70
	.125	43	EHN	65.0	.75	14.0	68.0	69	.90
	.187	44	EHW	200.0	.23	60.0	600.0	63	1.4
1¼	.094	45	EHM	21.0	2.6	2.8	6.8	74	.94
	.125	46	EHO	47.0	1.2	9.0	31.0	69	1.3
	.187	47	EHZ	148.0	.3	40.0	290.0	63	1.8
1½	.125	48	EHS	39.0	1.9	6.0	17.0	69	1.4
	.187	49	EHAA	122.0	.6	33.0	150.0	63	2.2
	.250	50	EHAC	290.0	.27	90.0	720.0	60	2.6
2	.187	51	EHAD	90.0	1.3	20.0	54.0	63	3.1
	.250	52	EHAG	210.0	.6	55.0	260.0	60	3.7

Carried in stock in 3-foot lengths - cut to length and looped to order.

Hardware Products Company, Inc.

191 WILLIAMS STREET • CHELSEA, MA 02150

A

ATLAS OF GEARED FIVEBAR LINKAGE COUPLER CURVES

C. Zhang, R. L. Norton, T. Hammond

The following pages of coupler curve data are excerpted from the complete work. See Sections 3.6 (p. 103), 4.8 (p. 164), 6.8 (p. 278), and 7.4 (p.319) for more information on the geared fivebar linkage. Use program FIVEBAR to investigate other linkage geometries.

Alpha = Coupler Link 3 / Link 2

Beta = Ground Link 1 / Link 2

Lambda = Gear Ratio = Gear 5 / Gear 2.[*]

Phase angle is noted on each plot of a coupler curve.

The dots along curves are at every 10 degrees of Link 2's rotation.

Linkage is symmetrical: Link 2 = Link 5 and Link 3 = Link 4

---

[*] Note that this lambda is the **inverse** of the λ which is defined in Sections 4.8, 6.8, and 7.4. See also Figures P4-4 (p. 139), P6-4 (p. 229), and P7-4 (p. 263). For example, a gear ratio λ of 2 in this atlas corresponds to a λ of 0.5 in the text and in program FIVEBAR. (The difference merely corresponds to a mirroring of the linkage from left to right.)

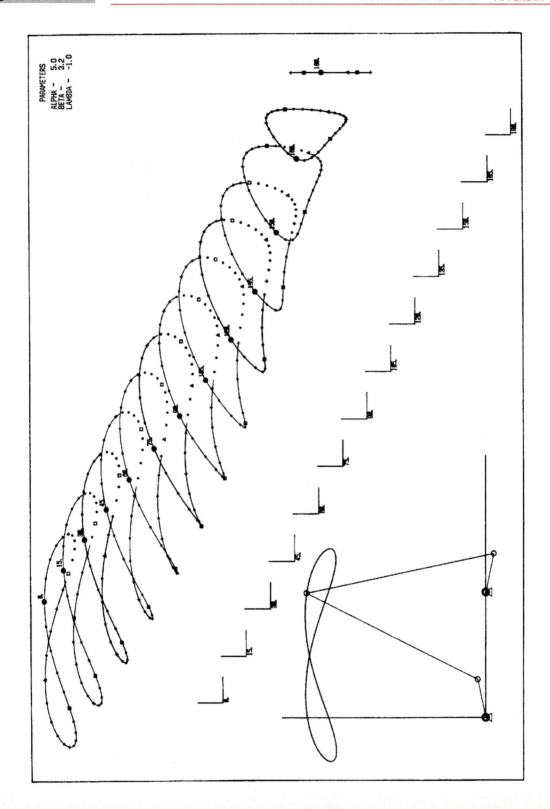

PARAMETERS
ALPHA –   5.0
BETA –   3.2
LAMBDA – –1.0

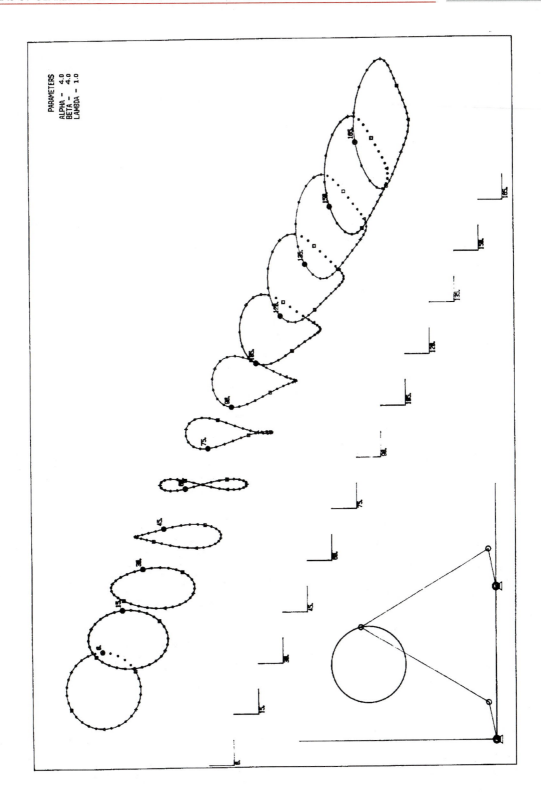

PARAMETERS
ALPHA – 4.0
BETA – 4.0
LAMBDA – 1.0

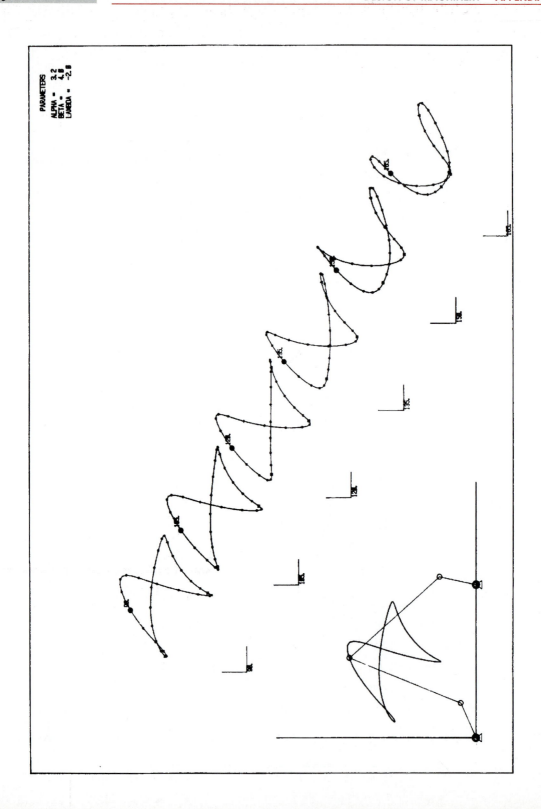

PARAMETERS

ALPHA = 3.2
BETA = 4.8
LAMBDA = -2.8

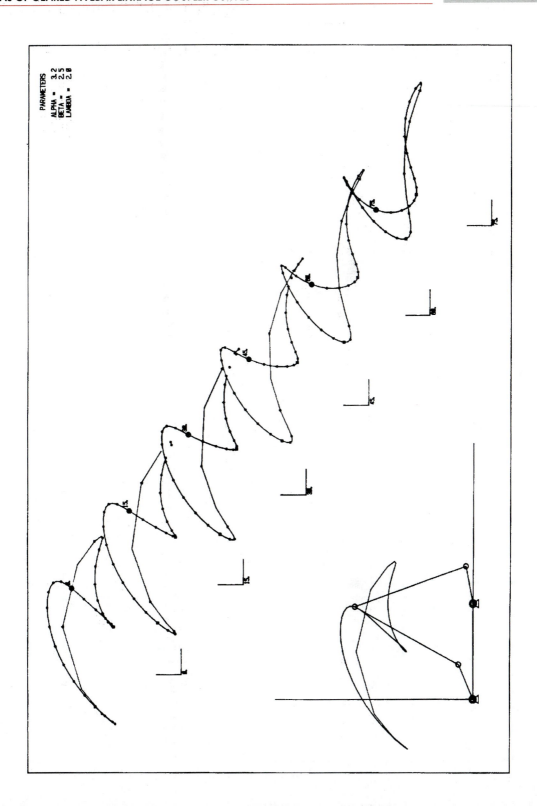

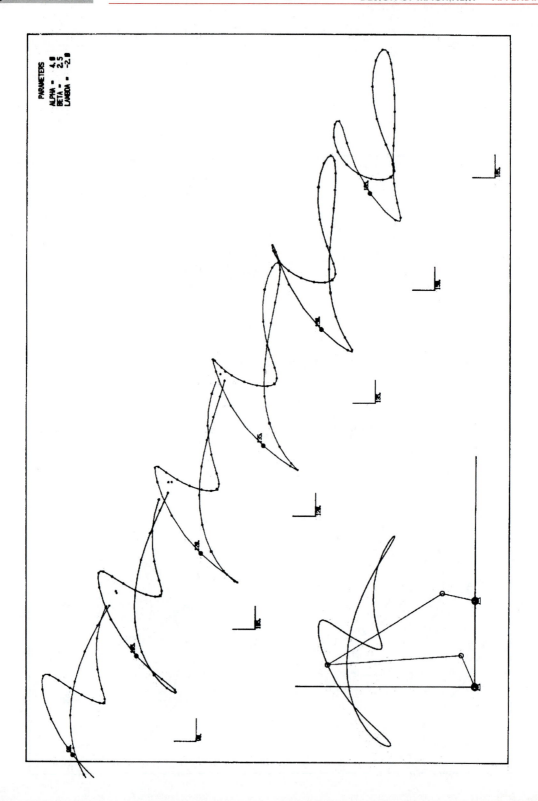

PARAMETERS
ALPHA = 4.0
BETA = 2.5
LAMBDA = -2.0

Appendix F

ANSWERS TO SELECTED PROBLEMS

CHAPTER 2 KINEMATICS FUNDAMENTALS

2-1

a. 1	b. 1	c. 2	d. 1	e. 7	f. 1	g. 4
h. 4	i. 4	j. 2	k. 1	l. 1	m. 2	n. 2
o. 4	p. as many as it has sections less one				q. 3	

2-3 a. 1 b. 3 c. 3 d. 3 e. 2

2-4 a. 6 b. 6 c. 5 d. 4, but 2 are dynamically coupled e. 10 f. 3

2-5 force closed

2-6

a. pure rotation
b. complex planar motion
c. pure translation
d. pure translation
e. pure rotation
f. complex planar motion
g. pure translation
h. pure translation
i. complex planar motion

2-7 a. 0 b. 1 c. 1 d. 3

2-8

a. structure - $DOF = 0$
b. mechanism - $DOF = 1$
c. mechanism - $DOF = 1$
d. mechanism - $DOF = 3$

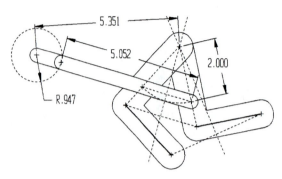
(a) One possible solution to Problem 3-3

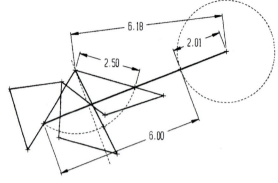
(b) One possible solution to Problem 3-5

FIGURE S3-1

Solutions to Problems 3-3 and 3-5

2-15 a. Grashof   b. non-Grashof   c. special-case Grashof

2-21

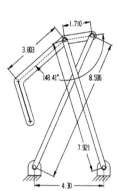

a. $M = 1$	b. $M = 1$	c. $M = 1$
d. $M = 1$	e. $M = -1$ (a paradox)	f. $M = 1$
g. $M = 1$	h. $M = 0$ (a paradox)	

2-24 a. $M = 1$ b. $M = 1$

2-26 $M = 1$

CHAPTER 3   GRAPHICAL LINKAGE SYNTHESIS

3-1

a.   Path generation
b.   Motion generation
c.   Function generation
d.   Path generation
e.   Path generation

Note that synthesis problems have many valid solutions. We cannot provide a "right answer" to all of these design problems. Check your solution with a cardboard model and/or by putting it into one of the programs supplied with the text.

3-3   See Figure S3-1.

3-5   See Figure S3-1.

3-6   See Figure S3-2.

3-8   See Figure S3-3.

FIGURE S3-2

Unique solution to
Problem 3-6

3-10   The solution using Figure 3-17 (p. 106) is shown in Figure S3-4. (Use program FOURBAR to check your solution.)

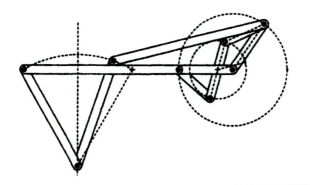

FIGURE S3-3

One possible solution to Problem 3-8

3-22 The transmission angle ranges from 31.5° to 89.9°.

3-23 Grashof crank-rocker. Transmission angle ranges from 58.1° to 89.8°.

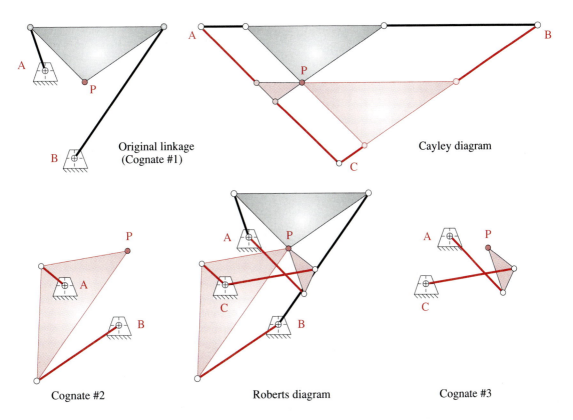

Original linkage (Cognate #1)

Cayley diagram

Cognate #2

Roberts diagram

Cognate #3

FIGURE S3-4

Solution to Problem 3-10. Finding the cognates of the fourbar linkage shown in Figure 3-17 (p. 106)

A

TABLE S4-1   Solutions for Problems 4-6, 4-7, and 4-13

Row	θ_3 open	θ_4 open	Trans Ang	θ_3 crossed	θ_4 crossed	Trans Ang
a	88.8	117.3	28.4	−115.2	−143.6	28.4
c	−53.1	16.5	69.6	173.3	103.6	69.6
e	7.5	78.2	70.7	− 79.0	− 149.7	70.7
g	−16.3	7.2	23.5	155.7	132.2	23.5
i	−1.5	103.1	75.4	−113.5	141.8	75.4
k	−13.2	31.9	45.2	−102.1	−147.3	45.2
m	−3.5	35.9	39.4	−96.5	−135.9	39.4

TABLE S4-2   Solutions for Problems 4-9 to 4-10

Row	θ_3 open	Slider open	θ_3 crossed	Slider crossed
a	180.1	5.0	−0.14	−3.0
c	205.9	9.8	−25.90	−4.6
e	175.0	16.4	4.20	−23.5
g	212.7	27.1	−32.70	−14.9

TABLE S4-3   Solutions for Problems 4-11 to 4-12

Row	θ_3 open	θ_4 open	R_B open	θ_3 crossed	θ_4 crossed	R_B crossed
a	232.7	142.7	1.79	−79.0	−169.0	−1.79
c	91.4	46.4	2.72	208.7	163.7	−11.20
e	158.2	128.2	6.17	−36.2	−66.2	−9.63

TABLE S4-4   Solutions for Problems 4-16 to 4-17

Row	θ_3 open	θ_4 open	θ_3 crossed	θ_4 crossed
a	173.6	−177.7	−115.2	− 124.0
c	17.6	64.0	−133.7	180.0
e	−164.0	−94.4	111.2	41.6
g	44.2	124.4	−69.1	−149.3
i	37.1	120.2	−67.4	−150.5

A

3-36   Grashof double-rocker. Works from 56° to 158° and from 202° to 310°. Transmission angle ranges from 0° to 90°.

3-42   Non-Grashof triple-rocker. Toggles at ± 55.4°. Transmission angle ranges from 0° to 88.8°.

CHAPTER 4   POSITION ANALYSIS

4-6   and   **4-7**   See Table S4-1 and the file P07-04*row*.4br.

4-9   and   **4-10**   See Table S4-2.

4-11   and   **4-12**   See Table S4-3.

4-13   See Table S4-1.

4-14   Open the file P07-04*row*.4br[†] in Program FOURBAR to see this solution.[*]

4-15   Open the file P07-04*row*.4br[†] in Program FOURBAR to see this solution.[*]

4-16   See Table S4-4.

4-17   See Table S4-4.

4-21   Open the file P04-21.4br in Program FOURBAR to see this solution.[*]

4-23   Open the file P04-23.4br in Program FOURBAR to see this solution.[*]

4-25   Open the file P04-25.4br in Program FOURBAR to see this solution.[*]

4-26   Open the file P04-26.4br in Program FOURBAR to see this solution.[*]

4-29   Open the file P04-29.4br in Program FOURBAR to see this solution.[*]

4-30   Open the file P04-30.4br in Program FOURBAR to see this solution.[*]

CHAPTER 5   ANALYTICAL LINKAGE SYNTHESIS

5-8   Given: $\alpha_2 = -62.5°$, $P_{21} = 2.47$, $\delta_2 = 120°$

For left dyad: Assume: $z = 1.075$, $\phi = 204°$ $\beta_2 = -27°$

Calculate: $\mathbf{W} = 3.67$ @ $-113.5°$

For right dyad: Assume: $s = 1.24$, $\psi = 74°$ $\gamma_2 = -40°$

Calculate: $\mathbf{U} = 5.46$ @ $-125.6°$

5-11   See Figure S5-1 for the solution. The link lengths are:

Link 1 = 4.35,   Link 2 = 3.39,   Link 3 = 1.94,   Link 4 = 3.87

5-15   See Figure S5-2 for the solution. The link lengths are:

Link 1 = 3.95,   Link 2 = 1.68,   Link 3 = 3.05,   Link 4 = 0.89

5-19   See Figure S5-3 for the solution. The link lengths are:

Link 1 = 2,   Link 2 = 2.5,   Link 3 = 1,   Link 4 = 2.5

* These files can be found in the PROBLEM SOLUTIONS folder on the CD-ROM included with this text.

A

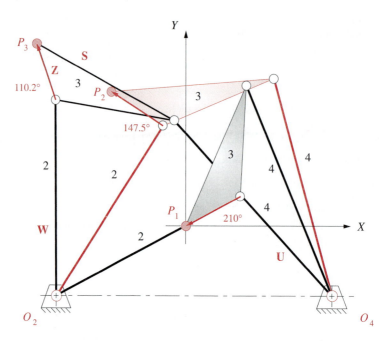

FIGURE S5-1

Solution to Problem 5-11. Open the file P05-11 in program FOURBAR for more information

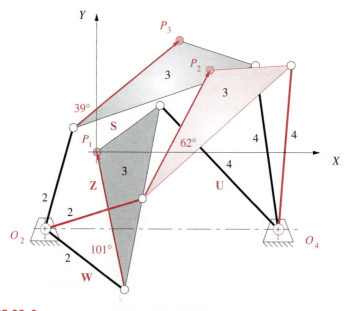

FIGURE S5-2

Solution to Problem 5-15. Open the file P05-15 in program FOURBAR for more information.

A

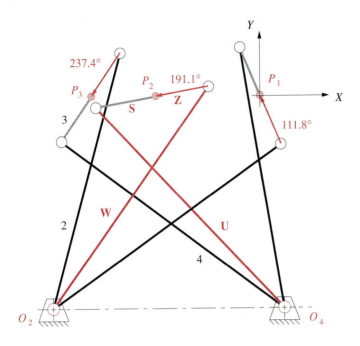

FIGURE S5-3

Solution to Problem 5-19. Open the file P05-19 in program FOURBAR for more information.

5-26 Given:

$\alpha_2 = -45°$,	$P_{21} = 184.78$ mm,	$\delta_2 = -5.28°$
$\alpha_3 = -90°$,	$P_{31} = 277.35$ mm,	$\delta_3 = -40.47°$
$O_{2x} = 86$ mm	$O_{2y} = -132$ mm	
$O_{4x} = 104$ mm	$O_{4y} = -155$ mm	

For left dyad: Calculate: $\beta_2 = -85.24°$ $\beta_3 = -164.47°$

Calculate: $W = 110.88$ mm $\theta = 124.24°$

Calculate: $Z = 46.74$ mm $\phi = 120.34°$

For right dyad: Calculate: $\gamma_2 = -75.25°$ $\gamma_3 = -159.53°$

Calculate: $U = 120.70$ mm $\sigma = 104.35°$

Calculate: $S = 83.29$ mm $\psi = 152.80°$

CHAPTER 6   VELOCITY ANALYSIS

6-4   and   **6-5**   See Table S6-1 and the file P07-04*row*.4br.

6-6   and   **6-7**   See Table S6-2.

6-8   and   **6-9**   See Table S6-3.

6-10   and   **6-11**   See Table S6-4.

6-47   Open the file P06-47.4br in Program FOURBAR to see this solution.[*]

6-48   Open the file P06-48.4br in Program FOURBAR to see this solution.[*]

6-49   Open the file P06-49.4br in Program FOURBAR to see this solution.[*]

6-51   Open the file P06-51.4br in Program FOURBAR to see this solution.[*]

6-62   Open the file P06-62.4br in Program FOURBAR to see this solution.[*]

CHAPTER 7   ACCELERATION ANALYSIS

7-3   and   **7-4**   See Table S7-1 and the file P07-04row.4br.

7-5   and   **7-6**   See Table S7-2.

7-7   and   **7-8**   See Table S7-3.

7-9   See Table S7-4.

7-39   Open the file P07-39.4br in Program FOURBAR to see this solution.[*]

7-40   Open the file P07-40.4br in Program FOURBAR to see this solution.[*]

7-41   Open the file P07-41.4br in Program FOURBAR to see this solution.[*]

7-42   Open the file P07-42.4br in Program FOURBAR to see this solution.[*]

7-44   Open the file P07-44.4br in Program FOURBAR to see this solution.[*]

7-56   Tipover at 19.0 to 20.3 mph; load slides at 16.2 to 19.5 mph.

CHAPTER 8   CAM DESIGN

Most of the problems in this cam chapter are design problems with more than one correct solution. Use program DYNACAM to check your solution obtained with *Mathcad* or *TKSolver* and also to explore various solutions and compare them to find the best one for the constraints given in each problem.

8-1   See Figure S8-1.

CHAPTER 9   GEAR TRAINS

9-1   Pitch diameter = 5.5, circular pitch = 0.785, addendum = 0.25, dedendum = 0.313, tooth thickness = 0.393, and clearance = 0.063.

9-5   a. $p_d = 4$,   b. $p_d = 2.67$

---

[*] These files can be found in the PROBLEM SOLUTIONS folder on the CD-ROM included with this text.

A

TABLE S6-1   Solutions for Problems 6-4 to 6-5

Row	ω_3 open	ω_4 open	V_P mag	V_P ang	ω_3 crossed	ω_4 crossed	V_P mag	V_P ang
a	−6.0	−4.0	40.8	58.2	−0.66	−2.66	22.0	129.4
c	−12.7	−19.8	273.8	−53.3	−22.7	−15.7	119.1	199.9
e	1.85	−40.8	260.5	−12.1	−23.3	19.3	139.9	42.0
g	76.4	146.8	798.4	92.9	239.0	168.6	1435.3	153.9
i	−25.3	25.6	103.1	−13.4	56.9	6.0	476.5	70.4
k	−56.2	−94.8	436.0	−77.4	−55.6	−16.9	362.7	79.3
m	18.3	83.0	680.8	149.2	7.73	−57.0	571.3	133.5

TABLE S6-2   Solutions for Problems 6-6 to 6-7

Row	V_A mag	V_A ang	ω_3 open	V_B mag open	ω_3 crossed	V_B mag crossed
a	14	135	−2.47	−9.9	2.47	−9.92
c	45	−120	5.42	−41.5	−5.42	−3.54
e	250	135	−8.86	−189.7	8.86	−163.80
g	700	60	−28.80	738.9	28.80	−38.90

TABLE S6-3   Solutions for Problems 6-8 to 6-9

Row	V_A mag	V_A ang	ω_3 open	V_{slip} open	V_B mag open	ω_3 crossed	V_{slip} crossed	V_B mag crossed
a	20.0	120.0	-10.3	33.5	41.2	3.6	−33.5	14.6
c	240.0	135.0	23.7	73.0	142.5	6.5	257.8	38.8
e	−180.0	165.0	-2.7	−176.0	5.4	−44.5	−17.1	89.0

TABLE S6-4   Solutions for Problems 6-10 to 6-11

Row	ω_3 open	ω_4 open	ω_3 crossed	ω_4 crossed
a	32.6	16.9	−75.2	−59.6
c	10.7	-2.6	−8.2	5.1
e	−158.3	−81.3	−116.8	−193.9
g	−8.9	−40.9	−48.5	−16.5
i	−40.1	47.9	59.6	−28.4

A

TABLE S7-1   Solutions for Problems 7-3 to 7-4

Row	α_3 open	α_4 open	A_P mag	A_P ang	α_3 crossed	α_4 crossed	A_P mag	A_P ang
a	26.1	53.3	419	240.4	77.9	50.7	298	−11.3
c	−154.4	−71.6	4400	238.9	−65.2	−148.0	3554	100.6
e	331.9	275.6	10 260	264.8	1287.7	1344.1	19 340	−65.5
g	−23 510.	−19 783.	172 688	191.0	−43 709.	−47 436.	273 634	−63.0
i	−344.6	505.3	9492	−81.1	121.9	−728.0	27 871	150.0
k	−2693.	−4054.	56 271	220.2	311.0	1672.1	27 759	−39.1
m	680.8	149.2	35 149	261.5	9266.1	10 303.	63 831	103.9

TABLE S7-2   Solutions for Problems 7-5 to 7-6

Row	A_A mag	A_A ang	α_3 open	A_B mag open	A_B ang open	α_3 cross	A_B mag cross	A_B ang cross
a	140	−135	25	124	180	−25	74	180
c	676	153	−29	709	180	29	490	180
e	12 500	45	−447	6653	0	447	11 095	0
g	70 000	150	−1136	62 688	180	1136	58 429	180

TABLE S7-3   Solutions for Problems 7-7 to 7-8

Row	α_3 open	α_4 open	A_{slip} open	α_3 crossed	α_4 crossed	A_{slip} crossed
a	130.5	130.5	−3111.7	−9.9	−9.9	1125.2
c	−212.9	−212.9	3327.9	−217.8	−217.8	6776.4
e	896.3	896.3	2020.1	595.6	595.6	−1044.4

TABLE S7-4   Solutions for Problem 7-9

Row	α_3 open	α_4 open	α_3 crossed	α_4 crossed
a	3191	2492	−6648	−5949
c	314	228	87	147
e	2171	−6524	7781	5414
g	−22 064	−23 717	−5529	−29 133
i	−5697	−3380	−2593	−7184

A

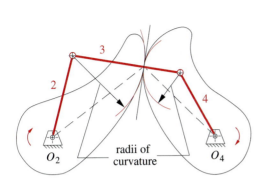

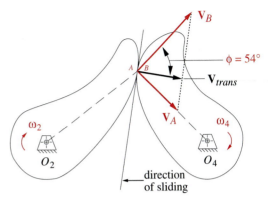

(a) Effective linkage for Problem 8-1

(b) Pressure angle φ for Problem 8-2

FIGURE S8-1

Solutions to Problems 8-1 and 8-2

9-6 Assume a minimum no. of teeth = 17, then: pinion = 17t and 2.125-in pitch dia. Gear = 153t and 19.125-in pitch dia. Contact ratio = 1.7.

9-7 Assume a minimum no. of teeth = 17, then: pinion = 17t and 2.83-in pitch dia. Gear = 136t and 22.67-in pitch dia. An idler gear of any dia. is needed to get the positive ratio. Contact ratio = 1.7.

9-10 Three stages of 4.2:1, 4:1, and 4.167:1 give –70:1. Stage 1 = 20t ($d = 2.0$ in.) to 84t ($d = 8.4$ in.). Stage 2 = 20t ($d = 2.0$ in) to 80t ($d = 8.0$in). Stage 3 = 18t ($d = 1.8$ in) to 75t ($d = 7.5$ in).

9-12 The square root of 150 is > 10 so will need three stages. 5 × 5 × 6 = 150. Using a minimum no. of teeth = 18 gives 18:90, 18:90 and 18:108 teeth. Pitch dias. are 3.0, 15 and 18 in. An idler (18t) is needed to make the overall ratio positive.

9-14 The factors 5 × 6 = 30. The ratios 14:70 and 12:72 revert to same center distance of 4.2. Pitch dias. are 1.4, 1.2, 7, and 7.2.

9-16 The factors 7.5 × 10 = 75. The ratios 22:165 and 17:170 revert to same center distance of 6.234. Pitch dias. are 1.833, 1.4167, 13.75, and 14.167.

9-19 The factors 2 × 1.5 = 3. The ratios 15:30 and 18:27 revert to the same center distance of 3.75. Pitch dias. are 2.5, 5, 3, and 4.5. The reverse train uses the same 1:2 first stage as the forward train, so it needs a second stage of 1:2.25 which is obtained with a 12:27 gearset. The center distance of the 12:27 reverse stage is 3.25 which is less than that of the forward stage. This allows the reverse gears to engage through an idler of any suitable diameter to reverse output direction.

9-21 For the low speed of 6:1, the factors 2.333 × 2.571 = 6. The ratios 15:35 and 14:36 revert to the same center distance of 3.125. Pitch dias. are 1.875, 4.375, 1.75, and 4.5. The second speed train uses the same 1:2.333 first stage as the low speed train, so it needs a second stage of 1:1.5 which is obtained with a 20:30 gearset which reverts to the same center distance of 3.125. The additional pitch dias. are 2.5 and 3.75. The reverse train also uses the same 1:2.333 first stage as both forward trains, so it needs a second stage of 1:1.714 which is obtained with a 14:24 gearset. The center distance of the 14:24 reverse stage is 2.375 which is less than that of the forward stages. This allows the reverse gears to engage through an idler of any suitable diameter to reverse output direction.

TABLE S9-1   Solution to Problem 9-29

Possible ratios for two-stage compound gear train to give the ratio 2.71828

Pinion1	Gear 1	Ratio1	Pinion 2	Gear 2	Ratio 2	Train ratio	Abs error
25	67	2.68	70	71	1.014	2.71828571	5.71E-06
29	57	1.966	47	65	1.383	2.71826853	1.15E-05
30	32	1.067	31	79	2.548	2.71827957	4.30E-07
30	64	2.133	62	79	1.274	2.71827957	4.30E-07
31	48	1.548	45	79	1.756	2.71827957	4.30E-07
31	64	2.065	60	79	1.317	2.71827957	4.30E-07
31	79	2.548	75	80	1.067	2.71827957	4.30E-07
35	67	1.914	50	71	1.420	2.71828571	5.71E-06

9-25 a. $\omega_2 = 790$,   c. $\omega_{arm} = -4.544$,   e. $\omega_6 = -61.98$

9-26 a. $\omega_2 = -59$,   c. $\omega_{arm} = 61.54$,   e. $\omega_6 = -63.33$

9-27 a. 560.2 rpm and 3.57 to 1,   b. $x = 560.2 \times 2 - 800 = 320.4$ rpm

9-29 See Table S9-1 for solution. The third row has the smallest error and smallest gears.

CHAPTER 10   DYNAMICS FUNDAMENTALS

10-1 *CG* @ 8.36 in. from handle end, $I_{zz} = 0.177$ in-lb-sec^2, $k = 8.99$ in.

10-2 *CG* @ 7.61 in. from handle end, $I_{zz} = 0.096$ in-lb-sec^2, $k = 8.48$ in.

10-4

    a. $x = 3.547$, $y = 4.8835$, $z = 1.4308$, $w = -1.3341$

    b. $x = -62.029$, $y = 0.2353$, $z = 17.897$, $w = 24.397$

10-6

    a.   In series: $k_{eff} = 3.09$,   Softer spring dominates

    b.   In parallel: $k_{eff} = 37.4$,   Stiffer spring dominates

10-9

    a.   In series: $c_{eff} = 1.09$,   Softer damper dominates

    b.   In parallel: $c_{eff} = 13.7$,   Stiffer damper dominates

10-12 $k_{eff} = 4.667$, $m_{eff} = 0.278$

10-20 Effective mass in 1st gear = 0.054 bl, 2nd gear = 0.096 bl, 3rd gear = 0.216 bl, 4th gear = 0.863 bl.

10-21 Effective spring constant at follower = 308.35 lb/in.

* These files can be found in the PROBLEM SOLUTIONS folder on the CD-ROM included with this text.

CHAPTER 11   DYNAMIC FORCE ANALYSIS

11-3 Open file P11-03*row*.sld in program SLIDER to check your solution.[*]

11-4 Open file P11-03*row*.sld in program SLIDER to check your solution.[*]

A

11-5 Open file P11-05*row*.4br in program FOURBAR to check your solution.[*]

11-6 Open file P11-05*row*.4br in program FOURBAR to check your solution.[*]

11-7 Open file P11-07*row*.4br in program FOURBAR to check your solution.[*]

11-12 $F_{12x} = -1246$ N, $F_{12y} = 940$ N; $F_{14x} = 735$ N, $F_{14y} = -2\ 219$ N;
$F_{32x} = 306$ N, $F_{32y} = -183$ N; $F_{43x} = 45.1$ N, $F_{43y} = -782$ N; $T_{12} = 7.14$ N-m

11-13 Open file P11-13.4br in program FOURBAR to check your solution.[*]

CHAPTER 12   BALANCING

12-1

    a. $m_b r_b = 0.934$, $\theta_b = -75.5°$

    c. $m_b r_b = 5.932$, $\theta_b = 152.3°$

    e. $m_b r_b = 7.448$, $\theta_b = -80.76°$

12-5

    a. $m_a r_a = 0.814$, $\theta_a = -175.2°$, $m_b r_b = 5.50$, $\theta_b = 152.1°$

    c. $m_a r_a = 7.482$, $\theta_a = -154.4°$, $m_b r_b = 7.993$, $\theta_b = 176.3°$

    e. $m_a r_a = 6.254$, $\theta_a = -84.5°$, $m_b r_b = 3.671$, $\theta_b = -73.9°$

12-6 $W_a = 3.56$ lb, $\theta_a = 44.44°$, $W_b = 2.13$ lb, $\theta_b = -129.4°$

12-7 $W_a = 4.2$ lb, $\theta_a = -61.8°$, $W_b = 3.11$ lb, $\theta_b = 135°$

12-8 These are the same linkages as in Problem 11-5. Open the file P11-05*row*.4br in program FOURBAR to check your solution.[*] Then use the program to calculate the flywheel data.

12-9 Open the file P12-09.4br in program FOURBAR to check your solution.[*]

CHAPTER 13   ENGINE DYNAMICS

13-1 Exact solution $= -\ 42\ 679.272$ in/sec @ 299.156° and 200 rad/sec

    Fourier series approximation $= -\ 42\ 703.631$ in/sec @ 299.156° and 200 rad/sec

    Error $= -0.0571\%$ $(-0.000\ 571)$

13-3 Gas torque $=\ 2\ 040$ (approx.),   Gas force $=\ 3\ 142$

13-5 Gas torque $=\ 2\ 039.53$ (approx.),   Gas torque $=\ 2\ 039.91$ (exact)

    Error $= 0.0186\%$ $(0.000\ 186)$

13-7

    a. $m_b = 0.00748$ at $l_b = 7.2$, $m_p = 0.01251$ at $l_p = 4.31$
    b. $m_b = 0.00800$ at $l_b = 7.2$, $m_a = 0.01200$ at $l_a = 4.80$
    c. $I_{model} = 0.6912$,   Error $= 11.48\%$ (0.1148)

13-9 $m_{2a} = 0.018$ at $r_a = 3.5$, $I_{model} = 0.2205$,   Error $= -26.5\%$ (-0.265)

[*] These files can be found in the PROBLEM SOLUTIONS folder on the CD-ROM included with this text.

TABLE S15-1
Solutions to Problem 15-6

	ω_n	ω_d	c_c
a.	3.42	3.38	8.2
b.	4.68	4.65	19.7
c.	0.26	0.26	15.5
d.	2.36	2.33	21.2
e.	5.18	5.02	29.0
f.	2.04	1.96	49.0

13-11 Open the file P13-11.eng in program ENGINE to check your solution.[*]

13-14 Open the file P13-14.eng in program ENGINE to check your solution.[*]

13-19 Open the file P13-19.eng in program ENGINE to check your solution.[*]

CHAPTER 14   MULTICYLINDER ENGINES

Use program ENGINE to check your solutions.[*]

CHAPTER 15   CAM DYNAMICS

15-1 to **15-5** Use program DYNACAM to solve these problems. There is not any *one right answer* to these design problems.

15-6 See Table S15-1

---

[*] These files can be found in the PROBLEM SOLUTIONS folder on the CD-ROM included with this text.

A

INDEX

A

acceleration 4, 144, 300
  absolute 302, 307, 308
  analysis
    analytical 308
    graphical 303, 304
  angular 522, 531, 541
    as free vector 308
    definition 300
    inverted slider-crank 318
  cam 352
  centripetal 301, 307, 514
  coriolis 313
  difference
    analytic solution 308, 309
    definition 303
    equation 302
    graphical solution 307
    in slider-crank 312
  discontinuities in 358
  human tolerance of 322
  linear 522, 531
  modified sine 367, 369
  modified trapezoidal 364
  multiplier factor (cams) 420
  normal 307
  of a valve train 709
  of any point on link 320
  of geared fivebar 319
  of piston 324
  of slip 318
  relative 303, 307
    peak 418

  sinusoidal 362
  tangential 301, 307
  tolerance 323
accelerometer 322
across variable 504
actuator 28, 517
addendum 435, 440
  circle 440
  modification coefficients 444
AGMA 440, 444
air
  cylinder 28, 77
  motor 65
all-wheel-drive 478
Ampere, Andre Marie 5
amplitude ratio 695
analog duplication 416
analogies 11, 685
analogs 504
analysis 10, 11, 76, 77, 521, 721
  definition 8
  of mechanisms 3, 22
analytical linkage
    synthesis 78, 80, 188, 189, 199
  compare to graphical 196
angle
  of a vector
    definition 156
  of approach 437
  of recess 437

angular velocity ratio 258, 261, 434
  definition 257
animate (in programs) 738, 745
antiparallelogram 48
  linkage 265
apparent position 148
applications
  assembly machines 385
  automobile engine 345
  automobile suspension 107, 262
  automobile transmission 435
  crank-shaper linkage 103
  engine valves 374
  indexing table drive 371
  movie camera 105
  of air motors 65
  of fluid power cylinders 65
  of hydraulic motors 65
  of kinematics 6
  of solenoids 66
  optical adjusting mechanism 263
  steam locomotives 48
  toggle linkages 81
approximate
  circle arc 126, 128
  dwell 130
  straight line 103, 120, 128
arc of action 437
arm (epicyclic) 462
Artobolevsky 6
asperities 25

CD-ROM INDEX